Foodborne and Waterborne Bacterial Pathogens

Epidemiology, Evolution and Molecular Biology

Edited by

Shah M. Faruque

Molecular Genetics Laboratory
International Centre for Diarrhoeal Disease Research,
Bangladesh;
Department of Genetic Engineering and Biotechnology
University of Dhaka
Dhaka
Bangladesh

Caister Academic Press

Caister Academic Press
Norfolk, UK

www.caister.com

British Library Cataloguing-in-Publication Data
A catalogue record for this book is available from the British Library

ISBN: 978-1-908230-06-5

Cover design adapted from Figure 10.3A

Printed and bound in Great Britain

Contents

Contributors

M. John Albert
Department of Microbiology
Faculty of Medicine
Kuwait University
Jabriya
Kuwait

john@hsc.edu.kw

Ishrat Jahan Azmi
Enteric Bacteriology Laboratory
International Centre for Diarrhoeal Disease Research, Bangladesh
Dhaka
Bangladesh

ishratazmi@icddrb.org

Patit Paban Bhowmick
Department of Microbiology
Karnataka Veterinary, Animal and Fisheries Sciences University
College of Fisheries
Mangalore
India

patit_micro@yahoo.co.in

Kathryn J. Boor
Food Science Department
Cornell University
Ithaca, NY
USA

kjb4@cornell.edu

Lieneke I. Bouwman
Department of Infectious Diseases & Immunology
Utrecht University
Utrecht
Netherlands

L.I.Bouwman@uu.nl

E. Fidelma Boyd
Department of Biological Sciences
University of Delaware
Newark, DE
USA

fboyd@udel.edu

Shah M. Faruque
Molecular Genetics Laboratory
International Centre for Diarrhoeal Disease Research, Bangladesh
Dhaka
Bangladesh

faruque@icddrb.org

Maria Fredriksson-Ahomaa
Department of Food Hygiene and Environmental Health
Faculty of Veterinary Medicine
University of Helsinki
Helsinki
Finland

maria.fredriksson-ahomaa@helsinki.fi

Iddya Karunasagar
Products, Trade and Marketing Service
Food and Agriculture Organisation
Rome
Italy

Iddya.Karunasagar@fao.org

Indrani Karunasagar
Department of Microbiology
Karnataka Veterinary, Animal and Fisheries Sciences University
College of Fisheries
Mangalore
India

karuna8sagar@yahoo.com

Sophia Kathariou
Department of Food Science
North Carolina State University
Raleigh, NC
USA

sophia_kathariou@ncsu.edu

Ballamoole Krishna Kumar
Department of Microbiology
Karnataka Veterinary, Animal and Fisheries Sciences University
College of Fisheries
Mangalore
India

findkkumar@gmail.com

Deekshit Vijaya Kumar
Department of Microbiology
Karnataka Veterinary, Animal and Fisheries Sciences University
College of Fisheries
Mangalore
India

dixith_14@yahoo.co.in

Riikka Laukkanen-Ninios
Department of Food Hygiene and Environmental Health
Faculty of Veterinary Medicine
University of Helsinki
Helsinki
Finland

riikka.laukkanen-ninios@helsinki.fi

Sangmi Lee
Department of Food Science
North Carolina State University
Raleigh, NC
USA

slee19@ncsu.edu

John J. Mekalanos
Department of Microbiology and Molecular Genetics
Harvard Medical School
Boston, MA
USA

John_mekalanos@hms.harvard.edu

G. Balakrish Nair
Translational Health Science and Technology Institute
Gurgaon
India

gbnair_2000@yahoo.com

Gururaja P. Pazhani
National Institute of Cholera and Enteric Diseases
Kolkata
India

pazhanigp@gmail.com

Jos P.M. van Putten
Department of Infectious Diseases & Immunology
Utrecht University
Utrecht
Netherlands

j.vanputten@uu.nl

M. Mahbubur Rahman
Acute Respiratory Infections Laboratory
International Centre for Diarrhoeal Disease Research, Bangladesh
Dhaka
Bangladesh

mahbubur@icddrb.org

Thandavaryan Ramamurthy
National Institute of Cholera and Enteric Diseases
Kolkata
India

rama1murthy@yahoo.com

Anusha Rohit
Department of Microbiology
Madras Medical Mission
Chennai
India

anusha3rohit@gmail.com

Robin M. Siletzky
Department of Food Science
North Carolina State University
Raleigh, NC
USA

rmsiletz@ncsu.edu

Kaisar Ali Talukder
Enteric Bacteriology Laboratory
International Centre for Diarrhoeal Disease Research, Bangladesh
Dhaka
Bangladesh

kaisar@icddrb.org

Marie Yeung
Biological Sciences Department
California Polytechnic State University
San Luis Obispo, CA
USA

pmyeung@calpoly.edu

Current books of interest

Bionanotechnology: Biological Self-assembly and its Applications	2013
Real-Time PCR in Food Science: Current Technology and Applications	2013
Bacterial Gene Regulation and Transcriptional Networks	2013
Bioremediation of Mercury: Current Research and Industrial Applications	2012
Neurospora: Genomics and Molecular Biology	2013
Rhabdoviruses	2012
Horizontal Gene Transfer in Microorganisms	2012
Microbial Ecological Theory: Current Perspectives	2012
Two-Component Systems in Bacteria	2012
Malaria Parasites: Comparative Genomics, Evolution and Molecular Biology	2013
Foodborne and Waterborne Bacterial Pathogens	2012
Yersinia: Systems Biology and Control	2012
Stress Response in Microbiology	2012
Bacterial Regulatory Networks	2012
Systems Microbiology: Current Topics and Applications	2012
Quantitative Real-time PCR in Applied Microbiology	2012
Bacterial Spores: Current Research and Applications	2012
Small DNA Tumour Viruses	2012
Extremophiles: Microbiology and Biotechnology	2012
Bacillus: Cellular and Molecular Biology (Second edition)	2012
Microbial Biofilms: Current Research and Applications	2012
Bacterial Glycomics: Current Research, Technology and Applications	2012
Non-coding RNAs and Epigenetic Regulation of Gene Expression	2012
Brucella: Molecular Microbiology and Genomics	2012
Molecular Virology and Control of Flaviviruses	2012
Bacterial Pathogenesis: Molecular and Cellular Mechanisms	2012
Bunyaviridae: Molecular and Cellular Biology	2011
Emerging Trends in Antibacterial Discovery: Answering the Call to Arms	2011
Epigenetics: A Reference Manual	2011
Metagenomics: Current Innovations and Future Trends	2011
Nitrogen Cycling in Bacteria: Molecular Analysis	2011
The Biology of Paramyxoviruses	2011
Helicobacter pylori	2011
Microbial Bioremediation of Non-metals: Current Research	2011
Lactic Acid Bacteria and Bifidobacteria: Current Progress in Advanced Research	2011

Full details at www.caister.com

Preface

The developing scenario of global changes in climate with increasing frequency of natural disasters linked to the threat of enhanced epidemic spread of foodborne and waterborne diseases warrants an update of knowledge on these infectious diseases and the causative agents. Gastrointestinal diseases of which a substantial proportion are due to food- or waterborne bacterial pathogens is the second most common cause of morbidity throughout the world after acute respiratory tract illness. In children alone there may be between 3.5 and 18 million deaths per year worldwide from diarrhoeal diseases.

The past 10–15 years have witnessed the unravelling of remarkable new insights on the biology of the pathogens that spread through water or contaminated foodstuff. These include the discovery of the filamentous phage that encodes cholera toxin in *Vibrio cholerae*, the existence of two chromosomes in *Vibrio* species and the whole genome sequencing of a number of these and related pathogens. The discovery of new mobile and integrative genetic elements, such as conjugative transposons, integrons, pathogenicity islands and phages, involved in horizontal transfer of virulence genes have provided new insights on the emergence of new pathogenic strains and their variants. Several foodborne and waterborne diseases have emerged in the past two decades as a consequence of changes in aetiological agents, hosts and the environment. Despite tremendous recent development in our understanding of the emergence, infectivity and spread of bacterial pathogens, such as *Vibrio cholerae*, pathogenic *Escherichia coli*, *Salmonella*, *Shigella* and *Campylobacter* spp., updated compilations of this knowledge are rare. It is therefore, timely to compile a volume comprising the current status of research on the biology of these and related pathogens.

A tremendous amount of pioneering work in recent times and the voluminous information flow has provided the impetus for putting this book together. At this time a whole set of new research has been directed towards understanding how new pathogenic strains emerge, newly emerged strains replace old strains, their survival strategy and competition in the environment, the molecular mechanism of cell to cell communication, and biofilm formation, as well as the genetic determinants of epidemic and pandemic properties. In this book, we have attempted to capture the essence of this mighty deluge of information regarding bacterial pathogens that spread through food and water. The scope of this book, as emphasized by the publishers, is on the epidemiology and molecular biology of these pathogens. Leading experts in the fields have described various bacterial pathogens, their pathogenic properties, population genetics, virulence genes, epidemiology, identification and control strategies. Other topics include the molecular basis for enhanced transmissibility of waterborne pathogens, their mode of survival in the environment, and above all the evolution of the different species to attain increased fitness both as a pathogen and an environmental organism. I hope that the readers find this book interesting and comprehensive.

Shah M. Faruque

Introduction

1

Shah M. Faruque

Abstract

Foodborne and waterborne bacterial pathogens are a major cause of mortality in developing countries and cause significant morbidity in developed nations. Some countries carry a disproportionately heavy burden of these infectious diseases because of inadequate resources to provide sanitation and hygienic facilities, and safe water. The most important bacterial pathogens transmitted through contaminated water and food include species or strains of *Salmonella*, *Vibrio* (e.g. *V. cholerae*, *V. parahaemolyticus*, *V. vulnificus*), *Shigella* (*S. dysenteriae*, *S. flexneri*, *S. sonnei*, *S. boydii*), *Escherichia coli*, *Yersinia*, *Staphylococcus* and *Campylobacter*. The pathogenic mechanisms of these bacteria involve synergistic actions of multiple virulence factors produced by the pathogen after infecting the host. Besides carrying sets of virulence genes which are often horizontally transferred between strains, many of these bacteria may also carry precise genetic programmes that allow them to adapt and survive in water leading to enhanced transmission or prolonged persistence in the aquatic environment. Understanding the epidemiology, pathogenesis and evolution of these pathogens can contribute significantly to control foodborne and waterborne diseases.

Introduction

Illness attributed to contaminated food or drink containing pathogenic microbes can be devastating, and in addition to death or suffering can cause negative economic impact, e.g. unwanted medical expenses and wage loss. Some of the major diseases transmitted through contaminated food and water include cholera, salmonellosis, shigellosis, amoebic dysentery, persistent diarrhoea, hepatitis A, viral gastroenteritis, and botulism or *Staphylococcus* food poisoning (Greer *et al.*, 2008). Pneumonic pathogens such as *Legionella* species are also transmitted through water whereas other foodborne pathogens include *Listeria monocytogenes*, which can cause meningitis. Several of these diseases have emerged or re-emerged in the past two decades as a consequence of changes in aetiological agents, hosts and the environment (Bartlett, 1996). These newly recognized pathogens which cause foodborne and waterborne diseases include *Campylobacter* spp., *Cryptosporidium parvum*, Vero cytotoxin-producing *Escherichia coli*, Norwalk virus, and hepatitis E virus. Thus, foodborne and waterborne diseases are caused by a variety of bacterial, viral and protozoal pathogens or the toxic substances they produce. The purpose of this book is to present a compilation of available information and scientific data derived from studies in bacterial pathogens which spread through food and water, with particular emphasis on the epidemiology, pathogenesis, molecular biology and prevention strategies.

Common symptoms of foodborne illness include vomiting, abdominal cramps, moderate to severe diarrhoea, and often fever. The causative microorganisms or their toxins can be found in the raw food products (such as *Salmonella* in raw chicken, beef or eggs; *V. parahaemolyticus* and *V. vulnificus* in seafood), or can be present in the people who handle the food (such as *Salmonella*, Norwalk virus, or hepatitis A virus in faeces or *Staphylococcus* in a cut or boil on the skin). Common waterborne pathogens such as *Vibrio cholerae*, diarrhoeagenic *E. coli* or *Shigella* spp., present in polluted water can also contaminate foods. In developing countries where safe water is inadequate, often waterborne bacteria contaminate food stuff and the pathogens may grow in the food prior to being consumed. Because of the inequitable distribution of resources, the burden of these infectious diseases is not uniformly distributed globally. Instead, some countries carry a disproportionately heavy burden of the foodborne and waterborne diseases (Bartlett, 1996). In many parts of the world, subsets of the population do not have access to clean potable water. In others, where public water supplies are available, system failures can place entire communities at risk.

The pathogens

The most severe form of waterborne and foodborne diarrheal disease, cholera, is caused by toxigenic strains of *V. cholerae* and remains an important cause of death in the developing world (Faruque *et al.*, 1998). These bacteria are part of the normal aquatic flora in estuarine and brackish waters (Colwell and Spira, 1992). The epidemics are not only initiated as a result of ingesting naturally occurring toxigenic *V. cholerae* in water but are also sustained by rapid transmission of the pathogen through water (Faruque *et al.*, 2006). Often foodstuff are also contaminated with the polluted water, and this adds to further epidemic spread. The genus *Vibrio* contains several bacterial species, of which *V. cholerae*, *V. parahaemolyticus* and *V. vulnificus* are the most important pathogens of humans (Janda *et al.*, 1988; West, 1989). Whereas *V. parahaemolyticus* is an important cause of diarrheal illness, *V. vulnificus* infections can range from self-limiting gastroenteritis and wound infections to severe necrotizing infections of soft tissues and fatal septicaemia (Tacket *et al.*, 1984). Of the various *Vibrio* species associated with human disease, *V. cholerae* is the most significant in terms of severity of disease, and number of people affected. However, of more than 200 known serogroups of *V. cholerae*, strains belonging to serogroups O1 and O139 are the causative agents of cholera epidemics, and have the ability to cause pandemic spread of the disease.

Other major foodborne and/or waterborne acute diarrheal agents include different groups of diarrhoeagenic *E. coli*, *Shigella*, *Salmonella* and *Campylobacter* species. *Shigella* is a cause of classic dysentery. The organism is less susceptible to acid than many other bacterial pathogens, and thus, as few as 10–100 organisms can cause disease, in part because the organisms can survive transit through the stomach (Bennish, 1991). All *Shigella* species, namely *S. dysenteriae*, *S. flexneri*, *S. sonnei* and *S. boydii* are pathogenic to humans, but the epidemiology varies in different parts of the world with different species and subgroups of *Shigella* affecting different subsets of the population. Intermediate bacterial replication in contaminated food or water is not required to achieve the low infectious dose. As a result, *Shigella* transmission can occur through direct person-to-person spread as well as from contaminated food and water.

Different categories of pathogenic *E. coli*, namely enterotoxigenic (ETEC), enteropathogenic (EPEC), enterohaemorrhagic (EHEC), enteroinvasive (EIEC), and enteroaggregative (EAEC) are a major cause of diarrhoea in infants and young children in developing countries and a leading cause of travellers' diarrhoea and haemorrhagic colitis, in developed countries (Levine and Edelman, 1984; Levine, 1987). Both symptomatic and asymptomatic human carriers are presumed to be the primary reservoirs for these pathogens. The enterohaemorrhagic (EHEC) group of *E. coli* contains an emerging pathogen, Vero cytotoxin-producing *E. coli* O157, which has its reservoir in cattle. Transmission of *E. coli* O157 can occur through a variety of routes including undercooked ground beef and unpasteurized milk, contaminated water which in turn contaminates vegetables and other foodstuffs, through person-to-person spread, and from animals to people.

Campylobacter spp. are a part of the bacterial flora in the gastrointestinal tract of animals, especially poultry, and hence animal-to-human transmission of infections occurs frequently. Campylobacter belongs to a distinct group of specialized Gram-negative bacteria designated rRNA superfamily VI (Vandamme *et al.*, 1991), which includes *Arcobacter* and *Helicobacter* in addition to the genus *Campylobacter*. Some Arcobacters cause intestinal infection in humans, whereas *Helicobacter pylori* causes gastritis and peptic ulcer disease, but there are other *Helicobacter* species which cause intestinal infection. The combination of spiral shape and long polar flagella of these bacteria allow rapid motility enabling these organisms to move through mucus. Thus this group of bacteria are adapted to colonize the surface of the mucous membranes of the alimentary tracts. *Campylobacter* species are commonly reported bacterial cause of intestinal diseases in many industrialized countries. *Campylobacter jejuni* is one of the most widespread bacterial causes of diarrhoea with an estimated 400 million cases worldwide. In developing countries, infection is very common in early childhood and an important cause of foodborne illness in young children, including less than 1-year-old infants (Coker *et al.*, 2002; Fullerton, *et al.*, 2007). In addition, its reported association with life-threatening cases of Guillain–Barré syndrome (GBS) is of great concern. Transmission generally occurs through consumption of contaminated water, raw milk or undercooked meat, especially poultry meat and meat products. Transmission also can occur through contaminated recreational waters.

Salmonella species continue to be one of the main causes of foodborne illness worldwide. Currently, there are three recognized species: *S. enterica*, the type species, and *S. bongori*, which was formerly subspecies V, and *S. subterranea* (Brenner *et al.*, 2000; Shelobolina, *et al.*, 2004). *S. enterica* is divided into six subspecies, each designated with a roman numeral and a name: (I) *enterica*, (II) *salamae*, (IIIa) *arizonae*, (IIIb) *diarizonae*, (IV) *houtenae* and (VI) *indica*. The Centers for Disease Control (CDC) uses names for serotypes in subspecies I (for example, serotypes Enteritidis, Typhimurium, Typhi and Choleraesuis) and uses antigenic formulas for unnamed serotypes described after 1966 in subspecies II, IV, and VI and in *S. bongori*. The name usually refers to the geographic location where the serotype was first isolated. Within *S. enterica* subsp. I, the most common O-antigen serogroups are A, B, C1, C2, D and E. These serogroups cause approximately 99% of *Salmonella* infections in humans and warm-blooded animals (Popoff and Le Minor, 1997).

Salmonella infections in humans are divided into typhoid fever caused by the *Salmonella enterica* serovar Typhi (commonly referred to as *S. typhi*), and a range of diarrhoeal diseases caused by a large number of non-typhoidal *Salmonella* serovars. The disease is transmitted by ingestion of food, including dairy products, or water contaminated by excreta from patients or chronic carriers or handled by infected persons (Parry, 2004). The highest incidence usually occurs where water supplies serving large populations are contaminated by faecal matter. *Salmonella* can adapt to different domestic and wild warm-blooded animals and take advantage of changing environmental factors. The intensive rearing of poultry using strains of poultry selected for food conversion efficiency and egg production, rather than disease resistance, provide fertile ground for *S. enteritidis* in recent times. Contamination of eggs and poultry meat as well as an increased consumption of poultry meat have led to the predominance of *S. enteritidis* as a cause of human salmonellosis in many countries.

Yersinia enterocolitica and *Listeria monocytogenes* are two other important foodborne pathogens. *Y. enterocolitica* has been a cause of acute gastroenteritis in many countries including Belgium, Canada, the Netherlands, Australia, New Zealand and parts of Germany (Cover and Aber, 1989; Wright *et al.*, 1995). Different serotypes of *Y. enterocolitica* are associated with human infections in different regions of the world. In Europe, Canada and Japan, occasional infections caused by *Y. enterocolitica* serovar O:3 and to a lesser extent O:9 are predominant whereas in the United States rare sporadic cases, are associated with multiple serovars, and outbreaks, which are more common, mainly involve serovar O:8. These bacteria can grow at low temperatures, and the increase in prevalence of these organisms in food may in part be due to the increased use of refrigeration in the food chain (Doyle, 1990).

L. monocytogenes causes severe, potentially lethal illness referred to as listeriosis in animals and humans. Like *Y. enterocolitica*, this organism can also multiply in food at low temperatures (Schlech *et al.*, 1983). Out of 12 serotypes of *L. monocytogenes*, only three serotypes, i.e. serotypes 1/2a, 1/2b and 4b, are responsible for over 95% of clinical cases. The incidence of listeriosis is uncommon in the general population, but it can cause severe disease in certain groups including pregnant women, persons with malignancy and those with immune deficiency.

Virulence genes and characteristics

The ability of pathogenic bacteria to cause severe infection depends on their virulence gene content, which varies between pathogenic and non-pathogenic strains. Virulence factors include bacterial toxins, cell surface proteins that mediate bacterial attachment, cell surface carbohydrates and proteins that protect a bacterium, and hydrolytic enzymes that may contribute to the pathogenicity of the bacterium. These virulence factors manipulate cellular processes inside the host cell, and often the host cells are forced to behave in ways that promote growth of the pathogen and make defence against infection more difficult. An important feature of most enteric bacterial pathogens seem to be their ability to adhere and colonize the intestine despite peristalsis and competition for nutrients by the indigenous flora of the gut. Virulence factors responsible for pathogenicity in enteric bacteria, such as *E. coli*, *Salmonella*, *Yersinia*, and *Shigella* spp. are often encoded by horizontally transferable genetic elements which include plasmids, transposons, genomic pathogenicity islands or bacteriophages.

Like most mucosal pathogens, *E. coli* follows a strategy of infection which includes colonization of a mucosal site, evasion of host defences, multiplication, and host damage. The presence of surface adherence fimbriae is a property of all *E. coli* strains, including non-pathogenic varieties. However, diarrhoeagenic *E. coli* strains possess specific fimbrial antigens that enhance their intestinal colonizing ability and allow adherence to the small bowel mucosa (Nataro and Kaper, 1998). For example, Typical EPEC isolates produce a type IV pilus called the bundle-forming pilus (BFP) (Giron *et al.*, 1991), which mediates inter-bacterial adherence and adherence to epithelial cells. ETEC of human and animal origin also have defined fimbriae that help ETEC to adhere to small bowel enterocytes prior to expression of one or more enterotoxins. Once colonization is established, the different categories of diarrhoeagenic *E. coli* strains exhibit a variety of pathogenetic strategies. Diarrhoeagenic *E. coli* strains are indeed distinguished and defined on the basis of pathogenic mechanisms, and their virulence gene contents. Three general paradigms by which *E. coli* may cause diarrhoea include enterotoxin production (ETEC and EAEC), invasion (EIEC), and/or intimate adherence with membrane signalling (EPEC and EHEC).

The virulence properties of *E. coli* are encoded by virulence-related plasmids and chromosomal pathogenicity islands. EIEC, EHEC, EAEC, and EPEC strains typically harbour highly conserved plasmid families, each encoding multiple virulence factors (Baldini *et al.* 1986; Nataro *et al.*, 1987; Hales *et al.*, 1992). The genes for colonization factors of ETEC are located either in plasmids or chromosome and are typically a polycistronic operon that includes the fimbrial subunit genes, chaperones, and ushers for biogenesis of the fimbriae. These genes may be horizontally acquired via flanking insertion sequences (IS) and transposons. The chromosomal virulence genes of EPEC and EHEC are organized as a cluster referred to as a pathogenicity island (PAI) (McDaniel *et al.*, 1995; McDaniel and Kaper, 1997). Such islands may represent a common way in which the genomes of pathogenic and non-pathogenic *E. coli* strains diverge genetically. Plasmids and pathogenicity islands carry clusters of virulence traits, but certain traits, e.g. ST may be transposon encoded (So and McCarthy, 1980), or Shiga toxin may be phage encoded (O'Brien *et al.*, 1992). Occasionally, the pathogenicity islands may also carry genes for type III secretion systems (T3SS), which are effectively a 'nano-injection syringe', with the help of which the bacteria pump virulence factors into the target cells (Heuck *et al.*, 1998). The genes encoding T3SS components in Gram-negative bacteria are usually located on pathogenicity islands, e.g. the locus of enterocyte effacement region of EHEC O157:H7 and EPEC, which carry genes encoding T3SS and other genes involved in pathogenesis (Clarke *et al.*, 2003). EPEC colonizes epithelial cells in the human small intestine by provoking effacement of the microvilli and intimating attachment to the host cells, a prominent pathogenic feature called attaching and effacing. EPEC normally cannot invade epithelial cells and rather induces the formation of an actin pedestal structure beneath the bacterium attached to the host cell surface. To achieve an intimate attachment, EPEC delivers a set of effector proteins such as Tir, EspB, and EspD into the host cytoplasm via the T3SS machinery.

The presence of haemolysins, namely thermostable direct haemolysin (TDH) and TDH-related protein (TRH) has been closely associated with the diarrheal illness caused by *V. parahaemolyticus*. TDH and TRH have various biological activities including haemolytic activity, cardiotoxicity and enterotoxicity. The *tdh* gene encoding TDH is associated with T3SS2 and an 80-kb pathogenicity island located in chromosome 2. Whereas T3SS1 appears to be common in all *V. parahaemolyticus* isolates, including those isolated from food and environmental sources and not implicated in human diseases, the T3SS2 is associated with enterotoxicity of *V. parahaemolyticus* (Shimohata and Takahashi, 2010).

The two most important virulence factors in *V. cholerae* include the cholera toxin (CT) which is primarily responsible for the acute diarrhoea, and a colonization factor known as toxin coregulated pilus (TCP) (Faruque *et al.*, 1998). The major virulence genes in *V. cholerae* are clustered in several regions of the bacterial chromosome. The *ctxAB* genes encoding the A and B subunits of CT are part of the genome of a lysogenic filamentous bacteriophage designated as CTXφ (Waldor and Mekalanos, 1996), whereas the genes for TCP are part of a larger cluster of genes known as *Vibrio* pathogenicity island (VPI), alternatively referred to as the TCP pathogenicity island, (Faruque and Mekalanos, 2003). *V. cholerae* colonizes the small intestine using TCP, and produces CT, which is mainly responsible for the manifestation of the disease. CT acts as a classical A–B type toxin, leading to ADP-ribosylation of a small G protein, and constitutive activation of adenylate cyclase, thus giving rise to increased levels of cyclic AMP within the host cell. This results in rapid efflux of chloride ions and water from host intestinal cells, and leads to diarrhoea and vomiting. Massive outpouring of fluid and electrolytes leads to severe dehydration, electrolyte abnormalities, and metabolic acidosis (Kaper *et al.*, 1995).

Salmonella species cause disease through their ability to adapt and grow in hostile environments within the body, simultaneously evading the immune system. During the disease process, salmonellae exist in both extracellular and intracellular environments, and the intracellular component plays a significant role in the disease. *S. enterica* serovar Typhi has a combination of characteristics that make it an effective pathogen. This bacterium contains an endotoxin typical of Gram-negative organisms, as well as the Vi antigen which is thought to increase virulence. It also produces and excretes a protein known as 'invasin' that allows non-phagocytic cells to take up the bacterium, which is able to live intracellularly. It is also able to inhibit the oxidative burst of leucocytes, making innate immune response ineffective. *Salmonella* pathogenesis involves a number of different bacterial gene products that are expressed in a coordinated fashion. Many of the *Salmonella* virulence genes are located in pathogenicity islands, which are large segments of DNA acquired by horizontal gene transfer. At least 21 different *Salmonella* pathogenicity islands (SPIs) are known to play a role in virulence, host specificity and the degree of invasiveness. Some of the SPIs encode type III secretion systems (T3SS) with which the bacteria pump a cocktail of virulence factors into the intestinal cells (Hansen-Wester and Hensel, 2001). The pathogenesis is ultimately based on the mode of action of this virulence factor cocktail, which may consist of a large number of different virulence factors.

Shigella spp. (*S. boydii, S. dysenteriae, S. flexneri and S. sonnei*) and enteroinvasive *E. coli* (EIEC) are responsible for a disease characterized by the destruction of the colonic mucosa that is induced upon bacterial invasion. *Shigella* and EIEC strains contain a virulence plasmid of approximately 220 kb that encodes determinants for entry into epithelial cells and dissemination from cell to cell. The ability of *Shigella* spp., to continuously spread intra- and intercellularly, is an essential component of the pathogenesis leading to bacillary dysentery. The *virG gene* (also called *icsA*) located on a plasmid of *S. flexneri* referred to as the large plasmid is required for cell-to-cell spreading and actin-based motility in mammalian cells. A similar intracellular actin-based movement has been reported for *Listeria monocytogenes*.

A type III secretion system (T3SS) encoded on the large plasmid is also a key virulence machinery of *S. flexneri*. The T3SS determines the interactions of *S. flexneri* with intestinal cells by consecutively translocating two sets of effector proteins into the target cells. Thus, *S. flexneri* controls invasion, intra- and intercellular spread, macrophage cell death, as well as host inflammatory responses. Some of the translocated effector proteins also show novel biochemical activities by which they intercept host cell signal transduction pathways.

Enteropathogenic *Yersinia* strains carry several plasmid and chromosomal virulence genes. A 70-kb virulence plasmid, pYV encodes factors that are required for survival and multiplication of *Yersinia* in the host (Cornelis *et al.*, 1998). The pYV carries genes for a type III secretion system called Ysc (Cornelis, 2002) and its secreted proteins called Yops, as well as another outer membrane protein YadA (*Yersinia* adhesin A). Extracellularly located *Yersinia* strains, come in close contact with the eukaryotic cell, and deliver Yops into the cytosol of the target cell using the Ysc. These Yops effector proteins protect *Yersinia* strains from the macrophages by destroying the phagocytic and signalling capacities of the macrophages, and eventually leading to induction of apoptosis.

Enteropathogenic *Yersinia* strains also require chromosomally encoded virulence factors to express full virulence. An attachment invasion locus (Ail) of the chromosome is involved in the attachment and invasion of the host cell by *Y. enterocolitica* and in the serum resistance of both *Y. enterocolitica* and *Y. pseudotuberculosis* (Miller *et al.*, 2001). At least two chromosomal genes, *inv* and *ail*, are needed for the invasion, and Inv promotes the entry of enteropathogenic *Yersinia* strains into epithelial cells of the ileum (Isberg and Barnes, 2001). Other virulence associated factors including thermostable enterotoxin (Yst) and mucoid *Yersinia* factor (Myf) of *Y. enterocolitica* and the pH6 antigen of *Y. pseudotuberculosis* are also chromosomally encoded. The pH6 antigen is a fimbrial structure that mediates agglutination of erythrocytes (Yang and Isberg, 1997).

Overview of detection methods and preventive measures

Pathogenic bacteria that are transmitted through water and/or food often have the ability to survive for a long time in the environment, and the potential for transmission may be positively correlated with the ability to survive outside the host. Some of these bacteria are part of the normal flora in the aquatic environments (Colwell *et al.*, 1984). In most cases bacteria that normally exist in water at a low level are concentrated in the foods, for example oysters may contain higher concentration of *V. parahaemolyticus* or *V. vulnificus* than the concentration in the surrounding water (Parvathy *et al.*, 2004; Drake *et al.*, 2007). Similarly, inanimate foodstuff contaminated with bacteria such as *V. cholerae*, *E. coli* or shigellae may support multiplication of the bacteria in the food before being consumed, thus increasing the risk of disease. Therefore, a key to preventing foodborne infections is to ascertain microbiological safety of the food and water. Available phenotypic detection methods relies on laboratory assays that correlate with certain pathogenic mechanism exhibited by the organism *in vivo*. For example, EPEC isolates possess the EAF plasmid encoding a type IV pilus called the bundle-forming pilus (BFP) (Giron *et al.*, 1991), which mediates adherence to epithelial cells. This localized adherence of EPEC as microcolonies on the intestine *in vivo* correlates with its adherence on mammalian cells such as HeLa and HEp-2 cells under *in vitro* conditions. Similarly, production of enterotoxin can be demonstrated by cell culture assays or animal models. Although available culture and phenotypic detection methods can identify the pathogens most of the time, there are instances when the phenotypic techniques become inadequate. One of these circumstances is when bacteria enter into a dormant state to promote persistence in the environment, and become difficult to grow in standard culture media.

Conditionally viable bacteria in the environment

Bacterial pathogens tightly regulate the expression of genes required for virulence, as well as their prolonged survival in water and this regulation is often in response to very specific environmental conditions. Under conditions that do not support optimal physiological functions, some bacteria enter into a metabolic state in which the cells are referred to as viable but non-culturable (VBNC), alternatively known as 'conditionally viable environmental cells' (CVEC) (Faruque *et al.*, 2006; Kamruzzaman *et al.*, 2010; Oliver, 2010). In this state the bacteria fail to grow on standard culture media, despite the fact that they remain alive and show some metabolic activities. Entering this physiological state is probably a survival mechanism for non-sporulating bacteria under unfavourable conditions (Oliver, 2010). Over 25 bacterial genera have been documented to exhibit this dormant metabolic state. The VBNC state of *Vibrio* species was first reported for *V. cholerae* by Colwell and co-workers (Xu *et al.*, 1982; Colwell and Spira, 1992). Subsequently, *V. parahaemolyticus* and *V. vulnificus* have also been reported to enter the VBNC state under starvation and/or temperature stress (Oliver *et al.*, 1995; Jiang and Chai, 1996; Mizunoe *et al.*, 2000; Johnston and Brown, 2002).

Recent studies with *V. cholerae* have established the pathogenic potential of this dormant form of waterborne bacteria (Faruque *et al.*, 2006; Kamruzzaman *et al.*, 2010). Efforts to understand the genetic regulations involved in the formation of the CVEC state have revealed that CVEC are in fact derived from bacterial biofilms which are dense aggregates of bacteria embedded in exopolysaccharide materials (Kamruzzaman *et al.*, 2010). The formation and dispersion of the biofilms are regulated through 'quorum sensing' which is a regulatory mechanism that occurs in bacterial communities in response to cell density, and through the sensing of extracellular signal molecules called autoinducers which the members of the community produce (Higgins *et al.*, 2007). Thus, waterborne bacterial pathogens possess appropriate genetic mechanisms to persist and survive in water. Furthermore, inability to culture does not rule out the existence of such pathogens in water.

Molecular diagnostic methods

Since many pathogenic bacteria also have non-pathogenic counterparts of the same species and subtypes that only lack the virulence genes, development of phenotypic diagnostic methods to discriminate between pathogenic and non-pathogenic strains within the species or subtypes is challenging. Although there are certain animal models to test toxicity and colonization or cell culture assays to test toxicity as well as adhesion, the methods are diverse and not applicable in general to all pathogens. Often the unreliability of culture methods to isolate pathogens and phenotypic methods to accurately determine their pathogenic potential, have led to the development of a variety of molecular diagnostic techniques. These molecular diagnostic methods have largely addressed the difficulties faced with the phenotypic methods and detection of pathogenic genes seem more convenient than the biological assays. PCR assays have been widely used to detect the presence of small number of pathogenic bacteria, in food samples or in water. These assays are also useful in detecting organisms that are not readily culturable by routine laboratory methods. A variety of genetic fingerprinting techniques have also allowed to track the spread and transmission of these pathogens. The most common examples of these techniques are pulsed-field gel electrophoresis (PFGE), ribotyping, multilocus sequence typing (MLST), amplified-fragment-length polymorphism (AFLP), and multiple-locus variable-number tandem-repeats analysis (MLVA). However, the improvement of phenotypic methods and culture methods to isolate live bacteria remain an essential step in detailed characterization of the bacteria.

Control measures

The most effective means of reducing food and waterborne diseases is to improve the sanitation and hygienic conditions, and the main reason for disproportionate occurrence of the diseases in certain parts of the world is inadequate resources to improve these conditions. Vaccination may be an interim way of reducing disease incidences, until sanitation, safe water and food safety are ensured for all. However, there are very few available vaccines yet to combat these diseases.

There are certain measures that can directly have an effect on reducing the incidence and transmission of food and waterborne diseases. For example, pathogens that spread through animal faeces, are to be prevented through adequate steps to prevent faecal contamination of meat during slaughter and processing, and through safe disposal of animal wastes. For seafood borne infections, such as *V. parahaemolyticus* and *V. vulnificus* the post-harvest management is important to control growth of the organism in the seafood.

An active surveillance system for timely identification of outbreak cases should lead to identify the source of contamination and track the spread of disease. Monitoring the emergence of new pathogens or the new appearance of a pathogen in an area is also important to prevent the spread of the organism.

A worrying development is the emergence of increased antibiotic resistance among these pathogens, particularly in *Salmonella* infections when treatment with an appropriate antibiotic is essential. On the other hand, antibiotics are not beneficial for patients with HUS from EHEC infection, since antibiotics may induce expression and release of Stx (Walterspiel *et al.*, 1992; Zhang *et al.*, 2000). Therefore, judicious use of antibiotics is advised for maximum effectiveness and to reduce the emergence of resistance.

Evolution of waterborne and foodborne pathogens have often been driven by the need to acquire the ability to switch between the two habitats, the host and the environment. Evolutionary changes have been promoted by the close interactions with other organisms that they come in contact in these habitats, often by horizontal transfer of genes. Mobility of virulence associated genes, islands or phage genomes have been documented or proposed for a variety of food- and waterborne pathogens (Faruque *et al.*, 1998, 2003; Ochi *et al.*, 2009). Since evolution is a continuous process, to detect the emergence of new waterborne and foodborne pathogens or strains with new drug resistance phenotype, it is important to monitor these organisms both in the environment and in clinical cases.

The following chapters will present more detailed description of the most important foodborne and water borne bacterial pathogens, in terms of their epidemiology, clinical presentations, genetics and evolution. Furthermore, methods for their identification, typing and possible ways of prevention will be discussed.

References

Baldini, M.M., Nataro, J.P., and Kaper, J.B. (1986). Localization of a determinant for HEp-2 adherence by enteropathogenic *Escherichia coli*. Infect. Immun. *52*, 334–336.

Bartlett, C.L.R. (1996). An overview of emerging foodborne and waterborne diseases. Eastern Mediterranean Health J. *2*, 51–60.

Bennish, M.L. (1991). Potentially lethal complications of shigellosis. Rev. Infect Dis. *13*, S319.

Brenner, F.W., Villar, R.G., Angulo, F.J., Tauxe, R., and Swaminathan, B. (2000). *Salmonella* Nomenclature. J. Clin. Microbiol. *38*, 2465–2467.

Clarke, S.C., Haigh, R.D., Freestone, P.P., and Williams, P.H. (2003). Virulence of enteropathogenic *Escherichia coli*, a global pathogen. Clin. Microbiol. Rev. *16*, 365–378.

Coker, A.O., Isokpehi, R.D., Thomas, B.N., Amisu, K.O., and Obi, C.L. (2002). Human campylobacteriosis in developing countries. Emerg. Infect. Dis. *8*, 237–44.

Colwell, R.R., and W.M.Spira. (1992). The ecology of *Vibrio cholerae*. In Cholera, Barua, D., and Greenough, W.B., III, ed. (Plenum Medical Book Co., New York), pp. 107–127.

Colwell, R.R., West, A.P.A., Maneval, D., Remmers, E.F., Elliot, E.L., and Carlson, N.E. (1984). Ecology of pathogenic vibrios in Chesapeake Bay. In Vibrios in the Environment, Colwell, R.R., ed. (John Wiley & Sons, Inc, Hoboken, NJ), pp. 367–387.

Cornelis, G.R. (2002). *Yersinia* type III secretion: Send in the effectors. J. Cell Biol. *158*, 401–408.

Cornelis, G.R., Boland, A., Boyd, A.P., Geuijen, C., Iriarte, M., Neyt, C., Sory, M.P., and Stainier, I. (1998). The virulence plasmid of *Yersinia*, an antihost genome. Microbiol. Mol. Biol. Rev. *62*, 1315.

Cover, T.L., and Aber, R.C. (1989). *Yersinia enterocolitica*. New Engl. J. Med. *321*, 16–24.

Doyle, M.P. (1990). Pathogenic *Escherichia coli*, *Yersinia enterocolitica* and *Vibrio parahaemolyticus*. Lancet *336*, 1111–5.

Drake, S.L., DePaola, A. and Jaykus, L. (2007). An overview of *Vibrio vulnificus* and *Vibrio parahaemolyticus*. Comprehensive Rev. Food Sci. Food Safety *6*, 120–144.

Faruque, S.M., Albert, M.J., and Mekalanos, J.J. (1998). Epidemiology, genetics, and ecology of toxigenic *Vibrio cholerae*. Microbiol. Mol. Biol. Rev. *62*, 1301–1314.

Faruque, S.M., and Mekalanos, J.J. (2003). Pathogenicity islands and phages in *Vibrio cholerae* evolution. Trends. Microbiol. *11*, 505–510.

Faruque, S.M., Biswas, K., Udden, S.M., Ahmad, Q.S., Sack, D.A., Nair, G.B., and Mekalanos, J.J. (2006). Transmissibility of cholera: *in vivo*-formed biofilms and their relationship to infectivity and persistence in the environment. Proc. Natl. Acad. Sci. U.S.A. *103*, 6350–6355.

Fullerton, K.E., Ingram, L.A., Jones, T.F., Anderson, B.J., McCarthy, P.V., Hurd, S., Shiferaw, B., Vugia, D., Haubert, N., Hayes, T., *et al.* (2007). Sporadic campylobacter infection in infants: a population-based surveillance case–control study. Pediatr. Infect. Dis. J. *26*, 19–24.

Giron, J.A., Ho, A.S.Y., and Schoolnik, G K. (1991). An inducible bundle-forming pilus of enteropathogenic *Escherichia coli*. Science *254*, 710–713.

Greer, A., Ng, V., and Fisman, D. (2008). Climate change and infectious diseases in North America: the road ahead. Canadian Med. Assoc. J. *178*, 715–722.

Hales, B.A., Hart, C.A., Batt, R.M., and Saunders, J.R. (1992). The large plasmids found in enterohemorrhagic and enteropathogenic *Escherichia coli* constitute a related series of transfer-defective Inc F-IIA replicons. Plasmid *28*, 183–193.

Hansen-Wester, I., and Hensel, M. (2001). *Salmonella* pathogenicity islands encoding type III secretion systems. Microbes Infect. *3*, 549–59.

Heuck, C.J. (1998). Type III protein secretion systems in bacterial pathogens of animals and plants. Microbiol. Mol. Biol. Rev. 62:379–433.

Higgins, D.A., Pomianek, M.E., Kraml, C.M., Taylor, R.K., Semmelhack, M.F. and Bassler, B.L. (2007). The major *Vibrio cholerae* autoinducer and its role in virulence factor production. Nature *450*, 883–885

Isberg, R.R., and Barnes, P. (2001). Subversion of integrins by enteropathogenic *Yersinia*. J. Cell. Sci. *114*, 21–28.

Janda, J.M., Powers, C., Bryant, R.G., and Abbott, S.L. (1988). Current perspective on the epidemiology and pathogenesis of clinically significant *Vibrio* spp. Clin Microbiol Rev. *1*, 245–267.

Kamruzzaman, M., Udden, S.M., Cameron, D.E., Calderwood, S.B., Nair, G.B., Mekalanos, J.J., and Faruque, S.M. (2010). Quorum-regulated biofilms enhance the development of conditionally viable, environmental *Vibrio cholerae*. Proc. Natl. Acad. Sci. U.S.A. *107*, 1588–93.

Kaper, J.B., Morris, J.G., Jr, and Levine, M.M. (1995). Cholera. Clin. Microbiol. Rev. *8*, 48–86.

Levine, M.M. (1987). *Escherichia coli* that cause diarrhoea: enterotoxigenic, enteropathogenic, enteroinvasive, enterohaemorrhagic, and enteroadherent. J. Infect. Dis. *155*, 377–389.

Levine, M.M., and Edelman, R. (1984). Enteropathogenic *Escherichia coli* of classic serotypes associated with infant diarrhoea: epidemiology and pathogenesis. Epidemiol. Rev. *6*, 31–51.

McDaniel, T.K., and Kaper, J.B. (1997). A cloned pathogenicity island from enteropathogenic *Escherichia coli* confers the attaching and effacing phenotype on *E. coli* K-12. Mol. Microbiol. *23*, 399–407.

McDaniel, T.K., Jarvis, K.G., Donnenberg, M.S., and Kaper, J.B. (1995). A genetic locus of enterocyte effacement conserved among diverse enterobacterial pathogens. Proc. Natl. Acad. Sci. U.S.A. *92*, 1664–1668.

Miller, V.L., Beer, K.B., Heusipp, G., Young, B.M., and Wachtel, M.R. (2001). Identification of regions of Ail required for the invasion and serum resistance phenotypes. Mol. Microbiol. *41*, 1053–1062.

Nataro, J.P., and Kaper, J.B. (1998). Diarrheagenic *Escherichia coli*. Clin Microbiol. Rev. *11*, 142–201.

Nataro, J.P., Maher, K.O., Mackie, P., and Kaper, J.B. (1987). Characterization of plasmids encoding the adherence factor of enteropathogenic *Escherichia coli*. Infect. Immun. *55*, 2370–2377.

O'Brien, A.D., Tesh, V.L., Donohue-Rolfe, A , Jackson, M.P., Olsnes, S., Sandvig, K., Lindberg, A.A., and Keusch, G.T. (1992). Shiga toxin: biochemistry, genetics, mode of action, and role in pathogenesis. Curr. Top. Microbiol. Immunol. *180*, 65–94

Ochi, S., Shimizu, T., Ohtani, K., Ichinose, Y., Arimitsu, H., Tsukamoto, K., Kato, M., and Tsuji, T. (2009). Nucleotide sequence analysis of the enterotoxigenic *Escherichia coli* Ent plasmid. DNA Res. *16*, 299–309.

Oliver, J.D. (1995). The viable but non-culturable state in the human pathogen Vibrio vulnificus. FEMS Microbiol. Lett. *133*, 203–208.

Oliver, J.D. (2010). Recent findings on the viable but nonculturable state in pathogenic bacteria. FEMS Microbiol. Rev. *34*, 415–425.

Parvathi, A., Kumar, H.S., Karunasagar, I. and Karunasagar, I. (2004). Detection and enumeration of *Vibrio vulnificus* in oysters from two estuaries along the Southwest coast of India using molecular methods. Appl. Environ. Microbiol. *70*, 6909–6913.

Parry, C.M. (2004). Typhoid Fever. Curr. Infect. Dis. Rep. *6*, 27–33.

Popoff, M.Y., and Le Minor, L. (1997). Antigenic formulas of the *Salmonella* serovars, 7th revision. World Health Organization Collaborating Centre for Reference and Research on *Salmonella* Pasteur Institute, Paris, France.

Schlech, W.F. 3rd, Lavigne, P.M., Bortolussi, R A., Allen, A.C., Haldane, E.V., Wort, A.J., Hightower, A.W., Johnson, S.E., King, S.H., Nicholls, E.S., *et al.* (1983). Epidemic listeriosis-evidence for transmission by food. New Engl. J. Med. *308*, 203–6.

Shelobolina, E.S., Sullivan, S.A., O'Neill, K.R., Nevin, K.P., and Lovley, D.R. (2004). Isolation, characterization, and U(VI)-reducing potential of a facultatively anaerobic, acid-resistant bacterium from low-pH, Nitrate- and U(VI)-contaminated subsurface sediment and description of *Salmonella subterranea* sp. nov. Appl. Environ. Microbiol. *70*, 2959–2965.

Shimohata, T., and Takahashi, A. (2010). Diarrhea induced by infection of *Vibrio parahaemolyticus*. J. Med. Invest. *57*, 179–182.

So, M., and McCarthy, B.J. (1980). Nucleotide sequence of transposon Tn 1681 encoding a heat-stable toxin (ST) and its identification in enterotoxigenic *Escherichia coli* strains. Proc. Natl. Acad. Sci. U.S.A. *77*, 4011–4015.

Tacket, C.O., Brenner, F., and Blake, P.A. (1984). Clinical features and an epidemiological study of *Vibrio vulnificus* infections. J Infect Dis. *149*, 558–561.

Vandamme, P., Falsen, E., Rossau, R., Hoste, B., Segers, P., Tytgat, R., and De Ley, J.(1991). Revision of Campylobacter, Helicobacter, and Wolinella taxonomy: emendation of generic descriptions and proposal of Arcobacter gen. nov. Int. J. Syst. Bacteriol. *41*, 88–103.

Waldor, M.K., and Mekalanos, J.J. (1996). Lysogenic conversion by a filamentous bacteriophage encoding cholera toxin. Science. *272*, 1910–1914.

Walterspiel, J.N., Ashkenazi, S., Morrow, A.L., and Cleary, T.G. (1992). Effect of subinhibitory concentrations of antibiotics on extracellular Shiga-like toxin I. Infection *20*, 25–29.

West, P.A. (1989). The human pathogenic vibrios-A public health update with environmental perspectives. Epidemiol. Inf. *103*, 1–34.

Wright, J., Fenwick, S., and McCarthy, M. (1995). Yersiniosis: an emerging problem in New Zealand. New Zealand Public Health Report 2, 65–6.

Xu, H.S., Roberts, N., Singleton, F.L., Attwell, R.W., Grimes, D., and Colwel, R.R. (1982) Survival and viability of nonculturable *Escherichia coli* and *Vibrio cholerae* in the estuarine and marine environment. Microb. Ecol. *8*, 313–323.

Yang, Y.X., and Isberg, R.R. (1997). Transcriptional regulation of the Yersinia pseudotuberculosis pH 6 antigen adhesin by two envelope-associated components. Mol. Microbiol. *24*, 499–510.

Zhang, X., McDaniel, A.D., Wolfe, L.E., Keusch, G.T, Waldor, M.K., and Acheson, D.W. (2000). Quinolone antibiotics induce Shiga toxin-encoding bacteriophages, toxin production, and death in mice. J. Infect. Dis. *181*, 664–670.

Epidemiology, Pathogenesis and Genetics of Diarrhoeagenic *Escherichia coli* Infections

2

Thandavarayan Ramamurthy and M. John Albert

Abstract

There are five categories of diarrhoeagenic *Escherichia coli* (DEC), namely enterotoxigenic, enteropathogenic, enterohaemorrhagic, enteroinvasive and enteroaggregative. They have evolved from non-pathogenic commensal strains by acquisition of specific virulence genes through mobile genetic elements. Their pathogenesis differs and they produce distinct clinical syndromes and pathological lesions and have different epidemiological characteristics. The virulence genes are carried on plasmids, bacteriophages, transposons or pathogenicity islands. DEC produce an array of virulence factors which include colonization factors, enterotoxins, cytotoxins, haemolysins, invasins etc. The diseases they produce range from acute watery diarrhoea to dysentery to bloody diarrhoea with haemolytic uraemic syndrome. Even though the major burden of the disease is in the developing world, no part of the world is free from them, and EHEC infections are predominant in developed countries. A variety of molecular tools have been developed to study the diversity and transmission of these pathogens. Even though attempts are being made, no ideal vaccine exists against any category of DEC, therefore maintaining appropriate food and water hygiene is the only way to keep the infections under control. We must also be on guard against the emergence of new pathogenic strains. The recent emergence of a hybrid enteroaggregative-haemorrhagic *E. coli* with the rare serotype of O104:H4 in Germany that caused high mortality rates is a case in point.

Introduction

Escherichia coli are normal flora of the large intestine. However, owing to the acquisition of specific virulence factors from other bacteria, specific pathogenic *E. coli* have evolved from the normal flora. These pathogenic *E. coli* cause diseases in the intestinal tract, urinary tract or the nervous system. The primary *E. coli* pathogens of the intestinal tract causing diarrhoeal diseases (also called pathovars or pathotypes) are divided into five categories: enteropathogenic *E. coli* (EPEC), enterotoxigenic *E. coli* (ETEC), enteroinvasive *E. coli* (EIEC), enterohaemorrhagic *E. coli* (EHEC) and enteroaggregative *E. coli* (EAEC). Although, there is a sixth category of *E. coli*, known as diffusively adherent *E. coli* (DAEC), since its pathogenicity is uncertain, it will not be discussed in this chapter. Different categories of diarrhoeagenic *E. coli* (DEC) have distinct characteristics with respect to virulence factors, clinical syndrome and epidemiological properties (Table 2.1). The route of transmission of these enteric pathogens is faecal–oral route, with food and water being the main vehicles of transmission (Hunter, 2003).

Enteropathogenic *E. coli* (EPEC)

EPEC was the first pathotype to be described when it caused outbreaks of diarrhoea in children in the United Kingdom in the early part of the twentieth century. They were recognized by isolation of serologically distinct strains from patients, and not from controls. Although the pathogenic mechanism of EPEC remained unknown for

Table 2.1 Clinical, pathological and epidemiological characteristics of disease caused by diarrhoeagenic *E. coli*

E. coli pathotype	Clinical syndrome	Intestinal pathology	Population at risk
ETEC	Watery diarrhoea	No notable pathology	Children in developing countries and travellers to these countries
EPEC	Non-specific diarrhoea	Attaching and effacing lesion in the small intestine	Children under 2 years of age in developing countries
EIEC	Dysentery	Inflammation of the large intestinal mucosa	Affect all age groups worldwide, but more common in developing countries
EHEC	Bloody diarrhoea and haemolytic uraemic syndrome (HUS)	Haemorrhagic colitis and attaching and effacing lesions in the large intestinal mucosa	Children and the elderly in developed countries
EAEC	Persistent diarrhoea	Thick mucus-containing biofilm encrusted with bacteria; cytotoxic change in enterocytes	Children in developing countries and travellers to these countries

ETEC, enterotoxigenic *E. coli*; EPEC, enteropathogenic *E. coli*; EIEC, enteroinvasive *E. coli*; EHEC, enterohaemorrhagic *E. coli*; EAEC, enteroaggregative *E. coli*.

several decades, EPEC strains were identified by O:H serotyping (Table 2.2). Because of numerous advancements made in the studies on the pathogenesis of EPEC in recent times, this pathotype is one of the best characterized of all pathogenic *E. coli*.

Clinical manifestations

The most common findings of EPEC infection are fever, diarrhoea and respiratory symptoms. Dehydration may be seen in infants under 1 year of age, and hypernatraemia and metabolic acidosis can occur (Kessner *et al.*, 1964). Illness lasts for about a week, yet protracted illness occurs with fatalities (Drucker *et al.*, 1970). Even though EPEC are not the agents of inflammatory diarrhoea, faecal leucocytes are seen occasionally in patients and faecal lactoferrin assay for inflammatory diarrhoea is more frequently positive in EPEC diarrhoea pointing to some degree of intestinal inflammation (Miller *et al.*, 1994).

Pathogenesis

The target site of EPEC pathogenesis is the small intestine. The virulence genes of EPEC are distributed in the chromosome and in a plasmid of 70–100 kb called the EAF (EPEC adherence factor) plasmid (Baldini *et al.*, 1986). Typical EPEC isolates only possess this plasmid. EAF plasmid encodes a type IV pilus called the bundle-forming pilus (BFP) (Giron *et al.*, 1991), which mediates interbacterial adherence and adherence to epithelial cells. This is manifested as localized adherence of EPEC as microcolonies on the intestine *in vivo* and on mammalian cells such as HeLa and HEp-2 cells *in vitro*. The plasmid also contains the *per* (plasmid-encoded regulator) locus, whose products regulate the *bfp* operon and most of the chromosomal genes in the locus for enterocyte effacement (LEE) found on a 35-kb pathogenicity island (PAI) in the chromosome (see below). The regulation is achieved through the LEE-encoded regulator (Ler) (Mellies *et al.*1999) and LuxS quorum sensing system (Sperandio *et al.*, 1999).

A characteristic intestinal pathology produced by EPEC infection is 'attaching and effacing' (A/E) lesion. After effacement of the intestinal microvilli, the bacteria intimately attach to the intestinal epithelial cells. The cytoplasmic membrane is drawn up like pedestals on which the bacteria sit. There is accumulation of polymerized actin directly beneath the adherent bacteria. The ability to induce this A/E histopathology is encoded by genes on the 35-kb LEE PAI mentioned above. Most of the 41 open reading frames (ORFs) of the LEE PAI encode a type III secretion system (T3SS) and the associated chaperons and effector proteins

Table 2.2 Important serotypes of diarrhoeagenic *E. coli*

E. coli pathotype	O : H serotype
EPEC	O18ac:H7; O20:H26/34; O25:H1; O26:H11/$^-$; O44:H34; O55;H6/7/$^-$;O86:H27/34/$^-$; O91:H7/$^-$; O111:H2/12/$^-$; O114:H10/32; O119:H6/$^-$; O125:H21/$^-$; O126:H2/$^-$; O127:H9/21/40/$^-$; O128:H2/7/8/12/$^-$; O142:H6; O158:H23
ETEC	O6:H16/$^-$; O8:H9/$^-$;O11:H27; O15:H11/$^-$; O20:H$^-$; O25:H42/$^-$; O27:H7/20; O63:H12/$^-$; O78:H11/12/20; O85:H7; O115:H11/40; O128:H7/21; O148:H28; O149:H10; O159: H20; O167:H5; O173:H$^-$
EHEC	O4:H$^-$; O5:H; O16:H6; O26:H11/21/32; O46:H31; O48:H21; O55:H7; O91:H10/21; O98:/H$^-$; O104:H4; O111:H2/8/$^-$; O113:H21; O117:H14; O118:H12; O119:H6; O125:H–; O126:H8; O126:H2; O145:H$^-$; O157:H7/$^-$; O172:H$^-$
EIEC	O28:H$^-$; O112:H$^-$; O115:H$^-$;O124:H30/32/$^-$; O136:H$^-$; O143:H$^-$; O144:H$^-$; O147:H$^-$; O152:H$^-$; O159:H$^-$; O164:H$^-$; O167:H4/5/$^-$
EAEC	O3:H2; O7:H$^-$; O15:H18; O44:H18; O77:H18; O86:H$^-$; O111:H21; O126:H27; O127:H2

EPEC, enteropathogenic *E. coli*; ETEC, enterotoxigenic *E. coli*; EHEC, enterohaemorrhagic *E. coli*; EIEC, enteroinvasive *E. coli*; EAEC, enteroaggregative *E. coli*.
Multiple H types found in certain serogroups are indicated. For example, H9/21/40/$^-$ indicates that H9,H21,H40 and H$^-$ (non-motile) strains have been identified in the indicated O group.

(EspA, EspB, EspD, EspF, EspH, Map, Tir). The effector proteins are injected into the host cell by T3SS. Many of the effector proteins have multiple functions and they subvert host cell processes. EspA forms a needle-like structure for injecting effector proteins into host cells. EspB and EspD form pores on the surface of host cells through which effector proteins are injected. Translocated intimin receptor (Tir) acts as the receptor for intimin, an outer membrane protein of EPEC encoded by the LEE, which mediates intimate attachment of EPEC to epithelial cells. Tir is phosphorylated by host tyrosine kinases. Phosphorylation of Tir recruits host cell adaptor protein, Nck, which activates neural Wiskott–Aldrich syndrome protein (N-WASP) and the actin-related protein 2/3 (ARP2/3) complex to mediate actin nucleation and pedestal formation. Mitochondrial-associated protein (Map) affects mitochondrial function and promotes filopodia (actin protrusion from cell surface) formation. EspF inhibits phagocytosis and disrupts tight junctions between enterocytes. EspB, a T3SS translocation protein, is also an inhibitor of phagocytosis. In addition, EspH also contributes to filopodia and pedestal structure formation. Non-LEE encoded (Nle) proteins (EspI, EspJ, Cif) also have roles in EPEC virulence. For example, EspI disrupts tight junctions, EspJ inhibits opsonophagocytosis by red blood cells, and cycle-inhibiting factor (Cif) induces apoptosis. There are several downstream effects of effector proteins on the cellular processes. Nuclear factor kappa B (NF-κB) is activated leading to the production of IL-8 and an inflammatory response.

Diarrhoea in EPEC infections probably results from multiple mechanisms that include active ion secretion, increased intestinal permeability, intestinal inflammation and loss of absorptive surface area resulting from microvillus effacement. The roles of effector proteins secreted via T3SS and their contribution to diarrhoea is somewhat known. For example, EspF decreases the activity of Na^+/H^+ exchanger 3 (NHE3), a mediator of Na^+ absorption (Hodges *et al.*, 2008); EspG-induced alterations of microtubules affect Cl^-/OH^-exchanger activity (Gill *et al.*, 2007). Moreover, both EspF and EspG mislocalize aquaporins (Guttman *et al.*, 2007), the channels responsible for rapid water transport, and inhibit the sodium–D-glucose cotransporter (SGLT1) (Dean *et al.*, 2006) and the intestinal serotonin transporter (SERT) (Esmaili *et al.*, 2009). The major water pump, SGLT1 is inhibited by a combination of Map, Tir and EspF. SERT is involved in the uptake of serotonin hormone that is involved in the intestinal transport of electrolytes and fluids. In addition, connexin 43 (Cx43) hemichannels are upregulated on the apical surface of the epithelium in response to bacterial infection, resulting in increased luminal water content. Collectively, the disruptions caused by these proteins in water and ion balance contribute to diarrhoea. A summary of the mechanisms is shown in Fig. 2.1.

Atypical EPEC

Typical EPEC are *E. coli* strains that are A/E positive, possess EAF plasmid and hence produce BFP, and negative for Shiga toxins (Stxs) (Kaper, 1996). On the other hand, atypical EPEC are positive for A/E lesion, but negative for both BFP (owing to absence of EAF plasmid) and Stxs (Kaper, 1996). Some studies have shown association of atypical EPEC with acute diarrhoea (Cohen *et al.*, 2005; Estrada-Garcia *et al.*, 2009), while others have not (Albert *et al.*, 1999; Nunes *et al.*, 2003). In developed countries, an association has been shown with persistent diarrhoea in children (Afset *et al.*, 2004; Nguyen *et al.*, 2006). Atypical EPEC have also been involved in some diarrhoeal outbreaks (Viljanen *et al.*, 1990; Hedberg *et al.*, 1997). The majority of atypical EPEC belongs to non-traditional EPEC serogroups, some to traditional EPEC serogroups and many are non-typeable. All outbreak isolates carried EAEC heat-stable enterotoxin1 (East-1) encoding gene (*astA*). Other studies have shown association with STEC (Shiga toxin producing *E. coli*) linked genes: OI-122 virulence genes (*efa1*/*lifA*, *nleB*, *nleE* and *ent*), *paa* (49.6%), *iha* (25.6%), *ehx*, *toxB*, *kat* and *espp* (6.0% each), and *ldaG* (2.6%). OI-122 is O island 122, a pathogenicity island. Atypical EPEC are present in domestic animals, and cross-infection to humans can occur (Hernandes *et al.*, 2009).

When atypical EPEC adhere to HEp-2 cells, they exhibit different patterns of adherence – localized – like diffuse or aggregative (Trabulsi *et al.*, 2002; Robins-Browne *et al.*, 2004). Since

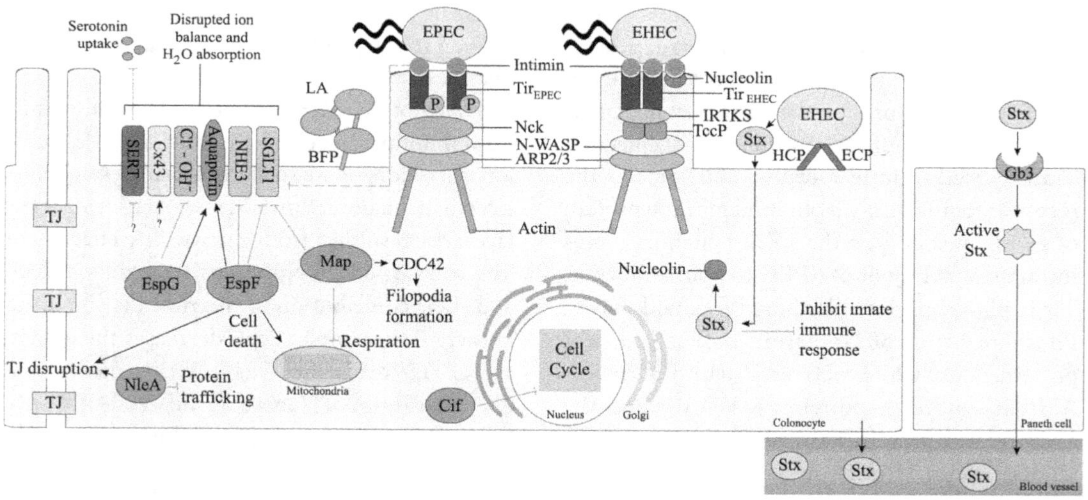

Figure 2.1 Pathogenic mechanisms of EPEC and EHEC. Both EPEC and EHEC are attaching and effacing pathogens, the former causing the lesion in the small intestine, and the latter in the colon. Effector proteins secreted by the type III secretion system can affect Cl^--OH^- and Na^+-H^+ exchanger (NHE3) activity, mislocalize aquaporins and inhibit sodium-D-glucose cotransporter 1 (SGLT1). Intimate attachment is mediated by the interaction between intimin and the translocated intimin receptor (Tir). There are slight differences in the mechanisms of pedestal formation between EPEC and EHEC. In EPEC, Tir is phosphorylated by host tyrosine kinases, and phosphorylated Tir recruits Nck, a host adaptor protein. Whereas in EHEC, Tir is not phosphorylated and pedestal formation is Nck-independent and actin rearrangements are mediated by Tir cytoskeleton-coupling protein (TccP). For full details, see the main text. Stx, Shiga toxin; Gb3, globotriaosylceramide receptor; HCP, haemorrhagic coli pilus; ECP, *E. coli* common pilus; IRTKS, insulin receptor tyrosine kinase substrate; Cif, cycle-inhibiting factor; N-WASP, neutral Wiskott–Aldrich syndrome protein; ARP2/3, actin-related protein 2/3; CDC42, cell division control protein 42; Map, mitochondrial-associated protein; LA, localized adherence; BFP, bundle-forming pilus; NleA, non-LEE-encoded effector A; Tj, tight junction; Cx43, connexin 43; SERT, serotonin transporter (adapted from Croxen and Finlay, 2010).

atypical EPEC lacks BFP, the search is on to identify alternative adherence factors. In one atypical EPEC strain, E128012 (serotype O114:H2) that causes diarrhoea, it was found that *pst-phoU*, a global regulatory network concerned with phosphate metabolism (Cheng *et al.*, 2009), contributes to virulence by causing increased expression of an unidentified adhesion.

Epidemiology of EPEC infection

Age distribution

EPEC infection is primarily a disease of infants younger than 2 years of age. In case–control studies, there is a strong correlation of isolation of EPEC from infants with diarrhoea compared with healthy controls. This correlation is strongest in infants under 6 months of age. In children older than 2 years, EPEC can be isolated with similar frequencies from diarrhoeal children as well as from control children, thereby showing no association with diarrhoea (Levine and Edelman, 1984). Even though diarrhoea can be demonstrated in adult volunteers after neutralization of gastric acidity with a heavy inoculum (Levine *et al.*, 1978), and outbreaks of diarrhoea in adults due to EPEC isolates have been reported (probably mediated by a heavy inoculum from a common source) (Costin *et al.*, 1964; Vijanen *et al.*, 1990), the resistance of older individuals to EPEC infection is probably due to loss of specific intestinal receptor.

Reservoirs and transmission

The reservoirs of EPEC infection is thought to be symptomatic and asymptomatic children, asymptomatic adult carriers, and mothers and persons handling infants (Levine and Edelman, 1984). EPEC infection spreads through hospital nurseries and day care centres from an index case (Bower *et al.*, 1989; Wu and Peng, 1992). As with other diarrhoeal pathogens, transmission of EPEC is faecal–oral with contaminated hands, infant foods or fomites serving as vehicles (Levine and Edelman, 1984). There is also potential for airborne transmission of EPEC as it has been isolated from dust and aerosols (Rogers, 1951). Waterborne (Yatsuyanga *et al.*, 2003) and foodborne (Hedberg *et al.*, 1997) outbreaks of EPEC infection have been reported in school children and adults. Epidemiological studies in different countries have shown that EPEC can be isolated from up to 18% of children with diarrhoea and from up to 10% of children without diarrhoea (Ochoa *et al.*, 2008). Symptomatic children excrete EPEC for up to 2 weeks after cessation of symptoms (Hill *et al.*, 1991).

EPEC in developed countries

EPEC caused explosive outbreaks of infantile diarrhoea in developed countries in the early part of the nineteenth century. They are no longer an important cause of diarrhoea in developed countries. However, they can still cause outbreaks in day care centres (Paulozzi *et al.*, 1986; Bower *et al.*, 1989) and children's wards (Rothbaum *et al.*, 1982). Moreover, in a study conducted in children with diarrhoea in Seattle, Washington, USA, *eae*-positive but EAF- and *stx*-negative isolates (atypical EPEC) were found in 5.6% of patients (Bokete *et al.*, 1997). This suggests that the current importance of EPEC infection in the US may be underestimated and further studies are required to assess its true contribution to diarrhoea.

EPEC in developing countries

EPEC is a major cause of infantile diarrhoea in developing countries. In many case–control studies, EPEC isolates have been found significantly more from children with diarrhoea than from controls (reviewed by Ochoa *et al.*, 2008). In some studies, 30–40% of infantile diarrhoea could be attributed to EPEC strains (Cravioto *et al.*, 1988; Gomes *et al.*, 1989). EPEC infections are present round the year with no seasonality (Albert *et al.*, 1999).

Molecular typing

Comparison of the amino acid sequences of the outer membrane virulence associated protein intimin from different EPEC strains have revealed that the N-terminal regions are highly conserved, while the C-terminal end shows variability (Frankel *et al.*, 1994). The cell binding activity is localized to the C-terminal 280 amino acids which determines tissue tropism (Frankel *et al.*, 1998). Up to 16 different intimin subtypes have been described (Lacher *et al.*, 2006). The nomenclature used to identify intimin subtypes is based on letters of the Greek alphabet. Of all the intimin subtypes, β, θ and ζ intimin subtypes seem to be most widely distributed subtypes (Jenkins *et al.*, 2003; Ramachandran *et al.*, 2003; Nguyen *et al.*, 2006; Albert *et al.*, 2009). The intimin genes are amplified by a set of universal primers and the products differentiated by agarose gel electrophoresis after digestion with a set of restriction enzymes (Ramachandran *et al.*, 2003). Further allelic subtyping of intimin locus (28 alleles are known) is possible by fluorescent restriction fragment length polymorphism (fRFLP) (Lacher *et al.*, 2006). Other methods used for molecular typing include ribotyping (Dalla-Costa *et al.*, 1998), pulsed-filed gel electrophoresis (Pitondo-Silva *et al.*, 2009), enterobacterial repetitive intergenic consensus-PCR (ERIC-PCR) (Rodrigues *et al.*, 2004), multilocus enzyme electrophoresis (MEE) (Dalla-Costa *et al.*, 1998) and multilocus sequence typing (MLST) (Pitondo-Silva *et al.*, 2009). By MLST, strains can be classified into certain clonal complexes, but no relation with serotypes and intimin subtypes is found (Pitondo-Silva *et al.*, 2009). On the other hand, good correlation has been found among ribotyping, MEE and ERIC-PCR (Dalla-Costa *et al.*, 1998).

Diagnosis

The important diagnostic tests for DEC are summarized in Table 2.3. There are serological tests, and phenotypic and genotypic tests for diagnosis of EPEC infection. Many routine laboratories still practise serogrouping for identification of classical serotypes of EPEC (Table 2.2). Previously, research studies used localized adherence assay for BFP and fluorescent actin staining assay for A/E lesion using HEp-2 or HeLa cells for diagnosis (Knutton *et al.*, 1991; Rosa *et al.*, 1998). Previous genotypic tests included DNA probe assays using EAF probe or *bfp* sequence and *eae* gene sequence (Albert *et al.*, 1995; Knutton *et al.*, 2001). These assays have been superseded by multiplex PCR assays (Aranda *et al.*, 2004; Vidal *et al.*, 2005). Typical EPEC is diagnosed by the presence of both *bfp* and *eae* genes and atypical EPEC by the presence of *eae* gene sequence and absence of *bfp* and *stx* genes. Absence of *stx* gene in an *eae* gene positive strain is to rule out the derivation of the strain from an EHEC strain (Muller *et al.*, 2006).

Prevention and treatment

Standard public health and infection control measures are important in interrupting transmission of the organism. These include food and water hygiene, personal hygienic measures such as hand washing and prevention of exposure to equipment, bedding or solutions. There is no

Table 2.3 Summary of tests employed for laboratory detection of diarrhoeagenic *E. coli*

E. coli pathotype	Phenotypic test	Genotypic test
ETEC	LT or ST detection by ELISA	Genes corresponding to phenotypic tests detected by DNA hybridization or PCR
EPEC	Localized adherence to tissue culture cells (for Bfp);fluorescent actin staining (for LEE pathogenicity island)	Genes corresponding to the phenotypic tests detected by DNA hybridization or PCR
EIEC	Sereny test; invasion of tissue culture cells	Genes associated with invasion (*ial*, *ipaH*) detected by DNA hybridization or PCR
EHEC	Production of Shiga toxin; production of EHEC haemolysin, presence of LEE	Genes corresponding to phenotypic test detected by DNA hybridization or PCR
EAEC	Aggregative adherence to tissue culture cells	Gene encoding adhesive fimbriae, global regulator of virulence or cytotoxin detected by DNA hybridization or PCR

ETEC, enterotoxigenic *E. coli*; EPEC, enteropathogenic *E. coli*; EIEC, enteroinvasive *E. coli*; EHEC, enterohaemorrhagic *E. coli*; EAEC, enteroaggregative *E. coli*; Bfp, bundle- forming pilus; LEE, locus for enterocyte effacement; LT, heat-labile enterotoxin; ST, heat-stable enterotoxin.

vaccine to prevent EPEC diarrhoea. In case of dehydration as with any diarrhoeal disease, it should be treated appropriately with parenteral or oral rehydration. Antibiotics have been found to be useful for treatment, but antibiotic therapy is complicated by resistance in EPEC (Donnenberg, 1995). Oral non-absorbable antibiotics, such as neomycin, gentamicin and trimethoprim-sulphamethoxazole have been proven to be effective in reducing the severity of diarrhoea due to EPEC (Nelson, 1971; Thoren *et al.*, 1980). In addition, prolonged diarrhoea due to EPEC is treated with appropriate nutritional supplements.

Enterohaemorrhagic *Escherichia coli* (EHEC)

Enterohaemorrhagic *E. coli* (EHEC) or Shiga toxin-producing *E. coli* (STEC) was first recognized when it caused large outbreaks of bloody diarrhoea (haemorrhagic colitis, HC) in the US in 1982. This was associated with ingestion of undercooked hamburgers at a fast food hamburger chain. Its initial recognition was by its rare serotype, O157:H7 (Riley *et al.*, 1983). At the same time, another report described an association of sporadic case of haemolytic uraemic syndrome (HUS) with enteric infection by an *E. coli* producing cytotoxin on Vero cells (Karmali *et al* 1983). Subsequently, it was found that Shiga toxin [Stx or verocytotoxin (VT)] production is the key virulence factor of this pathotype. Many studies have since shown that EHEC encompasses a number other serotypes (Table 2.2). Another important EHEC serotype that emerged recently is O104:H4 that caused an extensive outbreak of HC and HUS in Germany and other European countries and was linked to consumption of contaminated vegetable (Buchholz *et al.*, 2011).

Clinical manifestations

EHEC causes asymptomatic infection and diarrhoea. Some cases of diarrhoea may be non-bloody. Bloody diarrhoea (HC) may be preceded by non-bloody diarrhoea. Usually, there is no fever, but a peripheral white blood cell count of more than 10,000/µl, and abdominal tenderness are present (Slutsker *et al.*, 1997; Tarr *et al.*, 2005). HUS is a major systemic complication of EHEC infection which develops in 6–9% of cases about 5–10 days after the onset of diarrhoea (Tarr *et al.*, 2005). HUS is characterized by a triad of acute renal failure, microangiopathic haemolytic anaemia and thrombocytopenia. Some patients may also develop the related disorder, thrombotic thrombocytopenic purpura (TTP), characterized by fever and neurological symptoms (Su and Brandt, 1995). In the EHEC outbreak due to serotype O104:H4 in Germany, a higher proportion of individuals developed HC and HUS than disease caused by non-O157:H7 serotypes suggesting that this serotype is highly virulent (Frank *et al.*, 2011).

Pathogenesis

The key virulence factor of EHEC is the phage-encoded Stx. The name of the toxin is derived from the toxin produced by *Shigella dysenteriae* type 1 to which it is closely related. There are two subgroups of Stx produced by EHEC, Stx1 and Stx2, which can be found in different combinations in EHEC isolates. Stx2 has different subtypes as well. Stxs are responsible for the vascular damage (HC) and systemic effects of infection such as HUS. Stxs have an A:B5 structural motif. A subunit has two domains – A1 and A2 – and the A1 domain is enzymatically active. The B subunits form a pentamer that binds the toxin to its receptor on eukaryotic cells, the glycolipid globotriaosylceramide (Gb3). The bound toxin is internalized through clathrin-coated pits. It is then processed by retrograde transport through the Golgi apparatus and endoplasmic reticulum to release the enzymatically active A1 domain in the cytosol. It cleaves a specific adenine residue from the 28S rRNA in the 60S ribosomal subunit, arresting protein synthesis and causing cell death. Stxs enter the systemic circulation and are bound to polymorphonuclear leucocytes (Te Loo *et al.*, 2001). In the systemic circulation, Stxs target endothelial cells, causing vascular damage and bloody diarrhoea, and in some patients, a prothrombotic state that precedes HUS (Chandler *et al.*, 2002). Expression of toxin receptors on endothelial cells is upregulated by tumour necrosis factor-alpha (TNF-α) and interleukin (IL) 1b. These cytokines are in turn released by the interaction of Stxs with macrophages and monocytes (Sakiri *et al.*, 1998).

EHEC O157:H7 isolates and some other EHEC strains possess a 92-kb virulence plasmid called pO157. It encodes a haemolysin, a bifunctional catalase–peroxidase, a type II secretion system, and a secreted serine protease, EspP, a member of the autotransporter family that cleaves human coagulation factor V (Brunder *et al.*, 1997). It also encodes the StcE protein that cleaves the C1 esterase inhibitor (C1–1NH) of the complement pathway and could potentially contribute to the tissue damage, intestinal oedema and thrombotic abnormalities seen in EHEC infections (Lathem *et al.*, 2002). However, the role of this virulence plasmid in EHEC pathogenesis is not clear, since several EHEC isolates do not possess this plasmid and yet cause disease. Perhaps isolates which possess this plasmid may have enhanced virulence.

A small dose of about 10–100 organisms is sufficient to initiate infection (Schmid-Hempel and Frank, 2007). Being a low-dose pathogen, EHEC isolates are highly acid resistant and able to withstand the high acidity of stomach and the weak acidity of the intestinal tract. Several acid-resistant mechanisms including the alternative sigma factor, RpoS may contribute to this acid resistance. Once acid-resistance is induced, it remains active for prolonged periods of cold storage at 4°C (Lin *et al.*, 1996). After ingestion, EHEC isolates predominantly colonize the terminal ileum and the colon. The initial attachment of EHEC to intestinal epithelial cells is not well defined. A type IV pilus, called the haemorrhagic coli pilus is involved in adherence and biofilm formation (Xicohtencatl-Cortes *et al.*, 2009). Flagella and the *E. coli* common pilus might also be involved in the attachment to the host cells (Erdem *et al.*, 2007; Rendon *et al.*, 2007). The EHEC genome has the same LEE PAI as the EPEC genome, and secretes the same effector products as the EPEC via the type III secretory apparatus producing similar A/E lesions on the epithelial cells. As with EPEC, intimate attachment of EHEC to host cells occurs through interaction between intimin and Tir (Fig. 2.1). In addition to Tir, intimin recognizes the eukaryotic protein, nucleolin as a receptor on the epithelial cell surface, the expression of which is increased by Stx (Robinson *et al.*, 2006). However, the mechanism of pedestal formation by EHEC is slightly different from that of EPEC in that Tir is not tyrosine phosphorylated by the host cells (DeVinney *et al.*, 1999) and pedestal formation is Nck independent (Campellone and Leong, 2003). Subversion of the host cell actin cytoskeleton is mediated by a homologue of EspF called Tir cytoskeleton-coupling protein (Tccp synonymous with $EspF_u$) (Campellone *et al.*, 2004; Garmendia *et al.*, 2004). This protein is linked to Tir by a host protein, insulin receptor tyrosine kinase substrate (IRTKS) (Vingadassalom *et al.*, 2009). TccP interacts with N-WASP to activate ARP2/3 complex-mediated actin assembly (Cheng *et al.*, 2008). As in the EPEC, expression of proteins in the LEE is dependent upon the Ler regulatory protein (Mellies *et al.*, 1999) and the LuxS quorum sensing system (Sperandio *et al.*, 1999). In spite of the low inoculum infection of EHEC, an autoinducing signal produced by non-pathogenic *E. coli* in the lower intestinal tract can trigger expression of genes in the LEE.

The above mechanisms of pedestal formation are representative of prototypical EPEC and EHEC isolates. Lineage 2 EPEC isolates and non-O157:H7 EHEC isolates use a combination of Nck-dependent and Nck-independent mechanisms. The contribution of the A/E lesion to EHEC pathogenesis not very clear since not all EHEC isolates possess the pathogenicity island or produce the A/E lesion. It is likely that there are other mechanisms that allow EHEC to interact with the intestinal epithelium and initiate the disease.

Epidemiology of EHEC infection

Incidence and prevalence

EHEC infection can occur as outbreaks or as sporadic cases. In the FoodNet surveys of laboratory-confirmed cases of acute foodborne illnesses in many US states, the overall frequency of *E. coli* O157:H7 infections fell from 2.3 per 100,000 in 1996–1998 to 1.6 per 100,000 in 2009 (CDC, 2010). In a review of stool cultures performed in several hospitals in the United States in the early 1990s, a bacterial pathogen was identified in 5.6% cases, and *E. coli* O157: H7 was isolated in 0.4% of the stool samples (Slutsker *et al.*, 1997).

Developed versus developing countries

EHEC infections are a major problem of developed countries of northern hemisphere such as North America, Europe and Japan, although some middle-income countries of the southern hemisphere such as Argentina (Rivas *et al.*, 2008) and Chile (Prado *et al.*, 1997) have reported this pathogen. Many studies in developing countries have shown that EHEC infections are not important in these countries (Cravioto *et al.*, 1990; Rosa *et al.*, 1998; Albert *et al.*, 1999; Nguyen *et al.*, 2006). However, survey of cattle in developing countries has shown that they do carry Stx-producing *E. coli* in their intestine (Wani *et al.*, 2003; Vu-Khac *et al.*, 2008). The reason for the low level or lack of EHEC infection in developing countries is rather intriguing. In developing countries, EPEC infections are endemic and EPEC and EHEC share common antigens encoded by the LEE pathogenicity island and other O islands. It is postulated that EPEC infection might immunize individuals in developing countries against EHEC infection. Such a possibility has been shown in a mouse model of infection (Toledo *et al.*, 2011).

Age distribution

While EHEC infection can affect any age group, it is most common in children. HUS following EHEC infection characteristically occurs in children and the elderly.

Reservoirs and transmission

The source of human infection is animals. Stx-producing *E. coli* can be found in the faecal flora of a variety of animals including cattle, sheep, goats, cats, dogs, pigs, chickens and gulls. However, the majority of isolates harboured by animals are of serotypes other than O157:H7 and their pathogenicity is doubtful. Cattle are the most important species for human infection, and 10–25% of these animals carry Stx-producing *E. coli*. A variety of retail meats sold in grocery stores are also contaminated with Stx-producing *E. coli* (reviewed by Nataro and Kaper, 1998). EHEC can be transmitted by food and water and from person to person. The majority of cases are caused by contaminated meat, especially undercooked hamburger meat. Other food items linked to outbreaks are lettuce, radish sprouts, alfalfa sprouts, mayonnaise, unpasteurized apple juice, raw milk and fermented salami (reviewed by Nataro and Kaper, 1998). In the May 2011 outbreak of EHEC infection due to O104:H4 serotype in Germany, the outbreak was traced to contaminated bean sprout (Struelens *et al.*, 2011). A number of *E. coli* O157:H7 outbreaks have been associated with exposure to animals in county fairs, farms, and petting zoos, in association with poor hand hygiene (Steinmuller *et al.*, 2006).

Also, outbreaks of EHEC infection have been linked to recreational water and drinking water with the affected people developing bloody diarrhoea, HUS or both. The first outbreak associated with drinking water was in a township in Missouri, USA, possibly contaminated with sewage outflow (Geldriech *et al.*, 1992). The outbreaks in Scotland (Dev *et al.*, 1991) and South Africa (Isaacson *et al.*, 1993) have been linked to drinking water contamination possibly by cattle dung. An international outbreak affecting holiday makers while returning from Canary Islands was linked to contamination of well water (Pebody *et al.*, 1999). A faecally contaminated municipal water supply possibly from deer or elk affected a large number of people in Wyoming, USA (Olson *et al.*, 2002). The largest outbreak affecting more than 1000 people occurred in Ontario, Canada, and was due to contamination of drinking water from wells possibly by cattle faeces (Anon, 2000). There have been outbreaks linked to swimming pool (Friedman *et al.*, 1999), paddling pool (Brewster *et al.*, 1994) and natural lake (Keene *et al.*, 1994). The general assumption is that outbreaks have generally followed faecal accidents from bathers.

Non-O157:H7 serotypes

Most outbreaks of EHEC infection have been caused by O157:H7 isolates, suggesting that this serotype may be more virulent and more transmissible than other serotypes. Nevertheless, other serotypes of Stx-producing *E. coli* have been implicated in sporadic cases and outbreaks, and the incidence of disease due to other serotypes is on the rise (Johnson *et al.*, 1996). More than 200 serotypes of *E. coli* can express Stx, and within these serotypes both Stx-positive and Stx-negative isolates can be found (Johnson *et al.*, 1996). Some important EHEC serotypes are listed in Table 2.2. The significance of isolating a non-O157:H7 Stx-producing *E. coli* from human stool is not as certain as that of isolating an O157:H7 isolate. Additional virulence factors such as A/E phenotype and the pO157 plasmid expressing enterohaemolysin may be able to distinguish pathogenic from non-pathogenic EHEC isolates. However, many non-O157:H7 isolates that are clearly associated with outbreaks or sporadic human disease do not possess either *eae* gene or the pO157 plasmid (Willshaw *et al.*, 1992; Bokete *et al.*, 1993). This suggests that there must be additional, as yet unknown, virulence factors that distinguish pathogenic from non-pathogenic non-O157:H7 EHEC isolates.

Molecular typing

A number of genotyping methods have been developed to track EHEC infections and determine diversity and evolutionary relationships among these organisms.

Pulsed-field gel electrophoresis (PFGE)

The gold standard for subtyping O157:H7 serotype and other EHEC serotypes appears to be PFGE. This method classifies bacteria into subtypes (pulsotypes). Images of the PFGE gels are captured, which are then analysed by suitable software for similarity coefficients and dendrograms. The Centers for Disease Control and Prevention (CDC, Atlanta, GA, USA) has created a standardized PFGE protocol and an electronic database (PulseNet) for PFGE profiles of a number of foodborne bacterial pathogens including EHEC. PFGE fingerprint data can be submitted to the database by laboratories for comparison with the existing profiles to determine the novelty or otherwise of the submitted strain (Swaminathan *et al.*, 2006). PFGE applied to EHEC isolates has supplied valuable information such as outbreaks involving distant sites (Wetzel and LeJune, 2006) and spread of EHEC isolates from a common source (Shere *et al.*, 1998).

Ribotyping

There are conflicting reports on the discriminatory power of ribotyping for subtyping of EHEC isolates. While some studies have found it to be less discriminatory than PFGE (Martin *et al.*, 1996; Avery *et al.*, 2002), another study has found it to have a higher discriminatory power than PFGE (Richards *et al.*, 2006). It has been suggested that the discriminatory power of ribotyping depends on the inherent diversity of the strains.

Multilocus sequence typing (MLST)

MLST was unable to differentiate EHEC O157:H7 isolates for which a high diversity has been demonstrated by PFGE (Noller *et al.*, 2003; Gilmour *et al.*, 2005). Housekeeping genes that are highly conserved do not have sufficient polymorphisms to discriminate EHEC isolates.

Amplified-fragment-length polymorphism (AFLP)

A number of studies have applied AFLP to differentiate EHEC isolates. In one study, AFLP was capable of typing isolates that were untypeable by PFGE (Heir *et al.*, 2000). Therefore, it could be used to subtype isolates that are untypeable by PFGE. In other studies, AFLP clustered isolates according to geographical origins (Tsai *et al.*, 2005) and even showed a higher discriminatory power than PFGE (Zhao *et al.*, 2000).

Random amplification of polymorphic DNA (RAPD)

This technique has been used to subtype isolates by many investigators (Kim *et al.*, 2005; Birch *et al.*, 1996; Grif *et al.*, 1998). But it has a low discriminatory power than PFGE. RAPD's main disadvantage is its lack of reproducibility, making it difficult to compare results among laboratories. Because of low annealing temperatures, there are fluctuations in banding patterns. Therefore, RAPD is not useful for studying genetic diversity among isolates.

Multiple-locus variable-number tandem-repeats analysis (MLVA)

This method involves PCR amplification of multicopy DNA tandem repeats that are dispersed across the genome. Organisms have variable number of tandem repeats (VNTR), that are short repetitive DNA sequences that are organized in tandem, variable in size, location and copy number (Nakamura *et al.*, 1987). MLVA searches for VNTR loci presenting polymorphisms that can allow bacterial discrimination. One study showed the high discriminatory power of MLVA comparable to PFGE and there was a good correlation between MLVA and PFGE clusters (Lindstedt *et al.*, 2003). In another study involving O157:H7 isolates, MLVA showed a higher discriminatory power than PFGE (Hyytia-Trees *et al.*, 2006). In yet another study, isolates that were untypeable by PFGE could be typed by MLVA (Lindstedt *et al.*, 2007). Thus, MLVA shows great promise to subtype EHEC isolates, mainly because of its ability to differentiate isolates that cannot be discriminated by PFGE.

Octamer-based genome scanning and lineage-specific polymorphism assay

Octamer-based genome scanning (OBGS) has been used to determine the genetic diversity of EHEC isolates. OBGS is based on octamers skewed to the leading strand of replicating DNA molecule. These octamers are amplified by fluorescent-labelled primers and the isolates are discriminated based on the size distribution of OBGS amplicons. When applied to a diverse collection of EHEC isolates, it achieved a level of discrimination higher than that of PFGE (Kim *et al.*, 1999). An improved version of OBGS is lineage-specific polymorphism that amplifies six markers (LSPA-6) (Yang *et al.*, 2004). LSPA is simple, rapid and comparable among different laboratories.

Plasmid analysis

EHEC isolates carry a number of plasmids and it is possible to differentiate isolates based on their plasmid content. In fact plasmid analysis was done to trace the first reported O157:H7 outbreak to hamburger (Riley *et al.*, 1983). A number of studies since have shown that plasmid analysis has a lower discriminatory power than PFGE (Meng *et al.*, 1995; Radu *et al.*, 2001). Besides, some isolates lack plasmids and therefore, cannot be analysed by this technique.

Microarray analysis

Microarray based on 'tiling strategy' has identified single nucleotide polymorphism (SNP), multiple nucleotide polymorphism (MNPs) and copy number polymorphism (CNP) (Mukherjee *et al.*, 2006). This technique has also identified isolates as belonging to different lineages (Zhang *et al.*, 2007).

Single nucleotide polymorphisms

Single nucleotide polymorphisms (SNPs) can be identified by sequencing PCR or oligonucleotide arrays combined with conventional sequencing or pyrosequencing. With this technique, it was possible to identify emergence of new strains with a high virulence (Manning *et al.*, 2008).

Optical mapping

This method is based on construction of physical map of chromosome based on DNA restriction digestion fragments. Distinct DNA molecules are assembled into partial maps, followed by filling in of overlapping sequences resulting in a whole genome map. With this technique, discrimination among EHEC isolates was achieved (Kotewicz *et al.*, 2007). The discrimination was due to inversions associated with phage sequences, insertions, substitutions and restriction fragment length polymorphisms (RFLPs).

In summary, PFGE and MVLA have been shown to be highly discriminatory for studying EHEC diversity. Because of the superior discriminatory power of MVLA, it has the potential to replace PFGE as the preferred subtyping method. With sequence-based methods, it has become possible to identify location of SNPs or stretches of DNA sequences involved in genetic diversity. However, sequence based methods need further refinement, and issues regarding data interpretation guidelines and intra- and inter-laboratory reproducibility have to be resolved.

Diagnosis

The diagnostic tests for EHEC have been reviewed by Nataro and Kaper (1998). These can be classified into two categories: one aimed at detecting specifically O157:H7 serotype and the other, all EHEC isolates. The diagnostic techniques rely on the detection of a specific antigen (e.g. O157 or H7) and/or virulence factors (e.g. Stx1, Stx2 or *ehxA* gene). There are various tests available for specific detection of O157:H7 serotype. One such test is Sorbitol–MacConkey agar. This serotype does not ferment sorbitol and as a result appears as pale colony, whereas most of normal *E. coli* in stool ferment Sorbitol and appear as pink colony. There are also immune assays in ELISA format, latex agglutination format and dipstick format for detection of O157:H7 serotype and Stx1 and Stx2. Other assays include DNA probe assays and PCR assays for single (e.g. *stx1 or stx2* gene) or multiple genes (e.g. *eae, stx1, stx2* genes). The diagnostic sample is either stool itself or colonies after stool culture. In many studies, the highest yield of EHEC has been obtained while screening stool samples containing visible blood. Retrospective diagnosis is made by measurement of specific serum antibodies to lipopolysaccharide (LPS) antigen.

Prevention and treatment

Prevention of EHEC infection requires practices that reduce or eliminate faecal contamination of meat during slaughter and processing. Hazard analysis and critical control point (HACCP) system is the universally accepted management system for delivering microbiologically safe food. Identification and elimination of super shedding cattle of EHEC isolates is also another important step to reduce the risk of contamination of meat. Feeding cattle with hay instead of grain for a brief period before slaughter has been reported to dramatically reduce the number of EHEC in the intestinal tract by one study (Diez-Gonzalez *et al.*, 1998), while another study has disputed this finding (Hovde *et al.*, 1999). Food items should be cooked properly to excess of 68.3°C to eliminate EHEC contamination. Cross-contamination of cooked food with raw food should be avoided. Another important measure is appropriate hand washing after handling raw meat. Also, patients suffering from EHEC infection should be handled with attention to universal precautions to prevent person-to-person transmission of organisms. An active surveillance system for timely identification of outbreak cases should identify contaminated food items and appropriate handling of patients to prevent further spread of infection. Water hygiene is also an important aspect of prevention of outbreaks due to drinking water and recreational water contamination with EHEC.

Treatment of EHEC infection consists of supportive care and monitoring for the development of HUS. Antibiotics are not beneficial for patients with HUS. Experimental studies have shown that antibiotics induce expression and release of Stx (Walterspiel *et al.*, 1992; Zhang *et al.*, 2000). The risk of development of HUS with the use of antibiotics such as trimethoprim-sulphamethoxazole and β-lactam antibiotics has been shown in some studies (Pavia *et al.*, 1990; Wong *et al.*, 2000). Neutralizing toxin by hyperimmune antitoxin antiserum or oral toxin-binding agents to reduce absorption of the toxin by the gut is yet another approach which remains experimental at this stage (Tzipori *et al.*, 2004; Watanabe *et al.*, 2004). Supportive care in patients developing HUS includes control of fluid and electrolyte imbalance, dialysis, control of hypertension, or blood transfusion (Michael *et al.*, 2009). Vaccines are being investigated to promote immunity against specific serotypes (e.g. O157:H7) and/or Stxs either in humans or cattle. Vaccines are also being studied to reduce shedding and carriage rates of EHEC in ruminants (Smith *et al.*, 2009).

Enterotoxigenic *Escherichia coli* (ETEC)

ETEC is one of the larger pathotypes among DEC and is responsible for a majority of the episodes of infantile diarrhoea and deaths in developing countries. ETEC produces one or both types of enterotoxins – heat-labile (LT) and heat-stable (ST) enterotoxins. In this chapter, the LT- and ST-producing ETEC are named as LT-ETEC and ST-ETEC, respectively. In ETEC, adhesion to the intestinal epithelium is mediated by colonization factor antigens (CFs), which are filamentous protein structures on the bacterial surface. Apart from the enterotoxins, many novel virulence factors have been demonstrated in ETEC. The most common serogroups of human ETEC are shown in Table 2.2. About 60–70% of the isolates examined worldwide falls under specific serogroups and the remaining 30–40% belong to a large number of different serogroups or untypeable ones (Wolf, 1997). Epidemiological studies conducted in diarrhoea- endemic regions indicate that ETEC is mostly associated with diarrhoeal children than controls.

Enterotoxins of ETEC

Heat-labile enterotoxin (LT)

LT is a large (84,000 Da) immunogenic oligotoxin related to cholera toxin (CT) of *Vibrio cholerae* in sequence, structure and mechanism (Svennerholm and Tobias, 2008). The gene encoding LT (*elt*) resides on plasmids that may contain genes encoding ST and/or CFs. LT has two subtypes, LT-I and LT-II, which are divided respectively into LT-Ih (human) and LT-Ip (pig), and LT-IIa, LT-IIb and LTIIc. LT-I is expressed by ETEC isolates that are pathogenic for both human and animal, whereas, LT-II is associated with ETEC of animal origin and rarely with humans isolates. LT-I and LT-II have similar biological activities, but are immunologically distinct. Secretion of LT via vesicles composed of outer membrane and periplasm and its endocytosis is dependent on cholesterol-rich lipid rafts (Kesty *et al.*, 2004). These vesicles act as specifically targeted transport system that mediates entry of active LT and other bacterial envelope components into the host cells. The marked protein composition of vesicles and their ability to transfer LT may contribute to the pathogenicity of ETEC (Horstman and Kuehn, 2000). LT is translocated as folded protein complexes from the periplasm across the outer membrane through the type II secretion channel i.e. secretin GspD. Horstman and Kuehn (2002) proposed a model in which LT is secreted by a general secretory pathway across the outer membrane, followed by binding of LT to LPS via a GM_1-independent binding region on its B subunit.

LT produced by ETEC recognizes certain blood group epitopes. Based on the crystal structure analysis, the blood group A antigen analogue GalNAcα3-(Fucα2)-Galβ4-(Fucα3)-Glcβ was identified as a receptor for the B subunit of the native LT (Holmner *et al.*, 2007). Studies conducted in Bangladesh also showed that ETEC was more prevalent in diarrhoeal children in the A and AB groups than in the O blood group (Qadri *et al.*, 2007).

Heat-stable enterotoxin (ST)

ST produced by ETEC is a small polypeptide (2000 Da) which is non-immunogenic in its natural form. ST has unique properties such as heat-stability, low molecular weight and ability to induce fluid secretion from the intestine of 3–4 days old suckling mice. The gene encoding ST (*est*) resides on plasmids that also may contain genes encoding LT and/or CFs. Two groups of STs have been described- methanol-soluble (STI or STa) and methanol-insoluble (STII or STb). Although STb contains four cysteine residues which form disulfide bonds, it has no homology with STa. STa has two variants – STp (ST porcine or STIa) and STh (ST human or STIb). STs are expressed as 72-amino-acid pre-pro-peptides (Okamoto and Takahara, 1990). As a signal peptide, the pre-sequences guide translocation of the pre-pro-polypeptide across the inner membrane (Rasheed *et al.*, 1990). Outer membrane translocation is accompanied by cleavage of the pre-pro-peptide, generating the mature ST (Yamanaka *et al.*, 1997). Epidemiological investigations suggest that ETEC isolates secreting ST with or without LT induce the most severe disease among children in developing countries (Walker *et al.*, 2007). ST monomeric toxins that comprise many cysteine residues and their disulfide bonds account for the heat-stability of ST. Mutagenesis at one disulfide-bridge-forming cysteine in STa led to a dramatic reduction in the STa toxicity.

Colonization factors (CFs)

The fimbriae (also known as pili) help ETEC to adhere to small bowel enterocytes that follows expression of either or both LT and ST. ETEC of human and animal origin have defined fimbriae. Based on the morphological characteristics, the CFs can be subdivided into three types; rigid rods (CFA/I), bundle-forming flexible rods (CFA/II), and thin flexible wiry structures. Till date, more than 20 CFs have been recognized among human ETEC. The nomenclature for the CFs was simplified in 1996 designating all the fimbriae antigens as coli surface antigen (CS) except CFA/I (Gaastra and Svennerholm, 1996). Among ETEC isolates, CFA/1, CS1 to CS7, CS14, CS17, and CS21 are the most commonly detected CFs. However, in about 30–50% of the ETEC isolates from all over the world, no known CFs could be detected. Longus, a type IV pilus, now classified as CS21 participates in the colonization of human colonic cells and shares similarities with toxin-coregulated pilus of *V. cholerae* and the bundle-forming pilus of EPEC.

The genes for CFs located either in plasmid or chromosome are typically a polycistronic operon that includes the fimbrial subunit genes, chaperones, and ushers. These genes are horizontally acquired via flanking insertion sequences (IS) and transposons. Typical fimbrial gene clusters may have genes encoding the primary fimbrial subunit protein and accessory proteins. The N termini of the subunit proteins and the accessory proteins, are normally conserved and hence reflect structure–functional requirements of the CFs. Presence of different allelic combinations in the subunits (CssA and CssB) of CS6 has been reported (Sabui *et al.*, 2010) and a specific combination of these alleles were shown to enhance cellular binding and has direct influence on the disease outcome. Moreover, many CFs require specific genetic background to acquire and retain plasmids carrying CF encoding genes (Nada *et al.*, 2011).

As reported in many countries, prevalence of CFA was more in ST-ETEC isolates (Subekti *et al.*, 2003). In Bangladesh, ST-ETEC was detected at higher frequency with CFs (66%) whereas the LT-ETEC in controls lacked any of the CFs (92%) (Qadri *et al.*, 2007). Nada *et al.* (2010) observed that LT-ETEC has unique CFs which was not previously described in ST or LT/ST-ETEC. Children with blood group A and Lewis blood group 'a' antigen (Le^a) seem susceptible to ETEC diarrhoea, as the CFA/I binds to glycosphingolipds that are associated with Le^a expressed on epithelial cells (Jansson *et al.*, 2006). Birth cohort study conducted in Bangladesh also showed that children with the Le^a blood type had higher incidence of diarrhoea caused by ETEC expressing CFA/I group fimbriae (Ahmed *et al.*, 2009).

Other virulence factors

The important virulence factors of ETEC are LT/ST and CFs. In addition, several other putative virulence factors such as Tia, an outer membrane for adhesin and invasion; LeoA, associated with LT secretion; TibA, a glycosylated autotransporter for adhesion to epithelial cells; ClyA, a haemolysin for positive transcriptional activation; EatA, a serine protease autotransporter for fluid accumulation in the ileal loop model have been identified (Fleckenstein *et al.*, 1996, 2000, 2002; Westermark *et al.*, 2000; Ludwig *et al.*, 2004; Patel *et al.*, 2004; Sherlock *et al.*, 2005). In many enteric pathogens, flagella play an important role in adherence, secretion of virulence proteins and stimulation of innate immunity. The importance of flagella was demonstrated in a non-motile variant of ETEC with a mutation in the flagellar *fliD* that was unable to deliver LT (Dorsey *et al.*, 2006). Wang *et al.* (2002) verified that more than 30% of ETEC and EPEC harbour genes *irp2*, *fyua* and *asn-intB*, which were reported in the high-pathogenicity island (HPI) of *Yersinia enterocolitica*. The presence of HPI in ETEC and EPEC is also crucial for virulence with reference to their variations, evolution and regulation. Dual infections caused by ETEC and EPEC have often been reported in many diarrhoea-endemic regions. *In vitro* findings demonstrated that concomitant infection caused by ETEC and EPEC in the same host may cause cellular damage and this in turn might enhance the severity of diarrhoea (Crane *et al.*, 2006).

Apart from secreting the enterotoxins, ETEC is capable of invading the epithelial cells of human ileum and colon using two separate invasion loci, *tia* and *tib* (Lindenthal and Elsinghorst, 2001). Pore-forming cytolysin *α-hly* has also been reported in certain ETEC, which is located in a transferable 48–157 kb plasmid (Burgos and Beutin, 2010). Epidemiological significance of invasive characteristics of ETEC was demonstrated in a Peruvian study, demonstrating the association of ETEC with the presence of faecal leucocytes (Mercado *et al.*, 2011).

Pathogenesis

Mechanisms that contribute to the pathogenesis of ETEC are the expression of CFs and LT and/or ST. Pathogen–host cell contact seems obligatory for efficient toxin delivery and activation of adenylate cyclase (Dorsey *et al.*, 2006). The mechanisms of action of ST and LT toxins were reviewed extensively in several publications. An outline of pathogenesis mediated by these toxins is given below.

Heat-labile toxin (LT)

LT is an oligomeric protein composed of an A-polypeptide, non-covalently coupled to a pentameric B-polypeptides. LT and CT have been designated as members of the large family of toxins known as the A_1B_5 ADP-ribosylating heat-labile enterotoxins (HLTs). Binding of LT and CT to ganglioside receptors is mediated by the B polypeptides. Gangliosides are members of a heterogeneous family of sialylated glycosphingolipids expressed on the surface of eukaryotic cells. LT-I binds preferentially to ganglioside GM1 and also to glycoproteins. LT-IIa binds with greatest affinity to ganglioside GDlb, and LT-IIb binds preferentially to ganglioside GDla. The A subunit possesses an ADP-ribosylation activity that covalently modifies the α subunit of the Gs, GTP-binding protein, resulting in activation of adenylate cyclase and production of 3′,5′-cyclic AMP (cAMP) (Fig. 2.2). Intracellular accumulation of cAMP modulates cAMP-dependent protein kinase A (PKA), which phosphorylates the R domain of the cystic fibrosis transmembrane conductance regulator (CFTR). Accumulation of cAMP induces the cell to secrete chloride ions, electrolytes and water efflux into the intestinal lumen leading to significant volumes of watery diarrhoea. In addition to the above physiological mechanism, LT preconditions the host intestinal epithelium with host-derived cAMP for an enhanced adherence of ETEC in the early stages of infection (Johnson *et al.*, 2009). Expression of the LT increased intestinal epithelial cell adherence ~28-fold, whereas expression of the LT-B subunit increased adherence by ~6-fold.

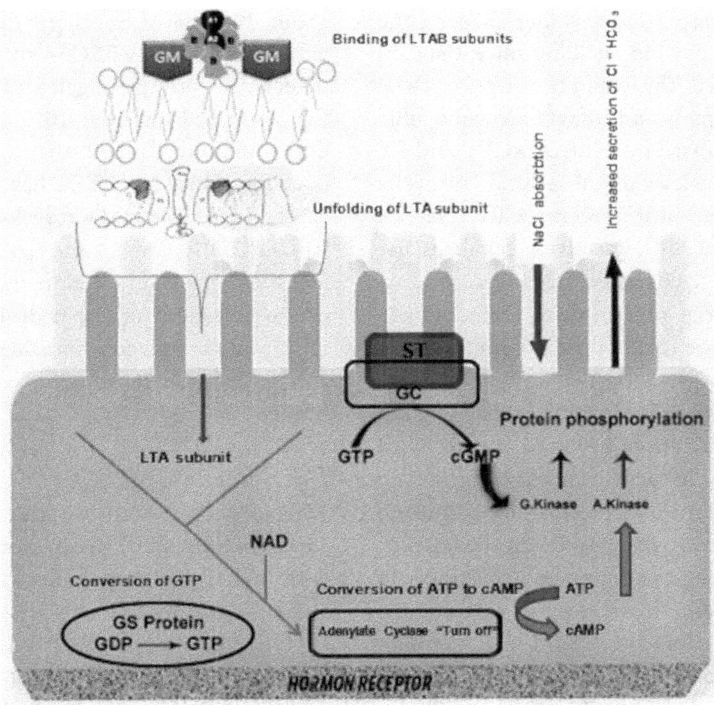

Figure 2.2 Pathogenic mechanisms of ETEC. After initial colonization, ETEC produces LT and ST LT is an oligomeric protein composed of an A-polypeptide and pentameric B-polypeptides. Binding of LT to GM1 receptors of eukaryotic cells is mediated by the B polypeptides. The A subunit possesses an ADP-ribosylation activity that covalently modifies the α subunit of the Gs GTP-binding protein, resulting in activation of adenylate cyclase and production of cAMP. Intracellular accumulation of cAMP modulates cAMP-dependent PKA. Accumulation of cAMP induces the cell to secrete chloride ions, electrolytes and water into the intestinal lumen that leads to watery diarrhoea. The other enterotoxin, ST binds to GC-C located in the brush border. This binding causes an increase in the concentration of intracellular cGMP that activates CFTR through cGMP-dependent G-kinase. This activation results in enhanced Cl^- and H_2O secretion and inhibition of Na^+ absorption in the small intestine.

Heat-stable toxin

STa binds in the lumen of intestine to specific transmembrane protein located in the brush border (guanylate cyclase C), which is the receptor of the endogenous ligands guanylin and uroguanylin (Uzzau and Fasano, 2000). This binding causes an increase in the concentration of intracellular cyclic GMP (cGMP) level. Increased cGMP activates the CFTR through cGMP-dependent protein kinase A phosphorylation that induces intestinal fluid secretion by the stimulation of chloride secretion. CFTR, which is a chloride channel, might be involved in chloride secretion. This activation results in enhanced Cl^- and H_2O secretion and inhibition of Na^+ absorption in the small intestine (Uzzau and Fasano, 2000) (Fig. 2.2). al-Majali *et al.* (1999) demonstrated an age-dependent variation in the density and affinity of the mouse enterocyte receptors specific for STa. Greater interaction of STa was identified on the enterocytes of 2-day-old suckling mice than on those of older mice. This could be the reason why the newborn animals and infants are vulnerable to STa-mediated diarrhoeal infection. LT and ST are regulated at the level of transcription by cyclic AMP (cAMP) receptor protein (CRP). It was demonstrated that CRP positively regulates the expression of the glycosylated TibA, an autotransporter and afimbrial adhesin using a binding site within an intergenic region upstream of *tibDBCA* (Espert *et al.*, 2011).

Compared with STa, the steps that lead to intestinal fluid secretion by STb are not well established. STb induces cell damage in the intestinal epithelium and these lesions indicate a loss of absorptive cells, weakening the absorption and stimulating the secretion of bicarbonate from intestinal cells. However, the receptor for STb is unidentified and the toxin may combine non-specifically to the plasma membrane prior to endocytosis. STb does not stimulate intracellular cAMP or cGMP concentrations, although it does stimulate increase in the intracellular calcium level from extracellular sources.

Epidemiology of ETEC

ETEC is a major cause of diarrhoea among children less than 3 years of age and in some travellers to developing countries. ETEC infection leads to profuse watery diarrhoea, often leading to severe dehydration or death. Worldwide, ETEC infections are estimated to result in 650 million cases of diarrhoea and 380,000 deaths in children below 5 years of age (Sizemore *et al.*, 2004) and close to 50 million children are asymptomatic ETEC carriers (Wennerås and Erling, 2004). Population-based studies conducted in developing countries from 1984 to 2005 reported that ETEC incidence ranges from 39 to 4460 infections/1000 persons per year and children between ages 6 and 18 months are the high risk groups (Gupta *et al.*, 2008). An average of 14% (2–36%) specimens were positive for ETEC in all age groups and 13% (3–39%) in children.

In developing countries, children are likely to experience 0.5 ETEC diarrhoeal episodes per year until the age of five years, after which the yearly incidence drops to 0.1. It has been estimated that about 30–70% of travellers' diarrhoea is due to bacteria, of which ETEC was the most frequently detected enteric pathogen and in this LT-ETEC was identified in about 20% of the cases (Wiedermann and Kollaritsch, 2006). Nosocomial diarrhoea caused by ETEC has also been reported (Dong *et al.*, 2010). Outbreaks of diarrhoea due to ETEC infection have been reported in many investigations from developed countries. In most of the investigations from developing countries, DEC was not given importance and hence the epidemiological information still remains inadequate.

ETEC in the Americas

A retrospective study conducted from January 1990 to June 2005 with morbidity associated with diarrhoea among deployed US military and similar populations showed it to be 38% of the infected individuals from the Middle East, 29% from South-East Asia, 27% from Latin America/Caribbean, and 6% from sub-Saharan Africa. ETEC, *Campylobacter*, and *Shigella* were identified to cause 38–45% of diarrhoea in these regions (Riddle *et al.*, 2006). In the USA, of the 16 outbreaks caused by ETEC from 1996 to 2003, 10 outbreaks were cased by the serotype O169:H41 (Beatty *et al.*, 2004). There have been seven cruise ship ETEC outbreaks in the USA (Sizemore *et al.*, 2004). A cohort study with US adult students travelling to Cuernavaca, Mexico showed that temperature variations significantly influenced the weekly attack rate of ETEC (Paredes-Paredes *et al.*, 2011). During 1994, ST-ETEC belonging to serotype O153:H45 was reported in a foodborne outbreak among attendees of a banquet in Milwaukee (Roels *et al.*, 1998). ETEC was associated in (62% of the cases) a large diarrhoeal outbreak due to overflow of sewage water in Valle de Chalco, Mexico (Cortés-Ortiz *et al.*, 2002). In Brazil and Nicaragua, LT-ETEC was identified more often than the ST or LT/ST-ETEC isolates from the diarrhoeal patients (Regua-Mangia *et al.*, 2004; Reyes *et al.*, 2009). Peruvian studies indicated that ETEC infected children aged >6 months had almost 5-fold increased risk of diarrhoea compared with younger children (Ochoa *et al.*, 2009a, 2011). The frequency of having at least one DEC was significantly high in HIV-positive (42%) patients with a high prevalence of ETEC (13%) (Garcia *et al.*, 2010). Occurrence of ETEC among diarrhoeal children from birth to <12 years of age was 22.3% in Bosnia and Herzegovina with the predominance of serotypes O78:K80 and O25:K11, especially during the first 12 months of life (Dedeić-Ljubović *et al.*, 2009).

Asia

ETEC isolates belonging to different serogroups harbouring *elt* gene encoding LT was associated with an outbreak of diarrhoea in Ahmedabad, India (Chakraborty *et al.*, 2001). However, genetically they belonged to different clones. Frequency of ETEC among hospitalized patients in Dhaka with acute diarrhoea was 11%, among them 36% were below 5 years of age (Chowdhury *et al.*, 2011). In 2004, epidemic diarrhoea caused by ST-ETEC was reported in Dhaka, Bangladesh (Qadri *et al.*, 2005a). In the Asian setting, malnourished children are prone to have ETEC-mediated diarrhoea (Mondal *et al.*, 2009). ETEC was the most common (7–18%) enteric pathogen in community and hospitalized patients in Jakarta, Indonesia, and children aged 1–15 years were found to be the most affected group (Oyofo *et al.*, 2001, 2002). The incidence of ST-ETEC was generally high ranging from 69–72% in Indonesia (Oyofo *et al.*, 2001; Subekti *et al.*, 2003). In Thailand, children with diarrhoea were infected with ETEC (3%), of which 5% of the cases were children aged between 2 and 12 months (Ratchtrachenchai *et al.*, 2004). Predominant CFs in Thailand included CS6 (29%) followed by CS13 (12%), CFA/I, CS2 and CS3 and CS17/CS19 (6% each), while 41% of the strains had unrecognizable CFs (Puiprom *et al.*, 2010).

Middle East

Studies conducted in Egypt showed that 14–50% children under 3 years of age with diarrhoea were positive for ETEC with a high frequency of ST-ETEC (Peruski *et al.*, 1999; Shaheen *et al.*, 2004; Wierzba *et al.*, 2006). The most common CF phenotypes in Egypt were CFA/I (10%), CS6 (9%), CS14 (6%), CS1 plus CS3 (4%), and unknown CFs (59%). In rural Egyptian infants, ETEC diarrhoea was 1.5 episodes/child/year and the incidence increased from 1.7 episodes in the first 6 months to 2.3 in the second 6 months and declined in higher age groups (Rao *et al.*, 2003). In Malatya, Turkey, ETEC was detected in 5.2% of diarrhoeal patients including children (10%) (Ozerol *et al.*, 2005). In the above investigation, no LT+ST-ETEC was detected. In Tehran, 33% of the hospitalized acute diarrhoeal patients were infected with ETEC (Jafari *et al.*, 2008). In a subsequent finding, ETEC was detected in 6.8% children >5 years of age (Jafari *et al.*, 2009). Among Israeli army soldiers, CFA was characterized in 68% of the ETEC isolates, in which CFAs of the CFA/II group and CS6 were the most prevalent (Cohen *et al.*, 2010). A decade long study clearly showed the changing trends in the incidence of ETEC in Tehran with the dominance of ST-ETEC since 1988, but a significant shift towards LT-gene carriage was detected during 1998 (Shahrokhi *et al.*, 2011).

Africa

ST-ETEC was high (51%) among travellers in Kenya with the expression of CFA/II (35%) and CFA/IV (46%) (Shaheen *et al.*, 2003). Interestingly, antigens other than CFs seem to contribute substantially to the acquisition of anticolonizing immunity in Guinea-Bissau (Valentiner-Branth *et al.*, 2003). Among hospitalized children in Dar es Salaam, Tanzania, the prevalence of ETEC was low (3.6%) and most of the isolates were LT-ETEC or LT/ST-ETEC (Moyo *et al.*, 2007). In Mozambique also, the prevalence of ETEC among children with diarrhoea was low (6.8%) (Mandomando *et al.*, 2007). In Tunisia, ST-ETEC was more frequently isolated (~72%) than LT-ETEC (21.5%) and ST/LT-ETEC (6.1%), and CFA/I (44.6%) and CS6 (11%) were the most common CFs (Al-Gallas *et al.*, 2007a).

Case–control studies

Case–control studies estimate the disease burden and validate the importance of aetiological agents for their overall role in causing the infections. In many of the case–control studies, ETEC has been found more in cases than controls though the proportions are not uniform. In Nicaragua, ETEC was identified more in diarrhoeal children than the asymptomatic controls (Paniagua *et al.*, 1997; Vilchez *et al.*, 2009). In addition, CFA/I was significantly higher in cases with diarrhoea and mostly associated with ST or ST/LT-ETEC. These *E. coli* were isolated more frequently from symptomatic children (18.3%) than asymptomatic children (13.3%) from north-eastern Argentina (Viboud *et al.*, 1999). ETEC was also detected more frequently (15.5%) in diarrhoeal children > 5 years of age in Tehran but not in controls (Salmanzadeh-Ahrabi *et al.*, 2005). ETEC was one of the pathogens significantly associated with diarrhoea in Bangladeshi children with a major burden in the first year of life (Albert *et al.*, 1999) and an annual incidence of 0.5 episode/child (Qadri *et al.*, 2007). In Vietnam, a case–control study has confirmed the association of ETEC with diarrhoeal cases (Hien *et al.*, 2008). In Nigeria, ETEC and other DEC were significantly associated with diarrhoeal cases as compared with controls (Nweze, 2010). Recent studies conducted in Thailand, India and Brazil also showed more isolation of ETEC in cases than in controls (Bodhidatta *et al.*, 2010; Rajendran *et al.*, 2010; Nunes *et al.*, 2011).

A few studies conducted in Brazil and Tunisia showed that there was no difference in the incidence of ETEC between cases and controls (Regua-Mangia *et al.*, 2004; Bueris *et al* 2007; Al-Gallas *et al.*, 2007b). In Peruvian children aged 2–24 months, the association of LT-ETEC was more with controls, whereas, ST + LT-ETEC, and ST-ETEC were detected more in cases (Rivera *et al.*, 2010). In Jordanian children, there was no difference in the incidence of ETEC infection in diarrhoeal children (10%) and control children (11%) (Shehabi *et al.*, 2003).

Sources of infection

Presence of *E. coli* that constitutes total and faecal coliforms is considered as an indicator of faecal pollution of water bodies. However, monitoring of DEC groups in the water bodies is not being made as a routine practice. In San Juan Sacatepequez, Guatemala, a seroepidemiological study revealed a high prevalence of ETEC antibodies (48–83%) among 6–36 months old children as an indirect evidence for waterborne infection (Steinberg *et al.*, 2004). Several diarrhoea outbreaks due to contaminated potable water supply in passenger and military ships have been reported. In most of these outbreaks, ETEC was frequently identified as a causative agent (Rooney *et al.*, 2004). The high risk factors for water contamination in ships include contamination of source water while loading, improper loading techniques, poor design and maintenance of storage tanks, cross-connections, back siphonage, and insufficient residual disinfectants. ETEC was isolated (30%) from waters of Bangladesh and epidemiological studies with clinical isolates indicated that surface waters act as a reservoir (Begum *et al.*, 2007). Natural aquatic flora and cultivated fauna in the Gangetic riverine system were reported as reservoirs of ETEC (Singh *et al.*, 2010). Owing to its long survival periods (~3 months) in waters without losing most of their virulence genes (Lothigius *et al.*, 2010), ETEC can be considered as one of the water-borne pathogens.

Contaminated foods also serve as a source of ETEC infection. ETEC was detected in street-vended chilli sauces in Mexico, indicating its potential as a source of infectious diarrhoea (Estrada-Garcia *et al.*, 2002). Independent risk factors for ETEC diarrhoea in Sao Paulo, Brazil include preparation of rice, beans, or soup in the morning and serving it to children in the afternoon (Sobel *et al.*, 2004). Large ETEC-mediated foodborne outbreaks have also been reported in the US (Beatty *et al.*, 2006). In these findings, ETEC isolates that produce LT and ST (serotype O6:H16) were reported. The Nevada ETEC outbreak in 2004 was associated with poor food-handling practices and infected food handlers in the restaurant (Jain *et al.*, 2008). Compared with other foodborne pathogens such as STEC, *Salmonella* and *Campylobacter* spp., ETEC-mediated infections are relatively less frequent.

Molecular typing

Genome analysis

In ETEC, the virulence-associated genes are mostly carried by plasmids which encode CFs, toxins, and other adhesins. The ETEC chromosome has the genomic background of *E. coli* K-12, a non-pathogenic strain and shares about 2.7 million bases of its core sequence. Complete genomic sequence of a prototype ETEC strain H10407 revealed that the chromosome is closely related to that of the non-pathogenic commensal *E. coli* strains (Crossman *et al.*, 2010). DNA subtractive hybridization analysis of ETEC strain H10407 and optical-map-based restriction site similarity also proved about 96% identity with the *E. coli* K-12 strain MG1655 (Chen *et al.*, 2006). However, there were no chromosomally encoded factors unique to ETEC strains. On the other hand, plasmids of the strain H10407 and other ETEC strains revealed mosaic structures. By verifying the distribution of virulence factors of DEC and the phylogenetic tree, Escobar-Páramo *et al.* (2004) hypothesized that with the different genetic backgrounds, EAEC and DAEC would have acquired the virulence factors. In the case of ETEC, EHEC and EIEC, specific genetic background is a prerequisite for the acquisition of each of their respective virulence factor.

The flanking regions of LT-encoding gene *eltAB* are highly conserved with 236- and 280-bp regions, followed by highly variable DNA sequences which consist mainly the partial insertion sequence (IS) elements such as IS*600*, IS*91*, IS*1162*, IS2, and IS*1294* (Schlör *et al.*, 2000). The operons encoding LT-IIa and LT-IIc are flanked by genes having significant homology to genes associated with lysogenic bacteriophage (Nawar *et al.*, 2010). Single-nucleotide polymorphic analysis of the LT-encoding genes of several ETEC strains revealed 21 polymorphic sites in *eltA* and nine in *eltB* indicating the diversity of LTs that may have impact on the pathogenesis of the isolates and the epidemiology of the disease (Lasaro *et al.*, 2008). One-third of the ETEC isolates harbour both *elt* and *est* genes and combination of these genes seems to increase the virulence. In one such isolate, these genes were found in a 13.5-kb DNA fragment with three IS*1*s (IS*1A*, *B* and *C*), IS*600* and an IS*Ec8*-like element (Ochi *et al.*, 2009). The IS*Ec8*-like element is located between *eltAB* and *estIa* and there were seven transposase genes between *eltA* and *estIa*, and two transposase genes were found upstream of *estIa* (Fig. 2.3). These observations suggest that the toxin region has evolved through multiple transposition events and may have been transferred horizontally.

In different epidemiological findings, it was proved that ETEC appears with different genetic backgrounds indicating enormous diversity among the isolates from the same geographical location or during outbreaks (Kahali *et al.*, 2004a). In multilocus sequence typing (MLST), ETEC isolates were distributed in all the phylogenetic lineages of *E. coli* thereby indicating that the acquisition of toxin genes is sufficient to generate an ETEC strain (Turner *et al.*, 2006a). In addition, the polyphyletic distribution of ETEC strains indicated that they have arisen several times without any common clonal lineage. MLST of human ETEC and non-ETEC from different countries showed 42 different lineages and were estimated to have emerged more than 900 years ago (Steinsland *et al.*, 2010). From

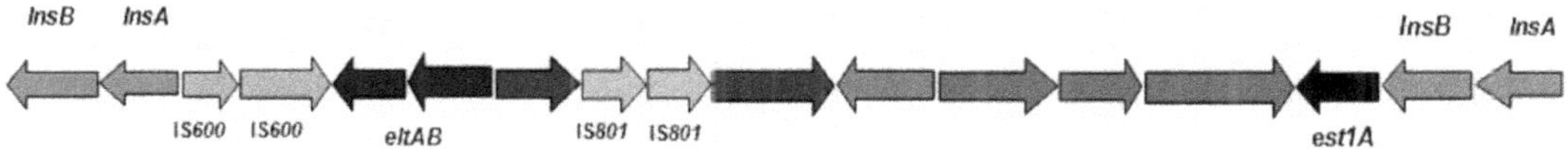

Figure 2.3 Genetic organization of *eltAB* and *estIa*, if present in the same isolate. *eltAB* and *estIa* are separated by seven transposase genes and two transposase genes are located upstream of *estIa*.

these results it appears that there is extensive exchange of enterotoxin and CF genes between different lineages with a common ancestral lineage for human and porcine ETEC. Sequence analysis of *mdh* using multilocus enzyme electrophoresis indicated that DEC has originated from commensal *E. coli* strains by acquisition of suitable virulence factors (Pupo *et al.*, 1997). Phylogenetic analysis also confirmed genetic variability in the known virulence associated genes of ETEC (Sahl *et al.*, 2011).

It is interesting to note that host genomic susceptibility in the form of variations in genes that encode molecules that mediate attachment, pathogen recognition, inflammatory cytokine response, and innate and acquired immunity are specific to certain enteric pathogens. Genetic polymorphisms in the CD14, lactoferrin and osteoprotegerin promoter genes seem associated with diarrhoea in travellers and SNPs in the IL-8 genes are also associated with increased risk for EAEC infection, whereas IL-10 haplotypes may have influence on ETEC infection (Flores and Okhuysen, 2009a).

Diagnosis

Serology-based pathogrouping of DEC has become an obsolete method since many of the serotypes are not specific to any pathogroup. Identification of ETEC is directly related to the number of typical colonies considered for screening (Galbadage *et al.*, 2009). Hence, it is mandatory to include many typical colonies (3–5 per sample) from the selective agar plates to increase the probability of getting a few pathogenic *E. coli* against the background of non-pathogenic ones. Commonly used assays for diagnosis are summarized in Table 2.3. ST can be detected by the suckling mouse or radioimmunoassays. For the detection of LT, either tissue culture assay using mouse adrenal tumour Y1/Chinese hamster ovarian (CHO) cell lines or rabbit ileal loop model is being used in many laboratories. Filter sterilized culture supernatants of LT-ETEC cause elongation of CHO cells *in vitro* or fluid accumulation in the rabbit ileal loop model. GM1 ganglioside based enzyme-linked immunosorbent assay is a widely used method for the detection of LT (Svennerholm and Wiklund, 1983). Use of monoclonal antibodies for the detection of CFs among ETEC was first demonstrated in 1997 (Helander *et al.*, 1997). DNA probes for detection of LT and ST encoding genes led to the study of ETEC in a much more detailed manner (Moseley *et al.*, 1982). Colony hybridization assay with pooled DNA probes was useful for the identification of toxin genes in ETEC along with 18 of 21 CF encoding genes (Steinsland *et al.*, 2003).

In many investigations, a PCR-based method was adopted for the detection of ETEC either directly from the stool specimens or using pure cultures. Several simplex and multiplex PCRs were designed for the detection of the toxin genes of ETEC or the structural genes encoding CFs (Ghosal *et al.*, 2007; Rodas *et al.*, 2009; Nada *et al.*, 2010). Common PCR targeting *gadAB* encoding the enzyme glutamate decarboxylase was found to be useful in the detection of ETEC, EPEC and STEC (Grant *et al.*, 2001) as this gene is conserved among DEC as well as *Shigella* spp. Loop-mediated isothermal amplification (LAMP) PCR for the detection of LT/ST-ETEC has been developed recently (Yano *et al.*, 2007). The detection limit of the LAMP-PCR was 4 and 40 cfu for LT and ST respectively, which is about 10-fold higher than that of conventional PCR assays. In terms of rapidity, the LAMP-PCR assay can be completed within 35 minutes and amplification can be visually seen based on the turbidity caused by magnesium pyrophosphate, a by-product of the reaction.

Gold nanoparticle based colourimetric detection system has been developed for diagnosis of *elt* harbouring ETEC (Jyotil *et al.*, 2011). ETEC or EAEC can be efficiently identified by PCR with DNA extracted from the cards with patients' faecal occult blood and this method is suitable for field studies where collection and transportation of stool specimens are difficult (Grimes *et al.*, 2008). Characterization of *fliC* genes from ETEC isolates by PCR-RFLP was shown to give better results than classical serotyping of H antigens (Ramos Moreno *et al.*, 2006). In addition, this method discriminated the non-motile ETEC isolates.

A highly sensitive and specific real-time fluorescence-based multiplex PCR for the detection of DEC has also been developed (Guion *et *al.*, 2008). Total time between preparation of DNA from the test colonies and completion of this PCR and melting-curve analysis was less than 90 minutes. A DNA microarray based detection of enterotoxin and common O serogroup encoding genes has been developed for the clinical diagnosis and epidemiological surveillance of ETEC (Wang *et al.*, 2010). This identification system can accommodate thousands of test isolates at a time.

Treatment

The treatment of ETEC-mediated diarrhoea is mainly based on the correction of dehydration and maintenance of hydration. Early rehydration using intravenous fluids such as Ringer's lactate is obligatory for patients with severe dehydration. Patients with mild and moderate diarrhoea can be given oral rehydration solutions.

Antimicrobials are useful only when the diagnosis or suspicion of ETEC-related diarrhoea or cholera is made. Empirical use of ciprofloxacin for adults and azithromycin for children has been described (Qadri *et al.*, 2005b). Travellers' diarrhoea caused by ETEC can be treated with non-absorbable antibiotics such as rifaximin. Combination of antimicrobials such as azithromycin and loperamide or levofloxacin and loperamide were found to be useful in the treatment of travellers with diarrhoea (Sanders *et al.*, 2007). Apart from shortening the duration of diarrhoea, rifaximin was found to suppress the expression of LT/ST and CS2/CS3 or CS6 (Jiang and DuPont, 2010).

Exclusive breastfeeding seems to protect infants against severe ETEC diarrhoea, but this practice was not associated with long-term protection as the child grows. Human milk contains α1,2-linked fucosylated glycans (α-FG), which include free and conjugated forms of oligosaccharides. *In vitro* and *in vivo* studies demonstrated that α-FG inhibits the binding of several bacterial toxins including the enterotoxins from ETEC (Morrow *et al.*, 2005). A study in Mexico on the impact of dietary risk factors on patterns of infection by LT-ETEC showed that breast milk antibodies and liquid infusions reduce diarrhoeal disease and the duration of infection (Long *et al.*, 1999). Since the human colostrum prevents CF expression, breast-feeding could protect children against ETEC-mediated diarrhoea (Corrêa *et al.*, 2006). The main constituent, calcium in milk seems to inhibit ETEC-mediated diarrhoea. When ETEC patients were provided with high dietary calcium (1100 mg/day), the duration of diarrhoea was shortened (1–2 days) compared with those who received low calcium (60 mg/day, 3 or more days) (Bovee-Oudenhoven *et al.*, 2003).

Provision of adequate nutrition is a critical factor in children in developing countries, where the diarrhoeal disease burden is very high. In the dietary management of persistent diarrhoea among hospitalized children, green banana and pectin were found to be useful. When the therapeutic effects were evaluated, green banana or pectin was found to reduce volumes of stool, oral rehydration solution, intravenous fluid, and numbers of vomiting episodes as well as duration of diarrhoea (Rabbani *et al.*, 2001). The protective effect of egg-yolk antibodies obtained from hens immunized with fimbrial antigens of ETEC was promising when tested with neonatal and early-weaned piglets (Marquardt *et al.*, 1999). Recently, compounds that inhibit the activity of the ETEC toxins and affect the colonization capabilities were identified. One such fluorenone-based compound, DC5 was shown to inhibit the accumulation of cAMP and colonization abilities of ETEC and similar adenylyl cyclase toxin-producing bacteria (Moen *et al.*, 2010). Vitamin A supplementation modified involvements of gut-cytokine immune response due to infection caused by ETEC and EPEC (Long *et al.*, 2011). In vitamin A-supplemented children, the faecal concentrations of IL-6 and IL-10 were very low when they were infected with ETEC and EPEC, respectively, thereby indicating the reduction in inflammatory response.

Probiotics are described as microorganisms that have a beneficial effect in the prevention and treatment of specific illness when they are ingested. Currently, the best-studied probiotics against ETEC are *Lactobacillus* species and *Bifidobacterium* species. Purified adhesin of *B. adolescentis* was shown to significantly inhibit the adhesion of ETEC to intestinal epithelial cell line, Lovo (Zhong *et al.*, 2004). In tissue culture studies, the human isolate of *L. casei* subsp. *rhamnosus* strain exhibited adhesive properties and prevented colonization of many enteric pathogens including ETEC (Forestier *et al.*, 2001). Toxin-binding properties of probiotics were also demonstrated. A recombinant probiotic strain of *E. coli* (CWG308) that expresses glycosyltransferase was shown to bind with LT-ETEC with a high avidity (Paton *et al.*, 2005). In *in vitro* and animal model studies, this genetic construct acted like lacto-*N*-neotetraose, and neutralized about 93.8% of the LT.

An update on ETEC vaccines

Even though the pathogenesis of ETEC infection has been well known for many years, an approved ETEC vaccine is not available especially for infants. An ideal ETEC vaccine should contain the common fimbrial antigens (CF antigen I and CS1-CS6), and the LT toxoid. The existing vaccine formulations are based on LT-B subunit and predominant CFs. This combination was estimated to protect about 90% of ETEC infections, with the expectation of strains of unknown CFs and those that express poorly immunogenic ST. Considering the larger numbers of unrecognisable CFs and ST-ETEC in diarrhoea-endemic settings, the current strategy for ETEC vaccine formulation has to be amended. The ST-specific vaccine is difficult to formulate because of cross-reactivity between anti-ST antibodies with structurally similar endogenous ligands (guanylin and uroguanylin) and non-immunogenic nature of ST (Taxt *et al.*, 2010). In many ETEC vaccine studies, the candidate strains gave poor immune response against the target CFs or had the outcome of mild to severe diarrhoea. Systematic review of 27 volunteer challenge studies with ETEC showed that the diarrhoeal outcome varied with strain, dose and type of enterotoxin (Porter *et al.*, 2011). Risk of nausea, abdominal cramps and headaches was mostly seen in females. Baseline anti-LT IgG titres were associated with decreased risk of diarrhoea but not with anti-LT IgA or anti-CF antibodies.

Different strategies have been attempted to deliver fimbriae and toxin antigens to the human immune system to elicit strong mucosal, in particular, intestinal immune responses. There has been some promise when testing different ETEC candidate vaccines for protection against diarrhoea in adult travellers. Some of the recent vaccine candidates such as a trivalent STa–LTB-STb fusion protein (You *et al.*, 2011), a recombinant strain co-expressing CFA/I along with CS2 (Tobias *et al.*, 2010) showed strong immunization potential in animal models. In healthy volunteers, oral delivery of attenuated ETEC that expresses CFA/I with deletions of the *aroC, ompC, ompF,* and toxin genes (*est* and *east*) caused greater than 2-fold increase in the levels of CFA/I specific IgA and IgG antibody-secreting cells (Turner *et al.*, 2006b). This vaccine was found to be safe with a high colonization ability. Similarly, a live attenuated ETEC strain PTL003, which expresses the CFA/II and its components, CS1 and CS3 was found to be highly immunogenic (McKenzie *et al.*, 2006).

A randomized, placebo-controlled trial with two doses of vaccine with CFs and CT-B subunit reduced the rate of severe episodes of travellers' diarrhoea due to ETEC infection and stimulated anti-LT antibodies, but did not reduce the overall rate of ETEC diarrhoea (Sack *et al.*, 2007). ETEC-LT vaccine delivered to the skin by patch has been tested in travellers to Mexico and Guatemala. This multidose vaccine showed that LT-patch recipients were protected against moderate to severe diarrhoea with protective efficacy from 75–84% (Frech *et al.*, 2008). In a double-blind, placebo-controlled trial with 50 μg of LT or placebo in a patch applied on days 0, 21, and 42, a 4-fold increase in anti-LT IgG (100%) and IgA (97%) antibodies was recorded. When challenged orally with 6×10^8 cfu of ETEC that produces both the toxins, this vaccine did not prevent but reduced the typical symptoms (McKenzie *et al.*, 2007). Disruption of stratum corneum using an electrocardiogram prep-pad before the LT vaccine patch application resulted in increased immunity (24-fold) in anti-LT IgG titre (Glenn *et al.*, 2007).

MucoRice-CTB is a cold chain-free rice-based oral vaccine that induces CTB-specific secretory IgA that gave a long-lasting protection against *V. cholerae* and LT-ETEC-induced diarrhoea (Tokuhara *et al.*, 2010). Transgenic plant-derived LT antigens offer a new strategy for development of a safe and inexpensive vaccine that is immunogenic without the need of any buffer to neutralize gastric acidity.

Enteroaggregative *Escherichia coli* (EAEC)

EAEC is considered as one of the emerging aetiological agents of diarrhoea worldwide. EAEC forms a characteristic aggregative adherence (AA) on cultured HEp-2 cells with a 'stacked-brick' configuration. Since its discovery, an increasing number of reports have implicated EAEC in a variety of settings as a causative agent of diarrhoea in both industrialized and developing countries. Initial studies from developing countries showed that EAEC was linked to persistent diarrhoea. However, recent surveys revealed that EAEC is also associated with acute/mild inflammatory diarrhoea, community outbreaks, travellers' diarrhoea and an important pathogen in malnourished children and human immunodeficiency virus-infected persons. EAEC is heterogeneous, and comprised diverse serotypes with a variety of putative virulence factors.

Clinical manifestations

The clinical characteristics of EAEC-mediated diarrhoea are described in many sporadic infections, outbreaks and human volunteer studies. Generally, the infection is accompanied by watery diarrhoea with or without mucus or vomiting and low-grade fever. Occult blood, faecal leucocytes and IL-8 were also recorded in the stools of diarrhoeal patients infected with EAEC indicating mucosal inflammation. In malnourished children, the EAEC causes severe mucosal damage, which leads to persistent or chronic diarrhoea (Opintan *et al.*, 2010). Recently, several other variations of EAEC-mediated clinical symptoms have been described mainly owing to host genetic susceptibility and host immune response.

Pathogenesis

Adherence to the intestinal epithelial cells

Typical EAEC carry AggR regulon which is encoded on a 113,346-bp virulence plasmid called, pAA (widely known as CVD432). For the proper functioning of fimbrial structure (mainly the aggregative adherence fimbriae, AAF) the *aatA* gene encoding a protein for surface translocation of dispersin is required. AAF is distantly related to the Dr family of adhesins and belongs to the IncFIIA family. A type VI secretion system (T6SS) within a chromosomal island called AAI was identified in a cluster of AggR-activated genes on the chromosome of the EAEC prototype strain, 042. The pathogenicity of EAEC includes colonization of the intestinal mucosa and elaboration of one or more cytotoxins such as *Shigella* enterotoxin 1 (ShET1), the plasmid-encoded toxin (Pet), and the enteroaggregative *E. coli* heat-stable toxin (EAST1) that are under the control of the AggR regulator.

Different animal models such as rat, rabbit and gnotobiotic piglets; tissue culture assay using cell lines such as T84, Caco 2, HT29, HeLa, and HEp-2; and *in vitro* organ culture with intestinal fragments have been employed to study the pathogenicity of EAEC. Adhere to epithelial cells is facilitated by the following three mechanisms.

1 *Fimbrial adhesins.* Based on the sequences of the major fimbrial subunits, three distinct AAF variants AAF/I, AAF/II, and AAF/III encoded by the *aggA*, *aafA*, and *agg3A* genes, respectively, have been described (Fig. 2.4). In most of the epidemiologic studies, the majority of EAEC do not express these AAF variants, despite the presence of AggR regulator and other pAA-borne genes. AAF/I fimbriae mediate adherence to HEp-2 cells and haemagglutination of human erythrocytes. The genes encoding AAF/I are separated into two regions. Region 1 contains a cluster of four genes, *aggDCBA*, coding for the chaperone, usher, invasion, and adhesin/fimbrial subunit, respectively and the *astA*. Region 2 encodes an AraC-like regulator, AggR that is required for the AAF/I expression (Nataro *et al.*, 1992, 1994). AAF/II is different from AAF/I and its fimbrial structural subunit (AafA) and has only 25% similarity to AggA of AAF/I. The genes encoding AAF/II are also separated into two regions with genes encoding the chaperone (*aafD*), the major fimbrial subunit (*aafA*), and the transcriptional activator (*aggR*) are present in region 1 and the other genes encoding the silent chaperone (*aafD'*), usher

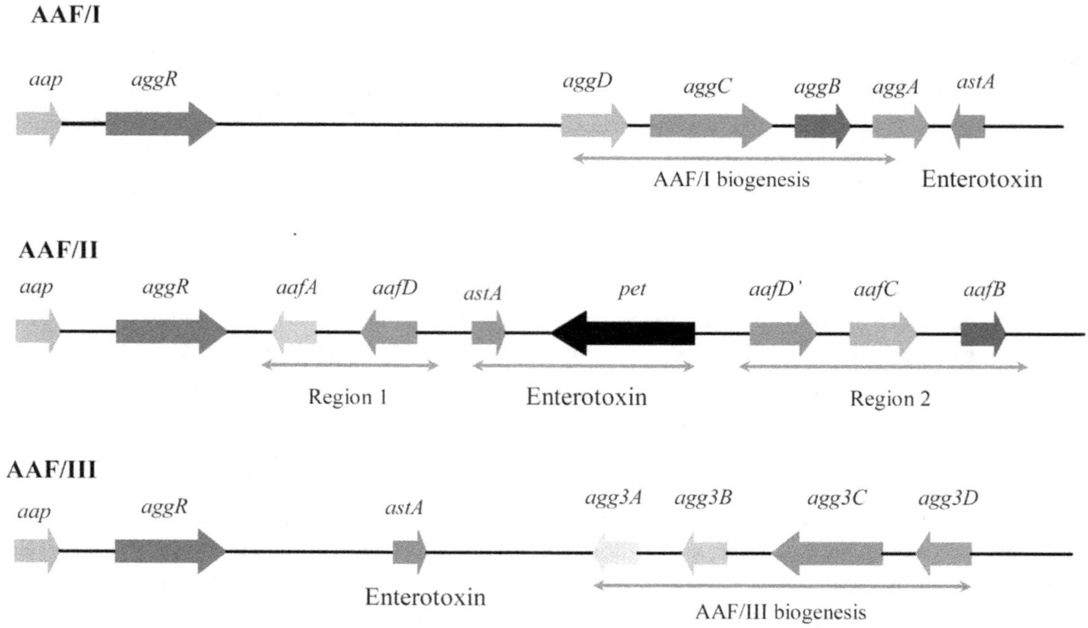

Figure 2.4 Schematic representation of AAF alleles. In the AAF/I allele, region 1contains a cluster of four genes, *aggDCBA* and region 2 encodes an AraC-like regulator, AggR. The genes encoding AAF/II allele are also separated into two regions with *aafD* and *aafA* in region 1 and genes *aafD*, *aafC* and *aafB* in region 2. The distinct AAF/III allele has *agg3ABCD* in the biogenesis region, which is comparable to that of AAF/I and the Agg-3BCD region has 63–67% identity with the corresponding proteins of AAF/I and AAF/II.

(*aafC*), and invasin (*aafB*) are located in region 2. The distinct AAF/III fimbrial structural protein Agg-3A was identified which functions as an adhesin to intestinal cells using individual filaments, which is different from bundles formed by AAF/I or AAF/II. Overall, the gene organization of the AAF/III cluster is comparable to that of AAF/I, and the Agg-3BCD region has 63–67% identity with the corresponding proteins of AAF/I and AAF/II.

2. *Afimbrial adhesin AfaD-III.* The afimbrial adhesin proteins AggB and AafB are encoded in the adhesin operons *agg* and *aaf*, respectively. AggB belongs to the family of afimbrial invasins. The genes *afaD* and *afaE* are located in the afimbrial adhesion (*afa-3*) gene cluster, and the respective proteins AfaD and AfaE are required for the internalization and binding of EAEC to the intestinal cell lines.
3. *Dispersin and biofilm formation.* EAEC adheres to the small and large intestinal mucosal surfaces and form a biofilm which may support colonization and make a strong barrier for the host antibacterial factors. The genetic regulation of this biofilm is mediated by AAF/II. In addition, a novel virulence factor, dispersin encoded by *aap* was reported in EAEC that may be important in promoting bacterial spread across the mucosa (Sheikh *et al.*, 2002).

The pAA plasmid carries the plasmid encoded toxin (Pet) and EAST-1. The heat-labile 108-kDa Pet is secreted using the T5SS (Navarro-Garcia, 2010) and is a serine protease autotransporter of *Enterobacteriaceae* (SPATE) that promotes crypt openings, cell rounding and extrusion of enterocytes in human intestinal tissues. The toxic effects of Pet include internalization into the host cells followed by cleavage of host cytoskeletal protein spectrin. The *pet* promoter is co-dependent on transcription factors CRP and Fis, and is a key virulence regulator in EAEC 042 (Rossiter *et al.*, 2011). It was also shown that Pet is highly homologous to the EspP protease of EHEC and to EspC of EPEC (Eslava *et al.*, 1998).

EAST-1 is a 4.1-kDa heat-stable toxin similar to the STa-produced by ETEC. STa requires six cysteine residues and three disulphide linkages for full activity whereas EAST-1 has four residues. Both these toxins activate guanylate cyclase. EAST-1 (encoded in the *astA*) located adjacent to *pet* is not only present in EAEC, but also in other categories of DEC. EAST1 has been shown to induce diarrhoea principally in humans, although this gene is reported in *E. coli* from many types of farm animals (Veilleux and Dubreuil, 2006). In a rabbit model, Savarino *et al.* (1991) showed that isolates of EAEC induce fluid secretion. The toxic effect of EAST-1 was demonstrated to increase short circuit current in Ussing chamber, an indication for the net anion secretion.

The chromosomally encoded Pic (protein involved in colonization) in EAEC has 100% homology with Pet of *Shigella flexneri*. However, the flanking sequence of *pic* in EAEC differs from that of *S. flexneri*, suggesting that this gene was acquired by horizontal transfer. The *pic* gene lies in the DNA coding strand, whereas *set1A* and *set1B* encoding ShET1 are carried on the opposite strand, but completely within the *pic* gene (Fig. 2.5). Pic in EAEC is a mucinase involved in intestinal colonization, serum resistance and haemagglutination (Henderson *et al.*, 1999). The immune response to Pet and Pic autotransporter proteins was confirmed in convalescent children after infection with EAEC thereby indicating their specific roles in the pathogenesis (Bellini *et al.*, 2005). *Shigella* enterotoxin 1 (ShET1) and haemolysin-E (HlyE) are chromosomally encoded toxins. The ShET1 consist of a single 20-kDa SetA protein associated with a pentamer of SetB subunits, which are thought to establish an oligomeric

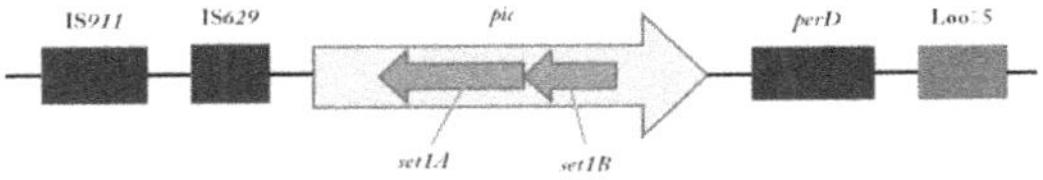

Figure 2.5 Schematic representation of *pic* gene of EAEC. The gene *pic* lies in the DNA coding strand, whereas the ShET1encoding genes *set1A* and *set1B* are placed on the opposite strand but fully within the *pic* gene.

toxin. Based on this archetype, ShET1 appears to induce intestinal secretion via cAMP and cGMP. HlyE oligomerizes and the oligomer makes a pore-forming toxin mediating *in vitro* cytolytic and cytopathic effects on human cells. But the role of *hlyE* is not well established as this gene was found in many non-pathogenic EAEC strains as well. The other chromosomally mediated *E. coli* common pilus gene (*ecpA*) was reported in the majority of the EAEC regardless of the presence or absence of AAF fimbrial genes (*aggA* or *aafA*). An *ecpA* mutant obtained from the strain 042 (with AAF/II) was not altered in adherence suggesting the role of AAF/II (Avelino *et al.*, 2010). The importance of *ecpA* was demonstrated in EAEC isolates lacking AAFs, as they adhered to cultured epithelial cells.

EAEC induces a steady loss of epithelial integrity supplemented by delocalization of tight junction proteins, leading to decrease in transepithelial electrical resistance (TER) and support in the expression AAF/II adhesion (Strauman *et al.*, 2010). The disruption of epithelial barrier function leads to the secretion of electrolytes, loss of mucosal protein, malabsorption and helps the pathogens to have access to the basolateral compartment. The extracellular matrix proteins fibronectin, laminin, and collagen IV were identified as receptors for adherence fimbriae (Flores and Okhuysen, 2009b). The genes encoding AggR are responsible for diffuse adherence, heat-resistant agglutinin-expressed colonization, aggregative adherence and biofilm formation (Boisen *et al.*, 2008; Bhargava *et al.*, 2009). The gene *hra1* that encodes an outer membrane protein was also found to confer autoaggregation, biofilm formation, and aggregative adherence (Bhargava *et al.*, 2009).

The invasive nature of EAEC was demonstrated in Caco-2 and T84 cells and the internalized cells multiplied and survived up to 72 hours *in vitro* (Pereira *et al.*, 2008). This strategy seems advantageous for survival and proliferation evading the host clearance mechanisms, immune defence and antimicrobial actions. p38 mitogen-activating protein kinase (MAPK) is a member of a family of stress-related kinases that impacts a varied range of cellular functions including host inflammatory responses to infectious bacterial products. EAEC surface protein, flagellin (FliC) produces interleukin 8 (IL-8) from several epithelial cell lines. The host response to FliC is mediated by Toll-like receptor 5 (TLR5), which signals through NF-κB to induce transcription of proinflammatory cytokines. TLR5 facilitates p38MAPK-dependent IL-8 secretion from EAEC infected epithelial cells and this process requires IL-8 promoter activation independent of NF-κB nuclear migration. EAEC-induced activation of MAPK in intestinal epithelial cells leads to *in vitro* induction of DNA-binding activity of NF-κB and AP-1, resulting in IL-8 secretion (Khan *et al.*, 2004, 2010). Pretreatment of cells with inhibitors for MEK, JNK and p38MAPK resulted in inhibition of the IL-8 secretion, confirming the specific stimulation of MAPK signalling pathways leading to nuclear responses. Apart from IL-8, the genes upregulated in strain 042-infected non-polarized T84 cells includes those for IL-6, TNF-α, GRO-α GRO-γ, ICAM-1, GM-CSF and IL-1α (Harrington *et al.*, 2005). Increased concentration of faecal IL-8 and other cytokines such as IL-1β and IL-1ra were significantly associated with the presence of *aggR* alone or with other virulence factors in EAEC (Jiang *et al.*, 2002).

Considering all the above-described virulence factors, the following three stage model of EAEC pathogenesis was proposed (Fig. 2.6): (i) adherence to the small and large bowel mucosal surfaces in a characteristic aggregative or 'stacked brick' adherence pattern; (ii) production and deposition of a large mucus biofilm encrusted with EAEC; and (iii) mucosal toxicity induced by EAST1, Pet and Pic, inflammation and induction of cytokines.

Epidemiology

Case–control studies

Studies conducted in Brazil, Congo, Jordan, Poland, Iran, Vietnam and the US showed that the incidence of EAEC was more in diarrhoeal children than the control children (Jalaluddin *et al.*, 1998; Shehabi *et al.*, 2003; Sobieszczańska and Osek, 2003; Regua-Mangia *et al.*, 2004; Cohen *et al.*, 2005; Nguyen *et al.*, 2005; Salmanzadeh-Ahrabi *et al.*, 2005; Hien *et al.*, 2008). Some of the virulence-specific genes were detected in EAEC from cases and not from controls. In Nigeria, EAEC isolates with genes encoding haem transport (*chu*) and synthesis (*iucA*) were found mostly in diarrhoeal cases and not in controls (Okeke *et al.*, 2004). EAEC virulence gene markers *aggR*, *aap*, *aat*, *astA* and/or *set1A* were identified more frequently in diarrhoeal cases (6%) than controls (2%) in the US (Cennimo *et al.*, 2009). The stool from cases was characterized by gross mucus with leucocytes, typical for inflammatory diarrhoea. In HCT-8 cell line, the *aggR* gene in EAEC expressed higher levels of IL-1ra, IL-6, IL-8 and TNF-α compared with *aggR*$^-$ EAEC. Positive correlation of EAEC with virulence gene markers such as *astA*, *pic* and CVD432 were found in diarrhoeal infants from Brazil (Pereira *et al.*, 2007). Persistent diarrhoea was related to CVD432-positive isolates, and the EAST1-positive isolates were statistically associated with acute diarrhoea in children > 6 months of age. EAEC harbouring the pAA plasmid were isolated at similar frequencies in Brazilian children, with (19.6%) and without (10.8%) diarrhoea (Piva *et al.*, 2003). However, the genes *astA* and *pic*, and the combination of *aap* with *pic* genes were significantly higher in children with diarrhoea than controls. Another study from Brazil fund that the virulence marker, Aap was significantly associated with infants with diarrhoea (22%) than with controls (3%) (Zamboni *et al.*, 2004). In addition, EAST-1 and Shf markers were most frequently detected (61%) in pAA probe-negative EAEC isolates.

In some studies, the EAEC was reported more in controls than cases. In Vellore, India, EAEC was the most common pathotypes more frequently detected in controls (24%) than in diarrhoeal cases (15%) (Rajendran *et al.*, 2010). Similarly, the proportion of EAEC was slightly higher in Brazilian control children (7%) than cases (5%) (Spano *et al.*, 2008). EAEC was detected in lower frequency in Nicaraguan children with diarrhoea (27.8%) than without diarrhoea (33%) (Vilchez *et al.*, 2009).

Travellers' diarrhoea (TD)

Among travellers returning from developing countries, EAEC was significantly higher among diarrhoeal cases than controls indicating the importance of this pathogen detection scheme.

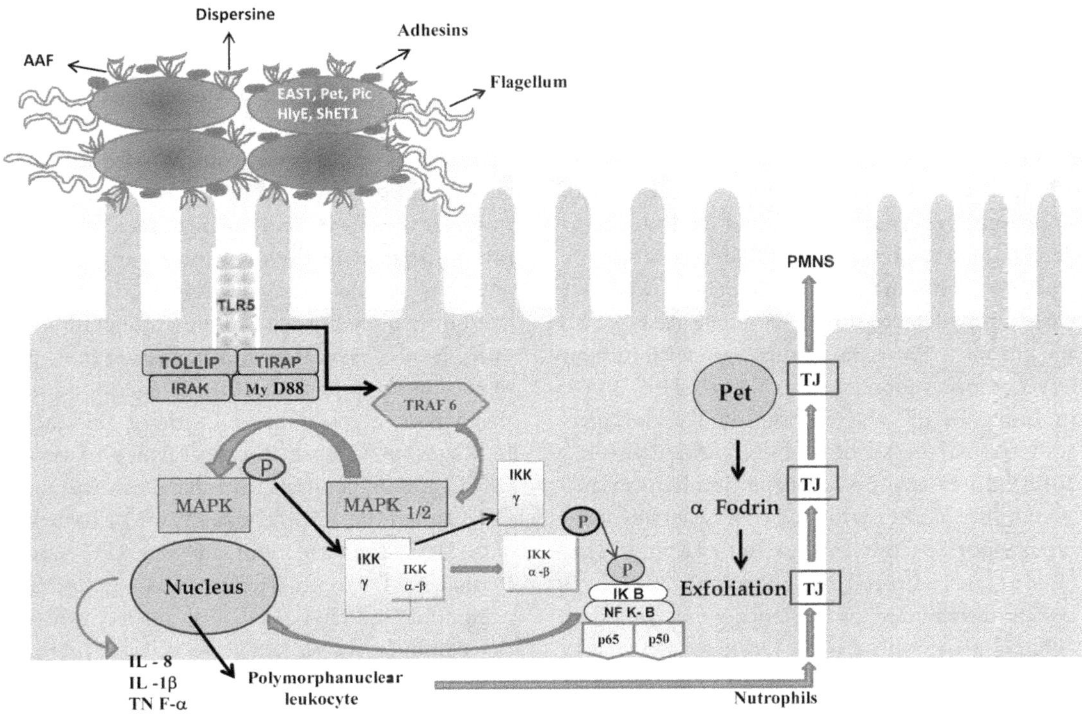

Figure 2.6 Pathogenic mechanisms of EAEC. During the initial stage, EAEC attaches to the intestinal mucosa and/or the mucus layer. The adherence process requires AAF fimbrial structures. The second stage includes formation of biofilm and expression of antiaggregation by dispersin across the intestinal mucosa. The final stage of infection involves release of EAEC toxins, and elicitation of an inflammatory response, mucosal toxicity and intestinal secretion. The toxins, Pet and EAST1 are associated with increased levels of faecal cytokines and inflammatory markers such as interleukins, interferon (INF)-α and IL-6, lactoferrin, faecal leucocytes, and occult blood. IL-8 is responsible for recruiting neutrophils to the epithelial mucosa and facilitates intestinal fluid secretion

The proportion of EAEC associated with diarrhoea was high (20–60%) among travellers visiting South Asia, sub-Saharan Africa and South America. A study conducted in three different geographical areas designated EAEC as a major cause of TD (26%) along with ETEC (30%) (Adachi *et al.*, 2001). Even among children > 5 years of age, EAEC was an aetiological agent of diarrhoea associated with international travel (Pabst *et al.*, 2003). From more than 50 published articles on TD (1973 to 2008), EAEC was found to be implicated with the infection next to ETEC (Shah *et al.*, 2009). Up to 60% of the US visitors to Mexico develop TD due to EAEC infection with a peak during winter months than summer diarrhoea caused by ETEC (Paredes-Paredes *et al.*, 2011). Travellers from the US and Europe with diarrhoea were often infected with EAEC having *aggR, aap, astA* and *set1A* genes (Huang *et al.*, 2007). Any two or three of these genes were associated with diarrhoea with gross mucus and faecal leucocytes.

Incidence and virulence gene distribution in EAEC

EAEC was commonly found in Argentinian children (31%) when they were tested since their birth and during the first 20 months of their lives (Quiroga *et al.*, 2004). Introduction of weaning foods and complete termination of breastfeeding were associated with an increase of asymptomatic infections. In many developing countries, it was shown that EAEC was appreciably detected in diarrhoeal children (8–20%) (Knutton *et al.*, 2001; Rodrigues *et al.*, 2002; Okeke *et al.*, 2003; Kahali *et al.*, 2004b; Araujo *et al.*, 2007; Mandomando *et al.*, 2007; Takahashi *et al.*, 2008; Ochoa *et al.*, 2009a; Usein *et al.*, 2009; Moreno *et al.*, 2010). In indigenous diarrhoeal cases from Japan, the US and Nigeria, prevalence of EAEC was next to EPEC (Nishikawa *et al.*, 2002; Vernacchio *et al.*, 2006; Nweze, 2010). A prospective study conducted in the US to determine the aetiology of diarrhoeal illness among patients and control subjects showed that EAEC was detected in 4.5% of the cases of all age groups without any association with foreign travel or immunodeficiency (Nataro *et al.*, 2006). In India, EAEC was predominantly identified among tribal pre-school children (Anvikar *et al.*, 2008). There was no serotype domination and most of the EAEC belonged to a variety of serotypes reported from Kolkata, India (Kahali *et al.*, 2004b) and in Brazil (Franzolin *et al.*, 2005).

The distribution of virulence genes varied in different geographical areas. Virulence gene markers such as CVD432 plasmid (19%), *astA* (34%) and *pet* (42%) were commonly detected in EAEC isolates from diarrhoeal children in Lambaréné, Gabon (Presterl *et al.*, 2003). The incidence of EAEC in children with diarrhoea from Mongolia was 15% and the AggR-positive EAEC in the diarrhoeal group was significantly higher with other virulence genes such as *aatA* (30.5%), *capU* (25.0%), *aap* (22.2%), *pet* (16.6%), *agg-3a*, (13.8%), *aafA* (11.1%), *set* (8.3%) and *aggA* (5.5%) (Sarantuya *et al.*, 2004). The detection rate of EAEC in Ifakara, Tanzania, from DEC positive diarrhoeal stool samples (35–63%) varied from 35% to 63% with a peak incidence during dry season (Vargas *et al.*, 2004). Except for AA, *aap*, and *aggR*, the rest of the virulence related genes such as *pet*, *shf*, *aggA*, and *aafA* were absent in EAEC from Iran (Bouzari *et al.*, 2005). Most of the EAEC from Tanzania harboured the *aggR* gene (61.6%) and *aat* gene alone (13%) (Mendez-Arancibia *et al.*, 2008). EAEC has also been implicated in the development of irritable bowel syndrome (IBS). The majority of the EAEC from adults with IBS carried the genes encoding different iron acquisition systems especially the *chuA* coding for a haem receptor and yersiniabactin-encoding genes (Sobieszczańska, 2008).

In Dar es Salaam, Tanzania, the prevalence of EAEC among diarrhoeal patients was high (14.6%) and most of them (82.9%) were classified as typical EAEC possessing the *aggR* gene, and 92.6% carried the *aat* gene (Moyo *et al.*, 2007). EAEC isolated from Brazilian children with diarrhoea as well as controls harboured the genes *fim*, *aggR* and *aap* in almost equal proportions. However, *hly*, *aero* and *aggA* sequences were more prevalent in the diarrhoeal group and the gene *kps* was detected exclusively in strains isolated from symptomatic children (Regua-Mangia *et al.*, 2009). Typical EAEC expressing the AggR regulon is an important cause of childhood diarrhoea than atypical EAEC without this factor. In Kagoshima, Japan, the prevalence of typical EAEC among children with diarrhoea was more (74%) than the atypical ones (26%) (Tokuda *et al.*, 2010). The detection rate of EAEC among Iranian children with diarrhoea was 10% and 87% using PCR and He La cell assay, respectively, and the combination virulence genes *aggR-aap-astA* was detected in 20% of the isolates (Aslani *et al.*, 2011).

Host genetic factors have established that SNP in the lactoferrin, osteoprotegerin, and IL-8/IL-10 genes are related with increased risks for diarrhoea caused by enteric pathogens (Flores and Okhuysen, 2009b; Hill and Beeching, 2010). SNP in the IL-8 gene was detected among EAEC-related infection in the US students who remained in Mexico for a long time. The probability of getting the infection was high with AA and AT genotypes at the –251 position compared with TT genotype (Jiang *et al.*, 2003). The faecal IL-8 was high among those who had the AA genotype.

Association with diarrhoea among HIV-infected individuals

Persistent diarrhoea is a main cause of morbidity in HIV-infected individuals. Wanke *et al.* (1998) showed that EAEC was significantly found in HIV-infected patients with diarrhoea compared with the non-diarrhoeal group. In Zambia, EAEC isolated from AIDS-related patients with persistent diarrhoea exhibited cytotoxic phenotype than isolates from the control group (Kelly *et al.*, 2003). Investigations made in Venda, South Africa on EAEC and its virulence-related genes with HIV-positive diarrhoeal cases and diarrhoea-negative controls indicated that the EAEC was significantly associated with cases (Samie *et al.*, 2007). In addition, the presence of EAEC-specific virulence genes was significantly associated with occult blood in the stool samples. In some studies, HIV infected individuals with and without diarrhoea showed no appreciable difference between these two groups either by PCR or cell adherence assay in detecting EAEC (Durrer *et al.*, 2000). In HIV-positive Peruvian children with and without diarrhoea, all the DEC including EAEC was found in equal frequencies and more in controls (Medina *et al.*, 2010).

Waterborne and foodborne outbreaks

Following an experimental challenge, virulence of the EAEC strain 042 has been established in volunteers with an inoculum of 10^{10} cfu (Nataro *et al.*, 1995a). The incubation period of EAEC in diarrhoeal illness ranges between 8 and 18 hours. EAEC was responsible for many diarrhoeal outbreaks due to consumption of water from open wells in India (Pai *et al.*, 1997), school lunch and party foods in Japan (Itoh *et al.*, 1997; Yatsuyanagi *et al.*, 2002) and restaurant foods in the UK (Smith *et al.*, 1997). EAEC serotype O92:H33 harbouring the virulence-related genes *aat, aggR, aap* and *set1A* was reported from a foodborne outbreak of gastroenteritis in Italy with suspected source from the pecorino cheese made with unpasteurized sheep's milk (Scavia *et al.*, 2008). During May-June 2011, more than 800 HUS and bloody diarrhoeal cases with about 40 deaths caused by a unique *E. coli* strain O104: H4 were reported from Germany and France. The suspected source of this infection was the German sprouts (Struelens *et al.*, 2011). HUS caused by this serotype of *E. coli* has already been reported in Korea (Bae *et al.*, 2006). The genome analysis of two isolates from German outbreak indicated that they belonged to the same MLST of EAEC (Brzuszkiewicz *et al.*, 2011). In addition, these isolates carried virulence-related genes of EAEC such as *aggA, aggR, set1, pic, aap* and a prophage encoding the Shiga toxin (Stx2), which is characteristic for EHEC (Bielaszewska *et al.*, 2011). Adherence of these isolates to intestinal epithelium similar to EAEC and expression of Stx2 could explain the high progression to HUS in a susceptible population. Unique genomic features of these O104:H4 isolates suggested that it may represent a new pathotype, enteroaggregative-haemorrhagic *E. coli* (EAHEC) (Brzuszkiewicz *et al.*, 2011).

Antimicrobial resistance

Compared with other diarrhoeagenic *E. coli*, EAEC was found to be highly resistant to many commonly used antimicrobials used in the treatment of diarrhoea (Vila *et al.*, 2001). The majority of the EAEC from diarrhoeal patients in Kolkata, India, exhibited multidrug resistance including resistance to fluoroquinolones (Kahali *et al.*, 2004b). Peruvian EAEC isolates were also multidrug resistant ,especially to ampicillin, cotrimoxazole, tetracycline and nalidixic acid (Ochoa *et al.*, 2009b). EAEC from Iran was highly resistant to ampicillin, erythromycin, cephalothin, co-trimoxazole, tetracycline and nalidixic acid (Aslani *et al.*, 2011). Generally, the typical EAEC seems to exhibit multidrug resistance to ampicillin, cefotaxime and tetracycline (Tokuda *et al.*, 2010). Surface hydrophobicity of EAEC seems to affect the response to ciprofloxacin. Mortensen *et al.* (2011) found that ciprofloxacin had an inhibitory effect on EAEC adhesion to HEp-2 cells due to the presence of hydrophobic positively charged surface protein, dispersin.

Molecular typing

In several molecular typing methods and in outer membrane protein profiles, EAEC was found to be heterogeneous irrespective of their serotypes (Suzart *et al.*, 1999; Kahali *et al.*, 2004b). These typing methods may be useful in outbreak situation but not in general surveillance. In several recent findings, specific targets were used to differentiate the pathogenic EAEC from the non-pathogenic counterpart. However, such techniques were not tested in larger epidemiological studies. In the MLST analysis, EAEC from case and controls of Nigerian children were identified as multiple pathogenic lineages with 40% of the isolates belonging to double locus variants of STs 10, 31, and 394 (Okeke *et al.*, 2010). The ST10-complex was associated with weaned children with diarrhoea and the other two clonal complexes possessed conserved chromosomal loci. The majority of EAEC isolates do not produce the aggregative adherence fimbriae I-III (AAFs), but harbour the *E. coli* common pilus (ECP) gene *ecpA* (Avelino *et al.*, 2010).

In a comparative *ecpA* mutational studies, ECP was shown to be important in strain 278-1 (AAF^-) for adherence to cultured epithelial cells than the prototype strain 042 (AAF/II^+). This indicates a potential role of ECP in adherence for EAEC strains lacking known AAFs and thus useful in case and control studies. In dual infection caused by EAEC harbouring F-pili gene (*traA*) and aggregative *Citrobacter freundii*, the putative F pili expressed by EAEC strains increased the biofilm formation 8-fold in the presence of *C. freundii* (Pereira *et al.*, 2010).

In EAEC, AraC-like regulator AggR is an important virulence regulator. AggR-regulated genes comprise those encoding the AAF, the dispersin protein, and a T6SS. Transcriptional initiation site of the *aggR* promoter was identified 40 nucleotides upstream of the translational start (Morin *et al.*, 2010). Regulation cascade controlling the *aggR* expression during EAEC infection is very complex. The *aggR* promoter (P) is positively regulated by the DNA-binding protein FIS and negatively regulated by the global regulator H-NS. Compared with *in vitro* tests, assays in mouse proved that EAEC colonized the intestine and possessed relatively high levels of P(*fis*) and P(*aggR*) activity and a low level of P(*hns*) (Morin *et al.*, 2010). Full genome sequence of the EAEC prototype strain 042 has been done along with a comparative metabolic profiling (Chaudhuri *et al.*, 2010) that has helped to generate information on proteomes and genomes.

Detection methods

The commonly used detection methods are presented in Table 2.3. Classical diagnosis of EAEC is the characteristic 'stacked-brick' like adhesion pattern when co-cultured with HEp-2 cells. However, this method is laborious while handling a large number of EAEC isolates. The clump-formation assay in Mueller–Hinton broth for the presumptive identification of EAEC was first proposed by Albert *et al.* (1993), and when this assay was used in combination with cell culture and *aggR* based PCR, the specificity and sensitivity was found to be 90% (Iwanaga *et al.*, 2002). Biofilm formation can also be detected and quantified by using a microtitre plate assay by crystal violet staining method. The aggregative property of EAEC interferes in serotyping by causing auto-agglutination (Weintraub, 2007) and this may be the reason why many isolates of EAEC remain serologically unidentifiable compared with other DEC.

EAEC-specific DNA probes made from the AA plasmid have been used in many epidemiological studies. Several PCR methods were established for the specific detection of EAEC individually or with other DEC. PCR assays targeting the chromosomally encoded gene, *aaiA* can be used for the detection of typical and atypical EAEC. The plasmid-encoded genes such as *aat, aggR,* and *aap,* are useful in the detection of typical EAEC. Since *astA* is not specific for EAEC, this gene can be used in combination with other EAEC genes. Expression of Pet can be directly detected from EAEC culture supernatants by slot blot immunoassay and can be used in clinical laboratories (Vilhena-Costa *et al.*, 2006). Faecal cards are generally used in the detection of occult blood. PCR assay using the DNA extracted from these cards significantly identified (49%) EAEC cases among the US travellers to Mexico with diarrhoea (Grimes *et al.*, 2008). This method is pertinent for large field studies that do not require immediate processing of stool samples for the detection of DEC. Existing methods for the detection of DEC are too expensive and labour-intensive in routine practice. A highly specific real-time fluorescence-based multiplex PCR for the six categories of DEC targeting eight different virulence genes has been developed (Guion *et al.*, 2008) with the completion time of about 90 min.

Treatment

EAEC infections are generally self-limiting and adequate oral rehydration still remains the first line of therapy. In developing countries, the risk factors for the development of persistent EAEC-mediated diarrhoea include young age, malnutrition, and immunosuppression. In patients with persistent diarrhoea and severe dehydration, antimicrobials such as azithromycin and rifaximin may be given to shorten the duration of diarrhoea (Huang and Dupont, 2004). For EAEC infection among travellers, rifaximin was shown to be very effective (Infante *et al.*, 2004). This antimicrobial agent shortens the duration of diarrhoea without altering the colonic flora and also reduces the expression of the matrix metalloproteinase-9 (Jiang *et al.*, 2010). However, the antimicrobial therapy depends on individual basis and remains largely as an empirical treatment. In infants, breastfeeding gives protection against persistent diarrhoea caused by EAEC. IgA from human colostrum inhibits adhesion of EAEC to HEp-2 cells mediated by AAF/II and suggests that these antibodies may play a role in protection of infants against diarrhoea caused by EAEC (Fernandes *et al.*, 2001). Studies on the broad-spectrum antiparasitic and antidiarrhoeal drug nitazoxanide with EAEC strains JM221 and/or 042 (expressing AAF/I and AAF/II, respectively) showed inhibition of biofilm production and haemagglutination of red blood cells thereby indicating its influence on fimbrial assembly (Shamir *et al.*, 2010).

In the Hep-2 cell assay, probiotic metabolites from *B. breve*, *L. casei* subsp. *casei* and *Enterococcus faecium* inhibited colonization of EAEC or induced changes in the adherence pattern, reducing the frequency and size of bacterial aggregates after the pre-adherence period (Lazăr *et al.*, 2009). Micronutrient supplementation in the form of zinc has several beneficiary effects. Zinc inhibited biofilm formation *in vitro* by the typical EAEC isolated from children, indicating its efficient use in the management of EAEC-mediated diarrhoea (Pereira *et al.*, 2010).

Enteroinvasive *Escherichia coli* (EIEC)

Enteroinvasive *Escherichia coli* (EIEC) causes invasive diarrhoea and shares virulence mechanisms and clinical manifestations with *Shigella*. EIEC strains are lysine decarboxylase negative, non-motile, lactose negative and resemble *Shigella* spp. during primary isolation. Phylogenetically, *Shigella* spp. and EIEC arose from independent clonal lines of *E. coli* by horizontal acquisition of virulence genes, forming a single pathotype. However, EIEC is less commonly reported compared with other members of DEC. EIEC was first reported during 1943–1945 among US soldiers in the Mediterranean region (Ewing and Gravatti, 1947) followed by a gastroenteritis outbreak among school children and hospitalized patients in the UK (Hobbs *et al.*, 1949; Rowe *et al.*, 1974) and waterborne outbreaks in Hungary (Lanyi *et al.*, 1959). In these reports, *E. coli* isolates belonging to serogroup O124 was identified.

Clinical manifestations

Though in the majority of the EIEC infected cases watery diarrhoea is most common, in few instances, patients may have typical dysentery syndrome (Marier *et al.*, 1973; Snyder *et al.*, 1984). Passage of stools is very frequent, but of low volume in contrast to the watery stools seen with other DEC infections.

Pathogenesis

In several publications, the basics of EIEC pathogenesis have been described that include penetration of intestinal epithelial cells, endocytic vacuole lysis, intracellular multiplication, relocation across the cytoplasm and migration into adjacent epithelial cells. The hallmark of the infection is the invasiveness of EIEC, which is mediated by a 120-MDa virulence plasmid, pInv with several virulence associated loci such as *mxi* and *spa* loci (Fig. 2.7). Both plasmid and chromosomal genes are involved in conferring pathogenicity to EIEC. T3SS mechanism is obligatory for the secretion of many proteins that are necessary for expression of maximum virulence.

The Ipa proteins (IpaBCDA) are secreted proteins, in which IpaB, IpaC, and IpaD are effectors of the invasion of EIEC/*Shigella*. IpaC promotes the uptake of the pathogen into the epithelial cell, whereas IpaB helps in the lysis of the phagocytic vacuole and stimulation of apoptosis in macrophages. The progression of the pathogen in the cytoplasm and to other epithelial cells is facilitated by the nucleation of cellular actin that is regulated by VirG. VirR is a chromosomally encoded histone-like protein and may act with a pInv encoded transcriptional activator VirF. The expression of Inv-encoded virulence factors is globally regulated by VirB and MxiE. Gross comparisons and sequence alignments of eight Inv plasmids suggest that the *Shigella* Inv plasmids have evolved from an EIEC Inv plasmid predecessor via successive

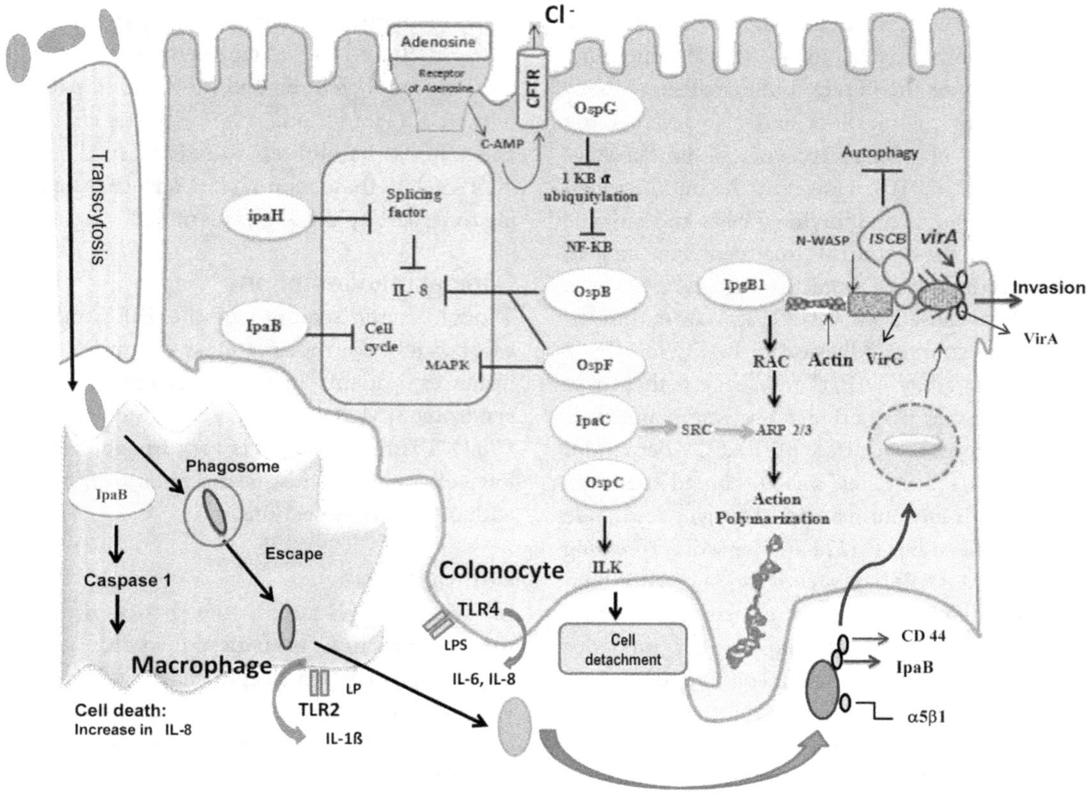

Figure 2.7 Pathogenic mechanism of EIEC. The basics of EIEC pathogenesis include penetration of intestinal epithelial cells, endocytic vacuole lysis, intracellular multiplication, relocation across the cytoplasm and migration into adjacent epithelial cells. The hallmark of the infection is the invasiveness of EIEC, which is mediated by a 120-MDa virulence plasmid, pInv with several virulence associated loci such as *mxi* and *spa* loci. Both plasmid and chromosomal genes are involved in conferring pathogenicity to EIEC. T3SS mechanism is obligatory for the secretion of many proteins that are necessary for expression of maximum virulence. See text for details.

deletions and additions of several DNA blocks (Johnson and Nolan, 2009). Another plasmid-borne *Shigella* enterotoxin encoding gene (*sen*) with a predicted size of 63-kDa activated Isc levels in Ussing chamber assays (Nataro *et al.*, 1995b). Presence of *sen* in majority of the isolates was correlated to the watery diarrhoea among EIEC-infected cases.

Pathogenesis of EIEC is complex as the pathogen has to survive in the host cell and trigger several mechanisms to initiate infection. Pathogenesis of EIEC begins in the colon, where the bacteria traverse the microfold cells (M cells) by transcytosis and move into the submucosa (Fig. 2.7). Since EIEC does not have flagella, it exploits the host's machinery for movement in the cytosol and cell-to-cell dissemination. Release of the pathogen from the M-cells causes an initial release of IL-1β that attracts polymorphonucleocytes (PMNs), which is a precursor to the secretagogue adenosine and activates Cl^- secretion through CFTR. EIEC is freed from dead macrophages into the submucosa, which follows invasion into the basolateral side of colonocytes with the aid of T3SS. This secretion system injects a number of effector proteins required for initial invasion, escape from the vacuole and movement within the cell. In the epithelial cell cytoplasm, EIEC promotes its survival using effectors to weaken the host cell processes. For this, IpaB mediates cell cycle arrest and OspE interacts with integrin-linked kinase (ILK) to prevent epithelial cell detachment. IpgD prevents apoptosis by stimulating the phosphoinositide 3-kinase and activates Akt proteins.

To actively stay inside the colonocytes, EIEC must also evade innate immune responses using many effectors. One of these effectors, OspG, binds ubiquitylated E2 proteins, which prevent degradation of inhibitor of nuclear factor-κB (NF-κB) subunit-α (IκBα) and thus inhibiting NF-κB activation. OspF is targeted to the nucleus, where it permanently dephosphorylates mitogen- stimulated protein kinases that are needed for the transcription of genes regulated by NF-κB79. IpaH is also targeted to the nucleus, where it interacts with a splicing factor implicated in the expression of inflammatory cytokines. The OspB, acts with OspF to reduce IL-8 levels by recruiting host factors that remodel chromatin.

Three effector proteins, OspF, OspG and IpaH modulate the host immune response and differentially modulate the host diarrhoeal response. OspF has a unique way of inhibiting MAPK signalling. These kinases trigger transcription factors influencing production of proinflammatory cytokines including IL-1, IL-6 and TNFα. MAP kinases are also targeted by IpaH. In contrast, OspG actively interferes with the ubiquitination of phospho-IκBα, preventing its degradation. IκBα generally binds to NF-κB and averts its translocation to the nucleus. OspF, IpaH and OspG affect both the MAP kinase cascade and NF-κB translocation, blocking many of the essential pathways required for transcriptional activation of host cytokines. Effectors like IpaC, activate SRC kinases at the site of bacterial contact to recruit the ARP2/3 complex and cause actin polymerization and ruffle formation for bacterial entry. RAC1, which can promote membrane ruffle formation, may be activated by IpgB1. The other effectors, such as IpgD, IpaA and VirA, are involved in the destabilization of actin and microtubules to promote invasion into a phagosome. The escape of EIEC from phagosomes is dependent on IpaB, IpaC, IpaD and IpaH. VirG localizes to a single pole and recruits and activates N-WASP and the ARP2/3 complex for actin polymerization, which drives the pathogen through the cell. VirA helps in the intracellular spread of EIEC by providing a channel that enables them to move throughout the cell.

Epidemiology

The pathogenic potential of EIEC was first demonstrated by DuPont *et al.* (1971) in human volunteer studies. Initial studies on outbreaks and sporadic cases caused by EIEC were reported in Australia (Riley, 1968) and Israel (Shmilovitz *et al.*, 1974). EIEC is one of the important causes of diarrhoeal disease and dysentery in children from less developed countries (Taylor *et al.*, 1986; Levine, 1987). In developed countries, EIEC is mostly associated with cruise ship (Snyder *et al.*, 1984) or water- and foodborne outbreaks (Lanyi *et al.*, 1959; Marier *et al.*, 1973; Tulloch *et al.*, 1973). EIEC outbreak due to person-to-person transmission was investigated in 1981 in Columbia, Missouri (Harris *et al.*, 1985). In this outbreak, EIEC serogroup O124 was implicated. During 1985, EIEC serogroup O143 outbreak was recorded in Houston, USA, due to consumption of contaminated guacamole (Gordillo *et al.*, 1992). A Taiwanese study identified EIEC in as high as 20% of patients, which was next to ETEC (67%) (Yang *et al.*, 2007). Case–control studies conducted in 22 rural communities in north-western Ecuador showed that the incidence of EIEC (3.2 cases/100 persons) was more than those of other diarrhoeal pathogens (Vieira *et al.*, 2007). In addition, EIEC displayed comparable risk factor relationships with other pathotypes analysed, but different age-specific infection rates were shown. Studies conducted in Nicaragua showed that the isolation rates of EIEC were similar in both the cases and controls (0.8% and 1.4%, respectively) (Vilchez *et al.*, 2009). Of the 32 diarrhoea episodes observed among Mexican children, 9% were associated with EIEC (Estrada-Garcia *et al.*, 2009). In Costa Rica, the incidence of EIEC infection was high (19%) among children with diarrhoea (Pérez *et al.*, 2010).

Detection methods

The detection methods for EIEC are summarized in Table 2.3. For the conventional culture, xylose–lysine–deoxycholate, in combination with MacConkey or eosine–methylene blue agar, seems to provide the best isolation rate for EIEC (Szakál and Pál, 2003). Sereney test using guinea pigs is considered as the gold standard to confirm the invasive nature of EIEC isolates (Sereny,

1957). A mouse model was also suggested in assessing the invasive property of EIEC (Murayama *et al.*, 1986). Plaque formation in HeLa cell monolayer was shown to be associated with virulence of EIEC/*Shigella* isolates (Ménard and Sansonetti, 1994).

An ELISA technique for the detection of IpaC and pro- and anti-inflammatory cytokines induced by whole bacterial cultures of EIEC as well as by their soluble substances have been reported to obviate the need for an animal model (Pal *et al.*, 1997; Chifiriuc *et al.*, 2010). Several DNA probes and PCR assays targeting the virulence genes such as *inv* and *ial* have been used for the detection of EIEC using either the isolates or directly in stool specimens (Gomes *et al.*, 1987; Small and Falknow 1988; Frankel *et al.*, 1989, 1990). Combinations of IpaC protein-specific monoclonal antibody (Mab) for colony blot assay, and *Shigella*/EIEC specific *ipaH* gene PCR assay have increasingly detected EIEC in stool specimens (Szakál *et al.*, 2003). A loop-mediated isothermal amplification (LAMP) method efficiently detected the *ipaH* gene of EIEC within 2 hours at a minimal count of 8 cfu per reaction (Song *et al.*, 2005).

Treatment

Antimicrobial therapy is mandatory for EIEC/*Shigella* infected cases as the dehydration status is less severe with modest requirement of ORS. In some countries, trimethoprim–sulfamethoxazole (TMP-SMX) is still useful for the treatment of dysentery cases. Ceftibuten was found to be as effective as TMP-SMX in the treatment of diarrhoea caused by EIEC in children (Prado *et al.*, 1992) and hence can be used in areas where TMP-SMX is less effective. Addition of EIEC to human intestinal epithelial cell lines along with the live probiotic strains of *Streptococcus thermophilus* and *Lactobacillus acidophilus* significantly limited adhesion, invasion, and physiological dysfunction induced by the pathogen. Thus, the live probiotics presumably protect intestinal epithelial cells from the deleterious effects of infection with EIEC (Resta-Lenert and Barrett, 2003). *In vitro* studies showed that upregulation of nitric oxide synthase and cyclooxygenase-2 by EIEC can modulate chloride secretion and barrier function in human intestinal epithelial cell lines. Thus, these enzymes represent possible therapeutic targets in infectious diarrhoea (Resta-Lenert and Barrett, 2002).

Concluding remarks

E. coli was considered as harmless commensal of the intestinal tract until the discovery of EPEC in the early part of the 20th century. In the following several decades, with the advent of techniques in the study of virulence and pathogenesis, other categories of DEC were discovered. Each pathotype of *E. coli* has evolved by acquiring specific set of virulence genes through mobile genetic elements-plasmids, transposons and bacteriophages. The factors encoded by these virulence genes affect a variety of cellular processes including protein synthesis, cell division, cytoskeletal function, apoptosis, and water and electrolyte transport. Our improved understanding of the molecular mechanisms of *E. coli* pathogenesis has uncovered new aspects of host response to infecting pathogens. However, our understanding of the interactions between virulence factors and host components remains incomplete. It will be interesting to further study the interactions to uncover how they contribute to the progression of the disease. This knowledge could be exploited in better treatment and prevention strategies. Because of the genome plasticity of *E. coli*, new pathogens will probably continue to evolve. The recent emergence of the highly virulent EHEC, O104:H4 in Germany is a case in point. We should be vigilant to the threats posed by new pathotypes and variants of the existing pathotypes of *E. coli* that may emerge in future. Water hygiene is an important aspect of the prevention strategy of DEC infections.

References

Adachi, J.A., Jiang, Z.D., Mathewson, J.J., Verenkar, M.P., Thompson, S., Martinez-Sandoval, F., Steffen, R., Ericsson, C.D., and DuPont H.L. (2001). Enteroaggregative *Escherichia coli* as a major etiologic agent in traveler's diarrhea in 3 regions of the world. Clin. Infect. Dis. 32, 1706–1709.

Afset, J.E., Bevanger, L., Romundstad, P., and Bergh, K. (2004). Association of atypical enteropathogenic *Escherichia coli* (EPEC) with prolonged diarrhea. J. Med. Microbiol. 53, 1137–1144.

Ahmed, T., Lundgren, A., Arifuzzaman, M., Qadri, F., Teneberg, S., and Svennerholm, A.M. (2009). Children with the Le(a+b⁻) blood group have increased susceptibility to diarrhea caused by enterotoxigenic *Escherichia coli* expressing colonization factor I group fimbriae. Infect. Immun. *77*, 2059–2064.

Albert, M.J., Qadri, F., Haque, A., and Bhuiyan, N.A. (1993). Bacterial clump formation at the surface of liquid culture as a rapid test for identification of enteroaggregative *Escherichia coli*. J. Clin. Microbiol. *31*, 1397–1399.

Albert, M.J., Faruque, SM., Faruque, A.S.G., Neogi, P.K., Ansaruzzaman, M., Bhuiyan, N.A., Alam, K., and Akbar, M.S. (1995). Controlled study of *Escherichia coli* diarrheal infections in Bangladeshi children. J. Clin. Microbiol. *33*, 973–977.

Albert, M.J., Faruque, A.S.G., Faruque, S.M., Sack, R.B., and Mahalanabis, D. (1999). Case–control study of enteropathogens associated with childhood diarrhea in Dhaka, Bangladesh. J. Clin. Microbiol. *37*, 3458–3464.

Albert, M.J., Rotimi, V.O., Dhar, R., Silpikurian, S., Pacsa, A.S., Molla, A.M., and Szucs, G. (2009). Diarrheagenic *Escherichia coli* are not a significant cause of diarrhoea in hospitalised children in Kuwait. BMC Microbiol. *9*, 62.

Al-Gallas, N., Abbassi, S.M., Hassan, A.B., and Aissa, R.B. (2007a). Genotypic and phenotypic profiles of enterotoxigenic *Escherichia coli* associated with acute diarrhea in Tunis, Tunisia. Curr. Microbiol. *55*, 47–55.

Al-Gallas, N., Bahri, O., Bouratbeen, A., Hassan, A.B., and Aissa, R.B. (2007b). Etiology of acute diarrhea in children and adults in Tunis, Tunisia, with emphasis on diarrheagenic *Escherichia coli*: prevalence, phenotyping, and molecular epidemiology. Am. J. Trop. Med. Hyg. *77*, 571–582.

Anon. (2000). Waterborne outbreak of gastroenteritis associated with a contaminated municipal water supply, Walkerton, Ontario, May–June 2000. Can. Commun. Dis. Rep. *26*, 170–173.

Anvikar, A.R., Dolla, C., Dutta, S., Rao, V.G., Gadge, V.S., Shukla, G.P., Rao, S., and Karforma, C. (2008). Role of *Escherichia coli* in acute diarrhoea in tribal preschool children of central India. Paediatr. Perinat. Epidemiol. *22*, 40–46.

Aranda, K.R.S., Fagundes-Neto, U., and Scaleksky, I.C.A. (2004). Evaluation of multiplex PCRs for diagnosis of infection with diarrheagenic *Escherichia coli* and *Shigella* spp. J. Clin. Microbiol. *42*, 5849–5853.

Araujo, J.M., Tabarelli, G.F., Aranda, K.R., Fabbricotti, S.H., Fagundes-Neto, U., Mendes, C.M., and Scaletsky, I.C. (2007). Typical enteroaggregative and atypical enteropathogenic types of *Escherichia coli* are the most prevalent diarrhea-associated pathotypes among Brazilian children. J. Clin. Microbiol. *45*, 3396–3399.

Aslani, M.M., Ahrabi, S.S., Alikhani, Y.M., Jafari, F., Zali, R.M., and Mani, M. (2008). Molecular detection and antimicrobial resistance of diarrheagenic *Escherichia coli* strains isolated from diarrheal cases. Saudi. Med. J. *29*, 388–392.

Aslani, M.M., Alikhani, M.Y., Zavari, A., Yousefi, R., and Zamani, A.R. (2011). Characterization of enteroaggregative *Escherichia coli* (EAEC) clinical isolates and their antibiotic resistance pattern. Int. J. Infect. Dis. *15*, 136–139.

Avelino, F., Saldaña, Z., Islam, S., Monteiro-Neto, V., Dall'Agnol, M., Eslava, C.A., and Girón, J.A. (2010). The majority of enteroaggregative *Escherichia coli* strains produce the *E. coli* common pilus when adhering to cultured epithelial cells. Int. J. Med. Microbiol. *300*, 440–448.

Avery, S.M., Liebana, E., Reid, C.A., Woodward, M.J., and Buncic, S. (2002). Combined use of two genetic fingerprinting methods, pulsed-field gel electrophoresis and ribotyping, for characterization of *Escherichia coli* O157 isolates from food animals, retail meats, and cases of human disease. J. Clin. Microbiol. *40*, 2806–2812.

Bae, W.K., Lee, Y.K., Cho, M.S., Ma, S.K., Kim, S.W., Kim, N.H., and Choi, K.C. (2006). A case of hemolytic uremic syndrome caused by *Escherichia coli* O104:H4. Yonsei. Med. J. *47*, 437–439.

Baldini, M.M., Nataro, J.P., and Kaper, J.B. (1986). Localization of a determinant for HEp-2 adherence by enteropathogenic *Escherichia coli*. Infect. Immun. *52*, 334–336.

Beatty, M.E., Bopp, C.A., Wells, J.G., Greene, K.D., Puhr, N.D., and Mintz, E.D. (2004). Enterotoxin-producing *Escherichia coli* O169:H41, United States. Emerg. Infect. Dis. *10*, 518–521.

Beatty, M.E., Adcock, P.M., Smith, S.W., Quinlan, K., Kamimoto, L.A., Rowe, S.Y., Scott, K., Conover, C., Varchmin, T., Bopp, C.A., *et al.* (2006). Epidemic diarrhea due to enterotoxigenic *Escherichia coli*. Clin. Infect. Dis. *42*, 329–334.

Begum, Y.A., Talukder, K.A., Nair, G.B., Khan, S.I., Svennerholm, A.M., Sack, R.B., and Qadri, F. (2007). Comparison of enterotoxigenic *Escherichia coli* isolated from surface water and diarrhoeal stool samples in Bangladesh. Can. J. Microbiol. *53*, 19–26.

Bellini, E.M., Elias, W.P., Gomes, T.A., Tanaka, T.L., Taddei, C.R., Huerta, R., Navarro-Garcia, F., and Martinez, M.B. (2005). Antibody response against plasmid-encoded toxin (Pet) and the protein involved in intestinal colonization (Pic) in children with diarrhea produced by enteroaggregative *Escherichia coli*. FEMS. Immunol. Med. Microbiol. *43*, 259–264.

Bhargava, S., Johnson, B.B., Hwang, J., Harris, T.A., George, A.S., Muir, A., Dorff, J., and Okeke, I.N. (2009). Heat-resistant agglutinin 1 is an accessory enteroaggregative *Escherichia coli* colonization factor. J. Bacteriol. *191*, 4934–4942.

Bielaszewska, M., Mellmann, A., Zhang, W., Köck, R., Fruth, A., Bauwens, A., Peters, G., and Karch, H. (2011). Characterisation of the *Escherichia coli* strain associated with an outbreak of haemolytic uraemic syndrome in Germany, a microbiological study. Lancet Infect. Dis. *11*, 671–676.

Birch, M., Denning, D.W., and Law, D. (1996). Rapid genotyping of *Escherichia coli* O157 isolates by random amplification of polymorphic DNA. Eur. J Clin. Microbiol. Infect. Dis. *15*, 297–302.

Bodhidatta, L., McDaniel, P., Sornsakrin, S., Srijan, A., Serichantalergs, O., and Mason, C.J. (2010). Case–control study of diarrheal disease etiology in a remote rural area in Western Thailand. Am. J. Trop. Med. Hyg. *83*, 1106–1109.

Boisen, N., Struve, C., Scheutz, F., Krogfelt, K.A., and Nataro, J.P. (2008). New adhesin of enteroaggregative *Escherichia coli* related to the Afa/Dr/AAF family. Infect. Immun. *76*, 3281–3292.

Bokete, T.N., O'Callahan, C.M., Clausen, C.R., Tang, N.M., Tran, N., Moseley, S.L., Fritsche, T.R., and Parr, P.I. (1993). Shiga-like toxin-producing *Escherichia coli* in Seattle children: a prospective study. Gastroenterology *105*, 1724–1731.

Bokete, T.N., Whittam, T.S., Wilson, R.A., Clausen, C.R., O'Callahan, C.M., Moseley, S.L., Fritsche, T.R., and Tarr, P.I. (1997). Genetic and phenotypic analysis of *Escherichia coli* with enteropathogenic characteristics isolated from Seattle children. J. Infect. Dis. *175*, 1382–1389.

Bouzari, S., Jafari, A., and Zarepour, M. (2005). Distribution of virulence related genes among enteroaggregative *Escherichia coli* isolates: using multiplex PCR and hybridization. Infect. Genet. Evol. *5*, 79–83.

Bovee-Oudenhoven, I.M., Lettink-Wissink, M.L., Doesburg. V.W., Witteman, B.J., and Van Der Meer, R. (2003). Diarrhea caused by enterotoxigenic *Escherichia coli* infection of humans is inhibited by dietary calcium. Gastroenterology *125*, 469–476.

Bower, J.R., Congeni, B.L., Cleary, T.G., Stone, R.T., Wanger, A., Murray, B.E., Mathewson, J.J., and Pickering, L.K. (1989). *Escherichia coli* O111:nonmotile as a pathogen in an outbreak of severe diarrhea associated with a day care center. J. Infect. Dis. *160*, 243–247.

Brewster, D.H., Brown, M.I., Robertson, D., Houghton, G.L., Bimson, J., and Sharp, J.C.(1994). An outbreak of *Escherichia coli* O157 associated with a children's paddling pool. Epidemiol. Infect. *112*, 441–447.

Brunder, W., Schmidt, H., and Karch, H. (1997). EspP, a novel extracellular serine protease of enterohaemorrhagic *Escherichia coli* O157:H7 cleaves human coagulation factor V. Mol. Microbiol. *24*, 767–778.

Brzuszkiewicz, E., Thürmer, A., Schuldes, J., Leimbach, A., Liesegang, H., Meyer, F.D., Boelter, J., Petersen, H., Gottschalk, G., and Daniel, R. (2011). Genome sequence analyses of two isolates from the recent *Escherichia coli* outbreak in Germany reveal the emergence of a new pathotype: Entero-Aggregative-Haemorrhagic *Escherichia coli* (EAHEC). Arch. Microbiol. *193*, 883–891.

Buchholz, U., Bernard, H., Werber, D., Böhmer, M.M., Remschmidt, C., Wilking, H., Deleré, Y., an der Heiden, M., Adlhoch, C., Dreesman, J., *et al.* (2011). German outbreak of *Escherichia coli* O104:H4 associated with sprouts. N. Engl. J. Med. *365*, 1763–1770.

Bueris, V., Sircili, M.P., Taddei, C.R., DosSantos, M.F., Franzolin, M.R., Martinez, M.B., Ferrer, S.R., Barreto, M.L., and Trabulsi, L.R. (2007). Detection of diarrheagenic *Escherichia coli* from children with and without diarrhea in Salvador, Bahia, Brazil. Mem. Inst. Oswaldo. Cruz. *102*, 839–844.

Burgos, Y., and Beutin, L. (2010). Common origin of plasmid encoded alpha-hemolysin genes in *Escherichia coli*. BMC. Microbiol. *10*, 193.

Campellone, K.G., and Leong, J.M. (2003). Tails of two Tirs: actin pedestal formation by enteropathogenic *E. coli* and enterohemorrhagic *E. coli* O157:H7. Curr. Opin. Microbiol. *6*, 82–90.

Campellone, K.G., Robbins, D., and Leong, J.M. (2004). EspFu is a translocated EHEC effector that interacts with Tir and N-WASP and promotes Nck-independent actin assembly. Dev. Cell *7*, 217–228.

Cennimo, D., Abbas, A., Huang, D.B., and Chiang, T. (2009). The prevalence and virulence characteristics of enteroaggregative *Escherichia coli* at an urgent-care clinic in the USA: a case–control study. J. Med. Microbiol. *58*, 403–407.

Centers for Disease Control and Prevention. (2010). Preliminary FoodNet data on the incidence of infection with pathogens transmitted commonly through food-10 states. MMWR Morb. Mortal. Wkly Rep. *59*, 418–422.

Chakraborty, S., Deokule, J.S., Garg, P., Bhattacharya, S.K., Nandy, R.K., Nair, G.B., Yamasaki, S., Takeda, Y., and Ramamurthy, T. (2001). Concomitant infection of enterotoxigenic *Escherichia coli* in an outbreak of cholera caused by *Vibrio cholerae* O1 and O139 in Ahmedabad, India. J. Clin. Microbiol. *39*, 3241–3246.

Chandler, W.L., Jelacic, S., Boster, D.R., Ciol, M.A., Williams, G.D., Watkins, S.L., Igarashi, T., and Tarr, P.I. (2002). Prothrombotic coagulation abnormalities preceding the hemolytic-uremic syndrome. N. Engl. J. Med. *346*, 23–32.

Chaudhuri, R.R., Sebaihia, M., Hobman, J.L., Webber, M.A., Leyton, D.L., Goldberg, M.D., Cunningham, A.F., Scott-Tucker, A., Ferguson, P.R., Thomas, C.M., *et al.* (2010). Complete genome sequence and comparative metabolic profiling of the prototypical enteroaggregative *Escherichia coli* strain 042. PLoS One *5*, e8801.

Chen, Q., Savarino, S.J., and Venkatesan, M.M. (2006). Subtractive hybridization and optical mapping of the enterotoxigenic *Escherichia coli* H10407 chromosome: isolation of unique sequences and demonstration of significant similarity to the chromosome of *E. coli* K-12. Microbiology *152*, 1041–1054.

Cheng, C., Tennant, S.M., Azzopardi, K.I., Bennett-Wood, V., Hartland, E.L., Robins-Browne, R.M., and Tauschek, M. (2009). Contribution of the *pst-phoU* operon to cell adherence by atypical enteropathogenic *Escherichia coli* and virulence of *Citrobacter rodentium*. Infect. Immun. *77*, 1936–1944.

Cheng, H., Skehan, B.M., Campellone, K.G., Leong, J.M., and Rosen, M.K. (2008). Structural mechanism of WASP activation by the enterohaemorrhagic *E. coli* effector EspF$_u$. Nature *454*, 1009–1013.

Chifiriuc, M.C., Bleotu, C., Măruţescu, L., Cristea, D., and Lazăr, V. (2010). The modulation of HeLa cells secretory patterns by invasive *Shigella* spp. and enteroinvasive *E. coli* bacterial cells and their soluble components. Roum. Arch. Microbiol. Immunol. *69*, 139–144.

Chowdhury, F., Rahman, M.A., Begum, Y.A., Khan, A.I., Faruque, A.S., Saha, N.C., Baby, N.I., Malek, M.A., Kumar, A.R., Svennerholm, A.M., *et al.* (2011). Impact of rapid urbanization on the rates of infection by *Vibrio cholerae* O1 and enterotoxigenic *Escherichia coli* in Dhaka, Bangladesh. PLoS Negl. Trop. Dis. *5*, e999.

Cohen, D., Tobias, J., Spungin-Bialik, A., Sela, T., Kayouf, R., Volovik, Y., Yavzori, M., and Ephros, M. (2010). Phenotypic characteristics of enterotoxigenic *Escherichia coli* associated with acute diarrhea among Israeli young adults. Foodborne. Pathog. Dis. *7*, 1159–1164.

Cohen, M.B., Nataro, J.P., Bernstein, D.I., Hawkins, J., Roberts, N., and Staat, M.A. (2005). Prevalence of diarrheagenic *Escherichia coli* in acute childhood enteritis: a prospective controlled study. J. Pediatr. *146*, 54–61.

Corrêa, S., Palmeira, P., Carneiro-Sampaio, M.M., Nishimura, S.L., and Guth, B.E. (2006). Human colostrum contains IgA antibodies reactive to colonization factors I and II of enterotoxigenic *Escherichia coli*. FEMS Immunol. Med. Microbiol. *47*, 199–206.

Cortés-Ortiz, I.A., Rodríguez-Angeles, G., Moreno-Escobar, E.A., Tenorio-Lara, J.M., Torres-Mazadiego, B.P., and Montiel-Vázquez, E. (2002). Outbreak caused by *Escherichia coli* in Chalco, Mexico. Salud. Publica. Mex. *44*, 297–302 (in Spanish)

Costin, I.D., Voiculescu, D., and Gorcea, V. (1964). An outbreak of food poisoning in adults associated with *Escherichia coli* serotype 86:B7:H34. Pathol. Microbiol. *27*, 68–78.

Crane, J.K., Choudhari, S.S., Naeher, T.M., and Duffey, M.E. (2006). Mutual enhancement of virulence by enterotoxigenic and enteropathogenic *Escherichia coli*. Infect. Immun. *74*, 1505–1515.

Cravioto, A., Reyes, R.E., and Ortega, R., Fernandez, G., Hernandez, R., and Lopez, D. (1988). Prospective study of diarrhoeal disease in a cohort of rural Mexican children: incidence and isolated pathogens during the first two years of life. Epidemiol. Rev. *101*, 123–134.

Cravioto, A., Reyes, R.E., Trujillo, F., Uribe, F., Navarro, A., de la Roca, J.M., Hernandez, J.M, Perez, G., and Vazquez, V. (1990). Risk of diarrhea during the first year of life associated with initial and subsequent colonization by specific enteropathogens. Am. J. Epidemiol. *131*, 886–904.

Crossman, L.C., Chaudhuri, R.R., Beason, S.A., Wells, T.J., Desvaux, M., Cunningham, A.F., Petty, N.K., Mahon, V., Brinkley, C., Hobman, J.L., *et al.* (2010). A commensal gone bad: complete genome sequence of the prototypal enterotoxigenic *Escherichia coli* strain H10407. J. Bacteriol. *192*, 5822–5831..

Croxen, M.A., and Finlay, B.B. (2010). Molecular mechanisms of *Escherichia coli* pathogenicity. Nat. Rev. Microbiol. *8*, 26–38.

Dalla-Costa, L.M., Irino, K., Rodrigues, J., Rivera, I.N.G., and Trabulsi, L.R. (2008). Characterisation of diarrhoeagenic *Escherichia coli* clones by ribotyping and ERIC-PCR. J. Med. Microbiol. *47*, 227–234.

Dean, P., Maresca, M., Schuller, S., Phillips, A.D., and Kenny, B. (2006). Potent diarrheagenic mechanism mediated by the cooperative action of three enteropathogenic *Escherichia coli*-injected effector proteins. Proc. Natl. Acad. Sci. U.S.A. *103*, 1876–1881.

Dedeić-Ljubović, A., Hukić, M., Bekić, D., and Zvizdić, A. (2009). Frequency and distribution of diarrhoeagenic *Escherichia coli* strains isolated from pediatric patients with diarrhoea in Bosnia and Herzegovina. Bosn. J. Basic. Med. Sci. *9*, 9148–9155.

Dev, V.J., Main, M., and Gould, I. (1991). Waterborne outbreak of *Escherichia coli* O157. Lancet *337*, 1412.

DeVinney, R., Stein, M., Reinscheid, D., Abe, A., Ruschkowski, S., and Finlay, B.B. (1999). Enterohemorrhagic *Escherichia coli* O157:H7 produces Tir, which is translocated to the host cell membrane but is not tyrosine phosphorylated. Infect. Immun. *67*, 2389–2398.

Diez-Gonzalez, F., Callaway, T.R., Kizoulis, M.G., and Russell, J.B. (1998). Grain feeding and the dissemination of acid-resistant *Escherichia coli* from cattle. Science *281*, 1666–1668.

Dong, Y.L., Yang, Q.Z., and Chen, S.J. (2010). Outbreak of nosocomial diarrhoea caused by enterotoxigenic *Escherichia coli* in a geriatric ward. J. Hosp. Infect. *74*, 405–406.

Donnenberg, M.S. (1995). Enteropathogenic *Escherichia coli*. In Infections of the Gastrointestinal Tract, Blaser, M.J., Smith, P.D., Ravdin, J.I., Greenberg, H.B., and Guerrant, R.L., eds. (Raven Press, New York), pp. 709–726.

Dorsey, F.C., Fischer, J.F., and Fleckenstein, J.M. (2006). Directed delivery of heat-labile enterotoxin by enterotoxigenic *Escherichia coli*. Cell. Microbiol. *8*, 1516–1527.

Drucker, M.M., Poliack, A., Yevien, R., and Sacks, T.G. 1970. Immunofluorescent demonstration of enteropathogenic, *Escherichia coli* in tissue of infants dying with enteritis. Pediatrics *46*, 855–864.

DuPont, H.L., Formal, S.B., Hornick, R.B., Snyder, M.J., Libonati, J.P., Sheahan, D.G., LaBrec, E.H., and Kalas, J.P. (1971). Pathogenesis of *Escherichia coli* diarrhea. N. Engl. J. Med. *285*, 1–9.

Durrer, P., Zbinden, R., Fleisch, F., Altwegg, M., Ledergerber, B., Karch, H., and Weber, R. (2000). Intestinal infection due to enteroaggregative *Escherichia coli* among human immunodeficiency virus-infected persons. J. Infect. Dis. *182*, 1540–1544.

Erdem, A.L., Avelino, F., Xicohtencatl-Cortes, J., and Giron, J.A. (2007). Host binding and adhesive properties of H6 and H7 flagella of attaching and effacing *Escherichia coli*. J. Bacteriol. *189*, 7426–7435.

Escobar-Páramo, P., Clermont, O., Blanc-Potard, A.B., Bui, H., Le Bouguénec, C., and Denamur, E. (2004). A specific genetic background is required for acquisition and expression of virulence factors in *Escherichia coli*. Mol. Biol. Evol. *2*, 1085–1094.

Eslava, C., Navarro-García, F., Czeczulin, J.R., Henderson, I.R., Cravioto, A., and Nataro, J.P. (1998). Pet, an autotransporter enterotoxin from enteroaggregative *Escherichia coli*. Infect. Immun. *66*, 3155–3163.

Esmaili, A., Nazir, S.F., Bothakur, A., Yu, D., Turner, J.R., Saksena, S., Singla, A., Hecht, G.A., Alrefai, W.A., and Gill, R.K. (2009). Enteropathogenic *Escherichia coli* infection inhibits intestinal serotonin transporter (SERT) function and expression. Gastroenterology *137*, 2074–2083.

Espert, S.M., Elsinghorst, E.A., and Munson, G.P. (2011). The *tib* adherence locus of enterotoxigenic *Escherichia coli* is regulated by cyclic AMP receptor protein. J. Bacteriol. *193*, 1369–1376.

Estrada-Garcia, T., Cerna, J.F., Thompson, M.R., and Lopez-Saucedo, C. (2002). Faecal contamination and enterotoxigenic *Escherichia coli* in street-vended chili sauces in Mexico and its public health relevance. Epidemiol. Infect. *129*, 223–226.

Estrada-Garcia, T., Lopez-Saucedo, C., Thompson-Bonilla, R., Abonce, M., Lopez-Hernandez, D., Santos, J.I., Rosado, J.L., DPont, H.L., and Long, K.Z. (2009). Association of diarrheagenic *Escherichia coli* pathotypes with infection and diarrhea among Mexican children and association of atypical enteropathogenic *E. coli* with acute diarrhea. J. Clin. Microbiol. *47*, 93–98.

Ewing, W.H., and Gravatti, J.L. (1947). *Shigella* types encountered in the Mediterranean area. J. Bacteriol. *53*, 191–195.

Fernandes, R.M., Carbonare, S.B., Carneiro-Sampaio, M.M., and Trabulsi, L.R. (2001). Inhibition of enteroaggregative *Escherichia coli* adhesion to HEp-2 cells by secretory immunoglobulin A from human colostrum. Pediatr. Infect. Dis. J. *20*, 672–678.

Fleckenstein, J.M., Kopecko, D.J., Warren, R.L., and Elsinghorst, E.A. (1996). Molecular characterization of the *tia* invasion locus from enterotoxigenic *Escherichia coli*. Infect. Immun. *64*, 2256–2265.

Fleckenstein, J.M., Lindler, L.E., Elsinghorst, E.A., and Dale, J.B. (2000). Identification of a gene within a pathogenicity island of enterotoxigenic *Escherichia coli* H10407 required for maximal secretion of the heat-labile enterotoxin. Infect. Immun. *68*, 2766–2774.

Fleckenstein, J.M., Holland J.T., and Hasty, D.L. (2002). Interaction of an outer membrane protein of enterotoxigenic *Escherichia coli* with cell surface heparan sulfate proteoglycans. Infect. Immun. *70*, 1530–1537.

Flores, J., and Okhuysen, P.C. (2009a). Genetics of susceptibility to infection with enteric pathogens. Curr. Opin. Infect. Dis. *22*, 471–476.

Flores, J., and Okhuysen, P.C. (2009b). Enteroaggregative *Escherichia coli* infection. Curr. Opin. Gastroenterol. *25*, 8–11.

Forestier, C., De Champs, C., Vatoux, C., and Joly, B. (2001). Probiotic activities of *Lactobacillus casei rhamnosus: in vitro* adherence to intestinal cells and antimicrobial properties. Res. Microbiol. *152*, 167–173.

Frank, C., Werber, D., Cramer, J.P., Askar, M., Faber, M., an der Heiden, M., Bernard, H., Fruth, A., Prager, R., Spode, A., *et al.* (2011). Epidemic profile of Shiga-toxin-producing *Escherichia coli* O104:N4 outbreak in Germany. N. Engl. J. Med. *365*, 1772–1780.

Frankel, G., Giron, J.A., Valmassoi, J., and Schoolnik, G.K. (1989). Multi-gene amplification: simultaneous detection of three virulence genes in diarrhoeal stool. Mol. Microbiol. *3*, 1729–1734.

Frankel, G., Riley, L., Giron, J.A., Valmassoi, J., Friedmann, A., Strockbine, N., Falkow, S., and Schoolnik, G.K. (1990). Detection of *Shigella* in feces using DNA amplification. J. Infect. Dis. *161*, 1252–1256.

Frankel, G., Candy, D.C.A., Everest, P., and Dougan, G. (1994). Characterization of the C-terminal domains of intimin-like proteins of enteropathogenic and enterohemorrhagic *Escherichia coli*, *Citrobacter freundii*, and *Hafnia alvei*. Infect. Immun. *62*, 1835–1842.

Frankel, G., Philips, A.D., Novakova, M., Batchelor, M., Hicks, S., and Dougan, G. (1998). Generation of *Escherichia coli* intimin derivatives with differing biological activities using site-directed mutagenesisof the intimin C terminus domain. Mol. Microbiol. *29*, 559–570.

Franzolin, M.R., Alves, R.C., Keller, R., Gomes, T.A., Beutin, L., Barreto, M.L., Milroy, C., Strina, A., Ribeiro, H., and Trabulsi, L.R. (2005). Prevalence of diarrheagenic *Escherichia coli* in children with diarrhea in Salvador, Bahia, Brazil. Mem. Inst. Oswaldo. Cruz. *100*, 359–363.

Frech, S.A., Dupont, H.L., Bourgeois, A.L., McKenzie, R., Belkind-Gerson, J., Figueroa, J.F., Okhuysen, P.C., Guerrero, N.H., Martinez-Sandoval, F.G., *et al.* (2008). Use of a patch containing heat-labile toxin from *Escherichia coli* against travellers' diarrhoea: a phase II, randomised, double-blind, placebo-controlled field trial. Lancet *371*, 2019–25.

Friedman, M.S., Roels, T., Koehler, J.E., Feldman, L., Bibb, W.F., and Blake, P. (1999). *Escherichia coli* O157:H7 outbreak associated with an improperly chlorinated swimming pool. Clin. Infect. Dis. *29*, 298–303.

Gaastra, W., and Svennerholm, A.M. (1996). Colonization factors of human enterotoxigenic *Escherichia coli* (ETEC). Trends. Microbiol. *4*, 444–452.

Galbadage, T., Jiang, Z.D., and DuPont, H.L. (2009). Improvement in detection of enterotoxigenic *Escherichia coli* in patients with travelers' diarrhea by increasing the number of *E. coli* colonies tested. Am. J. Trop. Med. Hyg. *80*, 20–23.

Garcia, C., Chincha, O., Leon, M., Iglesias, D., Barletta, F., Mercado, E., and Ochoa, T. (2010). High frequency of diarrheagenic *Escherichia coli* in human immunodeficiency virus (HIV) patients with and without diarrhea in Lima, Peru. Am. J. Trop. Med. Hyg. *82*, 1118–1120.

Garmendia, J., Phillips, A.D., Carlier, M.F., Chong, Y., Schuller, S., Marches, O., Dahan, S., Oswald, E., Shaw, R.K., Knutton, S., *et al.* (2004). TccP is an enterohaemorrhagic *Escherichia coli* O157:H7 type III effector protein that couples Tir to the actin-cytoskeleton. Cell. Microbiol. *6*, 1167–1183.

Geldreich, E.E., Fox, K.R., Goodrich, J.A., Rice, E.W., Clark, R.M., and Swerdlow, D.L. (1992). Searching for a water supply connectionin the Cabool, Missouri, disease outbreak of *Escherichia coli* O157:H7. Water Res. *26*, 1127–1137.

Ghosal, A., Bhowmick, R., Nandy, R.K., Ramamurthy, T., and Chatterjee, N.S. (2007). PCR-based identification of common colonization factor antigens of enterotoxigenic *Escherichia coli*. J. Clin. Microbiol. *45*, 3068–3071.

Gill, R.K., Borthakur, A., Hodges, K., Turner, J.R., Clayburgh, D.R., Saksena, S., Zaheer, A., Ramaswamy, K., Hecht, G., and Dudeja, P.K. (2007). Mechanism underlying inhibition of intestinal apical Cl/OH exchange following infection with enteropathogenic *E. coli*. J. Clin. Invest. *117*, 428–437.

Gilmour, M.W., Cote, T., Munro, J., Chui, L., Wylie, J., Isaac-Renton, J., Horsman, G., Tracz, D.M., Andrysiak, A., and Ng, L.K. (2005). Multilocus sequence typing of *Escherichia coli* O26:H11 isolates carrying *stx* in Canada does not identify genetic diversity. J. Clin. Microbiol. *43*, 5319–5323.

Giron, J.A., Ho, A.S.Y., and Schoolnik, G.K. (1991). An inducible bundle-forming pilus of enteropathogenic *Escherichia coli*. Science *254*, 710–713.

Glenn, G.M., Villar, C.P., Flyer, D.C., Bourgeois, A.L., McKenzie, R., Lavker, R.M., and Frech, S.A. (2007). Safety and immunogenicity of an enterotoxigenic *Escherichia coli* vaccine patch containing heat-labile toxin: use of skin pretreatment to disrupt the stratum corneum. Infect. Immun. *75*, 2163–2170.

Gomes, T.A., Toledo, M.R., Trabulsi, L.R., Wood, P.K., Morris, J.G.Jr. (1987). DNA probes for identification of enteroinvasive *Escherichia coli*. J. Clin. Microbiol. *25*, 2025–2027.

Gomes, T.A.T., Blake, P.A., and Trabulsi, L.R. (1989). Prevalence of *Escherichia coli* strains with localised, diffuse, and aggregative adherence to HeLa cells in infants with diarrhea and matched controls. J. Clin. Microbiol. *27*, 266–269.

Gordillo, M.E., Reeve, G.R., Pappas, J., Mathewson, J.J., DuPont, H.L., and Murray, B.E. (1992). Molecular characterization of strains of enteroinvasive *Escherichia coli* O143, including isolates from a large outbreak in Houston, Texas. J. Clin. Microbiol. *30*, 889–893.

Grant, M.A., Weagant, S.D., and Feng, P. (2001). Glutamate decarboxylase genes as a prescreening marker for detection of pathogenic *Escherichia coli* groups. Appl. Environ. Microbiol. *67*, 3110–3114.

Grif, K., Karch, H., Schneider, C., Daschner, F.D., Beutin, L., Cheasty, T., Smith, H., Rowe, B., Dierich, M.P., and Allerberger, F. (1998). Comparative study of five different techniques for epidemiological typing of *Escherichia coli* O157. Diagn. Microbiol. Infect. Dis. *32*, 165–176.

Grimes, K.A., Mohamed, J.A., Dupont, H.L., Padda, R.S., Jiang, Z.D., Flores, J., Belkind-Gerson, J., Martinez-Sandoval, F.G., and Okhuysen, P.C. (2008). PCR-based assay using occult blood detection cards for detection of diarrheagenic *Escherichia coli* in specimens from U.S. travelers to Mexico with acute diarrhea. J. Clin. Microbiol. *46*, 2227–2230.

Guion, C.E., Ochoa, T.J., Walker, C.M., Barletta, F., and Cleary, T.G. (2008). Detection of diarrheagenic *Escherichia coli* by use of melting-curve analysis and real-time multiplex PCR. J. Clin. Microbiol. *46*, 1752–1757.

Gupta, S.K., Keck, J., Ram, P.K., Crump, J.A., Miller, M.A., and Mintz, E.D. (2008). Part III. Analysis of data gaps pertaining to enterotoxigenic *Escherichia coli* infections in low and medium human development index countries, 1984–2005. Epidemiol. Infect. *136*, 721–738.

Guttman, J.A., Samji, F.N., Li, Y., Deng, W., Lin, A., and Finlay, B.B. (2007). Aquaporins contribute to diarrhea caused by attaching and effacing bacterial pathogens. Cell. Microbiol. *9*, 131–141.

Harrington, S.M., Strauman, M.C., Abe, C.M., and Nataro, J.P. (2005). Aggregative adherence fimbriae contribute to the inflammatory response of epithelial cells infected with enteroaggregative *Escherichia coli*. Cell. Microbiol. *7*, 1565–1578.

Harris, J.R., Mariano, J., Wells, J.G., Payn, B.J., Donnell, H.D., and Cohen, M.L. (1985). Person-to-person transmission in an outbreak of enteroinvasive *Escherichia coli*. Am. J. Epidemiol. *122*, 245–52.

Hedberg, C.W., Savarino, S.J., Besser, J.M., Paulus, C.J., Thelen, V.M., Myers, L.J., Cameron, D.N., Barrett, T.J., Kaper, J.B., and Osterholm, M.T. (1997). An outbreak of foodborne illness caused by *Escherichia coli* O39:NM, an agent not fitting into the existing scheme for classifying diarrhoeagenic *E. coli*. J. Infect. Dis. *176*, 1625–1628.

Heir, E., Lindstedt, B.A., Vardund, T., Wasteson, Y., and Kapperud, G. (2000). Genomic fingerprinting of Shiga toxin-producing *Escherichia coli* (STEC) strains: comparison of pulsed-field gel electrophoresis (PFGE) and fluorescent amplified-fragment-length polymorphism (FAFLP). Epdemiol. Infect. *125*, 537–548.

Helander, A., Grewal, H.M., Gaastra, W., and Svennerholm, A.M. (1997). Detection and characterization of the coli surface antigen 6 of enterotoxigenic *Escherichia coli* strains by using monoclonal antibodies. J. Clin. Microbiol. *35*, 867–872.

Henderson, I.R., Czeczulin, J., Eslava, C., Noriega, F., and Nataro, J.P. (1999). Characterization of Pic, a secreted protease of *Shigella flexneri* and enteroaggregative *Escherichia coli*. Infect. Immun. *67*, 5587–5596.

Hernandes, R.T., Elias, W.P., Vieira, M.A.M., and Gomes, T.A.T. (2009). An overview of atypical enteropathogenic *Escherichia coli*. FEMS Microbiol. Lett. *297*, 137–149.

Hien, B.T., Scheutz, F., Cam, P.D., Serichantalergs, O., Huong, T.T., Thu, T.M., and Dalsgaard, A. (2008). Diarrheagenic *Escherichia coli* and *Shigella* strains isolated from children in a hospital case–control study in Hanoi, Vietnam. J. Clin. Microbiol. *46*, 996–1004.

Hill, D.R., and Beeching, N.J. (2010). Travelers' diarrhea. Curr. Opin. Infect. Dis. *23*, 481–487.

Hill, S.M., Philips, A.D., and Walker-Smith, J.A. (1991). Enteropathogenic *Escherichia coli* and life threatening chronic diarrhoea. Gut 32, 154–158.

Hobbs, B.C., Thomas, M.E., and Taylor, J. (1949). School outbreak of gastro-enteritis associated with a pathogenic paracolon bacillus. Lancet *2(6577)*, 530–532.

Hodges, K., Alto, N.M., Ramaswamy, K., Dudeja, P.K., and Hecht, G. (2008). The enteropathogenic *Escherichia coli* effector protein EspF decreases sodium hydrogen exchanger 3 activity. Cell. Microbiol. *10*, 1735–1745.

Holmner, A., Askarieh, G., Okvist, M., and Krengel, U. (2007). Blood group antigen recognition by *Escherichia coli* heat-labile enterotoxin. J. Mol. Biol. *371*, 754–764.

Horstman, A.L., and Kuehn, M.J. (2000). Enterotoxigenic *Escherichia coli* secretes active heat-labile enterotoxin via outer membrane vesicles. J. Biol. Chem. *275*, 12489–12496.

Horstman, A.L., and Kuehn, M.J. (2002). Bacterial surface association of heat-labile enterotoxin through lipopolysaccharide after secretion via the general secretory pathway. J. Biol. Chem. 277, 32538–32545.

Hovde, C.J., Austin, P.R., Cloud, K.A., Williams, C.J., and Hunt, C.W. (1999). Effect of cattle diet on *Escherichia coli* O157:H7 acid resistance. Appl. Environ. Microbiol. *65*, 3233–3235.

Huang, D.B., and Dupont, H.L. (2004). Enteroaggregative *Escherichia coli:* an emerging pathogen in children. Semin. Pediatr. Infect. Dis. *15*, 266–271.

Huang, D.B., Mohamed, J.A., Nataro, J.P., DuPont, H.L., Jiang, Z.D., and Okhuysen, P.C. (2007). Virulence characteristics and the molecular epidemiology of enteroaggregative *Escherichia coli* isolates from travellers to developing countries. J. Med. Microbiol. *56*, 1386–1392.

Hunter, P.R. (2003). Drinking water and diarrhoeal disease due to *Escherichia coli*. J. Water Health *1*, 65–72.

Hyytia-Trees, E., Smole, S.C., Fields, P.A., Swaminathan, B., and Ribot, E.M. (2006). Second generation subtyping: a proposed PulseNet protocol for multiple-locus variable-number tandem repeat analysis of Shiga toxin-producing *Escherichia coli* O157 (STEC O157). Foodborne Pathog. Dis. *3*, 118–131.

Infante, R.M., Ericsson, C.D., Jiang, Z.D., Ke, S., Steffen, R., Riopel, L., Sack, D.A., and DuPont. H.L. (2004). Enteroaggregative *Escherichia coli* diarrhea in travelers: response to rifaximin therapy. Clin. Gastroenterol. Hepatol. *2*, 135–138.

Isaacson, M., Canter, P.H., Effler, P., Arntzen, L., Bomans, P., and Heenan, R. (1993). Haemorrhagic colitis epidemic in Africa. *Lancet 341*, 961.

Itoh, Y., Nagano, I., Kunishima, M., and Ezaki, T. (1997). Laboratory investigation of enteroaggregative *Escherichia coli* O untypeable: H10 associated with a massive outbreak of gastrointestinal illness. J. Clin. Microbiol. 35, 2546–2550.

Iwanaga, M., Song, T., Higa, N., Kakinohana, S., Toma, C., and Nakasone, N. (2002). Enteroaggregative *Escherichia coli*: incidence in Japan and usefulness of the clump-formation test. J. Infect. Chemother. *8*, 345–348.

Jafari, F., Shokrzadeh, L., Hamidian, M., Salmanzadeh-Ahrabi, S., and Zali, M.R. (2008). Acute diarrhea due to enteropathogenic bacteria in patients at hospitals in Tehran. Jpn. J. Infect. Dis. *61*, 269–273.

Jafari, F., Garcia-Gil, L.J., Salmanzadeh-Ahrabi, S., Shokrzadeh, L., Aslani, M.M., Pourhoseingholi, M.A., Derakhshan, F., Zali, M.R. (2009). Diagnosis and prevalence of enteropathogenic bacteria in children less than 5 years of age with acute diarrhea in Tehran children's hospitals. J. Infect. *58*, 21–27.

Jain, S., Chen, L., Dechet, A., Hertz, A.T., Brus, D.L., Hanley, K., Wilson, B., Frank, J., Greene, K.D., Parsons, M., *et al.* (2008). An outbreak of enterotoxigenic *Escherichia coli* associated with sushi restaurants in Nevada, 2004. Clin. Infect. Dis. *47*, 1–7.

Jalaluddin, S., De Mol, P., Hemelhof, W., Bauma, N., Brasseur, D., Hennart, P., Lomoyo, R.E., Rowe, B., and Butzler, J.P. (1998). Isolation and characterization of enteroaggregative *Escherichia coli* (EAggEC) by genotypic and phenotypic markers, isolated from diarrheal children in Congo. Clin. Microbiol. Infect. *4*, 213–219.

Jansson, L., Tobias, J., Lebens, M., Svennerholm, A.M., and Teneberg, S. (2006). The major subunit, CfaB, of colonization factor antigen i from enterotoxigenic Escherichia coli is a glycosphingolipid binding protein. Infect. Immun. 74, 3488–3497.

Jenkins, C., Lawson, A.J., Cheasty, T., Willshaw, G.A., Wright, P., Dougan, D., Frankel, G., and Smith, H.R. (2003). Subtyping intimin genes from enteropathogenic *Escherichia coli* associated with outbreaks and sporadic cases in the United Kingdom and Eire. Mol. Cell. Prob. *17*, 149–156.

Jiang, Z.D., Greenberg, D., Nataro, J.P., Steffen, R., and DuPont, H.L. (2002). Rate of occurrence and pathogenic effect of enteroaggregative *E.cherichia coli* virulence factors in international travelers. J. Clin. Microbiol. *40*, 4185–4190.

Jiang, Z.D., Okhuysen, P.C., Guo, D.C., He. R., King, T.M., DuPont. H.L., and Milewicz, D.M. (2003). Genetic susceptibility to enteroaggregative *Escherichia coli* diarrhea: polymorphhism in the interleukin-8 promotor region. J. Infect. Dis. *188*, 506–511.

Jiang, Z.D., Ke, S., and Dupont, H.L. (2010). Rifaximin-induced alteration of virulence of diarrhoea-producing *Escherichia coli* and *Shigella sonnei*. Int. J. Antimicrob. Agents 35, 278–281.

Johnson, A.M., Kaushik, R.S., Francis, D.H., Fleckenstein, J.M., and Hardwidge, P.R. (2009). Heat-labile enterotoxin promotes *Escherichia coli* adherence to intestinal epithelial cells. J. Bact. *191*, 178–186.

Johnson, R.P., Clarke, R.C., Wilson, J.B., Read, S.C., Rahn, K., Renwick, S.A., Sandhu, K.A., Alves, D., Karmali, M.A., Lior, H., *et al.* (1996). Growing concerns and recent outbreaks involving non-O157:h7 serotypes of verotoxigenic *Escherichia coli*. J. Food Prot. *59*, 112–1122.

Johnson, T.J., and Nolan, L.K. (2009). Pathogenomics of the virulence plasmids of *Escherichia coli*. Microbiol. Mol. Biol. Rev. 73, 750–774.

Jyotil, A., Singh, S.P., Yashpal, M., Dwivedi, P.D., and Shanker, R. (2011). Rapid detection of enterotoxigenic *Escherichia coli* gene using bio-conjugated gold nano-particles. J. Biomed. Nanotechnol. 7, 170–171.

Kahali, S., Sarkar, B., Chakraborty, S., Macaden, R., Deokule, J.S., Ballal, M., Nandy, R.K., Bhattacharya., S.K., Takeda, Y., and Ramamurthy, T. (2004a). Molecular epidemiology of diarrhoeagenic *Escherichia coli* associated with sporadic cases and outbreaks of diarrhoea between 2000 and 2001 in India. Eur. J. Epidemiol. *19*, 473–479.

Kahali, S., Sarkar, B., Rajendran, K., Khanam, J., Yamasaki, S., Nandy, R.K., Bhattacharya, S.K., and Ramamurthy, T. (2004b). Virulence characteristics and molecular epidemiology of enteroaggregative *Escherichia coli* isolates from hospitalized diarrheal patients in Kolkata, India. J. Clin. Microbiol. *42*, 4111–4120.

Kaper, J.B. (1996). Defining EPEC. Rev. Microbiol. Sao Paulo *27*, 130–133.

Karmali, M.A., Steele, B.T., Petric, M., and Lim, C. (1983). Sporadic cases of haemolytic uraemic syndrome associated with faecal cytotoxin and cytotoxin-producing *Escherichia coli* in stools. Lancet *1(8325)*, 619–620.

Keene, W.E., McAnulty, J.M., Hoesly, F.C., Williams, L.P.,Hedberg, K., Oxman, G.L., Barrett, T.J., Pfaller, M.A., and Fleming, D.W. (1994). A swimming-associated outbreak of hemorrhagic colitis caused by *Escherichia coli* O157:H7 and *Shigella sonnei*. New Engl. J. Med. *331*, 579–584.

Kelly, P., Hicks, S., Oloya, J., Mwansa, J, Sikakwa, L., Zulu, I., and Phillips, A. (2003). *Escherichia coli* enterovirulent phenotypes in Zambians with AIDS-related diarrhoea. Trans. R. Soc. Trop. Med. Hyg. *97*, 573–576.

Kessner, D.M., Shaughnessy, H.J., Googins, J., Rasmussen, C.M., Rose, M.J., Marshall, A.L.Jr., Andelman, S.L., Hall, J.B., and Rosenblom, P.J. (1964). An extensive community outbreak of diarrhea due to enteropathogenic *Escherichia coli* O111:B4. I.Epidemiologic studies. Am. J. Hyg. *76*, 27–43.

Kesty, N.C., Mason, K.M., Reedy, M., Miller, S.E., and Kuehn. M.J. (2004). Enterotoxigenic *Escherichia coli* vesicles target toxin delivery into mammalian cells. EMBO. J. *23*, 4538–4549.

Khan, M.A., Kang, J., and Steiner, T.S. (2004). Enteroaggregative *Escherichia coli* flagellin-induced interleukin-8 secretion requires Toll-like receptor 5-dependent p38 MAP kinase activation. Immunology *112*, 651–660.

Khan, K., Konarm, M., Goyalm, A., and Ghoshm, S. (2010). Enteroaggregative *Escherichia coli* infection induces IL-8 production via activation of mitogen-activated protein kinases and the transcription factors NF-kappaB and AP-1 in INT-407 cells. Mol. Cell. Biochem. *337*, 17–24.

Kim, J., Nietfeldt, J., and Benson, A.K. (1999). Octamer-based genome scanning distinguishes a unique subpopulation of *Escherichia coli* O157:H7 strains in cattle. Proc. Natl. Acad. Sci. U.S.A. *96*, 13288–13293.

Kim, J.Y., Kim, S.H., Kwon, N.H., Bae, W.K., Lim, J.Y., Koo, H.C., Kim, J.M., Noh, K.M., Jung, W.K., Park, K.T., *et al.* (2005). Isolation and identification of *Escherichia coli* O157:H7 using different detection methods and molecular determination by multiplex PCR and RAPD. J. Vet. Sci. *6*, 7–19.

Knutton, S., Phillips, A.D., Smith, H.R., Gross, R.J., Shaw, R., Watson, P., and Price, E. (1991). Screening for EPEC in infants with diarrhea by the fluorescent-actin staining test. Infect. Immun. *59*, 365–371.

Knutton, S., Shaw, R., Phillips, A.D., Smith, H.R., Willshaw, G.A., Watson, P., and Price, E. (2001). Phenotypic and genetic analysis of diarrhea-associated *Escherichia coli* isolated from children in the United Kingdom. J. Paediatr. Gastroenterol. Nutr. *33*, 32–40.

Kotewicz, M.L., Jackson, S.A., LeClerc, J.E., and Cebula, T.A. (2007). Optical maps distinguish individual strains of *Escherichia coli* O157:H7. Microbiology *153*, 1720–1733.

Lacher, D.W., Steinsland H., and Whittam, T.S. (2006). Allelic subtyping of the intimin locus (*eae*) of pathogenic *Escherichia coli* by fluorescent RFLP. FEMS Microbiol. Lett. *261*, 80–87.

Lanyi, B., Szita, J., Ringelhann, B., and Kovach, K. (1959). Waterborne epidemic caused by *Escherichia coli* serotype 124:72–32 in Paradsasvar. Orv.Hetil. *100*, 1075–1077. [In Hungarian.]

Lasaro, M.A., Rodrigues, J.F., Mathias-Santos, C., Guth, B.E., Balan, A., Sbrogio-Almeida, M.E., and Ferreira, L.C. (2008). Genetic diversity of heat-labile toxin expressed by enterotoxigenic *Escherichia coli* strains isolated from humans. J. Bacteriol. *190*, 2400–2410.

Lathem, W.W., Grys, T.E., Witowski, S.E., Torres, A.G., Kaper, J.B., Tarr, P.I., and Welch. (2002). StcE, a metalloprotease secreted by *Escherichia coli* O157:H7, specifically cleaves C1 esterase inhibitor. Mol. Microbiol. *45*, 277–288.

Lazăr, V., Miyazaki, Y., Hanawa, T., Chifiriuc, M.C., Dițu, L.M., Mărușescu, L., Bleotu, C., and Kamiya, S (2009). The influence of some probiotic supernatants on the growth and virulence features expression of several selected enteroaggregative *E. coli* clinical strains. Roum. Arch. Microbiol. Immunol. *68*, 207–214.

Levine, M.M. (1987). Escherichia coli that cause diarrhea: enterotoxigenic, enteropathogenic, enteroinvasive, enterohemorrhagic, and enteroadherent. J. Infect. Dis. 155, 377–389.

Levine, M.M., and Edelman, R. (1984). Enteropathogenic *Escherichia coli* of classic serotypes associated with infant diarrhea: epidemiology and pathogenesis. Epidemiol. Rev. *6*, 31–51.

Levine, M.M., Bergquist, E.J., Nalin, D.R., Waterman, D.H., Hornick, R.B., Young, C.R., and Sotman, S. (1978). *Escherichia coli* strains that cause diarrhea, but do not produce heat-labile or heat-stable enterotoxins and are non-invasive. Lancet *1(8074)*, 1119–1122.

Lin, J., Smith, M.P., Chapin, K.C., Baik, H.S., Bennett, G.N., and Foster, J.W. (1996). Mechanisms of acid resistance in enterohemorrhagic *Escherichia coli*. Appl. Environ. Microbiol. *62*, 3094–3100.

Lindenthal, C., and Elsinghorst, E.A. (2001). Enterotoxigenic *Escherichia coli* TibA glycoprotein adheres to human intestine epithelial cells. Infect. Immun. *69*, 52–57.

Lindstedt, B.A., Heir, E., Gjernes, E., Vardund, T., and Kapperud, G. (2003). DNA fingerprinting of Shiga-toxin producing *Escherichia coli* O157 based on Multiple-Locus Variable-Number Tandem-Repeats Analysis (MLVA). Ann. Clin. Microbiol. Antimicrob. *2*, 12.

Lindstedt, B.A., Brandal, L.T., Aas, L., Vardund, T., and Kapperud, G. (2007). Study of polymorphic variable-number of tandem repeats loci in the ECOR collection and in a set of pathogenic *Escherichia coli* and *Shigella* isolates for use in a genotyping assay. J. Microbiol. Methods *69*, 197–205.

Long, K., Vasquez-Garibay, E., Mathewson, J., de la Cabada, J., and DuPont, H. (1999). The impact of infant feeding patterns on infection and diarrheal disease due to enterotoxigenic *Escherichia coli*. Salud Publica Mex. *41*, 263–267.

Long, K.Z., Santos, J.I., Rosado, J.L., Estrada-Garcia, T., Haas, M., Al Mamun, A., DuPont, HL., and Nanthakumar, N.N. (2011). Vitamin A supplementation modifies the association between mucosal innate and adaptive immune responses and resolution of enteric pathogen infections. Am. J. Clin. Nutr. *93*, 578–585.

Lothigius, A., Sjöling, A., Svennerholm, A.M., and Bölin, I. (2010). Survival and gene expression of enterotoxigenic *Escherichia coli* during long-term incubation in sea water and freshwater. J. Appl. Microbiol. *108*, 1441–1449.

Ludwig, A., von Rhein, C., Bauer, S., Huttinger, C., and Goebel, W. (2004). Molecular analysis of cytolysin A (ClyA) in pathogenic *Escherichia coli* strains. J. Bacteriol. *186*, 5311–5320.

McKenzie, R., Bourgeois, A.L., Engstrom, F., Hall, E., Chang, H.S., Gomes, J.G., Kyle, J.L., Cassels, F., Turner, A.K., Randall, R., *et al.* (2006). Comparative safety and immunogenicity of two attenuated enterotoxigenic *Escherichia coli* vaccine strains in healthy adults. Infect. Immun. *74*, 994–1000.

McKenzie, R., Bourgeoism, A.L., Frechm, S.A, Flyer, D.C., Bloom, A., Kazempour, K., and Glenn, G.M. (2007). Transcutaneous immunization with the heat-labile toxin (LT) of enterotoxigenic *Escherichia coli* (ETEC): protective efficacy in a double-blind, placebo-controlled challenge study. Vaccine *25*, 3684–3691.

al-Majali, A.M., Robinson, J.P., Asem, E.K., Lamar, C., Freeman, M.J., and Saeed, A.M. (1999). Age-dependent variation in the density and affinity of *Escherichia coli* heat-stable enterotoxin receptors in mice. Adv. Exp. Med. Biol. *473*, 137–145.

Mandomando, I.M., Macete, E.V., Ruiz, J., Sanz, S., Abacassamo, F., Vallès, X., Sacarlal, J., Navia, M.M., Vila, J., Alonso, P.L., *et al.* (2007). Etiology of diarrhea in children younger than 5 years of age admitted in a rural hospital of southern Mozambique. Am. J. Trop. Med. Hyg. *76*, 522–527.

Manning, S.D., Motivala, A.S., Springman, A.C., Qi, W., Lacher, D.W., Ouellette, L.M., Mladonicky, J.M., Somsel, P., Rudrik, J.T., Dietrich, S.D., *et al.* (2008). Variation in virulence among clades of *Escherichia coli* O157:H7 associated with disease outbreaks. Proc. Natl. Acad. Sci. U.S.A. *105*, 4868–4873.

Marier, R., Wells, J.G., Swanson, R.C., Callahan, W., and Mehlman, I.J. (1973). An outbreak of enteropathogenic *Escherichia coli* foodborne disease traced to imported French cheese. Lancet *302*, 1376–1378.

Marquardt, R.R., Jin, L.Z., Kim, J.W., Fang, L., Frohlich, A.A., and Baidoo, S.K. (1999). Passive protective effect of egg-yolk antibodies against enterotoxigenic *Escherichia coli* K88+ infection in neonatal and early-weaned piglets. FEMS. Immunol. Med. Microbiol. *23*, 283–288.

Martin, I.E. Tyler, S.D., Tyler, K.D., Khakhria, R., and Johnson, W.M. (1996). Evaluation of ribotyping as epidemiological tool for typing *Escherichia coli* serogroup O157 isolates. J. Clin. Microbiol. *34*, 720–723.

Medina, A.M., Rivera, F.P., Romero, L.M., Kolevic, L.A., Castillo, M.E., Verne, E., Hernandez, R., Mayor, Y.E., Barletta, F., Mercado, E., *et al.* (2010). Diarrheagenic *Escherichia coli* in human immunodeficiency virus (HIV) pediatric patients in Lima, Peru. Am. J. Trop. Med. Hyg. *83*, 158–163.

Mellies, J.L., Elliot, S.J., Sperandio, V., Donnenberg, M.S., and Kaper, J.B. (1999). The Per regulon of enteropathogenic *Escherichia coli*: identification of a regulatory cascade and a novel transcriptional activator, the locus of enterocyte effacement (LEE)-encoded regulator (Ler). Mol. Microbiol. *33*, 296–306.

Ménard, R., and Sansonetti, P.J. (1994). *Shigella flexneri*: isolation of noninvasive mutants of Gram-negative pathogens. Methods Enzymol. *236*, 493–509.

Mendez-Arancibia, E., Vargas, M., Soto, S., Ruiz, J., Kahigwa, E., Schellenberg, D., Urassa, H., Gascón, J., and Vila, J. (2008). Prevalence of different virulence factors and biofilm production in enteroaggregative *Escherichia coli* isolates causing diarrhea in children in Ifakara (Tanzania). Am. J. Trop. Med. Hyg. *78*, 985–989.

Meng, J., Zhao, S., Zhao, T., and Doyle, M.P. (1995). Molecular characterisation of *Escherichia coli* O157:H7 isolates by pulsed-field gel electrophoresis and plasmid DNA analysis. J. Med. Microbiol. *42*, 258–263.

Mercado, E.H., Ochoa, T.J., Ecker, L., Cabello, M., Durand, D., Barletta, F., Molina, M., Gil, A.I., Huicho, L., Lanata, C.F., and Cleary, T.G. (2011). Fecal Leukocytes in children infected with diarrheagenic *Escherichia coli*. J. Clin. Microbiol. *49*, 1376–1381.

Michael, M., Elliot, E.J., Ridley, G.F., Hodson, E.M., and Craig, J.C. (2009). Interventions for haemolytic syndrome and thrombocytopaenic purpura. Cochrane Database Syst. Rev. *1*, CD003595.

Miller, J.R., Barrett, L., Kotloff, K., and Guerrant, R.L. (1994). A rapid diagnostic test for infectious and inflammatory enteritis. Arch. Intern. Med. *154*, 2660–2664.

Moen, S.T., Blumentritt, C.A., Slater, T.M., Patel, S.D., Tutt, C.B., Estrella-Jimenez, M.E., Pawlik, J., Sower, L., Popov, V.L., Schein, C.H., *et al.* (2010). Testing the efficacy and toxicity of adenylyl cyclase inhibitors against enteric pathogens using *in vitro* and *in vivo* models of infection. Infect. Immun. *78*, 1740–1749.

Mondal, D., Haque, R., Sack, R.B., Kirkpatrick, B.D., and Petri, W.A., Jr. (2009). Attribution of malnutrition to cause-specific diarrheal illness: evidence from a prospective study of preschool children in Mirpur, Dhaka, Bangladesh. Am. J. Trop. Med. Hyg. *80*, 824–826.

Moreno, A.C., Filho, A.F., Gomes, Tdo. A., Ramos, S.T., Montemor, L.P., Tavares, V.C., Filho Ldos, S., Irino, K., and Martinez, M.B. (2010). Etiology of childhood diarrhea in the northeast of Brazil: significant emergent diarrheal pathogens. Diagn. Microbiol. Infect. Dis. *66*, 50–57.

Morin, N., Tirling, C., Ivison, S.M., Kaur, A.P., Nataro, J.P., and Steiner, T.S. (2010). Autoactivation of the AggR regulator of enteroaggregative *Escherichia coli in vitro* and *in vivo*. FEMS. Immunol. Med. Microbiol. *58*, 344–355.

Morrow, A.L., Ruiz-Palacios, G.M., Jiang, X., and Newburg, D.S. (2005). Human-milk glycans that inhibit pathogen binding protect breast-feeding infants against infectious diarrhea. J. Nutr. *135*, 1304–1307.

Mortensen, N.P., Fowlkesm, J.D., Maggart, M., Doktycz, M.J., Nataro, J.P., Drusano, G., Allison, D.P. (2011). Effects of sub-minimum inhibitory concentrations of ciprofloxacin on enteroaggregative *Escherichia coli* and the role of the surface protein dispersin. Int. J. Antimicrob. Agents *38*, 27–34.

Moseley, S.L., Echeverria, P., Seriwatana, J., Tirapat, C., Chaicumpa, W., Sakuldaipeara, T., and Falkow, S. (1982). Identification of enterotoxigenic *Escherichia coli* by colony hybridization using three enterotoxin gene probes. J. Infect. Dis. *145*, 863–869.

Moyo, S.J., Maselle, S.Y., Matee, M.I., Langeland N., and Mylvaganam, H. (2007). Identification of diarrheagenic *Escherichia coli* isolated from infants and children in Dar es Salaam, Tanzania. BMC. Infect. Dis. *7*, 92.

Mukherjee, A., Jackson, S.A., LeClerc, J.E., and Cebula, T.A. (2006). Exploring genotypic and phenotypic diversity of microbes using microarray approaches. Toxicol. Mech. Methods *16*, 121–123.

Muller, D., Hagedorn, P., Brast, S., Heusipp, G., Bielaszewska, M., Friedrich, A.W., Karch, H, and Schmidt, M.A. (2006). Rapid identification and differentiation of clinical isolates of enteropathogenic *Escherichia coli* (EPEC), atypical EPEC, and Shiga toxin-producing *Escherichia coli* by a one-step multiplex PCR method. J. Clin. Microbiol. *44*, 2626–2629.

Murayama, S.Y., Sakai, T., Makino, S., Kurata, T., Sasakawa, C., and Yoshikawa, M. (1986). The use of mice in the Sereny test as a virulence assay of shigellae and enteroinvasive *Escherichia coli*. Infect. Immun. *51*, 696–698.

Nada, R.A., Shaheen, H.I., Touni, I., Fahmy, D., Armstrong, A.W., Weiner, M., and Klena, J.D. (2010). Design and validation of a multiplex polymerase chain reaction for the identification of enterotoxigenic *Escherichia coli* and associated colonization factor antigens. Diagn. Microbiol. Infect. Dis. *67*, 134–142.

Nada, R.A., Shaheen, H.I., Khalil, S.B., Mansour, A., El-Sayed, N., Touni, I., Weiner, M., Armstrong, A.W., and Klena, J.D. (2011). Discovery and phylogenetic analysis of novel members of class b enterotoxigenic *Escherichia coli* adhesive fimbriae. J. Clin. Microbiol. *49*, 1403–1410.

Nakamura, Y., Leppert, M., O'Connell, P., Wolff, R., Holm, T., Culver, M., Martin, C., Fujimoto, E., Hoff, M., and Kumlin, E. (1987). Variable number of tandem repeat (VNTR) markers for human gene mapping. Science *235*, 1616–1622.

Nataro, J., and Kaper, J.B. (1998). Diarrheagenic *Escherichia coli*. Clin. Microbiol. Rev. *11*, 142–201.

Nataro, J.P., Deng, Y., Maneval, D.R., German, A.L., Martin, W.C., and Levine, M.M. (1992). Aggregative adherence fimbriae I of enteroaggregative *Escherichia coli* mediate adherence to HEp-2 cells and hemagglutination of human erythrocytes. Infect. Immun. *60*, 2297–2304.

Nataro, J.P., Yikang, D., Yingkang, D., and Walker, K. (1994). AggR, a transcriptional activator of aggregative adherence fimbria I expression in enteroaggregative *Escherichia coli*. J. Bacteriol. *176*, 4691–4699.

Nataro, J.P., Deng, Y., Cookson, S., Cravioto, A., Savarino, S.J., Guers, L.D., Levine, M.M., and Tacket, C.O. (1995a). Heterogeneity of enteroaggregative *Escherichia coli* virulence demonstrated in volunteers. J. Infect. Dis. *171*, 465–468.

Nataro, J.P., Seriwatana, J., Fasano, A., Maneval, D.R., Guers, L.D., Noriega, F., Dubovsky, F., Levine, M.M., Morris, J.G., Jr. (1995b). Identification and cloning of a novel plasmid-encoded enterotoxin of enteroinvasive *Escherichia coli* and *Shigella* strains. Infect. Immun. *63*, 4721–4728.

Nataro, J.P., Mai, V., Johnson, J., Blackwelder, W.C., Heimer, R., Tirrell, S., Edberg, S.C., Braden, C.R., GlennMorris, J.Jr., and Hirshon, J.M. (2006). Diarrheagenic *Escherichia coli* infection in Baltimore, Maryland, and New Haven, Connecticut. Clin. Infect. Dis. *43*, 402–407.

Navarro-Garcia, F. (2010). Enteroaggregative *Escherichia coli* plasmid-encoded toxin. Future Microbiol. *5*, 1005–1013.

Nawar, H.F., King-Lyons, N.D., Hu, J.C., Pasek, R.C., and Connell, T.D. (2010). LT-IIc, a new member of the type II heat-labile enterotoxin family encoded by an *Escherichia coli* strain obtained from a nonmammalian host. Infect. Immun. 78, 4705–4713.

Nelson, J.D. (1971). Duration of neomycin for enteropathogenic *Escherichia coli* diarrheal disease. A comparative study of 113 cases. Pediatrics *48*, 248–258.

Nguyen, R.N., Taylor, L.S., Tauschek, M., and Robins-Browne, R.M. (2006). Atypical enteropathogenic *Escherichia coli* infection and prolonged diarrhea in children. Emerg. Infect. Dis. *12*, 597–603.

Nguyen, T.V., Le Van, P., Le Huy, C., Gia, K.N., and Weintraub, A. (2005). Detection and characterization of diarrheagenic *Escherichia coli* from young children in Hanoi, Vietnam. J. Clin. Microbiol. *43*, 755–760.

Nishikawa, Y., Zhou, Z., Hase, A., Ogasawara, J., Kitase, T., Abe, N., Nakamura, H., Wada, T., Ishii, E., Haruki, K., *et al.* (2002). Diarrheagenic *Escherichia coli* isolated from stools of sporadic cases of diarrheal illness in Osaka City, Japan between 1997 and 2000: prevalence of enteroaggregative *E. coli* heat-stable enterotoxin 1 gene-possessing *E. coli*. Jpn. J. Infect. Dis. *55*, 183–190.

Noller, A.C., McEllistrem, M.C., Stine, O.C., Morris, J.G., Jr., Boxrud, D.J., Dixon, B., and Harrison, L.H. (2003). Multilocus sequence typing reveals a lack of diversity among *Escherichia coli* O157:H7 isolates that are distinct by pulsed-field gel electrophoresis. J. Clin. Microbiol. *41*, 675–679.

Nunes, E.B., Saridakis, H.O., Irino, K., and Pelayo, J.S. (2003). Genotypic and phenotypic characterization of attaching and effacing *Escherichia coli* (AEEC) isolated from children with and without diarrhoea in Londrina, Brazil. J. Med. Microbiol. 52, 499–504.

Nunes, M.D., Penna, F.J., Franco, R.T., Mendes, E.N., and Magalhães, P.P. (2011). Enterotoxigenic *Escherichia coli* in children with acute diarrhea and controls in

Teresina/PI, Brazil: distribution of enterotoxin and colonization factor genes. J. Appl. Microbiol. *111*, 224–232.

Nweze, E.I. (2010). Aetiology of diarrhoea and virulence properties of diarrhoeagenic *Escherichia coli* among patients and healthy subjects in southeast Nigeria. J. Health. Popul. Nutr. *28*, 245–252.

Ochi, S., Shimizu, T., Ohtani, K., Ichinose, Y., Arimitsu, H., Tsukamoto, K., Kato, M., and Tsuji, T. (2009). Nucleotide sequence analysis of the enterotoxigenic *Escherichia coli* Ent plasmid. DNA Res. *16*, 299–309.

Ochoa, T.J., Barletta, F., Contreras, C., and Mercadoa, E. (2008). New insights into the epidemiology of enteropathogenic *Escherichia coli* infection. Trans. R. Soc. Trop. Med. Hyg. *102*, 852–856.

Ochoa, T.J., Ecker, L., Barletta, F., Mispireta, M.L., Gil, A.I., Contreras, C., Molina, M., Amemiya, I., Verastegui, H., Hall, E.R., *et al.* (2009a). Age-related susceptibility to infection with diarrheagenic *Escherichia coli* among infants from Periurban areas in Lima, Peru. Clin. Infect. Dis. *49*, 1694–1702.

Ochoa, T.J., Ruiz, J., Molina, M., Del Valle, L.J., Vargas, M., Gil, A.I., Ecker, L., Barletta, F., Hall, E., Cleary, T.G., and Lanata, C.F. (2009b). High frequency of antimicrobial drug resistance of diarrheagenic *Escherichia coli* in infants in Peru. Am. J. Trop. Med. Hyg. *81*, 296–301.

Ochoa, T.J., Mercado, E.H., Durand D., Rivera, F.P., Mosquito, S., Contreras, C., Riveros, M., Lluque, A., Barletta, F., Prada, A., *et al.* (2011). Frequency and pathotypes of diarrheagenic *Escherichia coli* in Peruvian children with and without diarrhea. Rev. Peru. Med. Exp. Salud. Publica. *28*, 13–20. [In Spanish.]

Okamoto, K., and M.Takahara. (1990). Synthesis of *Escherichia coli* heatstable enterotoxin STp as a pre-pro form and role of the pro sequence in secretion. J. Bacteriol. *172*, 5260–5265.

Okeke, I.N., Ojo, O., Lamikanra, A., and Kaper, J.B. (2003). Etiology of acute diarrhea in adults in southwestern Nigeria. J. Clin. Microbiol. *41*, 4525–4230.

Okeke, I.N., Scaletsky, I.C., Soars, E.H., Macfarlane, L.R., and Torres, A.G. (2004). Molecular epidemiology of the iron utilization genes of enteroaggregative *Escherichia coli*. J. Clin. Microbiol. 42, 36–44.

Okeke, I.N., Wallace-Gadsden, F., Simons, H.R., Matthews, N., Labar, A.S., Hwang, J., and Wain, J. (2010). Multi-locus sequence typing of enteroaggregative *Escherichia coli* isolates from Nigerian children uncovers multiple lineages. PLoS One *5*, e14093.

Olsen, S.J., Miller, G., Kennedy, M., Higgins, C., Walford, J., McKee, G., Fox, K., Bibb, W., and Mead, P. (2002). A waterborneoutbreak of *Escherichia coli* O157:H7 infections and hemolytic uremic syndrome: implications for rural water systems. Emerg. Infect. Dis. *8*, 370–375.

Opintan, J.A., Newman, M.J., Ayeh-Kumi, P.F., Affrim, R., Gepi-Attee, R., Sevilleja, J.E., Roche, J.K., Nataro, J.P., Warren, C.A., and Guerrant, R.L. (2010). Pediatric diarrhea in southern Ghana: etiology and association with intestinal inflammation and malnutrition. Am. J. Trop. Med. Hyg. *83*, 936–943.

Oyofo, B.A., Subekti, D.S., Svennerholm, A.M., Machpud, N.N., Tjaniadi, P., Komalarini, T.S., Setiawan, B., Campbell, J.R., Corwin, A.L., and Lesmana, M. (2001). Toxins and colonization factor antigens of enterotoxigenic *Escherichia coli* among residents of Jakarta, Indonesia. Am. J. Trop. Med. Hyg. *65*, 120–124.

Oyofo, B.A., Subekti, D., Tjaniadi, P., Machpud, N., Komalarini, S., Setiawan, B., Simanjuntak, C., Punjabi, N., Corwin, A.L., Wasfy, M., *et al.* (2002). Enteropathogens associated with acute diarrhea in community and hospital patients in Jakarta, Indonesia. FEMS. Immunol. Med. Microbiol. *34*, 139–146.

Ozerol, I.H., Bayraktar, M.R., Iseri, L., Otlu, B., and Durmaz, R. (2005). The prevalence and molecular typing of enterotoxigenic *Escherichia coli* strains isolated from diarrheic stools in Malatya, Turkey. New. Microbiol. *28*, 237–243.

Pabst, W.L., Altwegg, M., Kind, C., Mirjanic, S., Hardegger, D., and Nadal, D. (2003). Prevalence of enteroaggregative *Escherichia coli* among children with and without diarrhea in Switzerland. J. Clin. Microbiol. *41*, 2289–2293.

Pai, M., Kang, G., Ramakrishna, B.S., Venkataraman, A., and Muliyil, J. (1997). An epidemic of diarrhoea in South India caused by enteroaggregative *Escherichia coli*. Indian J. Med. Res. *106*, 7–12.

Pal, T., Al-Sweih, N.A., Herpay, M., and Chugh, T.D. (1997). Identification of enteroinvasive *Escherichia coli* and *Shigella* strains in pediatric patients by an IpaC-specific enzyme-linked immunosorbent assay. J. Clin. Microbiol. *35*, 1757–1760.

Paniagua, M., Espinoza, F., Ringman, M., Reizenstein, E., Svennerholm, A.M., and Hallander, H. (1997). Analysis of incidence of infection with enterotoxigenic *Escherichia coli* in a prospective cohort study of infant diarrhea in Nicaragua. J. Clin. Microbiol. *35*, 1404–1410.

Paredes-Paredes, M., Okhuysen, P.C., Flores, J., Mohamed, J.A., Padda, R.S., Gonzalez-Estrada, A., Haley, C.A., Carlin, L.G., Nair, P., and DuPont, H.L. (2011). Seasonality of diarrheagenic *Escherichia coli* pathotypes in the US students acquiring diarrhea in Mexico. J. Travel. Med. *18*, 121–125.

Patel, S.K., Dotson, J., Allen, K.P., and Fleckenstein, J.M. (2004). Identification and molecular characterization of EatA, an autotransporter protein of enterotoxigenic *Escherichia coli*. Infect. Immun. *72*, 1786–1794.

Paton, A.W., Jennings, M.P., Morona, R., Wang, H., Focareta, A., Roddam, L.F., and Paton, J.C. (2005). Recombinant probiotics for treatment and prevention of enterotoxigenic *Escherichia coli* diarrhea. Gastroenterology *128*, 1219–1228.

Paulozzi, L.J., Johnson, K.E., Kamahele, L.M., Clausen, C.R., Riley, L.W., and Helgerson, S.D. (1986). Diarrhea associated with adherent enteropathogenic *Escherichia coli* in an infant and toddler center, Seattle, Washington. Peditrics *77*, 296–300.

Pavia, A.T., Nichols, C.R., Green, D.P., Tauxe, R.V., Mottice, S., Greene, K.D., Wells, J.G., Seigler, R.L., Brewer, E.D., and Hannon, D. (1990).

Hemolytic-uremic syndrome during an outbreak of *Escherichia coli* O157:H7 infections for mentally retarded persons: clinical and epidemiological observations. J. Pediatr. *116*, 544–551.

Pebody, R.G., Furtado, C., Rojas, A., McCarthy, N., Nylen, G., Ruutu, P., Leino, T., Chalmers, R., de Jong, B., Donnelly, M., *et al.* (1999). An international outbreak of Vero cytotoxin-producing *Escherichia coli* O157 infection amongst tourists; a challenge for the European infectious disease surveillance network. Epidemiol. Infect. *123*, 217–223.

Pereira, A.C., Britto-Filho, J.D., José de Carvalho, J., de Luna, M.G., and Rosa, A.C. (2008). Enteroaggregative *Escherichia coli* (EAEC) strains enter and survive within cultured intestinal epithelial cells. Microb. Pathog. *45*, 310–314.

Pereira, A.L., Ferraz, L.R., Silva, R.S., and Giugliano, L.G. (2007). Enteroaggregative *Escherichia coli* virulence markers: positive association with distinct clinical characteristics and segregation into 3 enteropathogenic *E. coli* serogroups. J. Infect. Dis. *195*, 366–374.

Pereira, A.L., Silva, T.N., Gomes, A.C., Araújo, A.C., and Giugliano, L.G. (2010). Diarrhea-associated biofilm formed by enteroaggregative *Escherichia coli* and aggregative *Citrobacter freundii*: a consortium mediated by putative F pili. BMC. Microbiol. *10*, 57.

Pérez, C., Gómez-Duarte, O.G., and Arias, M.L. (2010). Diarrheagenic *Escherichia coli* in children from Costa Rica. Am. J. Trop. Med. Hyg. *83*, 292–297.

Peruski, L.F., Jr., Kay, B.A., El-Yazeed, R.A., El-Etr, S.H., Cravioto, A., Wierzba, T.F., Rao, M., El-Ghorab, N., Shaheen, H., Khalil, S.B., *et al.* (1999). Phenotypic diversity of enterotoxigenic *Escherichia coli* strains from a community-based study of pediatric diarrhea in periurban Egypt. J. Clin. Microbiol. *37*, 2974–2978.

Pitondo-Silva, A., Minarini, L.A.R., Camargo, I.L.B.C., and Darini, A.L.C. (2009). Clonal relationships determined by multilocus sequence typing among enteropathogenic *Escherichia coli* isolated in Brazil. Can. J. Microbiol. *55*, 672–679.

Piva, I.C., Pereira, A.L., Ferraz, L.R., Silva, R.S., Vieira, A.C., Blanco, J.E., Blanco, M., Blanco, J., and Giugliano, L.G. (2003). Virulence markers of enteroaggregative *Escherichia coli* isolated from children and adults with diarrhea in Brasília, Brazil. J. Clin. Microbiol. *41*, 1827–1832.

Porter, C.K., Riddle, M.S., Tribble, D.R., Louis Bougeois, A., McKenzie, R., Isidean, S.D., Sebeny, P., and Savarino, S.J. (2011). A systematic review of experimental infections with enterotoxigenic *Escherichia coli* (ETEC). Vaccine *29*, 5869–5895.

Prado, D., López, E., Liu, H., Devoto, S., Woloj, M., Contrini, M., Murray, B.E., Gómez, H., and Cleary, T.G. (1992). Ceftibuten and trimethoprim–sulfamethoxazole for treatment of *Shigella* and enteroinvasive *Escherichia coli* disease. Pediatr. Infect. Dis J. *11*, 644–647.

Prado, V., Martinez, J., Arellano, C., and Levine, M.M. (1997). Temporal variation of genotypes and serotypes of enterohemorrhagic *E. coli* isolated from Chilean children with asymptomatic infections or hemolytic uremic syndrome. Rev. Med. Chile *125*, 291–297.

Presterl, E., Zwick, R.H., Reichmann, S., Aichelburg, A., Winkler, S., Kremsner, P.G., and Graninger, W. (2003). Frequency and virulence properties of diarrheagenic *Escherichia coli* in children with diarrhea in Gabon. Am. J. Trop. Med. Hyg. *69*, 406–410.

Puiprom, O., Chantaroj, S., Gangnonngiw, W., Okada, K., Honda, T., Taniguchi, T., and Sawanpanyalert, P. (2010). Identification of colonization factors of enterotoxigenic *Escherichia coli* with PCR-based technique. Epidemiol. Infect. *138*, 519–524.

Pupo, G.M., Karaolis, D.K., Lan, R., and Reeves, P.R. (1997). Evolutionary relationships among pathogenic and nonpathogenic *Escherichia coli* strains inferred from multilocus enzyme electrophoresis and *mdh* sequence studies. Infect. Immun. *65*, 2685–2692.

Qadri, F., Khan, A.I., Faruque, A.S., Begum, Y.A., Chowdhury, F., Nair, G.B., Salam, M.A., Sack, D.A., and Svennerholm, A.M. (2005a). Enterotoxigenic *Escherichia coli* and *Vibrio cholerae* diarrhea, Bangladesh, 2004. Emerg. Infect. Dis. *11*, 1104–1107.

Qadri, F., Svennerholm, A.M., Faruque, A.S., and Sack, R.B. (2005b). Enterotoxigenic *Escherichia coli* in developing countries: epidemiology, microbiology, clinical features, treatment, and prevention. Clin. Microbiol. Rev. *18*, 465–483.

Qadri, F., Saha, A., Ahmed, T., Al Tarique, A., Begum, Y.A., and Svennerholm, A.M. (2007). Disease burden due to enterotoxigenic *Escherichia coli* in the first 2 years of life in an urban community in Bangladesh. Infect. Immun. *75*, 3961–3968.

Quiroga, M., Oviedo, P., Chinen, I., Pegels, E., Husulak, E., Binztein, N., Rivas, M., Schiavoni, L., and Vergara, M. (2004). Asymptomatic infections by diarrheagenic *Escherichia coli* in children from Misiones, Argentina, during the first twenty months of their lives. Rev. Inst. Med. Trop. Sao Paulo. *42*, 9–15.

Rabbani, G.H., Teka, T., Zaman, B., Majid, N., Khatun, M., and Fuchs, G.J. (2001). Clinical studies in persistent diarrhea: dietary management with green banana or pectin in Bangladeshi children. Gastroenterology *121*, 554–560.

Radu, S., Ling, O.W., Rusul, G., Karim, M.I., and Nishibuchi, M. (2001). Detection of *Escherichia coli* O157:H7 by multiplex PCR and their characterization by plasmid profiling, antimicrobial resistance, RAPD and PFGE analyses. J. Microbiol. Methods *46*, 131–139.

Rajendran, P., Ajjampur, S.S., Chidambaram, D., Chandrabose, G., Thangaraj, B., Sarkar, R., Samuelm, P., Rajan, D.P., and Kang, G. (2010). Pathotypes of diarrheagenic *Escherichia coli* in children attending a tertiary care hospital in South India. Diagn. Microbiol. Infect. Dis. *68*, 117–122.

Ramachandren, V., Brett, K., Hornitzky, M.A., Dowton, M., Bettelheim, K.A., Walker, M.J., and Djordjevic, S.P. (2003). Distribution of intimin subtypes among *Escherichia coli* isolates from ruminant and human sources. J. Clin. Microbiol. *41*, 5022–5032.

Ramos Moreno, A.C., Cabilio, Guth, B.E., and Baquerizo, Martinez, M. (2006). Can the *fliC* PCR-restriction fragment length polymorphism technique replace classic serotyping methods for characterizing the H antigen of enterotoxigenic *Escherichia coli* strains? J. Clin. Microbiol. *44*, 1453–1458.

Rao, M.R., Abu-Elyazeed, R., Savarino, S.J., Naficy, A.B., Wierzba, T.F., Abdel-Messih, I., Shaheen, H., Frenck, R.W., Jr., Svennerholm, A.M., and Clemens, J.D. (2003). High disease burden of diarrhea due to enterotoxigenic *Escherichia coli* among rural Egyptian infants and young children. J. Clin. Microbiol. *41*, 4862–4864.

Rasheed, J.K., Guzman-Verduzco, L.M., and Kupersztoch, Y.M. (1990). Two precursors of the heat-stable enterotoxin of *Escherichia coli*: evidence of extracellular processing. Mol. Microbiol. *4*, 265–273.

Ratchtrachenchai, O.A., Subpasu, S., Hayashi, H., and Ba-Thein, W. (2004). Prevalence of childhood diarrhoea-associated *Escherichia coli* in Thailand. J. Med. Microbiol. *53*, 237–243.

Regua-Mangia, A.H., Gomes, T.A., Vieira, M.A., Andrade, J.R., Irino, K., and Teixeira, L.M. (2004). Frequency and characteristics of diarrhoeagenic *Escherichia coli* strains isolated from children with and without diarrhoea in Rio de Janeiro, Brazil. J. Infect. *48*, 161–167.

Regua-Mangia, A.H., Gomes, T.A., Vieira, M.A., Irino, K., and Teixeira, L.M. (2009). Molecular typing and virulence of enteroaggregative *Escherichia coli* strains isolated from children with and without diarrhoea in Rio de Janeiro city, Brazil. J. Med. Microbiol. *58*, 414–422.

Rendon, M.A., Saldana, Z., Erdem, A.L., Monteiro-Neto, V., Vasquez, A., Kaper, J.B., Puente, J.L., and Giron, J.A. (2007). Commensal and pathogenic *Escherichia coli* use a common pilus adherence factor for epithelial cell colonization. Proc. Natl. Acad. Sci. U.S.A. *104*, 10637–10642.

Resta-Lenert, S., and Barrett, K.E. (2002). Enteroinvasive bacteria alter barrier and transport properties of human intestinal epithelium: role of iNOS and COX-2. Gastroenterology *122*, 1070–1087.

Resta-Lenert, S., and Barrett, K.E. (2003). Live probiotics protect intestinal epithelial cells from the effects of infection with enteroinvasive *Escherichia coli* (EIEC). Gut *52*, 988–997.

Reyes, D., Vilchez, S., Paniagua, M., Colque, P., Weintraub, A., Möllby, R., and Kühn, I. (2009). Diversity of intestinal *Escherichia coli* populations in Nicaraguan children with and without diarrhoea. J. Med. Microbiol. *58*, 1593–1600.

Richards, H.A., Perez-Conesa, D., Doane, C.A., Gillespie, B.E., Mount, J.R., Oliver, S.P., Pangloli, P., and Draughon, F.A. (2006). Genetic characterization of a diverse *Escherichia coli* O157:H7 population from a variety of farm environments. Foodborne Pathog. Dis. *3*, 259–265.

Riddle, M.S., Sanders, J.W., Putnam, S.D., and Tribble, D.R. (2006). Incidence, etiology, and impact of diarrhea among long-term travelers (US military and similar populations): a systematic review. Am. J. Trop. Med. Hyg. *74*, 891–900.

Riley, L.W., Remis, R.S., Helgerson, S.D., McGee, H.B., Wells, J.G., Davis, B.R., Herbert, R.J., Olcott, E.S., Johnson, L.M., Hargrett, N.T., *et al.* (1983). Hemorrhagic colitis associated with a rare *Escherichia coli* serotype. N. Engl. J. Med. *308*, 681–685.

Riley, W.B. (1968). An outbreak of diarrhoea in an infants' home due to a *Shigella*-like organism. Med. J. Aust. *2*, 1175–1176.

Rivas, M., Sosa-Estani, S., Rangel, J., Caletti, M.G., Valles, P., Roldan, C.D., Balbi, L., de Mollar, M.C.M., Amoedo, D., Miliwebsky, E., *et al.* (2008). Risk factors for sporadic Shiga toxin-producing *Escherichia coli* infections in children, Argentina. Emerg. Infect. Dis. *14*, 763–771.

Rivera, F.P., Ochoa, T.J., Maves, R.C., Bernal, M., Medina, A.M., Meza, R., Barletta, F., Mercado, E., Ecker, L., Gil, A.I., *et al.* (2010). Genotypic and phenotypic characterization of enterotoxigenic *Escherichia coli* strains isolated from Peruvian children. J. Clin. Microbiol. *48*, 3198–203.

Robins-Browne, R.M., Bordun, A.-M., Tauschek, M., Bennett-Wood, V., Russell, J., Oppedisano, F., Lister, N.A., Bettelheim, K.A., Fairley, C.K., Sinclair, M.I., *et al.* (2004). Atypical enteropathogenic, *Escherichia coli*: a leading cause of community-acquired gastroenteritis in Melbourne, Australia. Emerg. Infect. Dis. *10*,1797–1805.

Robinson, C.M., Sinclair, J.F., Smith, M.J., and Obrien, A.D. (2006). Shiga toxin of enterohemorrhagic *Escherichia coli* type O157:H7 promotes intestinal colonization. Proc. Natl. Acad. Sci. U.S.A. *103*, 9667–9672.

Rodas, C., Iniguez, V., Qadri, F., Wiklund, G., Svennerholm, A.M., and Sjöling, A. (2009). Development of multiplex PCR assays for detection of enterotoxigenic *Escherichia coli* colonization factors and toxins. J. Clin. Microbiol. *47*, 1218–1220.

Rodrigues, J., Acosta, V.C., Candeias, J.M., Souza, L.O., and Filho, F.J. (2002). Prevalence of diarrheogenic *Escherichia coli* and rotavirus among children from Botucatu, São Paulo State, Brazil. Braz. J. Med. Biol. Res. *35*, 1311–1318.

Rodrigues, J., Thomazini, C.M., Lopes, C.A.M., and Dantas, L.O. (2004). Concurrent infection in a dog and colonization in a child with a human enteropathogenic *Escherichia coli* clone. J. Clin. Microbiol. *42*, 1388–1389.

Roels, T.H., Proctorm, M.E., Robinsonm, L.C., Hulbert, K., Bopp, C.A., and Davis, J.P. (1998). Clinical features of infections due to *Escherichia coli* producing heat-stable toxin during an outbreak in Wisconsin: a rarely suspected cause of diarrhea in the United States. Clin. Infect. Dis. *26*, 898–902.

Rogers, K.B. (1951). The spread of infantile gastroenteritis in a cubicled ward. J. Hyg. *49*, 140–151.

Rooney, R.M., Bartram, J.K., Cramer, E.H., Mantha, S., Nichols, G., Suraj, R., and Todd, E.C. (2004). A review of outbreaks of waterborne disease associated with ships: evidence for risk management. Public Health Rep. *119*, 435–442.

Rosa, A.C., Mariano, A.T., Pereira, A.M., Tibana, A., Gomes, T.A., and Andrade, J.R. (1998). Enteropathogenicity markers in *Escherichia coli* isolated from infants with acute diarrhea and healthy controls in Rio de Janeiro, Brazil. J. Med. Microbiol. 47, 781–790.

Rossiter, A.E., Browning, D.F., Leyton, D.L., Johnson, M.D., Godfrey, R.E., Wardius, C.A., Desvaux, M., Cunningham, A.F., Ruiz-Perez, F., Nataro, J.P., *et al.* (2011). Transcription of the plasmid-encoded toxin gene from Enteroaggregative *Escherichia coli* is regulated by a novel co-activation mechanism involving CRP and Fis. Mol. Microbiol. *81*, 179–191.

Rothbaum, R., McAdams, A.J., Giannella, R., and Partin, J.C. (1982). A clinicopathological study of enterocyte-adherent *Escherichia coli*: a cause of protracted diarrhea in infants. Gastroenterology *83*, 441–454.

Rowe, B., Gross, R.J., and Allen, H.A. (1974). Enteropathogenic *E. coli* O 124 in the United Kingdom. Lancet *1(7850)*, 224–225.

Sabui, S., Ghosal, A., Dutta, S., Ghosh, A., Ramamurthy, T., Nataro, J.P., Hamabata, T., and Chatterjee, N.S. (2010). Allelic variation in colonization factor CS6 of enterotoxigenic *Escherichia coli* isolated from patients with acute diarrhoea and controls. J. Med. Microbiol. *59*, 770–779.

Sack, D.A., Shimko, J., Torres, O., Bourgeois, A.L., Francia, D.S., Gustafsson, B., Kärnell, A., Nyquist, I., and Svennerholm, A.M. (2007). Randomised, double-blind, safety and efficacy of a killed oral vaccine for enterotoxigenic *E. coli* diarrhoea of travellers to Guatemala and Mexico. Vaccine. *25*, 4392–4400.

Sahl, J.W., Steinsland, H., Redman, J.C., Angiuoli, S.V., Nataro, J.P., Sommerfelt, H., and Rasko, D.A. (2011). A comparative genomic analysis of diverse clonal types of enterotoxigenic *Escherichia coli* reveals pathovar-specific conservation. Infect. Immun. *79*, 950–960.

Sakiri, R., Ramegowda, B., and Tesh, V.L. (1998). Shiga toxin type 1 activates tumor necrosis factor-alpha gene transcription and nuclear translocation of the transcriptional activators nuclear factor-kappaB and activator protein-1. Blood *92*, 558–566.

Salmanzadeh-Ahrabi, S., Habibi, E., Jaafari, F., and Zali, M.R. (2005). Molecular epidemiology of *Escherichia coli* diarrhoea in children in Tehran. Ann. Trop. Paediatr. *25*, 35–39.

Samie, A., Obi, C.L., Dillingham, R., Pinkerton, R.C., and Guerrant, R.L. (2007). Enteroaggregative *Escherichia coli* in Venda, South Africa: distribution of virulence-related genes by multiplex polymerase chain reaction in stool samples of human immunodeficiency virus (HIV)-positive and HIV-negative individuals and primary school children. Am. J. Trop. Med. Hyg. *77*, 142–150.

Sanders, J.W., Frenck, R.W., Putnam, S.D., Riddle, M.S., Johnston, J.R., Ulukan, S., Rockabrand D.M., Monteville, M.R., and Tribble, D.R. (2007). Azithromycin and loperamide are comparable to levofloxacin and loperamide for the treatment of traveler's diarrhea in United States military personnel in Turkey. Clin. Infect. Dis. *45*, 294–301.

Sarantuya, J., Nishi, J., Wakimoto, N., Erdene, S., Nataro, J.P., Sheikh, J., Iwashita, M., Manago, K., Tokuda, K., Yoshinaga, M., *et al.* (2004). Typical enteroaggregative *Escherichia coli* is the most prevalent pathotype among *E. coli* strains causing diarrhea in Mongolian children. J. Clin. Microbiol. *42*, 133–139.

Savarino, S.J., Fasano, A., Robertson, D.C., and Levine, M.M. (1991). Enteroaggregative *Escherichia coli* elaborate a heat-stable enterotoxin demonstrable in an '*in vitro*' rabbit intestinal model. J. Clin. Invest. *87*, 1450–1455.

Scavia, G., Staffolani, M., Fisichella, S., Striano, G., Colletta, S., Ferri, G., Escher, M., Minelli, F., and Caprioli, A. (2008). Enteroaggregative *Escherichia coli* associated with a foodborne outbreak of gastroenteritis. J. Med. Microbiol. *57*, 1141–1146.

Schlör, S., Riedl, S., Blass, J., and Reidl, J. (2000). Genetic rearrangements of the regions adjacent to genes encoding heat-labile enterotoxins (*eltAB*) of enterotoxigenic *Escherichia coli* strains. Appl. Environ. Microbiol. *66*, 352–358.

Schmid-Hempel, P., and Frank, S.A. (2007). Pathogenesis, virulence and infective dose. PLoS Pathog. *3*, e147.

Sereny, B. (1957). Experimental keratoconjunctivitis shigellosa. Acta. Microbiol. Hung. *4*, 367–376.

Shah, N., DuPont, H.L., and Ramsey, D.J. (2009). Global etiology of travelers' diarrhea: systematic review from 1973 to the present. Am. J. Trop. Med. Hyg. *80*, 609–614.

Shaheen, H.I., Kamal, K.A., Wasfy, M.O., El-Ghorab, N.M., Lowe, B., Steffen, R., Kodkani, N., Amsler, L., Waiyaki, P., David, J.C., *et al.* (2003). Phenotypic diversity of enterotoxigenic *Escherichia coli* (ETEC) isolated from cases of travelers' diarrhea in Kenya. Int. J. Infect. Dis. *7*, 35–38.

Shaheen, H.I., Khalil, S.B., Rao, M.R., Abu Elyazeed, R., Wierzba, T.F., Peruski, L.F., Jr, Putnam, S., Navarro, A., Morsy, B.Z., Cravioto, A., *et al.* (2004). Phenotypic profiles of enterotoxigenic *Escherichia coli* associated with early childhood diarrhea in rural Egypt. J. Clin. Microbiol. *42*, 5588–5595.

Shahrokhi, N., Bouzari, S., and Jafari, A. (2011). Comparison of virulence markers and antibiotic resistance in enterotoxigenic *Escherichia coli* isolated ten years apart in Tehran. J. Infect. Dev. Ctries. *5*, 248–254.

Shamir, E.R., Warthan, M., Brown, S.P., Nataro, J.P., Guerrant, R.L., and Hoffman, P.S. (2010). Nitazoxanide inhibits biofilm production and hemagglutination by enteroaggregative *Escherichia coli* strains by blocking assembly of AafA fimbriae. Antimicrob. Agents. Chemother. *54*, 1526–1533.

Sheikh, J., Czeczulin, J.R., Harrington, S., Hicks, S., Henderson, I.R., Le Bouguenec, C., Gounon, P., Phillips, A., and Nataro, J.P. (2002). A novel dispersin protein in enteroaggregative *Escherichia coli*. J. Clin. Investig. *110*, 1329–1337.

Shehabi, A.A, Bulosm, N.K, and Hajjajm, K.G. (2003). Characterization of diarrhoeagenic *Escherichia coli* isolates in Jordanian children. Scand. J. Infect. Dis. *35*, 368–371.

Shere, J.A., Bartlett, K.J., and Kasper, C.W. (1998). Longitudinal study of *Escherichia coli* O157:H7 dissemination on four dairy farms in Wisconsin. Appl. Environ. Microbiol. *64*, 1390–1399.

Sherlock, O., Vejborg, R.M., and Klemm, P. (2005). The TibA adhesin/invasion from enterotoxigenic *Escherichia coli* is self-recognizing and induces bacterial aggregation and biofilm formation. Infect. Immun. *73*, 1954–1963.

Shmilovitz, M., Kretzer, B., and Levy, E. (1974). The anaerogenic serotype 147 as an etiologic agent of dysentery in Israel. Isr. J. Med. Sci. *10*, 1425–1429.

Singh, G., Vajpayee, P., Ram, S., and Shanker, R. (2010). Environmental reservoirs for enterotoxigenic *Escherichia coli* in south Asian Gangetic riverine system. Environ. Sci. Technol. *44*, 6475–6480.

Sizemore, D.R., Roland, K.L., and Ryan, U.S. (2004). Enterotoxigenic *Escherichia coli* virulence factors and vaccine approaches. Expert. Rev. Vaccines 3, 585–95.

Slutsker, L., Ries, A.A., and Greene, K.D. (1997). *Escherichia coli* O157:H7 diarrhea in the United States: clinical and epidemiologic features. Ann. Intern. Med. *126*, 505- 513.

Small, P.L., and Falkow, S. (1988). Identification of regions on a 230-kilobase plasmid from enteroinvasive *Escherichia coli* that are required for entry into HEp-2 cells. Infect. Immun. *56*, 225–229.

Smith, D.R., Moxley, R.A., and Peterson, R.E. (2009). A two-dose regime of a vaccine against type III secreted proteins reduced *Escherichia coli* O157:H7 colonization of the terminal rectum in commercial feed lots. Foodborne Pathog. Dis. *6*, 155–161.

Smith, H.R., Cheasty, T., and Rowe, B. (1997). Enteroaggregative *Escherichia coli* and outbreaks of gastroenteritis in UK. Lancet *350*, 814–815.

Snyder, J.D., Wells, J.G., Yashuk, J., Puhr, N., and Blake., P.A. (1984). Outbreak of invasive *Escherichia coli* gastroenteritis on a cruise ship. Am. J. Trop. Med. Hyg. *33*, 281–284.

Sobel, J., Gomes, T.A., Ramos, R.T., Hoekstra, M., Rodrigue, D., Rassi, V., and Griffin, P.M. (2004). Pathogen-specific risk factors and protective factors for acute diarrheal illness in children aged 12–59 months in São Paulo, Brazil. Clin. Infect. Dis. *38*, 1545–1551.

Sobieszczańska, B.M. (2008). Distribution of genes encoding iron uptake systems among enteroaggregative *Escherichia coli* strains isolated from adults with irritable bowel syndrome. Clin. Microbiol. Infect. *14*, 1083–1086.

Sobieszczańska, B.M., and Osek, J. (2003). Enteroaggregative and cell-detaching *Escherichia coli* strains among Polish children with and without diarrhea. Folia. Microbiol. (Praha) *48*, 823–8.

Song, T., Toma, C., Nakasone, N., and Iwanaga, M. (2005). Sensitive and rapid detection of *Shigella* and enteroinvasive *Escherichia coli* by a loop-mediated isothermal amplification method. FEMS. Microbiol. Lett. *243*, 259–263.

Spano, L.C., Sadovsky, A.D., Segui, P.N., Saick, K.W., Kitagawa, S.M., Pereira, F.E., Fagundes-Neto, U., and Scaletsky, I.C. (2008). Age-specific prevalence of diffusely adherent *Escherichia coli* in Brazilian children with acute diarrhoea. J. Med. Microbiol. *57*, 359–363.

Sperandio, V., Mellies, J.L., Nguyen, W., Shin., S and Kaper, J.B. (1999). Quorum sensing controls expression of the type III secretion gene transcription and protein secretion in enterohemorrhagic and enteropathogenic *Escherichia coli*. Proc. Natl. Acad. Sci. U.S.A. *96*, 15196–15201.

Steinberg, E.B., Mendoza, C.E., Glass, R., Arana, B., Lopez, M.B., Mejia, M., Gold, B.D., Priest, J.W., Bibb, W., Monroe, S.S., *et al.* (2004). Prevalence of infection with waterborne pathogens: a seroepidemiologic study in children 6–36 months old in San Juan Sacatepequez, Guatemala. Am J. Trop. Med. Hyg. *70*, 83–88.

Steinmuller, N., Demma, L., Bender, J.B., Edison, M., and Angulo, F.J. (2006). Outbreaks of enteric disease associated with animal contact: not just a foodborne problem anymore. Clin. Infect. Dis. *43*, 1596–1602.

Steinsland, H., Valentiner-Branth, P., Grewal, H.M., Gaastra, W., Mølbak, K.K., and Sommerfelt, H. (2003). Development and evaluation of genotypic assays for the detection and characterization of enterotoxigenic *Escherichia coli*. Diagn. Microbiol. Infect. Dis. *45*, 97–105.

Steinsland, H., Lacher, D.W., Sommerfelt, H., and Whittam, T.S. (2010). Ancestral lineages of human enterotoxigenic *Escherichia coli*. J. Clin. Microbiol. *48*, 2916–2924.

Strauman, M.C., Harper, J.M., Harrington, S.M., Boll, E.J., and Nataro, J.P. (2010). Enteroaggregative *Escherichia coli* disrupts epithelial cell tight junctions. Infect. Immun. 78, 4958–4964.

Struelens, M., Palm, D., and Takkinen, J. (2011). Enteroaggregative, Shiga toxin-producing *Escherichia coli* O104:H4 outbreak: new microbiological findings boost coordinated investigations by European public health laboratories. Euro Surveill. *16*, pii: 19890.

Su, C., and Brandt, L.J. (1995). *Escherichia coli* O157: H7 infection in humans. Ann. Intern. Med. *123*, 698–714.

Subekti, D.S., Lesmana, M., Tjaniadi, P., Machpud, N., Sriwati, Sukarma, Daniel, J.C., Alexander, W.K., Campbell, J.R., Corwin, A.L., *et al.* (2003). Prevalence of enterotoxigenic *Escherichia coli* (ETEC) in hospitalized acute diarrhea patients in Denpasar, Bali, Indonesia. Diagn. Microbiol. Infect. Dis. *47*, 399–405.

Suzart, S., Aparecida, T., Gomes, T., and Guth, B.E. (1999). Characterization of serotypes and outer membrane protein profiles in enteroaggregative *Escherichia coli* strains. Microbiol. Immunol. *43*, 201–205.

Svennerholm, A.M., and Wiklund, G. (1983). Rapid GM1-enzyme-linked immunosorbent assay with visual reading for identification of *Escherichia coli* heat-labile enterotoxin. J. Clin. Microbiol. *17*, 596–600.

Svennerholm, A.-M., and Tobias., J. (2008). Vaccines against enterotoxigenic *Escherichia coli*. Expert. Rev. Vaccines 7, 795–804.

Swaminathan, B., Gerner-Smidt, P., Ng, L.K., Lukinmaa, S., Kam, K.M., Rolando, S., Gutierrez, E.P., and Binsztein, N. (2006). Building PulseNet International: an interconnected system of laboratory networks to facilitate timely public health recognition and response to foodborne disease outbreaks and emerging foodborne disease. Foodborne Pathog. Dis. 3, 36–50.

Szakál, D., and Pál, T. (2003). Comparison of media for the selective culture of enteroinvasive *Escherichia coli*. Eur. J. Clin. Microbiol. Infect. Dis. 22, 235–241.

Szakál, D.D., Schneider, G., and Pál, T. (2003). A colony blot immune assay to identify enteroinvasive *Escherichia coli* and *Shigella* in stool samples. Diagn. Microbiol. Infect. Dis. *45*, 165–171.

Takahashi, E., Sultan, Z., Shimada, S., Aung, W.W., Nyein, M.M., Oo, K.N., Nair, G.B., Takeda, Y., and Okamoto, K. (2008). Studies on diarrheagenic *Escherichia coli* isolated from children with diarrhea in Myanmar. Microbiol. Immunol. *52*, 2–8.

Tarr, P.I., Gordon, C.A., and Chandler, W.L. (2005). Shiga-toxin-producing *Escherichia coli* and haemolytic uraemic syndrome. Lancet *365*, 1073–1086.

Taxt, A., Aasland, R., Sommerfelt, H., Nataro, J., and Puntervoll, P. (2010). Heat-stable enterotoxin of enterotoxigenic *Escherichia coli* as a vaccine target. Infect. Immun. *78*, 1824–1831.

Taylor, D.N., Echeverria, P., Pál, T., Sethabutr, O., Saiborisuth, S., Sricharmorn, S., Rowe, B., and Cross, J. (1986). The role of *Shigella* spp., enteroinvasive *Escherichia coli*, and other enteropathogens as causes of childhood dysentery in Thailand. J. Infect. Dis. *153*, 1132–1138.

Te Loo, D.M., van Hinsbergh, V.W., van den Heuvel, L.P., and Monnens, L.A. (2001). Detection of verocytotoxin bound to circulating ploymorphonuclear leukocytes of patients with hemolytic uremic syndrome. J. Am. Soc. Nephrol. *12*, 800–806.

Thoren, A., Wolde-Mariam, T., Stintzing, G., Wadstrom, T., and Habte, D. (1980). Antibiotics in the treatment of gastroenteritis caused by enteropathogenic *Escherichia coli*. J. Infect. Dis. *141*, 27–31.

Tobias, J., Holmgren, J., Hellman, M., Nygren, E., Lebens, M., and Svennerholm, A.M. (2010). Over-expression of major colonization factors of enterotoxigenic *Escherichia coli*, alone or together, on non-toxigenic *E. coli* bacteria. Vaccine *28*, 6977–6984.

Tokuda, K., Nishi, J., Imuta, N., Fujiyama, R., Kamenosono, A., Manago, K., Kawano, Y. (2010). Characterization of typical and atypical enteroaggregative *Escherichia coli* in Kagoshima, Japan: biofilm formation and acid resistance. Microbiol. Immunol. *54*, 320–329.

Tokuhara, D., Yuki, Y., Nochi, T., Kodama, T., Mejima, M., Kurokawa, S., Takahashi, Y., Nanno, M., Nakanishi, U., Takaiwa, F., *et al.* (2010). Secretory IgA-mediated protection against *V. cholerae* and heat-labile enterotoxin-producing enterotoxigenic *Escherichia coli* by rice-based vaccine. Proc. Natl. Acad. Sci. U.S.A. *107*, 8794–8799.

Toledo, C.C., Arvidsson, I., and Karpman, D. (2011). Cross-reactive protection against enterohemorrhagic *Escherichia coli* infection by enteropathogenic *E. coli* in a mouse model. Infect. Immun. *79*, 2224–2233.

Trabulsi, L.R., Keller, R., and Gomes, T.A.T. (2002). Typical and atypical enteropathogenic *Escherichia coli*. Emerg. Infect. Dis. *8*, 508–513.

Tsai, T.Y., Luo, W.C., Wu, F.T., and Pan, T.M. (2005). Molecular subtyping for *Escherichia coli* O157:H7 isolated in Taiwan. Microbiol. Immunol. *49*, 579–588.

Tulloch, E.F., Jr., Ryan, K.J., Formal, S.B., and Franklin, F.A. (1973). Invasive enteropathic *Escherichia coli* dysentery. An outbreak in 28 adults. Ann. Intern. Med. *79*, 13–17.

Turner, S.M., Chaudhuri, R.R., Jiang, Z.D., DuPont, H., Gyles, C., Penn, C.W., Pallen, M.J., and Henderson, I.R. (2006a). Phylogenetic comparisons reveal multiple acquisitions of the toxin genes by enterotoxigenic *Escherichia coli* strains of different evolutionary lineages. J. Clin. Microbiol. *44*, 4528–4536.

Turner, A.K., Beavis, J.C., Stephens, J.C., Greenwood, J., Gewert, C., Thomas, N., Deary, A., Casula, G., Daley, A., Kelly, P., *et al.* (2006b). Construction and phase I clinical evaluation of the safety and immunogenicity of a candidate enterotoxigenic *Escherichia coli* vaccine strain expressing colonization factor antigen CFA/I. Infect. Immun. *74*, 1062–1071.

Tzipori, S., Sheoran, A., Akiyoshi, D., Donohue-Rolfe, A., and Trachtman, H. (2004). Antibody therapy in the management of shiga toxin-induced hemolytic uremic syndrome. Clin. Microbiol. Rev. *17*, 926–941.

Usein, C.R., Tatu-Chitoiu, D., Ciontea, S., Condei, M., and Damian, M. (2009). *Escherichia coli* pathotypes associated with diarrhea in Romanian children younger than 5 years of age. Jpn. J. Infect. Dis. *62*, 289–293.

Uzzau, S., and Fasano, A. (2000). Cross-talk between enteric pathogens and the intestine. Cell. Microbiol. *2*, 83–89.

Valentiner-Branth, P., Steinsland, H., Fischer, T.K., Perch, M., Scheutz, F., Dias, F., Aaby, P., Mølbak, K., and Sommerfelt, H. (2003). Cohort study of Guinean children: incidence, pathogenicity, conferred protection, and attributable risk for enteropathogens during the first 2 years of life. J. Clin. Microbiol. *41*, 4238–4245.

Vargas, M., Gascon, J., Casals, C., Schellenberg, D., Urassa, H., Kahigwa, E., Ruiz, J., and Vila, J. (2004). Etiology of diarrhea in children less than five years of age in Ifakara, Tanzania. Am. J. Trop. Med. Hyg. *70*, 536–539.

Veilleux, S., and Dubreuil, J.D. (2006). Presence of *Escherichia coli* carrying the EAST1 toxin gene in farm animals. Vet. Res. *37*, 3–13.

Vernacchio, L., Vezina, R.M., Mitchell, A.A., Lesko, S.M., Plaut, A.G., and Acheson, D.W. (2006). Diarrhea in American infants and young children in the community setting: incidence, clinical presentation and microbiology. Pediatr. Infect. Dis. J. *25*, 2–7.

Viboud, G.I., Jouve, M.J., Binsztein, N., Vergara, M., Rivas, M., Quiroga, M., and Svennerholm, A.M. (1999). Prospective cohort study of enterotoxigenic *Escherichia coli* infections in Argentinean children. J. Clin. Microbiol. *37*, 2829–2833.

Vidal, M., Kruger, E., Duran, C., Lagos, R., Levine, M., Prado, V., Toro, C, and Vidal, R. (2005). Single multiplex PCR assay to identify simultaneously the six categories of diarrheagenic *Escherichia coli* associated with enteric infections. J. Clin. Microbiol. *43*, 5362–5365.

Vieira, N., Bates, S.J., Solberg, O.D., Ponce, K., Howsmon, R., Cevallos, W., Trueba, G., Riley, L., and Eisenberg, J.N. (2007). High prevalence of enteroinvasive *Escherichia coli* isolated in a remote region of northern coastal Ecuador. Am. J. Trop. Med. Hyg. *76*, 528–533.

Vila, J., Vargas, M., Ruiz, J., Espasa, M., Pujol, M., Corachán, M., Jiménez de, Anta, M.T., and Gascón, J. (2001). Susceptibility patterns of enteroaggregative *Escherichia coli* associated with traveller's diarrhoea: emergence of quinolone resistance. J. Med. Microbiol. *50*, 996–1000.

Vilchez, S., Reyes, D., Paniagua, M., Bucardo, F., Möllby, R., and Weintraub, A. (2009). Prevalence of diarrhoeagenic *Escherichia coli* in children from León, Nicaragua. J. Med. Microbiol. *58*, 630–637.

Vilhena-Costa, A.B., Piazza, R.M., Nara, J.M., Trabulsi, L.R., and Martinez, M.B. (2006). Slot blot immunoassay as a tool for plasmid-encoded toxin detection in enteroaggregative *Escherichia coli* culture supernatants. Diagn. Microbiol. Infect. Dis. *55*, 101–106.

Viljanen, M.K., Peltola, T., Junnila, S.Y.T., Olkkonen, L., Jarvinen, H., Kuistila, M., and Huovinen. (1990). Outbreak of diarrhoea due to *Escherichia coli* O111:B4 in school children and adults: association of Vi antigen-like reactivity. Lancet *336*, 831–834.

Vingadassalom, D., Kazlauskas, A., Skehan, B., Cheng, H.C., Magoun, L., Robbins, D., Rosen, M.K., Saksela, K., and Leong, J.M. (2009). Insulin receptor tyrosine kinase substrate links the *E. coli* O157:H7 actin assembly effectors Tir and EspF(U) during pedestal formation. Proc. Natl. Acad. Sci. U.S.A. *106*, 6754–6759.

Vu-Khac, H. and Cornick, N.A. (2008). Prevalence and genetic profiles of Shiga toxin-producing *Escherichia coli* strains isolated from buffaloes, cattle, and goats in central Vietnam. Vet. Microbiol. *126*, 356–363.

Walker, R.I., Steele, D., Aguado, T., and the Ad Hoc ETEC Technical Expert Committee. (2007). Analysis of strategies to successfully vaccinate infants in developing countries against enterotoxigenic *E. coli* (ETEC) disease. Vaccine *25*, 2545–2566.

Walterspiel, J.N., Ashkenazi, S., Morrow, A.L., and Cleary, T.G. (1992). Effect of subinhibitory concentrations of antibiotics on extracellular Shiga-like toxin I. Infection *20*, 25–29.

Wang, Q., Wang, S., Beutin, L., Cao, B., Feng, L., and Wang, L. (2010). Development of a DNA microarray for detection and serotyping of enterotoxigenic *Escherichia coli*. J. Clin. Microbiol. *48*, 2066–2074.

Wang, Y., Wang, H., Xiang, Q., Sun, S.X., and Yu, S.Y. (2002). Detection of the high-pathogenicity island of *Yersinia enterocolitica* in enterotoxigenic and enteropathogenic *E. coli* strains. Di Yi Jun Yi Da Xue Xue Bao *22*, 580–583.

Wani, S.A., Bhat, M.A., Samanta, I., Nishikawa, Y., and Buchh, A.A. (2003). Isolation and characterization of Shiga toxin-producing *Escherichia coli* (STEC) and enteropathogenic *Escherichia coli* (EPEC) from calves and lambs with diarrhea in India. Lett. Appl. Microbiol. *37*, 121–126.

Wanke, C.A., Mayer, H., Weber, R., Zbinden, R., Watson, D.A., and Acheson, D. (1998). Enteroaggregative *Escherichia coli* as a potential cause of diarrheal disease in adults infected with human immunodeficiency virus. J. Infect. Dis. *178*, 185–190.

Watanabe, M., Matsuoka, K, Kita, E., Igai, K., Higashi, N., Miyagawa, A., Watanabe, T., Yanoshita, R., Samejima, Y., Terunuma, D., *et al.* (2004). Oral therapeutic agents with highly clustered globotriose for treatment of Shiga toxigenic *Escherichia coli* infections. J. Infect. Dis. *189*, 360–368.

Weintraub, A. (2007). Enteroaggregative *Escherichia coli*: epidemiology, virulence and detection. J. Med. Microbiol. *56*, 4–8.

Wennerås, C., and Erling, V. (2004). Prevalence of enterotoxigenic *Escherichia coli*-associated diarrhoea and carrier state in the developing world. J. Health. Popul. Nutr. *22*, 370–382.

Westermark, M.J., Mizunoe, O.Y., Urbonaviciene, J., and Uhlin, B.E. (2000). Silencing and activation of ClyA cytotoxin expression in *Escherichia coli*. J. Bacteriol. *182*, 6347–6357.

Wetzel, A.N., and LeJeune, J.T. (2006). Clonal dissemination of *Escherichia coli* O157:H7 subtypes among dairy farms in northeast Ohio. Appl. Environ. Microbiol. *72*, 2621–2626.

Wiedermann, U., and Kollaritsch, H. (2006). Vaccines against traveler's diarrhoea and rotavirus disease – a review. Wien Klin Wochenschr. *118*, 19–20.

Wierzba, T.F., Abdel-Messih, I.A., Abu-Elyazeed, R., Putnam, S.D., Kamal, K.A., Rozmajzl, P., Ahmed, S.F., Fatah, A., Zabedy, K., Shaheen, H.I., *et al.* (2006). Clinic-based surveillance for bacterial- and rotavirus-associated diarrhea in Egyptian children. Am. J. Trop. Med. Hyg. *74*, 148–153.

Willshaw, G.A., Scotland S.M., Smith, H.R., and Rowe, B. (1992). Properties of verocytotoxin-producing *Escherichia coli* of human origin of O serogroups other than O157. J. Infect. Dis. *166*, 797–802.

Wolf, M.K. (1997). Occurrence, distribution, and associations of O and H serogroups, colonization factor antigens, and toxins of enterotoxigenic *Escherichia coli*. Clin. Microbiol. Rev. *10*, 569–584.

Wong, C.S., Jelacic, S., Habeeb, R.L., Watkins, S.L., and Tarr, P.I. (2000). The risk of the hemolytic-uremic syndrome after antibiotic treatment of *Escherichia coli* O157:H7 infections. N. Engl. J. Med. *342*, 1930–1936.

Wu, S.-X., and Peng, R.-Q. (1992). Studies on an outbreak of neonatal diarrhea caused by EPEC O127:H6 with plasmid analysis, restriction analysis and outer membrane protein determination. Acta Paediatr. Scand. *81*, 217–221.

Xicohtencatl-Cortes, J., Monteiro-Neto, V., Saldana, Z., Ledesma, M.A., Puente, J.L., and Giron, J.A. (2009). The type 4 pili of enterohemorrhagic *Escherichia coli* O157: H7 are multipurpose structures with pathogenic attributes. J. Bacteriol. *191*, 411–421.

Yamanaka, H., Nomura, T., Fujii, Y., and Okamoto. K. (1997). Extracellular secretion of *Escherichia coli* heat-stable enterotoxin I across the outer membrane. J. Bacteriol. *179*, 3383–3390.

Yang, J.R., Wu, F.T., Tsai, J.L., Mu, J.J., Lin, L.F., Chen, K.L., Kuo, S.H., Chiang, C.S., and Wu, H.S. (2007). Comparison between O serotyping method and multiplex real-time PCR to identify diarrheagenic *Escherichia coli* in Taiwan. J. Clin. Microbiol. *45*, 3620–3625.

Yang, Z., Kovar, J., Kim, J., Nietfeldt, J., Smith, D.R., Moxley, R.A., Olson, M.E., Fey, P.D., and Benson, A.K. (2004). Identification of common subpopulations of non-sorbitol-fermenting, beta-glucuronidase-negative

Escherichia coli O157:H7 from bovine production environments and human clinical samples. Appl. Environ. Microbiol. *70*, 6846–6854.

Yano, A., Ishimaru, R., and Hujikata, R. (2007). Rapid and sensitive detection of heat-labile I and heat-stable I enterotoxin genes of enterotoxigenic *Escherichia coli* by loop-mediated isothermal amplification. J. Microbiol. Methods *68*, 414–420.

Yatsuyanagi, J., Saito, S., Sato, H., Miyajima, Y., Amano, K., and Enomoto, K. (2002). Characterization of enteropathogenic and enteroaggregative *Escherichia coli* isolated from diarrheal outbreaks. J. Clin. Microbiol. *40*, 294–297.

Yatsuyanga, Y., Saito, S., Miyajima, Y., Amano, K-I., Enomoto, K. (2003). Characterization of atypical enteropathogenic *Escherichia coli* strains harboring the *astA* gene that were associated with a waterborne outbreak of diarrhea in Japan. J. Clin. Microbiol. *41*, 2033–2039.

You, J., Xu, Y., He, M., McAllister, T.A., Thacker, P.A., Li, X., Wang, T., and Jin, L. (2011). Protection of mice against enterotoxigenic *E. coli* by immunization with a polyvalent enterotoxin comprising a combination of LTB, STa, and STb. Appl. Microbiol. Biotechnol. *891*, 885–893.

Zamboni, A., Fabbricotti, S.H., Fagundes-Neto, U., and Scaletsky, I.C. (2004). Enteroaggregative *Escherichia coli* virulence factors are found to be associated with infantile diarrhea in Brazil. J. Clin. Microbiol. *42*, 1058–1063.

Zhang, X., McDaniel, A.D., Wolfe, L.E., Keusch, G.T, Waldor, M.K., and Acheson, D.W (2000). Quinolone antibiotics induce Shiga toxin-encoding bacteriophages, toxin production, and death in mice. J. Infect. Dis. *181*, 664–670.

Zhang, Y., Laing, C., Steele, M., Ziebell, K., Johnson, R., Benson, A.K., Taboada, E., and Gannon, V.P. (2007). Genomic evolution in major *Escherichia coli* O157:H7 lineages. BMC Genomics *8*, 121.

Zhao, S., Mitchell, S.E., Meng, J., Kresovich, S., Doyle, M.P., Dean, R.E., Casa, A.M., and Weller, J.W. (2000). Genomic typing of *Escherichia coli* O157:H7 by semi-automated fluorescent AFLP analysis. Microbes Infect. *2*, 107–113.

Zhong, S.S., Zhang, Z.S., Wang, J.D., Lai, Z.S., Wang, Q.Y., Pan, L.J., and Ren, Y.X. (2004). Competitive inhibition of adherence of enterotoxigenic *Escherichia coli*, enteropathogenic *Escherichia coli* and *Clostridium difficile* to intestinal epithelial cell line Lovo by purified adhesin of *Bifidobacterium adolescentis* 1027. World. J. Gastroenterol. *10*, 1630–1633.

Population Genetics and Molecular Epidemiology of *Shigella* Species

3

Kaisar Ali Talukder and Ishrat Jahan Azmi

Abstract

Shigellosis, also known as acute bacillary dysentery, produces inflammatory reactions and ulceration on the intestinal epithelium followed by bloody or mucoid diarrhoea. Shigellosis is caused by any one of the four species or groups of *Shigella*, namely, *S. dysenteriae*, *S. flexneri*, *S. boydii* and *S. sonnei*. At least 54 serotypes or subtypes of *Shigella* are currently recognized, of which *S. dysenteriae* has 16 serotypes, *S. flexneri* has 17 serotypes and subserotypes, *S. boydii* has 20 and *S. sonnei* has a single serotype. Shigellosis can occur in sporadic, epidemic and pandemic forms. To develop an effective vaccine it is important to monitor the prevalent serotypes and their changes around the world because immunity to *Shigella* is serotype specific. The genetic variability between serotypes and emergence of atypical strains accentuates the problems to the development of an effective vaccine. In view of the emergence of new strains with altered characteristics than the ones established globally for many years, there is also a need for revising the nomenclature for the three groups of *Shigella: flexneri, boydii* and *dysenteriae*. This chapter discusses the recent serotyping scheme of *Shigella* with special focus on the emergence of new variants and the necessity to make an updated scheme. Furthermore, the epidemiology, phenotypic and molecular characteristics, population genetics and clinical impact of these variants have been described.

Introduction

Shigella is a well-known human pathogen that causes diseases such as diarrhoea and bacillary dysentery called shigellosis (Wachsmuth and Morris, 1989). Since its discovery in 1897 by Kiyoshi Shiga, this genus and its associated disease has become endemic throughout the world. The global morbidity and mortality caused by infection with *Shigella* is a major public health concern, especially in developing countries where sanitation is poor. Approximately 165 million cases of *Shigella* infection occur annually worldwide with 1.1 million deaths. Most of the causality are children under 5 years of age (Kotloff *et al.*, 1999). However, a recent study in six Asian countries showed that the disease burden due to *Shigella* infection was 2.1 per 1000 population per year in all ages and 13.2 per 1000 population per year in children under 5 years (von Seidlein *et al.*, 2006). This figure was much higher than that in industrialized countries, since the estimated shigellosis incidence in all ages was 3.7 and 3.2 per 100,000 per year in the United States in 1999 (Gupta *et al.*, 2004) and in the Netherlands between 1996 and 2000 (Van Pelt *et al.*, 2003) respectively.

Shigella spread by direct faecal–hand–oral contact under poor hygienic conditions (Weissman *et al.*, 1975). Clinical infection can be transmitted by as few as 10 *Shigella* organisms (DuPont *et al.*, 1989) even without neutralization of gastric acid. *Shigella* being non-motile cannot penetrate mucus-covered intestinal cells. Instead, using its unique molecular mechanisms, *Shigella* initially traverse the mucosa through non-mucin-covered specialized antigen-processing microfold (M) cells, without disrupting them. Shigellae encounter the underlying lymphoid follicles within the mucosa, are ingested by resident macrophages, lyse the phagosomal vacuole and enter the cytoplasm where they multiply. It produces inflammatory reactions and ulceration on the intestinal epithelium followed by bloody diarrhoea and mucus in the stool. Complications in shigellosis include leukaemoid reaction, haemolytic uraemic syndrome (HUS), seizure, encephalopathy, abdominal cramps, rectal pain and septicaemia, which significantly increase morbidity and mortality.

Owing to increasing resistance of *Shigella* species to commonly used antibiotics treatment/intervention strategies have been hindered. Failure to provide clean water and sanitation and reduce overcrowding, have necessitated the development of effective vaccines against shigellosis. However, immunity to *Shigella* is serotype specific and vaccine protection will therefore depend on the representation of each serotype in the vaccine (Formal and Baker, 1953; Fontaine *et al.*, 1990; Noriega *et al.*, 1994, 1996). But the genetic variability between serotypes and emergence of atypical strains (El-Gendy *et al.*, 1999; Talukder *et al.*, 2001, 2002, 2003b) and novel *Shigella* serovars (Talukder *et al.*, 2007) accentuates the problems to the development of an effective vaccine.

There is a need for establishing new nomenclature for certain groups of *Shigella*, because of the emergence of new strains with altered characteristics than the ones established globally for many years. These new strains include novel *S. boydii* (Grimont *et al.*, 2007), *S. dysenteriae* (Talukder *et al.*, 2007), and new subserotypes of *S. flexneri* (Talukder *et al.*, 2001, 2002, 2003b). To develop a new effective vaccine it is also important to know the prevalence of different serotypes and their variants.

Biochemical characteristics

The genus *Shigella* of γ-proteobacteria belonging to the family *Enterobacteriaceae* are Gram-negative, non-motile, non-spore forming bacteria. The genus *Shigella* comprises four species (or groups), namely *S. dysenteriae*, *S. flexneri*, *S. boydii* and *S. sonnei*, and these species can be further divided into serotypes based on the structure of the O-antigen component of LPS present on the outer membrane of the cell wall (Simmons and Romanowska, 1987). The subgroups and serotypes are differentiated from one another by their biochemical traits (e.g. ability to ferment D-mannitol) and antigenic properties (Ewing, 1954). Shigellae are oxidase negative, and ferment glucose and other carbohydrates without producing gas with the exception of *S. flexneri* 6. By definition, lysine is not decarboxylated by all *Shigella* spp. and ornithine is decarboxylated by most isolates of *S. sonnei* and rare isolates of *S. boydii*. Generally all serotypes of *S. flexneri*, *boydii* and *sonnei* ferment mannitol except *S. flexneri* type 4 (Talukder *et al.*, 2002), *S. flexneri* type 6 and *S. boydii* 14 (Altwegg and Bockemühl, 1998). In addition, *Shigella* is Voges–Proskauer negative and methyl-red positive, do not utilize Simmons citrate, nor produce H_2S and are arginine dihydrolase and urease negative, and thus are very closely related to *Escherichia coli*. *Shigella* can be isolated from a number of commercially prepared plating media. Of these, MacConkey, XLD and *Shigella–Salmonella* agar (SS agar) are more common selective/differential agar media that are used for the isolation of *Shigella* species.

Serotype

Shigella dysenteriae (group A) has 15 serotypes (Bopp *et al.*, 2003); *S. flexneri* (group B) has 14 classical serotypes and subserotypes (Bopp *et al.*, 2003), *S. boydii* (group C) has 20 serotypes and *S. sonnei* has a single serotype (Levine *et al.*, 2007). *S. flexneri* serotypes (with the exception of serotype 6) have some degree of antigenic relatedness attributable to a common repeating tetrasaccharide unit, to which α-D-glucopyranosyl and *O*-acetyl groups are added, providing the basis for their 'type' (i.e. I to VI) and 'group' (i.e. 3,4, 6 and 7,8) antigenic factors (Edwards and Ewing, 1972; Carlin *et al.*, 1984). The common repeating tetrasaccharide unit is identical to *S. flexneri* serotype Y which have the 3.4 group antigen. One strain can thus contain one type or one or more group antigens (Table 3.1). Rabbit antisera raised against the specific type and group antigenic factors are routinely used in agglutination reactions to identify the *S. flexneri* serotypes (Table 3.1). Current taxonomy recognizes 14 classical *S. flexneri* serotypes or subserotypes defined by type (I, II, III, IV, V, VI) and group antigens (3, 4, 6, 7, 8) using absorbed antisera. When *S. flexneri* strains are not

Table 3.1 Serotyping of *S. flexneri* using monoclonal/polyclonal antisera

	Type-specific antisera						Group-specific antisera				Provisional antisera
Serotype	I	II	III[d]	IV	V	VI	3, 4	6	7, 8	E1037 (or IV–I)[b]	1c[b]
S. flexneri 1a	+	–	–	–	–	–	+	–	–	–	–
S. flexneri 1b	+	–	–	–	–	–	+	+	–	–	–
S. flexneri 2a	–	+	–	–	–	–	+	–	–	–	–
S. flexneri 2b	–	+	–	–	–	–	–	–	+	–	–
S. flexneri 3a	–	–	+	–	–	–	+/–	+	+	–	–
S. flexneri 3b	–	–	+	–	–	–	+/–	+	–	–	–
S. flexneri 4a	–	–	–	+	–	–	+	–	–	+/–	–
S. flexneri 4b	–	–	–	+	–	–	–	+	–	–	–
S. flexneri 4c[a]	–	–	–	+	–	–	–	–	+	–	–
S. flexneri 5a	–	–	–	–	+	–	+	–	–	–	–
S. flexneri 5b	–	–	–	–	+	–	–	–	+	–	–
S. flexneri 6	–	–	–	–	–	+	+/–	–	–	+/–	–
S. flexneri Variant X	–	–	–	–	–	–	–	–	+	+/–	–
S. flexneri Variant *Y*	–	–	–	–	–	–	+	–	–	+/–	–
Atypical *S. flexneri*[c]											
S. flexneri Type 4	–	–	–	+	–	–	–	–	–	+	–
S. flexneri 4X	–	–	–	–	–	–	–	–	–	+	–
S. flexneri 1c	–	–	–	–	–	–	–	–	–	–	+

Current taxonomy recognizes 14 classical *S. flexneri* serotypes or subserotypes defined by type (I, II, III, IV, V, VI) and group antigens (3.4, 6, 7.8) using absorbed antisera. When *S. flexneri* are not typed using absorbed antisera monoclonal antibody (IV-I, 1c) is used to type these strains following the scheme shown in Table 3.1.

[a]The subserotype 4c was proposed to the International Committee on Systematic Bacteriology Subcommittee on the Taxonomy of Enterobacteriaceae, for its incorporation in the *S. flexneri* scheme (Pryamukhina NS, Khomenko NA –1988); however, this subserotype has not yet been officially recognized.

[b]Monoclonal antibody. IV-I (or E1037): it is designated as new group factor since this antigen is shared among different serotypes. Monoclonal antibody 1c is only present in atypical *S. flexneri* designated as serotype 1c (Carline *et al.*, 1989; Talukder *et al.*, 2001, 2002, 2003).

[c]Strains could not be definitely serotyped with absorbed antisera following the current serotyping scheme. These could be serotyped using monoclonal antibody.

[d]*S. flexneri* 3a and 3b can be readily identified using absorbed or monoclonal antibody to group 6 specific antigen. Absorbed rabbit type 'III' antigen binds to group 6 antigen, indicating that type III antigen is identical to group antigen, which is present in other serotypes. Therefore, now we can omit this type factor from typing scheme (Labine *et al.*, 2007).

typed using absorbed antisera, monoclonal antibody (MASF IV-l, 1c) is used to type these strains following the scheme shown in the Table 3.1.

Until recently at least 50 serotypes of *Shigella* are recognized (Levine *et al.*, 2007). Beside, there are also some provisional (Talukder *et al.*, 2007) and new subserotypes of *S. flexneri* (Talukder *et al.*, 2001, 2002, 2003b). In 2011, Foster *et al.* proposed that provisional serotypes Y394 and 88-893 could be designated as two new serotypes 7a and 7b, respectively, in the *S. flexneri* typing scheme since Y394 agglutinated only with monoclonal antibody MASF1c (raised against Y394) and 88-893 agglutinated with group factor 6 and MASF1c. However, using monoclonal antibody MASF1c, *S. flexneri* 1c was detected from diverse geographical regions. These 1c strains are identical to 7a shown in Table 3.1. *S. flexneri* 1c monoclonal antibody raised against provisional serotype Y394 has also been used to confirm similar strains of *S. flexneri* isolated in Bangladesh, Egypt, Indonesia, Pakistan and Vietnam, and accounted for from 2% to up to 40% of *S. flexneri* isolates. The recent isolation of *S. flexneri* designated as 4s (Qiu *et al.*,

2011) from China is controversial. The antigenic property is of this strain is likely to be same as 4X earlier reported by Talukder *et al.* (2001).

S. flexneri type 6 has been placed as single serotype in the present classification scheme. Biochemical analysis of *S flexneri* type 6 showed two different biotypes (Edwards) and they are also differentiated by O-antigenic determinants since one group agglutinated only with type factor 6 and another group agglutinated with type factor 6 and group factor E1037 (IV-l). These two group strains were different in their PFGE and Ribotyping patterns. Interestingly, these strains were also different in the MLST profile, which suggested that they might have different origin. These findings suggest that type 6 variant could be classified as a different subserotype of type 6 (Talukder *et. al.*, unpublished data). Therefore, we propose to include strains of type 6 which agglutinate without and with group factor E1037 (IV-l) as 6a and 6b respectively in the classification scheme.

Until 1958, the time of the last major classification of the genus *Shigella* (Ent. sub. 1958), *S. dysenteriae* consisted of 10 serotypes. Subsequently, Ewing *et al.* (1958) described provisional serotypes 3873-50 and 3341-55, which were later proposed for addition to the *Shigella* scheme as *S. dysenteriae* serotypes 11 and 12, respectively (Brenner, 1984; Wathen-Grady *et al.*, 1990). In 1985, Shmilovitz *et al.*, identified a new provisional serotype represented by strain I9809–73. This provisional serotype was included in the *S. dysenteriae* scheme as serotype 13. Latter provisional serotypes E22383 and E23507 were originally reported in 1989 (Gross *et al.*, 1989). After extensive characterization and isolation from various geographical locations, Ansaruzzaman *et al.* (1995) suggested that these provisional serotypes be designated as serotypes 14 and 15, respectively. Recently, a group of non-agglutinating strains have been reported as novel serotype of *S. dysenteriae* isolated from diarrhoea patients (Talukder *et al.*, 2007). This potentially new serotypes were represented by strain KIVI 162. This novel serotype is one of the dominant serotypes among *S. dysenteriae* strains recently isolated in Bangladesh (Talukder *et. al.*, unpublished data). In addition, serotypic validation of these strains was done in reference laboratories including the National Public Health Laboratory at Colindale, UK (Talukder *et al.*, 2007), and French National Reference Centre for *E. coli* and *Shigella*, Institute of Pasteur, Paris, France (Personal communication). *S. boydii* comprises 19 serogroups. Ewing (1986) acknowledged 17 serogroups, which included two provisional serogroups represented by reference strain 2710-54 as *S. boydii* 16, and 3615-53 as *S. boydii* 17. In 1980 Gross *et al.* reported a new provisional serogroup, isolated in Bangladesh and designated serogroup 18 with the reference strain E10163. Recently provisional *S. boydii* serovar E 16553, designated serogroup 19, has been reported by several groups from diverse geographical origins (Gross *et al.*, 1982; Ansaruzzaman *et al.*, 2005) and serogroup 20 was recently proposed and confirmed as novel (Woodward *et al.*, 2005; Grimont *et al.*, 2007). Considering the prevalence and clinical impact of the new serotypes/subserotypes of *Shigella* species, and the proposal to include them in the serotyping scheme, at least 54 serotypes or subserotypes of *Shigella* are now recognized of which *S. dysenteriae* has 16 serotypes, *S. flexneri* has 17 classical serotypes and subserotypes, *S. boydii* has 20 and *S. sonnei* has a single serotype.

Global distribution and epidemiology

Although *Shigella* is endemic worldwide, it affects certain populations more than others. In developing countries, high rates of morbidity and mortality are known to occur among displaced populations (Kotloff *et al.*, 1999). In general, *S. sonnei* is more common in developed countries and *S. flexneri* and *S. dysenteriae* are more prevalent in developing countries. The proportions of each species vary from country to country.

Shigella dysenteriae serotype 1 was the dominant species after its discovery in 1918, but was replaced by *S. flexneri* after the First World War and virtually disappeared. Following the Second World War, *S. sonnei* gradually became the dominant species in the industrialized countries, while *S. flexneri* has persistent in developing countries especially in six Asian countries. There are few exceptions such as Thailand, where *S. sonnei* has also replaced *S. flexneri* as the most common isolate (von Seidlein *et al.*, 2006). *S. boydii* is present primarily in the Indian subcontinent whereas it was the second most common species accounting 23% in Bangladesh. In all these countries 4% of isolates were of *S. dysenteriae* spp., but none was *S. dysenteriae* type 1 (von Seidlein *et al.*, 2006).

Shigella dysenteriae 1 re-emerged during 1969 to 1972 in Mexico and Central America causing high mortality (Mendizabal-Morris *et al.*, 1971). Since then, epidemics due to *S. dysenteriae* 1 have been reported in many Asian countries such as Bangladesh (1973, 1984, 1993, 2003), Sri Lanka (1976), The Maldives (1982), Nepal (1984–85), Bhutan (1984–85), Myanmar (1984–85) (Rahaman *et al.*, 1975; Sheriff, 1982; Pal *et al.*, 1989; Naheed *et al.*, 2004; Talukder *et al.*, 2004) and Africa (Guerin *et al.*, 2003). In Bangladesh, each epidemic was associated with a high rate of mortality, particularly in children < 5 years of age. In each epidemic, *S. dysenteriae* 1 strains emerged acquiring resistance to the latest drug of choice, together with other effective drugs, and this was evident in epidemics of 1973–1974, 1983–1984 and 1993–1994, where strains were resistant to ampicillin, sulfamethoxazole/trimethoprim and nalidixic acid, respectively (Talukder *et al.*, 2004). Ciprofloxacin-resistant strains of *S. dysenteriae* 1 are found in different countries within a short duration, suggesting that the resistant strain may spread very rapidly, probably traversing a similar path as previous epidemic strains. In 2003 few Ciprofloxacin-resistant strain was isolated and since then *S. dysenteriae* 1 was not isolated at all in Bangladesh. A comparative genomic analysis was carried out between the strains of *S. dysenteriae* 1 collected from previous epidemics and the recent ciprofloxacin-**resistant** strains isolated from different countries. Pulsed-field gel electrophoresis (PFGE) and mutation analysis of *gyr* genes showed that all the Cip^r strains isolated in Bangladesh, India and Nepal, were completely different from that of the previous ciprofloxacin sensitive epidemic strains (Talukder *et al.*, 2004, 2006).

In contrast, there are no reports of outbreaks caused by serotypes of *S. dysenteriae* other than type 1 except for an outbreak of *S. dysenteriae* 2 (Kolavic *et al.*, 1997). In recent time, *S. dysenteriae* 1 was replaced by *S. dysenteriae* type 2 and type 4 in Bangladesh (Talukder *et al.*, 2003a), and an outbreak caused by *S. dysenteriae* type 4 occurred between June and December 2000 (Talukder *et al.*, 2005). *S. dysenteriae* type 2 and type 4 serotypes are becoming more prevalent in both urban and rural area of Bangladesh (Talukder *et al.*, unpublished data). Recently an outbreak of *S. dysenteriae* type 2 occurred during May–June 2009 in Sweden (Löfdahl *et al.*, 2009).

S. sonnei is the predominant species in industrialized countries (Alcoba-Florez *et al.*, 2005). A large outbreak of *S. sonnei* gastroenteritis occurred in Murcia region (south-east Spain) in 1995–1996. (García-Fulgueiras *et al.*, 2001) and in 2008, three cases of *S. sonnei* infection were identified among the Orthodox Jewish community of Antwerp, Belgium. Data from Israel, Spain, and the USA consistently demonstrate that *S. sonnei* is the most common serogroup found in industrialized countries (Kotloff *et al.*, 1999). In developing countries, the predominant serotype of *S. flexneri* is 2a, followed by 1b, 3a, 4a, and 6. In industrialized countries, most isolates are *S. flexneri* 2a or other unspecified type 2 strains (Kotloff *et al.*, 1999).

Of 25,010 patients enrolled in the 2% diarrhoeal disease surveillance system at Dhaka Hospital of ICDDR,B between 2000 and 2010, *Shigella* was isolated from 1098 (4.4%) patients. Of the 1098 *Shigella*-infected patients, 568 (51.7%) were aged < 5 years, 82 (7.5%) were aged 5–14 years, and 448 (40.8%) were aged over 15 years. The age distribution of patients infected with *Shigella* remained stable over time. Since 2007 *S. flexneri* has begun to show resistance to ciprofloxacin and mecillinam and by 2010, 45% of *S. flexneri* was resistant to ciprofloxacin and 42% to mecillinam. Other strains of *Shigella* such as *S. sonnei*, *S. boydii* and *S. dysenteriae* together had a 20% resistance to mecillinam and 10% to ciprofloxacin in 2010. In addition, isolation rates of *Shigella* were higher in rural Matlab (9%) hospital of ICDDR,B and another rural community-based study in Mirzapur (15%) (Faruque *et al.*, personal communication).

Of 14,560 strains of *Shigella* species isolated at the clinical microbiology laboratory of ICDDR,B Dhaka hospital between 1997 and 2010, 6349 strains were available for characterization using various phenotypic and genotypic techniques. Of

6349 strains, *S. flexneri* was found to be the most predominant species (61%) followed by *S. boydii* (17%), *S. sonnei* (10%), *S. dysenteriae* (8%), and *Shigella*-like organism (SLO) which did not agglutinate with any antisera of the established *Shigella* serovars (3%) (Fig. 3.1). During this study period, 23.4% of *S. flexneri*, 8% of *S. dysenteriae*, and 23% of *S. boydii* isolates were identified as atypical (Talukder *et al.*, 2003a) since these isolates did not react with either type or group antigen-specific known antisera. The prevalence of *S. sonnei* has surprisingly increased (27% in 2010 from 8% in 2000) and the prevalence of *S. dysenteriae* (3%) has drastically reduced in 2010 (Fig. 3.1). It is interesting to note that since 2004, *S. dysenteriae* type 1 was not found at all in Bangladesh. This indicates that improved sanitation system might have an impact on the epidemiology of *Shigella* spp.

Over the last 35 years, *S. flexneri* was always dominant in Bangladesh except during 1993–95 when it was replaced by *S. dysenteriae*. Within *S. flexneri*, *S. flexneri* 2a was the predominant serotype except in 1997 to 1999 when it was replaced by 2b and atypical *S. flexneri* (1c, type 4 and 4X) followed by *S. flexneri* 3a, 2b and 1b. It is interesting to note that atypical strain of *S. flexneri* increased from 0% during 1978–1984 to 23.8% during 1997–2000 (Fig. 3.2) (Talukder *et al.*, 2003b). In case of *S. boydii* and *S. dysenteriae* atypical strain increased from 0% in 1978–1984 to 23.% and 8% in 1997–2000 respectively (Talukder *et al* unpublished data).

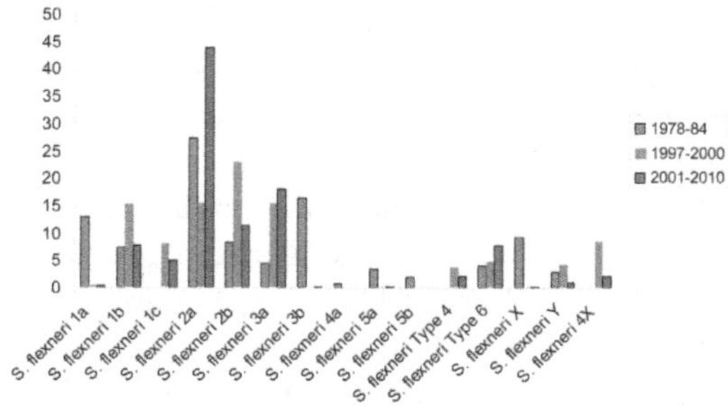

Figure 3.2 Distribution of different serotypes of *S. flexneri* during 1978 to 2010 in Dhaka, Bangladesh.

Molecular epidemiology

Atypical strains or newer subserotypes are being isolated in different parts of the world including Bangladesh. This is mainly because, commercially available antisera specific for each type and group factor antigens currently in use for typing are not able to cover all possible epitopes of the O-antigen. To overcome these constraints, these strains were typed using monoclonal antibody and finally these atypical strains were discriminated by molecular methods. Atypical *Shigella* and SLO

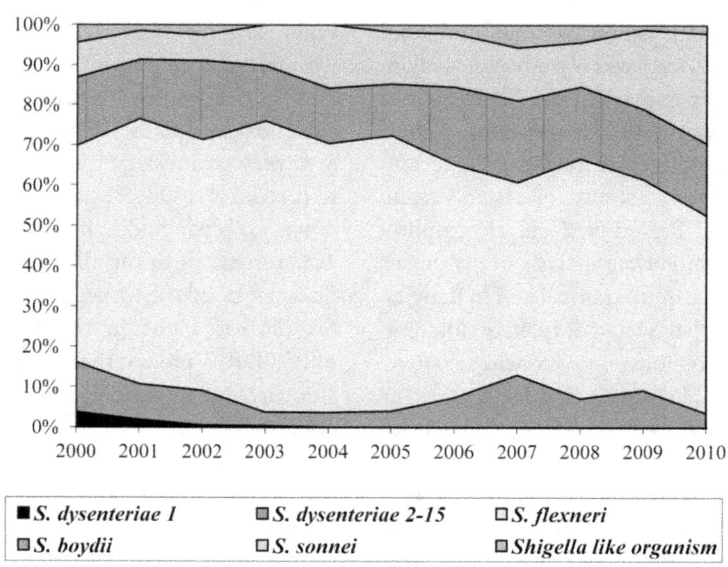

Figure 3.1 Prevalence of different species of *Shigella* in Dhaka, Bangladesh, during 2000 to 2010.

had biochemical characteristics typical of *Shigella*, and hence may be designated as new serotypes or subserotypes of *Shigella*. Recently 12% ($n = 1228$) of *Shigella* isolates from public health laboratories in 50 states of USA were not identified as belonging to a specific subgroup. A recent study in Bangladesh from January 1999 to December 2010 showed that 3% of the total isolates were SLOs, which did not react to any antisera of the established *Shigella* serovars. During this period, a large percentage of atypical strains were isolated which account for 23%, 8% and 23% among *S. flexneri*, *S. dysenteriae*, and *S. boydii* respectively. Trends of isolation of atypical serotypes and SLOs were found in Pakistan, Indonesia, Thailand and Vietnam (von Seidlein *et al.*, 2006).

In 1985, a new antigen called E1037 was detected in *S. flexneri* serotypes 4a, X, Y and 6 associated with other type- or group-specific antigens. Later monoclonal antibody against this antigen (E1037) was developed, which is designated as MASF IV-1. Another provisional serotype of *S. flexneri* 1c (strain # Y394), was first identified in Bangladesh by Carlin *et al.* (1989). This strain did not agglutinate with any type and group factor antigen but agglutinated only with monoclonal antibody MASF 1c. In Bangladesh (1997–2000) 23.4% of *S. flexneri* isolates were untypeable (Talukder *et al.*, 2001). Of these, one group is identified as 1c (8.2%). Second group agglutinated with type antigen factor IV-2 but did not agglutinate with group factor Y-5 or -6 thereby excluding their characterization into serotype 4a or 4b, and these strains were designated as *S. flexneri* type 4 (3.8%) (Talukder *et al.*, 2002). The remaining atypical strains agglutinated only with the MASF IV-1 specific for a new antigenic determinant (E1037) but did not react with any type and group antigen-specific antisera tested, and these were designated as 4X (Talukder *et al.*, 2001). A population-based study in six Asian countries found high prevalence of these atypical strains (von Seidlein *et al.*, 2006). Recently, Stagg *et al.* (2008) found that *S. flexneri* serotype 1c was the most prevalent (40%) among *S. flexneri* serotype isolated in Vietnam. A total of 272 atypical strains of *S. flexneri* were collected from diverse geographical locations, of which 1c (46%), type 4 (26%) and 4X (28%) were extensively characterized using different molecular techniques including plasmid profile, PFGE, ribotyping, and MLST. In Bangladesh two different PFGE types were observed among atypical *S. flexneri* type 4 (Talukder *et al.*, 2002) while three distinct types were found among the strains of *S. flexneri* 1c, of which PFGE pattern of two types were identical to that of 1a and 1b respectively and that of the remaining type was completely different from 1a and 1b strains (Talukder *et al.*, 2003b). Among *S. flexneri* 1c strains isolated in Bangladesh, only the predominant type (A) is found in Pakistan, Thailand, and Indonesia (Fig. 3.3A) (Talukder *et al.*, unpublished data). The two different types of *S. flexneri* type 4 were found in Bangladesh, of which the predominant clone (A) was also found in Australia, Vietnam and Pakistan whereas Indonesian strains were heterogeneous (Fig. 3.3B).

Plasmid profiling has long been used as molecular tools for the investigations of outbreak as well as epidemiological study of *Shigella* species (Haider *et al.*, 1988; Matsumoto *et al.*, 1998; Talukder *et al.*, 1999, 2003b; Terajima *et al.*, 2004). This technique is very simple, rapid and sensitive. Serotype specific plasmid was found in all most all species of *Shigella* that facilitates the identification of pathogen even in the case of cross reacting bacteria where traditional serotyping method is ineffective. When used as a fingerprint for strains, the plasmid profile may aid in differentiating strains, identifying a source of infection, or evaluating the efficacy of control measures (Tacket and Cohen, 1983). Several studies also showed that a number core plasmid are associated with the specificity of individual serotypes within *Shigella* species (Haider *et al.*, 1988, 1989; Talukder *et al.*, 2002, 2003a,b). It has been found that the core plasmids of *S. dysenteriae* type 1 are 140, 6, and 2 MDa and *S. flexneri* 2a are 140, 2.7, and 2.1 MDa irrespective of their drug resistance pattern in both the epidemic and non-epidemic isolates of *S. dysenteriae* type 1 (Haider *et al.*, 1988). Plasmid patterns of different serotypes of *S. dysenteriae* were found to be different (Talukder *et al.*, 2003a). Within a single serotype multiple patterns were observed, indicating their clonal heterogeneity. In a previous study, numerous plasmid patterns were found in *S. dysenteriae* type 1 isolated from scattered as well as defined geographical origins.

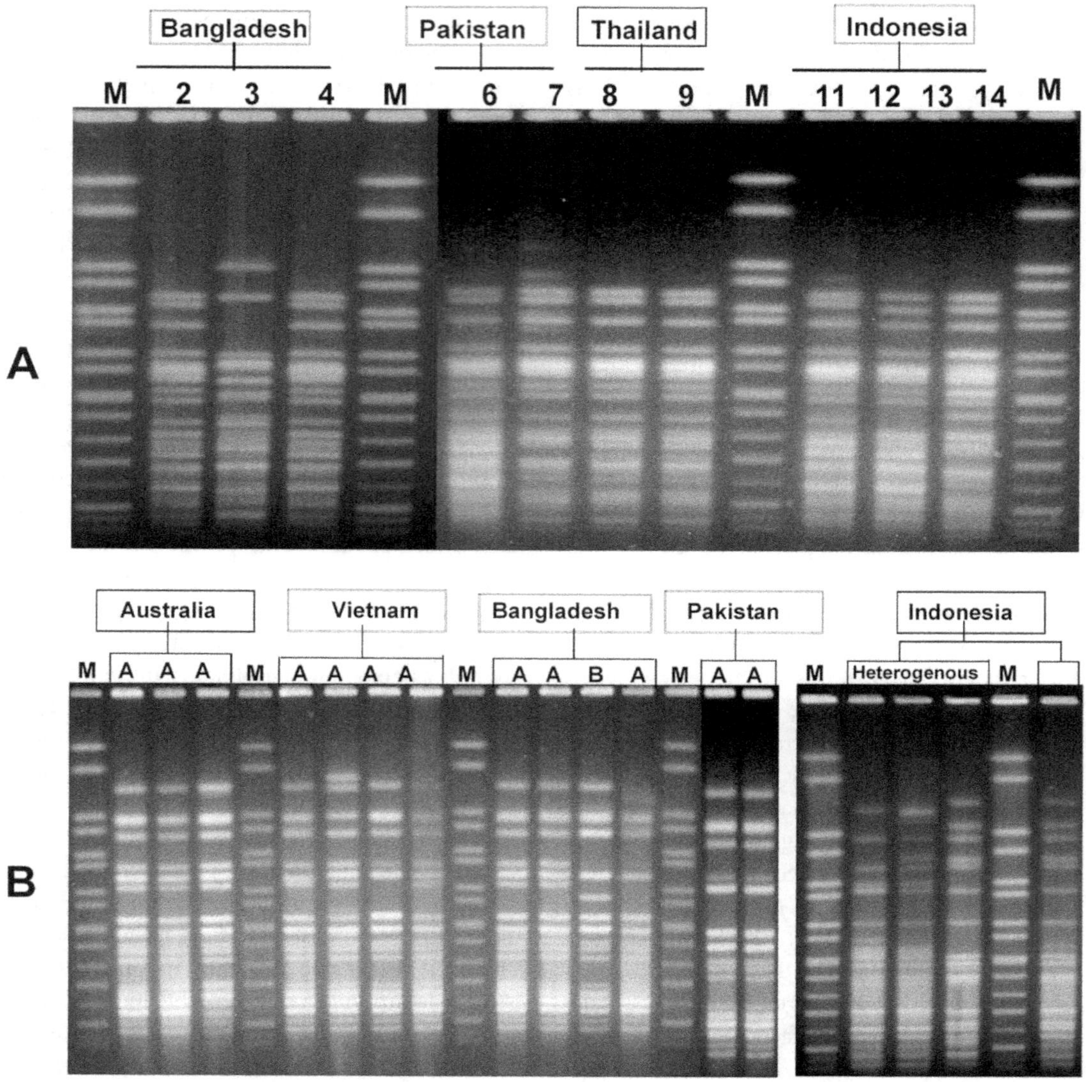

Figure 3.3 PFGE patterns of *Xba*I-digested chromosomal DNA from representative strains of (A) *S. flexneri* subserotype 1c and (B) *S. flexneri* type 4 from diverse geographical locations. (A) Lanes 2 through 4, strains from Bangladesh; 6 and 7, strains from Pakistan; 8 and 9 strains from Thailand; 11–14 strains from Indonesian. Two different types of *S. flexneri* type 4 (A and B) were found among strains isolated from Bangladesh. Predominant type (A) of Bangladeshi strains was also found in Australia, Vietnam and Pakistan whereas Indonesian strains were heterogeneous. Lanes marked with M contain *Salmonella Braenderup* DNA (marker).

Within serotype 4, plasmid analysis showed that 140, 70-60, 4 and 1.6 MDa plasmids were commonly present in all the strains in Bangladesh (Talukder *et al.*, 2005). It has been demonstrated that a middle range plasmid (35–80 MDa) are more likely to carry the antibiotic resistance determinants for ampicillin, tetracycline and trimethoprim–sulfamethoxazole (Talukder *et al.*, 2002, 2003a, 2003b). The presence of additional plasmids in particular serotypes suggested that plasmid profiles may be a useful tool in distinguishing between serotypes of *S. flexneri* (Haider *et al.*, 1989). It may also be possible to document the appearance of any new strain in a community by these patterns (Haider *et al.*, 1989; Talukder *et al.*, 2002, 2003a,b).

Plasmid profiling had been used as a useful method to identify and characterize the atypical as well as new variant of *Shigella* species. A good correlation was observed between plasmid profiles and the other typing procedures like biotyping, PFGE and ribotyping. It was also suggested that there were some small plasmids present in the strains which maintain their existence as a stable gene pool so these could be used as extrachromosomal markers for identification of new strains and differentiating the existing serotypes (Talukder *et al.*, 2002, 2003b, 2007). Analysis of plasmid DNA of *S. flexneri* serotype 1 (serotypes 1a, 1b and 1c) has shown that all the strains studied harboured 2.8, 2.1, 1.8, and 1.0 MDa core plasmids. These four core plasmids were unique to the strains of *S. flexneri* serotype 1, perhaps reflecting their dissemination from a single origin. However, an additional plasmid of 4.0 and/or 3.0 MDa was found in all strains of 1c isolated from Pakistan and Indonesia. This plasmids were found infrequently in Bangladeshi strains (Talukder *et al.*, unpublished data) (Fig. 3.4).

It is suggested by a number studies (Sansonetti *et al.*, 1982, 1983; Maurelli *et al.*, 1984) that the genetic determinants for the invasiveness of *Shigella* are located on the 140-MDa large plasmid. Strains of *Shigella* species which carry the *ipaH* gene on the 140-MDa plasmid, have the capacity to absorb Congo red dye, and provoke keratoconjunctivitis in the guinea pig's eye, a test for their invasiveness. It is now known that the *sen* gene, for *Shigella* enterotoxin is also located on the 140-MDa invasive plasmid and is present in all strains of *S. flexneri* that harbour this plasmid (Nataro *et al.*, 1995).

Among all diarrhoea episodes that occur every year in Bangladesh, 20% are caused by unidentified agent. We have been investigating the properties of these unidentified organisms and found that 3% of total isolates are biochemically related to *Shigella* but could not be serologically typed and hence were termed as *Shigella*-like organism (SLO) (Talukder *et al.*, 2007). In addition >4% *Shigella* were detected in culture negative stool samples collected in a rural population based study at Mirzapur, Bangladesh during January to December 2010 using a combination of molecular methods and culture. One cluster of SLO has finally been designated as a provisional serotype of *S. dysenteriae*. All these strains were non-lactose fermenters, with an identical biochemical pattern typical of *S. dysenteriae*. All strains were considered as invasive since they showed all the virulence properties of *Shigella* described above. These isolates had a unique PFGE type, which seemed to be related to *S. dysenteriae* distinguishable from all other *Shigella* and *E. coli* patterns. An antiserum prepared against one of these strains reacted with the remaining isolates of the cluster but did not show any cross reactivity with any of the other *Shigella* and *E. coli* reference strains. We proposed that these isolates represent a new provisional/novel serovar of *S. dysenteriae* (NSD) type strain KIVI 162 (Talukder *et al.*, 2007). It is interesting to note that among the *S. dysenteriae*

Country	Core plasmids (MDa)
Bangladesh	2.8, 2.1, 1.8, 1.0
Indonesia	3.0, 2.8, 2.1, 1.8, 1.0
Thailand	2.8, 2.1, 1.8, 1.0
Pakistan	4.0, 2.8, 2.1, 1.8, 1.0

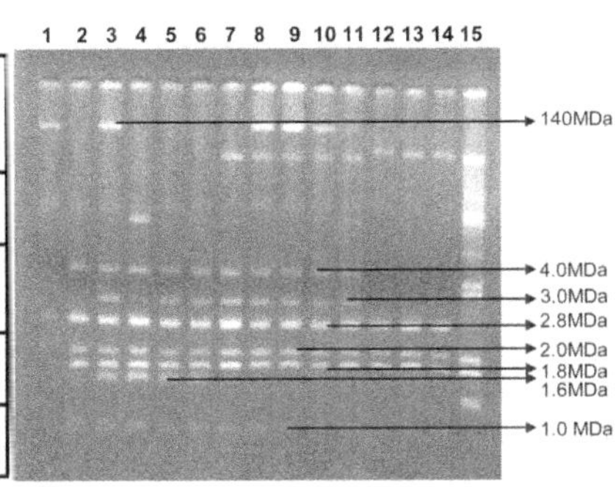

Figure 3.4 Agarose gel electrophoresis of plasmid DNA showing the representative patterns among different serotype of *S. flexneri* 1c isolated from diverse geographical location: Lanes: 1, *E. coli* PDK-9 (Marker); 2, *S. flexneri* 1c from Pakistan; 3–7, 14, *S. flexneri* 1c from Bangladesh; 8 and 9, *S. flexneri* 1c from Thailand; 10 and 11, *S. flexneri* 1c from Indonesia; 12, *S. flexneri* 1a from Bangladesh; 13, *S. flexneri* 1b from Bangladesh; 15, *E. coli* V-517.

strain isolated between 2008 and 2010, NSD (23%) was the second prevalent serotype after serotype 2 (39%). Therefore we suggest including strains of this new provisional/novel serovar of *S. dysenteriae* in the classification scheme as *S. dysenteriae* 16.

A second cluster of isolates showing biochemical properties typical of *S. boydii* harboured the 140 MDa plasmid, and exhibited invasive and toxigenic phenotype as showed by other NSD. These isolates were identical based on their PFGE and ribotyping patterns and were indistinguishable from other *S. boydii* serotypes. However, antiserum raised against K-730 a representative isolate, reacted with all the isolates of these cluster but not with other *Shigella* and *E. coli* reference strains (Talukder *et al.*, 2002b).

In 2004, Kalluri *et al.* reported a new provisional serogroup represented by strain 99–4528 in an epidemiological investigation. Very recently Woodward *et al.* (2005) reported the same new serogroup and later Grimont *et al.* (2007) characterized isolates of the provisional serogroup including the K-730 strain, phenotypically and genotypically and designated these as *S. boydii* 20 (Grimont *et al.*, 2007).

Multiple types of toxin and their corresponding protein products have been discovered in *Shigella* spp. and EIEC (O'Brien *et al.*, 1992; Sansonetti, 1998); a clear understanding of the role of toxin in Shigellosis remains to be studied. A large number of variants of Shiga toxin (STX) e.g. STX1 and STX 2 have been isolated illustrating diversity of the STX family (Gannon and Gyles, 1990; Ito *et al.*, 1990; Lin *et al.*, 1993). Shiga toxin (STX) belongs to the large family of ribosome-inactivating proteins (RIPs). Shiga toxin has also been shown to induce nuclear DNA damage in human endothelial cells and releases single-stranded DNA (Brigotti *et al.*, 2002). This toxin is mostly present in *S. dysenteriae* 1 and Shiga-toxin producing *E. coli*. In addition, two enterotoxins have been identified in *S. flexneri*: *Shigella* enterotoxin 1 (ShET-1) encoded by the set1 chromosome gene and *Shigella* enterotoxin 2 (ShET-2) encoded by the sen gene (140 MDa plasmid) (Vargas *et al.*, 1999). Interestingly, we have identified a group of clinical isolates of *S. flexneri* strains that do not contain any of *stx*, *set*, or *sen* genes. However, these strains exhibit very strong cytotoxic activity against HeLa cells. We have developed a neuron culture model to study if these strains exhibit neurotoxic activity. Using rat brain (cerebellar) neurons culture model we have found that these strains bear strong neurotoxic activity. We are studying the neurotoxic principle by biochemical fractionation. Our preliminary studies indicate that the neurotoxic activity is not due to *stx*, *set*, or *sen*, indicating that some *Shigella* strains may have additional cytotoxic factor(s).

Phylogenetic analysis

Phylogenetically, the *Shigella* spp. (except *S. boydii* 13) are distributed in seven clusters within the *E. coli branch* (Lan and Reeves, 2002). Various studies have shown that *Shigella* strains fall within *E. coli*, including studies using MLEE (Ochman *et al.*, 1983), ribotyping (Rolland *et al.*, 1998), MLST or a combination of MLEE and sequencing (Pupo *et al.*, 1997). The evolutionary relationships of all 46 serotypes of *Shigella* were elucidated recently by analysis of sequence variation in eight housekeeping genes (Pupo *et al.*, 2000) showing that most fall into three clusters of *E. coli*, each cluster representing a group of related *E. coli* clones. Interestingly, each cluster includes strains from different *Shigella* 'species'. Cluster 1 contains one *flexneri* (F) and the majority of *boydii* (B) and *dysenteriae* (D) strains (B1 to B4, B6, B8, B10, B14, B18, D3 to D7, D9, D11 to D13 and F6). Cluster 2 contain one *dysenteriae* and seven *Boydii* strain (B5, B7, B9, B11, B15, B16, B17 and D2). Cluster 3 contains one *boydii* strain (B12) and *flexneri* serotypes 1–5. *Sonnei* (S) and three *dysenteriae* strains (D1, D8 and D10) are outside of the three main clusters, but nonetheless clearly within *E. coli*, while B13 is distantly related to *E. coli* and represents an unnamed species. However, it has recently been shown that B13 fall into two groups, one of which is distant from *E. coli* but is within the species *Escherichia albertii*, while the other is within *E. coli* (Hyma *et al.*, 2005).

MLST analysis conducted by our group in the ICDDRB in collaboration with Whitman's lab of Michigan State University, USA, analysis of new *S. boydii* 20 and the *S. flexneri* 6 and 6a showed that these isolates fall within *Shigella* Group 1 that includes (according to Reeves) Boydii serotypes 1–6, 8, 10, 14, 18 and *dysenteriae* D3–7, 11, 12, 13 and *flexneri* 6 and 6a. Atypical *S. flexneri* 1c and *S. flexneri* type 4 belongs to *Shigella* Group 3 with *flexneri* 1 and F3a, F3c, F4a, F4b, F5, and Boydii 12 within *E. coli* (Fig. 3.5) (unpublished data). These cluster analyses further emphasizes that the atypical *Shigella* strains should be included in the serotype scheme of *Shigella*.

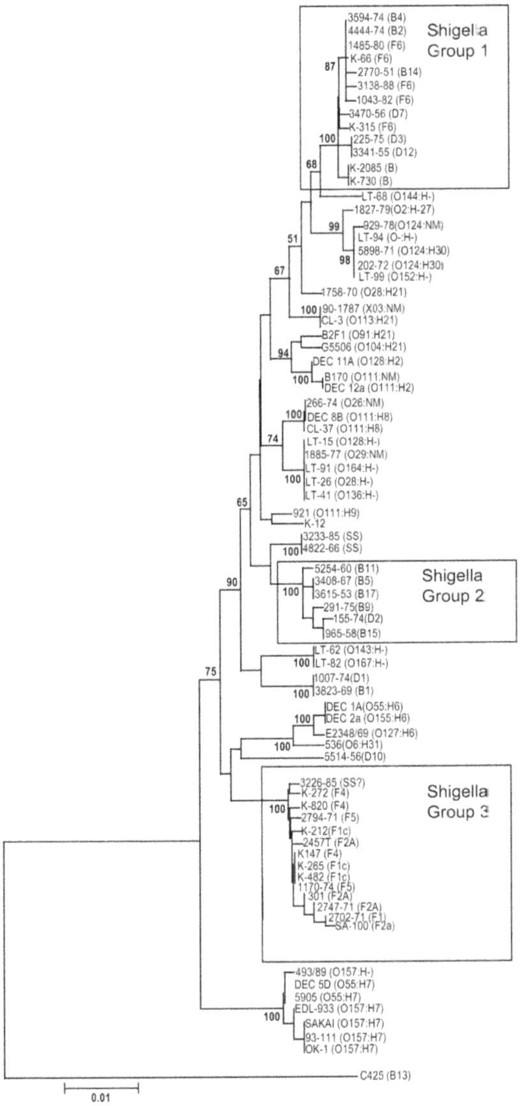

Figure 3.5 MLST analysis of new variants and atypical *S. flexneri* strains and existing *Shigella* serotypes: New *S. boydii* 20 (K-730, K-2085) and the atypical *S. flexneri* strains serotype 6a, K-66 and 6b, K-315 fall into *Shigella* Group 1 that includes *S. boydii* (B) serotypes B2, B4, B14 and *S. dysenteriae* (D) serotypes D3, D7 and D12. Atypical *S. flexneri* (F) type 4(F4), K-272; K-820 and K-147; *S. flexneri* 1c, (F1c), K-265 and K-482 fall into *Shigella* Group 3 with *S. flexneri* serotypes F1, F2a and F5.

References

Alcoba-Florez, J., Perz-Roth, E., Gonzalez-Linares, S., and Mendez-Alvarez, S. (2005). Outbreak of *Shigella sonnei* in a rural hotel in La Gomera, Canary Islands, Spain. Int. Microbiol. *8*, 133–136.

Altweg, M., and Bockemuhl, J. (1998). *Escherichia* and *Shigella*. In Topley & Wilson's Microbiology and Microbial Infections, Collier, L., Balows, A., and Sussman, M., eds. (Arnold, London), pp. 935–967.

Ansaruzzaman, M., Kibriya, A.K., Rahman, A., Neogi, P.K., Faruque, A.S., Rowe, B., and Albert, M.J. (1995). Detection of provisional serovars of *Shigella dysenteriae* and designation as *S. dysenteriae* serotypes 14 and 15. J. Clin. Microbiol. 33, 1423–1425.

Ansaruzzaman, M., Sultana, M., Talukder, K.A., Alam, K., Matsushita, S., Safa, A., Khajanchi, B.K., Dutta, D.K., Islam, Z., Albert, M.J., *et al.* (2005). Isolation and characterization of provisional serovar *Shigella* boydii E16553 from diarrhoeal patients in Bangladesh. J. Med. Microbiol. *54*, 477–480.

Bopp, C.A., Brenner, F.W., Fields, P.I., Wells, J.G., Strockbine, N.A. (2003). *Escherichia*, *Shigella*, and *Salmonella*. In Manual of clinical microbiology, 8th edition, Murray, P.R., Baron, E.J., Jorgensen, J.H., Pfaller, M.A., Yolken, R.H., eds. (ASM Press, Washington, DC), pp. 654–671.

Brenner, D.J. (1984). Recommendations on recent proposals for the classification of shigellae. Int. J. Syst. Bacteriol. *34*, 87–88.

Brigotti, M., Alfieri, R., Sestili, P., Bonelli, M., Petronini, P.G., Guidarelli, A., Barbieri, L., Stirpe, F., and Sperti, S. (2002). Damage to nuclear DNA induced by Shiga toxin 1 and ricin in human endothelial cells. Faseb J. *16*, 365–372.

Carlin, N.I., Lindberg, A.A., Bock, K., and Bundle, D.R. (1984). The *Shigella flexneri* O-antigenic polysaccharide chain. Nature of the biological repeating unit. Eur. J. Biochem. *139*, 189–194.

Carlin, N.I., Rahman, M., Sack, D.A., Zaman, A., Kay, B., and Lindberg, A.A. (1989). Use of monoclonal antibodies to type *Shigella flexneri* in Bangladesh. J. Clin. Microbiol. *27*, 1163–1166.

De Schrijver, K., Bertrand, S., Gutierrez Garitano, I., Van den Branden, D., and Van Schaeren, J. (2011). Outbreak of *Shigella sonnei* infections in the Orthodox Jewish community of Antwerp, Belgium, April to August 2008. Euro. Surveill. *16*, 15–20.

DuPont, H.L., Levine, M.M., Hornick, R.B., and Formal, S.B. (1989). Inoculum size in shigellosis and implications for expected mode of transmission. J. Infect. Dis. *159*, 1126–1128.

Edwards, P.R., and Ewing, W.H. (1972). Identification of Enterobacteriaceae. (Burgess Publishing Co., Minneapolis, MN), pp. 126–131.

El-Gendy, A., El-Ghorab, N., Lane, E.M., Elyazeed, R.A., Carlin, N.I., Mitry, M.M., Kay, B.A., Savarino, S.J., and Peruski, L.F., Jr. (1999). Identification of *Shigella flexneri* subserotype 1c in rural Egypt. J. Clin. Microbiol. *37*, 873–874.

Ewing, W.H. (1954). Mannitol negative varieties of *Shigella flexneri* serotypes. J. Immunol. 72, 404–410.

Ewing, W.H., Reavis, R.W., and Davis, B.R. (1958). Provisional *Shigella* serotypes. Can. J. Microbiol. *4*, 89–107.

Fontaine, A., Arondel, J., and Sansonetti, P.J. (1990). Construction and evaluation of live attenuated vaccine strains of *Shigella flexneri* and *Shigella dysenteriae* 1. Res. Microbiol. *141*, 907–912.

Formal, S.B., and Baker, E.E. (1953). Quantitative studies of cross reactions between *Shigella flexneri* types 1a, 1b, and 31. J. Immunol. *70*, 260–266.

Gannon, V.P., and Gyles, C.L. (1990). Characteristics of the Shiga-like toxin produced by *Escherichia coli* associated with porcine edema disease. Vet. Microbiol. *24*, 89–100.

Garcia-Fulgueiras, A., Sanchez, S., Guillen, J.J., Marsilla, B., Aladuena, A., and Navarro, C. (2001). A large outbreak of *Shigella sonnei* gastroenteritis associated with consumption of fresh pasteurised milk cheese. Eur. J. Epidemiol. *17*, 533–538.

Grimont, F., Lejay-Collin, M., Talukder, K.A., Carle, I., Issenhuth, S., Le Roux, K., and Grimont, P.A. (2007). Identification of a group of shigella-like isolates as *Shigella* boydii 20. J. Med. Microbiol. *56*, 749–754.

Gross, R.J., Thomas, L.V., Cheasty, T., Rowe, B., and Lindberg, A.A. (1989). Four new provisional serovars of *Shigella*. J. Clin. Microbiol. *27*, 829–831.

Gross, R.J., Thomas, L.V., Day, N.P., Cheasty, T., and Rowe, B. (1982). New provisional serovar of *Shigella boydii*. J. Clin. Microbiol. *16*, 1000–1002.

Gross, R.J., Thomas, L.V., and Rowe, B. (1980). New provisional serovar (E10163) of *Shigella boydii*. J. Clin. Microbiol. *12*, 167–169.

Guerin, P.J., Brasher, C., Baron, E., Mic, D., Grimont, F., Ryan, M., Aavitsland, P., and Legros, D. (2003). *Shigella dysenteriae* serotype 1 in west Africa: intervention strategy for an outbreak in Sierra Leone. Lancet *362*, 705–706.

Gupta, A., Polyak, C.S., Bishop, R.D., Sobel, J., and Mintz, E.D. (2004). Laboratory-confirmed shigellosis in the United States, 1989–2002: epidemiologic trends and patterns. Clin. Infect. Dis. *38*, 1372–1377.

Haider, K., Kay, B.A., Talukder, K.A., and Huq, M.I. (1988). Plasmid analysis of *Shigella dysenteriae* type 1 isolates obtained from widely scattered geographical locations. J. Clin. Microbiol. *26*, 2083–2086.

Haider, K., Huq, M.I., Talukder, K.A., and Ahmad, Q.S. (1989). Electropherotyping of plasmid DNA of different serotypes of *Shigella flexneri* isolated in Bangladesh. Epidemiol. Infect. *102*, 421–428.

Hyma, K.E., Lacher, D.W., Nelson, A.M., Bumbaugh, A.C., Janda, J.M., Strockbine, N.A., Young, V.B., and Whittam, T.S. (2005). Evolutionary genetics of a new pathogenic Escherichia species: *Escherichia albertii* and related *Shigella boydii* strains. J. Bacteriol. *187*, 619–628.

Ito, H., Terai, A., Kurazono, H., Takeda, Y., and Nishibuchi, M. (1990). Cloning and nucleotide sequencing of Vero toxin 2 variant genes from *Escherichia coli* O91:H21 isolated from a patient with the hemolytic uremic syndrome. Microb. Pathog. *8*, 47–60.

Kalluri, P., Cummings, K.C., Abbott, S., Malcolm, G.B., Hutcheson, K., Beall, A., Joyce, K., Polyak, C., Woodward, D., Caldeira, R., *et al.* (2004). Epidemiological features of a newly described serotype of *Shigella boydii*. Epidemiol. Infect. *132*, 579–583.

Kolavic, S.A., Kimura, A., Simons, S.L., Slutsker, L., Barth, S., and Haley, C.E. (1997). An outbreak of *Shigella dysenteriae* type 2 among laboratory workers due to intentional food contamination. JAMA *278*, 396–398.

Kotloff, K.L., Winickoff, J.P., Ivanoff, B., Clemens, J.D., Swerdlow, D.L., Sansonetti, P.J., Adak, G.K., and Levine, M.M. (1999). Global burden of *Shigella* infections: implications for vaccine development and implementation of control strategies. Bull. World Health Organ. *77*, 651–666.

Lan, R., and Reeves, P.R. (2002). *Escherichia coli* in disguise: molecular origins of *Shigella*. Microbes Infect. *4*, 1125–1132.

Levine, M.M., Kotloff, K.L., Barry, E.M., Pasetti, M.F., and Sztein, M.B. (2007). Clinical trials of *Shigella* vaccines: two steps forward and one step back on a long, hard road. Nat. Rev. Microbiol. *5*, 540–553.

Lin, Z., Kurazono, H., Yamasaki, S., and Takeda, Y. (1993). Detection of various variant verotoxin genes in *Escherichia coli* by polymerase chain reaction. Microbiol. Immunol. *37*, 543–548.

Lofdahl, M., Ivarsson, S., Andersson, S., Langmark, J., and Plym-Forshell, L. (2009). An outbreak of *Shigella dysenteriae* in Sweden, May–June 2009, with sugar snaps as the suspected source. Euro Surveill. *14*.

Matsumoto, M., Suzuki, Y., Saito, M., Ishikawa, N., and Ohta, M. (1998). Epidemiologic study of *Shigella sonnei* from sequential outbreaks and sporadic cases using different typing techniques. Microbiol. Immunol. *42*, 259–264.

Maurelli, A.T., Blackmon, B., and Curtiss, R., 3rd (1984). Loss of pigmentation in *Shigella flexneri* 2a is correlated with loss of virulence and virulence-associated plasmid. Infect. Immun. *43*, 397–401.

Mendizabal-Morris, C.A., Mata, L.J., Gangarosa, E.J., and Guzman, G. (1971). Epidemic Shiga-bacillus dysentery in Central America. Derivation of the epidemic and its progression in Guatemala, 1968–69. Am. J. Trop. Med. Hyg. *20*, 927–933.

Naheed, A., Kalluri, P., Talukder, K.A., Faruque, A.S., Khatun, F., Nair, G.B., Mintz, E.D., and Breiman, R.F. (2004). Fluoroquinolone-resistant *Shigella dysenteriae* type 1 in northeastern Bangladesh. Lancet Infect. Dis. *4*, 607–608.

Nataro, J.P., Seriwatana, J., Fasano, A., Maneval, D.R., Guers, L.D., Noriega, F., Dubovsky, F., Levine, M.M., and Morris, J.G., Jr. (1995). Identification and cloning of a novel plasmid-encoded enterotoxin of enteroinvasive *Escherichia coli* and *Shigella* strains. Infect. Immun. *63*, 4721–4728.

Noriega, F.R., Wang, J.Y., Losonsky, G., Maneval, D.R., Hone, D.M., and Levine, M.M. (1994). Construction and characterization of attenuated delta aroA delta virG *Shigella flexneri* 2a strain CVD 1203, a prototype live oral vaccine. Infect. Immun. *62*, 5168–5172.

Noriega, F.R., Losonsky, G., Lauderbaugh, C., Liao, F.M., Wang, J.Y., and Levine, M.M. (1996). Engineered deltaguaB-A deltavirG *Shigella flexneri* 2a strain CVD 1205: construction, safety, immunogenicity, and potential efficacy as a mucosal vaccine. Infect. Immun. *64*, 3055–3061.

O'Brien, A.D., Tesh, V.L., Donohue-Rolfe, A., Jackson, M.P., Olsnes, S., Sandvig, K., Lindberg, A.A., and Keusch, G.T. (1992). Shiga toxin: biochemistry, genetics, mode of action, and role in pathogenesis. Curr. Top Microbiol. Immunol. *180*, 65–94.

Ochman, H., Whittam, T.S., Caugant, D.A., and Selander, R.K. (1983). Enzyme polymorphism and genetic population structure in *Escherichia coli* and Shigella. J. Gen. Microbiol. *129*, 2715–2726.

Pal, S.C., Sengupta, P.G., Sen, D., Bhattacharya, S.K., and Deb, B.C. (1989). Epidemic shigellosis due to *Shigella dysenteriae* type 1 in south Asia. Indian J. Med. Res. *89*, 57–64.

van Pelt, W., de Wit, M.A., Wannet, W.J., Ligtvoet, E.J., Widdowson, M.A., and van Duynhoven, Y.T. (2003). Laboratory surveillance of bacterial gastroenteric pathogens in The Netherlands, 1991–2001. Epidemiol. Infect. *130*, 431–441.

Pupo, G.M., Karaolis, D.K., Lan, R., and Reeves, P.R. (1997). Evolutionary relationships among pathogenic and nonpathogenic *Escherichia coli* strains inferred from multilocus enzyme electrophoresis and mdh sequence studies. Infect. Immun. *65*, 2685–2692.

Pupo, G.M., Lan, R., and Reeves, P.R. (2000). Multiple independent origins of *Shigella* clones of *Escherichia coli* and convergent evolution of many of their characteristics. Proc. Natl. Acad. Sci. U.S.A. *97*, 10567–10572.

Qiu, S., Wang, Z., Chen, C., Liu, N., Jia, L., Liu, W., Wang, L., Hao, R., Zhang, L., Wang, Y., *et al.* (2011). Emergence of a novel *Shigella flexneri* serotype 4s strain that evolved from a serotype X variant in China. J. Clin. Microbiol. *49*, 1148–1150.

Rahaman, M.M., Khan, M.M., Aziz, K.M., Islam, M.S., and Kibriya, A.K. (1975). An outbreak of dysentery caused by *Shigella dysenteriae* type 1 on a coral island in the Bay of Bengal. J. Infect. Dis. *132*, 15–19.

Rolland, K., Lambert-Zechovsky, N., Picard, B., and Denamur, E. (1998). *Shigella* and enteroinvasive *Escherichia coli* strains are derived from distinct ancestral strains of *E. coli*. Microbiology *144*, 2667–2672.

Sansonetti, P.J. (1998). Molecular and cellular mechanisms of invasion of the intestinal barrier by enteric pathogens. The paradigm of *Shigella*. Folia Microbiol. (Praha) *43*, 239–246.

Sansonetti, P.J., Kopecko, D.J., and Formal, S.B. (1982). Involvement of a plasmid in the invasive ability of *Shigella flexneri*. Infect. Immun. *35*, 852–860.

Sansonetti, P.J., d'Hauteville, H., Ecobichon, C., and Pourcel, C. (1983). Molecular comparison of virulence plasmids in *Shigella* and enteroinvasive *Escherichia coli*. Ann. Microbiol. (Paris) *134*, 295–318.

Sheriff, M.H.R.L.t.l.f.m.M.e.w.s.d.S.L.F.P., 117–120. (1982). Lessons to learn from my Maldivian experience with *Shigella* dysentery. Sri Lankan Family Physician *5*, 117–120.

Shmilovitz, M., Kretzer, B., and Levi, S. (1985). A new provisional serovar of *Shigella dysenteriae*. J. Clin. Microbiol. *21*, 240–242.

Simmons, D.A., and Romanowska, E. (1987). Structure and biology of *Shigella flexneri* O antigens. J. Med. Microbiol. *23*, 289–302.

Stagg, R.M., Cam, P.D., and Verma, N.K. (2008). Identification of newly recognized serotype 1c as the most prevalent *Shigella flexneri* serotype in northern rural Vietnam. Epidemiol. Infect. *136*, 1134–1140.

Tacket, C.O., and Cohen, M.L. (1983). Shigellosis in day care centers: use of plasmid analysis to assess control measures. Pediatr. Infect. Dis. *2*, 127–130.

Talukder, K.A. (2002). Detection and characterization of new *Shigella boydii* isolated from diarrhoeal patients in Bangladesh. Paper presented at: 6th International Meeting on Molecular Epidemiology and Evolutionary Genetics in Infectious Diseases (MEEGID–VI) (Pasteur Institute, Paris, France).

Talukder, K.A., Dutta, D.K., and Albert, M.J. (1999). Evaluation of pulsed-field gel electrophoresis for typing of *Shigella dysenteriae* type 1. J. Med. Microbiol. *48*, 781–784.

Talukder, K.A., Dutta, D.K., Safa, A., Ansaruzzaman, M., Hassan, F., Alam, K., Islam, K.M., Carlin, N.I., Nair, G.B., and Sack, D.A. (2001). Altering trends in the dominance of *Shigella flexneri* serotypes and emergence of serologically atypical *S. flexneri* strains in Dhaka, Bangladesh. J. Clin. Microbiol. *39*, 3757–3759.

Talukder, K.A., Islam, M.A., Dutta, D.K., Hassan, F., Safa, A., Nair, G.B., and Sack, D.A. (2002). Phenotypic and genotypic characterization of serologically atypical strains of *Shigella flexneri* type 4 isolated in Dhaka, Bangladesh. J. Clin. Microbiol. *40*, 2490–2497.

Talukder, K.A., Islam, M.A., Khajanchi, B.K., Dutta, D.K., Islam, Z., Safa, A., Alam, K., Hossain, A., Nair, G.B., and Sack, D.A. (2003a). Temporal shifts in the dominance of serotypes of *Shigella dysenteriae* from 1999 to 2002 in Dhaka, Bangladesh. J. Clin. Microbiol. *41*, 5053–5058.

Talukder, K.A., Islam, Z., Islam, M.A., Dutta, D.K., Safa, A., Ansaruzzaman, M., Faruque, A.S., Shahed, S.N., Nair, G.B., and Sack, D.A. (2003b). Phenotypic and genotypic characterization of provisional serotype *Shigella flexneri* 1c and clonal relationships with 1a and 1b strains isolated in Bangladesh. J. Clin. Microbiol. *41*, 110–117.

Talukder, K.A., Khajanchi, B.K., Islam, M.A., Dutta, D.K., Islam, Z., Safa, A., Khan, G.Y., Alam, K., Hossain, M.A., Malla, S., *et al.* (2004). Genetic relatedness of ciprofloxacin-resistant *Shigella dysenteriae* type 1 strains isolated in south Asia. J. Antimicrob. Chemother. *54*, 730–734.

Talukder, K.A., Khajanchi, B.K., Dutta, D.K., Islam, Z., Islam, M.A., Iqbal, M.S., Nair, G.B., and Sack, D.A. (2005). An unusual cluster of dysentery due to *Shigella dysenteriae* type 4 in Dhaka, Bangladesh. J. Med. Microbiol. *54*, 511–513.

Talukder, K.A., Islam, Z., Dutta, D.K., Islam, M.A., Khajanchi, B.K., Azmi, I.J., Iqbal, M.S., Hossain, M.A., Faruque, A.S., Nair, G.B., *et al.* (2006). Antibiotic resistance and genetic diversity of *Shigella sonnei* isolated from patients with diarrhoea between 1999 and 2003 in Bangladesh. J. Med. Microbiol. *55*, 1257–1263.

Talukder, K.A., Mondol, A.S., Islam, M.A., Islam, Z., Dutta, D.K., Khajanchi, B.K., Azmi, I.J., Hossain, M.A., Rahman, M., Cheasty, T., *et al.* (2007). A novel serovar of *Shigella dysenteriae* from patients with diarrhoea in Bangladesh. J. Med. Microbiol. *56*, 654–658.

Terajima, J., Tamura, K., Hirose, K., Izumiya, H., Miyahara, M., Konuma, H., and Watanabe, H. (2004). A multi-prefectural outbreak of *Shigella sonnei* infections associated with eating oysters in Japan. Microbiol. Immunol. *48*, 49–52.

Vargas, M., Gascon, J., Jimenez, De Anta, M.T., and Vila, J. (1999). Prevalence of *Shigella* enterotoxins 1 and 2 among *Shigella* strains isolated from patients with traveler's diarrhea. J. Clin. Microbiol. *37*, 3608–3611.

von Seidlein, L., Kim, D.R., Ali, M., Lee, H., Wang, X., Thiem, V.D., Canh do, G., Chaicumpa, W., Agtini, M.D., Hossain, A., *et al.* (2006). A multicentre study of *Shigella* diarrhoea in six Asian countries: disease burden, clinical manifestations, and microbiology. PLoS Med. *3*, e353.

Wachsmuth, K., and Morris, G.K. (1989). *Shigella* foodborne bacterial pathogens. In Foodborne Bacterial Pathogens, Doyle M.P., ed. (Marcel Dekker, Inc., New York), pp. 447–462.

Wathen-Grady, H.G., Britt, L.E., Strockbine, N.A., and Wachsmuth, I.K. (1990). Characterization of *Shigella dysenteriae* serotypes 11, 12, and 13. J. Clin. Microbiol. *28*, 2580–2584.

Weissman, J.B., Gangorosa, E.J., Schmerler, A., Marier, R.L., and Lewis, J.N. (1975). Shigellosis in day-care centres. Lancet *1*, 88–90.

Woodward, D.L., Clark, C.G., Caldeira, R.A., Ahmed, R., Soule, G., Bryden, L., Tabor, H., Melito, P., Foster, R., Walsh, J., *et al.* (2005). Identification and characterization of *Shigella* boydii 20 serovar nov., a new and emerging *Shigella* serotype. J. Med. Microbiol. *54*, 741–748.

Molecular Pathogenesis, Epidemiology and Drug Resistance of *Shigella* species

4

Gururaja P. Pazhani and Thandavarayan Ramamurthy

Abstract

Shigella species are often associated with diarrhoea/dysentery among children in developing countries of Asia-African regions and also cause foodborne infections in developed countries. During infection, this pathogen secretes a number of effectors via the type III secretion system. Recently, *Shigella*-infected cases have increased considerably in Asia, but the death rate decreased substantially owing to nutritional and clinical interventions. Antimicrobials reduce the episodes of shigellosis. However, multiple antibiotic resistance in shigellae have increased over the years owing to improper use of antimicrobials in the treatment of diarrhoea. Several mechanisms such as plasmids and other mobile genetic elements are involved in the transmission of resistance in shigellae. Owing to poor efficacy of many of the existing *Shigella* vaccines, none has been licensed for use in endemic areas. This chapter reviews our current understanding of mechanism involved in the pathogenesis of *Shigella*, antimicrobial resistance and also its epidemiological importance and clinical management of shigellosis.

Introduction

The genus *Shigella* is a Gram-negative, facultative anaerobe and is the causative agent for shigellosis or 'bacillary dysentery'. Globally, shigellae cause over 164.7 million cases annually and 1.1 million deaths, of which 163.2 million occur in developing countries. In industrialized countries, this pathogen triggers many foodborne diarrhoeal outbreaks. An acute rectocolitis caused by *Shigella* reveals its ability to disrupt, invade, and cause the inflammatory damage of the intestinal epithelium. *Shigella* can be classified into four species *Shigella dysenteriae*, *Shigella flexneri*, *Shigella boydii* and *Shigella sonnei*, each consisting of several serotypes on the basis of O-specific antigens. This chapter reviews our current understanding of *Shigella* as a pathogen, its pathogenesis, epidemiology, antimicrobial resistance, and also the clinical management of shigellosis.

Pathogenesis

Several recent studies provide a paradigm shift in understanding the pathogenesis of *Shigella* spp. Colonization of the intestinal epithelium followed by invasion by this pathogen results in severe inflammatory colitis. *Shigella* possesses as many as 50 effectors that are injected via the type III secretion system (T3SS). Some of these effectors are delivered into the adjacent bacterial space and the others into the host cell cytoplasm and nucleus. The delivered effectors control host cell signalling and immune response, which ultimately promote the bacterial infection, eluding the host defence systems. The stepwise infection and mechanisms involved in the pathogenesis of *Shigella* includes; invasion of *Shigella* into the intestinal epithelial cells, replication within the epithelial cells and cell-to-cell spreading, cellular translocation of the epithelial barrier, host cell signal and inflammatory response, autophagy recognition and macrophage cell death. The details of pathogenesis of *Shigella* spp. has been explained as follows. Fig. 4.1 describes the epithelial invasion of *Shigella* spp. followed by activation of different bacterial and cellular machinery.

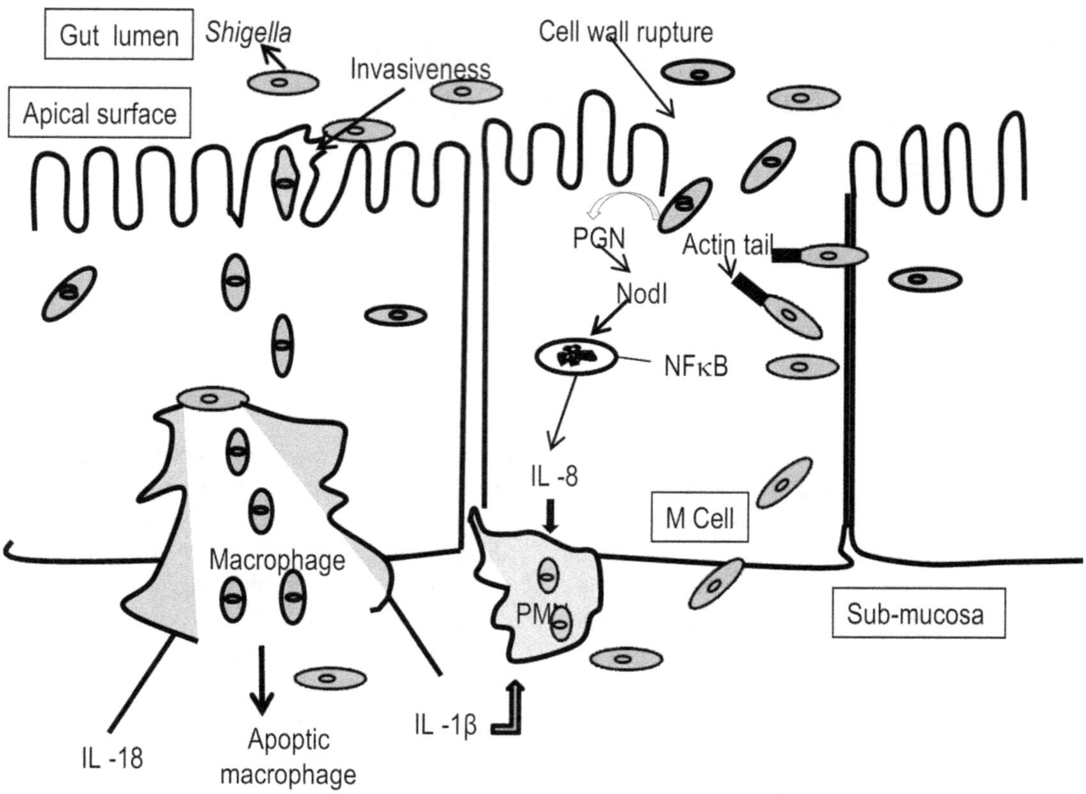

Figure 4.1 A model for *Shigella* infection of intestinal epithelium. Shigellae invade the epithelial barrier through M cell and encounters resident macrophages. The bacteria multiply within the macrophages and evade from them by inducing an apoptosis-like cell death followed by stimulation of proinflammatory responses. Released bacteria from macrophages invade epithelium from the basolateral side and move into the cytoplasm with the help of actin tail and spread into the adjacent cells. The intracellular bacteria cause release of IL-1 that evokes an influx of PMN, which transmigrate through the epithelium disrupts tight junctions and also destroy the integrity of the epithelial lining, thus enabling more luminal bacteria to reach submucosa. The cytoplasmic shigellae release PGN fragments and activate NF-κB, which triggers secretion of chemokine IL-8. This IL-8 plays major role in the recruitment of PMN. Finally, bacteria infected enterocytes are induced by endocytosis, as a result necrosis of epithelium.

Invasion into the intestinal epithelium

Human intestinal epithelium acts as a barrier against microbial invaders because epithelium consists of several layers of innate defence systems. *Shigella* has neither adherence factors nor flagella required for attaching or accessing the intestinal epithelium. These bacteria have the ability to colonize the intestinal epithelium by exploiting epithelial-cell functions, circumventing the host innate immune response and gaining accesses to deeper tissues (Ashida *et al.*, 2009). The lipid A component of lipopolysaccharide (LPS) molecule, the core and O-antigenic polysaccharide makes the pathogen resistant to host defence mechanisms such as opsonization, phagocytosis, and intracellular killing (Lindberg *et al.*, 1991). *OspG*, an effector protein initially identified in *Shigella* was shown to inhibit the host innate immune response. Kim *et al.* (2009) showed that *S. flexneri OspE*, a conserved type III effector, reinforces host cell adherence and prevents detachment of the infected cells by interacting with integrin-linked kinase. After this adherence, bacteria grow rapidly by utilizing available nutrients at the site of attachment.

Entry of bacteria into epithelial cells involves transient reorganization of host cell cytoskeleton at the site of bacterial interaction with the cell membrane, followed by various signalling pathways that elicit a macropinocytic and mutual exchanges of signals and responses. Interaction of bacteria with components of extracellular matrix, such as fibronectin, laminin or collagen type IV, stimulates the release of invasive plasmid antigen (Ipa) owing to activation of surface presentation antigen (Spa32) encoded by one of the *spa* genes on the large plasmid (Watarai *et al.*, 1995). In the anaerobic environment (gastrointestinal tract lumen), *spa* genes: *spa32* and *spa33* regulate secretion through the T3SS owing to the reduced secretions of Ipa. Marteyn *et al.* (2010) showed that oxygenation adjacent to the gastrointestinal mucosa occurs owing to diffusion from the capillary network at the tips of villi, which would reverse the anaerobic block of Ipa secretion thereby allowing T3SS activation at its precise site of action.

Major secretory proteins, the IpaA–D proteins and secretion apparatus, the Mxi-Spa translocon were identified *in vitro* for the entry of *S. flexneri* into epithelial cells (Parsot *et al.*, 1995). Both were activated upon contact of bacteria with epithelial cells. Shigellae are known to acquire pathogenicity islands (PIs), which consists of various genes encoding virulence traits and secretes effector molecules via T3SS. This system gets activated and delivers effector proteins into the host cytosol when bacteria contact with host cell (Winstanley and Hart, 2001). Electron microscopy studies identified three parts of T3SS; an external needle, transmembrane domain, and a cytoplasmic bulb, which are collectively termed as needle complex (NC). N-terminal sequencing of NC identified five major proteins, MxiD, MxiG, MxiJ, MxiH and MxiI. Of these, MxiG and MxiJ, MxiD and MxiH and MxiI, are recognized as inner membrane, outer membrane and small proteins, respectively (Blocker *et al.*, 2001). When in contact with host cell, *Shigella* induces protein secretion via T3SS. They inject secretary proteins through a 'needle' protruding from the bacterial surface directly into eukaryotic cells after an assembly of a 'translocator' pore in the host plasma membrane. The secretion turns off when the bacteria lose physical contact with a host cell (Martinez-Argudo *et al.*, 2010). The maximal expression of the genes encoding the T3SS and its substrates occur at high cell density. A stationary-phase culture derived from conditioned medium showed an increased expression of *ipa*, *mxi*, and *spa* invasion operons (Day *et al.*, 2001). T3SS in *Shigella* is regulated in response to changes in environmental osmolarity and temperature. Temperature-dependent regulation of *virF* is thought to occur at the transcriptional level. However, expression of *Shigella* virulence genes in response to both osmolarity and temperature involves the post-transcriptional regulation of *invE* (Mitobe *et al.*, 2009).

IpaA is essential for *S. flexneri* pathogenesis as it binds to the cytoskeletal protein vinculin and facilitates host cell entry (Ramarao *et al.*, 2007). Of the other secreted proteins, the Ipa complex; IpaD, IpaB, and IpaC, triggers a reorganization of the host-cell cytoskeleton leading to the formation of membrane ruffles, resulting a localized plasma membrane projection followed by long actin filaments, which engulf the bacterium. IpgC is a chaperone that binds to IpaB and C for effective host cell infection (Lokareddy *et al.*, 2010). At 37°C, the effectors molecules (Ipa protein) are synthesized and stored in the bacterial cytoplasm. N-terminus of IpaC binds with IpaB protein to form hydrophobic complex (Harrington *et al.*, 2003). This complex is inserted into the host cell membrane to form a pore for translocating the late effectors into the cell cytoplasm (Ménard *et al.*, 1996). C-terminus domain is exposed into the host cell cytoplasm in triggering actin nucleation/polymerization. This activation recruits tyrosine phosphorylated cortactin on the membrane, resulting in massive actin polymerization in the vicinity of the bacteria (Cossart and Sansonetti, 2004). IpgB1 effector, delivered from extracellular *Shigella* into the epithelial cytoplasm, which activate Rac1-WAVE-Arp2/3 pathway and induction of actin polymerization (Handa *et al.*, 2007). Finally, when bacteria are in contact with the surface of an epithelial cell induces cytoskeletal rearrangements resulting in phagocytosis and the secreted enzymes will degrade the phagosomal membrane, releasing the bacteria into the host cytoplasm (Theriot, 1995). Secretion and presentation of these proteins are controlled by IpaD at the pathogen–host interface (Espina *et al.*, 2006). Invasion of the colonic epithelium occurs from the basolateral surface followed by disruption of the colonic epithelial barrier. Both adherence and internalization into the basolateral surface of a polarized intestinal epithelium is dependent on the length of the LPS (i.e. rfc > rfaL > galU) and also plays an important role in mediating epithelial-derived signalling. This cell signalling leads to directed migration of polymorphonuclear leucocytes across intestinal epithelium (Köhler *et al.*, 2002).

Replication within the epithelial cells and cell-to-cell spreading

As a result of host inflammatory responses, *S. flexneri* can replicate inside the cytoplasm of epithelial cells (with a doubling time of 40 minutes), and cause epithelial cell lysis (Sansonetti *et al.*, 1986; Perdomo *et al.*, 1994). This process can also be accomplished without inducing major stress response pathways, or inducing host detachment or lysis (Mantis *et al.*, 1996). Different intracellular genes such as *alkA*, *citC*, and *wcaJ* were identified as required for the survival and replication within the cells (Shi *et al.*, 2004). The ability of *Shigella* to spread within the cytosol of epithelial cells and to infect adjacent cells is critical for the development of infection, which leads to mucosal abscesses. Owing to non-motile character, *Shigella* utilizes host cell microfilaments to generate intra as well as intercellular movements (Bernardini *et al.*, 1989). Ogawa *et al.* (1968) first reported the intracellular movement of *Shigella*. The movements in the cytoplasm are rapid in random directions, allowing the organism to come in contact with the plasma membrane.

Shigella exploits the host's actin-based polymerization machinery to assemble an actin tail that serves as the propulsive force required for spreading within and between the cells. IcsA (VirG) is responsible for these events in *Shigella* pathogenesis. The outer membrane protein IcsA belongs to the family of type V secreted (autotransported) virulence factors. Members of this family mediate their own translocation across the bacterial outer membrane (Brandon and Goldberg, 2001). This protein is localized at one pole of the bacterium and catalyses the polymerization of host cell actin. Within host cells, IcsA activates the host actin regulatory protein, neural Wiskott-Aldrich syndrome protein (N-WASP). This activation recruits the Arp2/3 complex, which nucleates host actin to form F-actin comet tails for the unidirectional movement of bacterium through the host-cell cytoplasm (Sandlin *et al.*, 1996; May and Morona, 2008). At the intermediate junction of the cells, long protrusions are formed with the bacteria at the tips followed by lysis of the two membranes, which leads to cell to cell spread (Theriot, 1995; Sansonetti *et al.*, 1999). VirF is an AraC-type transcriptional regulator. Expression of this gene controls normal life cycle of *Shigella* in the host, such as intracellular invasion and cell-to-cell spread. Gene knockout studies have shown their involvement in the cellular spread (Hurt *et al.*, 2010). The microtubules check the movement of *Shigella* within the cytoplasm. VirA protein that has alpha-tubulin-specific cysteine protease-like activity degrades microtubules (Yoshida *et al.*, 2006).

Cellular translocation of the epithelial barrier by *Shigella*

Shigellae invade the epithelial cells through their apical surface very poorly owing to the presence of glycolipids, which is a mucin layer that prevents shigellae to have an access to the apical membrane. Shigellae reach the basolateral region through the M cells (membranous epithelial cells) that lack mucous and glycocalyx facilities or by translocation to reach the subepithelial tissue. M cells sense and transport luminal antigens across the epithelial barrier, releasing it into an intraepithelial pocket. Once internalized in an endocytic vacuole of the M cell, shigellae move rapidly through the cell and are released into to the intraepithelial pocket. *Shigella* survives by inducing apoptosis of macrophages (Zychlinsky *et al.*, 1992), there by having an access to the basal side of epithelial cells, where bacteria can efficiently enter. A number of basolateral receptors capable of binding *Shigella* components have been identified. IpaB has an affinity to the transmembrane receptor CD44 (Skoudy *et al.*, 2000) and binding to this receptor induces cytoskeleton reform and promotes invasion of *S. flexneri* (Lafont *et al.*, 2002). Entry of *Shigella* is reduced after addition of cholesterol depletion using methyl-beta-cyclodextrin, which also inhibits macrophage cell death (Lafont *et al.*, 2002). IcsB helps to escape intracellular *Shigella* from autophagy by binding to cholesterol. Mutation or deletion in the cholesterol binding domain greatly impairs autophage escape of *Shigella* (Kayath *et al.*, 2010). Thus, cholesterol dependent signalling is required for the survival of intracellular *S. flexneri*.

Shigella-infected macrophages start releasing large quantities of proinflammatory cytokines and chemokines (Jung *et al.*, 1995). The inflammatory cytokine interleukin 1 (IL-1) triggers an inflammatory reaction characterized by extravasation of polymorphonuclear (PMN) leucocytes. Bacteria-infected cells initiate strongly proinflammatory cells through sustained activation of nuclear factor-kappaB (NF-κB). A major consequence of this is IL-8 production, which attracts PMNs. The LPS of intracellular *Shigella* also induce rapid and prolonged activation of NF-κB and terminus kinase through Nod1 causing production of IL-8 by epithelial cells. This is another signalling process that may locally subvert the villous epithelium and facilitate invasion of the pathogen (Girardin *et al.*, 2001). PMN cells migrate through the epithelium into the lumen of the colon and cause a chain of events such as disrupt the permeability of epithelium, destabilizing the integrity of the epithelial barrier and promote invasion of shigellae. Further colonization of the epithelium aggravates inflammation, which in turn causes extensive tissue destruction (Zychlinsky *et al.*, 1994).

S. flexneri LPS mediate epithelial-derived signalling, which leads to directed migration of PMNs across model intestinal epithelium. This signalling most likely involves in the activation of the mitogen-activated protein kinase (Köhler *et al.*, 2002). The other secreted products, ΔospF (OspF) and ΔospC1 (OspC1) of *S. flexneri* are essential for transcriptional response of infected epithelial cell to promote PMN transepithelial migration (Zurawski *et al.*, 2006). Recently, additional effectors, OspZ and NleE were detected to have similar roles in the upstream induction of host signalling pathways required for PMN transepithelial migration (Zurawski *et al.*, 2008).

Host cell signal and inflammatory response of *Shigella*

Bacterial pathogens need various factors such as LPS-A, capsule, outer membrane protein, effectors and toxins for initiation of host innate immune system for their survival in the host cell. During their multiplication, cytoplasm-invaded *Shigella* release peptidoglycans (PGNs), LPS and nucleic acids. These products are recognized by Toll-like receptors (TLRs) and Nod-like receptors (NLRs), which in turn stimulate inflammatory signalling cascades and activate the host immune system (Ashida *et al.*, 2011). LPS and LPS-binding protein (LBP) initiate signalling in monocytes via membrane-bound CD14. TLR4–CD14 complexes act through MyD88-dependent or -independent pathways and PGNs are mediated by the NOD/CARD family of proteins, which induces inflammation by activating the NF-κB (Guha and Mackman, 2001). Excessive activation of monocytes leads to intestinal bowel diseases and Crohn's disease (Kim *et al.*, 2011). The family of NF-κB transcription factors essentially regulate immune-related gene expression, which normally exists in an inactive state and bound to its inhibitor IκB in the cytoplasm. Upon activation, IκB undergoes phosphorylation, ubiquitinylation and subsequent degradation by the proteasome that allows NF-κB to translocate to the nucleus and to regulate the transcription of genes involved in the inflammatory response (Li and Verma, 2002). The effector OspG negatively controls the host innate response by binding to various ubiquitin-conjugating enzymes and promotes ubiquitination of phosphorylated inhibitor of NF-κB type alpha (phospho-IkappaBalpha) (Kim *et al.*, 2005). The injected IpaH9.8 via the T3SS plays an important role in modulating inflammatory response to *Shigella* infection. IpaH9.8 binds to a splicing factor U2AF, resulting reduction of U2AF level and modulates host immune responses by decreased expression of proinflammatory cytokines such as IL-8 or IL-1β (Okuda *et al.*, 2005).

Autophagy recognition of *Shigella*

Shigella enters the host cytosol to escape from the lytic compartment and gain a replicative niche. However, autophagy limits the growth of this pathogen in host cells and this mechanism protects the cytosol of eukaryotic cells from bacterial colonization. Autophagy is crucial for cell survival against invading bacteria, which targets unwanted cellular components or organelles, entrap them and finally get degraded via lysosomal fusion process. Effector IcsB helps in the protection of *Shigella* from this autophagy process. Intracellular actin-based motility protein VirG of *Shigella* induces autophagy by binding to the autophagy related protein, Atg5. However, IcsB competitively bind to VirG in place of Atg5, indicating some similarity with Atg5 and act as anti-Atg5-binding protein (Ogawa *et al.*, 2005, 2008). Phagocytosed *Shigella* breaks the vacuolar membrane for cytosolic access. In this process, proteins of vacuolar membranes remnants undergo polyubiquitination that trigger the autophagy activation by recruiting the autophagy markers LC3 and adaptor P62 (Dupont *et al.*, 2009).

Shigella-induced macrophage cell death

Shigella-induced macrophage cell death is an important hallmark in the pathogenesis of shigellosis. Once *Shigella* has been endocytosed by the M cells to submucosa, the indigenous macrophages and dendritic cells come into contact with the pathogen. An *in vitro* study has shown that *S. flexneri*-infected macrophages rapidly lose cell viability and lyse via a process that resembles apoptosis (Zychlinsky *et al.*, 1992). Macrophage death is mediated by Mxi-Spa-secreted effector protein IpaB. Deletion mutant of IpaB of *S. flexneri* was not cytotoxic but the purified IpaB infected with *Shigella* induced apoptosis. Receptor for this has been identified as interleukin-1β converting enzyme (ICE; also known as caspase-1) and this enzyme gets activated during *S. flexneri* infection (Chen *et al.*, 1996). Pretreatment of macrophages with specific caspase-1 inhibitors prevents apoptosis induced by *Shigella*, and macrophages isolated from caspase-1-deficient mice are resistant to *Shigella*-induced apoptosis (Hilbi *et al.*, 1998). These findings further confirm that delivery of IpaB to the macrophage cytosol stimulates the activation of caspase-1. Immunofluorescence and immunoelectron microscopy detected that IpaB gets disseminated throughout the cytoplasm of *Shigella*-infected macrophages and soon after secretion of IpaB, it binds to caspase-1 to initiate apoptosis and to promote the cleavage of IL-1β and IL-18 into biological active forms (Thirumalai *et al.*, 1997). The mechanism of cell death in the *Shigella*-induced macrophage killing was identified with specific morphological changes and fragmentation of nuclear DNA in their nuclear morphology (Guichon and Zychlinsky, 1997; Navarr and Zychlinsky, 2000).

Shigella-induced macrophage cells follow unique caspase-1-dependent apoptotic pathway, but not depend on other caspases (caspase-11/-3/-7) and other common apoptosis-regulating factor p53 (Navarr and Zychlinsky, 2000). Owing to non-involvement of caspases in the *Shigella*-induced cell death complicates the fact of apoptosis (Nonaka *et al.*, 2003). Macrophages infected *Shigella* can also induce cell death independent of caspase-1 or IpaB activity. It is known that major bacterial component of lipid A induces Toll-like receptor 4 (TLR-4) signalling pathways in apoptosis, and with the help of this it was shown that cell death induced by cytosolic bacteria was TLR-4 independent (Suzuki *et al.*, 2005). In the *Shigella* infected macrophages, Ipaf belong to NLR receptor family (also known as CLR) involved in activation of caspase-1 in response to pro-inflammation. Both Ipaf and caspase-1 were essential for pyroptosis, a specialized form of caspase-1-dependent cell death induced in macrophages by bacterial infection (Suzuki and Núñez, 2008). Cryopyrin (encodes CIAS1 gene) and ASC (Apoptotic Speck protein containing a CARD) are components of the inflammasome. These multiprotein complexes activate caspase-1 and then cleave procytokine IL-1β into active form.

The concept of *Shigella* induced cell death is very complex because of the involvement of multiple cell death pathways. Overall, the cell death is very important for early infection and befits *Shigella* out of apoptosis. Dying macrophages release the intact bacteria, which gains direct access to basolateral membrane of enterocysts and induces proinflammatory cytokines and chemokines, which promote spread of bacteria in the host tissue (Sansonetti, 2000).

Evolution of shigellae and their virulence

Although *Shigella* is closely related to *E. coli* this genus has traditionally been considered to cause bacillary dysentery. DNA hybridization study showed that *Shigella* exhibit >75% nucleotide similarity to different *E. coli* strains (Brenner *et al.*, 1972). Multilocus enzyme electrophoresis and comparative DNA sequence analysis of diverse strains of *Shigella* with enterohaemorrhagic *E. coli* (EHEC), enteroinvasive *E. coli* (EIEC), enteropathogenic *E. coli* (EPEC) and enterotoxigenic *E. coli* (ETEC) show its precise position in the phylogenetic tree (Pupo *et al.*, 1997). *Shigella*-like strains of *E. coli* that cause an invasive, dysenteric diarrhoeal illness were first described in 1971 (Levine, 1987). The biochemical properties and disease symptoms of EIEC are much similar to *Shigella* spp. However, EIEC is kept out of the genus *Shigella*, because some of them are motile and cause lysine decarboxylation. EIEC can be differentiated from *Shigella* only by a very limited number of tests, which include utilization of L-serine, D-xylose and/or sodium acetate, and mucate fermentation (Lan and Reeves, 2002). *Shigella* spp. and EIEC evolved from different *E. coli*, by horizontal acquisition of virulence genes, forming a single pathotype (Pupo *et al.*, 2000; Peng *et al*, 2006).

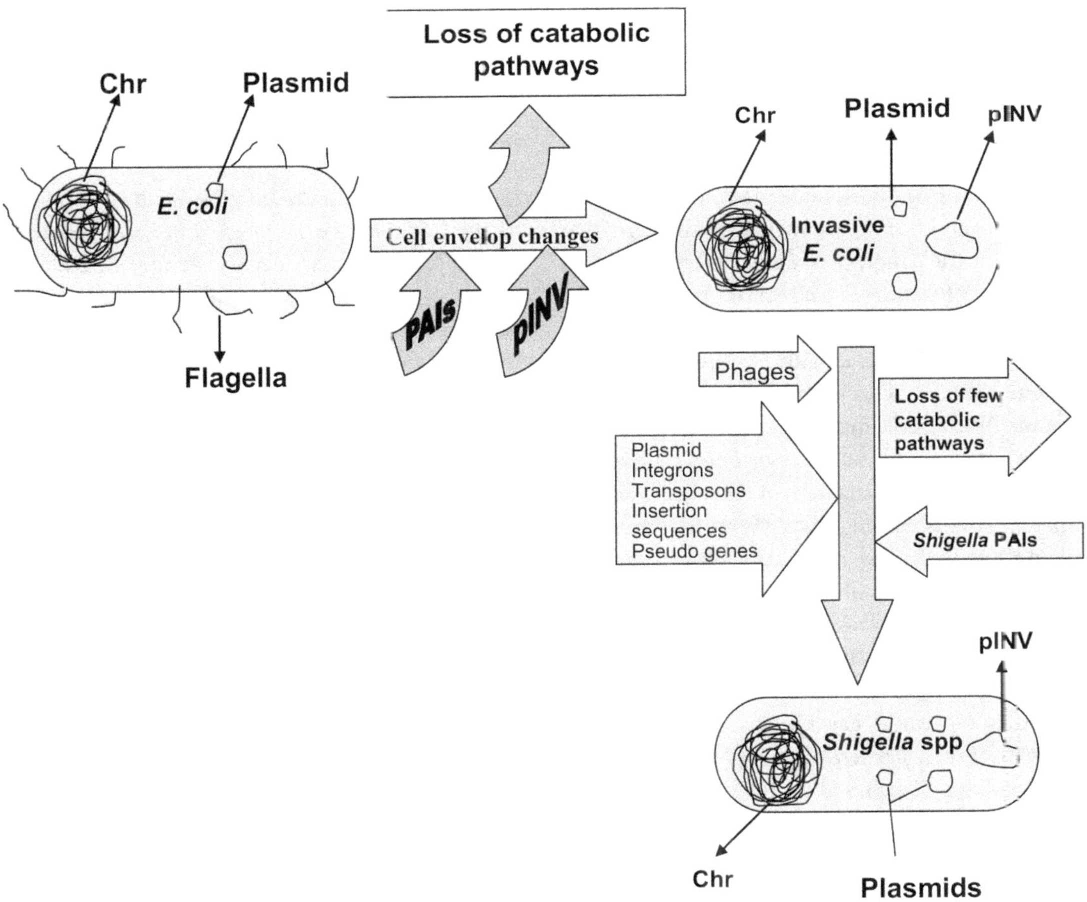

Figure 4.2 Multigenetic events are responsible for evolution of *Shigella* species from non-pathogenic *E. coli*. Abbreviations: Chr, chromosome; pINV, invasive plasmid; PAIs, pathogenicity islands.

Genetic events are responsible for the evolution of *Shigella* spp. from non-pathogenic *E. coli* as shown in Fig. 4.2. Molecular detection of *ipaH* is an alternative method for identifying non-agglutinable strains that are biochemically identical with *Shigella* spp. (Echeita Sarrionandia *et al.*, 2011).

EIEC and *Shigella* exhibit considerable antigenic cross-reactivity (Cheasty and Rowe, 1983). Some of the EIEC associated O antigens are identical to O antigens of *S. boydii* serotype 15/*S. dysenteriae* serotype 2, *S. boydii* serotype 3 and *S. dysenteriae* type 12. EIEC strains that shared O antigens with *Shigella* showed a higher metabolic activity than the *Shigella* strains (Lehmacher and Bockemühl, 2007). Acquisition of 220 kb pINV plasmid is a crucial step in the development of invasive forms of *E. coli* and *Shigella* spp. Plasmid pINV is frequently transferred into other *E. coli* by conjugation. However, in *S. flexneri* strains, pINV is unable to initiate the conjugation process (Sansonetti *et al.*, 1982). Furthermore, pINV was not transferred between the different *Shigella* and EIEC strains (Escobar-Paramo *et al.*, 2003).

Pylogenetic analysis using chromosomal housekeeping and virulence plasmid (VP) genes of *Shigella* and EIEC have suggested an alternative theory for the origin of *Shigella*. Entry of diverse ancestral VPs into *E. coli* in multiple transfer events might also contribute to the origin of *Shigella* spp. (Yang *et al.*, 2007). The transfer and chromosomal incorporation of pathogenicity islands (PAIs) are important steps in the evolution of pathogenic bacteria from non-pathogenic progenitors. The localization of PAIs varies among *Shigella* strains causing disparity in the virulence phenotypes (Yang *et al.*, 2005). As of now, five PAIs (SHI-0, SHI-1, SHI-2, SRL and SHI-3) have been identified in *Shigella* species (Rajakumar *et al.*, 1997a; Vokes *et al.*, 1999; Luck *et al.*, 2001; Purdy and Payne 2001; Ingersoll *et al.*, 2002). The pINV plasmid has specific names such as pWR100 for *S. flexneri* 5, and pMYSH6000 for *S. flexneri* type 2a. Based on the sequence, pINV plasmids are referred as pINVA and pINVB, and are detected in *Shigella* as well as in EIEC (Lan *et al.*, 2004). However, differences in pINV may play an important role in virulence. In another comparative study, pINVA of *S. flexneri* type 6 and pINVB of *S. flexneri* type 2a showed that majority of the virulence-associated genes are under strong selection pressure (Peng, 2009). Considerable variations in the genes encoding virulence in the plasmid pINV were observed including *sepA*, encoding for serine protease and *senA*, encoding an enterotoxin. These genes were mostly found in the *Shigella* rather then in EIEC (Lan *et al.*, 2004).

The A+T content of plasmid mediated virulence genes of *Shigella*, is higher (70–73%) than the chromosomal genes (50%) (Hale, 1991). These acquired genes also enhance the ability of *Shigella* species to adapt to the different host environments during the progress of infection (Peng *et al.*, 2006). In Southern hybridization analysis, virulence plasmids of *Shigella* and EIEC had different restriction endonuclease cleavage patterns (Sansonetti *et al.*, 1983). During the course of evolution, *Shigella* spp. not only acquired plasmid associated virulence genes but also lost some genes spontaneously and formed 'black holes'. This deletion of genes has been suggested to have facilitated the evolution of *Shigella* into its pathogenic mode of life (Maurelli *et al.*,1998; Peng *et al.*, 2006).

Genome sequence of *Shigella* has suggested that it might have originated from *E. coli* involving multiple evolutionary events (Peng *et al.*, 2006). *S. flexneri* genome has many pseudogenes, insertion sequences, numerous laterally acquired *S. flexneri*-specific sequences, as well as insertion sequence (IS)-mediated deletions, translocations and inversions (Wei *et al.*, 2003). Owing to these changes, *S. flexneri* strains are diverse in nature even within the same serogroup. The presence of large numbers of IS elements in the *Shigella* genome is likely to be the major cause of many of the genomic rearrangements. Large proportions of *S. flexneri* chromosomal genes are closely related to both non-pathogenic and enterohaemorrhagic *E. coli* (EHEC) strains (Yang *et al.*, 2005). A comparative study of the *S. flexneri* serotype 2a and *E. coli* K-12 genomes showed a large deletion of *cadA* region (encoding for lysine decarboxylation) in *Shigella*. When the *cadA* gene was experimentally introduced into *S. flexneri* serotype 2a, the strain became avirulent and its enterotoxin activity was inhibited (Maurelli *et al.*, 1998). This study has explained how the loss of certain metabolism associated genes have contributed to virulence of *Shigella* spp. Sequence analysis of four chromosomal genes (*trpA*, *trpB*, *pabB*, and *putP*) and three virulence plasmid genes (*ipaB*, *ipaD*, and *icsA*) of *Shigella* and EIEC showed that both the chromosomal and plasmid genes were similar. However, phylogenetic tree derived from chromosomal genes showed a distinctive phylogeny and fast diversification of *Shigella* and EIEC groups. (Escobar-Páramo *et al.*, 2003). It seems that EIEC is in an intermediate stage and are a potential precursor of 'full-blown' *Shigella* strains (Peng *et al.*, 2009).

EIEC and *Shigella* survive and escape from macrophages in a similar way. Owing to the differences in the expression of virulence genes, EIEC was less efficient in macrophage killing and stays longer in the infected cells (Bando *et al.*, 2010). In addition, compared with *Shigella*, the EIEC induces milder proinflammatory response, cell-to-cell dissemination and plaque formation during intercellular spreading (Moreno *et al.*, 2009). The other characteristics such as requirement of higher inoculum to cause infection, inability to produce Shiga toxins and less sever clinical complications differentiate EIEC from *Shigella* and perhaps owing to these reasons, EIEC is not associated with any large outbreaks (DuPont *et al.*, 1971; Hsia *et al.*, 1993; Johnson, 2000).

Epidemiology of *Shigella* species

Foodstuff are mostly contaminated owing to poor personal hygiene practices of *Shigella* infected food handlers, who were found to be responsible for many well studied outbreaks. Of 816 reports with 80,682 cases between 1927 and 2006, 33 outbreaks were caused by *Shigella* mostly in US, Canada, Europe and Australia (Greig *et al.*, 2007). During 2002, Centers for Disease Control and Prevention's emerging infections programme, Foodborne Diseases Active Surveillance Network (FoodNet) classified *Shigella* as the third most reported foodborne bacterial pathogen (Warren *et al.*, 2006). In addition, two recently published FoodNet reports showed that the incidence and trend of *Shigella* and other common pathogens remained as a leading cause of foodborne infections (Nyachuba, 2010). *Shigella* is acid resistant, salt tolerant, and can survive at infective levels in many types of foods including prepared and vacuum packed foods. At refrigerated temperatures, subsistence of *Shigella* is often enhanced. The other factors that influence the incidence of shigellosis are personal hygiene, nutritional status, safe disposal of faecal matters and storage/handling of food products. Incidence of *Shigella* spp. varies in different parts of the world and seasonality of infection also differs from one county to another. The major outbreaks caused by *Shigella* spp. are summarized in Table 4.1.

Seasonal variation in incidence of shigellosis

In some countries, *Shigella* infections occur throughout the year with peak in June and July (Srison and Pornpatkul, 1995). In the US and Europe, *Shigella* infections are most common during summer and early autumn. A study from Saudi Arabia shows that most of the *Shigella* strains have been isolated during April–May (Kagalwalla *et al.*, 1992) and also in the winter (Al-Eissa *et al.*, 1992). In Burkina Faso, *Shigella* species were detected during the cold dry period from December to February (Nitiema *et al.*, 2011). In Bangladesh, increase in the incidence of *Shigella* was reported during the post-monsoon season (August–November) and dry hot and dry pre-monsoon months (April and May). In Kolkata, India, *Shigella* infection was prevalent throughout year with high isolation rate during the summer and early monsoon months (Dutta *et al.*, 1992). In Laos and Guatemala, *Shigella* species were detected mainly during rainy seasons (June and July) (Phetsouvanh *et al.*, 1999). In Hungary, July to September was the peak period for the isolation *S. sonnei* and *S. flexneri* (Rudnai *et al.*,1981). Shigellosis occurs mainly during summer in Israel, but occasional winter outbreaks have also been reported (Ostroi *et al.*, 1996). In Malaysia, shigellosis occurred throughout the year, but the increased number of cases occur during May through August and also during the rainy season (November and December) (Banga Singh *et al.*, 2011). No clear explanation is still available for seasonal variations in different geographical regions. The seasonal peaks of dysentery may be due to complex social behaviours, climatic factors and the biological features of the organisms.

Table 4.1 Major outbreaks of shigellosis during 1954–2010

Year and country	Source	Outbreak information/serogroup	Reference
1954, USA	Person to person, faecal contamination, carriers in foodborne infections	19 outbreaks involving 1471 cases. *S. sonnei* and *S. flexneri*	Keller and Robbins (1956)
1958, UK	Imported infection	10 cases involving children and adult. *S. boydii* type 4	Browning and Boot (1960)
1964, USA	Person-to-person	28 cases caused by *S. flexneri*	Elsea *et al.* (1967)
1966, UK	Person-to-person	*S. sonnei*	McEvedy *et al.* (1966)
1967–68 USA	Food and water, potato salad and turkey	7 outbreaks and 51 cases. *S. flexneri*	Wheatley and Veer (1969)
1966, UK	Food handlers	200 cases. *S. sonnei*	Green *et al.* (1968)
1972, USA	Water	208 cases. *S. sonnei*	Baine *et al.* (1975)
1972, Mexico	Unknown	15 of 22 children in a TB ward, 5 died owing to *S. dysenteriae* type 1inection	Olarte *et al.* (1976)
1973, Bangladesh	Unknown	434 cases and 28 deaths. *S. dysenteriae* type 1	Rahaman *et al.* (1975)
1973, Caribbean cruise	Water and ice	585 passengers and 104 crew members. *S. flexneri* type 6 and *S. boydii* type 8	Merson *et al.* (1975)
1974, Canada	Contaminated lake water	62 persons out of 91 had shigellosis	White and Pedersen (1976)
1974, USA	Tap water	1200 cases, of whom 10 had culture-proven shigellosis	Weissman *et al.* (1976)
1977, USA	Person-to-person and contaminated water	11 deaths owing to diarrhoeal illness caused by *S. flexneri* 1b	Storch *et al.* (1980)
1976, Haiti	Unrefrigerated seafood	386 cases due to shigellosis and other diarrhoea	Berkelman *et al.* (1983)
1979, Zaire, Rwanda and Burundi	Not listed	Large outbreak due to *S. dysenteriae* type 1	Ebright *et al.* (1984)
1983, India	Person-to-person and contaminated water	Large outbreak due to *S. dysenteriae* type 1	Mathan *et al.* (1984)
1983, Israel	Vegetables	Community outbreak due to *S. sonnei*, *S. flexneri* and *S. boydii*	Porter *et al.* (1984)
1983, Rwanda	Not listed	524 cases. *S. dysenteriae* type 1	Vimont-Vicary and Rogerie (1985)
1983, USA	Tossed salads	Large outbreak due to *S. sonnei*	Martin *et al.* (1986)
1981–82, Rwanda	Not listed	11,801 cases of dysentery due to *S. dysenteriae* type 1	Huppertz (1986)
1984, India	Not listed	91 cases with attack rate of 9.7% and case fatality rate of 2.2. *S. dysentery* type 1	Sengupta *et al.* (1990)
1984, USA	Food handlers	1135 cases due to *S. flexneri* 4a	Spika *et al.* (1987)
1986–87 Thailand	Not listed	10,000 cases with case fatality rate of 0.9%. *S. dysenteriae* type 1	Taylor *et al.* (1989)
1986, USA	Shredded lettuce and food handlers	347 cases. *S. sonnei*	Davis *et al.* (1988)
1987, USA	Foodborne and person-to-person	342 cases with an attack rate range of 22–57%. *S. sonnei*	Coles *et al.* (1989)
1987, USA	Foodborne and person-to-person	Large outbreak among 12,700 attendees. *S. sonnei*	Wharton *et al.* (1990)

Year and country	Source	Outbreak information/serogroup	Reference
1986–88, USA	Water supply	50 outbreaks affecting 25,846 persons. *S. sonnei* was one of the aetiological agents	Levine *et al.* (1990)
1988, India	Not listed	62 cases *S. dysenteriae* type 1 and *S. flexneri*	Datta and Sen, (1990)
1988, USA	Uncooked tofu salad, food handlers	3175 cases *S. sonnei*	Lee *et al.* (1991)
1988, USA	Cold sandwich served on the flight	65 cases. *S. sonnei*	Hedberg *et al.* (1992)
1989, USA	Raw oysters	24 cases. *S. sonnei*	Reeve *et al.* (1989)
1989, Caribbean cruise	German potato salad, food handlers	818 passengers and 518 crew members. *S. flexneri 4a*	Lew *et al.* (1991)
1990, Burundi	Not listed	182 cases. *S. dysenteriae* type 1	Ries *et al.* (1994)
1990, Israel	Drinking water	8000 cases of *S. sonnei*	Egoz *et al.* (1991)
1991, Israel	Contaminated well water	1216 cases. *S. sonnei*	Simchen *et al.* (1991)
1991, Guatemala	Not listed	100 cases. *S. dysenteriae* type 1	Centers for Disease Control (1991)
1991, Spain	Water	26 cases. *S. sonnei*	Morera *et al.* (1995)
1991, USA	Moose soup	23 cases. *S. sonnei*	Gessner and Beller (1994)
1992, Japan	Familial, restaurant, and ark shell consumption	10 outbreaks, one caused by *S. flexneri* 3a, the remaining nine by *S. sonnei*	Yamada *et al.* (1993)
1992, USA	Tossed salad	46 cases. *S. flexneri*	Dunn *et al.* (1995)
1992, Thailand	Coconut milk dessert	33 cases. *S. dysenteriae* type 1	Hoge *et al.* (1995)
1992–93, U.S. troops in Somalia	Not listed	Of the 113 diarrhoeal patients 33% were infected with *Shigella* spp. (mainly *S. sonnei*)	Sharp *et al.* (1995)
1993, Taiwan	Contaminated water	Of the 207 dysentery cases *S. sonnei* was detected in 67 cases	Lai *et al.* (2000)
1993, Greece	Contaminated well water	138 cases. *S. sonnei*	Samonis *et al.* (1994)
1993, Mozambique	Not listed	47,483 cases and 199 deaths. *S. dysenteriae* type 1	Aragón *et al.* (1995)
1992–94 England and Wales	Foodborne	46 outbreaks were due to *S. sonnei*	Djuretic *et al* (1996)
1993–94, Zimbabwe	Not listed	Large dysentery outbreak. *S. dysenteriae* type 1	Mudzamiri *et al.* (1996)
1993–95, Rwanda, Tanzania and Democratic Republic of the Congo	Not listed	181,921 cases of bloody diarrhoea *S. dysenteriae* type	Kernéis *et al.* (2009)
1994, England and Wales	Iceberg lettuce	100 cases. *S. sonnei*	Frost *et al.* (1995)
1994, Norway	Iceberg lettuce	110 cases. *S. sonnei*	Kapperud *et al.* (1995)
1994, Zimbabwe	Borehole water	38 cases. *S. sonnei* and *S. boydii*	Tshimanga *et al.* (1997)
1995, Zaire	Restaurant food	77 military service men. *Shigella* spp.	Bruins *et al.* (1995)

Table 4.1 Continued

Year and country	Source	Outbreak information/serogroup	Reference
1995, Spain	Unchlorinated water	48 cases. *S. boydii* type 4	Jiménez t al. (1996)
1995, S. Arica	Not listed	158 cases. *S. dysenteriae* type 1	Chopra *et al.* (1997)
1995–96, Spain	Fresh pasteurized milk cheese	More than 200 cases. *S. sonnei*	Garcia-Fulgueiras *et al.* (2001)
1996, Kenya	Not listed	Large dysentery outbreak. *S. dysenteriae* type 1	Oundo *et al.* (1996)
1996, China	Not listed	40 cases. *S. flexneri* 2a	Chiou *et al.* (2001)
1997, Zimbabwe	Contamination of water	52 cases. *S. dysenteriae* type 1	Midzi *et al.* (2000)
1999, Saudi Arabia	Person to person	233 cases. *S. dysenteriae*	el Bushra and Bin Saeed (1999)
1999–2000, Chile	Not listed	450 outbreaks affecting 3022 individuals, in which *Shigella* spp. was next to *Salmonella* spp., *Staphylococcus aureus*	Prado *et al.* (2002)
1999–2000, Sierra Leone	Not listed	4218 cases. *S. dysenteriae* type 1	Guerin *et al.* (2003)
1999–2000, Poland	Foodborne	174 cases. *S. flexneri* and *S. sonnei*	Stypułkowska-Misiurewicz and Gonera (2003)
2000, Greece	Water samples	288 cases. *S. sonnei*	Alamanos *et al.* (2000)
2000, USA	Commercially prepared dip	406 cases. *S. sonnei*	Kimura *et al.* (2004)
2001, Japan	Imported oysters	Multiprefectural outbreaks. *S. sonnei*	Terajima *et al.* (2004)
1988–2002, Israel	Not listed	865 diarrhoeal outbreaks due to *Shigella*, *Salmonella* and *Staphylococcus* species	Schwaber *et al.* (2005)
2002, Spain	Contaminated water	181 cases *S. sonnei*	Arias *et al.* (2006)
2001–02, Taiwan	Not listed	122 children, *S. sonnei*	Chuang *et al.* (2006)
2002–03, India	Not listed	15 cases. *S. dysenteriae* type 1	Pazhani *et al.* (2004)
2002–03, Poland	Not listed	220 cases in 2002 and 75 cases 2003. *S. sonnei*, *S. flexneri*, and *S. dysenteriae* type 2	Stypułkowska-Misiurewicz (2005)
2003, Peru	Multisource including contaminated foods	172 cases. *Shigella* was one of the aetiological agents	Jones *et al.* (2006)
2004, airline passengers	Raw carrot	300–1500 passengers. *S. sonnei*	Gaynor *et al.* (2009)
2003–05, Poland	Not listed	154 cases *S. flexneri* and *S. sonnei*	Stypułkowska-Misiurewicz (2007)
2005, USA	Tomatoes	Multiple-restaurant outbreak. *S. flexneri* 2a	Reller *et al.* (2006)
2004–06, Poland	Mostly food handlers	189 cases *S. flexneri*, *S. boydii* and *S. sonnei*	Stypułkowska-Misiurewicz (2008)
2005, Spain	Potatoes with a sauce	146 cases. *S. sonnei*	Castell Monsalve *et al.* (2008)

Year and country	Source	Outbreak information/serogroup	Reference
2007, Austria	Foodborne	Four outbreaks *S. flexneri* and *S. sonnei*	Much *et al.* (2009)
2007, Iran	Raw vegetables	Prison outbreak with 701. *Shigella flexneri* serotype 3a	Ranjbar *et al.* (2010)
2007, India	Contaminated tap water	416 cases. *S. flexneri* 2a and 3a	Saha *et al.* (2009)
2007–08, Poland	Not listed	97 cases. *S. sonnei*, *S. boydii*	Stypułkowska-Misiurewicz and Baumann (2010)
2008, Denmark and Australia	Imported baby maize	215 cases Denmark and 12 in Australia. *S. sonnei*	Lewis *et al.* (2009)
2008, Austria	Salad and food handler	31 cases, *S. sonnei*	Kuo *et al.* (2009)
2009, Sweden	Sugar snaps from Kenya	35 confirmed cases. *S. dysenteriae* type 2	Löfdahl *et al.* (2009)
2009, Denmark	Imported fresh sugar peas	10 cases, *S. sonnei*	Muller *et al.* (2009)
2009, Belgium	Canteen-food	52 cases of shigellosis	Gutiérrez Garitano *et al.* (2011)
2009, Norway	Imported fresh sugar peas	Nine cases of *S. sonnei*	Heier *et al.* (2009)
2010, USA	Ice consumption	41 cases in dinner cruise. *S. sonnei*	Serdarevic *et al.* (2011)
2010, Spain	Drinking water	105 cases, *S. sonnei*	Godoy *et al.* (2011)

Early studies conducted in many *Shigella* endemic regions, identified *S. dysenteriae* as the predominant species which was replaced by *S. flexneri*. Thereafter, *S. flexneri* gradually became less prevalent in developed countries (Kostrzewski and Stypulkowska 1968; Green *et al.*, 1991). Cyclic pattern in the incidence of different species of *Shigella* suggest that it takes several years for herd immunity to reach a critical level in a given population before one species of *Shigella* replaces another (Dupont, 2000).

Risk groups

Shigellosis affects certain population more frequently than others. Generally, incidence of shigellosis is highest among children with the age group of 1–4 years. However, during *S. dysenteriae* type 1 epidemic, all the age groups have been found to be affected. A study conducted in the United States during 1989 through 2002 showed that children aged < 5 years had highest proportion of shigellosis, followed by those aged 5–9 years (Gupta *et al.*, 2004). Bacillary dysentery is primarily seen in children 6 months to 10 years of age and adults often acquire the illness from their children. In infants less than 6 months, shigellosis is not a common disease because most of them are breastfed and the higher concentration of IgA in human milk protects the children from diarrhoeal infections (Mata *et al.*, 1969; Hayani *et al.*, 1992). In industrialized countries, *Shigella* strains rarely cause severe illness in newborns (Haltalin, 1967). Children in day-care centres, patients in custodial institutions, homosexual men, international travellers and their household contacts were often identified as high-risk groups in industrialized countries. In developing countries, high rates of *Shigella*-mediated morbidity and mortality occur among displaced populations, HIV-associated immunodeficiency patients, malnourished infants and young children. Incidences of shigellosis among children are attributed to several factors. Young children generally do not follow good personal hygiene and may not have the acquired immunity against *Shigella* infection. Migrant workers in several developing countries and sudden mass displacement of people owing to war or disaster are also high-risk groups for shigellosis. In such settings, insufficient supplies of clean water, poor sanitation, and overcrowding and concomitant malnutrition are the prime factors for the infection. Epidemics of dysentery with high rates of morbidity and mortality among all age groups were reported among Bhutanese, Somalis, Burundians and Rwandans from 1991 to 1994 when their countries were in political turmoil (Kotloff *et al.*, 1999).

Worldwide incidence of shigellosis

Africa

A cohort study of children monitored from birth up to 2 years of age in Guinea-Bissau estimated that *Shigella* or EIEC incidence was 0.22 infection/child-year at risk (Valentiner-Branth *et al.*, 2003). In south-eastern Sierra Leone of Kenema district, the dysentery attack rate was 7.5% with higher rate of infection among children > 5 years (Guerin *et al.*, 2004). Between 2000 and 2002, a significant increase in *S. flexneri*, from 21.3% to 54.3%, and decrease in *S. dysenteriae*, from 21.3% to 1.1%, were recorded in Dakar, Senegal (Dromigny *et al.*, 2004). During the same period, the incidence of shigellosis due to *S. boydii* and *S. sonnei* remained stable. Stool samples analysed from randomly selected children from three primary schools in Calabar in Nigeria, indicated higher rates of shigellosis with *S. flexneri* and *S. sonnei* (12.0% and 8.8%, respectively) (Udo and Eja, 2004). Studies made in Ghana over a 50-year period indicated the predominance of *S. flexneri* with 4% change in the serodiversity (Opintan and Newman, 2007). HIV-1-related morbidity in adults was due to severe bacterial infections in Côte d'Ivoire and shigellosis accounted for 7% along with other bacterial infections (Attia *et al.*, 2001). In Burkina Faso, the incidence of *Shigella* spp. was 5.8% in all age groups (Nitiema *et al.*, 2011).

A retrospective cross-sectional study conducted in Gaborone, Botswana, from February 2003 through July 2008 showed that *Shigella* was the most common pathogen isolated among bacterial infection, which accounted for 4% (Rowe *et al.*, 2010). A survey conducted among diarrhoeal patients attending major public hospitals and children attending two public primary schools in Vhembe district of South Africa indicated that 8.5% patients were positive for shigellae (Samie *et al.*, 2009). In Egypt, *S. flexneri* was the dominant species in all age groups followed by *S. dysenteriae* and *S. sonnei* (El-Gendy *et al.*, 2011). Among different etiological agents of acute diarrhoea in Tunis, Tunisia, shigellae contributed 4%, mostly in adults (Al-Gallas *et al.*, 2007).

In Somalia, an average of 10–15% of the annual incidences of *Shigella* was recorded during 1983 through 1990 (Casalino *et al.*,1994). A clinic-based surveillance conducted in a rural community of Kenya from 1997 to 2003 estimated that *Shigella* spp. was the most frequently isolated bacteria with an incidence rate of 16% involving *S. flexneri* and *S. dysenteriae* (Brooks *et al.*, 2006). The aetiology of diarrhoea in children > 5 years of age group in Ifakara, Tanzania, during 1996 to 1997 revealed a very high incidence of *Shigella* spp. (24.14%) (Vargas *et al.*, 2004). Another cross-sectional study conducted among children ≤ 5 years of age in Dar es Salaam, Tanzania, between December 2005 and February 2006 indicated the predominance of *S. flexneri* (5.7%) than the other serogroups (Moyo *et al.*, 2011). In Compexe Pédiatrique of Bangui, Central Africa, the incidence of *Shigella* spp. was 9.6% during March–August 2007, which was high compared with a previous study conducted in in Bangui in 1983 (2.2%) (Manirakiza *et al.*, 2010).

Asia

A hospital based surveillance conducted in Dhaka, Bangladesh, showed that *Shigella* prevalence decreased steadily from 8–12% in the 1980s to 3% in 2008 (Khatun *et al.*, 2011). The same study showed that *S. flexneri* was mostly detected among endemic cases and the epidemic *S. dysenteriae* type 1 was detected with two peaks in 1984 and 1993 (Khatun *et al.*, 2011). A case–control study conducted among Cambodian children during 2004 through 2006 in Phnom Penh indicated that *Shigella* infection was high in cases (5.2%) than controls (1.7%) (Meng *et al.*, 2011). A cross-sectional study conducted at Henan Province in China during 2006 revealed that the overall prevalence of shigellosis was 13.2% and *S. flexneri* and *S. sonnei* were the more frequently isolated serogroups (Xia *et al.*, 2011). Multicentre population based study conducted in six Asian countries namely China, Vietnam, Thailand, Bangladesh, Pakistan and Indonesia indicated that the annual incidence of shigellosis was 13.2 per 1000 children > 5 years and 2.1 per 1000 in all age groups (von Seidlein *et al.*, 2006). The serogroup distribution analysis in the same study showed that *S. flexneri* was most frequently isolated in Bangladesh, China, Pakistan, Indonesia, and Vietnam, whereas in Thailand, *S. sonnei* was the most common serotype. The isolation rate of *Shigella* spp. among hospitalized children was almost 10% in Malaysia, which was the third most common bacterial agent responsible for childhood diarrhoea (Banga Singh *et al.*, 2011). In contrast to many Asian countries, *S. sonnei* was the predominant species in Malaysia followed by *S. flexneri* and *S. boydii* (Banga Singh *et al.*, 2011). During 2002–2004, prevalence of *Shigella* in Nepal was found to be 4.6% and its occurrence was detected throughout the year (Kansakar *et al.*, 2011).

In many parts of India, the distribution of *Shigella* spp. was mostly heterogeneous with predominance of *S. flexneri* in some regions. Increase in isolation rates of *S. sonnei* has been reported in Vellore (South India), Chandigarh (North India) and Kolkata (Eastern India) (Pazhani *et al.*, 2005). After the dysentery outbreaks in eastern India in the early 1980s, *S. dysenteriae* type 1 strains were replaced by various serotypes of *S. flexneri* (Pazhani *et al.*, 2005). After hiatus of 14 years, *S. dysenteriae* type 1 re-emerged in 2002 causing epidemics in several parts of India and after this year, isolation of this serotype has abruptly waned (Pazhani *et al.*, 2004). During 2001–2009, prevalence of *Shigella* spp. among hospitalized diarrhoeal cases in Kolkata was between 6% and 8% (Pazhani *et al.*, 2005; Ghosh *et al.*, 2011).

Middle East

In Iran, the incidence of *Shigellosis* during 2001–2002 was 16.8% (MoezArdalan *et al.*, 2003). In a study conducted in Tehran from 2001 to 2006 showed that the detection rate of *Shigella* spp. was about 4.5% with *S. flexneri* and *S. sonnei* as predominant serogroups. (Pourakbari *et al.*, 2010). In Kuwait, the incidence of *Shigella* spp. was 18% in 1990 with an upsurge in 1991 (24%) and 1996 (27%) (Jamal *et al.*, 1998). *Shigella* spp. was identified among 20% of the cases in Yemen, of which *S. flexneri* and *S. dysenteriae* were most frequently identified (Banajeh *et al.*, 2001). In Jordan, the incidence of shigellosis remained low (0.13%) during 1994–2000 (Battikhi, 2004). Among 20% of stool specimens examined in Israel, *Shigella* was identified in 7.1% of the diarrhoeal cases (Ziv *et al.*, 2011).

European Union

Compared with other regions, shigellosis is less frequent in Europe. A community-based study conducted in Denmark reported only 2% of *Shigella* incidence among children with diarrhoea (Molbak *et al.*, 1994). Studies conducted in northern Greece during 1971–96 showed that shigellosis was stable until 1990 and decreased during 1995–99 (Maraki *et al.*, 2003). Surveillance conducted by National Reference Centre for *Salmonella* and *Shigella* (NRCSS) in Belgium from 1990 assessed about 400 cases of *Shigella* infections per year. Shigellosis was the third leading bacterial gastrointestinal disease (68%) caused mainly by *S. sonnei* (Vrints *et al.*, 2009). In Sofia, Bulgaria, an average incidence of *Shigella* was 4.9% during 1987–2008 and this incidence rate was almost the same in developing countries (Ivanova *et al.*, 2010).

American continent

Laboratory surveillance for enteric pathogens in Canada during 2006 has shown that *Shigella* is a common cause of diarrhoeal illness in North America with an infection rate of 2.0 per 100,000 (Drews *et al.*, 2010). In Alberta, Canada, *S. sonnei* was frequently isolated followed by *S. flexneri*, *S. boydii* and *S. dysenteriae* during 2002–2007 (Drews *et al.*, 2010). *S. sonnei* and *S. flexneri* infected patients had a history of travel to the developing countries (Drews *et al.*, 2010). In the United States, 450,000 cases of shigellosis occur annually and majority of them are related to several outbreaks associated with food, water, child day-care centre, homosexual men and travel (Drews *et al.*, 2010). The US Foodborne disease surveillance programme reported that the *Shigella* incidence was 3.9/100,000, with highest frequencies among children aged 1–4 years (Haley *et al.*, 2010).

A community-based surveillance conducted among children >6 years old in Peruvian Amazon indicated that the incidence of diarrhoeal diseases was 4.4 episodes/child/year during 2002–2006 (Kosek *et al.*, 2008). A case–control study conducted in north-eastern Brazil to assess the epidemiological and clinical characteristics and changing nutritional status of infants suffering from acute diarrhoea showed that *Shigella* was one of pathogens commonly detected in children (Maranhão *et al.*, 2008). In another case–control study conducted in João Pessoa, north-east of Brazil, in diarrhoeal children, the detection rate of *Shigella* was about 4% (Moreno *et al.*, 2010). During 2004–05, *S. sonnei* and *S. flexneri* were found responsible for 95% episodes among the Chilean children aged 2–4 years (Hamilton-West *et al.*, 2007). A study conducted during 2006–2007 in Guadalajara, Mexico, showed that the prevalence of *Shigella* spp. among infants with acute diarrhoea was 2.8%, mostly due to contaminated food and water (Larrosa-Haro *et al.*, 2010).

Oceania

In Papua New Guinea, *Shigella* spp. was among the most common causes of enteric infections with the domination of *S. flexneri* serogroup. A retrospective analysis with the data collected from 2000–2009 revealed that 4% of the diarrhoeal cases were infected by *Shigella* spp. mainly by *S. boydii* and *S. dysenteriae* (Rosewell *et al.*, 2010).

Drug resistance in *Shigella species*

Excretion of profuse watery stool that leads to severe dehydration require adequate fluid replacement either by oral or intravenous rehydration. In addition to the rehydration therapy, shigellosis requires antimicrobial treatment to reduce the duration of the illness and to prevent transmission to close contacts. Fluoroquinolones, ceftriaxone and pivmecillinam are the antimicrobials currently recommended by the World Health Organization (WHO) for the treatment of dysentery in children, as these drugs are effective in reducing the mortality caused by *Shigella* spp. (Traa *et al.*, 2010). These drugs are also effective among immunocompromised children including neonates suffering from infections caused by multidrug-resistant strains of *Salmonella* and *Shigella* spp. (Leibovitz, 2006). The use of antimicrobials in the treatment of shigellosis varies from country to country. Pivmecillinam is not the drug of choice for the treatment of diarrhoea in many countries, but is being used in Bangladesh (Rahman *et al.*, 2007). Ceftriaxone is an ideal drug for the treatment of shigellosis in regions where fluoroquinolone resistance is common. The American Academy of Pediatrics and Infectious Diseases Society of America recommends azithromycin as an alternative drug for treatment of shigellosis (Jain *et al.*, 2005) and this drug was found effective against multidrug-resistant *Shigella* spp. in many countries.

Constant monitoring of susceptibility patterns of *Shigella* spp. is necessary for selection of appropriate antimicrobial agents for treatment. Antimicrobial therapy may begin based on the antimicrobial resistance profiles commonly found in *Shigella* strains in a particular region. Acquired resistance to antimicrobial drugs is becoming more prevalent among *Shigella* spp. in the developing countries. Antimicrobial resistance genes are often identified in plasmids, transposons, integrons, and constins. These resistance traits are disseminating mainly through horizontal gene transfer. Information on mechanisms involved in antimicrobial resistance is necessary to retard dissemination of antimicrobial resistance.

Shigella spp. have developed resistance to antimicrobials commonly used in paediatric practice such as ampicillin and trimethoprim–sulfamethoxazole during the past decade both in developing as well as developed countries. *Shigella* spp. has also acquired plasmid-mediated multiresistance to the commonly used antimicrobials as a first-line therapy (Niyogi *et al.*, 2000; Taneja *et al.*, 2005). Emergence of multidrug-resistant *Shigella* spp. to the new generation of antimicrobial agents has become a serious problem in developing countries.

S. dysenteriae type 1 is usually identified in human infection and cause epidemic bacillary dysentery. Among shigellae, this serotype causes the most severe form of disease. *S. dysenteriae* type 1 has a history of re-emergence after a span of decades in many developing countries (Talukder *et al.*, 2003). During each epidemic, *S. dysenteriae* type 1 alters its antibiotics resistance profile owing to several genetic modification and clonality (Pazhani *et al.*, 2004). Epidemiologically, *S. flexneri* in Asia and Africa and *S. sonnei* in Europe and US are important as these serogroups were predominantly found in these regions. Similar to the prevalence of different serotypes, antimicrobial resistance patterns also differ from country to country and even within the same country (Pazhani *et al.*, 2005; von Seidlein *et al.*, 2006). In Asia and Africa, antimicrobial resistance is an emerging problem among *Shigella* spp. and treatment options are becoming limited globally due its spread in several countries (Salam and Bennish, 1991; Kariuki and Hart, 2001, von Seidlein *et al.*, 2006). WHO has recommended that ciprofloxacin should be considered as a first-line antibiotic for the treatment of shigellosis and the use of nalidixic acid is not encouraged even in areas where it is still effective (WHO, 2004). Until 2007, ampicillin, nalidixic acid, fluoroquinolone, and a combination of trimethoprim and sulphamethoxazole represented the antimicrobials of choice to treat shigellosis. Recently, the therapies involved with these drugs are compromised owing to wide emergence of resistance to these commonly used antimicrobials. The rapid emergence of multidrug-resistant (MDR) *Shigella* strains is largely due to their ability to adapt under different antibiotic pressure. The bacteria acquire and disseminate exogenous genes associated with mobile genetic elements such as plasmids, transposons, integron and genomic islands (Rowe-Magnus *et al.*, 2002).

The increasing incidence of MDR microorganisms has led to tremendous interest in the genetics and mechanisms of resistance evolved by bacteria. In the majority of *Shigella* cases, resistance was found on conjugative R plasmids that has disseminated multiple antimicrobial resistance (Watanabe, 1963). In addition to plasmid-mediated resistance, an increasing evidence of chromosomally mediated drug resistance has been documented (Casalino *et al.*, 1994; Hens *et al.*, 2005). *Shigella* resistance locus (SRL), encoded in the chromosomal region was identified in a *S. flexneri* 2a strain (Rajakumar *et al.*, 1997a). This SRL may be involved in spreading of multiple antimicrobial resistance between different *Shigella* species (Turner *et al.*, 2001). Another route of acquisition and dissemination of resistance determinants is through mobile genetic elements including transposons and gene cassettes in integrons.

Horizontally transferable drug resistance

Shigella can acquire resistance when the antimicrobials are indiscriminately used. The most commonly encountered resistance mechanism among *Shigella* spp. is the acquisition of antimicrobial resistance genes through horizontal gene transfer, by which genetic elements move from one cell into another. These include R plasmids, integrons, conjugative resistance transposons and IS elements.

Plasmids

Plasmids are small, dispensable secondary replicons that exist independently as circular double stranded DNA molecule. When plasmids get into the host cell, their replication functions are supported by the host. Usually, the plasmids do not have core genes for cellular function such as growth and multiplication but harbour genes for bacterial survival in different environmental conditions. *Shigella* spp. are known to carry small as well as large plasmids and they often carry resistance genes encoding antimicrobial resistance in addition to other cellular functions. In *Shigella* spp., plasmid-encoded antibiotic resistance includes the most common antimicrobials such as β-lactams, sulfonamides, aminobenzoic acid derivatives, fluoroquinolones and aminoglycosides. Carriage of tetracycline resistance gene on conjugative R-plasmids was first reported in a *S. dysenteriae* strain isolated in Japan in 1953 (Watanabe, 1963). Resistance to trimethoprim in *Shigella* is known to be associated by different alleles of dihydrofolate reductase encoded in the *dfr* gene. In most of the trimethoprim resistance strains, the location of *dfr* has been found in a conjugative R-plasmid that can cotransfer other antimicrobial resistances as well. A similar-sized R-plasmid was reported in several *S. sonnei* strains from Bulgaria (Bratoeva *et al.*, 1989, 1992; Ling *et al.*, 1993; Barg *et al.*, 1995). Plasmids extracted from transconjugants of *S. dysenteriae* type 1 strains isolated in Kenya harboured multiple resistance genes encoding ampicillin, trimethoprim–sulfamethoxazole, tetracycline and chloramphenicol (Oundo *et al.*, 1996). In the western Ireland, an 8.4-kb non-transferable R-plasmid carrying *tetA*, *strA-strB* and *sul1* were found in 93% of the *S. sonnei* strains. Few strains carried transferable 5- to 60-kb plasmids and were responsible for resistance to ampicillin, sulfonamide, and tetracycline/trimethoprim (DeLappe *et al.*, 2003). In some countries, penicillin and other β-lactam antimicrobials (amoxicillin, ampicillin, piperacillin, and ticarcillin) were drugs of choice in the treatment of shigellosis. Ampicillin resistance determinants in *Shigella* were also reported within the mobile genetic elements (Navia *et al.*, 1999).

Most of the extended-spectrum β-lactamases (ESBLs) are derivatives of OXA, TEM and SHV enzymes, of which > 181 TEM-type β-lactamases and > 140 SHV-type enzymes were classified (http://www.lahey.org/studies). OXA type enzyme was mostly reported in *Shigella* species, which confer resistance to ampicillin and cephalothin but they are poorly inhibited by clavulanic acid (Bush *et al.*, 1995). In several investigations, the OXA-type β-lactamase is more prevalent than TEM-1 in ampicillin-resistant *S. flexneri* (Cavallo *et al.*, 1993; Navia *et al.*, 1999). Hyperproduction of TEM-1 enzyme was identified in *S. flexneri*, which was due to a single G → T transversion at position 1 of the −10 consensus sequence (Siu *et al.*, 1997). Ampicillin-resistant *S. sonnei* strains produce TEM-1 β-lactamase using the conjugally transferable plasmids (Lee *et al.*, 2001). In Korea, ESBL-encoding genes for TEM-19, TEM-15, and TEM-52 were identified in conjugal plasmids of *S. sonnei* strains (Kim *et al.*, 2004). A TEM type a TEM-116 derivative gene was also identified in a *S. flexneri* strain (Hu *et al.*, 2008).

To overcome resistance caused by β-lactamases, third generation cephalosporins such as ceftriaxone, cefotaxime and ceftazidime were introduced during 1990s. The cefotaximases (CTX-M) group of plasmid-mediated ESBL has emerged in many countries (Bradford *et al.*, 2001). CTX-M-enzyme is an expanding family of ESBLs that preferentially hydrolyse cefotaxime over ceftazidime and are inhibited by β-lactamase inhibitors such as clavulanic acid, sulbactam and tazobactam. Among *Shigella* spp., *S. sonnei* was the most frequently identified serogroup that exhibited ESBL production. During 1999, the ESBL-producing *S. dysenteriae* and *S. flexneri* were identified in India and Argentina, respectively (Acikgoz *et al.*, 2008).

Genetic background of ESBL has been reported in clinical strains of *S. sonnei* particularly for the bla-CTX-M-15 type, which was encoded on a 70-kb plasmid (Sabra *et al.*, 2009). This ESBL encoding gene was flanked by an insertion element (IS*Ecp1*) (Matar *et al.*, 2007; Sabra *et al.*, 2009). In *S. sonnei*, plasmid-mediated CTX-M-64 was identified and their deduced amino acid sequence revealed that this ESBL encoding gene may have originated by homologous recombination between CTX-M-15 and CTX-M-14-like β-lactamase. The carbapenem-hydrolysing β-lactamase encoded on a transferable plasmid pMS390 from a *S. flexneri* strain was first identified in Japan (O'Hara *et al.*, 1998). This β-lactamase hydrolysed oxyimino-cephalosporins also, but not the monobactams.

Owing to the emergence of ampicillin and co-trimoxazole-resistant *Shigella* strains, nalidixic acid, a quinolone derivative was used widely as the first-line antimicrobial agent in many countries. Clinical trials conducted during 1970–1980 showed its efficacy against *Shigella*. However, during late 1980s, nalidixic acid was also found to be less effective in the treatment of shigellosis (Sen *et al.*, 1988; Bennish and Salam, 1992). Until 2001, nalidixic acid reminded as drug of choice for the treatment of shigellosis in Pakistan, Ethiopia, Saudi Arabia and Kwazulu-Natal (South Africa) (Karas *et al.*, 1999, Mache, 2001; Mahmood, 2001; Panhotra *et al.*, 2004). Interestingly, *Shigella* spp. still remains susceptible to nalidixic acid in few countries such as Indonesia and Mozambique (Mandomando *et al.*, 2009; Herwana *et al.*, 2010). *S. flexneri* strains isolated from north-west Ethiopia were resistant to fluoroquinolones but susceptible to nalidixic acid (Tiruneh, 2009). During 1980s, *S. dysenteriae* type 1 became resistant to nalidixic acid followed by other serogroups of *Shigella* in the Indian subcontinent and other parts of the world (Rogerie *et al.*, 1986; Dutta *et al.*,1989; Bennish and Salam, 1992). In late 1990s, nalidixic acid was replaced by newer fluoroquinolone derivatives such as norfloxacin, ciprofloxacin and ofloxacin because of their simple treatment regime (two doses per day for 3 days, instead of four doses per day for 5 days with nalidixic acid) and more effectiveness. The first reduced susceptibility of *S. sonnei* to fluoroquinolone was documented in Japan (Horiuchi *et al.*, 1993). In 2002, appearance of *S. dysenteriae* type 1 strains resistant to fluoroquinolones was reported from India from several dysentery outbreaks (Bhattacharya *et al.*, 2003). From this point of time, fluoroquinolone resistance has increased in other countries and among other serogroups of *Shigella* as well.

The first appearance of ciprofloxacin-resistant *S. flexneri* type 2a was identified in Kolkata, India in 2003 and in the following years, the resistance increased from 41% to 91.6% during 2007–2010 (Pazhani *et al.*, 2005; Ghosh *et al.*, 2011). The same trend has also been reported in many South Asian countries (Kuo *et al.*, 2006; von Seidlein *et al.*, 2006). Till 2004, all the *S. sonnei* strains were susceptible for ciprofloxacin in the Indian subcontinent (Pazhani *et al.*, 2005, Talukder *et al.*, 2006). However, fluoroquinolone-resistant *S. sonnei* was predominantly seen in Kolkata, India (Ghosh *et al.*, 2011). In developed countries, though the nalidixic acid resistance is common among *S. sonnei*, the trend of fluoroquinolone resistance is slowly increasing (Kim *et al.*, 2008; Vrints *et al.*, 2009). *S. boydii* strains were more frequently found in South Asia and sub-Saharan Africa (Kotloff *et al.*, 1999). Since 2002 to 2004, quinolone-resistant *S. boydii* strains were not detected in India (Pazhani *et al.*, 2005). Currently, 60% of the strains have been found to be resistant to nalidixic acid and 28.5% of strains have shown reduced susceptibility to fluoroquinolones (Ghosh *et al.*, 2011). The emergence of fluoroquinolone resistance is rising year by year among all serogroups of *Shigella* spp.

Munshi *et al.* (1987) first attempted to characterize nalidixic acid-resistant *Shigella* strains isolated from epidemics from southern Bangladesh. It was thought that genes coding resistance for nalidixic acid was located on a conjugative 20-MDa plasmid (Munshi *et al.*, 1987). Later it was shown in Bangladesh that this plasmid in *S. dysenteriae* type 1 strain was not implicated in nalidixic acid, but produced nalidixic acid resistance mutants at a higher frequency. This resistance mechanism was not established with other antimicrobials such as chloramphenicol, tetracycline, or ciprofloxacin (Ashraf *et al.*, 1991). Emergence of plasmid-mediated quinolone resistance (PMQR) has been reported since 1998. PMQR genes such as *qnr* confer low-level resistance to fluoroquinolones and facilitate increase in the prevalence of quinolone resistance strains. The *qnr* genes encode pentapeptide repeat proteins, which protect DNA gyrase and topoisomerase IV from quinolone inhibition. To date, six *qnrA*, 20 *qnrB* and three *qnrS* alleles and also *qnrC* and *qnrD* genes have been reported mainly in southern and eastern Asia, North and South America and Europe (Hata *et al.*, 2005; Jacoby *et al.*, 2006). An update of different *qnr* alleles is available in the website www.lahey-clinic/qnrstudies.org. Plasmid mediated *qnrS* was first identified in a clinical isolate of *S. flexneri* 2b from Japan (Hata *et al.*, 2005). Subsequent to this finding, *qnrS* was identified in different serotypes of *S. flexneri* with a *qnrS2* allele from China (Pu *et al.*, 2009; Xiong *et al.*, 2010). Recent findings on origin of *qnr* genes specified that *qnrA* and *qnrS* originated from environmental *Shewanella algae* and *Vibrio splendidus*, respectively (Poirel *et al.*, 2005). Recently, two other PMQR mechanisms have been described that include aminoglycoside acetyltransferase gene (*aac(6')–1b-cr*), which is capable of modifying ciprofloxacin (Robicsek *et al.*, 2006) and the other is a quinolone efflux *qepA* gene (Yamane *et al.*, 2007). Except for *qnrS*, the other *qnr* alleles often carry antibiotic resistance genes that confer resistance to β-lactam, aminoglycosides, chloramphenicol, tetracycline, sulphonamides, trimethoprim and rifampin (Rodriguez-Martinez *et al.*, 2011). Owing to rapid emergence of resistance towards front-line antimicrobials and fluoroquinolones, the treatment option for shigellosis has been modified to cephalosporins and macrolide antibiotics.

Transposons and integrons

Transposons (Tn) carry diverse antibiotic resistance genes, and have the ability to move both inter and intra-molecularly and from one position into another. For example from one plasmid to another plasmid, or to a chromosome and thus carry the names 'conjugative transposons', or 'integrative conjugative element' (Roberts and Mullaney,

2009). These are small fragments of DNA that can transpose through the genome of their host via an RNA intermediate in the case of class I retrotransposon or via a 'cut-and-paste' mechanism for class II DNA transposons (Bessereau, 2006). Transposons generally fall into three groups. First one is insertion sequences (IS), which carry the gene for transposase enzyme and have inverted repeats (IRs). IS elements are generally found in many sites of chromosome and plasmids. Second group of transposons are composite transposons, which consist of two copies of same IS element in the terminal region, either as direct or inverted repeats, to the central section that encoded antibiotic resistance. Directions of these repeats play a vital role in their transcription and migrations into other specific sites. Third group of transposons are the non-composite transposons, which have IRs, transposase and antibiotic resistance genes. This assembly acts like both IS elements and composite transposon. Each transposon has its own IS and were designated as IS*1*, IS2, IS3, etc., whereas their corresponding transposons were designated with prefix of Tn and assigned a sequential number, e.g. Tn*1*, Tn2, Tn3, etc. Roberts *et al.* (2008) have described different types of transposons which are available at http://www.ucl.ac.uk/eastman/tn/.

In the bacterial pathogens, transposons are known to contribute to mobilization of integrons. Integrons are the gene expression system in which antimicrobial resistance genes are inserted. They are capable of incorporating any open reading frame (ORF) and convert them into functional genes like natural cloning vector. The basic components of integrons include integrase gene (*intI1*), the attachment site (*attI*) and the promoter region. Integrase is a member of the tyrosine site-specific recombinase family and are categorized into four commonly prevalent classes based on the homology of their proteins. Class 1 integrons are characterized by the presence of two conserved segments (CS), namely 5′-CS and 3′-CS. The 5′-CS consists of integrase gene (*intl1*), the attachment site (*attI*) and the promoter region, while 3′-CS carries sulphonamide resistance (*sul1*), quaternary ammonium compounds (*qac*Δ*E*) and disinfectants resistance determinants. Downstream region of 3′-CS is often associated with functionally unknown ORF. Antibiotic resistance determinants are embedded between these two conserved segments and number of resistance determinants may be single to multiple. This kind of unit structure of class 1 integrons is called In1 and the structure devoid of antimicrobial resistance genes in between CS is called In0 (Fig. 4.3a) (Bissonnette and Roy, 1992).

Class 2 integron was identified as a part of Tn7 transposons family, in which integrase (*intI2*) of 5′-CS is defective owing to a premature stop codon in this gene. The Tn7, *attl* site is placed between *intl2* and the first gene of antimicrobial resistance determinants. The 3′-CS of this class of integrons are encoded by promoter sequence of Tn7 transposons rather than *sul1* gene of class 1 integrons. Class 3 integron was identified in *Klebsiella* spp. harboured gene encoded for metallo-beta-lactams (bla_{IMP}) such as carbapenams. The flaking regions of class 3 integrons were identified with atypical recombination sites (Arakawa *et al.*,1995). A fourth class of integron was identified in the small chromosome of *V. cholerae*, in which a number of cassettes were associated with integrase (*intI4*) gene. Owing to the presence of large number of array cassettes, this class of integrons is called super-integrons. In bacteria, integrons not only disseminate antimicrobial resistance but also play an important role in acquisition of some of the virulence features.

Distribution of Integrons in *Shigella* species

Shigella has the ability to acquire mobile genetic elements such as plasmids and transposons. These mobile elements are often carried by the integrons. Antimicrobial resistances spread within the same species or across different species. Many studies prove that integrons also play an important role in the acquirement and spread of different antimicrobial resistance genes in *Shigella* species. However, integron analyses were predominantly made in a particular serotype or for a single class of integron. The location of transposon in the chromosome was identified in *S. sonnei* isolated from China (Pan *et al.*, 2006). It has been shown that the Tn7 was inserted in a single orientation into *att*Tn7, downstream of

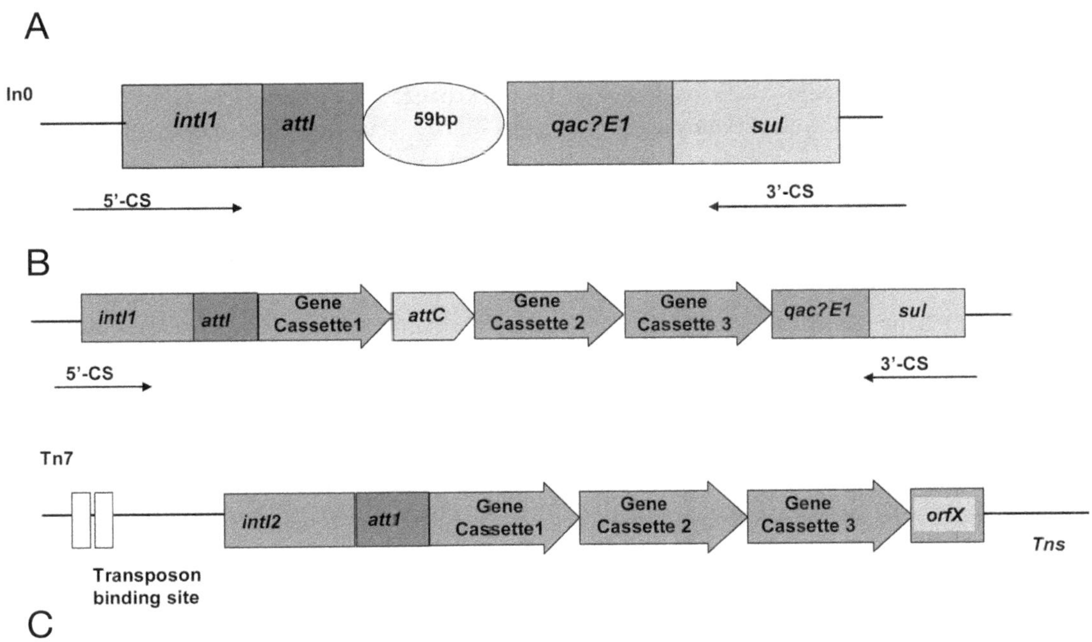

Figure 4.3 Structure of integrons. Structure of class 1 integron (A) devoid of resistance gene cassette and (B) with multiple resistance gene cassettes, and (C) Structure of class 2 integron as part of composite Tn7 transposon.

the *glmUS* operon (Pan *et al.*, 2006). In some instances, the integrase (*intI1*) gene of class 1 integron may not be present due truncation or absence of 3′-conversed segment (Fig. 4.3b). This kind of integron is known as atypical class 1 integron. Recently, the atypical class of integron is emerging among *S. flexneri* and *S. sonnei* isolated from China (Pan *et al.*, 2006). The atypical class 1 integron was detected only in the chromosome (Pan *et al.*, 2006). Atypical class 1 integron consisted of *intI1, attI,* with *bla0XA30* and *aadA1* gene cassettes and part of IS*1* sequence. The terminal gene of these gene cassettes was designated with superscript of SAI (*Shigella* atypical class 1 integron) (Pan *et al.*, 2006). However, atypical class 1 integron was first reported on the chromosome of *S. flexneri* 2a strain YSH6000 from Japan. The sequence of the gene cassettes of SAI had 99% homology with the corresponding sequence of strain YSH6000. Adjacent to this SAI, two resistance determinants for chloramphenicol and tetracycline were also identified and the whole fragment was described as SRL PAI (*Shigella* resistance locus, pathogenicity island) (Luck *et al.*, 2001). The atypical class 1 integrons were also reported in *Shigella* spp. isolated from Asia (Pan *et al.*, 2006), Africa, Europe, Central America (Dubois *et al.*, 2007) and Australia (Turner *et al.*, 2003). More recently, coexistence of the typical and atypical class 1 integron was identified in *S. flexneri* isolated from China in 2005 and 2006 (Zhu *et al.*, 2011). The atypical class 1 integron flanked by two copies of IS element and a typical class 1 integron was detected on a conjugative plasmid (Zhu *et al.*, 2011).

Class 2 integron was always identified within Tn7 transposon in majority of *S. sonnei* strains that harboured the gene cassette containing *dfrA1, sat1* and *aadA1*, conferring resistance to trimethoprim, streptothricin and streptomycin, respectively (Fig. 4.3c). Outbreak-related *S. sonnei* strains isolated from south-western Korea during 1999 harboured both class1 and class 2 integrons, in which class 1 integron comprised *dfrA12*, truncated *orfF*, and *aadA2* whereas the class 2 integron had *dfrA1* and *dfrA12* (Oh *et al.*, 2003). *Shigella* species isolated from travellers diarrhoea in Spain between 1995 and 2000 were identified with different resistance gene cassettes in class 1 integron and most of them confer resistance to trimethoprim at various levels (Navia *et al.*, 2004). The combination of class 1 and class 2 integrons are rarely reported in *Shigella* spp. A study from western Ireland showed that class 1 integron in *S. sonnei* strain contained *sat* and *aad* gene cassette and the class 2 integron harboured *dhfr1*, *sat* and *aad* genes (DeLappe *et al.*, 2003).

Chromosomally mediated drug resistance

Many members in the family Enterobacteriaceae including *E. coli* and *Shigella* carry a chromosomal *ampC* encoded β-lactamase. Generally, the expression of *ampC* is non-inducible and hence the β-lactamase activity remains at low levels. Mutations in their specific regions stimulate increased β-lactamase synthesis making the strains resistant to cephalosporins also (Normark *et al.*, 1986). Most of the acquired resistance to quinolones can be attributed to mutations in the genes encoding DNA gyrase or topoisomerase IV (Nakamura *et al.*, 1989). DNA gyrase consists of two A and B subunits, encoded by the *gyrA* and *gyrB*, respectively (Brown *et al.*, 1979). The majority of the quinolone resistance mutations are involved in a relatively small region at the N-terminus of *gyrA* called quinolone resistance-determining region (QRDR) (Yoshida *et al.*, 1990). The QRDR of GyrA and the corresponding ParC region (QRDR of ParC) are considered the mutational hotspots and thus responsible for determining resistance towards quinolone drugs.

In many clinical strains, high MIC levels for quinolone were detected with universal mutations in the *gyrA* at positions 83 and 87 (Heisig, 1996). DNA sequence analysis of N-terminal coding region of *gyrA* from nalidixic acid-resistant strains of *S. dysenteriae* type 1 revealed a C-to-T transition at nucleotide position 248 leading to a Ser-83 to Leu substitution, which was absent in susceptible strains (Rahman *et al.*, 1994; Jeong *et al.*, 2003). GyrA contains the QRDR near active site, between codons 67 and 106 and the changes of the polar amino acid serine to a hydrophobic leucine or isoleucine may alter the target structure and exhibit reduced susceptible to fluoroquinolones (Yoshida *et al.*, 1990). Rahman *et al.* (1994) have demonstrated that the amino acid change at position 83 of GyrA from serine to leucine leads to 8-fold increase in the MIC of ciprofloxacin in *S. dysenteriae*. *Shigella* strains with reduced susceptibility phenotype to fluoroquinolones had a single amino acid substitution either at position 83 or at 87 of GyrA (Mensa *et al.*, 2008), while the resistant strains showed the presence of at least one additional substitution in GyrA or ParC (Ruiz, 2003).

Mutations in the QRDR regions of *Shigella* species were reported from different geographical locations such as China, India, Bangladesh, Japan Korea, Spain and Israel (Chu *et al.*, 1998; Ahmed *et al.*, 1999; Jeong *et al.*, 2003; Talukder *et al.*, 2004; Pazhani *et al.*, 2004; Navia *et al.*, 2005; Mensa *et al.*, 2008). Fluoroquinolone-resistant *Shigella* species exhibited multiple mutations in GyrA and single mutation in ParC from India and southeast Asian countries (Pazhani *et al.*, 2004; Talukder *et al.*, 2004), single mutation in the GyrA and ParC from *S. flexneri* in China (Pu *et al.*, 2009) and double mutations at positions 83 and 87 in GyrA and double mutations at positions 80 and 91 in ParC in *S. flexneri* from Korea (Kim *et al.*, 2008). *S. flexneri* isolated from Spain related to travellers diarrhoea and sporadic cases from China showed resistance to nalidixic acid and reduced susceptibility to ciprofloxacin. However, these strains were devoid of any mutations in the quinolone target sites (Mensa *et al.*, 2008; Pu *et al.*, 2009). This result suggested that quinolone resistance not only involved mutation in the QRDR regions but also other active mechanisms.

Efflux mediated resistance in *Shigella* species

The complex membrane of Gram-negative bacteria contains of a variety of protein channels involved in the transport of different molecules such as nutrients and harmful substances including antibiotics, antiseptic compounds, dyes and detergents. Among these transporters, some are distinct energy dependent efflux pumps that recognize toxic substances and pump out from periplasm/cytoplasm to the exterior of the cell, thereby decreasing the intracellular accumulation of the harmful substances. The over expression of one of these pumps prevents the threshold intracellular concentration of the agent for effective inhibitory activity, which makes the bacterium resistant against a particular substance. Spread and emergence of multidrug resistance phenotypes may be due to overproduction of efflux pumps. Based on the sequence similarity, bacterial antibiotic efflux pumps belong to five superfamilies (www.biology.ucsd.edu/~msaier/transport). Efflux pumps for fluoroquinolones in Gram-negative bacteria have been grouped into three superfamilies namely, resistance-nodulation-division (RND) family, the major facilitator superfamily (MFS), and the multidrug and toxic compound extrusion (MATE) family (Nishino *et al.*, 2006). Efflux mediated resistance to fluoroquinolone has been identified in several bacteria. In *S. dysentery*, an efflux pump has been shown to confer a high-level resistance to quinolone (Ghosh *et al.*, 1998a; Ahamed *et al.*, 1999). To evaluate efflux pumps in fluoroquinolone resistance in *S. flexneri*, Kim *et al* (2008) conducted synergy experiments using efflux pumps inhibitor carbonyl cyanide *m*-chlorophenyl hydrazone (CCCP) with fluoroquinolones by agar dilution method. They found a decrease in the MIC of fluoroquinolones for *S. flexneri* of about 32- to 128-fold in agar plates containing CCCP when compared with those containing CCCP-free medium. An attempt was made to measure an expression of the efflux pump-encoding genes *mdfA*, *tolC* and *ydhE* belonging to MFS, RND and MATE superfamily transporters, respectively. These genes had increased expression in the presence of ciprofloxacin and accumulated low concentration of ciprofloxacin leading to reduced susceptibility towards this antibiotic (Kim *et al.*, 2008). In *S. dysenteriae*, resistance to ampicillin may be due to the alterations in the affinities or amounts of the penicillin-binding proteins for β-lactams, elaboration of β-lactamases and reduced permeability across the outer membrane (Ghosh *et al.*, 1998b).

Clinical management of shigellosis

The typical symptoms of shigellosis include diarrhoea/dysentery with frequent mucoid bloody stools, abdominal cramps and tenesmus. Since the inoculum required to transmit shigellosis is only $10–10^2$, the chances of person to person transmission or fly-borne infections are more. The major clinical intervention of diarrhoea consists of fluid and electrolyte rectification. Lactoferrin is an iron-binding protein present in mammalian milk and have antimicrobial, anti-inflammatory, and immunomodulatory functions. Several *in vitro* and *in vivo* studies have shown the influence of lactoferrin on enteric pathogens including *Shigella* by inhibiting the growth and impairing the function of surface expressed virulence factors (Ochoa and Cleary, 2009). Mother's milk may protect infants from many gastrointestinal infections and hence feeding during and after shigellosis is emphasized. In most of the pathogenic organisms, adhesion to host tissues is the prerequisite for the initiation of infectious diseases. This adhesion is mediated by lectins present on the surface of the pathogen that bind to complementary carbohydrates on the surface of the host tissues. To overcome this machinery, several anti-adhesions such as polyvalent saccharides (neoglycoproteins or dendrimers) and lectin-inhibitory saccharides have shown to protect several animal species against experimental infection by lectin-carrying pathogens (reviewed by Sharon, 2006). In addition, it is expected that the pathogenic strains resistant to anti-adhesive agents will emerge at a significantly lower rate than of strains that are resistant to antibiotics. These microbial lectin inhibitors may come up as future potential therapeutic agents of infectious diseases.

When antimicrobial therapy is essential, selection of a specific drug should be made based upon the susceptibility patterns of the pathogen. Antimicrobial agents are the mainstay of therapy for shigellosis and appropriate therapy shortens the duration of both clinical symptoms and faecal excretion of the pathogen. Ciprofloxacin, ceftriaxone and pivmecillinam are the antibiotics currently recommended by the World Health Organization (WHO) for the treatment of dysentery in children. Meta-analyses to determine the effect of treatment with ciprofloxacin, ceftriaxone and pivmecillinam showed that these antimicrobials were effective in reducing the clinical and bacteriological signs and symptoms of dysentery (Traa *et al.*, 2010). An assessment of antimicrobial susceptibility patterns to different species of *Shigella* are required to direct empiric therapy. Owing to its excellent spectrum of activity, penetration and convenient ways of administration, fluoroquinolones are widely used in the treatment of various infectious diseases in adults. However, owing to joint/cartilage toxicity in juvenile animal studies, use of fluoroquinolones in children is limited. With the exception of cystic fibrosis and life-endangering infections, the use of fluoroquinolones in paediatrics should be limited to Gram-negative pathogens (Leibovitz, 2006). Rifaximin is a poorly absorbed rifamycin antimicrobial drug with *in vitro* activity against Gram-positive, Gram-negative and anaerobic bacteria. Rifaximin shortens the duration of travellers' diarrhoea and non-dysenteric diarrhoeal illness owing to *S. sonnei* (Huang and DuPont, 2005). Selection of resistant mutants is unusual with rifaximin, which can alter the virulence of enteric pathogens in addition to organism's inhibition.

Owing to the wide range of *Shigella* serotypes and subtypes, there is a need for a multivalent vaccine representing prevalent species and serotypes. The main set back of *Shigella* vaccine development is the lack of a suitable animal model for testing vaccine efficacy. Several formulations such as subunit parenteral live attenuated oral, killed oral, conjugate, broad spectrum and proteosome-based vaccines for *Shigella* have been made and widely tested. However, the great obstacle to use these vaccines against shigellosis in children from developing areas is their poor immune responses due to their minimal maternal antibodies (Kweon, 2008). The efficacy of the live attenuated vaccine candidates vary with the colonization ability and immunogenic potential based on the non-endemic and endemic areas. Fermented soy sauce (shoyu) was found to possesses antimicrobial activity against many bacteria including *S. flexneri* (Kataoka, 2005). Probiotics have also been studied for their effectiveness in the prevention and treatment of diarrhoea caused by *Shigella* and other enteric pathogens (Rolfe, 2000). Probiotics inhibit intestinal bacterial enzymes, epithelial and mucosal adherence, epithelial invasion and production of antimicrobial substances. Preventative measures mainly include personal hygiene, especially promotion of hand washing.

Conclusion

Shigella has an extraordinary ability to acquire accessory genes, making them capable of causing infections in humans. Evidently, many genetic elements are responsible for genome rearrangements and some attributes to the antimicrobial resistance. Antimicrobial resistance genes have been shown to accumulate on mobile elements, leading to a situation in which multidrug resistance phenotypes can be transferred to a susceptible recipient via a single or multiple genetic events. Considering these mechanisms and rapid spread of antimicrobial resistance in *Shigella* spp., continuous monitoring of this pathogen for antimicrobial susceptibility is mandatory to ensure effective clinical management. Risk reduction strategies and control interventions at every step from the food preparation process to consumption point are crucial to reduce the burden of food-borne infections. Completion of whole genome sequencing of seven representative strains from each of the *Shigella* species has introduced an era of updating the knowledge on *Shigella* pathogenesis, and this may foster the development of safe and efficient vaccines. Evaluation of vaccine candidates against shigellosis are urgently needed in terms of safety, immunogenicity, and protection in volunteer challenge models.

References

Acikgoz, Z.C., Eser, O.K., and Kocagöz, S. (2008). CTX-M-3 type beta-lactamase producing *Shigella sonnei* isolates from pediatric bacillary dysentery cases. Jpn. J. Infect. Dis. *61*, 135–137.

Ahamed, J., Gangopadhyay, J., Kundu, M., and Sinha, A.K. (1999). Mechanisms of quinolone resistance in clinical isolates of *Shigella dysenteriae*. Antimicrob. Agents Chemother. *43*, 2333–2334.

Alamanos, Y., Maipa, V., Levidiotou, S., and Gessouli, E. (2000). A community waterborne outbreak of gastro-enteritis attributed to *Shigella sonnei*. Epidemiol Infect. *125*, 499–503.

Al-Eissa, Y., Al-Zami, F., Al-Kharashi, M., Kambal, A., Chowdhury, M., and Al-Ayed, I. (1992). The relative importance of *Shigella* in the aetiology of childhood gastroenteritis in Saudi Arabia. Scand. J. Infect. Dis. *24*, 347–351.

Al-Gallas, N., Bahri, O., Bouratbeen, A., Ben Haasen, A., and Ben, Aissa, R. (2007). Etiology of acute diarrhea in children and adults in Tunis, Tunisia, with emphasis on diarrheagenic *Escherichia coli*: prevalence, phenotyping, and molecular epidemiology. Am. J. Trop. Med. Hyg. 77, 571–582.

Aragón, M., Barreto, A., Chambule, J., Noya, A., and Tallarico. M. (1995). Shigellosis in Mozambique: the 1993 outbreak rehabilitation – a follow-up study. Trop. Doct. 25, 159–162.

Arakawa, Y., Murakami, M., Suzuki, K., Ito, H., Wacharotayankun, R., Ohsuka, S., Kato, N., and Ohta, M. (1995). A novel integron-like element carrying the metallo-beta-lactamase gene bla_{IMP}. Antimicrob. Agents Chemother. 39,1612–1615.

Arias, C., Sala, M.R., Domínguez, A., Bartolomé, R., Benavente, A., Veciana, P., Pedrol, A., and Hoyo, G; and Outbreak Working Group. (2006). Waterborne epidemic outbreak of *Shigella sonnei* gastroenteritis in Santa Maria de Palautordera, Catalonia, Spain. Epidemiol. Infect. *134*, 598–604.

Ashida, H., Ogawa, M., Mimuro, H., and Sasakawa, C. (2009). *Shigella* infection of intestinal epithelium and circumvention of the host innate defense system. Curr. Top. Microbiol. Immunol. *337*, 231–255.

Ashida, H., Kim, M., Schmidt-Supprian, M., Ma, A., Ogawa, M., and Sasakawa, C. (2010). A bacterial E3 ubiquitin ligase IpaH9.8 targets NEMO/IKKgamma to dampen the host NF-kappaB-mediated inflammatory response. Nat. Cell Biol. *12*, 66–73.

Ashida, H., Ogawa, M., Kim, M., Suzuki, S., Sanada, T., Punginelli, C., Mimuro, H., and Sasakawa, C. (2011). *Shigella* deploy multiple countermeasures against host innate immune responses. Curr. Opin. Microbiol. *14*, 16–23.

Ashraf, M.M., Ahmed, Z.U., and Sack, D.A. (1991). Unusual association of a plasmid with nalidixic acid resistance in an epidemic strain of *Shigella dysenteriae* type 1 from Asia. Can. J. Microbiol. *37*, 59–63.

Attia, A., Huët, C., Anglaret, X., Toure, S., Ouassa, T., Gourvellec, G., Menan, H., Dakoury-Dogbo, N., Combe, P., Chêne, G., *et al.* (2001). HIV-1-related morbidity in adults, Abidjan, Côte d'Ivoire: a nidus for bacterial diseases. J. Acquir. Immune. Defic. Syndr. *28*,478–486.

Baine, W.B., Herron, C.A., Bridson, K., Barker, W.H., Jr., Lindell, S., Mallison, G.F., Wells, J.G., Martin, W.T., Kosuri, M.R., Carr, F., *et al.*(1975). Waterborne shigellosis at a public school. Am. J. Epidemiol. *101*, 323–332.

Banajeh, S.M., Ba-Oum, N.H., and Al-Sanabani, R.M. (2001). Bacterial etiology and anti-microbial resistance of childhood diarrhea in Yemen. J. Trop. Pediatr. *47*, 301–303.

Bando, S.Y., Moreno, A.C., Albuquerque, J.A., Amhaz, J.M., Moreira-Filho, C.A., and Martinez, M.B. (2010). Expression of bacterial virulence factors and cytokines during *in vitro* macrophage infection by enteroinvasive *Escherichia coli* and *Shigella flexneri*: a comparative study. Mem. Inst. Oswaldo Cruz. *105*, 786–791.

Banga Singh, K.K., Ojha, S.C., Deris, Z.Z., and Rahman, R.A. (2011). A 9-year study of shigellosis in Northeast Malaysia: Antimicrobial susceptibility and shifting species dominance. Z. Gesundh. Wiss. *19*, 231–236.

Barg, N.L., Register, S., Thomson, C., and Amyes, S. (1995). Sequence identity with type VIII and association with IS*176* of type IIIc dihydrofolate reductase from *Shigella sonnei*. Antimicrob Agents Chemother. *39*, 112–116.

Battikhi, M.N. (2004). Bloody diarrhea cases caused by *Shigella* and amoeba in Jordan. New Microbiol. *27*, 37–47.

Bennish, M.L., and Salam, M.A. (1992). Rethinking options for the treatment of shigellosis. J. Antimicrob. Chemother. *30*, 243–247.

Berkelman, R.L., Cohen, M.L., Yashuk, J., Barrett, T., Wells, J.G., and Blake, P.A. (1983). Traveler's diarrhea at sea: two multi-pathogen outbreaks caused by food eaten on shore visits. Am. J. Public Health. *73*, 770–772.

Bernardini, M.L., Mounier, J., d'Hauteville, H., Coquis-Rondon, M., and Sansonetti, P.J. (1989). Identification of *icsA*, a plasmid locus of *Shigella flexneri* that governs bacterial intra- and intercellular spread through interaction with F-actin. Proc. Natl. Acad. Sci. U.S.A. *86*, 3867–3871.

Bessereau, J.L. (2006). Transposons in *C. elegans*. In Worm Book. Anderson, P., and Hodgkin, J., eds. Available at: http://www.wormbook.org/chapters/www_transposons/transposons.html/

Bhattacharya, S.K., Sarkar, K., Balakrish Nair, G., Faruque, A.S., and Sack, D.A. (2003). Multidrug-resistant *Shigella dysenteriae* type 1 in south Asia. Lancet Infect. Dis. 3,.755.

Bissonnette, L., and Roy, P.H. (1992). Characterization of In0 of *Pseudomonas aeruginosa* plasmid pVS1, an ancestor of integrons of multiresistance plasmids and transposons of Gram-negative bacteria. J. Bacteriol. *174*, 1248–1257.

Blocker, A., Jouihri, N., Larquet, E., Gounon, P., Ebel, F., Parsot, C., Sansonetti, P., and Allaoui, A. (2001). Structure and composition of the *Shigella flexneri* "needle complex", a part of its type III secreton. Mol. Microbiol. *39*, 652–663.

Bradford, P.A. (2001). Extended-spectrum beta-lactamases in the 21st century: characterization, epidemiology, and detection of this important resistance threat. Clin. Microbiol. Rev. *14*, 933–951.

Brandon, L.D., and Goldberg, M.B. (2001). Periplasmic transit and disulfide bond formation of the autotransported *Shigella* protein *IcsA*. J. Bacteriol. *183*, 951–958.

Bratoeva, M.P., and John, J.F, Jr. (1989). Dissemination of trimethoprim-resistant clones of *Shigella sonnei* in Bulgaria. J. Infect. Dis. *159*, 648–653.

Bratoeva, M.P., John, J.F., and Barg, N.L. (1992). Molecular epidemiology of trimethoprim-resistant *Shigella boydii* serotype 2 strains from Bulgaria. J. Clin. Microbiol. *30*, 1428–1431.

Brenner, D.J., Fanning, G.R., Steigerwalt, A.G., Orskov, I., and Orskov, F. (1972). Polynucleotide sequence relatedness among three groups of pathogenic *Escherichia coli* strains. Infect. Immun. *6*, 308–315.

Brooks, J.T., Ochieng, J.B., Kumar, L., Okoth, G., Shapiro, R.L., Wells, J.G., Bird, M., Bopp, C., Chege, W., Beatty, M.E., *et al.* (2006). Surveillance for bacterial diarrhea and antimicrobial resistance in rural western Kenya, 1997–2003. Clin. Infect. Dis. *43*, 393–401.

Brown, P.O., Peebles, C.L., and Cozzarelli, N.R. (1979). A topoisomerase from *Escherichia coli* related to DNA gyrase. Proc. Natl. Acad. Sci. U.S.A. *76*, 6110–6114.

Browning, P.M., Boot, P.A., and Stevenson, J.S. (1960). Outbreak of dysentery caused by *Shigella boydii* 4. Br. Med. J. *14*, 1483.

Bruins, J., Bwire, R., Slootman, E.J., and van Leusden, A.J. (1995). Diarrhea morbidity among Dutch military service men in Goma, Zaire. Mil. Med. *160*, 446–448.

Bush, K., Jacoby, G.A., and Medeiros, A.A. (1995). A functional classification scheme for beta-lactamases and its correlation with molecular structure. Antimicrob. Agents Chemother. *39*,1211–1233.

el Bushra, H.E., and Bin Saeed, A.A. (1999). Intrafamilial person-to-person spread of bacillary dysentery due to *Shigella dysenteriae* in southwestern Saudi Arabia. East Afr. Med. J. *76*, 255–259.

Casalino, M., Nicoletti, M., Salvia, A., Colonna, B., Pazzani, C., Calconi, A., Mohamud, K.A., and Maimone, F. (1994). Characterization of endemic *S. flexneri* strains in Somalia: antimicrobial resistance, plasmid profiles, and serotype correlation. J. Clin. Microbiol. *32*, 1179–1183.

Castell Monsalve, J., Gutiérrez Avila, G., Rodolfo Saavedra, R., and Santos Azorín, A. (2008). Shigellosis outbreak with 146 cases related to a fair. Gac. Sanit. *22*, 35–39. [Spanish]

Cavallo, J.D., Bercion, R., Baudet, J.M., Samson, T., France, M., and Meyran, M. (1993). Antibiotic sensitivity of 140 strains of *Shigella* isolated in Djibouti. Bull. Soc. Pathol. Exot. *86*, 35–40.

Centers for Disease Control. (1991). *Shigella dysenteriae* type 1 – Guatemala, 1991. MMWR Morb. Mortal. Wkly Rep. *40*, *421*, 427–428.

Cheasty, T., and Rowe, B. (1983). Antigenic relationships between the enteroinvasive *Escherichia coli* O antigens O28ac, O112ac, O124, O136, O143, O144, O152, and O164 and *Shigella* O antigens. J. Clin. Microbiol. *17*, 681–684.

Chen, Y., Smith, M.R., Thirumalai, K., and Zychlinsky, A. (1996). A bacterial invasin induces macrophage apoptosis by binding directly to ICE. EMBO J. *15*, 3853–3860.

Chiou, C.S., Hsu, W.B., Wei, H.L., and Chen, J.H. (2001). Molecular epidemiology of a *Shigella flexneri* outbreak in a mountainous township in Taiwan, Republic of China. J. Clin. Microbiol. *39*, 1048–1056.

Chopra, M., Wilkinson, D., and Stirling, S. (1997). Epidemic *Shigella* dysentery in children in northern KwaZulu-Natal. S. Afr. Med J. *87*, 48–51.

Chu, Y.W., Houang, E.T., and Cheng, A.F. (1998). Novel combination of mutations in the DNA gyrase and topoisomerase IV genes in laboratory-grown fluoroquinolone-resistant *Shigella flexneri* mutants. Antimicrob. Agents Chemother. *42*, 3051–3052.

Chuang, Y.Y., Huang, Y.C., and Lin, S.Y. (2006). Outbreak of *Shigella sonnei* gastroenteritis in Northeastern Taiwan. Pediatr. Infect. Dis. J. *25*, 92–94.

Coles, F.B., Kondracki, S.F., Gallo, R.J., Chalker, D., and Morse, D.L. (1989). Shigellosis outbreaks at summer camps for the mentally retarded in New York State. Am. J. Epidemiol. *130*, 966–975.

Cossart, P., and Sansonetti, P.J. (2004). Bacterial invasion: the paradigms of enteroinvasive pathogens. Science *304*, 242–248.

Datta, P., and Sen, D. (1990). Outbreak of dysentery due to nalidixic acid resistant *S. dysenteriae* 1 at Agartala, Tripura: a hospital based study. Indian J. Public Health. *34*, 11–14.

Davis, H., Taylor, J.P., Perdue, J.N., Stelma, G.N.Jr, Humphreys, J.M.Jr, Rowntree, R., 3rd., and Greene, K.D. (1988). A shigellosis outbreak traced to commercially distributed shredded lettuce. Am. J. Epidemiol. *128*, 1312–1321.

Day, W.A., Jr., and Maurelli, A.T. (2001). *Shigella flexneri* LuxS quorum-sensing system modulates *virB* expression but is not essential for virulence. Infect. Immun. *69*, 15–23.

DeLappe, N., O'Halloran, F., Fanning, S., Corbett-Feeney, G., Cheasty, T., and Cormican, M. (2003). Antimicrobial resistance and genetic diversity of *Shigella sonnei* isolates from western Ireland, an area of low incidence of infection. J. Clin. Microbiol. *41*, 1919–1924.

Djuretic, T., Wall, P.G., Ryan, M.J., Evans, H.S., Adak, G.K., and Cowden, J.M. (1996). General outbreaks of infectious intestinal disease in England and Wales 1992 to 1994. Commun. Dis. Rep. CDR Rev. *6*, 57–63.

Drews, S.J., Lau, C., Andersen, M., Ferrato, C., Simmonds, K., Stafford, L., Fisher, B., Everett, D., and Louie, M. (2010). Laboratory based surveillance of travel-related *Shigella sonnei* and *Shigella flexneri* in Alberta from 2002 to 2007. Global Health *6*, 20.

Dromigny, J.A., Macondo, E.A., Juergens-Behr, A., Siby, T., and Perrier-Gros-Claude, J.D. (2004). The distribution and antibiotic susceptibility of *Shigella* isolates in Dakar, Senegal (2000–2002). Int. J. Antimicrob. Agents. *24*, 307–308.

Dubois, V., Parizano, M.P., Arpin, C., Coulange, L., Bezian, M.C., and Quentin, C. (2007). High genetic stability of integrons in clinical isolates of *Shigella* spp. of worldwide origin. Antimicrob. Agents Chemother. *51*,1333–1340.

Dunn, R.A., Hall, W.N., Altamirano, J.V., Dietrich, S.E., Robinson-Dunn, B., and Johnson, D.R. (1995). Outbreak of *Shigella flexneri* linked to salad prepared at a central commissary in Michigan. Public Health Rep. *110*, 580–586.

Dupont, H.L. (2000). *Shigella* species (bacillary dysentery). In Principle and Practice of Infectious Diseases, Mandell, G.L., Dougles, R.G., and Bennet, J.E. eds. (Churichill Livingstone, New York), pp. 1716–1722.

Dupont, H.L., Formal, S.B., Hornick, R.B., Snyde, M.J., Libonati, J.P., Sheahan, D.G., LaBrec, E.H., and Kalas, J.P. (1971). Pathogenesis of *E. coli* diarrhea. N. Engl. J. Med. *285*, 1–9.

Dupont, N., Lacas-Gervais, S., Bertout, J., Paz, I., Freche, B., Van Nhieu, G.T., van der Goot, F.G., Sansonetti, P.J., and Lafont, F. (2009). *Shigella* phagocytic vacuolar membrane remnants participate in the cellular response to pathogen invasion and are regulated by autophagy. Cell Host Microbe *6*, 137–149.

Dutta, P., Dutta, D., Bhattacharya, S.K., Sen, D., Mitra, U., Ghosh, A.R., Lahiri, M., and Deb, B.C. (1989). Clinical & bacteriological profiles of shigellosis in Calcutta before & after an epidemic (1984–87). Indian J. Med. Res. *89*, 132–137.

Dutta, P., Bhattacharya, S.K., Sen, D., Bhattacharya, M.K., Mitra, U., Rasaily, R., Manna, B., Mukherjee, A., and Pal, S.C. (1992). Shigellosis in children: a prospective hospital based study. Indian Pediatr. *29*, 1125–1130.

Ebright, J.R., Moore, E.C., Sanborn, W.R., Schaberg, D., Kyle, J., and Ishida, K. (1984). Epidemic Shiga bacillus dysentery in Central Africa. Am. J. Trop. Med. Hyg. *33*,1192–1197.

Echeita Sarrionandia, M.A, León, S.H., and Baamonde, C.S. (2011). Invasive gastroenteritis, anything new?. Enferm. Infecc. Microbiol. Clin. *S3*, 55–60.

Egoz, N., Shmilovitz, M., Kretzer, B., Lucian, M., Porat, V., and Raz, R. (1991). An outbreak of *Shigella sonnei* infection due to contamination of a municipal water supply in northern Israel. J. Infect. *22*, 87–93.

El-Gendy, A.M., Mansour, A., Weiner, M.A., Pimentel, G., Armstrong, A.W., Young, S.Y., Elsayed, N., and Klena, J.D. (2011). Genetic diversity and antibiotic resistance in *Shigella dysenteriae* and *Shigella boydii* strains isolated from children aged < 5 years in Egypt. Epidemiol. Infect. *140*, 299–310.

Elsea, W.R., Partridge, R.A., and Neter, E. (1967). Epidemiologic and microbiological study of a *Shigella flexneri* outbreak. Public Health Rep. *82*, 347–352.

Escobar-Páramo, P., Giudicelli, C., Parsot, C., and Denamur, E. (2003). The evolutionary history of *Shigella* and enteroinvasive *Escherichia coli* revised. J. Mol. Evol. *57*, 140–148.

Espina, M., Olive, A.J., Kenjale, R., Moore, D.S., Ausar, S.F., Kaminski, R.W., Oaks, E.V., Middaugh, C.R., Picking, W.D., and Picking, W.L. (2006). IpaD localizes to the tip of the type III secretion system needle of *Shigella flexneri*. Infect. Immun. 74, 4391–4400.

Frost, J.A., McEvoy, M.B., Bentley, C.A., and Andersson, Y. (1995). An outbreak of *Shigella sonnei* infection associated with consumption of iceberg lettuce. Emerg. Infect. Dis. *1*, 6–9.

Garcia-Fulgueiras, A., Sánchez, S., Guillén, J.J., Marsilla, B., Aladueña, A., and Navarro, C. (2001). A large outbreak of *Shigella sonnei* gastroenteritis associated with consumption of fresh pasteurised milk cheese. Eur. J. Epidemiol. *17*, 533–538.

Gaynor, K., Park, S.Y., Kanenaka, R., Colindres, R., Mintz, E., Ram, P.K., Kitsutani, P., Nakata, M., Wedel, S., Boxrud, D., *et al.* (2009). International foodborne outbreak of *Shigella sonnei* infection in airline passengers. Epidemiol. Infect. *137*, 335–341.

Gessner, B.D., and Beller, M. (1994). Moose scup shigellosis in Alaska. West. J. Med. *160*, 430–433.

Ghosh, A.S., Ahamed, J., Chauhan, K.K., and Kundu, M. (1998a). Involvement of an efflux system in high-level fluoroquinolone resistance of *Shigella dysenteriae*. Biochem Biophys. Res. Commun. *242*, 54–56.

Ghosh, A.S., Kar, A.K., and Kundu, M. (1998b). Alterations in high molecular mass penicillin-binding protein 1 associated with beta-lactam resistance in *Shigella dysenteriae*. Biochem. Biophys. Res. Commun. *248*, 669–672.

Ghosh, S., Pazhani, G.P., Chowdhury, G., Guin, S., Dutta, S., Rajendran, K., Bhattacharya, M.K., Takeda, Y., Niyogi, S.K., Nair, G.B., *et al.* (2011). Genetic Characteristics and Changing Antimicrobial Resistance among *Shigella* spp. Isolated from Hospitalized Diarrhoeal Patients in Kolkata, India. J. Med. Microbiol. *60*, 1460–1466.

Girardin, S.E., Tournebize, R., Mavris, M., Page, A.L., Li, X., Stark, G.R., Bertin, J., DiStefano, P.S., Yaniv, M., Sansonetti, P.J., *et al.* (2001). CARD4/Nod1 mediates NF-kappaB and JNK activation by invasive *Shigella flexneri*. EMBO Rep. *2*, 736–742.

Godoy, P., Bartolomé, R., Torres, J., Espinet, L., Escobar, A., Nuin, C., and Domínguez, A. (2011) Waterborne outbreak of *Shigella sonnei* caused by consumption of public supply. Water. Gac. Sanit. *25*, 363–367. [In Spanish.]

Green, D.M., Scott, S.S, Mowat, D.A., Shearer, E.J., and Thomson, J.M. (1968). Water-borne outbreak of viral gastroenteritis and *Sonnei* dysentery. J. Hyg. (Lond). *66*, 383–392.

Green, M.S., Block, C., Cohen, D., and Slater, P.E. (1991). Four decades of shigellosis in Israel: epidemiology of a growing public health problem. Rev. Infect. Dis. *13*, 248–253.

Greig, J.D., Todd, E.C., Bartleson, C.A., and Michaels, B.S. (2007). Outbreaks where food workers have been implicated in the spread of foodborne disease. Part 1. Description of the problem, methods, and agents involved. J. Food. Prot. *70*, 1752–1761.

Guerin, P.J., Brasher, C., Baron, E., Mic, D., Grimont, F., Ryan, M., Aavitsland, P., and Legros, D. (2003). *S. dysenteriae* type 1 in West Africa: intervention strategy for an outbreak in Sierra. Leone. Lancet *362*, 705–706.

Guerin, P.J., Brasher, C., Baron, E., Mic, D., Grimont, F., Ryan, M., Aavitsland, P., and Legros, D. (2004). Case management of a multidrug-resistant *Shigella dysenteriae* serotype 1 outbreak in a crisis context in Sierra Leone, 1999–2000. Trans. R. Soc, Trop. Med. Hyg. *98*,635–643.

Guha, M., and Mackman, N. (2001). LPS induction of gene expression in human monocytes. Cell Signal. *13*, 85–94.

Guichon, A., and Zychlinsky, A. (1997). Clinical isolates of *Shigella* species induce apoptosis in macrophages. J. Infect. Dis. *175*, 470–473.

Gupta, A., Polyak, C.S., Bishop, R.D., Sobel, J., and Mintz, E.D. (2004). Laboratory-confimed shigellosis in the United States, 1989–2002: Epidemiological trends and patterns. Clin. Infect. Dis. *38*, 1372–1377.

Gutiérrez Garitano, I., Naranjo, M., Forier, A., Hendriks, R., DE Schrijver, K., Bertrand, S., Dierick, K., Robesyn, E., and Quoilin, S. (2011). Shigellosis outbreak linked to canteen-food consumption in a public institution: a matched case–control study. Epidemiol. Infect. *139*, 1956–1964.

Hale, T.L. (1991). Genetic basis of virulence in *Shigella* species. Microbiol. Rev. *55*, 206–224.

Haley, C.C., Ong, K.L., Hedberg, K., Cieslak, P.R., Scallan, E., Marcus, R., Shin, S., Cronquist, A., Gillespie, J., Jones, T.F., *et al.* (2010). Risk factors for sporadic shigellosis, FoodNet 2005. Foodborne Pathog. Dis. *7*, 741–747.

Haltalin, K.C. (1967). Neonatal shigellosis. Report of 16 cases and review of the literature. Am. J. Dis. Child. *114*, 603–611.

Hamilton-West, M.C., Prado, J.V., Hormazábal, O.J.C, Lagos, Z.R., Benadof, F.D., Mendoza, N.C., Elgueta, N.A., Tapia, P.C., C. Fuentes, D.M., Alvarez, A.I., *et al.* (2007). *Shigella* spp. infections in children living in the Metropolitan Region, Chile, during summer of 2004–2005. Rev. Med. Chil. *135*, 1388–1396.

Handa, Y., Suzuki, M., Ohya, K., Iwai, H., Ishijima, N., Koleske, A.J., Fukui, Y., and Sasakawa, C. (2007). *Shigella* IpgB1 promotes bacterial entry through the ELMO-Dock180 machinery. Nat. Cell Biol. *9*, 121–128.

Harrington, A.T., Hearn, P.D., Picking, W.L., Barker, J.R., Wessel, A., and Picking, W.D. (2003). Structural characterization of the N terminus of IpaC from *Shigella flexneri*. Infect. Immun. *71*, 1255–1264.

Hata, M., Suzuki, M., Matsumoto, M., Takahashi, M., Sato, K., Ibe, S., and Sakae, K. (2005). Cloning of a novel gene for quinolone resistance from a transferable plasmid in *Shigella flexneri* 2b. Antimicrob. Agents Chemother. *49*, 801–803.

Hayani, K.C., Guerrero, M.L., Morrow, A.L., Gomez, H.F., Winsor, D.K., Ruiz-Palacios, G.M., and Cleary, T.G. (1992). Concentration of milk secretory immunoglobulin A against *Shigella* virulence plasmid-associated antigens as predictor of symptom status in *Shigella*-infected breast-fed infants. J. Pediatr. *121*, 852–856.

Hedberg, C.W., Levine, W.C., White, K.E., Carlson, R.H., Winsor, D.K., Cameron, D.N., MacDonald, K.L., and Osterholm, M.T. (1992). An international foodborne outbreak of shigellosis associated with a commercial airline. JAMA *268*, 3208–3212.

Heier, B.T., Nygard, K., Kapperud, G., Lindstedt, B.A., Johannessen, G.S., and Blekkan, H. (2009). *Shigella sonnei* infections in Norway associated with sugar peas, May-June 2009. Euro Surveill. *14*, 19243.

Heisig, P. (1996). Genetic evidence for a role of *parC* mutations in development of high-level fluoroquinolone resistance in *Escherichia coli.* Antimicrob. Agents Chemother. *40*, 879–885.

Hens, D.K., Niyogi, S.K., and Kumar, R. (2005). Epidemic strain *Shigella dysenteriae* type 1 Dt66 encodes several drug resistances by chromosome. Arch. Med. Res. *36*,399–403.

Herwana, E., Surjawidjaja, J.E., Salim, O.C.h., Indriani, N., Bukitwetan, P., and Lesmana, M. (2010). *Shigella*-associated diarrhea in children in South Jakarta, Indonesia. Southeast Asian J. Trop. Med. Public Health. *41*, 418–425.

Hilbi, H., Moss, J.E., Hersh, D., Chen, Y., Arondel, J., Banerjee, S., Flavell, R.A., Yuan, J., Sansonetti, P.J., and Zychlinsky, A. (1998). *Shigella*-induced apoptosis is dependent on caspase-1 which binds to IpaB. J. Biol. Chem. *273*, 32895–32900.

Hoge, C.W., Bodhidatta, L., Tungtaem, C., and Echeverria, P. (1995). Emergence of nalidixic acid resistant *Shigella dysenteriae* type 1 in Thailand: an outbreak associated with consumption of a coconut milk dessert. Int. J. Epidemiol. *24*, 228–232.

Horiuchi, S., Inagaki, Y., Yamamoto, N., Okamura, N., Imagawa, Y., and Nakaya. R. (1993). Reduced susceptibilities of *Shigella sonnei* strains isolated from patients with dysentery to fluoroquinolones. Antimicrob. Agents Chemother. *37*, 2486–2489.

Hsia, R.C., Small, P.L., and Bavoil, P.M. (1993). Characterization of virulence genes of enteroinvasive *E. coli* by TnphoA mutagenesis: identification of *invX*, a gene required for entry intoHEp-2 cells. J. Bacteriol. *175*, 4817–4823.

Hu, G.Z., Chen, H.Y., Si, H.B., Deng, L.X., Wei, Z.Y., Yuan, L., and Kuang, X.H. (2008). Phenotypic and molecular characterization of TEM-116 extended-spectrum beta-lactamase produced by a *Shigella flexneri* clinical isolate from chickens. FEMS Microbiol. Lett. *279*,162–166.

Huang, D.B., and DuPont, H.L. (2005). Rifaximin – a novel antimicrobial for enteric infections. J. Infect. *50*, 97–106.

Huppertz, H.I. (1986). An epidemic of bacillary dysentery in western Rwanda 1981–1982. Cent. Afr. J. Med. *32*, 79–82.

Hurt, J.K., McQuade, T.J., Emanuele, A., Larsen, M.J., and Garcia, G.A. (2010). High-throughput screening of the virulence regulator VirF: a novel antibacterial target for shigellosis. J. Biomol. Screen. *15*, 379–387.

Ingersoll, M., Groisman, E.A., and Zychlinsky, A. (2002). Pathogenicity islands of *Shigella*. Curr. Top. Microbiol. Immunol. *264*, 49–65.

Ivanova, K., Marina, M., Petrov, P., and Kantardjiev, T. (2010). Campylobacteriosis and other bacterial gastrointestinal diseases in Sofia, Bulgaria, for the period 1987–2008. Euro Surveill. *15*, 9474.

Jacoby, G.A., Walsh, K.E., Mills, D.M., Walker, V.J., Oh, H., Robicsek, A., and Hooper, D.C. (2006). *qnrB*, another plasmid-mediated gene for quinolone resistance. Antimicrob. Agents Chemother. *50*, 1178–1182.

Jain, S.K., Gupta, A., Glanz, B., Dick, J., and Siberry, G.K. (2005). Antimicrobial-resistant *Shigella sonnei*: limited antimicrobial treatment options for children and challenges of interpreting *in vitro* azithromycin susceptibility. Pediatr. Infect. Dis J. *24*, 494–497.

Jamal, W.Y., Rotimi, V.O., Chugh, T.D., and Pal, T. (1998). Prevalence and susceptibility of *Shigella* species to 11 antibiotics in a Kuwait teaching hospital. J. Chemother. *10*, 285–290.

Jeong, Y.S., Lee, J.C., Kang, H.Y., Yu, H.S., Lee, E.Y., Choi, C.H., Tae, S.H., Lee, Y.C., Cho, D.T., and Seol, S.Y. (2003). Epidemiology of nalidixic acid resistance and TEM-1- and TEM-52-mediated ampicillin resistance of *Shigella sonnei* isolates obtained in Korea between 1980 and 2000. Antimicrob. Agents Chemother. *47*, 3719–3723.

Jiménez, M., Fuster, C., Raya, C., Díaz,A., and Carbajal, J. (1996). Outbreak of *Shigella boydii* dysentery in the county of El Bierz. Enferm. Infecc. Microbiol. Clin. *14*, 155–159. [Spanish]

Johnson, J.R. (2000). *Shigella* and *Escherichia coli* at the crossroads: machiavellian masqueraders or taxonomic treachery?. J. Med. Microbiol. *49*, 583–585.

Jones, F.R., Ortiz, M., Soriano, I., Utz, G., Saldarriaga, E., Cumpa, R., Collantes, V., Leandro, Y., Bernal, M.M., Ucanan, L.E., *et al.* (2006). Outbreak of gastroenteritis at a Peruvian naval base. Mil. Med. *17*, 1095–1099.

Jung, H.C., Eckmann, L., Yang, S.K., Panja, A., Fierer, J., Morzycka-Wroblewska, E., and Kagnoff, M.F. (1995). A distinct array of proinflammatory cytokines is expressed in human colon epithelial cells in response to bacterial invasion. J. Clin. Invest. *95*, 55–65.

Kagalwalla, A.F., Khan, S.N., Kagalwalla, Y.A., Alola, S., and Yaish, H. (1992). Childhood shigellosis in Saudi Arabia. Pediatr. Infect. Dis. J. *11*, 215–219.

Kansakar, P., Baral, P., Malla, S., and Ghimire, G.R. (2011). Antimicrobial susceptibilities of enteric bacterial pathogens isolated in Kathmandu, Nepal, during 2002–2004. J. Infect. Dev. Ctries. 5,163–168.

Kapperud, G., Rørvik, L.M., Hasseltvedt, V., Høiby, E.A., Iversen, B.G., Staveland, K., Johnsen, G., Leitao, J., Herikstad, H., Andersson, Y., *et al.* (1995). Outbreak of *Shigella sonnei* infection traced to imported iceberg lettuce. J. Clin. Microbiol. 33, 609–614.

Karas, J.A., Pillay, D.G., Naicker, T., and Sturm, A.W. (1999). Laboratory surveillance of *Shigella dysenteriae* type 1 in KwaZulu-Natal. S. Afr. Med. J. *89*, 59–63.

Kariuki, S., and Hart, C.A. (2001). Global aspects of antimicrobial-resistant enteric bacteria. Curr. Opin. Infect. Dis. *14*, 579–586.

Kataoka, S. (2005). Functional effects of Japanese style fermented soy sauce (shoyu) and its components. J. Biosci. Bioeng. *100*, 227–234.

Kayath, C.A., Hussey, S., El hajjami, N., Nagra, K., Philpott, D., and Allaoui, A. (2010). Escape of intracellular *Shigella* from autophagy requires binding to cholesterol through the type III effector, IcsB. Microbes. Infect. *12*, 956–966.

Keller, M.D., and Robbins, M.L. (1956). An outbreak of *Shigella* gastroenteritis. Public Health Rep. *71*, 856–862.

Kernéis, S., Guerin, P.J., von Seidlein, L., Legros, D., and Grais, R.F. (2009). A look back at an ongoing problem: *Shigella dysenteriae* type 1 epidemics in refugee settings in Central Africa (1993–1995). PLoS One *4*, e4494.

Khatun, F., Faruque, A.S., Koeck, J.L., Olliaro, P., Millet, P., Paris, N., Malek, M.A., Salam, M.A., and Luby, S. (2011). Changing species distribution and antimicrobial susceptibility pattern of *Shigella* over a 29-year period (1980–2008). Epidemiol. Infect. *139*, 446–452.

Kim, D.W., Lenzen, G., Page, A.L., Legrain, P., Sansonetti, P.J., and Parsot, C. (2005). The *Shigella flexneri* effector OspG interferes with innate immune responses by targeting ubiquitin-conjugating enzymes. Proc. Natl. Acad. Sci. U.S.A. *102*, 14046–14051.

Kim, H.G., Lee, S.Y., Kim, N.R., Lee, H.Y., Ko, M.Y., Jung, B.J., Kim, C.M., Lee, J.M., Park, J.H., Han, S.H., *et al.* (2011). *Lactobacillus plantarum* lipoteichoic acid down-regulated *Shigella flexneri* peptidoglycan-induced inflammation. Mol. Immunol. *48*, 382–391.

Kim, J.Y., Kim, S.H., Jeon, S.M., Park, M.S., Rhie, H.G., and Lee, B.K. (2008). Resistance to fluoroquinolones by the combination of target site mutations and enhanced expression of genes for efflux pumps in *Shigella flexneri* and *Shigella sonnei* strains isolated in Korea. Clin. Microbiol. Infect. *14*, 760–765.

Kim, M., Ogawa, M., Fujita, Y., Yoshikawa, Y., Nagai, T., Koyama, T., Nagai, S., Lange, A., Fässler, R., and Sasakawa, C. (2009). Bacteria hijack integrin-linked kinase to stabilize focal adhesions and block cell detachment. Nature *459*, 578–582.

Kim, S., Kim, J., Kang, Y., Park, Y., and Lee, B. (2004). Occurrence of extended-spectrum beta-lactamases in members of the genus *Shigella* in the Republic of Korea. J. Clin. Microbiol. *42*, 5264–5269.

Kimura, A.C., Johnson, K., Palumbo, M.S., Hopkins, J., Boase, J.C., Reporter, R., Goldoft, M., Stefonek, K.R., Farrar, J.A., Van Gilder, T.J., *et al.* (2004). Multistate shigellosis outbreak and commercially prepared food, United States. Emerg. Infect. Dis. *10*, 1147–1149.

Köhler, H., Rodrigues, S.P., and McCormick, B.A. (2002). *Shigella flexneri* interactions with the basolateral membrane domain of polarized model intestinal epithelium: role of lipopolysaccharide in cell invasion and in activation of the mitogen-activated protein kinase ERK. Infect. Immun. *70*, 1150–1158.

Kosek, M., Yori, P.P., Pan, W.K., Olortegui, M.P., Gilman, R.H., Perez, J., Chavez, C.B., Sanchez, G.M., Burga, R., and Hall, E. (2008). Epidemiology of highly endemic multiply antibiotic-resistant shigellosis in children in the Peruvian Amazon. Pediatrics. *122*, e541–549.

Kostrzewski, J., and Stypulkowska-Misiurewicz. (1968). Changes in the epidemiology of dysentery in Poland and the situation in Europe. Arch. Immunol. Ther. Exp. (Warsz). *16*, 429–451.

Kotloff, K.L., Winickoff, J.P., Ivanoff, B., Clemens, J.D., Swerdlow, D.L., Sansonetti, P.J., Adak, G.K., and Levine, M.M. (1999). Global burden of *Shigella* infections: implications for vaccine development and implementation of control strategies. Bull. World Health Organ. *77*, 651–666.

Kuo, C.Y., Su, L.H., Perera, J., Carlos, C., Tan, B.H., Kumarasinghe, G., So, T., Van, P.H., Chongthaleong, A., Song, J.H., *et al.* (2006). Antimicrobial susceptibility of *Shigella* isolates in eight Asian countries, 2001–2004. J. Microbiol. Immunol. Infect. *41*, 107–111.

Kuo, H.W., Kasper, S., Jelovcan, S., Höger, G., Lederer, I., König, C., Pridnig, G., Luckner-Hornischer, A., Allerberger, F., and Schmid, D. (2009). A food-borne outbreak of *Shigella sonnei* gastroenteritis, Austria, 2008. Wien. Klin. Wochenschr. *121*, 157–163.

Kweon, M.N. (2008). Shigellosis: the current status of vaccine development. Curr. Opin. Infect. Dis. *21*, 313–318.

Lafont, F., Tran Van Nhieu, G., Hanada, K., Sansonetti, P., and van der Goot, F.G. (2002). Initial steps of *Shigella* infection depend on the cholesterol/sphingolipid raft-mediated CD44–IpaB interaction. EMBO J. *21*, 4449–4457.

Lai, S.W., Lin, H.C., Lin, C.C., Liu, C.S., Shih, Y.T., Peng, C.T., Tsai, C.H., Tsai, F.J., and Ng, K.C. (2000). Clinical analysis of a dysentery outbreak in Taichung. Acta Paediatr. Taiwan. *41*, 18–21.

Lan, R., Alles, M.C., Donohoe, K., Martinez, M.B., and Reeves, P.R. (2004). Molecular evolutionary relationships of enteroinvasive *Escherichia coli* and *Shigella* spp. Infect. Immun. *72*, 5080–5088.

Lan, R., and Reeves, P.R. (2002). *Escherichia coli* in disguise: molecular origins of *Shigella*. Microbes. Infect. *4*, 1125–1132.

Larrosa-Haro, A., Macias-Rosales, R., Sánchez-Ramírez, C.A., Cortés-López, M.C., and Aguilar-Benavides, S. (2010). Seasonal variation of enteropathogens in infants and preschoolers with acute diarrhea in western Mexico. J. Pediatr. Gastroenterol. Nutr. *51*, 534–536.

Lee, J.C., Oh, J.Y., Kim, K.S., Jeong, Y.W., Cho, J.W., Park, J.C., Seol, S.Y., and Cho, D.T. (2001). Antimicrobial resistance of *Shigella sonnei* in Korea during the last two decades. APMIS *109*, 228–234.

Lee, L.A., Ostroff, S.M., McGee, H.B., Johnson, D.R., Downes, F.P., Cameron, D.N., Bean, N.H., and Griffin, P.M. (1991). An outbreak of shigellosis at an outdoor music festival. Am. J. Epidemiol. *133*, 608–615.

Lehmacher, A., and Bockemühl, J. (2007). L-Sorbose utilization by virulent *Escherichia coli* and *Shigella*: different metabolic adaptation of pathotypes. Int. J. Med. Microbiol. *297*,245–254.

Leibovitz, E. (2006). The use of fluoroquinolones in children. Curr. Opin. Pediatr. 18:64–70.

Levine, M.M. (1987). *Escherichia coli* that cause diarrhea: enterotoxigenic, enteropathogenic, enteroinvasive, enterohemorrhagic, and enteroadherent. J. Infect. Dis. *155*, 377–389.

Levine, W.C., Stephenson, W.T., and Craun, G.F. (1990). Waterborne disease outbreaks, 1986–1988. MMWR. CDC. Surveill. Summ. *39*, 1–13.

Lew, J.F., Swerdlow, D.L., Dance, M.E., Griffin, P.M., Bopp, C.A., Gillenwater, M.J., Mercatante, T., and Glass, R.I. (1991). An outbreak of shigellosis aboard a cruise ship caused by a multiple-antibiotic-resistant strain of *Shigella flexneri*. Am. J. Epidemiol. *134*, 413–420.

Lewis, H.C., Ethelberg, S., Olsen, K.E., Nielsen, E.M., Lisby, M., Madsen, S.B., Boel, J., Stafford, R., Kirk, M., Smith, H.V., *et al.* (2009). Outbreaks of *Shigella sonnei* infections in Denmark and Australia linked to consumption of imported raw baby corn. Epidemiol. Infect. *137*, 326–334.

Li, Q., and Verma, I.M. (2002). NF-kappaB regulation in the immune system. Nat. Rev. Immunol. *2*, 725–734.

Lindberg, A.A., Kärnell, A., and Weintraub, A. (1991). The lipopolysaccharide of *Shigella* bacteria as a virulence factor. Rev. Infect Dis. *S4*, 279–284.

Ling, J.M., Shaw, P.C., Kam, K.M., Cheng, A.F., and French, G.L. (1993). Molecular studies of plasmids of multiply-resistant *Shigella* spp. in Hong Kong. Epidemiol. Infect. *110*, 437–446.

Löfdahl, M., Ivarsson, S., Andersson, S., Långmark, J., and Plym-Forshell, L. (2009). An outbreak of *Shigella dysenteriae* in Sweden, May-June *2009*, with sugar snaps as the suspected source. Euro Surveill. *14*, pii: 19268.

Lokareddy, R.K., Lunelli, M., Eilers, B., Wolter, V., and Kolbe, M. (2010). Combination of two separate binding domains defines stoichiometry between type III secretion system chaperone *IpgC* and translocator protein IpaB. J. Biol. Chem. *285*, 39965–39975.

Luck, S.N., Turner, S.A., Rajakumar, K., Sakellaris, H., and Adler, B. (2001). Ferric dicitrate transport system (Fec) of *Shigella flexneri* 2a YSH6000 is encoded on a novel pathogenicity island carrying multiple antibiotic resistance genes. Infect. Immun. *69*, 6012–6021.

Mache, A. (2001). Antibiotic resistance and sero-groups of *Shigella* among paediatric out-patients in southwest Ethiopia. East Afr. Med. J. *78*, 296–299.

Mahmood, A. (2001). Nalidixic acid is still the drug of choice for shigellosis in Pakistan. J. Pak. Med. Assoc. *5*,101.

Mandomando, I., Jaintilal, D., Pons, M.J., Vallès, X., Espasa, M., Mensa, L., Sigaúque, B., Sanz, S., Sacarlal, J., Macete, E., *et al.* (2009). Antimicrobial susceptibility and mechanisms of resistance in *Shigella* and *Salmonella* isolates from children under five years of age with diarrhea in rural Mozambique. Antimicrob. Agents Chemother. *53*, 2450–2454.

Manirakiza, A., Megne-Boudjeka, P., Bobossi-Serengbe, G., Bercion, R., and Le Faou, A. (2010). Prevalence of shigellosis diarrhoea in a paediatrics population: hospital based survey in Bangui, Central African Republic. J. Infect. Dev. Ctries. *4*, 655–657.

Mantis, N., Prévost, M.C., and Sansonetti, P. (1996). Analysis of epithelial cell stress response during infection by *Shigella flexneri*. Infect. Immun. *64*, 2474–2482.

Maraki, S., Georgiladakis, A., Tselentis, Y., and Samonis, G. (2003). A 5-year study of the bacterial pathogens associated with acute diarrhea on the island of Crete, Greece, and their resistance to antibiotics. Eur. J. Epidemiol. *18*, 85–90.

Maranhão, H.S., Medeiros, M.C., Scaletsky, I.C., Fagundes-Neto, U., and Morais, M.B. (2008). The epidemiological and clinical characteristics and nutritional development of infants with acute diarrhoea, in north-eastern Brazil. Ann. Trop. Med. Parasitol. *102*, 357–365.

Marteyn, B., West, N.P., Browning, D.F., Cole, J.A., Shaw, J.G., Palm, F., Mounier, J., Prévost, M.C., Sansonetti, P., and Tang, C.M. (2010). Modulation of *Shigella* virulence in response to available oxygen *in vivo*. Nature *465*, 355–358.

Martin, D.L., Gustafson, T.L., Pelosi, J.W., Suarez, L., and Pierce, G.V. (1986). Contaminated produce – a common source for two outbreaks of *Shigella* gastroenteritis. Am. J. Epidemiol. *124*, 299–305.

Martinez-Argudo, I., and Blocker, A.J. (2010). The *Shigella* T3SS needle transmits a signal for MxiC release, which controls secretion of effectors. Mol. Microbiol. *78*, 1365–1378.

Mata, L.J., Urrutia, J.J., Garcia, B., Fernandez, R., and Behar, M. (1969). *Shigella* infection in breast-fed Guatemalan Indian neonates. Am. J. Dis. Child. *117*, 142–146.

Matar, G.M., Jaafar, R., Sabra, A., Hart, C.A., Corkill, J.E., Dbaibo, G.S., and Araj, G.F. (2007). First detection and sequence analysis of the *bla-CTX-M-15* gene in Lebanese isolates of extended-spectrum-beta-lactamase-producing *Shigella sonnei*. Ann. Trop. Med. Parasitol. *101*, 511–517.

Mathan, V.I., Bhat, P., Kapadia, C.R., Ponniah, J., and Baker, S.J. (1984). Epidemic dysentery caused by the Shiga bacillus in a southern Indian village. J. Diarrhoeal Dis. Res. *2*, 27–32.

Maurelli, A.T., Fernandez, R.E., Bloch, C.A., Rode, C.K., and Fasano, A. (1998). "Black holes" and bacterial pathogenicity: a large genomic deletion that enhances the virulence of *Shigella* spp. and enteroinvasive *Escherichia coli*. Proc. Natl. Acad. Sci. U.S.A. *95*, 3943–3948.

May, K.L., and Morona, R. (2008). Mutagenesis of the *Shigella flexneri* autotransporter *IcsA* reveals novel functional regions involved in *IcsA* biogenesis and recruitment of host neural Wiscott-Aldrich syndrome protein. J. Bacteriol. *190*, 4666–4676.

McEvedy, C.P., Griffith, A., and Hall, T. (1966). Two school epidemics. Br. Med. J. *26*, 1300–1302.

Ménard, R., Dehio, C., amd, Sansonetti, P.J. (1996). Bacterial entry into epithelial cells: the paradigm of *Shigella*. Trends Microbiol. *4*, 220–226.

Meng, C.Y., Smith, B.L., Bodhidatta, L., Richard, S.A., Vansith, K., Thy, B., Srijan, A., Serichantalergs, O., and Mason, C.J. (2011). Etiology of diarrhea in young children and patterns of antibiotic resistance in Cambodia. Pediatr. Infect. Dis. J. *30*, 331–335.

Mensa, L., Marco, F., Vila, J., Gascón, J., and Ruiz, J. (2008). Quinolone resistance among *Shigella* spp. isolated from travellers returning from India. Clin. Microbiol. Infect. *14*, 279–281.

Merson, M.H., Tenney, J.H., Meyers, J.D., Wood, B.T., Wells, J.G., Rymzo, W., Cline, B., DeWitt, W.E., Skaliy, P., Mallison, F. (1975). Shigellosis at sea: an outbreak aboard a passenger cruise ship. Am. J. Epidemiol. *101*, 165–175.

Midzi, S.M., Tshimanga, M., Siziya, S., Marufu, T., and Mabiza, E.T. (2000). An outbreak of dysentery in a rural district of Zimbabwe: the role of personal hygiene at public gatherings. Cent. Afr. J. Med. *46*, 150–153.

Mitobe, J., Morita-Ishihara, T., Ishihama, A., and Watanabe, H. (2009). Involvement of RNA-binding protein Hfq in the osmotic-response regulation of *invE* gene expression in *Shigella sonnei*. BMC Microbiol. *9*, 110.

MoezArdalan, K., Zali, M.R., Dallal, M.M., Hemami, M.R., and Salmanzadeh-Ahrabi, S. (2003). Prevalence and pattern of antimicrobial resistance of *Shigella* species among patients with acute diarrhea in Karaj, Tehran, Iran. J. Health Popul. Nutr. *21*, 96–102.

Molbak, K., Wested, N., Hojlyng, N., Scheutz, F., Gottschau, A., Aaby, P., and da, Silva, A.P. (1994). The etiology of early childhood diarrhea: a community study from Guinea-Bissau. J. Infect. Dis. *169*, 581–587.

Moreno, A.C., Ferreira, L.G., and Martinez, M.B. (2009). Enteroinvasive *Escherichia coli* vs. *Shigella flexneri*: how different patterns of gene expression affect virulence. FEMS Microbiol. Lett. *301*,156–163.

Moreno, A.C., Filho, A.F., Gomes Tdo, A., Ramos, S.T., Montemor, L.P., Tavares, V.C., Filho Ldos, S., Irino, K., and Martinez, M.B. (2010). Etiology of childhood diarrhea in the northeast of Brazil: significant emergent diarrheal pathogens. Diagn. Microbiol. Infect. Dis. *66*, 50–57.

Morera, M.A., Espejo, E., Coll, P., Simó, M., Uriz, M.S., Llovet, T., Rodríguez, M., Martínez, A., March, F., and Bella, F. (1995). Epidemic outbreak of shigellosis following water intake. Enferm. Infect. Microbiol. Clin. *13*, 160–165. [In Spanish.]

Moyo, S.J., Gro, N., Matee, M.I, Kitundu, J., Myrmel, H., Mylvaganam, H., Maselle, S.Y., and Langeland, N. (2011). Age specific aetiological agents of diarrhoea in hospitalized children aged less than five years in Dar es Salaam, Tanzania. BMC Pediatr. *11*, 19.

Much, P., Pichler, J., Kasper, S.S., and Allerberger, F. (2009). Foodborne outbreaks, Austria 2007. Wien. Klin. Wochenschr. *121*, 77–85.

Mudzamiri, W.S., L'Herminez, M., Peterson, D.E., and Nhindiri, P. (1996). Hospitalized dysentery cases during an outbreak of *Shigella dysenteriae* type I: Ndanga District Hospital, Zimbabwe. Cent. Afr. J. Med. *42*, 177–179

Muller, L., Jensen, T., Petersen, R.F., Mølbak, K., and Ethelberg, S. (2009). Imported fresh sugar peas as suspected source of an outbreak of *Shigella sonnei* in Denmark, April-May 2009. Euro Surveill. *14*, 19241.

Munshi, M.H., Sack, D.A., Haider, K., Ahmed, Z.U., Rahaman, M.M., and Morshed, M.G. (1987). Plasmid-mediated resistance to nalidixic acid in *Shigella dysenteriae* type 1. Lancet *2*, 419–421.

Nakamura, S., Nakamura, M., Kojima, T., and Yoshida, H. (1989). *gyrA* and *gyrB* mutations in quinolone-resistant strains of *Escherichia coli*. Antimicrob. Agents Chemother. *33*, 254–255.

Navarre, W.W., and Zychlinsky, A. (2000). Pathogen-induced apoptosis of macrophages: a common end for different pathogenic strategies. Cell Microbiol. *2*, 265–273.

Navia, M.M., Capitano, L., Ruiz, J., Vargas, M., Urassa, H., Schellemberg, D., Gascon, J., and Vila, J. (1999). Typing and characterization of mechanisms of resistance of *Shigella* spp. isolated from feces of children under 5 years of age from Ifakara, Tanzania. J. Clin. Microbiol. *37*, 3113–3117.

Navia, M.M., Ruiz, J., and Vila, J. (2004). Molecular characterization of the integrons in *Shigella* strains isolated from patients with traveler's diarrhea. Diagn. Microbiol. Infect. Dis. *48*, 175–179.

Navia, M.M., Gascón, J., and Vila, J. (2005). Analysis of the mechanisms of resistance to several antimicrobial agents in *Shigella* spp. causing travellers' diarrhoea. Clin. Microbiol. Infect. *11*, 1044–1047.

Nishino, K., Latifi, T., and Groisman, E.A. (2006). Virulence and drug resistance roles of multidrug efflux systems of *Salmonella enterica* serovar Typhimurium. Mol. Microbiol. *59*, 126–141.

Nitiema, L.W., Nordgren, J., Ouermi, D., Dianou, D., Traore, A.S., Svensson, L., and Simpore, J. (2011). Burden of rotavirus and other enteropathogens among children with diarrhea in Burkina Faso. Int. J. Infect. Dis. *15*, e646–652.

Niyogi, S.K., Dutta, P., and Mitra, U. (2000). Trends in antimicrobial resistance of *Shigella* species isolated from children with acute diarrhea. Indian Pediatr. *37*, 339–341.

Nonaka, T., Kuwabara, T., Mimuro, H., Kuwae, A., and Imajoh-Ohmi, S. (2003). *Shigella*-induced necrosis and apoptosis of U937 cells and J774 macrophages. Microbiology *149*, 2513–2527.

Normark, S., Lindquist, S., and Lindberg, F. (1986). Chromosomal beta-lactam resistance in enterobacteria. Scand J. Infect. Dis. Suppl. *49*, 38–45.

Nyachuba, D.G. (2010). Foodborne illness: is it on the rise? Nutr. Rev. *68*, 257–269.

Ochoa, T.J, and Cleary, T.G. (2009). Effect of lactoferrin on enteric pathogens. Biochimie *91*, 30–34.

Ogawa, H., Nakamura, A., and Nakaya, R. (1968). Cinemicrographic study of tissue cell cultures infected with *Shigella flexneri*. Jpn. J. Med. Sci. Biol. *21*, 259–273.

Ogawa, M., Yoshimori, T., Suzuki, T., Sagara, H., Mizushima, N., and Sasakawa, C. (2005). Escape of intracellular *Shigella* from autophagy. Science *307*, 727–731.

Ogawa, M., Handa, Y., Ashida, H., Suzuki, M., and Sasakawa, C. (2008). The versatility of *Shigella* effectors. Nat. Rev. Microbiol. *6*, 11–16.

Oh, J.Y., Yu, H.S., Kim, S.K., Seol, S.Y., Cho, D.T., Lee, J.C. (2003). Changes in patterns of antimicrobial susceptibility and integron carriage among *Shigella sonnei* isolates from southwestern Korea during epidemic periods. J. Clin. Microbiol. *41*, 421–423.

O'Hara, K., Haruta, S., Sawai, T., Tsunoda, M., and Iyobe, S. (1998). Novel metallo beta-lactamase mediated by a *Shigella flexneri* plasmid. FEMS Microbiol. Lett. *162*, 201–206.

Okuda, J., Toyotome, T., Kataoka, N., Ohno, M., Abe, H., Shimura, Y., Seyedarabi, A., Pickersgill, R., and Sasakawa, C. (2005). *Shigella* effector IpaH9.8 binds to a splicing factor U2AF(35) to modulate host immune responses. Biochem. Biophys. Res. Commun. 333, 531–539.

Olarte, J., Filloy, L., and Galindo, E. (1976). Resistance of *Shigella dysenteriae* type 1 to ampicillin and other antimicrobial agents: strains isolated during a dysentery outbreak in a hospital in Mexico City. J. Infect. Dis. *133*, 572–575.

Opintan, J., and Newman, M.J. (2007). Distribution of serogroups and serotypes of multiple drug resistant *Shigella* isolates. Ghana Med. J. *41*, 8–29.

Ostroi, P., Anis, E., and Green, M.S. (1996). Shigellosis in Isreal-update 1995. Public Health Rev. *24*, 213–225.

Oundo, J.O., Saidi, S.M., Mwituria, J.K., and Waiyaki, P.G. (1996). Outbreak of dysentery is Kenya due to multi-resistant polyclonal strain of *Shigella dysenteriae* type 1. Afr. J. Health Sci. *3*, 80–83.

Pan, J.C., Ye, R., Meng, D.M., Zhang, W., Wang, H.Q., and Liu, K.Z. (2006). Molecular characteristics of class 1 and class 2 integrons and their relationships to antibiotic resistance in clinical isolates of *Shigella sonnei* and *Shigella flexneri*. J. Antimicrob. Chemother. *58*, 288–296.

Panhotra, B.R., Saxena, A.K., and Al-Mulhim, K. (2004). Emergence of nalidixic acid resistance in *Shigella sonnei* isolated from patients having acute diarrheal disease: report from eastern province of Saudi Arabia. Jpn. J. Infect. Dis. *57*, 116–118.

Parsot, C., Ménard, R., Gounon, P., and Sansonetti, P.J. (1995). Enhanced secretion through the *Shigella flexneri* Mxi-Spa translocon leads to assembly of extracellular proteins into macromolecular structures. Mol. Microbiol. *16*, 291–300.

Pazhani, G.P., Sarkar, B., Ramamurthy, T., Bhattacharya, S.K., Takeda, Y., and Niyogi, S.K. (2004). Clonal multidrug-resistant *Shigella dysenteriae* type 1 strains associated with epidemic and sporadic dysenteries in eastern India. Antimicrob. Agents Chemother. *48*, 681–684.

Pazhani, G.P., Ramamurthy, T., Mitra, U., Bhattacharya, S.K., and Niyogi, S.K. (2005). Species diversity and antimicrobial resistance of *Shigella* spp. isolated between 2001 and 2004 from hospitalized children with diarrhoea in Kolkata (Calcutta), India. Epidemiol. Infect. *133*, 1089–1095.

Peng, J., Zhang, X., Yang, J., Wang, J., Yang, E., Bin, W., Wei, C., Sun, M., and Jin, Q. (2006). The use of comparative genomic hybridization to characterize genome dynamics and diversity among the serotypes of *Shigella*. BMC Genomics *7*, 218.

Peng, J., Yang, J., and Jin, Q. (2009) The molecular evolutionary history of *Shigella* spp. and enteroinvasive *Escherichia coli*. Infect. Gen. Evol. *9*, 147–152.

Perdomo, J.J., Cavaillon, J.M., Huerre, M., Ohayon, H., Gounon, P., and Sansonetti. P.J. (1994). Acute inflammation causes epithelial invasion and mucosal destruction in experimental shigellosis. J. Exp. Med. *180*, 1307–1319.

Phetsouvanh, R., Midorikawa, Y. and Nakamura, S. (1999). The seasonal variation in the microbial agents implicated in the etiology of diarrheal diseases among children in Lao People's Democratic Republic. Southeast Asian J. Trop. Med. Public Health. *30*, 319–323.

Poirel, L., Liard, A., Rodriguez-Martinez, J.M., and Nordmann, P. (2005). *Vibrionaceae* as a possible source of Qnr-like quinolone resistance determinants. J. Antimicrob Chemother. *56*, 1118–1121.

Porter, B., Schinder, E., Nagar H., Gilad, Y., and Torek, V. (1984). An outbreak of shigellosis in an ultra-orthodox Jewish community. Soc. Sci. Med. *18*, 1061–1062.

Pourakbari, B., Mamishi, S., Mashoori, N., Mahboobi, N., Ashtiani, M.H., Afsharpaiman, S., and Abedini, M. (2010). Frequency and antimicrobial susceptibility of *Shigella* species isolated in Children Medical Center Hospital, Tehran, Iran, 2001–2006. Braz. J. Infect. Dis. *14*, 153–7.

Prado, V., Solari, V., Alvarez, I.M., Arellano, C., Vidal, R., Carreño, M., Mamani, N., Fuentes, D., O'Ryan, M., and Muñoz, V. (2002). Epidemiological situation of foodborne diseases in Santiago, Chile in 1999–2000. Rev. Med. Chil. *130*, 495–501. [In Spanish.]

Pu, X.Y., Pan, J.C., Wang, H.Q., Zhang, W., Huang, Z.C., and Gu, Y.M. (2009). Characterization of fluoroquinolone-resistant *Shigella flexneri* in Hangzhou area of China. J. Antimicrob. Chemother. *63*, 917–920.

Pupo, G.M., Karaolis, D.K., Lan, R., and Reeves, P.R. (1997). Evolutionary relationships among pathogenic and nonpathogenic *Escherichia coli* strains inferred from multilocus enzyme electrophoresis and *mdh* sequence studies. Infect. Immun. *65*, 2685–2692.

Pupo, G.M., Lan, R., and Reeves, P.R. (2000). Multiple independent origins of *Shigella* clones of *Escherichia coli* and convergent evolution of many of their characteristics. Proc. Natl. Acad. Sci. U.S.A. *97*, 10567–10572.

Purdy, G.E., and Payne, S.M. (2001). The SHI-3 iron transport island of *Shigella boydii* 0–1392 carries the genes for aerobactin synthesis and transport. J. Bacteriol. *183*, 4176–4182.

Rahaman, M.M., Khan, M.M., Aziz, K.M., Islam, M.S., and Kibriya, A.K. (1975). An outbreak of dysentery caused by *Shigella dysenteriae* type 1 on a coral island in the Bay of Bengal. J. Infect. Dis. *132*, 15–19.

Rahman, M., Mauff, G., Levy, J., Couturier, M., Pulverer, G., Glasdorff, N., and Butzler, J.P. (1994). Detection of 4-quinolone resistance mutation in *gyrA* gene of *Shigella dysenteriae* type 1 by PCR. Antimicrob. Agents Chemother. *38*, 2488–2491.

Rahman, M., Shoma, S., Rashid, H., El Arifeen, S., Baqui, A.H., Siddique, A.K., Nair, G.B., and Sack, D.A. (2007). Increasing spectrum in antimicrobial resistance of *Shigella* isolates in Bangladesh: resistance to azithromycin and ceftriaxone and decreased susceptibility to ciprofloxacin. J. Health Popul. Nutr. *25*,158–167.

Rajakumar, K., Bulach, D., Davies, J., Ambrose, L., Sasakawa, C., and Adler, B. (1997a). Identification of a chromosomal *Shigella flexneri* multi-antibiotic resistance locus which shares sequence and organizational similarity with the resistance region of the plasmid NR1. Plasmid 37,159–168.

Rajakumar, K., Sasakawa, C., and Adler, B. (1997b). Use of a novel approach, termed island probing, identifies the *Shigella flexneri* she pathogenicity island which encodes a homolog of the immunoglobulin A protease-like family of proteins. Infect. Immun. *65*, 4606–4614.

Ramarao, N., Le Clainche, C., Izard, T., Bourdet-Sicard, R., Ageron, E., Sansonetti, P.J, Carlier, M.F., and Tran Van Nhieu, G. (2007). Capping of actin filaments by vinculin activated by the *Shigella* IpaA carboxyl-terminal domain. FEBS Lett. *581*, 853–857.

Ranjbar, R., Hosseini, M.J., Kaffashian, A.R., and Farshad, S. (2010). An outbreak of shigellosis due to *Shigella flexneri* serotype 3a in a prison in Iran. Arch. Iran. Med. *13*, 413–416.

Reeve, G., Martin, D.L., Pappas, J., Thompson, R.E., and Greene, K.D. (1989). An outbreak of shigellosis associated with the consumption of raw oysters. N. Engl. J. Med. *321*, 224–227.

Reller, M.E., Nelson, J.M., Mølbak, K., Ackman, D.M., Schoonmaker-Bopp, D.J., Root, T,P., and Mintz, E.D. (2006). A large, multiple-restaurant outbreak of infection with *Shigella flexneri* serotype 2a traced to tomatoes. Clin. Infect. Dis. *42*, 163–169.

Ries, A.A., Wells, J.G., Olivola, D., Ntakibirora, M., Nyandwi, S., Ntibakivayo, M., Ivey, C.B., Greene, K.D., Tenover, F.C., Wahlquist, S.P., *et al.* (1994). Epidemic *Shigella dysenteriae* type 1 in Burundi: panresistance and implications for prevention. J. Infect. Dis. *169*, 1035–1041.

Roberts, A.P., Chandler, M., Courvalin, P., Guédon, G., Mullany, P., Pembroke, T., Rood, J.I., Smith, C.J., Summers, A.O., Tsuda, M., and Berg, D.E. (2008). Revised nomenclature for transposable genetic elements. Plasmid *60*, 167–173.

Roberts, A.P., and Mullany, P. (2009). A modular master on the move: the Tn916 family of mobile genetic elements. Trends Microbiol. *17*, 251–258.

Robicsek, A., Strahilevitz, J., Jacoby, G.A., Macielag, M., Abbanat, D., Park, C.H., Bush, K., and Hooper, D.C. (2006). Fluoroquinolone-modifying enzyme: a new adaptation of a common aminoglycoside acetyltransferase. Nat. Med. *12*, 83–88.

Rodríguez-Martínez, J.M., Cano, M.E., Velasco, C., Martínez-Martínez, L., and Pascual, A. (2011). Plasmid-mediated quinolone resistance: an update. J. Infect. Chemother. *17*, 149–182.

Rogerie, F., Ott, D., Vandepitte, J., Verbist, L., Lemmens, P., and Habiyaremye, I. (1986). Comparison of norfloxacin and nalidixic acid for treatment of dysentery caused by *Shigella dysenteriae* type 1 in adults. Antimicrob. Agents Chemother. *29*, 883–886.

Rolfe, R.D. (2000). The role of probiotic cultures in the control of gastrointestinal health. J. Nutr. *130*, 396S–402S.

Rosewell, A., Ropa, B., Posanai, E., Dutta, S.R., Mola, G., Zwi, A., and MacIntyre, C.R. (2010). *Shigella* spp. antimicrobial drug resistance, Papua New Guinea, 2000–2009. Emerg. Infect. Dis. *16*, 1797–1799.

Rowe, J.S., Shah, S.S., Motlhagodi, S., Bafana, M., Tawanana, E., Truong, H.T., Wood, S.M., Zetola, N.M., and Steenhoff, A.P. (2010). An epidemiologic review of enteropathogens in Gaborone, Botswana: shifting patterns of resistance in an HIV endemic region. PLoS One 5, e10924.

Rowe-Magnus, A.D., Davies, J., and Mazel, D. (2002). Impact of integrons and transposons on the evolution of resistance and virulence. Curr. Top. Microbiol. Immunol. *264*,167–188.

Rudnai, O., Straub, I., Laszlo, V.G., Hajnal, A., and Lanyi, B. (1981). *Salmonella* and *Shigella* surveillance in Hungary. 1972–1976. II. *Shigella* surveillance. Acta. Microbiol. Acad. Sci. Hung. *28*, 53–65.

Ruiz, J. (2003). Mechanisms of resistance to quinolones: target alterations, decreased accumulation and DNA gyrase protection. J. Antimicrob. Chemother. *51*, 1109–1117.

Sabra, A.H., Araj, G.F., Kattar, M.M, Abi-Rached, R.Y., Khairallah, M.T., Klena, J.D., and Matar, G.M. (2009). Molecular characterization of ESBL-producing *Shigella sonnei* isolates from patients with bacilliary dysentery in Lebanon. J. Infect. Dev. Ctries. 3, 300–305.

Saha, T., Murhekar, M., Hutin, Y.J., and Ramamurthy, T. (2009). An urban, water-borne outbreak of diarrhoea and shigellosis in a district town in eastern India. Natl. Med. J. India. 22, 237–239.

Salam, M.A., and Bennish, M.L. (1991). Antimicrobial therapy for shigellosis. Rev. Infect. Dis. *13*, S332–341.

Samie, A., Guerrant, R.L., Barrett, L., Bessong, P.O., Igumbor, E.O., and Obi, C.L. (2009). Prevalence of intestinal parasitic and bacterial pathogens in diarrhoeal and non-diarrhoeal human stools from Vhembe District, South Africa. J. Health Popul. Nutr. *27*, 739–745.

Samonis, G., Elting, L., Skoulika, E., Maraki, S., and Tselentis, Y. (1994). An outbreak of diarrhoeal disease attributed to *Shigella sonnei*. Epidemiol. Infect. *112*, 235–245.

Sandlin, R.C., Goldberg, M.B., and Maurelli, A.T. (1996). Effect of O side-chain length and composition on the virulence of *Shigella flexneri* 2a. Mol. Microbiol. *22*, 63–73.

Sansonetti, P.J. (2000). Phagocytosis, a cell biology view. J. Cell Sci. *113*, 3355–3356.

Sansonetti, P.J, Kopecko, D.J., and Formal, S.B. (1982). Involvement of a plasmid in the invasive ability of *Shigella flexneri*. Infect. Immun. *35*, 852–860.

Sansonetti, P.J., Ryter, A., Clerc, P., Maurelli, A.T., and Mounier. J. (1986). Multiplication of *Shigella flexneri* within HeLa cells: lysis of the phagocytic vacuole and plasmid-mediated contact hemolysis. Infect. Immun. *51*, 461–469.

Sansonetti, P.J., Hale, T.L., Dammin, G.J., Kapfer,C., Collins, H.H.Jr., and Formal, S.B. (1983). Alterations in the pathogenicity of *Escherichia coli* K-12 after transfer of plasmid and chromosomal genes from *Shigella flexneri*. Infect. Immun. *39*, 1392–1402.

Sansonetti, P.J., Tran Van Nhieu, G., and Egile, C. (1999). Rupture of the intestinal epithelial barrier and mucosal invasion by *Shigella flexneri*. Clin. Infect. Dis. *28*, 466–475.

Schwaber, M.J., Grotto, I., Balicer, R.D., Davidovitch, N., Zelikovitch, Y., and Huerta, M. (2005). Infectious diarrheal outbreaks in the Israeli military, 1988–2002. Mil Med. *170*, 634–637.

von Seidlein, L., Kim, D.R., Ali, M., Lee, H., Wang, X., Thiem, V.D., Canh, do G., Chaicumpa, W., Agtini, M.D., Hossain, A., *et al.* (2006). A multicentre study of *Shigella* diarrhoea in six Asian countries: disease burden, clinical manifestations, and microbiology. PLoS Med. *3*, e353.

Sen, D., Dutta, P., Deb, B.C., and Pal, S.C. (1988). Nalidixic-acid resistant *Shigella dysenteriae* type 1 in eastern India. Lancet *2*, 911.

Sengupta, P.G., Mandal, S., Sen, D., Das, P., Deb, B.C., and Pal, S.C. (1990). Multidrug resistant epidemic shigellosis in a village in west Bengal, 1984. Indian J. Public Health. *34*,15–19.

Serdarevic, F., Jones, R.C., Weaver, K.N., Black, S.R., Ritger, K.A., Guichard, F., Dombroski, P., Emanuel, B.P., Miller, L., and Gerber, S.I. (2011). Multipathogen waterborne disease outbreak associated with a dinner cruise on Lake Michigan. Epidemiol. Infect. *140*, 1–5.

Sharon, N. (2006). Carbohydrates as future anti-adhesion drugs for infectious diseases. Biochim. Biophys. Acta *1760*, 527–537.

Sharp, T.W., Thornton, S.A., Wallace, M.R., Defraites, R.F., Sanchez, J.L., Batchelor, R.A., Rozmajzl, P.J., Hanson, R.K., Echeverria, P., Kapikian, A.Z., *et al.* (1995). Diarrheal disease among military personnel during Operation Restore Hope, Somalia, 1992–1993. Am. J. Trop. Med. Hyg. *52*, 188–193.

Shi, Z., Wang, H., Hu, K., Feng, E., Yao, X., Huang, L., Su, G., Hunag, P., and Huang, C. (2004). Screening and identification of *Shigella flexneri* 2a virulence-related genes induced after invasion of epithelial cells. Sci. China C. Life Sci. *47*, 494–502.

Simchen, E., Jeeraphat, S., Shihab, S., and Fattal, B. (1991). An epidemic of waterborne *Shigella* gastroenteritis in Kibbutzim of western Galilee in Israel. Int. J. Epidemiol. *20*, 1081–1087.

Siu, L.K., Ho, P.L., Yuen, K.Y., Wong, S.S., and Chau, P.Y. (1997). Transferable hyperproduction of TEM-1 beta-lactamase in *Shigella flexneri* due to a point mutation in the pribnow box. Antimicrob. Agents Chemother. *41*, 468–470.

Skoudy, A., Mounier, J., Aruffo, A., Ohayon, H., Gounon, P., Sansonetti, P., and Tran Van Nhieu, G. (2000). CD44 binds to the *Shigella* IpaB protein and participates in bacterial invasion of epithelial cells. Cell Microbiol. *2*, 19–33.

Spika, J.S., Dabis, F., Hargrett-Bean, N., Salcedo, J., Veillard, S., and Blake, P.A. (1987). Shigellosis at a Caribbean resort. Hamburger and North American origin as risk factors. Am. J. Epidemiol. *126*, 1173–1780.

Srison, D., and Pornpatkul, V. (1995). Shigellosis in Thai children: experience from a rural hospital 1985–1993. Southeast Asian J. Trop. Med. Public Health *26*, 347–349.

Storch, G.A., Gunn, R.A., Martin, W.T., Pollard, R.A., and Sinclair, S.P. (1980). Shigellosis in the Marshall Islands: epidemiologic aspects of an outbreak. Am. J. Trop. Med. Hyg. *29*, 456–463.

Stypułkowska-Misiurewicz, H. (2005). Dysentery in Poland in 2003. Przegl. Epidemiol. *59*, 263–268. [In Polish.]

Stypułkowska-Misiurewicz, H. (2007). Dysentery in Poland in 2005. Przegl. Epidemiol. *61*, 249–256. [In Polish.]

Stypułkowska-Misiurewicz, H. (2008). Dysentery and amoebiasis in Poland in 2006. Przegl. Epidemiol. *62*, 267–274. [In Polish.]

Stypułkowska-Misiurewicz, H., and Baumann, A. (2010). Dysentery and amoebiasis in Poland in 2008. Przegl. Epidemiol. *64*, 199–203. [In Polish.]

Stypułkowska-Misiurewicz, H., and Gonera, E. (2003). Dysentery in Poland in 2001. Przegl. Epidemiol. *57*, 77–84. [In Polish.]

Suzuki, T., Nakanishi, K., Tsutsui, H., Iwai, H., Akira, S., Inohara, N., Chamaillard, M., Nuñez, G., and Sasakawa, C. (2005). A novel caspase-1/toll-like receptor 4-independent pathway of cell death induced by cytosolic *Shigella* in infected macrophages. J. Biol. Chem. *280*, 14042–14050.

Suzuki, T., and Núñez, G. (2008). A role for Nod-like receptors in autophagy induced by *Shigella* infection. Autophagy *4*,73–75.

Talukder, K.A., Islam, M.A., Khajanchi, B.K., Dutta, D.K., Islam, Z., Safa, A., Alam, K., Hossain, A., Nair, G.B., and Sack, D.A. (2003). Temporal shifts in the dominance of serotypes of *Shigella dysenteriae* from 1999 to 2002 in Dhaka, Bangladesh. J. Clin. Microbiol. *41*, 5053–5058.

Talukder, K.A., Khajanchi, B.K., Islam, M.A., Dutta, D.K., Islam, Z., Safa, A., Khan, G.Y., Alam, K., Hossain, M.A., Malla, S., *et al.* (2004). Genetic relatedness of ciprofloxacin-resistant *Shigella dysenteriae* type 1 strains isolated in south Asia. J. Antimicrob. Chemother. *54*,730–734.

Talukder, K.A., Islam, Z., Dutta, D.K., Islam, M.A., Khajanchi, B.K., Azmi, I.J., Iqbal, M.S., Hossain, M.A., Faruque, A.S., Nair, G.B., *et al.* (2006). Antibiotic resistance and genetic diversity of *Shigella sonnei* isolated from patients with diarrhoea between 1999 and 2003 in Bangladesh. J. Med. Microbiol. *55*, 1257–1263

Taneja, N., Lyngdoh, V., Vermani, A., Mohan, B., Rao, P., Singh, M., Dogra, A., Singh, M.P., and Sharma, M. (2005). Re-emergence of multi-drug resistant *Shigella dysenteriae* with added resistance to ciprofloxacin in north India & their plasmid profiles. Indian J. Med. Res. *122*, 348–354.

Taylor, D.N., Bodhidatta, L., Brown, J.E., Echeverria, P., Kunanusont, C., Naigowit, P., Hanchalay, S., Chatkaeomorakot, A., and Lindberg, A.A. (1989). Introduction and spread of multi-resistant *Shigella dysenteriae* I in Thailand. Am. Trop. Med Hyg. *40*, 77–85.

Terajima, J., Tamura, K., Hirose, K., Izumiya, H., Miyahara, M., Konuma, H., and Watanabe, H. (2004). A multi-prefectural outbreak of *Shigella sonnei* infections associated with eating oysters in Japan. Microbiol. Immunol. *48*, 49–52.

Theriot, J.A. (1995). The cell biology of infection by intracellular bacterial pathogens. Annu. Rev. Cell Dev. Biol. *11*, 213–239.

Thirumalai, K., Kim, K.S., and Zychlinsky, A. (1997). IpaB, a *Shigella flexneri* invasin, colocalizes with interleukin-1 beta-converting enzyme in the cytoplasm of macrophages. Infect. Immun. *65*, 787–793.

Tiruneh, M. (2009). Serodiversity and antimicrobial resistance pattern of *Shigella* isolates at Gondar University teaching hospital, Northwest Ethiopia. Jpn. J. Infect. Dis. *62*, 93–97.

Traa, B.S., Walker, C.L., Munos, M., and Black, R.E. (2010). Antibiotics for the treatment of dysentery in children. Int. J. Epidemiol. *39*, 70–74.

Tshimanga, M., Peterson, D.E., and Dlodlo, R.A. (1997). Using epidemiologic tools to control an outbreak of diarrhoea in a textile factory, Bulawayo, Zimbabwe. East. Afr. Med. J. *74*, 719–722.

Turner, S.A., Luck, S.N., Sakellaris, H., Rajakumar, K., and Adler, B. (2001). Nested deletions of the SRL pathogenicity island of *Shigella flexneri* 2a. J. Bacteriol. *183*, 5535–5543.

Turner, S.A., Luck, S.N., Sakellaris, H., Rajakumar, K., and Adler, B. (2003). Molecular epidemiology of the SRL pathogenicity island. Antimicrob. Agents Chemother. *47*, 727–734.

Udo, S.M., and Eja, M.E. (2004). Prevalence and antibiotic resistant shigellae among primary school children in urban Calabar, Nigeria. Asia Pac. J. Public Health. *16*, 41–44.

Valentiner-Branth, P., Steinsland, H., Fischer, T.K., Perch, M., Scheutz, F., Dias, F., Aaby, P., Mølbak, K., and Sommerfelt, H. (2003). Cohort study of Guinean children: incidence, pathogenicity, conferred protection, and attributable risk for enteropathogens during the first 2 years of life. J. Clin. Microbiol. *41*, 4238–4245.

Vargas, M., Gascon, J., Casals, C., Schellenberg, D., Urassa, H., Kahigwa, E., Ruiz, J., and Vila, J. (2004). Etiology of diarrhoea in children less than five years of age in Ifakara, Tanzania. Am.J. Trop. Med. Hyg. *70*, 536–539.

Vimont-Vicary, P., and Rogerie, F. (1985). 3-year study of shigellosis epidemic in Rwanda, Central Africa. Problems of public health and bacteriological aspects. Me. Trop. *45*, 235–243. [French]

Vokes, S.A., Reeves, S.A., Torres, A.G., and Payne, S.M. (1999). The aerobactin iron trans-port system genes in *Shigella flexneri* are present within a pathogenicity island. Mol. Microbiol. *33*, 63–73.

Vrints, M., Mairiaux, E., Van Meervenne, E., Collard, J.M., and Bertrand, S. (2009). Surveillance of antibiotic susceptibility patterns among *Shigella sonnei* strains isolated in Belgium during the 18-year period 1990 to 2007. J. Clin. Microbiol. *47*, 1379–1385.

Warren, B.R., Parish, M.E., and Schneider, K.R. (2006). *Shigella* as a foodborne pathogen and current methods for detection in food. Crit. Rev. Food Sci. Nutr. *46*, 551–567.

Watanabe, T. (1963). Episome-mediated transfer of drug resistance in *Enterobacteriaceae*. VI. High-frequency resistance transfer system in *Escherichia coli*. J. Bacteriol. *85*,788–794.

Watarai, M., Tobe, T., Yoshikawa, M., and Sasakawa, C. (1995). Contact of *Shigella* with host cells triggers release of Ipa invasins and is an essential function of invasiveness. EMBO J. *14*, 2461–2470.

Wei, J., Goldberg, M.B., Burland, V., Venkatesan, M.M., Deng, W., Fournier, G., Mayhew, G.F., Plunkett G., 3rd, Rose, D.J., Darling, A., *et al.* (2003). Complete genome sequence and comparative genomics of *Shigella flexneri* serotype 2a strain 2457T. Infect. Immun. *71*, 2775–2786.

Weissman, J.B., Craun, G.F., Lawrence, D.N., Pollard, R.A., Saslaw, M.S., and Gangarosa, E.J. (1976). An epidemic of gastroenteritis traced to a contaminated public water supply. Am. J. Epidemiol. *103*, 391–398.

Wharton, M., Spiegel, R.A., Horan, J.M., Tauxe, R.V., Wells, J.G., Barg, N., Herndon, J., Meriwether, R.A., MacCormack, J.N., and Levine, R.H. (1990). A large outbreak of antibiotic-resistant shigellosis at a mass gathering. J. Infect. Dis. *162*, 1324–1328.

Wheatley, W., and Veer, J.B., Jr. (1969). A foodborne outbreak of shigellosis on an Indian reservation. Public Health Rep. *84*, 563–567.

White, F.M., and Pedersen, A.T. (1976). Epidemic shigellosis on a worktrain in Labrador. Can. Med. Assoc. J. *115*, 647–649.

Winstanley, C., and Hart, C.A. (2001). Type III secretion systems and pathogenicity islands. J. Med. Microbiol. *50*, 116–126.

World Health Organization. (2004). International note. Antibiotics in the management of shigellosis. Can. Commun. Dis. Rep. *30*, 202–203.

Xia, S., Xu, B., Huang, L., Zhao, J.Y., Ran, L., Zhang, J., Chen, H., Pulsrikarn, C., Pornruangwong, S., Aarestrup, F.M., *et al.* (2011). Prevalence and characterization of human *Shigella* infections in Henan Province, China, in 2006. J. Clin. Microbiol. *49*, 232–242.

Xiong, Z., Li, J., Li, T., Shen, J., Hu, F., and Wang, M. (2010). Prevalence of plasmid-mediated quinolone-resistance determinants in *Shigella flexneri* isolates from Anhui Province, China. J. Antibiot. (Tokyo) *63*, 187–189.

Yamada, S., Ohta, K., Obata, H., Matsushita, S., Hirata, I., and Kudoh, Y. (1993). On the outbreak of shigellosis in Tokyo, 1992. Kansenshogaku Zasshi *67*, 1183–1189. [Japanese]

Yamane, K., Wachino, J., Suzuki, S., Kimura, K., Shibata, N., Kato, H., Shibayama, K., Konda, T., and Arakawa, Y. (2007). New plasmid-mediated fluoroquinolone efflux pump, QepA, found in an *Escherichia coli* clinical isolate. Antimicrob. Agents Chemother. *51*,3354–3360.

Yang, F., Yang, J., Zhang, X., Chen, L., Jiang, Y., Yan, Y., Tang, X., Wang, J., Xiong, Z., Dong, J., *et al.* (2005). Genome dynamics and diversity of *Shigella* species, the etiologic agents of bacillary dysentery. Nucleic Acids Res. *33*, 6445–6458.

Yang, J., Nie, H., Chen, L., Zhang, X., Yang, F., Xu, X., Zhu, Y., Yu, J., and Jin, Q. (2007). Revisiting the molecular evolutionary history of *Shigella* spp. J. Mol. Evol. *64*,71–79.

Yoshida, H., Bogaki, M., Nakamura, M., and Nakamura, S. (1990). Quinolone resistance-determining region in the DNA gyrase *gyrA* gene of *Escherichia coli*. Antimicrob Agents Chemother. *34*, 1271–1272.

Yoshida, S., Handa, Y., Suzuki, T., Ogawa, M., Suzuki, M., Tamai, A., Abe, A., Katayama, E., and Sasakawa, C. (2006). Microtubule-severing activity of *Shigella* is pivotal for intercellular spreading. Science *314*, 985–989.

Zhu, J.Y., Duan, G.C., Yang, H.Y., Fan, Q.T., and Xi, Y.L. (2011). Atypical class 1 integron coexists with class 1 and class 2 integrons in multi-drug resistant *Shigella flexneri* isolates from China. Curr. Microbiol. *62*, 802–806.

Ziv, T., Heymann, A.D., Azuri, J., Leshno, M., and Cohen, D. (2011). Assessment of the underestimation of childhood diarrhoeal disease burden in Israel. Epidemiol. Infect. *139*, 1379–1387.

Zurawski, D.V., Mitsuhata, C., Mumy, K.L., McCormick, B.A., and Maurelli, A.T. (2006). OspF and OspC1 are *Shigella flexneri* type III secretion system effectors that are required for postinvasion aspects of virulence. Infect. Immun. *74*, 5964–5976.

Zurawski, D.V., Mumy, K.L., Badea, L., Prentice, J.A., Hartland, E.L., McCormick, B.A., and Maurelli, A.T. (2008). The NleE/OspZ family of effector proteins is required for polymorphonuclear transepithelial migration, a characteristic shared by enteropathogenic *Escherichia coli* and *Shigella flexneri* infections. Infect. Immun. *76*, 369–379.

Zychlinsky, A., Prevost, M.C., and Sansonetti, P.J. (1992). *Shigella flexneri* induces apoptosis in infected macrophages. Nature *358*, 167–169.

Zychlinsky, A., Perdomo, J.J., and Sansonetti, P.J. (1994). Molecular and cellular mechanisms of tissue invasion by *Shigella flexneri*. Ann. N. Y. Acad. Sci. *730*, 197–208.

Salmonellae: Taxonomy, Genomics and Antimicrobial Resistance

5

Mahbubur Rahman

Abstract

The genus *Salmonella* has three species, namely *Salmonella enterica, Salmonella bongori* and *Salmonella subterranea*. The type species *S. enterica* is further classified into six subspecies: *enterica* (subsp. I), *arizonae* (subsp. IIIa), *diarizonae* (subsp. IIIb), *houtenae* (subsp. IV), *indica* (subsp. VI) and *salamae* (subsp. II). *Salmonella* strains belong to over 50 serogroups based on the O antigen, and to over 2500 serovars (each having a unique combination of somatic O, flagellar H1 and H2 antigens). Most of these serovars (1531) belong to *Salmonella* subsp. *enterica,* and cause more than 99% of the diseases in humans including gastroenteritis and enteric fever (typhoid). Genome sequencing and comparative genomic analysis of 28 *S. enterica* serovars identified similarity of core regions of the genomes, together with evidence of recombination and rearrangement, genomic degradation, pseudogenes and clonal diversity both within and among the serovars. Genomic comparisons of host-restricted (*S.* Typhi, *S.* Paratyphi and *S.* Gallinarum) and host-adapted (*S.* Typhimurium and *S.* Enteridis) *S. enterica* serovars indicate that genomic degradation is a common evolutionary mechanism for host adaptation and increased pathogenicity of *Salmonella.* Drug resistances in *Salmonella* are mainly due to *Salmonella* genomic island 1 (an integrative mobile element) carrying various antibiotic resistance gene clusters, and to conjugative R plasmids which confer resistance to many antibiotics including extended-spectrum cephalosporins. Continuous genetic re-assortment in *Salmonella* leading to increased virulence and the emergence of resistance to multiple drugs are of significant public health concern.

Classification and nomenclature of *Salmonella*

The genus *Salmonella,* which belongs to the family *Enterobacteriaceae,* was named after Daniel E. Salmon, an American veterinarian who first isolated *S. choleraesuis* from pigs with hog cholera in 1884 (Tindall *et al.,* 2005). The genus *Salmonella* was designated by Lignières in 1900. Salmonellae are Gram-negative motile bacilli that are oxidase negative and catalase -positive, produce gas from D-glucose and utilize citrate as a sole carbon source. Salmonellae have several endotoxins: antigens O, H, and Vi. *Salmonella* is an important pathogen of both humans and animals.

The nomenclature and classification of *Salmonella* species have been modified and restructured many times during the last century (Ezaki *et al.,* 2000). The antigenic classification system of various *Salmonella* serovars used today has accumulated from many years of studies on antibody interactions with surface antigens of *Salmonella* organisms established originally by Kauffman and White (Popoff and Le Minor, 2001). All antigenic formulae of recognized *Salmonella* serotypes are listed in a document called the Kauffmann–White scheme. The WHO Collaborating Centre for Reference and Research on *Salmonella* at the Pasteur Institute, Paris, France, has been updating the scheme (Popoff *et al.,* 2004). In the latest report published in 2004, there were a total of 2541 serovars in the genus *Salmonella* (Table 5.1).

Traditionally, *Salmonella* species were named in accordance with the Kaufmann–White typing system, defined by different combinations of somatic O, surface Vi, and flagellar H antigens. Thus, *Salmonella* organisms are differentiated by

Table 5.1 *Salmonella* nomenclature

Taxonomic position and nomenclature				No. of serovers in each species or subspecies
Genus	Species	Subspecies	Serovers/serotypes	
Salmonella	*enterica*	*enterica* (subspecies I)	Enteritidis, Typhimurium, Typhi, Paratyphi, Choleraesuis	1531
		salamae (subspecies II)	9,46:z:z39	502
		arizonae (subspecies IIIa)	43:z29:–	95
		diarizonae (subspecies IIIb)	6,7:l,v:1,5,7	333
		houtenae (subspecies IV)	21:m,t:–	72
		indica (subspecies VI)	59:z36:–	13
	bongori	–	13,22:z39:–	23
	subterranea (6)	–	–	–

serotyping using antisera to two highly variable surface antigens, the O antigen reflecting variation in the exposed part of the lipopolysaccharide, and the H antigen, reflecting variation in flagellin, the major protein of the flagellum, with about 46 and 114 recognized antigenic forms respectively. Most *Salmonella* carry two flagellin genes that code for H1 and H2 variants of the H antigen. Isolates are identified by serotyping into more than 2500 reported serovars, each possessing a unique combination of O, H1 and H2 antigens (Grimont and Weill, 2007).

According to the Judicial Commission of the International Committee of Systematic Bacteriology the genus *Salmonella* is currently divided into three species: *Salmonella enterica, Salmonella bongori* and *Salmonella subterranean* (Reeves *et al.*, 1989; Shelobolina, *et al.*, 2004) based on DNA homology and numerical taxonomy. In 2005, '*Salmonella enterica*' gained official approval as the type species of the genus *Salmonella. S. enterica* is further divided into six distinct subspecies (subsp.): *Salmonella enterica* subsp. *enterica* (subsp. I), *Salmonella enterica* subsp. *salamae* (subsp. II), *Salmonella enterica* subsp. *arizonae* (subsp. IIIa), *Salmonella enterica* subsp. *diarizonae,* (subsp. IIIb), *Salmonella enterica* subsp. *houtenae* (subsp. IV), *Salmonella enterica* subsp. *indica* (subsp. VI). In subspecies I, serovars are designated by a name usually indicative of associated diseases, geographic origins, or usual habitats. In the remaining subspecies, as well as those of *S. bongori,* antigenic formulae determined according to the Kauffmann–White scheme are used for unnamed serovars. To avoid confusion between serovars and species, the serovar name is not italicized and starts with a capital letter. When cited the first time in a report, the genus and species and subspecies names are given followed by the word 'serovar/type' and then the serovar/type name, e.g. *Salmonella enterica* subsp. *enterica*/I serotype Typhimurium. Thereafter, the name may be shortened with the genus name followed directly by the serovar/type name, e.g. *Salmonella* Typhimurium. For designating strains by their antigenic formulae, the subspecies name is written in Roman letters (not italicized) followed by their antigenic formulae, including O (somatic) antigens, H (flagellar) antigens (phase 1), and H antigens (phase 2, if present). A colon is used between each antigen, e.g. *Salmonella* serotype II 39:z10:z6. For serotypes in *S. bongori* (which previously belonged to subgenus V), V is still used for consistency, e.g. *S.* V 13,22:z35:–.

Genomics of *Salmonella*

Salmonella emerged as an intracellular pathogen after diverging from a common ancestor with the genus *Escherichia* about 100–150 million years ago. *Salmonella* is closely related to *Escherichia coli,* but has an additional large number of virulence genes. Many virulence genes are located in genomic islands (GIs), which are large segments of DNA acquired by horizontal gene transfer and are usually located near tRNA genes, which are believed to facilitate the integration of the GIs into the chromosome. Many *Salmonella*-specific GIs known as *Salmonella* pathogenicity islands (SPIs) play a role in virulence, host specificity and degree of invasiveness (Marcus *et al.*, 2000). The islands were found to be typically well conserved in several, but not all, isolates. Twenty-one SPIs are known to date in *Salmonella*. *S.* Typhimurium and *S.* Typhi genomes contain 11 common SPIs (SPIs-1 to 6, 9, 11, 12, 13 and 16). SPIs-8 and 10 were initially found in *S.* Typhi, and were considered as absent in *S.* Typhimurium. However, at both of these chromosomal locations in *S.* Typhimurium, there is a completely different set of genes. There is only one SPI specific to *S.* Typhimurium, namely SPI-14, and four other SPIs are specific to *S.* Typhi (SPIs-7, 15, 17 and 18). SPIs-19, 20 and 21 are absent in both serovars (Hensel, 2004).

Salmonella strain has a large stable core genome (~2800 gene families), whilst there is an abundance of accessory genes, including SPIs, transposable elements, phages, and plasmid DNA. The pan-genome size is estimated to be around 10,000 gene families. In *Salmonella enterica*, ~10% of any individual genome is variable, including many horizontally acquired or mobile elements such as prophages, transposons, plasmids, and insertion elements. There is conservation in the genomes in relation to chromosomal location and DNA structure and variation in gene content in variable genomic regions or islands. These include the SPIs but also encompass phage insertion sites and transposable elements. Gene content is often, but not uniformly correlated to *Salmonella* serovar.

Chromosomes of *Salmonella* are dynamic, undergoing rearrangements that change the order and expression of genes. The complete sequence of 45 *Salmonella* genomes covering 21 serotypes of *Salmonella* subsp. *enterica* has been determined. These include S. Typhimurium strain LT2, S. Enteritidis PT4, *S.* Typhi strains CT18 and Ty2 and *S.* Paratyphi A strain ATCC9150 (McClelland *et al.*, 2001; Parkhill *et al.*, 2001; Thomson *et al.*, 2008; Holt *et al.*, 2009). The availability of these genome sequences and of newer sequencing technologies that make draft genome sequence simpler and more cost effective have provided new opportunities to understand the evolution of *Salmonella*.

The chromosome sizes range from ~4546 kbp (*S.* Javiana draft chromosome) to ~4951 kbp (*S.* Weltevreden draft chromosomes). In comparison, the two host-restricted *S.* Paratyphi A strains had the smallest chromosomes (~4582 kbp). The 13 other and two human host-restricted *S.* Typhi Ty2 256 and CT18 strains, however, carried chromosomes of intermediate sizes (4792 kbp and 4809 kbp), suggesting that host adaptation and increased pathogenicity in *S. enterica* serovars are not reflected in smaller genome sizes. The two Typhi and Paratyphi A strains are presumably at different stages of their evolution towards specialization to a host-adapted pathogenic lifestyle. Thus, differences in chromosomal size are reflective of the degree of genome degradation. In this case, the different evolutionary stages would, however, not be reflected in the pseudogene contents, which are similar in both serovars (Fig. 5.1).

The main features of some *Salmonella enterica* genomes are summarized in Table 5.2 to show the similarities and differences. The most striking feature of the analysis is the predominant similarity and synteny of core regions of the genomes, including many of the *Salmonella* pathogenicity islands (SPI). There is an extremely close relationship between the genomes of *S.* Enteritidis and *S.* Gallinarum, suggesting the latter is a direct evolutionary descendent of the former. In comparison to *S.* Enteritidis PT4, *S.* Gallinarum 287/91 harbours higher number of predicted pseudogenes. In contrast, the number of pseudogenes in *S.* Gallinarum 287/91 is slightly higher than that of the human-restricted *S.* Typhi CT18 and lower that *S. enterica* serovar Paratyphi A (173 pseudogenes).

Organization of *Salmonella* genomes helps us to understand the mechanisms by which different species or serovars have evolved. Differences in the *S.* Typhi genome occur mainly because of genomic rearrangements involving recombination between different rRNA operons or IS*200* elements. Each serovar evolves through the acquisition of genetic elements by horizontal gene transfer or by gene degradation. The genomes of *S.* Typhimurium strain LT2 and *S.* Typhi strain CT18 are composed of 4 857 432 and 4 809 037 bp, respectively. Both serovars share about 89% of genes (Ochman *et al.*, 2000; McClelland *et al.*, 2004; Sabbagh *et al.*, 2010).

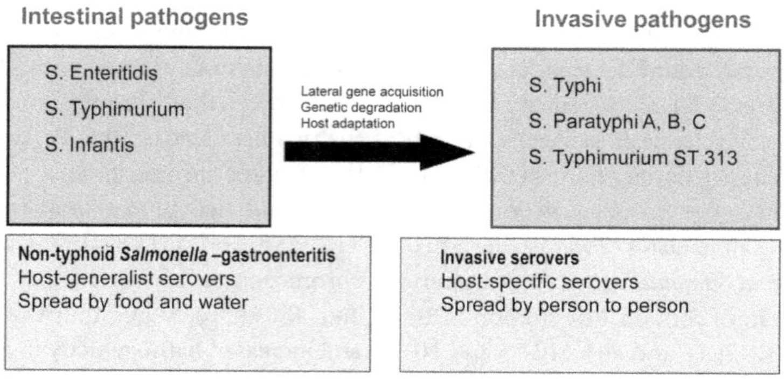

Figure 5.1 Genomic degradation, host adaptation and lateral gene transfer are associated with evolution of new *Salmonella* serovars with increased pathogenic potential.

Table 5.2 General properties genomes of some *Salmonella* enterica serovars

Salmonella enterica serovars				
Serovar	*S.* Enteritidis PT4	*S.* Typhimurium LT2	*S.* Gallinarum 287/91	*S.* Typhi CT18
Genome size (MB)	4.69	4.86	4.66	4.81
Genes	4363	4635	4466	5065
Average gene size	953	947	939	958
Unique gene families in genome (*n*)	29	43	135	349
Salmonella pathogenicity islands (*n*)	14	11	14	15
Pseudogenes	113	25	309	204
Percent G + C content (%)	52.17	52.22	52.20	52.09
No. of coding sequence	4318	4451	4274	4599
Coding density (%)	85.5	86.8	79.9	87.6
rRNA operons	7	7	7	7
tRNA	84	85	75	78

Differences between *S.* Typhimurium and *S.* Typhi include ~480 genes unique to *S.* Typhimurium and ~600 genes unique to *S.* Typhi (Sabbagh *et al.*, 2010). *Salmonella* pathogenicity islands (SPIs), plasmids, functional prophages and phage remnants contribute significantly to the genetic diversity among *S. enterica* strains.

Most *S.* Typhimurium strains contain a self-transmissible ~90 kb virulence plasmid (pSLT) carrying virulence genes such as the *spv* operon involved in intramacrophage survival, and the plasmid-encoded fimbriae (*pef*) operon. Additionally, *S.* Typhimurium may also carry multidrug-resistance plasmids of high molecular weight (up to 200 kb) as well as smaller plasmids (<20 kb) of unknown functions. *Salmonella enterica* serovar Typhi strain CT18 harbours pHCM1, an incHI1 plasmid of about 218 kb carrying genes for resistance to antibiotics and heavy metals.

Sequence diversity at multiple, conserved housekeeping genes by multilocus sequence typing suggests that *S.* Typhi emerged relatively recently, i.e. 15,000–150,000 years ago. Full sequence analysis suggests that *S.* Typhi and the evolution of known *S. enterica* sublineages is mostly mediated by two mechanisms: (i) the loss of coding sequences (gene degradation) with known metabolic functions and (ii) the acquisition of horizontally transferred genes, phage and plasmid (Fig. 5.1).

Antimicrobial resistance in *Salmonella*

Multidrug-resistant *S.* Typhimurium DT104 emerged in Europe and the United States and in the early 1990s (Threlfall *et al.*, 1998; Mølbak *et al.*, 1999; Baucheron *et al.*, 2004). This unique strain is characterized by concomitant resistance to five antimicrobial agents, ampicillin (A), chloramphenicol (C), streptomycin (S), sulfonamide (Su), and tetracycline (T) with an R-type referred to as ACSSuT. As of the mid-1990s, *S.* Typhimurium DT104 had become one of the most prevalent strains among human isolates of *Salmonella*, second only to *S.* Enteritidis strain PT4 in the United Kingdom. In the United States, the prevalence of *S.* Typhimurium isolates of R-type ACSSuT had increased to 34% in 1996. A study conducted by the National Antimicrobial Resistance Monitoring System (NARMS) in 1997–1998 indicated that 65% of *S.* Typhimurium isolates of R-type ACSSuT had similar phage type to DT104. Molecular typing analysis revealed that most isolates of *S.* Typhimurium DT104 had very similar genetic fingerprints, suggesting that clonal spread might have been responsible for the increasing prevalence of this unique resistant strain of *S.* Typhimurium. Problems arising owing to the increasing antimicrobial resistance of *Salmonella* include not only difficulties in antimicrobial therapy but also the apparent predilection of the organism to cause serious diseases. In the United States, a recent report from the NARMS indicates that the risk of bloodstream infections in patients infected with antimicrobial-resistant non-typhoid *Salmonella*, particularly *S.* Typhimurium, is 2-fold greater than in those infected with pan-susceptible strains. Information from the Foodborne Diseases Active Surveillance Network (FoodNet) also indicates a more than 4-fold risk of hospitalization in patients with resistant *Salmonella* infections. Similarly, a Danish study reported that infection with quinolone-resistant *S.* Typhimurium was associated with a 3-fold higher risk of invasive illness or death within 90 days of infection, compared with that observed for infection with pan-susceptible strains.

Multidrug-resistant *S.* Typhi as a global health problem

Salmonella enterica serovars Typhi and Paratyphi, although not prevalent in the developed world, continue to cause enteric fever in many developing countries, particularly in Asia and northern regions of Africa. The dissemination of multidrug-resistant *S.* Typhi in Asia and Africa pose an increased risk to developed countries via travellers, since its first emergence in 1987 in China (Rahman *et al.*, 2001, 2001, 2006; Crump and Mint, 2010). MDR typhoid fever continues to be a problem in the Indian subcontinent and North Africa. A recent survey of eight Asian countries revealed nalidixic acid resistance in 5–51% of *S.* Typhi isolates, and many nalidixic acid-resistant strains are resistant to multiple antibiotics. Isolates resistant to nalidixic acid are also less responsive to fluoroquinolone antibiotics. Nalidixic acid-susceptible *Salmonella* isolates with reduced susceptibility to fluoroquinolones are also increasingly being recognized, and the presence of these isolates may require modification in current clinical microbiology practices used to detect non-susceptible strains. The latter resistance pattern is caused by plasmid-borne *qnr* resistance determinants, which raise MICs for fluoroquinolones but not nalidixic acid. An unexpected recent finding is that mutations conferring antibiotic resistance do not necessarily appear to reduce *Salmonella* fitness in the absence of antibiotic selection. In south Asia and China, MDR *Salmonella enterica* serovar Paratyphi has emerged as an important health problem.

S. Typhimurium ST313: a unique MDR strain causing death in sub-Saharan Africa

Most non-typhoidal *Salmonella* (NTS) are associated with gastroenteritis. However, there has been a dramatic increase in reports of NTS-associated invasive disease in sub-Saharan Africa (Gordon *et al.*, 2008). S. Typhimurium ST313 (Kingsley *et al.*, 2009), a unique multidrug-resistant strain causes bacteraemia/

meningitis in children and HIV- and *Plasmodium falciparum*-infected adult in sub-Saharan Africa resulting in high morbidity and mortality (case fatality rates are around 20–25% for children and up to 50% in adults). *S.* Typhimurium isolates are responsible for a significant proportion of the reported invasive NTS in this region. Multilocus sequence analysis of invasive S. Typhimurium from Malawi and Kenya identified a dominant type, designated ST313, which currently is rarely reported outside of Africa. Whole-genome sequencing of a multiple drug-resistant (MDR) ST313 NTS isolate, D23580, identified a distinct prophage repertoire and a composite genetic element encoding MDR genes located on a virulence associated plasmid. Further, there was evidence of genome degradation, including pseudogene formation and chromosomal deletions, when compared with other S. Typhimurium genome sequences. Some of this genome degradation involved genes previously implicated in virulence of S. Typhimurium or genes for which the orthologues in S. Typhi are either pseudogenes or are absent. Genome analysis of other epidemic ST313 isolates from Malawi and Kenya provided evidence for microevolution and clonal replacement. ST313 clusters most closely with the host-specific *S.* Typhimurium strain ST128, which causes serious systemic disease in pigeons suggesting that ST313 may be in the very early stages of host adaptation. It is particularly intriguing that some of the genes lost by ST313 have been implicated in *S.* Typhimurium virulence or are also missing or non-functional in S. Typhi, suggesting that ST313 may be in the process of evolving from an intestinal pathogen spread by food and water to an invasive enteric fever pathogen spread from person to person.

References

Baucheron, S., Chaslus-Dancla, E., and Cloeckaert, A. (2004). Role of TolC and *parC* mutation in high-level fluoroquinolone resistance in *Salmonella enterica* serotype Typhimurium DT204. J. Antimicrob. Chemother. 53, 657–659.

Crump, J.A., and Mintz, E.D. (2010). Global trends in typhoid and paratyphoid fever. Clin. Infect. Dis. *50*, 241–246.

Ezaki, T., Kawamura, Y., and Yabuuchi, E. (2000). Recognition of nomenclatural standing of *Salmonella typhi* (Approved Lists 1980), *Salmonella enteritidis* (Approved Lists 1980) and *Salmonella typhimurium* (Approved Lists 1980), and conservation of the specific epithets enteritidis and typhimurium. Request for an opinion. Int. J. Syst. Evol. Microbiol. *50*, 945–947.

Gordon, M.A., Graham, S.M., Walsh, A.L., Wilson, L.K., Phiri, A., Molyneux, E.M., *et al.* (2008). Epidemics of invasive *Salmonella enterica* serovar Enteritidis and *Salmonella enterica* serovar Typhimurium infection associated with multidrug resistance among adults and children in Malawi. Clin. Infect. Dis. *46*, 963–969.

Grimont, P.A.D., and Weill, F.-X. (2007). Antigenic formulae of the *Salmonella* serovars, 9th edn. WHO Collaborating Center for Reference and Research on Salmonella. Institut Pasteur, Paris.

Hensel, M. (2004) Evolution of pathogenicity islands of *Salmonella* enterica. Int. J. Med. Microbiol. *294*, 95–102.

Holt, K.E., Thomson, N.R., Wain, J., Langridge, G.C., Hasan, R., Bhutta, Z.A., Quail, M.A., Norbertczak, H., Walker, D., Simmonds, M., *et al.* (2009). Pseudogene accumulation in the evolutionary histories of *Salmonella* enterica serovars Paratyphi A and Typhi. BMC Genomics *10*, 36.

Kingsley, R.A., Msefula, C.L., Thomson, N.R., Kariuki, S., Holt, K.E., Gordon, M.A., Harris, D., Clarke, L., Whitehead, S., Sangal, V., *et al.* (2009) Epidemic multiple drug resistant *Salmonella* Typhimurium causing invasive disease in sub-Saharan Africa have a distinct genotype. Genome. Res. *19*, 2279–2287.

McClelland, M., Sanderson, K.E., Spieth, J., Clifton, S.W., Latreille, P., Courtney, L., Porwollik, S., Ali, J., Dante, M., Du, F., *et al.* (2001).Complete genome sequence of *Salmonella enterica* serovar Typhimurium LT2. Nature 413, 852–856.

McClelland, M., Sanderson, K.E., Clifton, S.W., Latreille, P., Porwollik, S., Sabo, A., Meyer, R., Bieri, T., Ozersky, P., McLellan, M., *et al.* (2004). Comparison of genome degradation in Paratyphi A and Typhi, human-restricted serovars of *Salmonella enterica* that cause typhoid. Nat. Genet. *36*, 1268–1274.

Marcus, S.L., Brumell, J.H., Pfeifer, C.G., Finlay, B.B. (2000). *Salmonella* pathogenicity islands: big virulence in small packages. Microbes. Infect. 2, 145–156.

Mølbak, K., Baggesen, D.L., Aarestrup, F.M., Ebbesen, J.M., Engberg, J., Frydendahl, K., Gerner-Smidt, P., Petersen, A.M., and Wegener, H.C. (1999). An outbreak of multidrug-resistant, quinolone-resistant *Salmonella enterica* serotype Typhimurium DT104. N. Engl. J. Med. *341*, 1420–1425.

Ochman, H., Lawrence, J.G., and Groisman, E.A. (2000). Lateral gene transfer and the nature of bacterial innovation. Nature *405*, 299–304.

Parkhill, J., Dougan, G., James, K.D., Thomson, N.R., Pickard, D., Wain, J., Churcher, C., Mungall, K.L., Bentley, S.D., Holden, M.T., *et al.*(2001). Complete genome sequence of a multiple drug resistant *Salmonella enterica* serovar Typhi CT18. Nature *413*, 848–852.

Popoff, M.Y., and Le Minor. (2001). L. Antigenic formulas of the *Salmonella* serovars, 8th revision. Paris:

World Health Organization Collaborating Centre for Reference and Research on *Salmonella*, Pasteur Institute, Paris.

Popoff, M.Y., Bockemuhl, J., Gheesling, L.L. (2004) Supplement 2002 (no. 46) to the Kauffmann-White scheme. Res. Microbiol. *155*, 568–70.

Rahman, M., Islam, H., Ahmed, D. and Sack, R.B. (2001). Emergence of multiresistant *Salmonella* Gloucester and *Salmonella* Typhimurium in Bangladesh. J. Health Popul. Nutri. *19*, 191–198.

Rahman, M., Ahmad, A., and Shoma, S. (2002). Decline in epidemic of multidrug resistant *Salmonella* Typhi is not associated with emergence of susceptible strain in Bangladesh. Epidemiol. Infect. *129*, 29–34.

Rahman, M., Siddique, A.K., Shoma, S., Rashid, H., Salam, M.A. Ahmed, Q.S. Nair, G.B. and Breiman, R.F. (2006). Emergence of multidrug-resistant *Salmonella enterica* serotype Typhi with decreased ciprofloxacin susceptibility in Bangladesh. Epidemiol. Infect. *134*, 433–438.

Reeves, M.W., Evins, G.M., Heiba, A.A., Plikaytis, B.D., Farmer, J.J., III. (1989) Clonal nature of *Salmonella* typhi and its genetic relatedness to other salmonellae as shown by multilocus enzyme electrophoresis, and proposal of *Salmonella bongori* comb. nov. J. Clin. Microbiol. *27*, 313–320

Sabbagh, S.C., Forest, C.G., Lepage, C., Leclerc, J.M., and Daigle, F. (2010). So similar, yet so different: uncovering distinctive features in the genomes of *Salmonella* enterica serovars Typhimurium and Typhi. FEMS Microbiol. Lett. *305*, 1–13.

Shelobolina, E.S., Sullivan, S.A., O'Neill, K.R., Nevin, K.P., Lovley, D.R. (2004). Isolation, characterization, and U(VI)-reducing potential of a facultatively anaerobic, acid-resistant bacterium from low-pH, nitrate- and U(VI)-contaminated subsurface sediment and description of *Salmonella subterranean* sp. nov. Appl. Environ. Microbiol. *70*, 2959- 2965.

Thomson, N.R., Clayton, D.J., Windhorst, D., Vernikos, G., Davidson, S., Churcher, C., Quail, M.A., Stevens, M., Jones, M.A., Watson, M., *et al.* (2008). Comparative genome analysis of *Salmonella* Enteritidis PT4 and *Salmonella* Gallinarum 287/91 provides insights into evolutionary and host adaptation pathways. Genome Res. *18*, 1624–1637.

Threlfall, E.J., Ward, L.R., and Rowe, B. (1998). Multiresistant *Salmonella typhimurium* DT 104 and bacteraemia. Lancet 352, 287–288.

Tindall, B.J., Grimont, P.A., Garrity, G.M., and Euzeby, J.P. (2005). Nomenclature and taxonomy of the genus *Salmonella*. Int. J. Syst. Evol. Microbiol. *55*, 521–524.

Molecular Aspects of Pathogenesis and Drug Resistance in *Salmonella* Species

6

Indrani Karunasagar, Patit Paban Bhowmick and Deekshit Vijaya Kumar

Abstract

Foodborne and waterborne infections due to *Salmonella* species are a major concern worldwide. Several virulence genes have been identified in *Salmonella* and located in clusters called *Salmonella* pathogenicity island (SPI). There are 17 pathogenicity islands of *Salmonella* reported to date. *Salmonella* possess two distinct type three secretion system (T3SS) encoded by genes present in two different SPI *viz.* SPI-1 and SPI-2 that play an important role in adhesion, invasion and survival in the host cells. It has been also reported that a variety of *Salmonella* phenotypes associated with bioluminescence, biofilm formation, conjugation, motility, competence, and antibiotic production, are regulated in response to signalling molecules of quorum-sensing systems. The development of resistance to various antibiotics (particularly in Asian countries) including extended-spectrum cephalosporins worldwide is a cause of concern. Some variants of *Salmonella* have developed multidrug resistance as an integral part of the genetic material of the organism, and are therefore likely to retain their drug resistance genes even when antimicrobial drugs are no longer used. The role of plasmids, bacteriophages, transposons and integrons in the transfer of resistance genes is discussed.

Introduction

Salmonella are food/waterborne Gram-negative rods belonging to the family Enterobacteriaceae which cause clinical conditions ranging from mild gastrointestinal infections to systemic infections such as typhoid fever (Lacey, 1993). Over 4500 known serovars of *Salmonella* belong to either of the two species of *Salmonella*, i.e. *S. enterica* and *S. bongori*. Based on host adaptation, *Salmonella* serotypes are classified into three groups: (i) typhoidal (enteric) *Salmonella* responsible for typhoid and paratyphoid fever in humans and generally not pathogenic for animals; (ii) non-typhoidal *Salmonella* causing gastrointestinal illness in humans and a broad range of animals including other mammals, reptiles, birds and insects; and (iii) *Salmonella*-causing infections in specific host such as *S.* Abortovis and *S.* Gallinarum to sheep and poultry respectively. *Salmonella* infection is typically acquired by the intake of contaminated food like beef, fruit, vegetables and dairy products or water. However, exposure to pet reptiles and amphibians, which are often carriers of the bacteria, also cause salmonellosis in humans. Eggs and poultry are the most common source of infection (Linam and Gerber, 2007). Contaminated frozen potpies, puffed vegetables and snacks have been the main source of recent outbreaks (CDCP, 2008). Isolation of non typhoidal serovars from fish, shellfish and other seafood has been reported in several countries. Filter feeders such as oysters, shrimps and shellfish concentrate particulate matter including pathogenic bacteria present in the water which might explain the presence of *Salmonella* in seafood. *S.* Typhimurium and *S. enterica* serovar Enteritidis are the most common serovars isolated from human cases in many countries including Australia, New Zealand, Canada, the European Union (EU) and the United States (US). In developing countries such as Thailand,

Malaysia and India, human cases due to seafood associated *Salmonella* are frequently reported. Understanding the prevalence, activities and survival of seafood associated *Salmonella* is important for assessment of risk to human health. Volunteer studies point out that 10^5 to 10^{10} bacteria are required to initiate an infection (Glaser and Newman, 1982) but the exact number varies with the strain of the bacteria and the physiological condition of the host. Earlier studies have shown that even as low as 200 bacteria are enough for non-typhoidal gastroenteritis (Blacklow *et al.*, 1972). Although high gastric pH can reduce the infectious dose, *Salmonella* has an adaptive acid tolerance response which helps it to survive the low pH condition of the stomach (Garcia-del Portillo *et al.*, 1993). Variation in virulence among different serovars and ability to cause infections in different hosts is responsible for the evolution of virulence genes. *S.* Typhimurium harbours at least 80 different virulence genes, most of which are present in the chromosome in clusters called *Salmonella* pathogenicity island (SPI) (Table 6.1). Significant morbidity and mortality are associated with salmonellosis.

Table 6.1 Virulence factors responsible for *Salmonella* infection

Virulence factors	Examples	Functions	References
Adherence factors	Type 1 fimbriae (encoded by *fim* genes)	Adhesion to the murine small intestine	Baumler (1996)
	Plasmid-encoded fimbriae (*pef* genes)	Adhesion to the murine small intestine	Baumler (1996)
	Long polar fimbriae (*lpf* genes)	Adhesion to the macrophages	Edwards (2000)
	Thin aggregative fimbriae (*agf*, *stm1987*, *waaG* gene etc.)	Adhesion to abiotic surface	Garcia *et al.* (2004)
	S-fimbria-specific gene (*sfa*)	Binds to sialic acid-specific receptors on uroepithelial cells and on brain cells	Morschhauser *et al.* (2004)
	Flagella	Providing driving force to increase the penetrate rate	Khoramian-Falsafi *et al.* (1990)
Type III secretion	SPI-1 coded T3SS	Invasion of *Salmonella* inside the host	Galan and Collmer (1999)
System (T3SS)	SPI-2 coded T3SS	Survival of *Salmonella* inside the SCV	Kuhle and Hensel (2004)
Virulence plasmid	*spvRABCD* operon (8 kb)	Encodes the major virulence phenotype of the plasmid to multiply in the tissue macrophages	Williamson *et al.* (1988)
VI antigen	Capsular polysaccharide composed of N-acetylglucosamine uronic acid	Reduces the level of secretion of *S.* Typhi-induced tumour necrosis factor alpha and increases the level of resistance of the organism to oxidative killing	Hirose *et al.* (1997)
Gifsy-1 and Gifsy-2	Superoxide dismutase (*sod* gene)	Converts toxic superoxide to less toxic peroxide, may help the bacteria to survive inside the macrophage oxidative burst	Salyers and Whitt (2002)
Gene involved in metabolic adaptation	*aroA* gene	Involves in the synthesis of 5-enolpyruvylshikimate 3-phosphate, required for synthesis of aromatic amino acids	Chatfield (1990)
Toxins	Endotoxins (LPS)	Inhibit the activity of acetylcholinesterase	Van Asten and van Dijk (2005)
	Heat labile enterotoxin related to *E. coli*	Disrupt membrane permeability	
	Verotoxins related to Shiga-like toxins	The toxin acts on the lining of the blood vessels, the vascular endothelium	Ashkenazi and Cleary (1990)

Molecular aspects of pathogenesis

Adhesion of *Salmonella* with host cells

The preliminary step in pathogenesis by *Salmonella* is the adhesion of bacteria to specific host receptor. *S.* Typhimurium produces a number of adhesins. These include type 1 fimbriae (encoded by *fim* genes), plasmid-encoded fimbriae (*pef* genes), long polar fimbriae (*lpf* genes) and thin aggregative fimbriae (curli, encoded by *agf* genes). Generally bacteria attach to target cells via bacterial surface components (pili or fimbriae and flagella). For example, the PE (plasmid encoded) fimbriae of serovar Typhimurium is involved in bacterial adhesion to the murine small intestine (Baumler *et al.*, 1997), while SEF14 fimbriae in serovar Enteritidis, play a role in the bacterial uptake by macrophages (Edwards *et al.*, 2000). Flagella provide the driving force for increasing the penetration rate of the bacteria into the host mucus layer to reach the host cell surface. In many studies, bacterial motility was found to be essential for adherence or invasion (Khoramian-Falsafi, 1990; Moreira *et al.*, 2010). Generally, the initial interaction between *Salmonella* and the intestinal epithelium occur through M (Microfold) cells in which their basolateral surface is invaginated to form a pocket-like structure, to which macrophages and lymphocytes can migrate and where they interact with antigenic particles (dos Reis and Horn, 2010).

Cellular invasion pathway of *Salmonella*

Following adhesion, infection of *Salmonella* involves internalization, a process induced by the bacteria themselves and referred to as bacteria-mediated endocytosis. Microscopic studies reveal that *Salmonella* have the ability to enter the M cells of the intestine by bacterial-mediated endocytosis (Francis *et al.*, 1992). Shortly after the bacteria adhere to the apical epithelial surface, using various fimbrial adhesins, they disrupt the normal epithelial brush border to induce the membrane ruffling that engulf the organisms (Takeuchi, 1967; Baumler *et al.*, 1996). In this process bacteria use the type three secretion system (T3SS), a supramolecular needle-like structure encoded by *Salmonella* pathogenicity island 1 (SPI-1) (Kubori *et al.*, 1998).

Assembly of SPI-1 encoded T3SS

SPI-1 is a 40 kb region in the *Salmonella* genome, located at centisome 63 and flanked by the *flhA* and *mutS* genes (Mills *et al.*, 1995). It is believed that SPI-1 is acquired very early on in the evolution of *Salmonella* spp. at the point of divergence of *Salmonella* and *Escherichia* spp. from a common ancestor (Li *et al.*, 1995; Bäumler, 1997; Galán, 1999). Using electron microscopy, Kubori *et al.* (1998) have shown that this needle apparatus is multiringed with a cylindrical base that is anchored to both the inner and outer membranes and a slender, hollow, needle-like barrel projecting outward from the outer membrane. For the proper assembly of the needle complex, SPI-1 contains a group of genes present in *inv, spa, prg* and *org* operons called as SPI-1-associated T3SS apparatus protein (Table 6.2) (Eichelberg *et al.*, 1994; Collazo *et al.*, 1995; Jones and Falkow, 1996; Fu and Galan, 1998a,b). The *prgHIJK* operon is mainly responsible for construction of SPI-1 encoded T3SS. PrgH and PrgK are known to be major base component of the needle structure and PrgI and PrgJ are part of the external needle structure. The construction of T3SS is controlled by the operon *invJICBAEGFH*. While InvJ controls the length of the T3SS, InvA and InvC control assembly of T3SS and release of effector proteins (Eichelberg *et al.*, 1994) and InvI regulates expression of PrgI (Sukhan *et al.*, 2001). InvH increases the T3SS assembly process and a mutation in the gene reduces the normal secretion of several genes like, *prgH, prgK, prgI, invG* and *sipC* (Sukhan *et al.*, 2001). Another operon found to be involved with the T3SS apparatus is the *orgABC*. Initially it was identified as a single operon (*orgA*); later Klein *et al.* (2000) split this region into two ORFs *orgA* and *orgB* along with the identification of another ORF, *orgC*, overlapping the *orgB* gene. The proteins OrgA and OrgB are involved in invasion and secretion of other apparatus proteins (e.g. protein PrgI) (Klein *et al.*, 2000; Sukhan *et al.*, 2001). The *spa* locus (*spaOPQR*) has also been shown to be part of the T3SS

Table 6.2 Genes coded of *Salmonella* pathogenicity island, SPI-1 and their cellular functions

Gene	Cellular functions	References
Apparatus proteins		
prgH, *prgK* and *invG*	Components of the base of the needle structure	Kimbrough and Miller (2000)
prgI and *prgJ*	Parts of the external needle substructure	Sukhan *et al.* (2003)
invJ	Require for cell invasion	Sukhan *et al.* (2003)
invA and *invC*	Assembly of needle complex	Kubori *et al.* (2000)
invH	Increase the efficiency of T3SS assembly process	Sukhan *et al.* (2001)
orgA and *orgB*	Require in invasion and secretion	Jones and Falkow (1994)
orgC	Require in secretion	Jones and Falkow (1994)
Effector proteins		
avrA	Inhibits activation of the proinflammatory antiapoptotic NF-κB transcription factor, initiating and enhancing cell apoptosis	Hardt and Galán (1997)
sipA	Decreases the critical concentration of G-actin and increases the stability of F-actin; inducing membrane ruffling and bacterial internalization	Hersh *et al.* (1999)
sipC	Nucleates and bundles actin, contributing to bacterial invasion	Hayward and Koronakis (1999), Chang *et al.* (2005a,b)
sopA	Stimulates PMN transmigration by HECT-like E3 ubiquitin ligase activity	Wood *et al.* (2000)
sopB	Activates Cdc42, RhoG, AktA and chloride secretion through its inositol phosphatase activity and disrupts tight junctions	Galyov *et al.* (1997)
sopD	Stimulates fluid accumulation in bovine ligated ileal loops and contributes to diarrhoea in calves and systemic disease in mice	Jones *et al.* (1998)
sopE	Activates Cdc42, Rac1 and RhoG by its GEF activity and disrupts tight junctions	Wood *et al.* (1996)
sopE2	Activates Cdc42, Rac1 and RhoG by its GEF activity and disrupts tight junctions	Stender *et al.* (2000)
sptP	Disrupts the actin cytoskeleton of the cell	Fu and Galán (1998b), Murli *et al.* (2001)
Chaperone proteins		
invB	Acts as a chaperone for range of substrate proteins, including SipA, SopA, SopE and SopE2	Bronstein *et al.* (2000)
sicP	Acts as a specific chaperone for SptP	Fu and Galán (1998a)
sicA	Function as a putative chaperone for the T3SS associated Sip proteins	Kaniga *et al.* (1995b)
iagB	Involved in the passage of the needle complex through the bacterial peptidoglycan layer	Sukhan (2000)
lacP	Involved in the modification of effectors before translocation	Kaniga *et al.* (1995a), Collazo and Galán (1997)
Regulator proteins		
invE	Controls translocation process by regulating and interacting with the SipB-SipC-SicA protein complex	Kubori and Galán (2002)
invF	Controls the expression of T3SS-associated proteins encoded within and outside of SPI-1	Darwin and Miller (2001)
hilA	Central regulator of SPI-1	Ellermeier and Slauch (2008)

Gene	Cellular functions	References
hilC	Derepresses *hilA* expression by binding directly to the URS in the *hilA* promoter region	Schechter *et al.* (1999)
hilD	Essential for the expression of *hilA*	Boddicker *et al.* (2003)

apparatus and is essential for host invasion and secretion of proteins like InvJ, SipB and SipC (Collazo and Galán, 1996).

Role of effectors in cellular invasion of *Salmonella* spp.

Salmonella inject 12 effectors through the SPI-1 encoded T3SS to trigger their invasion into the host (Fig. 6.1) (Table 6.2). Probably the first proteins delivered through the T3SS to the host cells are *Salmonella* inner proteins (Sip) A and C. SipA and SipC promote bacterial internalization and modulation of actin dynamics. SipC is inserted in the host cell plasma membrane with a cytoplasmic domain and acts as a translocon for other effector proteins (Hayward and Koronaki, 1999). N terminal domain of the SipC binds with C terminal domain of the actin; as a result actin polymerization occurs on the host cell surface (Chang *et al.*, 2005a,b; Cain *et al.*, 2008). SipA helps to initiate actin polymerization by increasing the stability of filaments and blocking the action of an actin depolymerizing factor (ADF/confilin) (dos Reis and Horn, 2010). The *Salmonella* outer proteins SopE (SopE1 and SopE2), SopB and SptP are also delivered through the T3SS translocon into the host cells where SopE1, SopE2 and SopB activate the host cell's small GTPase proteins (e.g. Cdc42, Rac-1 and Rho). The GTPase proteins are involved in the regulation of the formation of F-actin filaments, the dynamics of the cytoskeleton, membrane ruffling and bacterial internalization by macropinocytosis (Stender *et al.*, 2000; Friebel *et al.*, 2001; Zhou *et*

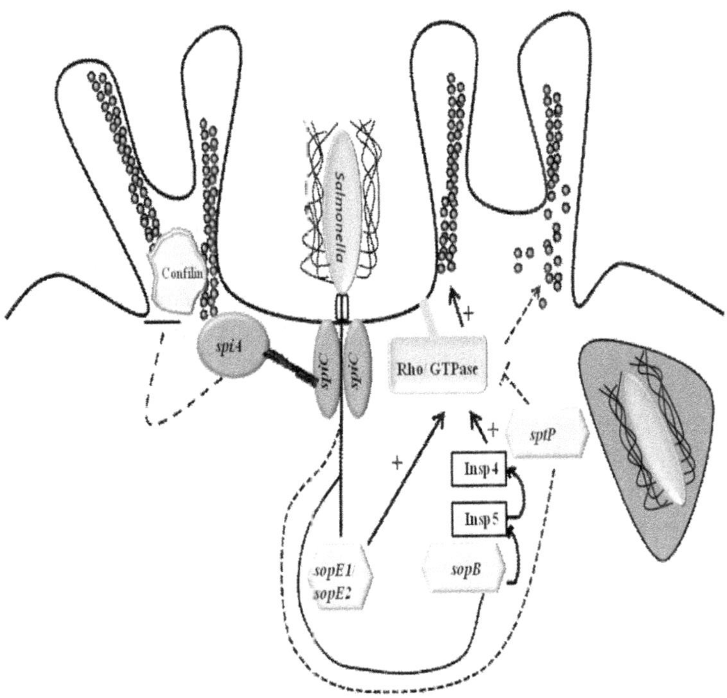

Figure 6.1 Role of effector proteins in cellular invasion of *Salmonella*.

al., 2001; Patel and Galan, 2006; Nagarajan *et al.*, 2009). SopE1 and SopE2 are guanine nucleotide exchange factors (GEFs) for the Rho monomeric GTPases, RhoG, Rac1 and Cdc42 (Friebel *et al.*, 2001), which are involved in the formation of filopodia extensions and lamellipodia structures. SopB has phosphoinositide phosphatase activity and stimulates Cdc42, a cell division control protein. This activation occurs by a cellular exchange factor called SGEF which has a phosphoinositide binding domain activated by SopB generated phosphoinositide fluxes (Ellerbroek *et al.*, 2004). This is followed by activation of the Wiskott–Aldrich syndrome protein family members, N- WASP and WAVE2, which provide actin-related protein 2/3 (Arp2/3) complex to sites of membrane ruffles and stimulate actin polymerization (Criss and Casanova, 2003; Unsworth *et al.*, 2004). Takeuchi (1967) had earlier observed that the epithelial cell soon recovers its normal shape following macropinocytosis of bacteria using another effector protein SptP, which acts as a GTPase-activating factor for the RhoGTPases targeted by the SopE and SopB proteins, thereby switching off the signal (Fu and Galán, 1999). Another effector protein, SipB is involved in macrophage apoptosis by activation of caspase-1 (Hersh *et al.*, 1999; Gong *et al.*, 2009). AvrA protein inhibits activation of the proinflammatory antiapoptotic NF-κB transcription initiation and enhances cell apoptosis. SopA protein stimulates polymorphonuclear transmigration by HECT-like E3 ubiquitin ligase activity.

Regulation of cellular invasion of *Salmonella*

Expression/repression of SPI-1 coded genes is tightly regulated by environmental conditions such as temperature, low or limited oxygen, high osmolarity, neutral to slightly basic pH and late log to stationary phase of bacterial growth (Lee and Falkow, 1990; Jones and Falkow, 1994; Bajaj *et al.*, 1996). SPI-1 codes for several transcriptional regulators like HilA, HilD, HilC and InvF which control expression of all SPI-1 coded genes (Maithreye *et al.*, 2007; Ellermeier and Slauch, 2008; Baek *et al.*, 2009) (Table 6.2). The *hilA* gene is considered a central regulator of SPI-1. It encodes an OmpR/ToxR family transcriptional regulator and binds to promoters of T3SS-associated genes like *prg/org* and *inv/spa* operons and activates them (Lee *et al.*, 1992; Bajaj *et al.*, 1996). Expression of the genes of the operon coding for apparatus proteins is followed by induction of the effector proteins coded by *sic/sip* operon, *sigD/sopB*, and *sopE* (Darwin and Miller, 2001; Boddicker *et al.*, 2003). Schechter *et al.* (1999) mapped an upstream repression sequence (URS) and was the first to identify it in the *hilA* promoter region. They also identified two SPI-1-associated genes encoding proteins, HilC and HilD that could counteract repression of *hilA* gene. HilC and HilD initially bind to URS sites within *hilA* and derepress *hilA* transcription. Following this, HilA binds to *invF* and *prgH* transcription start sites and activate expression. This finally results in expression of the genes encoding T3SS components (Akbar *et al.*, 2003). *hilA* gene is also regulated by another protein, RtsA, which is not coded by SPI-1 (Ellermeier and Slauch, 2003). Ellermeier *et al.* (2005) showed that HilC, HilD and RtsA can regulate expression of *hilA* gene either individually or together in a feed forward loop.

Intracellular life style of *Salmonella*

Once intracellular, either in epithelial cells or phagocytes, *Salmonella* is enclosed in an acidic compartment called a *Salmonella*-containing vacuole (SCV) (Dunlap *et al.*, 1991; Steele-Mortimer *et al.*, 1999). SCV biogenesis involves three stages, early (< 30 minutes), intermediate (30 minutes to 5 hours) and the late (> 5 hours), and is characterized by the presence of early endocytic members, namely early endosome antigen 1 (EEA1), the transferring receptor (*trf*) and several Rab GTPases such as Rab4, Rab5 and Rab11. SopE and the inositol phosphatase activity of SopB are required for SCV recruitment of Rab5 (Mukherjee *et al.*, 2001; Mallo *et al.*, 2008) which then binds with the phosphatidylinositol 3-kinase Vps34 required for the recruitment of EEA1 (Steele-Mortimer, 2008). 5–10 minutes post invasion, few of these proteins are replaced by lysosome associated membrane proteins (LAMPs) which accumulate further over many hours. After 30–60 minutes, Rab5 is replaced by Rab7, a transport regulator between late endosome and lysosome. Rab7 appears to recruit LAMP1 to the SCV and promoting the early juxtanuclear trafficking of the SCV, via the adaptor protein RILP (Rab7-interacting lysosomal protein) (Lahiri *et al.*, 2010). Following this, *sopB* activates host kinase protein (AKT1) on the cell membrane which in turn activates (phosphorylates) cytosolic AS160 (Rab14). Activated Rab14 fails to bind with phagosomal membrane and helps to recruit effectors that inhibit SCV fusion with lysosomes (Kuijl *et al.*, 2007). Eswarappa *et al.* (2010) suggested that *Salmonella* tackles the lysosomal degradation problem in an elegant manner by causing an imbalance in the ratio between SCVs and acidic lysosomes. LAMP1/2 and Rab7 are required for both SCV and lysosome production. When the number of SCVs increase, a deficiency of LAMP1/2 and Rab7 for the formation of lysosome is noted. So lysosome biogenesis may stop and there is likelihood of imbalance in the ratio. Lahiri *et al.* (2010) suggested that *Salmonella* tend to follow different mechanisms to avoid lysosome effect. Inside the host, each SCV contains only one bacterium since this is more difficult for the host cells to defend than a single or few SCVs containing several bacteria. This implies that there will be no competition for nutrients among bacteria within SCV. A single bacterium per SCV is a bacterial strategy to counteract host defence mechanisms and the same theory holds good for the decrease in the lysosome number. After 3 hours of invasion *Salmonella* express the unique T3SS to translocate effector proteins across the SCV which are encoded by *Salmonella* pathogenicity island 2 (SPI-2). Finally after 5 hours of invasion the SCV is characterized by the presence of long membrane structures, called *Salmonella*-induced filaments (Sifs) that extend from the SCV and thus provide space for *Salmonella* to replicate (dos Reis and Horn, 2010).

Assembly of SPI-2 coded T3SS

SPI-2 is 40 kb in size and is inserted adjacent to *valV* tRNA gene (Hensel *et al.*, 1998). A region of 25 kb containing 31 genes encodes the type III secretion system. These 31 genes are predicted to encode four types of proteins, namely apparatus, effector, regulatory and chaperone (Table 6.3). A second region of 15 kb encodes tetrathionate reductase (Ttr) and involved in anaerobic respiration dispensable for systemic virulence. Of the listed 31 genes, nine are involved in construction of SPI-2 encoded T3SS. SsaL, SsaM and SsaP are essential for forming the base of the needle appendages of T3SS (Chakravorty *et al.*, 2005). While the SsaC protein is predicted to form the outer membrane ring, SsaV and SsaT are predicted to be part of the T3SS apparatus in the inner membrane of the bacteria (Freeman *et al.*, 2002). SsaR, SsaG and SsaI are involved in the secretion of effector proteins.

Role of effectors in *Salmonella* survival inside the host

SPI-2 encodes three translocon proteins, SseBCD for translocation of SPI-2 coded effector proteins (Table 6.3). SseC and SseD are 24% and 27% identical to EspD and EspB of enteropathogenic *E. coli* (EPEC) respectively (Hensel *et al.*, 1998). Both SseC and SseD also have homology to YopB of *Yersinia pseudotuberculosis*, a protein that is required for delivery of Yop proteins into the host cell (Hakansson, 1996). The SseB protein is secreted on the bacterial cell surface and is loosely associated with the membrane It was predicted that SseB may be part of a structure through which proteins are translocated (Miao *et al.*, 1999). Secretion of SseC and SseD proteins occurs only in the presence of SseB. Our previous study showed that deletion of *sseC* gene leads to inability to survive in HeLa cells showing that this gene has an important role in intracellular survival of *S.* Weltevreden in human epithelial cells (Bhowmick *et al.*, 2011a). Other studies have shown that SseC protein is required for the proliferation of *S. enterica* inside host cells and thus is essential for the virulence (Klein and Jones, 2001; Nikolaus *et al.*, 2001; Eswarappa *et al.*, 2008). At present, 19 effector proteins have been identified (Table 6.3), of which only three genes are coded by SPI-2 and the remaining 16 not encoded by SPI-2 are associated with prophage or cryptic bacteriophages. Three effector proteins located on SPI-2 are SpiC, SseF and SseG. SseF and SseG are putative effector proteins implicated in the formation of *Salmonella*-induced aggregation of host endosomes (Guy *et al.*, 2000). SseG and SseF are responsible for placing the SCV in perinuclear

Table 6.3 Genes coded by *Salmonella* pathogenicity island, SPI-2 and their cellular functions

Gene	Cellular functions	References
Apparatus protein		
ssaL, *ssaM* and *ssaP*	Components of the base of the needle appendages of T3SS and helped in secretion of SPI-1 encoded protein SipC	Chakravortty *et al*. (2005)
ssaC	Form the outer membrane ring of the SPI-2 T3SS system	Freeman *et al*. (2002)
ssaV and *ssaT*	Part of the TTS apparatus in the inner membrane of the bacteria	Freeman *et al*. (2002)
ssaR, *ssaG* and *ssaI*	Part of T3SS apparatus involved in secretion of effectors proteins	Chakravortty *et al*. (2005)
Translocon protein		
sseB	Part of a structure through which proteins are translocated	Miao *et al*. (1999)
sseC and *sseD*	Translocon for various effector proteins	Hensel *et al*. (1998)
SPI-2 coded effector proteins		
ssaB/spiC	Inhibits the fusion of phagosome–lysosome vesicles and controls secretion process of translocons	Uchiya *et al*. (1999)
sseF and sseG	Contributes to Sif formation adn microtubule bundling	Hensel *et al*. (1998), Ramsden *et al*. (2007a,b)
sifA	Induces Sif formation, Association of SCV with the Golgi network is SifA	Brumell *et al*. (2001), Deiwick *et al*. (2006)
sifB	Not directly involved is *Salmonella* infection	Freeman *et al*. (2003)
sopD2	Contributes to Sif formation	Brumell *et al*. (2003)
sseL	Deubiquitinase	Coombes *et al*. (2007)
sseI	Contributes to host-cell dissemination	Miao and Miller (2000)
sseK1 and *sseK2*	Affect *Salmonella* virulence through the SPI-2 T3SS system	Kujat-Choy *et al*. (2004)
pipB2	Contributes to Sif formation	Knodler *et al*. (2003)
gogB, *pipB*, *steA*, *steB* and *steC*	Unknown	Knodler *et al*. (2003), Coombes *et al*. (2005)
Regulator proteins		
ssrA	Sensor kinase	Ochman *et al*. (1996), Deiwick *et al*. (1999)
ssrB	Transcriptional activator	Ochman *et al*. (1996), Deiwick *et al*. (1999)
phoP/phoQ	Regulates expression of SseK1 and SseK2	Kujat-Choy *et al*. (2004)
ompR	Regulates expression of SsrA	Feng *et al*. (2003, 2004)
envZ	Involved in the phosphorylation of OmpR and thereby converting it into its active form, OmpR-P	Feng *et al*. (2003, 2004)

region next to the Golgi apparatus of the host (Salcedo and Holden, 2003). SseF is required for survival by altering the trafficking of the SCV. This positioning facilitates interaction of the SCV with the endocytic and secretory pathways, allowing the pathogen to acquire extra membrane for expansion of the SCV and nutrients to promote intracellular survival. SscB, a protein encoded by region upstream of *sseF* acts as a chaperon for SseF and SseG to maintain their stability inside the bacteria. SpiC is involved in expression of the flagellin protein and mediates activation of the signal transduction pathway in macrophages (Uchiya *et al.*, 2008, 2009). Among 16 proteins not encoded by SPI-2, 14 proteins (SseL, SspH1, SspH2, SlrP, SifA, SifB, SopD2, SteA, SteB, SteA, SteB, SseJ, SteC and SseI) are characterized by a conserved N- terminal domain (Miao *et al.*, 1999; Miao and

Miller, 2000) and the remaining two proteins, PipB and PipB2 (Knodler and Steele-Mortimer, 2005) lack the conserved region. PipB is encoded by *Salmonella* pathogenicity island 5 together with the SPI-1 effector protein SopB. PipB2 is involved in modulation of phagosome tabulation (Ohlson *et al.*, 2008). The N-terminal domain of SseL, SspH1, SspH2, SlrP, SifA, SifB, SopD2, SteA, SteB, SteA, SteB, SseJ, SteC and SseI contain a translocation signal (Miao and Miller, 2000). The C-terminal region (amino acids 140–408) is 29% identical to several members of the GDSL lipase family (Brumell *et al.*, 2003). SteC is essential for the formation of SCV-associated F actin (Poh *et al.*, 2008). SseI and SspH2 bind the host actin cross-linking protein filamin (Miao *et al.*, 2003) and the latter also interacts with the cellular G-actin binding protein profilin and inhibits actin polymerization rates (Miao *et al.*, 2003). SseJ shows phospholipase A activity (Lossi *et al.*, 2008) and Ruiz-Albert *et al.* (2002) found that its translocation is responsible for the loss of vacuolar membrane integrity. Accumulation of cholesterol plays an important role in bacterial invasion and modification of the SCV (Kuhle and Hensel, 2004).

SPI-2 associated regulator proteins

The two-component regulon SsrAB present in SPI-2, (Ochman *et al.*, 1996) is observed to be stimulated by an acidic pH condition (Table 6.3) (Lee *et al.*, 2000; Miao *et al.*, 2002). SsrA is a putative sensor kinase and SsrB has been identified as a transcriptional activator that shares sequence homology with proteins like UvrY of *E. coli* and SirA of *S.* Typhimurium (Ochman *et al.*, 1996). Deiwick *et al.* (1999) reported that induction of SPI-2 gene expression is modulated by SsrAB and PhoPQ under Mg^{2+} and phosphate limitation. Transcription of SsrA and SsrB is uncoupled and overexpression of *ssrB* in the absence of SsrA is sufficient to activate SPI-2 gene expression (Feng *et al.*, 2003, 2004; Walthers *et al.*, 2007). SsrA has been recently shown to promote reproductive tract (but not intestinal) colonization of chicken by *S.* Enteritidis (Bohez *et al.*, 2008). SsrB is autoregulated and appears to have a dual function –activation of SPI-2-associated genes by binding to their promoters and to counteract gene silencing by the H-NS protein (Feng *et al.*, 2003, 2004; Walthers *et al.*, 2007).

Virulence genes present in SPI-3

Salmonella possesses another unique pathogenicity island, SPI-3 (Blanc-Potard and Groisman, 1997), harbouring 10 genes including eight ORFs, all of which contain potential Shine–Dalgarno sequences (Table 6.4). The SPI-3-encoded genes are organized in six transcriptional units. SugR comprises 519 amino acids and is encoded by the first gene present in SPI-3. It is a cytoplasmic protein exhibiting 44% similarity to the PgaA antigen of the *Porphyromonas gingivalis* and 34% with putative ATP binding protein from *E. coli* (Lim, 1992). RhuM with 215 aa and RmbA with 205 aa are predicted as cytoplasmic proteins (Lawley *et al.*, 2006). RmbA exhibits 39% similarity with *E. coli* ORF product. MisL with 955 aa is recognized as an immunoglobulin A1 protease of autotransported protein (Maurer *et al.*, 1997). The MarT (for membrane-associated regulator) protein has homology with ORF269 encoded protein of *E. coli* K-12 and exhibits similarity in its N-terminal domain to the ToxR protein from *V. cholerae.* FidL, 154 aa and SlsA, 226 aa is predicted as inner membrane protein (putative transmembrane domain and comprises residues 114–130) and code for putative signal sequences. CigR comprising 159 aa is predicted to be a membrane protein.

In *Salmonella* there are three transport systems for Mg^{2+} transport, CorA, MgtA and MgtB. The *corA* gene encodes a 40-kDa protein product which mediates both the influx and efflux of Mg^{2+} (Hmiel *et al.*, 1989). Two additional Mg^{2+} uptake systems are encoded by the *mgtA* and *mgtCB* loci present in SPI-3. The *mgtA* and *mgtCB* loci encode putative P-type ATPases (Hmiel *et al.*, 1989; Maguire *et al.*, 1992; Smith *et al.*, 1993a). The *mgtCB* locus encodes two proteins, MgtC and MgtB. MgtC is a hydrophobic protein with 5–6 transmembrane segments based on hydropathy analysis and a predicted molecular mass of 22.5 kDa (Smith *et al.*, 1995; Tao *et al.*, 1998). MgtB, the P-type ATPase, has a molecular mass of 102 kDa (Smith *et al.*, 1993a, 1995). Günkel *et al.* (2006) suggested that MgtC may be involved in the modulation of cellular ion homeostasis and thus, membrane

Table 6.4 Genes coded by *Salmonella* pathogenicity island SPI-3, SPI-4, and SPI-5 and their cellular functions

Gene	Cellular functions	References
Genes coded by SPI-3		
misL	Functions as autotransporter	Blanc-Potard *et al.* (1999)
marT	Acts as transcriptional activator of the misL promoter	Tükel *et al.* (2007)
rmbA	Requires for colonization of calves	Morgan *et al.* (2004)
slsA	Needed for colonization in both calves and chicks	Morgan *et al.* (2004)
sugR and *rhuM*	Cytoplasmic protein; involved in systemic disease process	Lawley *et al.* (2006)
cigR	Membrane protein; involved in systemic disease process	Lawley *et al.* (2006)
fidL	Inner membrane protein; involved in systemic disease process	Lawley *et al.* (2006)
mgtB	Magnesium transporter	Blanc-Potard *et al.* (1999)
mgtC	Magnesium transporter; also involves in the modulation of cellular ion homeostasis	Günkel *et al.* (2006)
Genes coded by SPI-4		
siiE	Mediates contact-dependent adhesion of *Salmonella* to epithelial cell surfaces	Gerlach *et al.* (2007)
siiCDF	Putative outer membrane component, a putative fusion protein spanning the inner and outer membranes	Gerlach *et al.* (2007)
siiA and *siiB*	May have a regulatory or chaperone function	Gerlach *et al.* (2007)
Genes coded by SPI-5		
sopB	Mediates chloride secretion in infected epithelial cells; leading to fluid influx, neutrophil accumulation in the gut and intestinal inflammation; all of which cause diarrhoea	Galyov *et al.* (1997), Norris *et al.* (1998)
pipC	Functions as a chaperone for SopB	Darwin *et al.* (2001)
pipB	Involved in formation and maintenance of SCV and Sif	Knodler *et al.* (2002)
pipA	Development of systemic infection and virulence	Knodler *et al.* (2002)
pipD and *orfX*	Function unknown	

potential through the activation of the host cell Na^+, K^+-ATPase. They are negatively regulated by two-component PhoPQ regulatory system. The *phoPQ* locus is an essential control element in virulence (Groisman *et al.*, 1989; Miller *et al.*, 1989; Gunn and Miller, 1996). Generally Mg^{2+} binds to PhoQ in the membrane and regulates its interaction with PhoP (Garcia-Vescovi *et al.*, 1996). Under normal laboratory growth conditions, there is sufficient Mg^{2+} in the medium to bind to PhoQ and inhibit PhoP activity. At low Mg^{2+} concentration, it dissociates from PhoQ, allowing the activation of PhoP which is an apparent transcription factor and activates at least 30 genes that appear to fall under the direct or indirect control of Mg^{2+} and PhoPQ (Rang *et al.*, 2007; Retamal *et al.*, 2009). In addition to low Mg^{2+} concentration, *S.* Typhimurium that invade epithelial and macrophage cells are also exposed to acidic pH (Riesenberg-Wilmes, 1996) which is also involved in regulation of MgtA and MgtCB (Bearson *et al.*, 1998; Tao *et al.*, 1998). Although *mgtA*, *mgtB* and *mgtC* are markedly induced upon invasion of a eukaryotic cell none is necessary either for invasion or for short term survival within the eukaryotic host (Blanc Potard and Groisman, 1997). The *mgtB* and *mgtC* genes are not always associated within enterobacteria members (Blanc Potard and Groisman, 1997). According to a study by Gunzel *et al.* (2006) MgtC would act as a subunit of the host cell's Na^+, K^+-ATPase either by replacing the endogenous subunit or by binding to the existing heterodimer. One of the main effects of invasion of enteric bacteria into mammalian cells is the induction of apoptosis (Monack *et al.*, 1996, 1997; Navarre and Zychlinsky, 2000).

Virulence genes present in SPI-4

SPI-4 was first described as a pathogenicity island by Wong *et al.* (1998). SPI-4 is an insertion of 27 kb and is located adjacent to *ssb* tRNA gene (Wong *et al.*, 1998). SPI-4 contains six open reading frames comprising a very large novel secreted protein (SiiE), component of a T1SS (SiiDCF) and two other putative secreted proteins (SiiAB) (Morgan *et al.*, 2004, 2007; Gerlach *et al.*, 2007; Main-Hester *et al.*, 2008) (Table 6.4). SiiE is a highly repetitive ~600 kDa non-fimbrial adhesin which mediates contact-dependent adhesion of *Salmonella* to epithelial cell surfaces (Gerlach *et al.*, 2007). SiiDCF, an ATP-binding cassette component, is predicted to be putative outer membrane protein (Morgan *et al.*, 2007) required for the secretion of SiiE protein. Gerlach *et al.* (2007) predicted that SiiA and SiiB may play the role of chaperone. An operon polarity suppressor (*ops*) sequence within the 5′untranslated region upstream of *siiA* is required for optimal SPI-4 expression (Main-Hester *et al.*, 2008).

Virulence genes present in SPI-5

SPI-5 is a small locus of 7.6 kb inserted adjacent to tRNA *serT* locus located on 24 centisomes of the *Salmonella* chromosome. SPI-5 encodes at least six genes, *pipD, sigD/sopB, sigE, pipB, pipC* and *pipA*, all of which contribute to enteropathogenesis as assessed in a calf model of infection (Wood *et al.*, 1998) (Table 6.4). SPI-5 encodes the effector proteins for both T3SS encoded by SPI-I and SPI-2 (Steele-Mortimer *et al.*, 2000; Zhou *et al.*, 2001). SigD/SopB is an inositol polyphosphate phosphatase (Norris *et al.*, 1998) that activates mammalian protooncogene Akt (Steele-Mortimer *et al.*, 2000) and promotes host cell membrane ruffling and invasion (Hong and Miller, 1998; Zhou *et al.*, 2001). It has been shown that, SopB gene is expressed upon bacterial invasion during late log/early stationary phase (Hong and Miller, 1998; Pfeifer *et al.*, 1999; Papezova *et al.*, 2007). PipB is translocated by the SPI-2 T3SS to the SCV and Sif (Kondler *et al.*, 2002). Knodler *et al.* (2002) in their study have shown the possibility that the reduction in virulence for the *pipB* mutant could have resulted from polar effects on the *pipA* gene. Expression of *pipB* gene occurs during stationary phage of the bacterial growth and is regulated by *ssrAB* of the SPI-2. Specific functions of other SPI-5-encoded genes are still unknown.

Virulence genes present in SPI-6

Salmonella pathogenicity island 6 is a locus of 47 kb inserted adjacent to *aspV* tRNA gene and contain the *saf* and *tcf* gene cluster for fimbriae (Parkhill *et al.*, 2001). Of the seven open reading frames of SPI-6, ORF 1 and 2 encode putative *Klebsiella aerogenes* AtsA and AtsB homologues that function in sulphate metabolism (Murooka *et al.*, 1996) and is regulated by the *moaR* gene. ORF 3 to 6 is a 4 kb region which has homology with a region of *Y. pestis* (Buchrieser *et al.*, 1998) and is known to be directly involved in virulence. Deletion of SPI-6 had no effect on system pathogenesis but reduced invasion of cultured cells was observed (Foleksson *et al.*, 2002). Mutagenesis of ORF2, ORF4 and ORF6 (*cat2*) resulted in impaired infection of chicken macrophages. The *cat2* gene helps to maintain the growth rate when succinate is the main carbon source (Zhao *et al.*, 2001). SPI-6 is involved in creating a phenotype that helps bacteria to go into a metabolic state that enables growth inside the host cell. Bioinformatics based studies reveal the presence of a unique T6SS encoded region in the SPI-6 of *Salmonella*, although there is no experimental evidence on the functionality of this system till date (Folkesson *et al.*, 2002). T6SS are known to be linked to a variety of functions such as adherence, cytotoxicity, host cell invasion, biofilm formation, survival within macrophages and persistence within the host (Bingle *et al.*, 2008; Cascales, 2008; Filloux *et al.*, 2008; Pukatzki *et al.*, 2009).

Virulence genes present in SPI-7

Salmonella pathogenicity island 7 is 134 kb in size and located between partially duplicated tRNA *pheU* sites. It was first discovered as a large insertion in the genome of the *S.* Typhi, and initially referred to as major pathogenic island (MPI) (Zhang *et al.*, 1997). SPI-7 contains 151 potential open reading frames including those for Vi exopolysaccharide, type IVB pili, putative conjugal transfer and *sopE* bacteriophage. There are two putative prophages integrated in the SPI-7 of *S.* Typhi whereas only one has been found in *S.* Dublin and *S.* Paratyphi C which may promote the excision and integration of the island. The important virulence factor encoded by SPI-7 is the Vi antigen, a capsular exopolysaccharide (Pickard *et al.*, 2003). Vi expression is associated with a cluster of ten genes located at position 4,409,519 bp on the *S.* Typhi chromosome, known as the *viaB* operon (Liu and Sanderson, 1995), regulated by two component regulator systems, *rcsB-rcsC* (Houng *et al.*, 1992) and *ompR-envZ* (Pickard *et al.*, 1994). The Vi capsular polysaccharide was first identified as being associated with virulence in mice. Structurally, it is a polymer of *N*-acetylaminohexuronic acid. Two genetic loci, *viaA* locus and *viaB* locus, are required for the production of Vi antigen.

Virulence genes present in SPI-8

Salmonella pathogenicity island 8 is a locus of 6.8 kb located adjacent to *pheV* tRNA gene (Parkhill *et al.*, 2001). Putative virulence factor include gene for bacteriocin but no functional data have been reported so far. Based on the genomic data available, Hensel (2004) predicted that SPI-8 would be restricted to *S.* Typhi but Saroj *et al.* (2008) predicted its presence also in *S.* Paratyphi A but absent in *S.* Worthington, Dublin, Paratyphi B and C.

Virulence genes present in other *Salmonella* pathogenicity islands

Pathogenicity island 9 (SPI-9) is a locus of 16 kb located near a lysogenic phage genome (Parkhill *et al.*, 2001). The putative virulence factor includes a type I secretion system and large repeats in toxins (RTX) like protein. SPI-10 locus is a large insertion of 32.8 kb located near tRNA *leuX* (Parkhill *et al.*, 2001). Townsend *et al.* (2001) reported SPI-10 to be restricted to *Salmonella* serovars Typhi, Paratyphi A, Dublin, Enteritidis and Gallinarum but Saroj *et al.* (2008) demonstrated it to be also present in serovars Typhimurium and Washington. A known virulence factor encoded by SPI-10 is *sef* fimbriae whose distribution is restricted to *S.* Typhi and *S.* Enteritidis and is considered one of the factors that determine host specificity (Townsend *et al.*, 2001). It has been predicted that two lysogenic phages, Gifsy-I and Gifsy-II are known to contribute to *Salmonella* virulence. Various virulence factors are associated with Gifsy phages (Figueroa-Bossi *et al.*, 2001; Lawley *et al.*, 2006), e.g. *gipA* of Gifsy-1 (Stanley *et al.*, 2000) and *sod* of Gifsy-2 (Salyers and Whitt, 2002). The *gipA* gene, carried on the lambdoid phage genome Gifsy-1, affects the ability of serovar Typhimurium to grow within Peyer's patches in the small intestine of mice. The *sod* gene of Gifsy-II encodes a superoxide dismutase, which help the bacteria to survive the oxidative burst within the macrophage (Salyers and Whitt, 2002). Gifsy-2 increased the LD_{50} of the strain significantly. SPI-11 was named following the sequencing of the *S.* Choleraesuis genome (Chiu *et al.*, 2005). This island which is 14 kb in length is located adjacent to the Gifsy-1 prophage. SPI-11 encoded *pagC* and *pagD* genes contribute to intramacrophage survival and systemic infection in mice (Miller *et al.*, 1989), regulated by SlyA and the PhoP/PhoQ (Gunn *et al.*, 1995; Navarre *et al.*, 2005). SPI-12 is 6.3 kb in length and is inserted next to the *proL* tRNA gene of *S.* Typhi and *S.* Choleraesuis. The only virulence factor present in SPI-12 is $SspH_2$. This is a secreted effector protein of T3SS that influences the rate of actin polymerization inside the infected cells (Miao *et al.*, 2003). SPI-13 is a 19.5-kb region located next to the *phe V* tRNA gene and of the 18 genes present in the island, only three are associated with virulence viz., *gacD*, *gtrA* and *gtrB* and have been shown to be responsible for the virulence of *S.* Gallinarum in a 1-day-old chick infection model (Shah *et al.*, 2005). SPI-14 carries two genes, namely *gpiAB*, which also have been found associated with the virulence of *S.* Gallinarum in 1-day-old chicks (Shah *et al.*, 2005). More recently, Blondel *et al.* (2009) discovered another three pathogenicity islands (SPI-19, 20 and 21) in *Salmonella*. SPI-19 is a locus of 45 kb, present in subset of serotypes belonging to *S.* Dublin, Weltevreden, Agona, Gallinarum and Enteritidis. It encodes two ORFs for Hcp1 and Hcp2 proteins. *hcp1* is a putative operon that includes most of the T6SS function and *hcp2* is found upstream of a VgrG homologue. SPI-20 is a locus of~ 34 kb located adjacent to *aspV* tRNA gene having 28 ORFs with 17 of them involved in T6SS function. SPI-21 locus is located adjacent to *thrW* tRNA gene and encodes a T6SS in the genome of serotype IIIa 62:z4,z23. It has 57 ORFs, 20 of which encode T6SS function. In addition SPI-21 includes four ORFs encoding putative colicin/pyocin immunity proteins.

Virulence plasmid

Plasmids involved in virulence of *Salmonella* vary in size (50–95 kb) and are less in copy number. Generally they are found in some specific serotypes of *Salmonella* such as *S.* Typhimurium, Abortusovis, Dublin, Enteritidis, Newport and Paratyphi C. These plasmids determine the infection of the spleen, the liver and perhaps the mesenteric lymph nodes and code for the ability to multiply in the tissue macrophages (Rychlik *et al.*, 2006). All virulence plasmid-containing *Salmonella* spp. share a highly conserved 8 kb operon encompassing five genes, designated *spvRABCD* (serovars associated plasmid-encoded virulence) (Rychlik *et al.*, 2006). Other loci of the plasmid involved in the biosynthesis of fimbriae and in conferring serum resistance, may play a role in other stages of the infection process. SpvA (28 kDa) is present in the outer membrane, SpvB (66 kDa) in the cytoplasm and inner membrane, SpvC (28 kDa) in the cytoplasm and SpvD (25 kDa) is exported outside the cells. SpvB is involved in actin degradation and apoptotic cell death (Tezcan-Merdol *et al.*, 2005). SpvC is required for the functional activity of SpvB and also interacts with spvD for its function (Matsui *et al.*, 2001). SpvR (33 kDa polypeptide) protein has the ability of self-activation. It binds with the operator region of *spvRABCD* operon to initiate the expression. It is known that RpoS protein encoding an alternative sigma factor σ^s, plays an important role in regulation of *spv* operon (Coynault *et al.*, 1992).

Toxins

Salmonella spp. are capable of producing both endotoxins and exotoxins (van Asten and van Dijk, 2005). The lipopolysaccharide (endotoxin) is a major component of the outer membrane of many enterobacteria, including *Salmonella* spp. (Hitchcock *et al.*, 1986; Raetz and Whitfield, 2002). Heat-labile enterotoxin related to the *E. coli* heat-labile enterotoxin-LT or cholera toxin-CT, may be produced by *Salmonella*. A histotoxin, which cleaved inter-enterocyte tight junctions, was found to be involved in the interaction of *Salmonella* serotype Typhimurium strain GM3 and serotype Dublin strain 3246 with explants of rabbit terminal ileum (Bolton *et al.*, 1999). Cytolethal distending toxin (CDT) is an important toxin which causes host cells to arrest during their cell cycle by inflicting DNA damage. Haghjoo and Galan (2004) found that *S.* Typhi encodes CDT activity which depends on the function of a CdtB homologous protein. Spano *et al.* (2008) suggested that PltA and PltB, homologues of the pertussis toxin, are required for the delivery of CdtB from an intracellular compartment to target cells via autocrine and paracrine pathways.

Quorum sensing in *Salmonella*

Quorum sensing, otherwise referred to as cell-to-cell communication, is the regulation of gene expression in response to changes in cell density, and is the mechanism in bacteria that involves the synthesis, secretion and detection of hormone-like molecules known as autoinducers (Elvers and Park, 2002; Reading and Sperandio, 2006). Once the signal molecule achieves the critical threshold concentration, it stimulates expression of several genes. Gram-negative bacteria use quorum sensing circuits to regulate many physiological activities such as symbiosis, virulence, competence, conjugation, antibiotic production, motility, sporulation and biofilm formation. In general, Gram-negative bacteria use acylated homoserine lactones as autoinducers (Miller and Bassler, 2001) which are signal molecules synthesized intracellularly and are either passively released or actively secreted to the outside of the cells. When autoinducers accumulate above the minimal threshold level required for detection, cognate receptors bind the autoinducers and trigger signal transduction cascades that result in population-wide changes in gene expression.

Quorum sensing in *Salmonella* involves LuxS/A1-2 system which is the most widespread quorum sensing system (Surette and Bassler, 1999). It has been seen that production of A1-2 increases in mid exponential phase and degrades in stationary phase. Several environmental factors like logarithmic growth rate, carbon source, neutral pH and low osmolarity regulate production and degradation of A1-2 signal (Surette and Bassler, 1998, 1999). LuxS and AI-2 regulate the expression of *lsr* (*luxS* regulated) operon encoding an ATP binding cassette (ABC)-type transporter. LsrB, directly binds with A1-2 signal but other genes in the *lsr* operon that encode LsrA, LsrC and LsrD proteins form an ABC transporter complex that internalizes the signal molecule. The internalized AI-2 is phosphorylated by LsrK and further modified by LsrF and LsrG (Taga *et al.*, 2001). DNA binding protein coded by a gene located adjacent to the *lsr* operon is required to link with AI-2 to express the *lsr* operon. The LuxS protein catalyses the final step in biosynthesis of AI-2, generating the signalling molecule 4,5-dihydroxy-2,3-pentanedione (DPD) (Meijler *et al.*, 2004). A study by Choi *et al.* (2007) provides a link between quorum sensing and regulatory cascade of virulence genes of SPI-1. They suggested that expression of *luxS* gene is required for cell density-dependent induction of *invF* gene and other genes regulated by *invF* such as *sicA*, *sigD*, and *sopE*. A recent study suggests that LuxS-based quorum sensing does not affect the ability of *S.* Typhimurium to express the SPI-1 coded type 3 secretion system which induce membrane ruffles, or invade epithelial cells (Perrett *et al.*, 2009). In the absence of AI-2, *lsr* operon is repressed by a protein coded by *lseR* gene and this operon is derepressed by the processing of phospho-AI-2 through LsrK (Taga *et al.*, 2001). Overexpression of *lseR* gene decreases the survival of *Salmonella* inside the macrophage by decreasing the expression of genes involved in the oxidative stress response, such as *sodA*, *sodCI* and *sodCII*.

Numerous species of Gram-negative bacteria use *N*-acylhomoserine lactone (AHL) signals to monitor their own population density. Though *Salmonella* species do not synthesize AHLs, they contain the AHL receptor, SdiA (Ahmer *et al.*, 1998), which can bind AHLs produced by other bacterial species like *Aeromonas hydrophila*, *Hafnia alvei* and *Yersinia enterocolitica* and regulate gene transcription (Michael *et al.*, 2001). The *sdiA* gene (DNA binding transcriptional activator) expresses under the control of the *araBAD* promoter on a multicopy plasmid (pJVR2) (Ahmer *et al.*, 1998); consequently expression of *sdiA* gene is dependent on the presence of arabinose. The *sdiA* gene regulates the *rck* (resistance to complement killing) gene on the *Salmonella* virulence plasmid (*pefI-srgD-srgA-srgB-rck-srgC*) and a single gene, *srgE* (*sdiA*-regulated gene E) acquired by horizontal gene transfer in the chromosome (Smith and Ahmer, 2003; Smith *et al.*, 2011).

Genetic determinants of biofilm production

It is now generally accepted that bacteria in the environment predominantly live as sessile communities called biofilms. The biofilm production in *Salmonella* is well documented as an adaptation mechanism. Typhoidal *Salmonella* spp. may survive within their asymptomatic human hosts by forming biofilms on the surface of gallstones which explains the failure of antibiotic treatment to clear the pathogen from asymptomatic hosts. Non-typhoidal *Salmonella* spp. can also colonize gallstones under laboratory conditions (Prouty *et al.*, 2002). It has been suggested that the thickness of biofilm brought about by production of exopolysaccharide and other extra-polymeric substance produce a diffusion barrier thereby preventing the uptake of antibiotics and biocides (Costerton *et al.*, 1999). A survival strategy for bacteria to adhere to surfaces leads to the development of biofilm by a sequential process of initial attachment, microcolony formation, production of extracellular polymeric substances (ECP) and maturation (Davey and O'Toole, 2000). The process is triggered by environmental signals and once the structure has developed, some bacteria from the outermost layer are continuously released into the liquid medium, enabling the biofilm to spread and establish over the surface (O'Toole *et al.*, 2000; Stoodley *et al.*, 2002; Lasa, 2006). In nutrient-rich condition, the main component that influences the biofilm formation is cellulose (β-1,4-glucan) which is the most abundant organic matter synthesized on earth (Choi and O'Day, 1984). Adhesion to surface stimulates the expression of biofilm specific genes such as *adrA* gene. The AdrA protein of *S.* Typhimurium controls the level of cyclic di-GMP which regulates cellulose production and biofilm production (Garcia *et al.*, 2004). OmpR234-dependent *csgD* and *csgBA* are activated during stationary phase of bacterial growth and stimulate the *adrA* gene expression. The *csgD* gene upregulates the production of Curlin, a major component of curli (Brombacher *et al.*, 2003). The *csgDEFG* operon (Hammar *et al.*, 1995) and STM1987 (GGDEF domain containing protein A, GcpA) are critical for biofilm formation in the nutrient- deficient medium (Garcia *et al.*, 2004). At 37°C cellulose production by S. Typhimurium in ATM (low nutrient medium), is seen to be independent of regulatory proteins such as RpoS, CsgD and OmpR but is dependent on the putative diguanylate cyclase, GcpA (Garcia *et al.*, 2004). c-di-GMP as a positive allosteric activator of the cellulose synthase enzyme has been reported in *Gluconacetobacter xylinus* accounting for the extracellular cellulose matrix (Aloni *et al.*, 1982). In a recent study on S. Weltevreden, we observed that, *gcpA* is responsible for biofilm production in both low and high nutrient conditions (Bhowmick *et al.*, 2011b). The study conclusively showed that in the absence of *adrA* expression, overexpressed *gcpA* is required for activation of cellulose synthesis and biofilm formation in both trypticase soy broth (TSB) and diluted TSB.

Drug resistance of *Salmonella* spp.

The introduction of antibiotics for treating and preventing infections in humans is the most fascinating development in medicine. However their indiscriminate use has led to the rapid evolution and spread of bacterial resistance, which poses a major threat for successful treatment of infectious diseases. Antimicrobial resistance is one of the growing concerns in many regions of the world. Unless a rational policy for its use is in place, there is a threat that infectious diseases will become gradually untreatable.

Salmonella within the subspecies 1 are amongst the most resistant bacterial pathogens and are wide spread in both developed and developing countries. Since the early 1990s, the multiresistant strains of S. Typhimurium definitive phage type 104, displayed resistance to up to six commonly used antibiotics. The development of resistance to key antimicrobials such as fluoroquinolones, particularly in Asian countries, and also extended-spectrum cephalosporins worldwide is a cause of concern (Miriagou, 2004). Since, *Salmonella* is an intracellular pathogen, the effective way to eradicate this organism is to use antibiotics that have intracellular activity. The earliest groups of drugs involved to treat *Salmonella* infections, were neomycin and colistin. This was followed by use of absorbable drugs such as ampicillin, amoxicillin, chloramphenicol, tetracycline, and co-trimoxazole which unfortunately do not have substantial intracellular activity. The new era of drugs began with the use of potent intracellular drugs with potent activity such as 5-fluoroquinolone compounds (norfloxacin, ofloxacin, fleroxacin, ciprofloxacin). A word of caution needs to be kept in mind since frequent use of these antibiotics has been found to lead to resistance in *Salmonella*.

Development of antibiotic resistance

Antibacterial agents used for the treatment of bacterial infections have four modes of action, namely (1) inhibition of cell wall synthesis by interfering with the enzymes required for the synthesis of the peptidoglycan layer (McManus, 1997; Martone, 1998); (2) inhibition of protein synthesis wherein the drug binds the 50S and 30S ribosomal subunit and inhibit the protein synthesis; (3) interference with nucleic acid synthesis by inhibiting DNA and RNA synthetase; and (4) inhibition of a metabolic pathway by targeting a metabolic enzyme. Susceptible bacteria can acquire resistance to new antimicrobial agent through many ways of which mutations in the genetic structure is one of the causes. Such mutations may cause resistance by (1) altering the target protein to which the antibacterial agent binds by modifying or eliminating the binding site; (2) upregulating the production of enzymes that inactivate the antimicrobial agent; (3) downregulating or altering an outer membrane protein channel that the drug requires for cell entry; and (4) upregulating pumps that expel the drug from the cell. In all these cases, the bacterial strains acquire resistance to new antibiotics owing to chromosomal mutation and such selection is termed as vertical evolution. Bacteria also develop resistance through the acquisition of new genetic material from other resistant organisms by several methods which include conjugation (via plasmids and conjugative transposons), transduction (via bacteriophages), and transformation (via uptake of chromosomal DNA, plasmids, and other DNAs from dying organisms) (Levy and Marshall, 2004) and these modes of transfer result in horizontal evolution. Other mechanisms of transferring resistance include virulence plasmids, transposons, integrons etc.

Antibiotic resistance pattern in *Salmonella*

Antibiotic resistance is an emerging trend and seen frequently in multiresistant strains of *S.* Typhimurium isolated from animals (Martel *et al.*, 1996) and humans (Breuil *et al.*, 1996). Strains of *Salmonella* resistant to several antimicrobial agents have been reported worldwide (Angulo and Griffin, 2000; Breuil *et al.*, 2000; Van Duijkeren *et al.*, 2003). In 1964, genes conferring resistance to multiple antibiotics were first identified in UK in *S.* Typhimurium and could be transferred between unrelated bacteria such as *Bacteroides* and also between related members of Enterobacteriaceae such as *Salmonella* and *E. coli* (O'Brien, 2002). Several studies have reported the prevalence of multiresistance genes in different serotypes of *Salmonella* (Aarts *et al.*, 2001).

Transmission of antibiotic resistance in *Salmonella*

The mobile genetic elements such as plasmids, transposons and integrons, which disseminate antibiotic resistance genes by horizontal or vertical transfer, play an important role in the evolution and dissemination of multidrug resistance (Gomez *et al.*, 1997; Boyd *et al.*, 2002).

Bacterial plasmids

Plasmid mediated resistance is the transfer of antimicrobial-resistant genes carried on plasmids and can be transferred between prokaryotes through horizontal gene transfer. They are important in bacterial evolution because they affect replication, metabolism, fertility as well as resistance to antibiotics, toxins (bacteriocins) and bacteriophages. The spread of multiple antimicrobial resistance has been enhanced by selective pressure from human and veterinary medicine (Carattoli, 2003) and plasmids often contain the information for resistance to antibiotics (Levy, 2002). A single R-plasmid can code for resistance of up to 10 different antibiotics simultaneously. Plasmids are able to replicate within a cell and are subject to mutations involving either the loss or gain of genes. Furthermore, they are capable of combining with other plasmids, thus conferring resistance to several antibiotics that can reside on one plasmid. Most importantly, bacteria are capable of transferring plasmids from one cell to another through a process termed conjugation, which is a mechanism of horizontal gene transfer and allows the transfer of plasmids coding for antibiotic resistance among an entire colony of bacterial cells (Levy, 2002). Many serovars of *Salmonella* harbour virulence plasmids which are important for systemic infection. They vary in their size depending on the serovars, e.g. 95 kb for *S.* Typhimurium, 60 kb for *S.* Enteritidis, 80 kb for *S.* Dublin, and ranging from 50 to 110 kb for *S.* Choleraesuis (Chu *et al.*, 1999). A highly conserved 8 kb operon of *Salmonella* virulence plasmid (*spv*) consisting of five genes, designated *spvRABCD* is present in all *Salmonella* serotypes and is found to contribute to colonization and resistance to complement killing by *rck* gene (Gulig *et al.*, 1993). The gene *spvB* encodes an ADP-ribosylating toxin that modifies actin directly and totally disrupts the cytoskeleton of the cell (Slayers and Whitt, 2002).

Bacteriophages

Bacteriophages are bacterial viruses discovered independently by Twort in 1915 in England and by d'Herelle at the Pasteur Institute in Paris in 1917. Like viruses, bacteriophages are incapable of independent growth in artificial media and are obligate intracellular parasites. They reproduce by the process of replication, a process in which many copies are made of each viral component and then assembled to produce many phage particles. Replication consists of the steps involving adsorption, penetration, transcription, assembly and finally release by burst.

Bacteriophages in the aquatic environment are very diverse and most that have been studied elaborately include the viruses that infect the enteric group such as *E. coli* and *Salmonella* Typhimurium. Those most commonly found in nature have ds DNA genome though there are others that have ss RNA genome, ds RNA genome and ss DNA genomes. Two types of viral life cycles, namely virulent and temperate, exist. Virulent also known as lytic phages lyse or kill the host after infection while the temperate or lysogenic as they are otherwise referred to, achieve a state where they get integrated into the genome and replicate along with the host genome without killing them. They provide a mechanism to transfer antibiotic resistance genes through lysogenic cycle, the process referred to as phage mediated transduction. When these viruses enter new hosts, they are able to integrate their DNA as well as the antibiotic resistance genes picked up from the previous host into the new host's chromosome (Levy, 2002). Zhang and Le Jeune (2008) showed that wild-type phage mediated transduction also plays an important role in the dissemination of antimicrobial resistance genes. According to Porwollik and McClelland (2003), phage genomes are commonly encountered in the genome of most *Salmonella*. To date about 95% of strains of *S.* Typhimurium examined contain complete inducible prophage genomes and of them 99% were shown to be capable of generalized transduction of chromosomal host markers and plasmids (Schicklmaier *et al.*, 1998). All these studies elucidate the importance of wild type phages in transmitting antibiotic resistance among *Salmonella* serovars.

Transposons

Besides plasmid-mediated mechanism, there are several other mechanisms of gene transfer. It is possible for resistance genes to reside on small pieces of DNA called transposons. In general, these pieces of DNA contain terminal regions that participate in recombination and express a protein(s) (e.g. transposase or recombinase) that facilitates incorporation into and movement from specific genomic regions. These pieces of DNA have the capability to 'jump' from one region of the chromosome, to another and vice versa. This way, a resistant gene can be directly incorporated into host chromosomal DNA and not be dependent on plasmid transfer for spread (Levy, 2002). Conjugative transposons are unique in having qualities of plasmids and facilitate the transfer of endogenous plasmids from one organism to another (Alekshun and Levy, 2007). Pezzella *et al.* (2004) showed that most of the *tetA* genes were localized within a deleted Tn*1721* transposon variant.

Integrons

Among genetic determinants responsible for capturing and expression of resistance genes, DNA elements called integrons have been described. Since late 1980s integrons have been recognized as naturally occurring gene expression elements responsible for the recruitment and assembly of antibiotic resistance genes in clusters. These integrons are associated with multidrug resistance and such multidrug-resistant integrons in *Salmonella* are a threat to empirical treatment. Integrons encode a site-specific recombinase, the integrase, which efficiently promote the acquisition of exogenous genes (Collis and Hall, 1995). There are two groups of integrons, resistance integrons and super integrons. Genes known to be present in resistance integrons encode resistance to antibiotics while gene cassettes within the super integrons encode a variety of functions. Four classes of integrons are defined based on the homology of the integrase proteins (40–60% amino acid identity) of which only two are present in *Salmonella* (Fluit and Schmitz, 2004), with most clinical isolates belonging to class 1 type of integron (Martínez-Freijo *et al.*, 1999). The region upstream of class 1 integrons consists of a 5′-conserved segment containing an *attI* site, a common promoter sequence and the opposite strand contains an integrase gene (*int*) (Collis and Hall, 1995; Recchia and Hall, 1995), whereas, the 3′-conserved segment contains a quaternary ammonium compound (antiseptic) resistance gene (*qacEΔ1*), a sulphonamide resistance gene (*sulI*), and an open reading frame (*orf5*) of unknown function (Hall and Stokes, 1993). About 60–100 different gene cassettes located on integrons and associated with resistance genes have been characterized, and these are responsible for bacterial resistance to a broad spectrum of antimicrobial agents (Recchia and Hall, 1995). The class 1 integrase catalyses site-specific recombination between *attI* of integron and a 59-base element of a mobile gene cassette containing various antibiotic-resistant genes and unknown open reading frames (Recchia and Hall, 1995). With this structure the class 1 integrons is able to express gene cassettes that generally code for antibiotic resistance (Sabate and Prats, 2002). In particular, the Tn*21* transposon, which is a member of the Tn3 family of transposable elements, has been associated with a class 1 integron, named In2. Tn*21* and In2 have been found widely distributed in Gram-negative bacteria (Liebert *et al.*, 2000). Several studies have documented the prevalence of integrons in multidrug-resistant serotypes of *Salmonella* and have found them to be widespread (Daly and Fanning, 2000; Guerra *et al.*, 2000; Khan *et al.*, 2006; Havlickova *et al.*, 2009). We have recently reported the presence of class 1 integrons (GenBank acc. no. HM769861, JF800673, JF800674) in seafood-associated non-typhoidal *Salmonella* (Deekshit *et al.*, 2012). Class 2 integrons have been described within the Tn7 family of transposons (Radstrom *et al.*, 1994). These integrons have a deletion rendering the integrase ineffective. Only a few class 2 integrons have been described in *Salmonella* (Fluit and Schmitz, 2004) while only one class 3 integron has been reported which is not associated with transposon but contains the bla_{IMP} gene cassette conferring resistance to carbapenems (Arakawa *et al.*, 1995).

Resistance of *Salmonella* to various groups of antibiotics

Resistance to aminoglycosides

There are three mechanisms of aminoglycoside resistance in bacteria – reduced uptake or decreased cell permeability, alteration at the ribosomal binding sites, or production of aminoglycoside-modifying enzymes leading to the enzymatic detoxification of drugs. In *Salmonella*, aminoglycoside resistance is mediated by modifying enzymes which attach certain groups to the aminoglycoside and destroy its antibacterial activity. Over 50 different modifying enzymes have been identified (Davies and Wright, 1997) and the genes encoding them are usually found on plasmids and transposons. There are three classes of aminoglycoside modifying enzymes: *O*-adenyltransferases (catalyses ATP-dependent adenylation of hydroxyl group), *N*-acetyltransferases (catalyses acetyl CoA-dependent acetylation of an amino group) and *O*-phosphotransferases (catalyses ATP-dependent phosphorylation of a hydroxyl group). There are two alternatively used nomenclatures for the genes encoding the same aminoglycoside modifying enzyme: one designation aph (3″)-1b refers to the type of modification (aminoglycoside phosphotransfer at the position 3″ and the subtype of the gene 1b) and the other *strA* refers to the streptomycin resistance and the subtype of the gene A.

Among the *aad* (aminogylcoside O-adenyltransferase) genes, only those which act at the position 3″ (*aadA*) and 2″ (*aadB*) are known in *Salmonella* (Table 6.5). Streptomycin resistance is mediated by *aadA* while *aadB* mediates gentamycin, kanamycin and tobramycin resistance. Various serotypes of *Salmonella* carrying *aadA1* (Chen *et al.*, 2004; Randall *et al.*, 2004), *aadA2* (Chen *et al.*, 2004; Randall *et al.*, 2004), *aadA7* and *aadA21* (Doublet *et al.*, 2004; Levings *et al.*, 2005) genes have been identified. Among the genes conferring resistance to streptomycin, the *aph (6)-Ia* gene (*strA*) and the *aph (6)-1d* gene (*strB*) is reported to be widely distributed in *Salmonella* (Sundin and Bender, 1996).

Resistance to β-lactam antibiotics

Resistance to β-lactam antibiotics is mainly due to inactivation by β-lactamase enzymes and also due to decreased ability to bind penicillin-binding protein (PBPs) and other mechanism such as permeability barrier in both Gram-positive and Gram-negative bacteria. About 400 different β-lactamases have been identified from both Gram-positive and Gram-negative bacteria. In *Salmonella* resistance is mainly mediated by β-lactamase enzymes (Mitsuhashi *et al.*, 1982) which inactivate the antibiotic. In *Salmonella*, there exists a wide variety of β- lactamases encoded by considerable number of genes (*bla*). At least 10 subgroups of β-lactamases have been identified encoding TEM-, SHV-, PSE-, OXA-, PER-, CTX-M-,CMY-, ACC-, DHA- or KPC-type β-lactamases (Table 6.5). In *Salmonella*,

Table 6.5 Antibiotic resistance determinants and their mechanism of action

Resistance phenotype	Resistance genes	Specific mechanism
Aminoglycosides	*aadA1,aadA2, aadA21, aadA7, aadB, aadD, aph(3')-IIa, aacC4,aac(6'),aac(3)-Ia aac(3)-IIa*	Alteration of ribosomal protein, protein synthesis inhibitor, inhibition of ribosomal translocation, disruption of the integrity of bacterial cell membrane
Tetracyclines	*tetA, tetB, tetC, tetD, tetE, tetG, tetL*	Energy-dependent efflux, ribosomal protection and enzymatic inactivation
Sulfonamides	*suII, suIII, suIIII*	Drug-resistant dihydropteroate synthase
Phenicols, chloramphenicols, florfenicols	*catA (catA1, catA2, catA3), cmlA, cmlB, floR, catB (catB1, catB2, catB3)*	Enzymatic inactivation of the type A or type B chloramphenicol acetyltransferases (CAT), export of chloramphenicol by the specific efflux proteins
β-Lactams	*bla*-TEM, *bla*-SHV, *bla*-PSE, *bla*-OXA, *bla*-PER, *bla*-CTX-M, *bla*-CMY, *bla*-ACC, *bla*-DHA, *bla*-KPC	Inactivation by β-lactamase enzymes, permeability barrier in both Gram-positive and Gram-negative bacteria
Trimethoprims	*dfrA (dfrA1,dfrA2,dfrA7,dfrA25), dfrB*	Mutations in the promoter region or in the dihydrofolate reductase structural gene
Fluoroquinolones	*aac(6')-IB-cr, qnrA, qnrB*	Mutations in bacterial DNA and topoisomerase IV, active efflux

there are several studies that report the presence of TEM- (Pasquali *et al.*, 2005; Dierikx *et al.*, 2010), SHV- type (Hasman *et al.*, 2005), PSE-1 (Briggs and Fratamico, 1999; Boyd *et al.*, 2002), OXA (Tosini *et al.*, 1998) CTX-M (Di Conza *et al.*, 2002; Hasman *et al.*, 2005; Dierikx *et al.*, 2010) and ACC- type β-lactamases (Dierikxa *et al.*, 2010).

Resistance to fluoroquinolones

High degree of fluoroquinolone resistance in *Salmonella* was first described in serovar Typhimurium DT204 in the early 1990s in Germany (Heisig *et al.*, 1993). More recently, outbreaks of fluoroquinolone-resistant *Salmonella* infections have been reported in the United States, Taiwan, and Japan (Chiu *et al.*, 2002). Bacterial resistance to fluoroquinolone is usually mediated by mutations in bacterial DNA gyrase (*gyrA* and *gyrB*) and topoisomerase IV (*parC* and *parE*) genes, as well as by active efflux (Hawkey, 2003). At least six major groups of active drug efflux pump transporters have been identified in prokaryotes to date: ATP-binding cassette (ABC), major facilitator superfamily (MFS), small multidrug resistance (SMR), multiantimicrobial resistance (MAR), resistance nodulation division (RND) and multidrug and toxic compound extrusion (MATE) (Van Bambeke *et al.*, 2000). Efflux pumps that contribute to antibiotic resistance have been described in a number of clinically important bacteria like *Salmonella* (AcrAB-TolC) (Buckley *et al.*, 2006), *Campylobacter jejuni* (CmeABC) and *Escherichia coli* (AcrAB-TolC, AcrEF-TolC, EmrB, and EmrD). Although target gene mutations and efflux pumps are two known mechanisms associated with fluoroquinolone resistance in bacteria, the additive or synergistic contribution of the two mechanisms in emerging fluoroquinolone resistance is not clear in *Salmonella*.

Resistance to phenicols

In *Salmonella*, the predominant resistance mechanism to chloramphenicol is by the enzymatic inactivation of the type A or type B chloramphenicol acetyltransferases (CAT) as well as export of chloramphenicol by the specific efflux proteins. Consequently there are two types of cat genes, *catA* and *catB* that encode two types of CatA protein (Table 6.5). The gene *catA1* was detected on Transposon 9 in different serotypes like *S.* Typhi, *S.* Typhimurium DT104, *S.* Agona and *S.* Derby. The *catA2* has been detected on a multiresistant plasmid from *S.* Choleraesuis (Chiu *et al.*, 2005) and *S.* Enteritidis (Chen *et al.*, 2004; Randall *et al.*, 2004). Three different types *catB* (*catB2, catB3* and *catB8*) have been found in *Salmonella*. All *catB* resistance genes are located upon gene cassettes identified as class I multiresistance integrons (Ahmed *et al.*, 2005). The gene *floR* encoding proteins that export florphenicol, also designated as flo, flost or pp-flo, plays an important role as a part of SGI-1 associated multiresistant gene cluster (Briggs *et al.*, 1999). The chloramphenicol exporter gene *cmlA* is a cassette-borne gene found in plasmid located class I integrons in *S.* Typhimurium. Recently, a new *cmlA4* variant has been identified in a plasmid-borne class I integron of *S.* Agona (Michael *et al.*, 2006).

Resistance to sulphonamides

Sulphonamides are the first antibiotics developed for large-scale clinical use which target dihydropteroate synthase (DHPS). Sulfonamide resistance in Gram-negative bacilli generally arises from the acquisition of either of the two genes *sul1* and *sul2*, encoding forms of dihydropteroate synthase that are not inhibited by the drug (Enne *et al.*, 2001). The *sul1* gene is normally found linked to other resistance genes in class 1 integrons, while *sul2* is usually located on small non-conjugative plasmids or large transmissible multiresistance plasmids (Enne *et al.*, 2001). The recently described *sul3* has been detected together with *sul1*on a large multiresistance plasmid from *S.* Choleraesuis (Chiu *et al.*, 2005).

In *Salmonella*, all three sulphonamide resistance genes – *sul1*, *sul2* and *sul3* – are reported (Table 6.5). PCR analysis has revealed the presence of *sul1* in many *Salmonella* serovars (Chen *et al.*, 2004; Randall *et al.*, 2004; Michael *et al.*, 2006). *Sul2* has been physically linked to streptomycin-resistant genes *strA-strB* found on plasmids and has been isolated from R-plasmids of *Salmonella* (Chu *et al.*, 2001). In more than 70 typical class 1 integrons isolated so far, the sequences of the *sul1* gene are identical. These integrons have been isolated from various bacteria of diverse ecological niches such as human *E. coli* (Enne *et al.*, 2001), *S.* Choleraesuis (Chu *et al.*, 2001) and serovar Typhimurium DT104 (Ng *et al.*, 1999).

Resistance to tetracyclines

The tetracyclines, first discovered in the 1940s, are the family of antibiotics that inhibit protein synthesis by preventing the attachment of aminoacyl-tRNA to the ribosomal acceptor (A) site. Mechanisms encoded by tetracycline-resistant genes include energy-dependent efflux, ribosomal protection and enzymatic inactivation (Roberts, 2005). Currently, at least 38 classes of tetracycline-resistant genes have been described from a variety of bacteria (Roberts, 2005) (Table 6.5). Of these, 23 genes code for efflux proteins, 11 for ribosomal protection proteins, three for inactivating enzyme and one for unknown function (Roberts, 2005; Zhang *et al.*, 2009). Among these genes classes, A to D and G are most frequently detected in the *Enterobacteriaceae* (Roberts, 1996; Chopra and Roberts, 2001). However, the detection rates of *tet* (A) and *tet* (B) was higher than *tet* (C) and *tet* (D) from the bacteria isolated from aquatic environments, which included *Salmonella* and other enterobacteriaceae (Tao *et al.*, 2010). Of the 35 *tet* genes, only five that are reported in *Salmonella* are, *tet* (A), *tet* (B) (Frech and Schwarz, 1998), *tet* (C) (Bernardi and Bernardi, 1984), *tet* (D) (Allard *et al.*, 1993) and *tet* (G) (Briggs and Fratamico, 1999). All these genes code for a membrane-associated efflux protein consisting of 12 transmembrane segments which can export tetracycline, oxytetracycline, chlortetracycline and doxycycline (Michael *et al.*, 2006). The *tet* (A) gene has been found on the plasmid and the chromosome, while *tet* (B), *tet* (C) and *tet* (D) have been detected on the chromosome of different *Salmonella* serotypes (Frech and Schwarz, 2000). These five different *tet* genes are associated with different elements: The *tet* (G) has been exclusively detected as a component of *Salmonella* genomic island (SGI-1) associated multiresistant gene cluster located upon the chromosome of *S.* Tyhimurium DT104 and other phage types (Pasquali *et al.*, 2004). The two genes *tet* (A) and *tet* (B) are associated with small non conjugative transposons Tn*1721* and Tn*10*, respectively. Plasmids conferring resistance to *tet* (C) and *tet* (D) have been identified only in single cases (Bernardi and Bernardi, 1984; Allard *et al.*, 1993).

Resistance to trimethoprim

Trimethoprim compound is a diaminopyrimidine that selectively inhibits bacterial dihydrofolate reductase (DHFR) by preventing the reduction of dihydrofolate to tetrahydrofolate. Tetrahydrofolate is an essential precursor in the *de novo* synthesis of the intermediate thymidine monophosphate (dTMP), precursor of DNA metabolite thymidine triphosphate. Bacteria are unable to take up folic acid from the host environment and are thus dependent on their own *de novo* synthesis. Inhibition of the enzyme starves the bacteria for nucleotides necessary for DNA replication (Skold, 2001). The most common trimethoprim resistance mechanism is the supplementation of a trimethoprim sensitive dihydrofolate reductase with a trimethoprim-resistant dihydrofolate reductase, resulting in high-level trimethoprim resistance in various bacteria. There are 30 *dfr* (dihydrofolate reductase) genes which are subdivided on the basis of their structure as *dfr1* and *dfr2* (Pattishall *et al.*, 1977), recently referred to as *dfrA* and *dfrB* (Skold, 2001). A total of 13 *dfrA* genes were detected in various *Salmonella* serovars with most of these located in class 1 and class 2 integrons (Daly and Fanning, 2000; Peirano, 2006) (Table 6.5). The association of these genes with mobile genetic elements such as transposons or plasmids has often been observed, and the presence of these genes in cassettes as part of integrons is also common (Peirano, 2006).

Impact of drug resistance on the epidemiology of *Salmonella* infections

In many countries the incidence of human *Salmonella* infections has increased markedly over the years. In the United States, non-typhoid *Salmonella* serotypes affect approximately 2–3 million persons and cause 500–2000 deaths each year (Altekruse *et al.*, 1997). It is also estimated that there are up to 16–33 million cases of typhoid fever causing more than 600,000 deaths every year (Edelman and Levine 1986; WHO, 1996). Although chloramphenicol has been used as a drug of choice for the treatment of typhoid fever, the emergence of its resistance has increased drastically in recent years. In the 1970s, chloramphenicol-resistant strains of typhoidal *Salmonella* emerged in the Indian subcontinent and Mexico (Paniker and Vimala, 1972). The plasmid encoded resistance was directed not only to chloramphenicol but also to ampicillin and trimethoprim–sulfamethoxazole, thereby leading to therapeutic complications and increased mortality (Row *et al.*, 1997). Most of the cases of typhoid fever in the United States occur secondarily owing to international travels with multidrug resistance showing an increasing trend from < 1 to > 10% (Ackers *et al.*, 2000). In a study in Europe, one third of the typhoidal isolates showed reduced susceptibility to fluoroquinolones, thus indicating a need for higher doses of this drug for treatment purposes (Threfall *et al.*, 2003). In some areas there is an increasing resistance of *S.* Typhi strains to fluoroquinolones while a decreasing resistance to chloramphenicol has been observed (Gautam *et al.*, 2002), which could influence the future treatment regimes/recommendations in developing countries where the economics/cost of treatment plays an important role.

Antimicrobial resistance is a major problem in the treatment of cases of non-typhoidal *Salmonella* (Saxena *et al.*, 1989; Tassios *et al.*, 1997; Su *et al.*, 2001; Chiu *et al.*, 2002). Until the 1980s, conventional antimicrobial agents such as ampicillin, chloramphenicol and trimethoprim–sulfamethoxazole had been the choice of drugs for the treatment of salmonellosis. However, in many areas of the world, multidrug resistance with more than 50% rates of resistance to these commonly used antimicrobial agents has been reported (Saxena *et al.*, 1989; Su *et al.*, 2001; Chiu *et al.*, 2002). Later extended-spectrum cephalosporins and fluoroquinolones came to be used as an alternative agent in the treatment of multidrug-resistant *Salmonella* infections (Goossens *et al.*, 1984; Barnass *et al.*, 1990). However, resistance of *Salmonella* serotypes to these antibiotics have been increasingly reported since 1991 (Poupart *et al.*, 1991; Bradford *et al.*, 1998; Baraniak *et al.*, 2002; Chiu *et al.*, 2002).

The association between antimicrobial resistance and salmonellosis has several repercussions (Barza, 2002; Barza and Travers, 2002; Travers and Barza, 2002), with the most important being its impact on the treatment of infections. Patients with invasive salmonellosis require effective antimicrobial therapy. Growing antimicrobial resistance may add to the difficulty in treating patients with such infections, thus leading to increased morbidity and mortality (Mead *et al.*,

1999). Many previous reports have provided evidence of development of antimicrobial-resistant *Salmonella* serotypes (Helms *et al.*, 2002; Su *et al.*, 2003). Epidemiological reports have indicated that antimicrobial-resistant strains of *Salmonella* could cause more prolonged or more severe illness than susceptible strains (Travers *et al.*, 2002) and could be due to the acquisition of virulence plasmid containing resistance genes. Such type of resistance characteristics mediated by plasmid could provide an advantage to the virulent *Salmonella* to cause infections in the unfavourable drug environment leading to increased mortality in infected humans and animals. In conclusion, plasmids, bacteriophages, transposons, integrons and even pathogenicity islands of *Salmonella* are continuously evolving with time. It appears likely that more antibiotic-resistant and virulent *Salmonella* will arise posing a serious problem in the treatment of *Salmonella* infections in humans and animals.

Acknowledgements

The authors are thankful to the Indian Council of Medical Research and the Department of Biotechnology, Government of India, and the Bioinformatics Centre, College of Fisheries, Mangalore, for the research funding and logistics support to carry out some of the research presented in this chapter.

References

Aarts, H.J., Boumedine, S.K., Nesme, X., and Cloeckaert, A. (2001). Molecular tools for the characterization of antibiotic-resistant bacteria. Vet. Res. *32*, 363–380.

Ackers, M.L., Phur, N.D., Tauxe, R.V., and Mintz, E.D. (2000). Laboratory-based surveillance of *Salmonella* serotype typhi infections in the United States; antimicrobial resistance on the rise. JAMA *283*, 2668–2673.

Ahmed, W., Neller, R., and Katouli, M. (2005). Host species-specific metabolic fingerprint database for *Enterococci* and *Escherichia coli* and its application to identify sources of fecal contamination in surface waters. Appl. Environ. Microbiol. *71*, 4461–4468.

Ahmer, B.M., van Reeuwijk, M.J., Timmers, C.D., Valentine, P.J., and Heffron, F. (1998). *Salmonella* Typhimurium encodes an SdiA homolog, a putative quorum sensor of the LuxR family that regulates genes on the virulence plasmid. J. Bacteriol. *180*, 1185–1193.

Akbar, S., Schechter, L.M., Lostroh, C.P., and Lee, C.A. (2003). AraC/XylS family members, HilD and HilC, directly activate virulence gene expression independently of HilA in *Salmonella* Typhimurium. Mol. Microbiol. *47*, 715–728.

Alekshun, M.N., and Levy, S.B. (2007). Molecular mechanisms of antibacterial multidrug resistance. Cell *128*, 1037–1050.

Allard, J.D., Gibson, M.L., Vu, L.H., Nguyen, T.T., and Bertrand K.P. (1993). Nucleotide sequence of class D tetracycline resistance genes from *Salmonella* Ordonez. Mol. Gen. Genet. *237*, 301–305.

Aloni, Y., Delmer, D.P., and Benziman, M. (1982). Achievement of high rates of *in vitro* synthesis of 1,4-D-D-glUCanA: activation bycooperative interaction of the acetobacterxylinum enzyme system with GTP, polyethylene glycol and a protein factor. Proc. Natl. Acad. Sci. U.S.A. *79*, 6448–6452.

Altekruse, S.F., Cohen, M.L., and Swerdlow, D.L. (1997). Emerging foodborne diseases. Emerg. Infect. Dis. *3*, 285–293.

Angulo, F.J., and Griffin, P.M. (2000). Changes in antimicrobial resistance in *Salmonella enterica* serovar Typhimurium. Emerg. Infect. Dis. *6*, 436–438.

Arakawa, Y., Murakami, M., Suzuki, K., Ito, H., Wacharotayankun, R., Ohsuka, S., Kato, N., and Ohta, M. (1995). A novel integron-like element carrying the metallo-beta-lactamase gene bla_{IMP}. Antimicrob. Agents Chemother. *39*, 1612–1615.

Ashkenazi, S., and Cleary, T.G. (1990). A method for detecting Shiga toxin and Shiga-like toxin-I in pure and mixed culture. J. Med. Microbiol. *32*, 255–261.

van Asten, A.J., and van Dijk, J.E. (2005). (MiniReview) Distribution of "classic" virulence factors among *Salmonella* spp. FEMS Immunol. Med. Microbiol. *44*, 251–259.

Baek, C.H., Wang, S., Roland, K.L., and Curtiss, R., 3rd. (2009). Leucine-responsive regulatory protein (Lrp) acts as a virulence repressor in *Salmonella enterica* serovar Typhimurium. J. Bacteriol. *191*, 1278–1292.

Bajaj, V., Lucas, R.L., Hwang, C., and Lee, C.A. (1996). Co-ordinate regulation of *Salmonella* Typhimurium invasion genes by environmental and regulatory factors is mediated by control of *hilA* expression. Mol. Microbiol. *22*, 703–714.

van Bambeke, F., Balzi, E., and Tulkens, P.M. (2000), Antibiotic efflux pumps. Biochem. Pharmacol. *60*, 457–470.

Baraniak, A., Sadowy, E., Hryniewicz, W., and Gniadkowski. M. (2002). Two different extended-spectrum beta-lactamases (ESBLs) in one of the first ESBL-producing *Salmonella* isolates in Poland. J. Clin. Microbiol. *40*, 1095–1097.

Barnass, S., Franklin, J., and Tabaqchali. S. (1990). The successful treatment of multiresistant non-enteric salmonellosis with seven day oral ciprofloxacin. J. Antimicrob. Chemother. *25*, 299–300.

Barza, M. (2002). Potential mechanisms of increased disease in humans from antimicrobial resistance in food animals. Clin. Infect. Dis. *34*, 123–125.

Barza, M., and Travers. K. (2002). Excess infections due to antimicrobial resistance: the "Attributable Fraction". Clin. Infect. Dis. *34*, 126–130.

Bäumler, A.J., Gilde, A.J., Tsolis, R.M., van der Velden, A.W.M., Ahmer, B.M.M., and Heffron, F. (1997).

Contribution of horizontal gene transfer and deletion events to development of distinctive patterns of fimbrial operons during evolution of *Salmonella* Serotypes. J. Bacteriol. *179*, 317–322.

Bäumler, A.J., Tsolis, R.M., Kusters, J.G., Hoffmann, S., and Heffron, F. (1996). The *pef* fimbrial operon of *Salmonella* Typhimurium mediates adhesion to murine small intestine and is necessary for fluid accumulation in the infant mouse. Infect. Immun. *64*, 61–68.

Bearson, B., Wilson, L., and Foster, J.W. (1998). A low pH-inducible, PhoPQ dependent acid tolerance response protects *Salmonella* Typhimurium against inorganic acid stress. J. Bacteriol. *180*, 2409–2417.

Bernardi, A., and Bernardi, F. (1984). Complete sequence of pSC101. Nucleic Acid Res. *12*, 9415–9426.

Bhowmick, P.P., Devananda, D., Ruwandeepika, H.A.D., Karunsagar, I., and Karunsagar, I. (2011a). Presence of *Salmonella* pathogenicity island 2 genes in seafood associated *Salmonella* serovars and the role of *sseC* gene in survival of *Salmonella enterica* serovar Weltevreden in epithelial cells. Microbiology *157*, 160–168.

Bhowmick, P.P., Devananda, D., Ruwandeepika, H.A.D., Thilo, M.F., Shabarinath, S., Karunasagar, I., and Karunasagar, I. (2011b). *gcpA* (*stm1987*) is critical for cellulose production and biofilm formation on polystyrene surface by *Salmonella enterica* serovar Weltevreden in both high and low nutrient medium. Microb. Pathog. *50*, 114–122.

Bingle, L.E., Bailey, C.M., and Pallen, M.J. (2008). Type VI secretion: a beginner's guide. Curr. Opin. Microbiol. *11*, 3–8.

Blanc-Potard, A.B., and Groisman, E.A. (1997). The *Salmonella selC* locus contains a Pathogenicity Island mediating intramacrophage survival. EMBO J. *16*, 5376–5385.

Blacklow, N.R., Dolin, R., Fedson, D.S., Dupont, H., Northrup, R.S., Hornock, R.B., and Chanock, R.M. (1972). Acute infectious nonbacterial gastroenteritis: etiology and pathogenesis. Ann. Intern. Med. *76*, 993.

Blondel, C.J., Jiménez, J.C., Contreras, I., and Santiviago, C.A. (2009). Comparative genomic analysis uncovers 3 novel loci encoding type six secretion systems differentially distributed in *Salmonella* serotypes. BMC Genomics *10*, 354.

Boddicker, J.D., Knosp, B.M., and Jones, B.D. (2003). Transcription of the *Salmonella* invasion gene activator, *hilA*, requires HilD activation in the absence of negative regulators. J. Bacteriol. *185*, 525–533.

Bohez, L., Gantois, I., Ducatelle, R., Pasmans, F., Dewulf, J., Haesebrouck, F., and Van Immerseel, F. (2008). The *Salmonella* Pathogenicity Island 2 regulator *ssrA* promotes reproductive tract but not intestinal colonization in chickens. Vet. Microbiol. *126*, 216–224.

Bolton, A.J., Martin, G.D., Osborne, M.P., Wallis, T.S., and Stephen, J. (1999) Invasiveness of *Salmonella* serotypes Typhimurium, Cholerae- suis and Dublin for rabbit terminal ileum *in vitro*. J. Med. Microbiol. *48*, 801–810.

Boyd, D., Cloeckaert, A., Chaslus-Dancla, E., and Mulvey, M.R. (2002). Characterization of variant *Salmonella* genomic island 1 multidrugresistance regions from serovars Typhimurium DT104 and Agona. Antimicrob. Agents. Chemother. *46*, 1714–1722.

Bradford, P.A., Yang, Y., Sahm, D., Grope, I., Gardovska, D., and Storch, G. (1998). CTX-M-5, a novel cefotaxime-hydrolyzing beta-lactamase from an outbreak of *Salmonella* Typhimurium in Latvia. Antimicrob. Agents Chemother. *42*, 1980–1984.

Breuil, J., Berger, N., Dublanchet, A., and le collège Bvh. (1996). Sensibilité aux antibiotiques de 2800 souches de Salmonelles et Shigelles isolées en France en 1994. Med. Mal. Infect. *26*, 420–425.

Breuil, J., Brisabois, A., Casin, I., Armand-Lefèvre, L., Fremy, S., and Collatz, E. (2000). Antibiotic resistance in Salmonellae isolated from humans and animals in France: comparative data from 1994 and 1997. J. Antimicrob. Chemother. *46*, 965–971.

Briggs, C.E., and Fratamico, P.M. (1999). Molecular characterization of an antibiotic resistance gene cluster of *Salmonella* Typhimurium DT104. Antimicrob. Agents Chemother. *43*, 846–849.

Brombacher, E., Dorel, C.A., Zehnder, J.B., and Landini, P. (2003). The curli biosynthesis regulator CSGD co-ordinates the expression of both positive and negative determinants for biofilm formation in *Escherichia coli*. Microbiology *49*, 2847–2857.

Bronstein, P.A., Miao, E.A., and Miller, S.I. (2000). InvB is a type III secretion chaperone specific for SspA. J. Bacteriol. *182*, 6638–6644.

Brumell, J.H., Rosenberger, C.M., Gotto, G.T., Marcus, S.L., and Finlay, B.B. (2001). SifA permits survival and replication of *Salmonella* Typhimurium in murine macrophages. Cell. Microbiol. 3, 75–84.

Brumell, J.H., Kujat-Choy, S., Brown, N.F., Vallance, B.A., Knodler, L.A., and Finlay, B.B. (2003). SopD2 is a novel type III secreted effector of *Salmonella* Typhimurium that targets late endocytic compartments upon delivery into host cells. Traffic 4, 36–48.

Buchrieser, C., Prentice, M., and Carniel, E. (1998). The 102-kilobase unstable region of *Yersinia pestis* comprises a high-pathogenicity island linked to a pigmentation segment which undergoes internal rearrangement. J. Bacteriol. *180*, 2321–2329.

Buchrieser, C., Rusniok, C., Frangeul, L., Couve, E., Billault, A., Kunst, F., Carniel, E., and Glaser, P. (1999). The 102-kilobase pgm locus of Yersinia pestis: sequence analysis and comparison of selected regions among different *Yersinia pestis* and *Yersinia pseudotuberculosis* strains. Infect. Immun. *67*, 4851–4861.

Buckley, A.M., Webber, M.A., Cooles, S., Randall, L.P., La Ragione, R.M., Woodward, M.J., and Piddock, L.J. (2006). The AcrAB–TolC efflux system of *Salmonella enterica* serovar Typhimurium plays a role in pathogenesis. Cell Microbiol. *8*, 847–856.

Cain, R.J., Hayward, R.D., and Koronakis, V. (2008). Deciphering interplay between *Salmonella* invasion effectors. PLoS Pathog. *4*, e1000037.

Carattoli, A. (2003). "Plasmid-mediated antimicrobial resistance in *Salmonella enterica*". Curr. Issues. Mol. Biol. *5*, 113–122.

Cascales, E. (2008). The type VI secretion toolkit. EMBO Rep. *9*, 735–741.

Centers for Disease Control and Prevention. (2008). Preliminary Food Net data on the incidence of infection with pathogens transmitted commonly through food 10 states, 2007. MMWR Morb. Mortal. Wkly Rep. *57*, 366–370.

Chakravortty, D., Rohde, M., Jäger, L., Deiwick, J., and Hensel, M. (2005). Formation of a novel surface structure encoded by *Salmonella* pathogenicity island 2. EMBO J. *24*, 2043–2052.

Chang, J., Chen, J., and Zhou, D. (2005a). Delineation and characterization of the actin nucleation and effector translocation activities of *Salmonella SipC*. Mol. Microbiol. *55*, 1379–1389.

Chang, J., Myeni, S.K., Lin, T.L., Wu, C.C., Staiger, C.J., and Zhou, D. (2005b). SipC multimerization promotes actin nucleation and contributes to *Salmonella*-induced inflammation. Mol. Microbiol. *66*, 1548–1556.

Chatfield, S., Dougan, G., and Charles, I. (1990). Complete nucleotide sequence of the *aroA* gene from *Salmonella* Typhi encoding 5-enolpyruvylshikimate 3-phosphate synthase. Nucleic. Acids Res. *18*, 6133.

Chen, S., Zhao, S., White, D.G., Schroeder, C.M., Lu, R., Yang, H., Mcdermott, P.F., Ayers, S., and Meng, J. (2004). Characterization of multiple antimicrobial-resistant *Salmonella* serovars isolated from retail meat. Appl. Environ. Microbiol. *70*, 1–7.

Chiu, C.H., Wu, T.L., Su, L.H., Chu, C., Kuo, A.J., Chien, M.S., and Lin, T.Y. (2002). The emergence in Taiwan of fluoroquinone resistance in *Salmonella enterica* serotype Cholerasuis. N. Engl. J. Med. *346*, 413–419.

Chiu, C.H., Tang, P., Chu, C., Hu, S., Bao, Q., Yu, J., Chou, Y.Y., Wang, H.S., and Lee, Y.S. (2005). The genome sequence of *Salmonella enterica* serovar Choleraesuis, a highly invasive and resistant zoonotic pathogen. Nucleic Acids Res. *33*, 1690–1698.

Choi, C.H.A., and O'Day, H.D. (1984). Calcofluor staining of cellulose during microcyst differentiation in wild-type and mutant strains of polysphondylium pallidum. J. Bacteriol. *157*, 291–296.

Choi, J., Shin, D., and Ryu, S. (2007). Implication of quorum sensing in *Salmonella enterica* serovar Typhimurium virulence: the *luxS* gene is necessary for expression of genes in Pathogenicity Island 1. Infect. Immun. *75*, 4885- 4890.

Chopra, I., and Roberts, M. (2001). Tetracycline antibiotics: Mode of action, applications, molecular biology, and epidemiology of bacterial resistance. Microbiol. Mol. Biol. Rev. *65*, 232–260.

Chu, C., Hong, S.F., Tsai, C., Lin, W.S., Liu, T.P., and Ou, J.T. (1999). Comparative physical and genetic maps of the virulence plasmids of *Salmonella enterica* serovars Typhimurium, Enteritidis, Choleraesuis, and Dublin. Infect. Immun. *67*, 2611–2614.

Chu, C., Chiu, C., Wu, W., Ch, C., Liu, T., and Ou, J. (2001). Large drug resistance virulence plasmids of clinical isolates of *Salmonella entrica* Serovar Choleaesuis. Antimicrob. Agents Chemother. *54*, 2299–2203.

Collazo, C.M., and Galán, J.E. (1996). Requirement for exported proteins in secretion through the invasion-associated type III system of *Salmonella* Typhimurium. Infect. Immun. *64*, 3524–3531.

Collazo, C.M., Zierler, M.K., and Galan, J.E. (1995). Functional analysis of the *Salmonella* Typhimurium invasion genes *invl* and *invJ* and identification of a target of the protein secretion apparatus encoded in the inv locus. Mol. Microbiol. *15*, 25–38.

Collis, C.M., and Hall, R.M. (1995). Expression of antibiotic resistance genes in the integrated cassette of integrons. Antimicrobiol. Agents Chemother. *39*, 155–162.

Coombes, B.K., Wickham, M.E., Lowden, M.J., Brown, N.F., and Finlay, B.B. (2005). Negative regulation of *Salmonella* Pathogenicity Island 2 is required for contextual control of virulence during typhoid. Proc. Natl. Acad. Sci. U.S.A. *102*, 17460- 17465.

Coombes, B.K., Lowden, M.J., Bishop, J.L., Wickham, M.E., Brown, N.F., Duong, N., Osborne, S., Gal-Mor, O., and Finlay, B.B. (2007). SseL Is a *Salmonella*-specific translocated effector integrated into the SsrB-controlled *Salmonella* pathogenicity island 2 type III secretion system. Infect. Immun. *75*, 574–580.

Costerton, J.W., Stewart, P.S., and Greenberg, E.P. (1999). Bacterial biofilms: a common cause of persistent infections. Science *284*, 1318–1322.

Coynault, C., Robbe-Saule, V., Popoff, M.Y., and Norel, F. (1992). Growth phase and SpvR regulation of transcription of *Salmonella* Typhimurium *spvABC* virulence genes. Microb. Pathog. *13*, 133–143.

Criss, A.K., and Casanova, J.E. (2003). Coordinate regulation of *Salmonella enterica* serovar Typhimurium invasion of epithelial cells by the Arp2/3 complex and Rho GTPases. Infect. Immun. *71*, 2885–2891.

Daly, M., and Fanning, S. (2000). Characterization and chromosomal mapping of antimicrobial resistance genes in *Salmonella enterica* serotype Typhimurium. Appl. Environ. Microbiol. *66*, 4842–4848.

Darwin, K.H., and Miller, V.L. (2001). Type III secretion chaperone-dependent regulation: activation of virulence genes by SicA and InvF in *Salmonella* Typhimurium. EMBO J. *20*, 1850–1862.

Darwin, K.H., Robinson, L.S., and Miller, V.I. (2001). SigE is a chaperone for the *Salmonella enterica* serovar Typhimurium invasion protein SigD. J. Bacteriol. *183*, 1452–1454.

Davey, M.E., and O'Toole, G.A. (2000). Microbial biofilms: from ecology to molecular genetics. Mol. Biol. Rev. *64*, 847–867.

Davies, J., and Wright, G.D. (1997). Bacterial resistance to aminoglycoside antibiotics. Trends. Microbiol. *5*, 234–240.

Deekshit, V.K., Kumar, K.B., Rai, P., Srikumar, S., Karunasagar, I., and Karunasagar, I. (2012). Detection of Class 1 integrons in *Salmonella* Wettevreden and silent antibiotic resistance genes in some seafood-associated non-typhoidal isolates of *Salmonella* in southwest coast of India. J. Appl. Microbiol. (accepted with minor revision).

Deiwick, J., Nikolaus, T., Erdogan, S., and Hensel, M. (1999). Environmental regulation of *Salmonella* pathogenicity island 2 gene expression. Mol. Microbiol. *31*, 1759- 1773.

Dierikxa, C., van, Essen-Zandbergen, A., Veldman, K., Smith, H., and Mevius, D. (2010). Increased detection of extended spectrum beta-lactamase producing *Salmonella* enterica and *Escherichia coli* isolates from poultry. Vet. Microbiol. *145*, 273–278.

Doublet, B., Weill, F.X., Fabre, L., Chaslus-Dancla, E., and Cloeckaert, A. (2004). Variant *Salmonella* genomic island 1 antibiotic resistance gene cluster containing a novel 30-N-aminoglycoside acetyltransferase gene cassette, aac (3)-Id, in *Salmonella enterica* serovar Newport. Antimicrob. Agents Chemother. *48*, 3806–3812.

van Duijkeren, E., Wannet, W.J., Houwers, D.J., and Van Pelt, W. (2003). Antimicrobial susceptibilities of *Salmonella* strains isolated from humans, cattle, pigs, and chickens in the Netherlands from 1984 to 2001. J. Clin. Microbiol. *41*, 3574–3578.

Dunlap, N.E., Benjamin, W.H., Jr., McCall, R.D., Jr., Tilden, A.B., and Briles, D.E. (1991). A "safe-site for *Salmonella* Typhimurium is within splenic cells during the early phase of infection in mice. Microb. Pathog. *10*, 297–310.

Edelman, R., and Levine, M.M. (1986). Summary of an international workshop on typhoid fever. Rev. Infect. Dis. *8*, 329–349.

Edwards, R.A., Schifferli, D.M., and Maloy, S.R. (2000). A role for *Salmonella* fimbriae in intraperitoneal infections. Proc. Natl. Acad. Sci. U.S.A. *97*, 1258–1262.

Eichelberg, K., Gindcchio, C.C., and Galan, J.E. (1994). Molecular and functional characterization of the *Salmonella* Typhimurium invasion genes *invB* and *invC*: homology of *invC* to the F0 F1 ATPase family of proteins. J. Bacteriol. *176*, 4501–4510.

Ellerbroek, S.M., Wennerberg, K., Arthur, W.T., Dunty, J.M., Bowman, D.R., DeMali, K.A., Der, C., and Burridge, K. (2004). SGEF, a RhoG guanine nucleotide exchange factor that stimulates macropinocytosis. Mol. Biol. Cell *15*, 3309–3319.

Ellermeier, J.R., and Slauch, J.M. (2003). RtsA and RtsB coordinately regulate expression of the invasion and flagellar genes in *Salmonella enterica* serovar Typhimurium. J. Bacteriol. *185*, 5096–5108.

Ellermeier, J.R., and Slauch, J.M. (2008). Fur regulates expression of the *Salmonella* pathogenicity island 1 type III secretion system through HilD. J. Bacteriol. *190*, 476–486.

Ellermeier, C.D., Ellermeier, J.R., and Slauch, J.M. (2005). HilD, HilC and RstA constitute a feed forward loop that controls expression of the SPI1 type three secretion system regulator *hilA* in *Salmonella enterica* serovar Typhimurium. Mol. Microbiol. *57*, 691–705.

Elvers, K.T., and Park, S.F. (2002). Quorum sensing in *Campylobacter jejuni*: detection of the *luxS* encoded signalling molecule. Microbiology *148*, 1475–1481

Enne, V.I., Livermore, D.M., Stephens, P., and Hall, L.M.C. (2001). Persistence of sulphonamide resistance in *Escherichia coli* in the UK despite national prescribing restriction. Lancet *357*, 1325–1328.

Eswarappa, S.M., Panguluri, K.K., Hensel, M., and Chakravortty, D. (2008). The yejABEF operon of *Salmonella* confers resistance to antimicrobial peptides and contributes to its virulence. Microbiology *154*, 666–678.

Eswarappa, S.M., Negi, V.D., Chakraborty, S., Chandrasekhar Sagar, B.K., and Chakravortty, D. (2010). Division of the *Salmonella*-containing vacuole and depletion of acidic lysosomes in *Salmonella*-infected host cells are novel strategies of *Salmonella enterica* to avoid lysosomes. Infect. Immun. *78*, 68–79.

Feng, X., Oropeza, R., and Kenney, L.J. (2003). Dual regulation by phospho-OmpR of *ssrA/B* gene expression in *Salmonella* pathogenicity island 2. Mol. Microbiol. *48*, 1131–1143.

Feng, X., Walthers, D., Oropeza, R., and Kenney, L.J. (2004). The response regulator SsrB activates transcription and binds to a region overlapping OmpR binding sites at *Salmonella* pathogenicity island 2. Mol. Microbiol. *54*, 823–835.

Figueroa-Bossi, N., Uzzau, S., Maloriol, D., and Bossi, L. (2001). Variable assortment of prophages provides a transferable repertoire of pathogenic determinants in *Salmonella*. Mol. Microbiol. 39, 260–271.

Filloux, A., Hachani, A., and Bleves, S. (2008). The bacterial type VI secretion machine: yet another player for protein transport across membranes. Microbiology *154*, 1570–1583.

Fluit, A.C., and Schmitz, F.J. (2004). Resistance integrons and super-integrons. Clin. Microbiol. Infect. *10*, 272–288.

Folkesson, A., Lofdahl, S., and Normark, S. (2002). The *Salmonella enterica* subspecies I specific centisome 7 genomic island encodes novel protein families present in bacteria living in close contact with eukaryotic cells. Res. Microbiol. *153*, 537–545.

Francis, C.L., Starnbach, M.N., and Falkow, S (1992). Morphological and cytoskeletal changes in epithelial cells occur immediately upon interaction with *Salmonella* Typhimurium grown under low-oxygen conditions. Mol. Microbiol. *6*, 3077–3087.

Frech, G., and Schwarz, S. (1998). Tetracycline resistance in *Salmonella enterica* subsp. *enterica* serovar Dublin isolates. Antimicrob. Agents Chemother. *42*, 1288–1289.

Frech, G., and Schwarz, S. (2000). Molecular analysis of tetracycline resistance in *Salmonella enterica* subsp. *enterica* serovars Typhimurium, Enteritidis, Dublin, Choleraesuis, Hadar, Saintpaul: construction and application of specific gene probes. J. Appl. Microbiol. *89*, 633–641.

Freeman, J.A., Rappl, C., Kuhle, V., Hensel, M., and Miller, S.I. (2002). SpiC is required for translocation of *Salmonella* pathogenicity island 2 effectors and secretion of translocon proteins SseB and SseC. J. Bacteriol. *184*, 4971–4980.

Friebel, A., Ilchmann, H., Aepfelbacher, M., Ehrbar, K., Machleidt, W., and Hardt, W.D. (2001). SopE and SopE2 from *Salmonella* Typhimurium activate different sets of RhoGTPases of the host cell. J. Biol. Chem. *276*, 34035–34040.

Fu, Y., and Galán, J.E. (1998a). Identification of a specific chaperone for SptP, a substrate of the centisome 63 type III secretion system of *Salmonella* Typhimurium. J. Bacteriol. *180*, 3393–3399.

Fu, Y., and Galan, J.E. (1998b). The *Salmonella* Typhimurium tyrosine phosphatase SptP is translocated into host cells and disrupts the actin ctyoskeleton. Mol. Microbiol. *27*, 359–368.

Galán, J.E. (1999). Interaction of *Salmonella* with host cells through the centisome 63 type III secertion system. Curr. Opin. Microbiol. 2, 46–50.

Galan, J.E., and Collmer, A. (1999). Type III secretion machines: bacterial devices for protein delivery into host cells. Science *284*, 1322–1328.

Galyov, E.E., Wood, M.W., Rosqvist, R., Mullan, P.B., Watson, P.R., Hedges, S., and Wallis, T.S. (1997). A secreted effector protein of *Salmonella* Dublin is translocated into eukaryotic cells and mediates inflammation and fluid secretion in infected ileal mucosa. Mol. Microbiol. *25*, 903–912.

Garcia, B., Latasa, C., Solano, C., Portillo, G.D.F., Gamazo, C., and Lasa, I. (2004). Role of the GGDEF protein family in *Salmonella* cellulose biosynthesis and biofilm formation. Mol. Microbiol. *54*, 264–277.

Garcia-del Purtillo, F., Zwick, M.B., Leung, K.Y., and Finlay, B.B. (1993). Intracellular replication of *Salmonella* within epithelial cells is associated with filamentous structures containing lysosomal membrane glycoproteins. Inf. Agents Dis. *2*, 227–231.

Garcia-Vescovi, E., Soncini, F.C., and Groisman, E.A. (1996). Mg^{2+} as an extracellular signal environmental regulation of *Salmonella* virulence. Cell *84*, 165–174.

Gautam, V., Gupta, N.K., Chaudhary, U., and Arora, D.R. (2002). Sensitivity pattern of *Salmonella* serotypes in Northern India. Braz. J. Infect. Dis. *6*, 281-287.

Gerlach, R.G., Jäckel, D., Stecher, B., Wagner, C., Lupas, A., Hardt, W.D., and Hensel, M. (2007). *Salmonella* pathogenicity island 4 encodes a giant non-fimbrial adhesin and the cognate type I secretion system. Cell Microbiol. 9, 1834–1850.

Glaser, M.J., and Newman, L.S. (1982). A review of human salmonellosis. I. Infective dose. Rev. Infect. Dis. *34*, 1096–1106.

Gomez, T.M., Motarjemi, Y., Miyagawa, S., Kaferstein, F.K., and Stohr, K. (1997). Foodborne salmonellosis. World Health Stat. Q. *50*, 81–89.

Gong, H., Su, J., Bai, Y., Miao, L., Kim, K., Yang, Y., Liu, F., and Lu, S. (2009). Characterization of the expression of *Salmonella* type III secretion system factor PrgI, SipA, SipB, SopE2, SpaO, and SptP in cultures and in mice. BMC Microbiology 9, 73.

Goossens, H., Vanhoof, R., de Mol, P., Grados, O., Ghysels, G., and Butzler. J.P. (1984). *In-vitro* susceptibility of salmonellae to antimicrobial agents. J. Antimicrob. Chemother. *13*, 559–565.

Groisman, E.A., Chiao, E., Lipps, C.J., and Heffron, F. (1989). *Salmonella* Typhimurium the influence of new mutations conferring co resistance on the *corA* Mg^{2+} transport system. Mol. Microbiol. *5*, 2753–2762.

Guerra, B., Soto, S., Cal, S., and Mendoza, M.C., (2000). Antimicrobial resistance and spread of class 1 integrons among *Salmonella* serotypes. Antimicrob. Agents Chemother. *44*, 2166–2169.

Gulig, P.A., Danbara, H., Guiney, D.G., Lax, A.J., Norel, F., and Rhen, M. (1993). Molecular analysis of *spv* virulence genes of the *Salmonella* virulence plasmids. Mol. Microbiol. *7*, 825–830.

Gunn, J.S., and Miller, S.I. (1996). PhoP-PhoQ activates transcription of *pmrAB*, encoding a two component regulatory system involved in *Salmonella* Typhimurium antimicrobial peptide resistance. J. Bacteriol. *178*, 6857–6864.

Gunn, J.S., Alpuche-Aranda, C.M., Loomis, W.P., Belden, W.J., and Miller, S.I. (1995). Characterization of the *Salmonella* Typhimurium *pagC/pagD* chromosomal region. J. Bacteriol. *177*, 5040–5047.

Gunzel, D., Kucharski, L.M., Kehres, D.G., Romero, M.F., and Maguire, M.E. (2006). The MgtC virulence factor of *Salmonella enterica* serovar Typhimurium activates Na+, K+ -ATPase. J. Bacteriol. *188*, 5586–5594.

Guy, R.L., Gonias, L.A., and Stein, M.A. (2000). Aggregation of host endosomes by *Salmonella* requires SPI-2 translocation of SseFG and involves SpvR and the *fms-aroE* intragenic region. Mol. Microbiol. *37*, 1417–1435

Haghjoo, E., and Galán, J.E. (2004). *Salmonella* Typhi encodes a functional cytolethal distending toxin that is delivered into host cells by a bacterial-internalization pathway. Proc. Natl. Acad. Sci. U.S.A. *101*, 4614–4619.

Hakansson, S., Schesser, K., Persson, C., Galyov, E.E., Rosqvist, R., Homble, F., and Wolf-Watz, H. (1996). The YopB protein of *Yersinia pseudotuberculosis* is essential for the translocation of Yop effector proteins across the target cell plasma membrane and displays a contact-dependent membrane disrupting activity. EMBO J. *15*, 5812–5823.

Hall, R.M., and Stokes, H.W. (1993). Novel DNA elements which capture genes by site-specific recombination. Genetica *90*, 115–132.

Hammar, M., Arnqvist, A., Bian, Z., Olsen, A., and Normark, S. (1995). Expression of two *csg* operons is required for production of fibronectin and Congo Red binding curli polymers in *Escherichia Coli* K-12. Mol. Microbiol. *18*, 661–670.

Hardt, W.D., and Galán, J.E. (1997). A secreted *Salmonella* protein with homology to an avirulence determinant of plant pathogenic bacteria. Proc. Natl. Acad. Sci. U.S.A. *94*, 9887–9892.

Hasman, H., Mevius, D., Veldman, K., Olesen, I., and Aarestrup, F.M. (2005). β-lactamases among extended-spectrum β-lactamase (ESBL)-resistant *Salmonella* from poultry, poultry products and human patients in the Netherlands. J. Antimicrob. Chemother. *56*, 115–121.

Havlickova, H., Hradecka, H., Bernardyova, I., and Rychlik, I. (2009). Distribution of integrons and SGI1 among antibiotic-resistant *Salmonella enterica* isolates of animal origin. Vet. Microbiol. *133*, 193–198.

Hawkey, M. (2003). Mechanisms of quinolone action and microbial response. J. Antimicrob. Chemother. *51*, 29–35.

Hayward, R.D., and Koronakis, V. (1999). Direct nucleation and bundling of actin by the SipC protein of invasive *Salmonella*. EMBO J. *18*, 4926–4934.

Heisig, P. (1993). High-level fiuroquinolone resistance in a *Salmonella* Typhimurium isolate due to alterations in both *gyrA* and *gyrB* genes. J. Antimicrob. Chemother. *32*, 367–377.

Helms, M., Vastrup, P., Gerner-Smidt, P., and Mølbak. K. (2002). Excess mortality associated with antimicrobial drug-resistant *Salmonella* Typhimurium. Emerg. Infect. Dis. *8*, 490–495.

Hensel, M. (2004). (REVIEW) Evolution of pathogenicity islands of *Salmonella enterica*. Int. J. Med. Microbiol. *294*, 95–102.

Hensel, M., Shea, J.E., Waterman, S.R., Mundy, R., Nikolaus, T., Banks, G., Vazquez-Torres, A., Gleeson, C., Fang, F.C., and Holden, D.W. (1998). Genes encoding putative effector proteins of the type III secretion system of *Salmonella* pathogenicity island 2 are required for bacterial virulence and proliferation in macrophages. Mol. Microbiol. *30*, 163–174.

Hersh, D., Monack, D.M., Smith, M.R., Ghori, N., Falkow, S., and Zychlinsky, A. (1999). The *Salmonella* invasion *sipB* induces macrophage apoptosis by binding to capase-1. Proc. Natl. Acad. Sci. U.S.A. *96*, 2396–2401.

Hirose, K., Ezaki, T., Miyake, M., Li, T., Khan, A.Q., Kawamura, Y., Yokoyama, H., and Takami, T. (1997). Survival of Vi-capsulated and Vi-deleted *Salmonella* Typhi strains in cultured macrophage expressing different levels of CD14 antigen. FEMS Microbiol. Lett. *147*, 259–265.

Hitchcock, P.J., Leive, L., Mäkelä, H., Rietschel, E.T., Strittmatter, W., and Morrison, D.C. (1986). Lipopolysaccharide nomenclature – past, present, and future. J. Bacteriol. *166*, 699–705.

Hmiel, S.P., Snavely, M.D., Florer, J.B., Maguire, M.E., and Miller, C.G. (1989). Magnesium transport in *Salmonella* Typhimurium: genetic characterization and cloning of three magnesium transport loci. J. Bacteriol. *171*, 4742–4751.

Hong, K.H., and Miller, V.L. (1998). Identification of a novel *Salmonella* invasion locus homologous to *Shigella ipgDE*. J. Bacteriol. *180*, 1793–1802.

Houng, H.H., Noon, K.F., Ou, J.T., and Baron, L. (1992). Expression of Vi antigen in *Escherichia coli* K-12: characterization of ViaB from *Citrobacter freundii* and identity of ViaA with RcsB. J. Bacteriol. *174*, 5910–5915.

Jones, B.D., and Falkow, S. (1996). Salmonellosis: Host immune responses and bacterial virulence determinants. Ann. Rev. Immunol. *14*, 533–561.

Jones, B.D., and Falkow, S. (1994). Identification and Characterization of a *Salmonella* Typhimurium oxygen-regulated gene required for bacterial internalization. Infect. Immun. *62*, 3745–3752.

Jones, M.A., Wood, M.W., Mullan, P.B., Watson, P.R., Wallis, T.S., and Galyov, E.E. (1998). Secreted effector proteins of *Salmonella* Dublin act in concert to induce enteritis. Infect. Immun. *66*, 5799–5804.

Kaniga, K., Trollinger, D. and Galán, J.E. (1995a). Identification of two targets of the Type III protein secretion system encoded by the *inv* and *spa* loci of *Salmonella* Typhimurium that have homology to the *Shigella* IpaD and IpaA Proteins. J. Bacteriol. *177*, 7078–7085.

Kaniga, K., Tucker, S., Trollinger, D., and Galán, J.E. (1995b). Homologues of the *Shigella* IpaB and IpaC invasins are required for *Salmonella* Typhimurium entry into cultured epithelial cells. J. Bacteriol. *177*, 3965–3971.

Khan, A.A., Cheng, C.M., Khanh, T.V., Summage-West, C., Nawaz, M.S., and Khan, S.A. (2006). Characterization of class 1 integron resistance gene cassettes in *Salmonella enterica* serovars Oslo and Bareily from imported seafood. J. Antimicrob. Chemother. *58*, 1308–1310.

Khoramian-Falsafi, T., Harayama, S., Kutsukake, K., and Pechere, J.C. (1990). Effect of motility and chemotaxis on the invasion of *Salmonella* Typhimurium into HeLa cells. Microb. Pathog. *9*, 47–53.

Kimbrough, T.G., and Miller, S.I. (2000). Contribution of *Salmonella* Typhimurium type III secretion components to needle complex formation. Proc. Natl. Acad. Sci. U.S.A. *97*, 11008–11013.

Klein, J.R., and Jones, B.D. (2001). *Salmonella* pathogenicity island 2-encoded proteins *SseC* and *SseD* are essential for virulence and are substrates of the type III secretion system. Infect. Immun. *69*, 737–743.

Klein, J.R., Fahlen, T.F., and Jones, B.D. (2000). Transcriptional organization and function of invasion genes within *Salmonella* enterica Serovar Typhimurium pathogenicity island 1, Including the *prgH*, *prgI*, *prgJ*, *prgK*, *orgA*, *orgB* and *orgC* genes. Infect. Immun. *68*, 3368–3376.

Knodler, L.A., and Steele-Mortimer, O. (2005). The *Salmonella* effector PipB2 affects late endosome/lysosome distribution to mediate Sif extension. Mol. Biol. Cell 16, 4108–4123.

Knodler, L.A., Celli, J., Hardt, W.D., Vallance, B.A., Yip, C., and Finlay, B.B. (2002). *Salmonella* effectors within a single pathogenicity island are differentially expressed and translocated by separate type III secretion systems. Mol. Microbiol. *43*, 1089–1103.

Kubori, T., and Galán, J.E. (2002). *Salmonella* type III secretion-associated protein InvE controls translocation of effector proteins into host cells. J. Bacteriol. *184*, 4699–4708.

Kubori, T., Matsushima, Y., Nakamura, D., Uralil, J., Lara-Tejero, M., Sukhan, A., Galán, J.E., and Aizawa, S.I. (1998). Supramolecular structure of the *Salmonella* Typhimurium type III protein secretion system. Science *280*, 602–605.

Kuhle, V., and Hensel, M. (2004). Cellular microbiology of intracellular *Salmonella enterica*: functions of the type III secretion system encoded by *Salmonella* pathogenicity island 2. Cell Mol. Life. Sci. *61*, 2812–2826.

Kuijl, C., Savage, N.D., Marsman, M., Tuin, A.W., Janssen, L., Egan, D.A, Ketema, M., van den Nieuwendijk, R., van den Eeden, S.J.F., Geluk, A., *et al.* (2007). Intracellular bacterial growth is controlled by a kinase network around PKB/AKT1. Nature *450*, 725–730.

Kujat-Choy, S.L., Boyle, E.C., Gal-Mor, O., Goode, D.L., Valdez, Y., Vallance, B.A. and Finlay, B.B (2004). SseK1 and SseK2 are novel translocated proteins of *Salmonella enterica* Serovar Typhimurium. Infect. Immun. 72, 5115–5125.

Lacey, R.W. (1993). Food borne bacterial infections. Parasitololgy *107*, 575–593.

Lahiri, A., Eswarappa, S.M., Das, P., and Chakravortty, D. (2010). Altering the balance between pathogen containing vacuoles and lysosomes: a lesson from *Salmonella*. Virelence, *1*, 325–329.

Lasa, I. (2006). Towards the identification of the common features of bacterial biofilm development. Int. Microbiol. *9*, 21–28.

Lawley, T.D., Chan, K., Thompson, L.J., Kim, C.C., Govoni, G.R., and Monack, D.M. (2006). Genome-Wide Screen for *Salmonella* genes required for long-term systemic infection of the mouse. PLoS Pathog. *2*, e11.

Lee, A.K., Detweiler, C.S., and Falkow, S. (2000). OmpR regulates the two-component system SsrA-SsrB in *Salmonella* pathogenicity island 2. J. Bacteriol. *182*, 771–781.

Lee, C.A., and Falkow, S. (1990). The ability of *Salmonella* to enter mammalian cells is affected by bacterial growth state. Proc. Natl. Acad. Sci. U.S.A. *87*, 4304–4308.

Lee, C.A., Jones, B.D., and Falkow, S. (1992). Identification of a *Salmonella* Typhimurium invasion locus by selection for hyperinvasive mutants. Proc. Natl. Acad. Sci. U.S.A. *89*, 1847–1851.

Levings, R.S., Partridge, S.R., Lightfoot, D., Hall, R.M., and Djordjevic, S.P. (2005). New integron-associated gene cassette encoding a 3-N-aminoglycoside acetyltransferase. Antimicrob. Agents Chemother. *49*, 1238–1241.

Levy, S.B., (2002). The Antibiotic Paradox: how the misuse of antibiotics destroys their curative powers. JAMA *288*, 2898–2899.

Levy, S.B., and Marshall, B. (2004). Antibacterial resistance worldwide: causes, challenges and responses. Nat. Med. *10*, 122--129.

Li, J., Ochman, H., Groisman, E.A., Boyd, E.F., Salomon, F., Nelson, K., and Selander, R.K. (1995). Relationship between evolutionary rate and cellular location among the Inv/Spa invasion protein of *Salmonella enterica*. Proc. Natl. Acad. Sci. U.S.A. *92*, 7252–7256.

Liebert, C.A., Hall, R.M., and Summers, A.O. (2000). Transposon Tn*21*, flagship of the floating genome. Microbiol. Mol. Biol. Rev. *63*, 507–522.

Lilic, M., Galkin, V.E., Orlova, A., VanLoock, M.S., Egelman, E.H., and Stebbins, C.E. (2003). *Salmonella* SipA polymerizes actin by stapling filaments with nonglobular protein arms. Science *301*, 1918–1921.

Lim, D. (1992). Structure and biosynthesis of unbranched multicopy single-stranded DNA by reverse transcriptase in a clinical *Escherichia coli* isolate. Mol. Microbiol. *6*, 3531–3542.

Linam, W.M., and Gerber, M.A. (2007). Changing epidemiology and prevention of *Salmonella* infections. Pediatr. Infect. Dis. J. *26*, 747–748.

Liu, S.L., and Sanderson, K.E. (1995). Genomic cleavage map of *Salmonella* Typhi TY2. J. Bacteriol. *177*, 5099–5107.

Lossi, N.S., Rolhion, N., Magee, A.I., Boyle, C., and Holden, D.W. (2008). The *Salmonella* SPI-2 effector SseJ exhibits eukaryotic activatordependent phospholipase A and glycerophospholipid: cholesterol acyltransferase activity. Microbiology *154*, 2680–2688.

Maguire, M.E., Snavely, M.D., Leizman, J.B., Gura, S., Bagga, D., Tao, T., and Smith, D.L. (1992). Mg^{2+} transporting P-type ATPases of *Salmonella* Typhimurium. Wrong way, wrong place enzymes. Ann. N. Y. Acad. Sci. *671*, 244–256.

Main-Hester, K.L., Colpitts, K.M., Thomas, G.A., Fang, F.C. and Libby, S.J. (2008). Coordinate regulation of *Salmonella* pathogenicity island 1 (SPI1) and SPI4 in *Salmonella enterica* serovar Typhimurium. Infect. Immun. *76*, 1024–1035.

Maithreye, R., Sharmila, A., and Mande, S. (2007). Modelling of the regulation of the *hilA* promoter of type three secretion system of *Salmonella enterica* serovar Typhimurium. Syst. Synth. Biol. *1*, 129–137.

Mallo, G.V., Espina, M., Smith, A.C., Terebiznik, M.R., Aleman, A., Finlay, B.B., Rameh, L.E., Grinstein, S., and Brumell, J.H. (2008). SopB promotes phosphatidylinositol 3-phosphate formation on *Salmonella* vacuoles by recruiting Rab5 and Vps34. J. Cell Biol. *182*, 741–752.

Martel, J.L., Chaslus-Dancla, E., Coudert, M., and Lafont, J.P., (1996). Evolution de la sensibilité aux antibiotiques des salmonelles d'origine bovine en France. Med. Mal. Infect. *26*, 415–419.

Marti´Nez-Freijo, P., Fluit, A.C., Schmitz, F.J., Verhoev, J., and Jones, M.E. (1999). Many class 1 integrons comprise stable structures occurring in different species of *Enterobacteriaceae* isolated from widespread geographic regions in Europe. Antimicrobiol. Agents Chemother. *43*, 686–689.

Martone, W.J. (1998). Spread of vancomycin-resistant enterococci: why did it happen in the United States? Infect. Control Hosp. Epidemiol. *19*, 539–545.

Matsui, H., Bacot, C.M., Garlington, W.A., Doyle, T.J., Roberts, S., and Gulig, P.A. (2001). Virulence plasmid-borne *spvB* and *spvC* genes can replace the 90-Kilobase plasmid in conferring virulence to *Salmonella enterica* serovar Typhimurium in subcutaneously inoculated mice. J. Bacteriol. *183*, 4652–4658.

Maurer, J., Jose, J., and Meyer, T.F. (1997). Autodisplay: one-component system for efficient surface display and release of soluble recombinant proteins from *Escherichia coli*. J. Bacteriol. *179*, 794–804.

McManus, M.C. (1997). Mechanisms of bacterial resistance to antimicrobial agents. Am. J. Health. Syst. Pharm. *54*, 1420–1433.

Mead, P.S., Slutsker, L., Dietz, V., McCaig, L.F., Bresse, J.S., Shapiro, C., Griffin, P.M., and Tauxe, R.V. (1999). Food-related illness and death in the United States. Emerg. Infect. Dis. *5*, 607–625.

Meijler, M.M., Hom, L.G., Kaufmann, G.F., McKenzie, K.M., Sun, C., Moss, J.A., Matsushita, M., and Janda, K.D. (2004). Synthesis and biological validation of a ubiquitous quorum-sensing molecule. Angew. Chem. Int. Ed. Engl. *43*, 2106–2108.

Miao, E.A., and Miller, S.I. (2000). A conserved amino acid sequence directing intracellular type III secretion by *Salmonella* Typhimurium. Proc. Natl. Acad. Sci. U.S.A. *97*, 7539–7544.

Miao, E.A., Scherer, C.A., Tsolis, R.M., Kingsley, R.A., Adams, L.G., Bäumler, A.J., and Miller, S.I. (1999). *Salmonella* Typhimurium leucine-rich repeat proteins are targeted to the SPI-1 and SPI-2 type III secretion systems. Mol. Microbiol. *34*, 850–864.

Miao, E.A., Brittnacher, M., Haraga, A., Jeng, R.L., Weich, M.D., and Miller, S.I. (2003). *Salmonella* effectors translocated across the vacuolar membrane interact with the actin cytoskeleton. Mol. Microbiol. *48*, 401–415.

Michael, B., Smith, J.N., Swift, S., Heffron, F., Brian, M.M., and Ahmer, B.M.M. (2001). SdiA of *Salmonella enterica* is a *luxR* homolog that detects mixed microbial communities. J. Bacteriol. *183*, 5733–5742.

Michael, G.B., Cardoso, M., Rabsch, W., and Schwarz, S. (2006). Phenotypic and genotypic differentiation of porcine *Salmonella enterica* subsp. *enterica* serovar Derby isolates. Vet. Microbiol. *118*, 312–318.

Miller, J.I., Kukral, A.M., and Mekalanos, J.J. (1989). A two component regulatory system (PhoP-PhoQ) controls *Salmonella* Typhimurium virulence. Proc. Natl. Acad. Sci. U.S.A. *86*, 5054–5058.

Miller, M.B., and Bassler, B.L. (2001) Quorum sensing in bacteria. Annu. Rev. Microbiol. *55*, 165–199.

Mills, D.M., Bajaj, V., and Lee, C.A. (1995). A 40 kb chromosomal fragment encoding *Salmonella* Typhimurium invasion genes is absent from the corresponding region of the *Escherichia coli* K-12 chromosome. Mol. Microbiol. *15*, 749–759.

Miriagou, V., Tassios, P.T., Legakis, N.J., and Tzouvelekis, L.S. (2004). Expanded spectrum cephalosporin resistance in non-typhoid *Salmonella*. Int J. Antimicrob. Agents 23, 547–555.

Mitsuhashi, S., Inoue, M., Saino, Y., and Ogashiwa, M. (1982). *In vitro* antibacterial activity of SCH29582, a new 1-lactam antibiotic. J. Antimicrob. Chemother. *9*, 91–95.

Monack, D.M., Raupach, B., Hromockyi, A.E., and Falkow, S. (1996). *Salmonella* Typhimurium invasion induces apoptosis in infected macrophages. Proc. Natl. Acad. Sci. U.S.A. *93*, 9833–9838.

Monack, D.M., Mecsas, J., Ghori, N., and Falkow, S. (1997). *Yersinia* signals macrophages to undergo apoptosis and YopJ is necessary for this cell death. Proc. Natl. Acad. Sci. U.S.A. *94*, 10385–10390.

Moreira, C.G., Weinshenker, D., and Sperandio, V. (2010). QseC mediates *Salmonella enterica* serovar Typhimurium virulence *in vitro* and *in vivo*. Infect. Immun. *78*, 914–926.

Morgan, E., Campbell, J.D., Rowe, S.C., Bispham, J., Stevens, M.P., Bowen, A.J., Barrow, P.A., Maskell, D.J., and Wallis, T.S. (2004). Identification of host-specific colonization factors of *Salmonella enterica* serovar Typhimurium. Mol. Microbiol. *54*, 994–1010.

Morgan, E., Bowen, A.J., Carnell, S.C., Wallis, T.S., and Stevens, M.P. (2007). SiiE is secreted by the *Salmonella enterica* serovar Typhimurium pathogenicity island 4-encoded secretion system and contributes to intestinal colonization in cattle. Infect. Immun. *75*, 1524–1533.

Morschhäuser, J., Vetter, V., Emödy, L., and Hacker, J. (1994). Adhesin regulatory genes within large, unstable DNA regions of pathogenic *Escherichia coli*: cross-talk between different adhesin gene clusters. Mol. Microbiol. *11*, 555–566.

Mukherjee, K., Parashuraman, S., Raje, M., and Mukhopadhyay, A. (2001). SopE acts as an Rab5-specific nucleotide exchange factor and recruits non-prenylated Rab5 on *Salmonella*-containing phagosomes to promote fusion with early endosomes. J. Biol. Chem. *276*, 23607–23615.

Murli, S., Watson, R.O., and Galán, J.E. (2001) Role of tyrosine kinases and the tyrosine phosphatase SptP in the interaction of *Salmonella* with host cells. Cell Microbiol. 3, 795–810.

Murooka, Y., Azakami, H., and Yamashita, M. (1996). The monoamine regulon including syntheses of arylsulfatase and monoamine oxidase in bacteria. Biosci. Biotechnol. Biochem. *60*, 935–941.

Nagarajan, A.G., Balasundaram, S.V., Janice, J., Karnam, G., Eswarappa, S.M., and Chakravortty, D. (2009). *sopB* of *Salmonella enterica* serovar Typhimurium is a potential DNA vaccine candidate in conjugation with live attenuated bacteria. Vaccine *27*, 2804–2811.

Navarre, W.W., and Zychlinsky, A. (2000). Pathogen-induced apoptosis of macrophages: a common end for different pathogenic strategies. Cell Microbiol. 2, 265–273.

Navarre, W.W., Halsey, T.A., Walthers, D., Frye, J., McClelland, M., Potter, J.L., Kenney, L.J., Gunn, J.S., Fang, F.C., and Libby, S.J. (2005). Co-regulation of *Salmonella enterica* genes required for virulence and resistance to antimicrobial peptides by SlyA and PhoP/PhoQ. Mol. Microbiol. *56*, 492–508.

Ng, I., Liu, S.L., and Sanderson, K.E. (1999). Role of genomic rearrangements in producing new ribotypes of *Salmonella* Typhi. J. Bacteriol. *181*, 3536–3541.

Nikolaus, T., Deiwick, J., Rappl, C., Freeman, J.A., Schröder, W., Miller, S.I., and Hensel, M. (2001). SseBCD proteins are secreted by the type III secretion system of *Salmonella* pathogenicity island 2 and function as a translocon. J. Bacteriol. *183*, 6036–6045.

Norris, F.A., Wilson, M.P., Wallis, T.S., Galyov, E.E., and Majerus, P.W. (1998). SopB, a protein required for virulence of *Salmonella* Dublin, is an inositol phosphate phosphatase. Proc. Natl. Acad. Sci. U.S.A. *95*, 14057–14059.

O'Brien, T.F. (2002). Emergence, spread and environmental effect of antimicrobial resistance: how use of an antimicrobial anywhere can increase resistance to any antimicrobial anywhere else. Clin. Infect. Dis. *34*, 78–84.

Ohlson, M.B., Huang, Z., Alto, N.M., Blanc, M.P., Dixon, J.E., Chai, J., and Miller, S.I. (2008). Structure and function of *Salmonella* SifA indicate that its interactions with SkiP, SseJ, and RhoA family GTPases induce endosomal tubulation. Cell. Host Microbe. 4, 434–446.

O'Toole, G., Kaplan, H.B., and Kolter, R. (2000). Biofilm formation as microbial development. Annu. Rev. Microbiol. *54*, 49–79.

Ochman, H., Soncini, F.C., Solomon, F., and Groisman, E.A. (1996). Identification of a pathogenicity island required for *Salmonella* survival in host cells. Proc. Natl. Acad. Sci. U.S.A. *93*, 7800–7804.

Paniker, C.K., and Vimala, C.N., (1972). Transferable chloramphenicol resistance in *Salmonella* Typhi. Nature *239*, 109–110.

Papezova, K., Gregorova, D., Jonuschies, J., and Rychlik, I. (2007): Ordered expression of virulence genes in *Salmonella enterica* serovar Typhimurium. Folia Microbiol. *52*, 107–114.

Parkhill, J., Dougan, G., James, K.D., Thomson, N.R., Pickard, D., wain, J., Churcher, C., Mungall, K.L., Bentley, S.D., Holden, M.T.G. (2001). Complete genome sequence of a multiple drug resistant *Salmonella enterica* serovar Typhi CT18. Nature *413*, 848–852.

Pasquali, F., de Cesare, A., Ricci, A., Kehrenberg, C., Schwarz, S., and Manfreda, G. (2004). Phage types, ribotypes and tetracycline resistance genes of *Salmonella enterica* subsp. *enterica* serovar Typhimurium strains isolated from different origins in Italy. Vet. Microbiol. *103*, 71–76.

Patel, J.C., and Galan, J.E. (2006). Differential activation and function of Rho GTPases during *Salmonella*-host cell interactions. J. Cell Biol. *175*, 453–463.

Pattishall, K.H., Acar, J., Burchall, J.J., Goldstein, F.W., and Harvey, R.J. (1977). Two distinct types of trimethoprim-resistant dihy- drofolate reductase specified by R-plasmids of different incompatibility groups J. Biol. Chem. *252*, 2319–2323.

Perrett, C.A., Karavolos, M.H., Humphrey, S., Mastroeni, P., Martinez-Argudo, I., Spencer, H., Bulmer, D., Winzer, K., McGhie, E., Koronakis, V., *et al.* (2009). LuxS-based quorum sensing does not affect the ability of *Salmonella enterica* serovar Typhimurium to express the SPI-1 type 3 secretion system, induce membrane ruffles, or invade epithelial cells. J. Bacteriol. *191*, 7253–7259.

Pezzella, C., Ricci, A., DiGiannatale, E., Luzzi, I., and Carattoli, A. (2004). Tetracycline and streptomycin resistance genes, transposons, and plasmids in *Salmonella* enterica isolates from animals in Italy. Antimicrob. Agents Chemother. *48*, 903–908.

Pfeifer, C.G., Marcus, S.L., Steele-Mortimer, O., Knodler, L.A., and Finlay, B.B. (1999). *Salmonella* Typhimurium virulence genes are induced upon bacterial invasion into phagocytic and nonphagocytic cells. Infect. Immun. *67*, 5690–5698.

Pickard, D., Wain, J., Baker, S., Line, A., Chohan, S., Fookes, M., Barron, A., Gaora, P.O., Chabalgoity, J.A., Thanky, N., *et al.* (2003). Composition, acquisition, and distribution of the Vi exopolysaccharide-encoding *Salmonella enterica* pathogenicity island SPI-7. J. Bacteriol. *185*, 5055–5065.

Pickard, D., Li, J., Roberts, M., Maskell, D., Hone, D., Levine, M., Dougan, G., and Chatfield, S. (1994). Characterization of defined *ompR* mutants of *Salmonella* Typhi: OmpR is involved in the regulation of Vi polysaccharide expression. Infect. Immun. *62*, 3984–3993.

Poh, J., Odendall, C., Spanos, A., Boyle, C., Liu, M., Freemont, P., and Holden, D.W. (2008). SteC is a *Salmonella* kinase required for SPI-2- dependent F-actin remodelling. Cell Microbiol. *10*, 20–30.

Porwollik, S., and McClelland, M. (2003). Lateral gene transfer in *Salmonella*. Microbes. Infect. *5*, 977–989.

Poupart, M.C., Chanal, C., Sirot, D., Labia, R., and Sirot, J. (1991). Identification of CTX-2, a novel cefotaximase from a *Salmonella* Mbandaka isolate. Antimicrob. Agents Chemother. *35*, 1498–1500.

Prouty, A.M., Schwesinger, W.H., and Gunn, J.S. (2002). Biofilm formation and interaction with the surfaces of gallstones by *Salmonella* spp. Infect. Immun. *70*, 2640–2649.

Pukatzki, S., McAuley, S.B., and Miyata, S.T. (2009). The type VI secretion system: translocation of effectors and effector-domains. Curr. Opin. Microbiol. *12*, 11–17.

Radstrom, P., Skold, O., Swedberg, G., Flensburg, J., Roy, P.H., and Sundstrom, L. (1994). Transposon Tn*5090* of plasmid R751, which carries an integron, is related to Tn7, Mu, and the retroelements. J. Bacteriol. *176*, 3257–3268.

Raetz, C.R.H., and Whitfield, C. (2002). Lipopolysaccharide endotoxins. Annu. Rev. Biochem. *71*, 635–700.

Ramsden, A.E., Mota, L.J., Münter, S., Shorte, S.L. and Holden, D.W. (2007). The SPI-2 type III secretion system restricts motility of *Salmonella*-containing vacuoles. Cell. Microbiol. *9*, 2517–2529.

Randall, L.P., Cooles, S.W., Osborn, M.K., Piddock, L.J.V., and Woodward, M.J. (2004). Antibiotic resistance genes, integrons, and multiple antibiotic resistances in thirty-five serotypes of *Salmonella enterica* isolated from humans and animals in the UK. J. Antimicrob. Chemother. *53*, 208–216.

Rang, C., Alix, E., Felix, C., Heitz, A., Tasse, L., and Blanc-Potard, A.B. (2007). Dual role of the MgtC virulence factor in host and non-host environments. Mol. Microbiol. *63*, 605–622.

Reading, N., and Sperandio, V. (2006). Quorum sensing: the many languages of bacteria. FEMS Microbiol. Lett. *254*, 1–11.

Recchia, G.D., and Hall, R.M. (1995). Gene cassettes: a new class of mobile element. Microbiology *141*, 3015–3027.

dos Reis, R.S., and Horn, F. (2010). Enteropathogenic *Escherichia coli*, *Salmonella*, *Shigella* and *Yersinia*: cellular aspects of host bacteria interactions in enteric diseases. Gut Pathog. *2*, 8.

Retamal, P., Castillo-Ruiz, M., and Mora, G.C. (2009). Characterization of MgtC, a virulence factor of *Salmonella enterica* serovar Typhi. PLoS One. 4, e5551.

Riesenberg-Wilmes, M.R., Bearson, B., Foster, J.W., and Curtiss, R. (1996). Role of the acid tolerance response in virulence of *Salmonella* Typhimurium. Infect. Immun. *64*, 1085–1092.

Roberts, M.C. (1996). Tetracycline resistance determinants: mechanisms of action, regulation of expression, genetic mobility, and distribution. FEMS Microbiol. Rev. *19*, 1–24.

Roberts, M.C. (2005). Update on acquired tetracycline resistance genes. FEMS Microbiol. Lett. *245*, 195–203

Row, B., Ward, L.R., and Threlfall, E.J. (1997). Multidrug-resistant *Salmonella* Typhi: a worldwide epidemic. Clin. Infect. Dis. *24*, 106–109.

Ruiz-Albert, J., Yu, X.J., Beuzon, C.R., Blakey, A.N., Galyov, E.E., and Holden, D.W. (2002). Complementary activities of SseJ and SifA regulate dynamics of the *Salmonella* Typhimurium vacuolar membrane. Mol. Microbiol. *44*, 645–661.

Rychlik, I., Gregorova, D., and Hradecka, H. (2006). Distribution and function of plasmids in *Salmonella enterica*. Vet. Microbiol. *112*, 1–10.

Sabate, M., and Prats, G. (2002). Estructuray función de los integrons. Enferm. Infect. Microbiol. Clin. *20*, 341–345.

Salcedo, S.P., and Holden, D.W. (2003). *SseG*, a virulence protein that targets *Salmonella* to the Golgi network. EMBO J. *22*, 5003–5014.

Salyers, A.A., and Whitt, D.D. (2002). *Salmonella* species. In Bacterial Pathogenesis: A Molecular Approach. Salyers, A.A., and Whitt D.D., eds (ASM Press, Washington, DC), pp. 381–397.

Saroj, S.D., Shashidhar, R., Karani, M., and Bandeka, J.R. (2008). Distribution of *Salmonella* pathogenicity island SPI-8 and SPI-10 among different serotypes of *Salmonella*. J. Med. Microbiol. *57*, 424–427.

Saxena, S.N., Kumari, N., Saini, S.S., Soni, D.V., Pahwa, R.K., and Jayasheela, M. (1989). Surveillance of Salmonellae in India for drug resistance. Indian J. Med. Sci. *43*, 145–150.

Schechter, L.M., Damrauer, S.M., and Lee, C.A. (1999). Two AraC/XylS family members can independently counteract the effect of repressing sequences upstream of the *hilA* promoter. Mol. Microbiol. *32*, 629–642.

Schicklmaier, P., Moser, E., Wieland, T., Rabsch, W., and Schmieger, H. (1998). A comparative study on the frequency of prophages among natural isolates of *Salmonella* and *Escherichia coli* with emphasis on generalized transducers. Antonie Van Leeuwenhoek *73*, 49–54.

Shah, D.H., Lee, M.J., Park, J.H., Lee, J.H., Eo, S.K., Kwon, J.T., and Chae, J.S. (2005). Identification of *Salmonella* Gallinarum virulence genes in a chicken infection model using PCR-based signature-tagged mutagenesis. Microbiology *151*, 3957–3968.

Skold, O. (2001). Resistance to trimethoprim and sulphonamides. Vet. Res. *32*, 261–273.

Smith, D.L., Tao, T., and Maguire. M.E. (1993). Membrane topology of a P-type ATPase: the MgtB transport protein of *Salmonella* Typhimurium J. Biol. Chem. *268*, 22469–22479.

Smith, J.N., and Ahmer, B.M. (2003). Detection of other microbial species by *Salmonella*: expression of the SdiA regulon. J. Bacteriol. *185*, 1357–1366.

Smith, J.L., Fratamico, P.M., and Yan, X. (2011). Eavesdropping by bacteria: the role of SdiA in *Escherichia coli* and *Salmonella* enterica serovar Typhimurium quorum sensing. Foodborne Pathog. Dis. *8*, 169–178.

Smith, R.L., Thompson, L.J., and Maguire, M.E. (1995). Cloning and characterization of *mgtE*, a putative new class of Mg^{2+} transporter from Bacillus firmus OF4. J. Bacteriol. *177*, 1233–1238.

Spano, S., Ugalde, J.E., and Galán, J.E. (2008). Delivery of a *Salmonella* Typhi exotoxin from a host intracellular compartment. Cell Host Microbe *3*, 30–38.

Stanley, T.L., Ellermeier, C.D., and Slauch, J.M. (2000). Tissue-specific gene expression identifies a gene in the lysogenic phage Gifsy-1 that affects *Salmonella enterica* serovar Typhimurium survival in Peyer's patches. J. Bacteriol. *182*, 4406–4413.

Steele-Mortimer, O. (2008). The *Salmonella*-containing vacuole: moving with the times. Curr. Opin. Microbiol. *11*, 38–45.

Steele-Mortimer, O., Meresse, S., Gorvel, J.P., Toh, B.H., and Finlay, B.B. (1999). Biogenesis of *Salmonella* Typhimurium containing vacuoles in epithelial cells involves interactions with the early endocytic pathway. Cell Microbiol. *1*, 33–49.

Steele-Mortimer, O., Knodler, L.A., Marcus, S.L., Scheid, M.P., Goh, B., Pfeifer, C.G., Duronio, V., and Finlay, B.B. (2000). Activation of Akt/protein kinase B in epithelial cells by the *Salmonella* Typhimurium effector *sigD*. J. Biol. Chem. *275*, 37718–37724.

Stender, S., Friebel, A., Linder, S., Rohde, M., Mirold, S., and Hardt, W.D. (2000). Identification of SopE2 from *Salmonella* Typhimurium, a conserved guanine nucleotide exchange factor for Cdc42 of the host cell. Mol. Microbiol. *36*, 1206–1221.

Stoodley, P., Sauer, K., Davies, D.G., and Costerton, J.W. (2002). Biofilm as complex differentiated communities. Annu. Rev. Microbiol. *56*, 187–209.

Su, L.H., Chiu, C.H., Kuo, A.J., Chia, J.H., Sun C.F., Leu, H.S., and Wu. T.L. (2001). Secular trends in incidence and antimicrobial resistance among clinical isolates of *Salmonella* at a university hospital in Taiwan, 1983–1999. Epidemiol. Infect. *127*, 207–213.

Su, L.H., Chiu, C.H., Chu, C., Wang, M.H., Chia, J.H., and Wu, T.L. (2003). *In vivo* acquisition of ceftriaxone resistance in *Salmonella enterica* serotype Anatum. Antimicrob. Agents Chemother. *47*, 563–567.

Sukhan, A. (2000). (Review) The invasion-associated type III secretion system of *Salmonella* Typhimurium: common and unique features. Cell. Mol. Life Sci. *57*, 1033–1049.

Sukhan, A. (2001). (Review) The invasion-associated type III secretion system of *Salmonella* Typhimurium: common and unique features. Cell Mol. Life. Sci. *57*, 1033–1049.

Sukhan, A., Kubori, T., and Galán, J.E. (2003). Synthesis and localization of the *Salmonella* SPI-1 type III secretion complex proteins PrgI and PrgJ. J. Bacteriol. *185*, 3480–3483.

Sundin, G.W., and Bender, C.L. (1996). Dissemination of the *strA-strB* streptomycin-resistance genes among commensal and pathogenic bacteria from humans, animals and plants. Mol. Ecol. *5*, 133–143.

Surette, M.G., and Bassler, B.L. (1998). Quorum sensing in *Escherichia coli* and *Salmonella* Typhimurium. Proc. Natl. Acad. Sci. U.S.A. *95*, 7046–7050.

Surette, M.G., and Bassler, B.L. (1999). Quorum sensing in *Escherichia coli, Salmonella* Typhimurium, and *Vibrio harveyi*: A new family of genes responsible for autoinducer production. Mol. Microbiol. *31*, 585–596.

Taga, M.E., Semmelhack, J.L., and Bassler, B.L. (2001). The LuxS-dependent autoinducer AI-2 controls the expression of an ABC transporter that functions in AI-2 uptake in *Salmonella* Typhimurium. Mol. Microbiol. *42*, 777–793.

Takeuchi, A. (1967). Electron microscope studies of experimental *Salmonella* infection. Penetration into the intestinal epithelium by *Salmonella* Typhimurium. Am. J. Pathol. *50*, 109–136.

Tao, R., Ying, G.G., Su, H.C., Zhou, H.W., and Sidhu, J.P.S. (2010). Detection of antibiotic resistance and tetracycline resistance genes in *Enterobacteriaceae* isolated from the Pearl rivers in South China. Environ. Pollut. *158*, 2101–2109.

Tao, T., Grulich, P.F., Kucharski, L.M., Smith, R.L., and Maguire, M.E. (1998). Magnesium transport in *Salmonella* Typhimurium: biphasic time and magnesium dependence of the transcription of the *mgtA* and *mgtCB* loci. Microbiology *144*, 655–664.

Tassios, P.T., Markogiannakis, A., Vatopoulos, A.C., Katsanikou, E., Velonakis, E.N., Kourea-Kremastinou, J., and Legakis, N.J. (1997). Molecular epidemiology of antibiotic resistance of *Salmonella* Enteritidis during a 7-year period in Greece. J. Clin. Microbiol. *35*, 1316–1321.

Tezcan-Merdol, D., Engstrand, L., and Rhen, M. (2005). *Salmonella enterica* SpvB mediated ADP-ribosylation as an activator for host cell actin degradation. Int. J. Med. Microbiol. *295*, 201–212.

Threlfall, E.J., Fisher, I.S., and Berghold, C. (2003). Trends in antimicrobial drug resistance in *Salmonella enterica* serotype Typhi and Paratyphi A isolated in Europe, 1999–2001. Int. J. Antimicrob. Agents *22*, 487–491.

Tosini, F., Visca, P., Luzzi, I., Dionisi, A.M., Pezzella, C., Petrucca, A., and Carattoli, A. (1998). Class 1 integron-borne multiple-antibiotic resistance carried by IncFI and IncL/M plasmids in *Salmonella enterica* serotype Typhimurium, Antimicrob. Agents Chemother. *42*, 3053–3058.

Townsend, S.M., Kramer, N.E., Edwards, R., Baker, S., Hamlin, N., Simmonds, M., Stevens, K., Malony, S., Parkhill, J., Dougan, G. (2001). *Salmonella enterica* serovar Typhi possesses a unique repertoire of fimbrial gene sequences. Infect. Immun. *69*, 2894–2901.

Travers, K., and Barza. M. (2002). Morbidity of infections caused by antimicrobial- resistant bacteria. Clin. Infect. Dis. *34*, 131–134.

Tükel, C., Akçelik, M., de Jong, M.F., Şimşek, Ö., Tsolis, R.M., and Bäumler, A.J. (2007). MarT activates expression of the MisL autotransporter protein of *Salmonella enterica* serotype Typhimurium. J. Bacteriol. *189*, 3922–3926.

Uchiya, K.I., Barbieri, M.A., Funato, K., Shah, A.H., Stahl, P.D., and Groisman, E.A. (1999). A *Salmonella* virulence protein that inhibits cellular trafficking. EMBO J. *18*, 3924- 3933.

Uchiya, K.I., Sugita, A., and Nikai, T. (2008). *Salmonella* virulence factor *SpiC* is involved in expression of flagellin protein and mediates activation of the signal transduction pathways in macrophages. Microbiology *154*, 3491–3502.

Uchiya, K.I., Sugita, A., and Nikai, T. (2009). Involvement of SPI-2-encoded SpiC in flagellum synthesis in *Salmonella* enterica serovar Typhimurium. BMC Microbiol. *9*, 179.

Unsworth, K.E., Way, M., McNiven, M., Machesky, L., and Holden, D.W. (2004). Analysis of the mechanisms of *Salmonella*-induced actin assembly during invasion of host cells and intracellular replication. Cell Microbiol. *6*, 1041–1055.

Walthers, D., Caroll, R.K., Navarre, W.W., Libby, S.J., Fang, F.C., and Kenney, L.J. (2007). The response regulator SsrB activate expression of diverse *Salmonella* pathogenicity island 2 promoters and counters silencing by the nucleoid-associated protein H-NS. Mol. Microbiol. *65*, 477–493.

Williamson, C.M., Pullinger, G.D. and Lax, A.J. (1988). Identification of an essential virulence region on *Salmonella* plasmids. Microb. Pathog. *5*, 469–473.

Wong, K., McCelland, M., Stillwell, L.S., Sisk, E.C., Thurston, S.J., and Saffer, J.D. (1998). Identification and sequence analysis of a 27-kilobase chromosomal fragment containing a *Salmonella* pathogenicity island located at 92 minutes on the chromosome map of *Salmonella enterica* Serovar Typhimurium LT2. Infect. Immun. *66*, 3365–3371.

Wood, M.W., Rosqvist, R., Mullan, P.B., Edwards, M.H., and Galyov, E.E. (1996). SopE, a secreted protein of *Salmonella* Dublin, is translocated into the target eukaryotic cells via a *sip*-dependent mechanism and promotes bacterial entry. Mol. Microbiol. *22*, 327–338.

Wood, M.W., Jones, M.A., Watson, P.R., Hedges, S., Wallis, T.S., and Galyov, E.E. (1998). Identification of a pathogenicity island required for *Salmonella* enteropathogenicity. Mol. Microbiol. *29*, 883–891.

Wood, M.W., Jones, M.A., Watson, P.R., Siber, A.M., McCormick, B.A., Hedges, S., Rosqvist, R., Wallis, T.S., and Galyov, E.E. (2000). The secreted effector protein of *Salmonella dublin*, SopA, is translocated into eukaryotic cells and influences the induction of enteritis. Cell. Microbiol. *2*, 293–303.

World Heath Organization (1996). The world heath heath report 1996- fighting disease fostering development. World health forum. *18*, 1–8.

Zhang, X.L., Morris, C., and Hackett, J. (1997). Molecular cloning, nucleotide sequence, and function of a site-specific recombinase encoded in the major Pathogenicity Island of *Salmonella* Typhi. Gene *202*, 139–146.

Zhang, X.X., Zhang, T., and Fang, H.H.P. (2009). Antibiotics resistance genes in water environment. Appl. Microbiol. Biotechnol. *82*, 397–414.

Zhang, Y., and LeJeune, J.T. (2008). Transduction of *bla* (CMY-2), *tet* (A), and *tet* (B) from *Salmonella enterica* subspecies *enterica* serovar Heidelberg to *S.* Typhimurium. Vet. Microbiol. *129*, 418–425.

Zhao, C., Ge, B., de Villena, J., Sudler, R., Yeh, E., Zhao, S., White, D.G., Wagner, D., and Meng, J. (2001). Prevalence of *Campylobacter* spp., *Escherichia coli*, and *Salmonella* serovars in retail chicken, turkey, pork, and beef from the greater Washington, D.C., area. Appl. Environ. Microbiol. *67*, 5431–5436.

Zhou, D., Chen, L.M., Hernandez, L., Shears, S.B., and Galan, J.E. (2001). A *Salmonella* inositol polyphosphatase acts in conjunction with other bacterial effectors to promote host cell actin cytoskeleton rearrangements and bacterial internalization. Mol. Microbiol. *39*, 248–259.

Epidemiology, Molecular Biology and Detection of Foodborne *Vibrio parahaemolyticus* Infections

7

Marie Yeung and Kathryn J. Boor

Abstract

As a natural inhabitant of the marine environment, *Vibrio parahaemolyticus* is frequently present in seafood, and particularly in oysters. *V. parahaemolyticus* can multiply rapidly under favourable conditions, but also may exist in a viable but non-culturable state under unfavourable conditions. A small subset of this species can cause human disease, with acute gastroenteritis as the predominant clinical manifestation. Two pore-forming haemolysins, thermostable direct haemolysin (TDH) and TDH-related haemolysin (TRH), are recognized as contributors to *V. parahaemolyticus* pathogenesis. Therefore, many detection methods for pathogenic strains focus on determining the presence of the genes encoding these haemolysins or on detection of beta-haemolysis on Wagatsuma agar (i.e. the Kanagawa Phenomenon). Serotyping and molecular fingerprinting techniques are also used to subtype *V. parahaemolyticus* isolates. One of the most effective strategies for preventing *V. parahaemolyticus* infection from consumption of raw or uncooked seafood is to reduce seafood post-harvest storage temperatures to prevent the growth of this species.

Introduction

As a member of the *Vibrio* genus, *V. parahaemolyticus* is a Gram-negative, non-spore-forming, curved, rod-shaped bacterium. It exhibits high motility in both liquid and semi-solid environments. *Vibrio* species contain two chromosomes, an uncommon characteristic among bacteria, with the larger chromosome (chromosome 1) and smaller chromosome (chromosome 2) in the size range of 3.0–3.3 Mb and 0.8–2.4 Mb, respectively (Okada *et al.*, 2005). The genome sequence of *V. parahaemolyticus* strain RIMD2210633 suggests that the chromosomes of this species are ~3.29 Mb and ~1.88 Mb respectively. The genome contains 11 copies of rRNA operons, with 10 on chromosome 1. The relatively large number of rRNA operons in *V. parahaemolyticus* may contribute to its fast-growing phenotype (Makino *et al.*, 2003).

V. parahaemolyticus is a natural inhabitant of marine waters and therefore commonly found in estuaries and seafood worldwide. Most *V. parahaemolyticus* strains are non-pathogenic to humans, yet the pathogenic subtypes have caused epidemics and pandemics, hence they are considered to be major foodborne pathogens in many countries. The incidence of *V. parahaemolyticus* infection in the US has shown an upwards trend since 2000. Human infections are most commonly associated with the consumption of raw, undercooked or contaminated shellfish. A number of individual virulence factors, including the thermostable direct haemolysin (TDH; encoded by the gene, *tdh*) and TDH-related haemolysin (TRH, encoded by the gene, *trh*), have been identified. Genome sequence data from *V. parahaemolyticus* RIMD2210633 also revealed the presence of two type three secretion systems (T3SS), with one present in each chromosome (Makino *et al.*, 2003).

V. parahaemolyticus is routinely isolated using thiosulfate-citrate-bile salts-sucrose (TCBS) agar as a selective and differential medium. Presumptive colonies on TCBS are then subjected to an array of biochemical tests and/or to molecular assays targeting the presence of species-specific genes. As *V. parahaemolyticus* appears able to persist in a viable-but-non-culturable (VBNC) state (Jiang and Chai, 1996; Mizunoe *et al.*, 2000; Johnston and Brown, 2002), application of culture-independent methods may be required to rule out the presence of this organism in marine-derived samples.

In this chapter, we will discuss the epidemiology and virulence characteristics of this important foodborne pathogen. Additionally, conventional and emerging techniques for isolation, identification, and subtyping *V. parahaemolyticus* will be presented. Mitigation strategies to control foodborne microorganisms in seafood, including *V. parahaemolyticus*, will also be described.

Worldwide distribution

As a natural inhabitant of estuarine marine water, *V. parahaemolyticus* is widely distributed in inshore marine waters throughout the world. In the US, the organism was first demonstrated in seawater, sediments, and shellfish in the Puget Sound region (Baross and Liston, 1968, 1970) and has been isolated consistently from such environments. Filter feeders in marine habitats, such as bivalve shellfish, can also harbour *V. parahaemolyticus* as part of their natural flora (Kueh and Chan, 1985).

The relatively recent increase in research studies in the US and around the globe on *V. parahaemolyticus* is at least partly a consequence of the emergence of the O3:K6 pandemic clone in the mid 1990s. This pandemic clone was characterized by apparent enhanced virulence and thereby a lower infectious dose than for previously identified *V. parahaemolyticus* serotypes (FDA, 2005). The epidemiology and genetics of this O3:K6 serovar and other related subtypes is discussed in more detail in Chapter 8.

To illustrate the ubiquity of *V. parahaemolyticus* and their versatility in adapting to different micro-environments, selected recent studies (published in or after 2010) reporting the prevalence of *V. parahaemolyticus* in retail oysters or the environment are described below. These studies provide a basis for assessing the risk of *V. parahaemolyticus* infection in the respective countries. Since the infectious dose of *V. parahaemolyticus* is considered high (10^7–10^8 organisms) (Sanyal and Sen, 1974; Hoashi *et al.*, 1990), and as the pathogenic subtypes represent only a fraction (in most cases, < 3%) of the total population, the risk of *V. parahaemolyticus* infection is thought to be substantial only when levels of this organism are high in seafood at retail. Specifically, the US Food and Drug Administration (FDA)/Interstate Shellfish Sanitation Conference (ISSC) recommends that oysters not harbour > 10,000 cells of *V. parahaemolyticus* cells/gram (FDA, 2005). However, as our understanding of the differential virulence potential among various subtypes is enhanced, this threshold may need to be adjusted seasonally and/or geographically, depending on the serotypes present in the seafood.

DePaola *et al.* (2010) reported that 77% of the retail oyster samples ($n = 396$) collected throughout 2007 were positive for *V. parahaemolyticus*. Close to one-third of the oyster samples from the Gulf Coast region had high levels of *V. parahaemolyticus*, as defined by > 4 log MPN/g. In this study, the occurrence of *V. parahaemolyticus* did not correlate with the presence of faecal pollution but, rather, with the presence of another *Vibrio* pathogen, *V. vulnificus*. Highest levels of *V. parahaemolyticus* were present in seafood during the summer months, while the lowest levels were present in the winter months. Pathogenic strains of *V. parahaemolyticus*, defined as harbouring *tdh* or *trh* genes, were isolated more frequently in Gulf Coast waters than from the Pacific Northwest. This outcome is probably due to the overall higher temperature in the Gulf Coast (FDA, 2005). A study conducted in 1998–1999 showed that the distribution and level of total *V. parahaemolyticus* in both regions remained consistent over a 10-year period (Cook *et al.*, 2002).

Sobrinho *et al.* (2011) detected *V. parahaemolyticus* in all 74 raw oysters collected in Sao Paulo State, Brazil, over a span of 11 months, with levels ranging from 1.8 to 6.0 log MPN/g. Samples collected during winter months harboured lower levels of *V. parahaemolyticus* (mean log MPN/g of 3.8) than those collected in fall or summer (mean log MPN/g of 4.6 and 5.0, respectively).

A previous study examining 20 oyster samples collected in the same region during the late 1990s found that all samples harboured *V. parahaemolyticus*, with densities varying from 3.6 to 44 MPN/g and from 12 to 2400 MPN/g in untreated and depurated oyster samples, respectively (Ristori *et al.*, 2007). Another previous study also showed that 100% of the oyster samples contained *V. parahaemolyticus* at mean log MPN/g of 3.51 ($n = 30$) (Gelli *et al.*, 1979).

An apparent upward trend in *V. parahaemolyticus* density could suggest that an oyster growing environment has become more favourable to *V. parahaemolyticus* viability, and/or that post-harvest practices in this region favour the growth of this microorganism. To illustrate, increased ocean water temperatures were linked to elevated *V. parahaemolyticus* densities and thus were deemed partly responsible for an illness outbreak (McLaughlin *et al.*, 2005). Failure to rapidly chill oysters also allows the proliferation of *V. parahaemolyticus*.

Of 122 South China retail oyster samples collected from July 2007 to June 2008, Chen *et al.* (2010) found that 109 (89.3%) were positive for *V. parahaemolyticus*, with 59 (48.4%) containing > 100 MPN/g. In New Zealand, *V. parahaemolyticus* was detected in 55 of 58 oysters harvested in representative growing areas in the North Island from December to April 2009, with numbers ranging from < 3 to 1500 MPN/g with a geometric mean density of 99.3 MPN/g (Kirs *et al.*, 2011). During the sampling period, only a narrow range of salinity(≥ 31 ppt) and temperature (18–24°C) was encountered, thus no correlation was found between *V. parahaemolyticus* density and these environmental factors. The authors of this study noted that the frequency of *V. parahaemolyticus* positive samples was ~ 1.7-fold higher compared with a study published in 1985, but no associations were established to explain the apparent increase.

For many reported prevalence studies, culture-dependent methods were the primary tool for isolation and/or quantification of *V. parahaemolyticus* (e.g. Cook *et al.*, 2002; DePaola *et al.*, 2003a, 2010; Hongping *et al.*, 2011; Oberbeckmann *et al.*, 2011; Sobrinho Pde *et al.*, 2011). As discussed below, *V. parahaemolyticus* can also exist in a non-culturable state. The impact and contributions of non-culturable *V. parahaemolyticus* to overall microbial ecology and human foodborne illnesses is poorly understood.

Influence of environmental factors

V. parahaemolyticus is a fast-growing bacterium and under optimal growth temperature (between 35°C and 39°C), its generation time is less than 20 minutes (Barrow and Miller, 1974; Jackson, 1974). In the laboratory, a pure culture of *V. parahaemolyticus* (1% vol/vol inoculum of a stationary phase culture into fresh medium) can re-enter stationary phase in less than 6 hours (Yeung and Boor, 2004). A doubling time of 1.8 hours was noted in live oysters that were held at ambient temperature of 26°C (Gooch *et al.*, 2002). The optimal growth range is consistent with the observation that *V. parahaemolyticus* is most prevalent in the marine environment during warm (> 25°C) summer months. Colwell *et al.* (1984) reported that *V. parahaemolyticus* was not detected in Chesapeake Bay seawater during the winter months (November to March), although a small number of *V. parahaemolyticus* was isolated from marine sediment samples. Thus, it was proposed that *V. parahaemolyticus* survives through the winter in marine sediments and then multiplies when the temperature rises, or following re-introduction of the organism into seawater when temperatures are permissive.

As a halophile, salinity is an absolute requirement for growth and survival of *V. parahaemolyticus*. Under laboratory conditions, *V. parahaemolyticus* can grow in sodium chloride concentrations ranging from 0.5 to 10%, with optimal levels between 1 and 3%. Salinity concentrations encountered by *V. parahaemolyticus* in marine environments typically range between 0.8–3% (DePaola *et al.*, 2000). A survey following outbreaks in Washington, Texas and New York in 1997 and 1998 found a slight negative correlation between *V. parahaemolyticus* numbers and a sample site's salinity (DePaola *et al*, 2000). However, statistically, salinity is not a strong factor in determining the level of *V. parahaemolyticus* in the US waters (FDA, 2005). The relatively small salinity differences commonly encountered in nature have only a small influence on bacterial numbers in seawater relative to the effects of ambient temperature (Cook *et al.*, 2002; FDA, 2005).

Viable but non-culturable state

Under some conditions that do not support optimal physiological functions, some bacteria do not grow on typical laboratory culturing media, despite the fact that they remain alive or viable and exhibit some metabolic activities. This so-called 'viable but non-culturable' (VBNC) state of *Vibrio* species was first reported for *V. cholerae* (Xu *et al.*, 1982). Subsequently, *V. parahaemolyticus* has also been reported to enter the VBNC state under starvation and/or temperature stress (Jiang and Chai, 1996; Mizunoe *et al.*, 2000; Johnston and Brown, 2002), regardless of the pathogenic potential of the strains under study (Bates and Oliver, 2004). For example, 6 *V. parahaemolyticus* strains entered the VBNC state after 12 days of starvation at 4°C (Mizunoe *et al.*, 2000). Direct microscopic observations supplemented with live/dead staining suggested that bacterial cells were viable, however, plating on Luria–Bertani broth with 3% NaCl did not produce visible colonies. Under electron microscopic observation, starved cells formed blebs on the surface of the rod-shaped cells. Other investigators reported that VBNC *V. parahaemolyticus* cells were coccoid or spheroid, rather than rod-shaped (Jiang and Chai, 1996; Johnston and Brown, 2002). MreB, the prokaryotic homologue of eukaryotic actin, has been identified as having a role in altered *V. parahaemolyticus* morphology in the VBNC state (Chiu *et al.*, 2008).

Over 25 bacterial genera have been documented to exhibit the physiological state referred to as VBNC, alternatively known as 'conditionally viable environmental cells' (CVEC) (Oliver, 1995; Faruque *et al.*, 2006; Kamruzzaman *et al.*, 2010). Entering this metabolic state is probably a survival mechanism for non-sporulating bacteria under unfavourable conditions (Oliver, 2010). *V. parahaemolyticus* cells are thought to exist as VBNC cells during the winter season, essentially as 'dormant' cells in marine sediments. In warmer months, these cells resuscitate or reproduce, which may enable their reintroduction into water columns. Whether the cells resuscitate or reproduce has been a subject of investigation, which remains to be resolved (Mizunoe *et al.*, 2000; Coutard *et al.*, 2007a,b). Nevertheless, the ability of *V. parahaemolyticus* to enter into a non-culturable state is a potential public health concern as resuscitated VBNC cells could regain virulence (Baffone *et al.*, 2003), yet conventional detection methods relying on typical growth media may not detect the presence of this pathogen in food.

Epidemiology of foodborne *V. parahaemolyticus* infections

Incidence

V. parahaemolyticus was first identified after an outbreak of food poisoning which caused 272 cases in Osaka, Japan, in 1950 (Fujino *et al.*, 1974). Although not all strains have the ability to cause human disease, *V. parahaemolyticus* is one of the leading causative agents of foodborne disease in some coastal countries and regions such as Japan, Taiwan and Hong Kong (Pan *et al.*, 1997; Wong *et al.*, 2000; Chan and Chan, 2008), where consumption of raw seafood is not uncommon. A systematic risk assessment of this enteric pathogen was conducted by the US government (FDA, 2005). In the US, the first recognized outbreak of *V. parahaemolyticus* infection occurred in Maryland in 1971, as a result of contaminated crabmeat consumption (Molenda *et al.*, 1972). Since then, an additional ~50 outbreaks of *V. parahaemolyticus* infection have been documented in the US (Fishbein *et al.*, 1974; Daniels *et al.*, 2000a), with a recent outbreak occurring in Washington state in summer 2010, which resulted in four cases of *V. parahaemolyticus* infection and the closing of two oyster-growing areas (ProMED-mail, 2010). Based on surveillance data collected from Florida, Alabama, Louisiana and Texas, there were 345 sporadic cases of *V. parahaemolyticus* infection in these states between 1988 and 1997 (Daniels *et al.*, 2000a). In addition to routine surveillance by the states participating in the Gulf Coast *Vibrio* Surveillance System, *Vibrio* spp. have also been included in the Foodborne Diseases Active Surveillance Network (FoodNet) since 1996. Maintained and operated by the Centers for Disease Control and Prevention (CDC), the Food Safety and Inspection Service (FSIS) of the United State Department of Agriculture (USDA), FDA and ten participating state health departments, FoodNet is an active, population-based surveillance system in ten US states for all laboratory-confirmed infections. Based on the FoodNet data, *Vibrio* spp. was responsible for 160 foodborne cases in 2009. *V. parahaemolyticus* is the predominant foodborne *Vibrio* spp. in the US, followed by *V. vulnificus and V. cholerae*. While other foodborne pathogens such as *Campylobacter* and *Escherichia coli* O157 have shown an overall decreasing trend of incidence, the incidence of *Vibrio* infection has been generally increasing since 2000 (CDC, 2010a). This apparent increase could reflect a number of circumstances, including a true increase in the number of *V. parahaemolyticus* illnesses, enhanced surveillance for this pathogen, or improved detection capabilities in the clinical laboratories. Despite the increased number of cases, the overall public health impact of *V. parahaemolyticus* infections remains less than that of some other foodborne pathogens (e.g. *Salmonella* spp., *Campylobacter* spp.), with an annual mean of five outbreaks and 114 outbreak-associated illnesses between 2002 and 2006 (CDC, 2010b). Cases of foodborne *V. parahaemolyticus* infections are likely to be under-reported, as most individuals with acute gastroenteritis do not seek medical help. After accounting for non-reported cases, CDC estimates there are 7880 cases of *Vibrio* illness in the US annually. Among these total *Vibrio* illnesses, approximately 4500 are attributable to *V. parahaemolyticus* (FDA, 2005).

Clinical manifestations and treatment

V. parahaemolyticus is the most frequent cause of foodborne *Vibrio*-associated gastroenteritis in the United States (Daniels *et al.*, 2000a). The most prevalent symptoms are diarrhoea and abdominal pain, followed by nausea, vomiting, and fever. Headache and chills are also reported. The median duration of the illness is 2.5 days, and the incubation period is 4–96 hours following ingestion of pathogenic strains, with an average of 15 hours (http://www.fda.gov/Food/FoodSafety/FoodborneIllness/FoodborneIllnessFoodbornePathogensNaturalToxins/BadBugBook/ucm070452.htm). While most affected individuals experience gastrointestinal symptoms through the consumption of contaminated food, some are infected through wound exposure, which may account for one-third of all *V. parahaemolyticus* cases (Daniels *et al.*, 2000a). Mortality rates are about 2% for gastroenteritis cases, but may be as high as 29% for those who develop septicaemia (Oliver and Kaper, 2007). Most cases of *V. parahaemolyticus* gastroenteritis are self-limiting and can be treated by oral rehydration, alone. Occasionally, when a patient does not respond to rehydration therapy, treatment with antibiotics such as doxycycline, ciprofloxacin, or erythromycin becomes necessary (Qadri *et al.*, 2003).

Susceptible population

While all seafood (and especially shellfish)-consuming populations appear equally susceptible to *V. parahaemolyticus* gastroenteritis, epidemiological investigations have identified a subpopulation that is more vulnerable to development of life-threatening septicaemia as a result of foodborne *V. parahaemolyticus* infections (FDA, 2005). This subpopulation includes patients with underlying medical conditions, such as liver disease, alcoholism, diabetes, malignancy, renal disease, immunodeficiency, haematological disease, and gastric surgery. This subpopulation also represents a high-risk group for contracting severe illnesses caused by *V. vulnificus*, a bacterium also commonly found in shellfish.

Associated foods

Shellfish, which live by filter feeding, can concentrate bacteria such as *V. parahaemolyticus* to levels higher than those in the surrounding water. *V. parahaemolyticus* can also multiply and colonize in the gut of filter feeders. *V. parahaemolyticus* is sensitive to heat, therefore, illnesses due to *V. parahaemolyticus* are usually associated with the consumption of contaminated raw or undercooked molluscan shellfish (e.g. oysters, clams and mussels). In the US, raw oysters are the predominantly incriminated food linked to *V. parahaemolyticus* outbreaks (Hlady, 1997; FDA, 2005). Cooked crustaceans (e.g. shrimp, crab, and lobsters) have also been implicated in *V. parahaemolyticus* infections (Blake *et al.*, 1980).

The presence of *V. parahaemolyticus* in cooked products results from either improper cooking or recontamination of the cooked product.

Outbreak patterns and recent trends

As *V. parahaemolyticus* is a mesophilic bacterium that is most prevalent during the summer months, foodborne outbreaks due to *V. parahaemolyticus* also show a seasonal pattern (Daniels *et al.*, 2000a; Lesmana *et al.*, 2001). While levels of *V. parahaemolyticus* in freshly harvested seafood are generally below the predicted infectious dose of 10^7 to 10^8 organisms, the ability of this organism to multiply rapidly at ambient temperatures can result in the presence of sufficient bacteria in foods to cause disease. To illustrate, 10^2 to 10^3 *V. parahaemolyticus* organisms/g in live American oysters have been shown to increase 50-fold (1.7 log cfu/g) and 790-fold (2.9 log cfu/g) at 26°C by 10 hours and 24 hours after harvest, respectively (Gooch *et al.*, 2002).

Historically, coliforms and faecal coliforms have been used as indicator organisms to screen for probable contamination of foods and water with enteric pathogens. However, the relationship between the presence of these indicator organisms and *V. parahaemolyticus* in the environment is not clearly established. Correlations were found between the numbers of *V. parahaemolyticus* and those of (faecal) coliforms in mussels, clams and cockles sampled in Spain (Martinez *et al.*, 2009). In contrast, other studies have shown no statistically significant association in oysters, water and/or sediment samples collected in the US (Shiaris *et al.*, 1987; Parveen *et al.*, 2008), or in mussels collected in Italy (Normanno *et al.*, 2006). The absence of a correlation is not surprising as *V. parahaemolyticus* is a natural inhabitant of the marine environment whereas the presence of pathogens expected to correlate with the presence of faecal coliforms (e.g. *Salmonella* spp.) reflects contamination originating from human and other land animal faecal matter. Therefore, *V. parahaemolyticus* infection linked to oyster consumption can occur even when implicated foods are harvested from growing areas harbouring acceptable levels of faecal coliforms (Daniels *et al.*, 2000b). Consequently, monitoring for the presence of *V. parahaemolyticus* in seafood harvesting areas, rather than the presence of indicator organisms of faecal contamination, is an important part of safeguarding the seafood supply.

Until the mid-1990s, the majority of *V. parahaemolyticus* infections worldwide were sporadic in nature, characterized by causal associations with multiple, diverse serotypes. One exception was the predominant association of O4:K12 strains with gastroenteritis cases on the US West Coast between 1979 and 1992 (Nolan *et al.*, 1984; Abbott *et al.*, 1989; Okuda *et al.*, 1997; DePaola *et al.*, 2000). Since 1996, a particular serotype, O3:K6, and a few other related strains, have become a predominant pathogenic subtype. Since the first cluster of cases occurred in Calcutta, this O3:K6 clonal group has been implicated in outbreaks worldwide, including in the US. Relative to typical *V. parahaemolyticus* outbreaks, the magnitude of some outbreaks caused by this pandemic clone has been exceptionally high, suggesting this pathogenic subtype has a lower infectious dose and, hence, increased virulence. In Chapter 8, the emergence of this pandemic clone and its pathogenesis will be discussed in more detail.

The incidence of foodborne *V. parahaemolyticus* infections in the US varies year by year but it is typically within the thousands range. Based on FoodNet data, the yearly estimates of foodborne *V. parahaemolyticus* infection (i.e. excluding wound infection) were approximately 2700, 9800 and 5600 cases for 1996, 1997 and 1998, respectively (FDA, 2005). When focusing on cases associated with raw oyster consumption only, which is predicted to be 62% of all *V. parahaemolyticus* illness cases, CDC estimates an annual burden of approximately 2790 cases of *V. parahaemolyticus* illness (FDA, 2005), based on more recent data from 1998 to 2002. The number of reported cases posted to the ProMED Mail voluntary reporting system, which is maintained by the International Society for Infectious Diseases, also corroborates the estimated total number of *V. parahaemolyticus* cases in this range (ProMED-mail, 2006a,b,c,d, 2007, 2008, 2010), after the number of confirmed and probable cases are multiplied by a factor of 20 (Mead *et al.*, 1999). The surge of cases in 1997 and 1998 was associated with the emergence of the O3:K6 clonal group. On the other hand, according to Scallan *et al.* (2011), who used multipliers of 1.1 and 142.4 to account for under-reporting and underdiagnosis respectively, the estimated range of episodes of *V. parahaemolyticus* infections in the US is 18,260–58,027 per year.

To assess the progress of the US effort to enhance food safety, one of the objectives listed in Healthy People 2020 (http://www.healthy-peoplegov/2020/default.aspx) is to reduce the incidence of laboratory-confirmed *Vibrio* species from 0.3 cases per 100,000 population per year (this baseline was the average number of cases from 2006 to 2008) to 0.2 by year 2020. Developed by the US Government, Healthy People provides science-based, 10-year national objectives for improving the health of all Americans. The food safety objective for the *Vibrio* spp. emerged for the first time in the Healthy People 2020 document; national targets had been set previously for reductions in *Campylobacter*, *Salmonella*, Shiga toxin-producing *E. coli* O157, and *Listeria* infections (http://www.healthypeople.gov/2010/).

Virulence factors and virulence genes of *V. parahaemolyticus*

Though it is known that *tdh*-negative *V. parahaemolyticus* strains may be virulent, e.g. such as when a strain has *trh* (Sochard and Colwell, 1977; Hondo *et al.*, 1987, 1988), *tdh* remains the most recognized virulence factor for this foodborne pathogen. Previous studies have shown that some *V. parahaemolyticus* strains also possess characteristics identified as important for virulence in other microorganisms (Yamamoto *et al.*, 1994; Suthienkul *et al.*, 1995; Osawa and Yamai, 1996; Okuda *et al.*, 1997; Yeung *et al.*, 2002). These include the presence of the urease gene (*ure*), the ability to produce the iron-binding protein (vibrioferrin), the ability to exhibit different motility patterns under liquid and semi-solid environments, etc. The recently identified type three secretion systems (T3SS) in *V. parahaemolyticus* RIMD2110633 (Makino *et al.*, 2003) has expanded the list of (putative) virulence factors. Some of these virulence factors were probably acquired through horizontal transfer, such as the pathogenicity island that was identified on chromosome 2 (Sugiyama *et al.*, 2008). A better understanding of the specific contributions of each of the different factors to overall *V. parahaemolyticus* virulence is needed to better understand disease risks associated with the pathogen (FDA, 2005). The factors described below have been either established or proposed to be associated with virulence of *V. parahaemolyticus*.

Haemolysins

Thermolabile haemolysin (TLH)

Virtually all *V. parahaemolyticus* strains but not other vibrios possess the thermolabile haemolysin gene (*tlh*); thus it can be used as a species-specific marker (Taniguchi *et al.*, 1985). Despite its haemolytic activity, epidemiological data does not support a role for TLH in human disease, hence the presence of *tlh* is exclusively used for the detection of total *V. parahaemolyticus*, while the presence of *tdh* and/or *trh* is used to indicate pathogenic strains. Haemolytic activities of the TLH isolated from *V. parahaemolyticus* T-3454 are abolished after heating at 95°C for 10 minutes (Sakurai *et al.*, 1974). Based on the nucleotide sequence, the preprotein and the mature protein are predicted to consist of 418 and 393 amino acids, respectively, with estimated molecular weights of 47.5 kDa and 45.3 kDa. The GC content of *tlh* gene is 47.6%, which is consistent with the average genomic GC content of *V. parahaemolyticus* (Taniguchi *et al.*, 1986). Genomic sequencing confirmed that the *tlh* gene (also called the lecithin-dependent haemolysin gene) is on the small chromosome (http://genome.bio.titech.ac.jp/bacteria/vpara/index.html). TLH does not exhibit beta-haemolysis on Wagatsuma agar and may not be produced at all under some experimental conditions (Sakurai *et al.*, 1974).

Thermostable direct haemolysin (TDH)

Many epidemiological studies indicate that the presence of the *tdh* gene, and thereby the production of TDH, is associated almost exclusively with clinical isolates, as compared with only < 5% of environmental isolates (Miyamoto *et al.*, 1969; Wong *et al.*, 2000). In addition, empirical studies showed that TDH induces cytotoxicity, haemolysis, cardiotoxicity, enterotoxicity and lethality in various experimental models (Raimondi *et al.*, 2000; Hiyoshi *et al.*, 2010; Matsuda *et al.*, 2010).

Early epidemiological studies identified an association between a *V. parahaemolyticus* strain's ability to cause beta-haemolysis on Wagatsuma blood agar, referred to as the Kanagawa phenomenon (KP), and its ability to cause gastroenteritis (Miyamoto *et al.*, 1969). TDH was later identified as the primary haemolysin responsible for the KP (Takeda, 1983). During the late 1980s to early 1990s, several groups raced to obtain the DNA and amino acid (aa) sequence of TDH present in clinical *V. parahaemolyticus* isolates. Nishibuchi and Kaper (1985) and Tsunasawa *et al.* (1987) were among the first to characterize *tdh* in *V. parahaemolyticus* WP1. The published sequences by these two groups differed by 7 aa residues. Later, Iida and Yamamoto (1990) cloned and sequenced *tdh* genes from strain T4750 and found that this strain contained two *tdh* copies, which were assigned *tdhA* and *tdhS*. These two copies shared 97.2% sequence homology (Iida and Yamamoto, 1990) and were predicted to differ in eight aa residues. The *tdhA* amino acid sequence is the same as that reported by Taniguchi *et al.* (1986) except in position 118 (Gln vs. Glu). More recently, genomic sequencing also identified *tdhA* (VPA1314) and *tdhS* (VPA1378) in strain RIMD2110633 (Makino *et al.*, 2003) with identical aa sequences to that reported by Nishibuchi and Kaper (1985, 1995). Duplication and variation of *tdh* is well documented (Nishibuchi and Kaper, 1990). Baba *et al.* (1991) found five sequence variants of the *tdh* gene (designated as *tdh1* through *tdh5*) sharing >97% similarity. Other variants were named as *tdh/I* (Honda *et al.*, 1991) and *tdh/II* (Nagayama *et al.*, 1995).

Between the two *tdh* copies, *tdhA* and *tdhS* (or *tdh2 and tdh1*), only one of them, *tdhA* (or *tdh2*), appears to be predominantly responsible for the KP (Nishibuchi *et al.*, 1991; Okuda and Nishibuchi, 1998). The differential haemolytic activities of these gene copies appear to be controlled at the transcription level (Okuda and Nishibuchi, 1998). Between *tdhA* and *tdhS*, the enhanced expression of the former gene variant appears to be due to the presence of a strong promoter. Compared with *tdhS*, the promoter of *tdhA* was characterized by a single-base difference in the –35 region (Okuda and Nishibuchi, 1998). Expression by *tdhS*, on the other hand, was partial and probably insignificant (Iida and Yamamoto, 1990). This variation in one nucleotide of the promoter binding site may influence the binding affinity of RNA polymerase and hence expression of the gene. Strains harbouring *tdh3*, *tdh4* or *tdh5* also did not exhibit significant expression of these genes (Wong *et al.*, 2000). In fact, their low levels of transcription were similar to that of *tdhS* (Wong *et al.*, 2000).

The primary target of TDH appears to be enterocytes that express the surface receptors GT1 and GD1a ganglioside (Takeda *et al.*, 1976; Shimohata and Takahashi, 2010; Takeda, 2011). TDH acts on cellular membranes as a pore-forming toxin that alters ion flux in intestinal cells, thereby leading to a secretory response and diarrhoea (Honda *et al.*, 1992; Fabbri *et al.*, 1999; Raimondi *et al.*, 2000; Takahashi *et al.*, 2000b). The process appears to be mediated by an increased intracellular concentration of Ca^{2+} in association with the phosphorylation of protein kinase C, which results in Cl^- secretion (Takahashi *et al.*, 2000b). To yield a haemolysis phenotype, a 25-kDa host protein must be phosphorylated after TDH binds to the erythrocyte membrane (Yoh *et al.*, 1996).

Expression of TDH is under the regulation of Hfq, an RNA-binding protein important for the fitness and virulence of many bacteria (Nakano *et al.*, 2008). As a global regulator, it is thought to act as an RNA chaperone by mediating interactions among many regulatory RNAs and their mRNA targets (Chao and Vogel, 2010). An *hfq* deletion mutant had an upregulated *tdhA* promoter, resulting in increased transcriptional activity, leading to higher TDH production (Nakano *et al.*, 2008). Other putative virulence factors (e.g. VopC, a homologue to *E. coli* cytotoxic necrotizing factor) are also upregulated in the absence of Hfq (Nakano *et al.*, 2008). In a study by Lin *et al.* (1993), another regulator, ToxR, was shown to play a role in TDH production. The transcriptional activity of *toxR* is not dependent on Hfq (Nakano *et al.*, 2008). More recently, two novel ToxR-like transcriptional regulatory proteins, VtrA and VtrB, were found to regulate the expression of the pathogenicity island on chromosome 2, which contains *tdh*. The gene *vtrB* was in turn under the control of *vtrA* (Kodama *et al.*, 2010).

Both the epidemiological and empirical research strongly support TDH as an important virulence factor of *V. parahaemolyticus*. Strains lacking TDH, on the other hand, should not be automatically regarded as non-pathogenic. Multiple *in vivo* and *in vitro* studies have been performed to examine the relationship between TDH and the pathogenesis of *V. parahaemolyticus*. For example, Hoashi *et al.* (1990) compared TDH expression patterns of 14 *V. parahaemolyticus* strains with their abilities to cause disease in mice following intraperitoneal or oral–gastric inoculations. Strain lethality and pathological effects developed similarly regardless of a strain's TDH-production capacity. Further, some strains bearing *tdh* were found to be KP negative. These results suggest that TDH expression may vary among strains and that the mere presence of the *tdh* gene may not reflect a strain's ability to cause disease (Hoashi *et al.*, 1990). Using 16 *V. parahaemolyticus* strains, Hackney *et al.* (1980) found that the intensity of *V. parahaemolyticus* adherence to human foetal epithelial cells was related more to a strain's ability to cause foodborne illness than to the KP. Iijima *et al.* (1981) also showed that *V. parahaemolyticus* adherence to two types of human epithelial cells was not related to the haemolysin-induced KP in 32 strains. A KP-negative strain isolated from a clinical source was shown to produce a non-TDH haemolysin-like substance that induced fluid accumulation (Honda *et al.*, 1988). This haemolysin, which is further described below, was termed TDH-related haemolysin (TRH). In 1991, Honda *et al.* reported the presence of yet another *V. parahaemolyticus* haemolysin that also is similar to TDH, but is not identical to TRH (Honda *et al.*, 1991). Taken together, these results suggest that (i) *V. parahaemolyticus* strains may produce several different, but related, haemolysins; (ii) TDH is not the sole contributor to *V. parahaemolyticus* pathogenicity; and, therefore, (iii) the presence of *tdh*, alone, cannot reliably indicate a virulent strain. Additional putative virulence factors appear to also contribute to the pathogenesis of *V. parahaemolyticus*.

TDH-related haemolysin (TRH)

Sochard and Colwell (1977) demonstrated that a KP-negative strain isolated from food produced a toxin-like proteinaceous substance that was lethal in mice at high concentrations and diarrhoeagenic at lower concentrations. The authors concluded that this toxin was unlikely to be TDH as the protein was heat labile. In 1985, an outbreak of gastroenteritis in the Maldives was linked to KP-negative isolates of *V. parahaemolyticus* shown to produce a TRH rather than TDH (Honda *et al.*, 1987, 1988).

Both TDH and TRH are reversible amyloid toxins, and are immunologically and biologically similar (Honda *et al.*, 1988; Burdette *et al.*, 2008). As with TDH, TRH induces Ca^{2+}-activated Cl^- channels in cultured human colonic epithelial cells which result in altered ion flux (Takahashi *et al.*, 2000a). It also stimulated fluid secretion in the rabbit ileal loop test, supporting the role of this haemolysin in the induction of diarrhoea (Honda *et al.*, 1988; Xu *et al.*, 1994b). Despite the similarity to TDH, TRH did not exhibit KP and was labile to heat treatment at 60 C for 10 minutes (Honda *et al.*, 1988).

Duplication of *trh* gene was noted in different studies, and many variations were detected from strain to strain but they could be clustered into two subgroups represented as *trh1* and *trh2*. These variants shared 84% sequence identity; both were ~70% identical to *tdh2* (Honda *et al.*, 1988; Nishibuchi *et al.*, 1989; Kishishita *et al.*, 1992). Both *trh* gene variants contained a 12-bp inverted repeat in the ribosome-binding site, immediately following the +1 position. Using an experimental design where the promoter and inverted repeat region were swapped between *tdh* and *trh* genes, Nakaguchi and Nishibuchi (2005) deduced that the low-level transcription of *trh* was dependent on the promoter, not the inverted repeat.

Most environmental isolates (not implicated in human illness) do not possess *tdh* or *trh*. On the other hand, *V. parahaemolyticus* gastroenteritis has been linked to the presence of either one of these genes. Few clinical isolates appear to contain both of them (Shirai *et al.*, 1990; Baba *et al.*, 1991; Kishishita *et al.*, 1992; Xu *et al.*, 1994a). Hence *tdh* and *trh* are now commonly recognized as the important virulence factors for *V. parahaemolyticus*.

Type III secretion systems and pathogenicity island

Functional and evolutionary aspects of the Type III secretion system (T3SS) in Gram-negative bacteria have been the subject of many reviews (DeVinney *et al.*, 1999; Schroeder and Hilbi, 2008; Marlovits and Stebbins, 2010). T3SS is thought to be triggered when a bacterium comes in close contact with host cells, and hence, this system is also called contact-dependent secretion. Delivered through the T3SS apparatus, bacterial proteins (termed 'effectors') including virulence determinants can be exported directly into the host cytosol. In the clinical strain RIMD2210633, two T3SS were found (Makino *et al.*, 2003) – the T3SS1 (located on chromosome 1) was structurally analogous to the Ysc secretion system observed in *Yersinia* spp. while the T3SS2 (located on chromosome 2) was analogous to the Inv-Mxi-Spa secretion system in *Salmonella* spp. and *Shigella* spp. (Troisfontaines and Cornelis, 2005).

Since the discovery of these T3SS, a surge of research has been conducted on the prevalence, genetic structure, and mechanisms of these putative virulence factors in *V. parahaemolyticus* (Liverman *et al.*, 2007; Burdette *et al.*, 2008; Abendroth *et al.*, 2009; Chao *et al.*, 2009; Hiyoshi *et al.*, 2010; Matlawska-Wasowska *et al.*, 2010; Shimohata *et al.*, 2011). T3SS1 appears to be common in all *V. parahaemolyticus* isolates, including those isolated from food and environmental sources not implicated in human diseases (Chao *et al.*, 2009). On the other hand, T3SS2 appears to be exclusively present in clinical isolates (Park *et al.*, 2004) – both pandemic and non-pandemic (Wang *et al.*, 2006). A closer look at the genetic region containing the T3SS2 led to the identification of a pathogenicity island (PAI) (Sugiyama *et al.*, 2008). Indeed, the genes encoding T3SS components in Gram-negative bacteria are usually located on PAIs; such as the locus of enterocyte effacement region of enterohaemorrhagic *E. coli* O157:H7 and enteropathogenic *E. coli*, which contains genes encoding T3SS and other genes involved in pathogenesis (Clarke *et al.*, 2003). The PAI on chromosome 2 of *V. parahaemolyticus* RIMD2110633 has a GC content of 39.8%, in contrast with the average GC content of 45.4% for the remainder of the genome, suggesting that the island was acquired by horizontal transfer. The mobile element with a transposase containing the DDE motif may have been involved in PAI formation in this strain (Sugiyama *et al.*, 2008). This PAI is approximately 80 kb (Izutsu *et al.*, 2008). A novel T3SS was found in a *tdh*-negative and *trh*-positive *V. parahaemolyticus* strain TH3996 upon sequencing the DNA region of approximately 100 kb surrounding the *trh* gene (Okada *et al.*, 2009). This novel T3SS was found to be phylogenetically similar to the T3SS2 in KP-positive *V. parahaemolyticus* strains, though it belonged to a distinctly different lineage (Okada *et al.*, 2009). Subsequently the T3SS2 found in KP-positive and KP-negative strains is sometimes referred to T3SS2α and T3SS2β, respectively.

The most recognized virulence factor of *V. parahaemolyticus*, *tdh*, is associated with T3SS2 (and thus the 80-kb PAI). With only one exception (out of 26 isolates), Wang *et al.* (2006) established a positive correlation between T3SS2 and *tdh*. Comparative genomic analysis also confirmed that the T3SS2-containing PAI was exclusively conserved in KP-positive but not in KP-negative strains (Izutsu *et al.*, 2008; Sugiyama *et al.*, 2008). In addition to the T3SS2 apparatus genes in the PAI, RIMD2110633 bears two *tdh* genes as well as other genes that have been associated with virulence in other bacteria, such as those encoding homologues of the *E. coli* cytotoxic necrotizing factor and of the *Pseudomonas* exoenzyme T (Makino *et al.*, 2003; Chao and Vogel, 2010).

In laboratory studies, all T3SS were found to show *in vivo* and *in vitro* virulence properties (Park *et al.*, 2004; Okada *et al.*, 2009; Hiyoshi *et al.*, 2010; Pineyro *et al.*, 2010). Specifically, 48-h-old piglets inoculated orogastrically with 10^{11} cfu of a genetically engineered *V. parahaemolyticus* strain NY-4 (serotype O3:K6) harbouring T3SS2 alone resulted in acute, self-limiting diarrhoea; whereas the same strain harbouring T3SS1 alone did not produce any clinical symptoms (Pineyro *et al.*, 2010). On the other hand, this study also showed that 5×10^5 cfu of *V. parahaemolyticus* NY-4 and a T3SS2 deletion mutant (i.e. harbouring T3SS1 alone) caused 80–100% mortality in C57BL/6 mice upon intrapulmonary inoculation.

Inoculation with a T3SS1 deletion mutant or a double deletion mutant caused no mortality. In another study, cytotoxic activity (as measured by lactate dehydrogenase release) of infected HeLa cells was greater in the presence of T3SS1 genes than with the T3SS2 genes (Park *et al.*, 2004). In contrast, T3SS2-related genes, but not T3SS1, were responsible for the intestinal fluid accumulation in a rabbit ileal loop test (Park *et al.*, 2004).

T3SS1, which is regulated by ExsA and ExsD (Zhou *et al.*, 2010a), is responsible for the pro-inflammatory cell death of HeLa cells and RAW 264.7 macrophages as a result of autophagy, cell rounding and cell lysis (Burdette *et al.*, 2008). The cytotoxicity was thought to be a necrotic process (Burdette *et al.*, 2008) even though earlier studies predicted a role for apoptosis in cell death (Bhattacharjee *et al.*, 2006). The cytotoxicities observed in the tissue culture model were in part caused by effectors VPA0450 (Broberg *et al.*, 2010) and VP1680 (Shimohata *et al.*, 2011). VPA0450 is located on chromosome 2 but this effector is delivered through T3SS1. It contains five of the six catalytic motifs that are present in the active site of inositol polyphosphate 5-phosphatases (IPP5Cs), which dephosphorylate the D5 phosphate on inositol rings; so this protein effector could act as an IPP5C altering the homeostasis of membrane-bound inositol-signalling molecules (Broberg *et al.*, 2010). Consequently, the adaptor protein-binding sites required for membrane and cytoskeleton assembly is disrupted, thereby compromising membrane integrity (Broberg *et al.*, 2010). The inflammation observed in T3SS1-mediated cytotoxicity in Caco-2 cells is probably associated with VP1680, which triggers interleukin-8 (IL-8) secretion through activating extracellular signal-regulated kinase 1/2 and p38 mitogen-activated protein kinases (MAPK) (Matlawska-Wasowska *et al.*, 2010; Shimohata *et al.*, 2011). Though these effectors appear to have the ability to directly manipulate host cells, expression of secretion apparatus proteins must precede their translocation. For example, VP1659 is important in the secretion of effectors responsible for autophagy and actin rearrangement in HeLa cells (Zhou *et al.*, 2010b). More effectors are likely to be involved in pathogenesis and they are summarized in Table 7.1.

The T3SS2α, which is regulated by two novel ToxR-like transcriptional regulatory proteins, VtrA and VtrB, was shown to induce *in vivo* enterotoxicity and *in vitro* cytotoxicity (Kodama *et al.*, 2010). T3SS2β (found in *trh*-positive but KP-negative *V. parahaemolyticus* strains) was also shown to induce fluid accumulation in a rabbit ileal loop test (Okada *et al.*, 2009) Mutant strains lacking the two effectors delivered by T3SS2α, VopB2 and VopD2, were incapable of causing fluid accumulation in a rabbit ileal loop model (Kodama *et al.*, 2008). VopA and VopL were shown to disrupt innate immunity and actin cytoskeleton in HeLa cells, respectively (Liverman *et al.*, 2007; Trosky *et al.*, 2007). Interestingly, while some T3SS1 effectors activated MAPK, VopA inhibited this important signalling pathway (Trosky *et al.*, 2007). T3SS2α was also shown to inhibit IL-8 mRNA transcription at early stages of interaction between the bacterium and the Caco-2 cell (Matlawska-Wasowska *et al.*, 2010). These results suggest that an interaction exists between T3SS1 and T3SS2 in eliciting host cell responses.

As mentioned previously, virtually all *V. parahaemolyticus* strains harbour T3SS1. Thus, its presence alone is not sufficient to cause human disease even though it has been demonstrated to have virulence properties in laboratory experiments. Infection dosages in tissue culture tests used a multiplicity of infection (MOI) of 10–100 (Burdette *et al.*, 2008; Matlawska-Wasowska *et al.*, 2010) while 10^8 cells were injected directly into the rabbit ileal loops (Park *et al.*, 2004) and 10^{11} cells were used in oral infection (Pineyro *et al.*, 2010). These high doses may not be realistic when compared with the numbers of *V. parahaemolyticus* to which humans are typically exposed (FDA, 2005). While there is a lack of association between T3SS1 and *V. parahaemolyticus* illnesses, there is a correlation between T3SS2 and clinical isolates. This discordant association may be due to the differential regulation and expression of these two secretion systems. For instance, Gode-Potratz (2010) found that, while T3SS1 gene expression was regulated by LafK and ExsA under the influence of calcium and iron levels, the upregulation of gene expression was not observed for T3SS2. The expression

Table 7.1 Putative virulence determinants delivered through Type III secretion systems (T3SS) in *V. parahaemolyticus*

Effector; and the cognate T3SS	Description/function	Reference
VopA/P (VPA1346); T3SS2	Acetyltransferase activity inhibiting the binding of ATP to the MAPK kinase Disrupts innate immunity Similar to YopP	Trosky *et al.* (2004, 2007), Shimohata and Takahashi (2010)
VopB2/D2; T3SS2	Deletion mutants lack the ability to cause fluid accumulation	Kodama *et al.* (2008), Shimohata and Takahashi (2010)
VopC (VPA1321); T3SS2	38% homology to *E. coli* cytotoxic necrotizing factor	Shimohata and Takahashi (2010)
VopD (VP1656), T3SS1	Induces toll-like receptor-independent apoptosis in macrophage through NF-κB inhibition	Bhattacharjee *et al.* (2006)
VopL (VPA1370); T3SS2	Influences the assembly of actin	Liverman *et al.* (2007), Trosky *et al.* (2007), Shimohata and Takahashi (2010)
VopQ; T3SS1	Necessary and sufficient to induce autophagy	Burdette *et al.* (2009)
VopS; T3SS1	AMPylator contributing to cell rounding by inhibiting Rho guanosine triphosphatases	Yarbrough *et al.* (2009)
VopT (VPA1327); T3SS2	ADP-ribosyltransferase activity and ribosylates Ras Partially responsible for the T3SS2-dependent cytotoxicity in Caco-2	Shimohata and Takahashi (2010)
VopX; T3SS2	Deletion mutant reduces the ability to colonize *in vivo*	Alam *et al.* (2011)
VPA0450; T3SS1	Inositol polyphosphate 5-phosphatase disrupting host cell membrane integrity	Broberg *et al.* (2010)
VP1680; T3SS1	Triggers IL-8 secretion in Caco-2 cells	Shimohata *et al.* (2011)

of T3SS1 was repressed under high cell density (Henke and Bassler, 2004). The role of quorum sensing on T3SS2 is not clear.

When comparing *V. parahaemolyticus* with another important human pathogen, *V. cholerae*, it is interesting to note that some clinical manifestations are similar (i.e. gastroenteritis), except for the characteristic rice water stool associated with *V. cholerae* infection. However, *V. parahaemolyticus* is generally considered to elicit inflammatory diarrhoea, as opposed to the non-inflammatory secretory diarrhoea caused by *V. cholerae* (Qadri *et al.*, 2003). Furthermore, the main virulence factor of *V. parahaemolyticus* and *V. cholerae* (TDH and cholera toxin, respectively) has different modes of action, such as modulating Ca^{2+} versus cAMP. It is tempting to speculate that a major difference in pathogenicity between these closely related organisms is, at least in part, due to the presence of T3SS (especially T3SS2) in *V. parahaemolyticus*. In general, T3SS are not present in *V. cholerae* although a T3SS2-related gene cluster has been found in some O39, non-O1 and non-O139 strains (Dziejman *et al.*, 2005; Alam *et al.*, 2011). It has yet to be established if the invasion capabilities of some *V. parahaemolyticus* strains can be attributed to the presence of T3SS. For other organisms, T3SS has been shown to enable a pathogen to establish an intimate interaction with the host and enhance intracellular infection, e.g. *Burkholderia pseudomallei* (Stevens *et al.*, 2002), *Salmonella* Typhimurium (Steele-Mortimer *et al.*, 2002) and *Shigella flexneri* (Schuch *et al.*, 1999).

Other factors associated with virulence

Swarming

In addition to the polar flagellum, *V. parahaemolyticus* also possess lateral flagella, which enable migration across semi-solid surfaces in a phenomenon called swarming (Fig. 7.1). Motility phenotype has been linked to virulence factor expression in *V. cholerae* (Gardel and Mekalanos, 1996). Specifically, *V. cholerae* mutants with greater motility showed increased expression of virulence-associated fucose-sensitive haemagglutination and HEp-2 adhesins. The lateral flagella in *Aeromonas* spp. appear to aid host cell invasion (Kirov, 2003). As an enhanced swarming ability might allow bacterial cells to more rapidly and effectively reach a target host, Yeung *et al.* (2002) compared the swarming ability of the pandemic O3:K6 strains with non-O3:K6 food and clinical strains. Though spreading could not be quantified accurately owing to burst formation (Fig. 7.1) with some strains, the pandemic O3:K6 group demonstrated an enhanced ability to swarm over agar surfaces. The presence of magnesium appeared to further stimulate swarming.

Lateral flagella genes (*laf*) were once thought to be under the sole control of a transcriptional activator LafK. However, by using a transcriptomic approach, a recent study by McCarter's group (Gode-Potratz *et al.*, 2011) showed that LafK is not the master regulator as some *laf* genes are controlled by other regulator(s). Upon encountering the agar surface, *V. parahaemolyticus* expresses ~70 surface-sensing genes, including *laf*. Some of these surface-sensing genes were shown empirically to lower the level of a second messenger, c-di-GMP, and increase cytotoxicity towards Chinese hamster ovary CCL-61 cells. Further characterization of *V. parahaemolyticus* motility behaviour is needed to clarify its role, if any, in pathogenesis.

Urea hydrolysis

Urea hydrolysis also has been proposed as an additional virulence marker for some pathogenic *V. parahaemolyticus* strains as a strong correlation has been demonstrated between the presence of the *trh* gene and urease production (Suthienkul *et al.*, 1995; Osawa *et al.*, 1996; Okuda *et al.*, 1997). In particular, urea hydrolysis has been shown to be a good marker for the O4 *V. parahaemolyticus* strains associated with sporadic illness in the Pacific Northwest (Kaysner *et al.*, 1994). This correlation is probably due to the presence of *trh* and *ure* in the same locus on the chromosome (Iida *et al.*, 1998). The production of urease, in theory, enhances the pathogen's survival in acidic environment such as those encountered by *Helicobacter pylori* in the human stomach. However, experimental data supporting its role in the fitness of *V. parahaemolyticus* is lacking. Further, urease activity is not a universal marker for pathogenic *V. parahaemolyticus*, as urease-negative strains, including the pandemic O3:K6 strains, have been isolated from a number of patients. Taken together, urea hydrolysis is not a reliable virulence marker for *V. parahaemolyticus*.

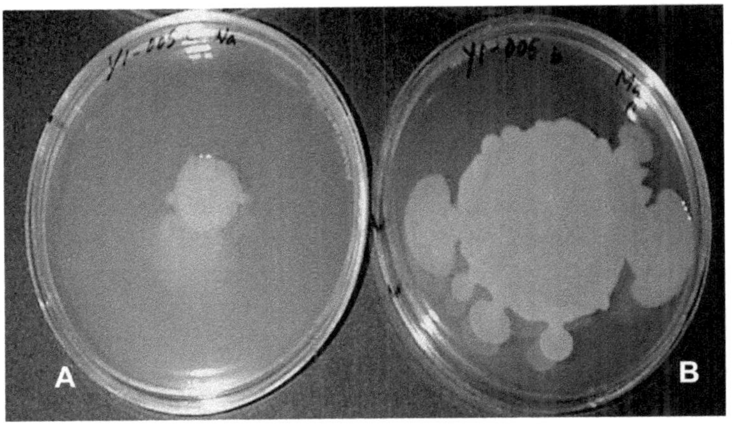

Figure 7.1 Representative swarming phenotype of a *V. parahaemolyticus* isolate on tryptic soy agar medium supplemented with 0.5% NaCl (left) or 0.5% NaCl plus 0.15% $MgSO_4$ (right). For this isolate, swarming is stimulated by the addition of magnesium. Cultures on the right also exhibit pronounced 'burst formation'.

DNA methyltransferase

In *V. cholerae,* mannose-fucose-resistant haemagglutinin (MFRHA) is considered a virulence factor as mutants lacking MFRHA activity had diminished ability to persist in infant mouse gut (Franzon *et al.,* 1993). Using a subtractive hybridization technique, Wang *et al.* (2006) identified a clone of clinical *V. parahaemolyticus* O3:K6 that contains a 22.79-kb pathogenicity island-like element carrying the DNA methyltransferase (MTase) gene sharing 72% aa identity to MFRHA. This gene is present in most, but not all, pandemic strains (Wang *et al.,* 2006). However, MTase is only a putative virulence factor as empirical data are lacking.

Summary of virulence factors

In summary, epidemiological studies suggest that *tdh* and *trh* are important virulence factors in *V. parahaemolyticus,* and this conclusion is well supported by multiple laboratory studies. The role of T3SS in *V. parahaemolyticus* pathogenesis is yet to be deciphered, even though their functions are well established for other pathogens. Evidence linking the presence of other factors (e.g. MTase, motility) and *V. parahaemolyticus* infection cases is currently limited. Thus, further research is necessary to gain an understanding of the role of these proposed factors in the ability of *V. parahaemolyticus* to cause human disease.

Isolation, identification and subtyping of *V. parahaemolyticus*

According to the FDA risk assessment document (FDA, 2005), the number of *V. parahaemolyticus* present in oysters is the most critical predictor of the risk of contracting *V. parahaemolyticus* infections. Given the ubiquitous presence of this organism in the marine environment, preventing contamination of shellfish with *V. parahaemolyticus* is virtually impossible. Therefore, to prevent *V. parahaemolyticus* infections and to determine the safety of reopening shellfish growing areas following an outbreak, regulatory agencies and the seafood industry need reliable methods for quantifying pathogenic *V. parahaemolyticus* present in a variety of matrices. Rapid methods are particularly desirable to facilitate timely decision-making on the need to close or ability to reopen shellfish-growing areas.

Methods for isolating and identifying *V. parahaemolyticus*

The standard method for *V. parahaemolyticus* detection described in the FDA's Bacteriological Analytical Manual (FDA, 2004) specifies a three-tube most probable number (MPN) enrichment in alkaline peptone water or alkaline peptone salt broth followed by isolation on TCBS agar. Hydrophobic grid membrane filtration may also be used in conjunction with tryptic soy agar containing magnesium sulfate plus 3% sodium chloride and *V. parahaemolyticus* sucrose agar media. Colonies typical of *V. parahaemolyticus* are then subjected to a number of biochemical tests to distinguish *V. parahaemolyticus* from other marine vibrios. Confirmation can also be performed using PCR to amplify the species-specific thermolabile haemolysin gene, *tlh*. Primer sequences for *tlh* amplification are shown in Table 7.2. Another species-specific marker, *tox*R, has been suggested to provide enhanced specificity over the sole use of *tlh* (Oberbeckmann *et al.,* 2011). Specifically, using 29 *tlh*-positive isolates collected from German Bight (North Sea), three were classified as *V. alginolyticus* based on *rpoB* sequencing. These three isolates were also only weakly positive for *tox*R. The authors concluded that if these weakly positive results were interpreted as negative, *toxR*-PCR is more specific than *tlh*-PCR in speciating *V. parahaemolyticus.* Similarly, *toxR*-PCR was also regarded by Croci *et al.* (2007) to be more reliable than *tlh*-PCR, as the specificity was 100% ($n = 16$), 44% higher than that for *tlh*-PCR. However, the sensitivity of *toxR*-PCR in this study was lower than for *tlh*-PCR, yielding 96% positive signals ($n = 14$) versus 100%. A putative new species-specific marker was identified by Yu *et al.* (2010) who demonstrated 100% sensitivity and specificity using the *irgB* gene from 293 *V. parahaemolyticus* strains representing ATCC, clinical, environment and seafood sources along with 46 strains representing 10 different *Vibrio* species and 17 non-*Vibrio* species. The detection limit was 0.17 pg purified genomic DNA (the equivalent cfu was not determined in this study). Presently,

Table 7.2 Reported sequences of primers used in PCR assays for the detection of total *V. parahaemolyticus* (*tlh*, *toxR*) and pathogenic *V. parahaemolyticus* (*tdh*, *trh*)

Sequence (F, forward direction; R, reverse direction; 5′ → 3′)	Product size (bp)	Reference
***tlh* (sometimes noted as *tl*)**		
F: aaa gcg gat tat gca gaa gca ctg R: gct act ttc tag cat ttt ctc tgc	450	Taniguchi *et al.* (1985, 1986), Bej *et al.* (1999)[a]
toxR		
F: tgt act gtt gaa cgc cta a R: cac gtt ctc ata cga gtg	503	Neogi *et al.* (2010)
F: gtc ttc tga cgc aat cgt tg R: ata cga gtg gtt gct gtc atg	366	Kim *et al.* (1999)
tdh		
F: cca tct gtc cct ttt cct gcc R: cca cta cca ctc tca tat gc	425	Wang *et al.* (2011)
F: gta aag gtc tct gac ttt tgg ac R: tgg aat aga acc ttc atc ttc acc	270	Nishibuchi and Kaper (1985), Bej *et al.* (1999)[a]
F: gca tta ctg tta ctt ata gag t R: att att att aat tat ttt caa ag	700	Yeung *et al.* (2002)
F: atc tgt ccc ttt tcc tgc cc R: ttc ttt gtt gga tat aca cat tac c	457	Williams *et al.* (2004)
trh		
F: cta ctt tgc att cag ttt gc R: ctc tga ttt tgt gaa gac cg	355	Kishishita *et al.* (1992)
F: ttg gct tcg ata ttt tca gta tct R: cat aac aaa cat atg ccc att tcc g	500	Bej *et al.* (1999) [a]

[a]Primer sets described in this study can be used in simultaneous identification of total and pathogenic *V. parahaemolyticus* by multiplex PCR.

isolation of putative *V. parahaemolyticus* colonies on TCBS agar followed by confirmation using biochemical methods or multiplex PCR targeting *tlh*, *tdh* and *trh* remains the standard approach (FDA, 2004).

Thiosulfate-citrate-bile salts-sucrose medium has routinely been used to differentiate species within the *Vibrio* genus according to their ability to ferment sucrose (FDA, 2004). More recent agar formulations incorporate chromogenic substrates to differentiate species based on their differential enzymatic release. For example, indoxyl-β-glucoside and indoxyl-3-galactoside have been used as the chromogenic substrates to differentiate *V. parahaemolyticus* colonies (which appear bluish-green) from those of *V. cholerae* (purple) owing to their differential abilities to produce β-glucosidase and β-galactosidase (Hara-Kudo *et al.*, 2001). Different formulations of chromogenic agar, e.g. Compact Dry VP (Kodaka *et al.*, 2009), Bio-Chrome *Vibrio* medium (Su *et al.*, 2005), and chromID™ (Eddabra *et al.*, 2011) have been evaluated and were reported to perform comparably to or better than TCBS. An advantage of these chromogenic media is that the colouring of the surrounding medium is minimal, thereby facilitating the isolation of particular colonies. However, universal application of chromogenic agar in *V. parahaemolyticus* detection will require rigorous standardization with robust results corroborated by multiple laboratories.

Traditional biochemical tests performed in separate test tubes (e.g. screening for arginine dihydrolase, cellobiose and lactose reactions) are time-consuming and labour intensive. More convenient formats (e.g. api® 20E diagnostic strip) are available to bypass the tedious substrate and glassware preparation. Commercial kits typically contain multiple substrates (generally, carbohydrates) in a dehydrated format, along with pH indicators, all packaged in disposable materials. Users inoculate bacterial cultures into separate wells or tubules, then incubate the inoculated media as directed. Positive and negative results are generally indicated by colour reactions. According to the manufacturers' websites, commercial systems such as the api® 20E and api® NE (BioMérieux), RapID™ NF PLUS System (Remel Inc.), Crystal Enteric/Non-fermenter ID kit (Becton Dickinson), and other semi-automated system for Enterobacteriaceae, e.g. ID32E (Monnet *et al.*, 1994), are claimed to identify *V. parahaemolyticus* and the tests require as little as 4 hours to 1–2 days.

Methods to detect VBNC form of *V. parahaemolyticus*

The existence of VBNC form of *V. parahaemolyticus* warrants application of detection methods that are independent of culturing to ensure the absence of pathogens in food and environmental samples. There is no standard method in the detection of VBNC *V. parahaemolyticus*. The methods described below can be applied for this purpose, although they should be further optimized for this species.

A direct viable count method was originally developed by Kogure's group (1979) for enumeration of viable bacteria in a natural marine environment. This method is based on incubation of samples in a nutrient broth supplemented with an antimicrobial agent – nalidixic acid. The addition of nalidixic acid inhibits DNA synthesis and therefore prevents cell division. Cells continue to grow in the absence of cell division, leading to an elongated and/or fattened shape. Viable elongated cells thus are enumerated by microscopic observation. Yokomaku *et al.* (2000) modified this method by adding glycine to compromise the walls of the growing cells. Relative to dead cells, these glycine-treated viable cells are more easily lysed by a single freeze–thaw treatment and therefore their number can be determined by subtraction. A more convenient method to distinguish live versus dead cells is through the use of commercial tools such as the Live/Dead BacLight™ bacterium viability kit coupled with fluorescence microscopy or flow cytometry (Falcioni *et al.*, 2008). The principle of these kits is based on the premise that the intact cell membrane of viable cells acts as a selective barrier prohibiting the translocation of some dyes such as SYBR Green while the compromised cell membrane of dead cells permits the diffusion of dyes such as propidium iodide into the cell. A potential significant drawback of the methods mentioned above is their inability to differentiate *V. parahaemolyticus* cells in a mixed culture. To circumvent this problem, Griffitt *et al.* (2011) developed a fluorescence visualization technique called recognition of individual gene fluorescence *in situ* hybridization (RING-FISH). This technique targets *tlh* for the enumeration of *V. parahaemolyticus*, coupled with the Kogure method to distinguish viable from dead cells.

Although routine growth media may not resuscitate the non-culturable form of *V. parahaemolyticus* (Mizunoe *et al.*, 2000), it may be possible to apply conditions to induce conversion from the non-culturable state to a normal state prior to culturing. However, the application of a single enrichment step in alkaline peptone water at 37°C may fail to recover VBNC or otherwise injured cells. The addition of supplements, shifting of growth temperature or the application of a multistep enrichment procedure, (e.g. culturing samples in nutrient broth at least twice) may be necessary to resuscitate or allow reproduction of VBNC or injured cells (Mizunoe *et al.*, 2000; Wong *et al.*, 2004). However, inclusion of a multistep enrichment procedure would inevitably prolong detection time in the laboratory. Microscopy (as described above) and molecular techniques such as reverse transcription-PCR probably are the most appropriate tools currently available for relatively rapid detection of the VBNC form of *V. parahaemolyticus*.

Methods to differentiate pathogenic and non-pathogenic *V. parahaemolyticus*

Kanagawa haemolytic test

Traditionally, beta haemolysis on Wagatsuma blood agar, i.e. KP, is the most important phenotypic test for pathogenic *V. parahaemolyticus* strains. The standard blood agar medium that is widely available for clinical purposes may not be suitable for this test because TDH, the *V. parahaemolyticus* haemolysin responsible for KP, binds specifically to GT1 and GD1a gangliosides that are not generally present in the horse blood typically present in standard blood agar medium (Takeda *et al.*, 2011). To make 1 l of the Wagatsuma blood agar, the following ingredients are needed: 70 g NaCl, 3 g yeast extract, 10 g peptone, 15 g agar, 5 g K_2HPO_4, 10 g mannitol, 0.001 g crystal violet, 20% fresh (within 24 hours of withdrawal) human or rabbit blood in anticoagulant (FDA, 2004; Wang *et al.*, 2011).

Although KP has been used as an important pathogenic criterion, this haemolysis assay is less able to accurately distinguish pathogenic *V. parahaemolyticus* strains from the more prevalent non-pathogenic ones relative to molecular methods. For example, Hongping *et al.* (2011) found that 88.66% of clinical isolates were positive for the presence of *tdh* gene, while 83.61% exhibited KP, suggesting the presence of *tdh* might be more indicative of virulence than measurement of haemolytic activity. Canizalez-Roman *et al.* (2011) also showed that KP exhibited low sensitivity (75.6%) and specificity (52.4%) for identifying *tdh*-positive *V. parahaemolyticus*. Indeed, relative to most genotypic methods, phenotypic characterization is more influenced by physical experimental conditions such as incubation time and temperature. The need for fresh blood for preparing the Wagatsuma agar is also inconvenient. Hence, when assessing the pathogenic potential of *V. parahaemolyticus*, many studies applied molecular techniques to detect the virulence gene(s) instead of conducting the Kanagawa haemolytic test (DePaola *et al.*, 2003b; Fabbro *et al.*, 2010; Sobrinho Pde *et al.*, 2011).

Detection of *tdh* and *trh* genes

The KP phenotype is strongly correlated with the *tdhA* (*tdh2*) or, simply, *tdh* gene. There is no *in vitro* assay for the haemolytic activity encoded by *trh*. Screening for the presence of both virulence factors is used to differentiate *V. parahaemolyticus* strains predicted to be pathogenic. To detect *tdh* or *trh*, *V. parahaemolyticus* isolates are subjected to PCR or colony hybridization using gene probes. Previously used primer sequences for the amplification of *tdh* and *trh* are shown in Table 7.2. Real-time PCR techniques which have the advantage of providing quantitative data while avoiding the need for electrophoresis, have been evaluated by several groups (Rizvi and Bej, 2010; Robert-Pillot *et al.*, 2010; Kang *et al.*, 2011). A newly developed technique, loop-mediated isothermal amplification, shows promise in replacing PCR to amplify *toxR*, *tdh*, and *trh* (Chen and Ge, 2010; Yamazaki *et al.*, 2010). With specially designed primer sets and the incorporation of chromogenic chemicals, this method eliminates both the heat denaturation process and the need for electrophoresis required with conventional PCR techniques.

Currently, the criterion used to determine whether a particular *V. parahaemolyticus* isolate belongs to the pathogenic subtype is based on the presence of *tdh* and/or *trh*. Through application of technologies that enable detection of multiple genes, such as multiplex PCR or DNA microarrays (Kim *et al.*, 2010), it is possible to simultaneously screen for additional virulence markers such as genes coding for the apparatus and effectors of the T3SS2.

Studies towards new phenotypic tests for pathogenic *V. parahaemolyticus*

Since some clinical *V. parahaemolyticus* strains posses neither *tdh* nor *trh*, other factors are predicted to be involved in pathogenesis. Given the relative ease and convenience of conducting phenotypic analyses as compared with genetic analysis or animal experiments, phenotypic characterization is often the first step in assessing the pathogenic potential of a microorganism. One classical test is the haemolysis assay which is an important virulence assay for some pathogens (Nizet, 2002; Wiles *et al.*, 2008). However, KP is an exception as it requires fresh blood. Another example may be to exploit the link between a foodborne pathogen's infectious dose with its ability to survive under acidic conditions such as those encountered in the human stomach (Gorden and Small, 1993; Waterman and Small, 1996; Audia *et al.*, 2001). In the quest to elucidate physiological differences between pathogenic and non-pathogenic *V. parahaemolyticus* strains, some assays have been conducted to compare the fitness (viability) and virulence between subgroups. The experimental designs used in the studies described below were chosen to represent typical

(i) conditions that may be experienced by *V. parahaemolyticus* in its natural habitat; (ii) conditions experienced during food processing or preservation treatment; and (iii) conditions experienced during infection of animal and human hosts.

Acid sensitivity

Generally *V. parahaemolyticus* is regarded to be sensitive to acid, a property which probably in part contributes to its high infectious dose. The emergence of the O3:K6 pandemic clone, which appeared to be more virulent than other previously identified *V. parahaemolyticus* serotypes, renewed interest in studying the survival of *V. parahaemolyticus* under acidic conditions. Wong *et al.* (1998) reported a 1-log reduction for a clinical *V. parahaemolyticus* strain following 60 minutes of exposure to pH 4.4. Koga *et al.* (1999) reported a 2-log reduction for another clinical isolate following 60 minutes of exposure to pH 2.0. Both groups tested *V. parahaemolyticus* cells that had been harvested in log phase. On the other hand, Yeung and Boor (2004) examined the effect of acid on survival capabilities and cytotoxicities of different *V. parahaemolyticus* strains at two growth phases. Survival of acid-challenged stationary-phase cells differed by up to 3 log cfu/ml among the 25 isolates tested. However, no differences in acid sensitivity were found between clinical and food isolates, suggesting response to acid cannot be reliably used to discriminate pathogenic *V. parahaemolyticus.*

Metal resistance

If a particular *V. parahaemolyticus* subtype has enhanced survival capabilities in an environment, the risk for human exposure to this subtype may be increased. Increased possibility of human exposure may underlie the emergence of pandemic clones. Ju *et al.* (2006) compared viabilities of 26 *V. parahaemolyticus* isolates recovered from clinical and food sources following exposure to high magnesium (300 mM) or iron-limiting (<0.96 mM) conditions. Within the diverse collection of strains examined in this study, viabilities under either condition did not uniformly differ between the clinical and food isolates, nor between *tdh*-positive and -negative isolates. Therefore, the authors concluded that neither phenotypic characteristic can reliably differentiate potentially pathogenic *V. parahaemolyticus* strains. Nonetheless, some environmental but not the clinical strains have been shown to tolerate arsenate (Chauhan *et al.*, 1997). It remains to be determined if resistance to this metal can be used as an indicator for environmental strains incapable of causing human diseases.

Siderophore production

Iron is essential for bacterial survival. Iron storage proteins, such as the haem-containing bacterioferritins and the haem-free ferritins, are widespread in bacteria (Andrews, 1998). In order to acquire iron from their surroundings, many pathogens, such as *E. coli* and *Klebsiella pneumoniae*, produce iron (particularly ferric iron) chelators called siderophores. Under iron limiting conditions, *V. parahaemolyticus* produces a novel siderophore called vibrioferrin (Yamamoto *et al.*, 1994). The vibrioferrin is produced when the medium contains a limited amount of iron (e.g. 0–1 μM $FeCl_3$), or when an iron chelator, EDDA, is added to iron rich (e.g. 5 μM) medium (Yamamoto *et al.*, 1999). These investigators showed that clinical isolates ($n=44$; 37 KP+) had higher vibrioferrin levels in spent supernatant than that secreted by either food or environmental isolates ($n=63$; all KP-negative) when grown in medium containing 0.2 μM of $FeCl_3$. However, no single 'cut-off' level of secreted vibrioferrin was defined to separate the clinical and non-clinical isolates into two distinct clusters.

Other *in vitro* methods studied

In the quest to find effective methods capable of differentiating pathogenic subtypes (as defined by capability of causing disease, i.e. clinical vs. non-clinical strains), many phenotypic methods have been evaluated. These include biofilm formation, capsule size, protease production, and motility (Yeung *et al.*, 2002; Mahoney *et al.*, 2010). Virulence characteristics of clinical and environmental isolates under the effect of temperature was also studied (Mahoney *et al.*, 2010). Overall, these methods were found to lack discriminatory power as clinical and non-clinical subpopulations tend to overlap, at least partially, under the experimental conditions. As with phenotypic characterization, Yeung *et al.* (2007) showed that measurement of lactate dehydrogenase release as an indicator for cytotoxicity could not definitively identify pathogenic strains, even when using five different cell lines. Therefore, genotypic methods (e.g. the presence of *tdh*) remain the standard for defining pathogenic *V. parahaemolyticus*.

Subtyping of *V. parahaemolyticus*

Serotyping

To track the transmission of foodborne outbreaks, detection at the species level is not sufficient. Therefore, various types of methods have been used to identify *V. parahaemolyticus* at the subspecies level. The traditional and most widely applied subtyping method for *V. parahaemolyticus* is serotyping. There are currently 76 recognized serotypes (FDA, 2004) based on the O and K antigens. The H antigen is common to all *V. parahaemolyticus* strains and therefore offers little value in differentiation. Interestingly, O3:K6 strains isolated after 1996 show different genetic patterns than those isolated before, based on results from some molecular methods, e.g. arbitrarily primed-PCR, ribotyping and pulsed-field gel electrophoresis (PFGE), (Chowdhury *et al.*, 2000; Matsumoto *et al.*, 2000). Using such methods, strains bearing serotypes O4:K68 and O1:KUT, which have been responsible for gastroenteritis in India and other Southeast Asian countries, are deemed genetically similar to the 'new' O3:K6 strains.

Molecular fingerprinting

Molecular methods have become important tools in taxonomic and epidemiologic studies. For *V. parahaemolyticus*, the applications of molecular subtyping methods have focused on the determination of the temporal and spatial relationships among *V. parahaemolyticus* subtypes, and the differentiation of pandemic versus non-pandemic clonal groups (reviewed by Staley and Harwood, 2010). The sequence-based methods, such as multilocus sequence typing (MLST), could also be used to evaluate the inter- and intra-specific phylogenetic relationship (Montieri *et al.*, 2010; Yan *et al.*, 2011).

According to a review summarizing 16 studies, examples of molecular subtyping methods applied to *V. parahaemolyticus* include genomic restriction enzyme analysis (DGREA), randomly amplified polymorphic DNA (RAPD)-PCR, group-specific (GS)-PCR, PCR on *orf8* of phage 237, repetitive-sequence-based PCRs, ribotyping, and PFGE (Staley and Harwood, 2010). Some methods generate a pattern derived from examination of the whole genome. For example, PFGE and DGREA are based on the principle of restriction fragment length polymorphism (RFLP), in which genomic DNA is cut with restriction enzyme(s) generating fragments of different sizes. Following electrophoresis, fragments are separated according to their sizes yielding a profile resembling a 'fingerprint' for each strain. Similarly, patterns generated by ribotyping reflect restriction fragment length polymorphisms in rRNA operons. RAPD-PCR and repetitive-sequence-based PCR generate profiles from amplified DNA fragments of different sizes.

To discriminate between pandemic and non-pandemic strains, it appears that PFGE of *NotI* digested genomic DNA is more appropriate than ribotyping using *Eco*RI, RAPD-PCR, GS-PCR and *orf8*-PCR because it produces more diverse patterns, and groups the pandemic strains in closely related clusters (Hlady and Klontz, 1996; Yeung *et al.*, 2002). Other studies also validated the reproducibility and discriminatory nature of PFGE to achieve this goal (Okura *et al.*, 2003; Staley and Harwood, 2010). PFGE also has the advantage of reflecting the genetic diversity among strains, an important piece of information not provided by GS-PCR or *orf8*-PCR. Analogous to employing *tdh*-PCR to identify pathogenic *V. parahaemolyticus*, GS-PCR or *orf8*-PCR was developed as a diagnostic tool to identify the pandemic clonal group (Nasu *et al.*, 2000; Okura *et al.*, 2004). In some studies, repetitive-sequence-based PCR was found to be slightly more discriminatory than PFGE as it generated greater numbers of different patterns and was less likely to yield untypeable results caused by DNA degradation (Marshall *et al.*, 1999; Wong and Lin, 2001).

Approaches to mitigate foodborne *V. parahaemolyticus* infections

At the pre-harvest level

Relaying and depuration

Common approaches to reduce or remove bacterial contaminants in shellfish include relaying and depuration (National Shellfish Sanitation Program, 2009). In the relaying process, contaminated shellfish are transferred from restricted areas to approved areas for natural biological purification. Depuration is similar to relaying, except that contaminated shellfish are transferred to a controlled aquatic environment instead of a natural ambient environment. The seawater of the depuration tank may be disinfected by UV irradiation, ozone, etc. These methods can be used to reduce natural toxins (Lassus *et al.*, 2000) and *E. coli* (Dore and Lees, 1995). Motes and DePaola (1996) also showed that offshore suspension relaying reduced 3–4 log of *V. vulnificus* in Gulf Coast oysters with less than 6% oyster mortality. Depuration with UV-sterilized seawater reduced *V. parahaemolyticus* populations in Pacific oysters by > 3 log MPN/g when the system was held at 5°C and operated for 96–144 hours (Su *et al.*, 2010). However, neither method is considered adequate to completely remove *V. parahaemolyticus* (Croci *et al.*, 2002; FDA, 2005; Wang *et al.*, 2010). The inadequacy of these conventional methods to remove *V. parahaemolyticus* has been hypothesized to result from interactions between a ligand of *V. parahaemolyticus* (e.g. flagella, other surface molecules) and corresponding receptor sites on oyster tissues. Wang *et al.* (2010) showed that *V. parahaemolyticus* tend to colonize the digestive glands and gills rather than in adductor muscle and mantle cilia, suggesting that some mechanisms exist to allow for better adhesion to the former sites. Le Guyader *et al.* (2006) demonstrated that a ligand–receptor interaction occurs between norovirus and oysters. It is yet to be determined if specific binding occurs for *V. parahaemolyticus*. Ren and Su (2006) evaluated the viability of *Vibrio* spp. in oysters treated with electrolysed oxidizing (EO) water (30 ppm chlorine, pH 2.82, and 1131 mV oxidation-reduction potential). While treatment with EO water for 4 to 6 hours reduced *V. parahaemolyticus* and *V. vulnificus* numbers in oyster samples by 1.05 to 1.13 log, a prolonged exposure with EO water led to oyster mortality. This study did not determine the effects of EO water on the sensory attributes of oysters.

Other innovative methods

Since conventional pre-harvest mitigation methods are not completely effective in eliminating *V. parahaemolyticus* from seafood, and given the ubiquitous nature of *V. parahaemolyticus* in the marine environment, methods to reduce their pre-harvest densities will require some innovative approaches. The methods described below have not been fully evaluated with *V. parahaemolyticus* in oysters, and more research is expected to follow in the near future.

The first approach targets the colonization step of *V. parahaemolyticus* on seafood tissues. If the ability to colonize is diminished, the initial count of *V. parahaemolyticus* in freshly harvested seafood is expected to be low. One way to disrupt the colonization abilities is to interfere with quorum sensing among bacterial cells (Zhang and Dong, 2004). Some bacteria are known to produce unique compounds called autoinducers (Nasser and Reverchon, 2007) that are believed to play an important role in cell-to-cell communication. Defoirdt *et al.* (2006) showed that both the natural and synthetic form of a compound, brominated furanone, disrupted the quorum sensing induced by autoinducers for several *Vibrio* spp. Subsequently, the possibility of using this compound to reduce the colonization of *Vibrio* spp. in gnotobiotic brine shrimps (*Artemia franiscana*) was tested. The addition of 5–20 mg/l furanone (5Z)-4-bromo-5-(bromomethylene)-3-butyl-2(5*H*)-furanone to the pond was found to enhance the survival of shrimps challenged with the shrimp pathogens *V. campbelli*, *V. harveyi* and *V. parahaemolyticus*. This innovative approach, though promising, is expensive owing to the need to extract, isolate and purify or synthesize the appropriate signal molecules.

Another approach capitalizes on the competitive exclusion (CE) nature among organisms. Initially described by Nurmi and Rantala (1973), the underlying principle of CE is the intentional addition of organisms to outcompete the target organisms. In the presence of these CE organisms, the target may be inhibited owing to shortage of nutrients and binding sites. In addition, some CE organisms may produce metabolic by-products that are inhibitory to their competitors, thereby further suppressing their growth. For example, nisin, a peptide produced by *Lactococcus lactis* subsp. *lactis*, can cause bactericidal and bacteriostatic effects towards some foodborne pathogens (Cleveland *et al.*, 2001). In theory, any organism that can effectively compete with the target organisms can be used as a CE organism. Thus, natural flora from food products can be exploited to control specific foodborne pathogens. Previous studies supported the concept of using CE organisms to outcompete foodborne pathogens in the pre-harvest stage for a variety of food products. For example, neonatal calves administered non-pathogenic *E. coli* had reduced faecal shedding of enterohaemorrhagic *E. coli* O111:NM and O26:H11 (Zhao *et al.*, 2003). Toro *et al.* (2005) showed that chicken treated with probiotic bacteria (*Lactobacillus rhamnosus, L. plantarum, L. delbrueckii* subsp. *bulgaricus, L. acidophilus, Bifidobacterium bifidum, Streptococcus thermophilus,* and *Enterococcus faecium*) exhibited enhanced resistance to *Salmonella* infections. Indigenous *Campylobacter jejuni* isolates (non-pathogenic to human) from chicks were more effective in colonizing chicken caeca than those from human origins (Chen and Stern, 2001), suggesting that it is possible to use chicken *C. jejuni* isolates as CE bacteria against human isolates in poultry.

CE has also emerged as an effective biocontrol method in seafood products. For example, antagonistic effects were found between *Pseudomonas fluorescens* AH2 and *V. anguillarum* in fish, when the concentration of the CE bacteria was 100 to 1000 times greater than that of the fish pathogen (Gram *et al.*, 1999). Three of five lactic acid bacteria reduced the ability of fish pathogens such as *Aeromonas salmonicida* and *Carnobacterium piscicola* to adhere on fish intestinal mucus (Balcazar *et al.*, 2007). Rico-Mora *et al.* (1998) isolated a native microorganism in diatom, designated as SK-05, which grew rapidly in organic-poor substrates. This microorganism helped prevent the establishment of *V. alginolyticus* in diatom. The authors concluded that the inhibitory effect was due solely to the growth capability of SK-05 on the substrate, as it did not show any bacteriostatic or bactericidal activity.

Though the aforementioned approaches have demonstrated different degrees of potential in food safety, research is lacking on their usefulness in preventing *V. parahaemolyticus* infections. As a result, detection of *V. parahaemolyticus* and prevention of movement of contaminated foods into the food supply are currently the most effective methods that can be applied at the pre-harvest stage for preventing human illness.

At the post-harvest level

Storage temperature control

Levels of *V. parahaemolyticus* in freshly harvested seafood are generally below the recommended level of 10,000 cfu/g (FDA, 2005). Yeung *et al.* (2003) determined that freshly harvested oysters from New York Long Island Sound contained 66–1700 *V. parahaemolyticus* MPN/g. The risk of individuals contracting *V. parahaemolyticus* infection through the consumption of these oysters should be low provided that the numbers of *V. parahaemolyticus* decline or remain steady along the 'farm-to-table' continuum. The length of time at growth permissive temperature is a decisive factor of the multiplication of the bacteria.

After storing live oysters at 26°C for 24 hours, Kaufman *et al.* (2003) reported 13- to 26-fold increases in *V. parahaemolyticus* numbers. Gooch *et al.* (2002) also reported as much as a 790-fold increase in *V. parahaemolyticus* numbers under these conditions. Since *V. parahaemolyticus* can multiply rapidly at ambient temperatures, temperature control is a critical post-harvest mitigation method to prevent pathogenic strains from reaching infectious dose levels. Under approved US guidelines, shellfish may remain unrefrigerated at an average temperature of ≥ 27°C for as long as 10 hours after harvest (National Shellfish Sanitation Program, 2009). Retail oysters that have been held at ambient temperatures for 10 hours could harbour 10–100 times more *V. parahaemolyticus* than present at harvest (Gooch *et al.*, 2001). On the other hand, cold storage or freezing of oysters offered a small reduction (up to 2 log reduction) of *V. parahaemolyticus* numbers (FDA, 2005).

Thermal and non-thermal treatments

Apart from temperature control, several post-harvest processing treatments can be employed to reduce *V. parahaemolyticus* to low or undetectable levels. If shellfish are consumed fully cooked, the risk of contracting *V. parahaemolyticus* infections will be drastically reduced because this organism is sensitive to heat. For example, a > 7-log reduction in viable cell count was reported when *V. parahaemolyticus* cells were heat treated at 55°C for 2 minutes (Nishikawa *et al.*, 1993). The time required to cause a 1-log reduction in viable cell count (D-value) at 47°C was 2.03 minutes (Wong *et al.*, 2002). Therefore, a mild heat treatment is effective in substantially reducing *V. parahaemolyticus* numbers. Andrews *et al.* (2000) and Johnston and Brown (2002) recommended using a low temperature pasteurization treatment (e.g. 50°C for up to 15 minutes) to reduce the risk of *Vibrio* infections from raw oyster consumption. Despite its killing effect on *V. parahaemolyticus*, heat treatment unavoidably alters oyster quality in a manner deemed unfavourable by many consumers. Therefore, to meet the demand for fresh, raw oysters, other post-harvest methods have been developed, as described below.

High-pressure processing (sometimes called high-hydrostatic-pressure processing) and irradiation have been evaluated against *V. parahaemolyticus* survival. It appears that *V. parahaemolyticus* is sensitive to both treatments. With oyster samples, a pressure of 300 MPa for 180 seconds is sufficient to achieve a > 5-log reduction in *V. parahaemolyticus* numbers, including the O3:K6 strains (Cook, 2003). Ma and Su (2011) validated that the treatment of 293 MPa for 120 s at 8 ± 1°C was sufficient to reduce the numbers by > 3.5 log; yielding a shelf life of Pacific oysters of 6–8 days when stored at 5°C or 16–18 days when stored in ice. Other non-thermal methods include ionizing irradiation and X-ray treatment. While high doses of gamma irradiation have the undesirable effect of killing oysters, lower, non-lethal doses (e.g. < 3.0 kGy) can substantially reduce *V. parahaemolyticus* numbers without significantly altering the textural and sensory properties of oysters (Jakabi *et al.*, 2003). The D_{10} value (the radiation dose required to eliminate 1 log of a bacterial population) was found to be 0.03–0.04 kGy, which was generally lower than for the Gram-positive bacteria (Dion *et al.*, 1994). Similarly, 5 kGy of X-rays did not kill oysters, yet this level was sufficient to reduce the numbers of *V. parahaemolyticus* residing in whole oyster by 6 log (Mahmoud and Burrage, 2009). The same study also showed that a lower dose (2 kGy) could achieve the same effect on half-shell oysters.

At the consumer level

Consumer behaviour also represents an integral component affecting numbers of *V. parahaemolyticus* infections. As with many other foodborne illnesses, the risk of *V. parahaemolyticus* infections can be reduced through application of hygienic food handling techniques. Such practices include holding seafood at a sufficiently low temperature to prevent *V. parahaemolyticus* from multiplying rapidly, avoiding cross-contamination of ready-to-eat foods with uncooked seafood, and, if possible, thorough cooking of raw seafood to destroy *V. parahaemolyticus*. These precautions are important at the consumer level, as mishandled raw or undercooked oysters can cause illness even if they had originally been harvested from approved sites (Daniels *et al.*, 2000a,b). Last but not least, notices or warnings, such as those displayed conspicuously in food service establishments that serve raw oysters in some US states, may help educate consumers about the risks associated with consuming raw oysters.

Conclusions

Vibrio parahaemolyticus is ubiquitous in the global marine environments. The ability of some strains to cause human diseases through food consumption is highly correlated to the presence of genes encoding thermostable direct haemolysin (*tdh*) and/or TDH-related haemolysin (*trh*). Recent studies also suggest the Type III secretion system on the smaller chromosome may have a role in *V. parahaemolyticus* infection. Though the larger chromosome also contains another Type III secretion system, its implication in *V.*

parahaemolyticus infection is unclear as virtually all *V. parahaemolyticus* strains, including the non-pathogenic ones, harbour this secretion system. Empirical and epidemiological studies indicate that a high infectious dose is required for *V. parahaemolyticus* infection. Therefore, maintaining the numbers of *V. parahaemolyticus* at low levels is critical to prevent *V. parahaemolyticus*-mediated foodborne illness. This can be achieved by chilling freshly harvested seafood immediately post-harvest, or by applying thermal and non-thermal treatments. Different detection methods can be readily adopted to determine the levels of total and pathogenic *V. parahaemolyticus*, including the non-culturable ones, in food and the environment. A better understanding of the prevalence, physiology, pathogenesis of *V. parahaemolyticus* and of consumers' behavioural patterns towards seafood consumption will enhance our ability to assess the risk of *V. parahaemolyticus* infection and to efficiently manage this foodborne pathogen.

References

Abbott, S.L., Powers, C., Kaysner, C.A., Takeda, Y., Ishibashi, M., Joseph, S.W., and Janda, J.M. (1989). Emergence of a restricted bioserovar of *Vibrio parahaemolyticus* as the predominant cause of *Vibrio*-associated gastroenteritis on the West Coast of the United States and Mexico. J. Clin. Microbiol. *27*, 2891–2893.

Abendroth, J., Kreger, A.C., and Hol, W.G. (2009). The dimer formed by the periplasmic domain of EpsL from the Type 2 Secretion System of *Vibrio parahaemolyticus*. J. Struct. Biol. *168*, 313–322.

Alam, A., Miller, K.A., Chaand, M., Butler, J.S., and Dziejman, M. (2011). Identification of *Vibrio cholerae* type III secretion system effector proteins. Infect. Immun. *79*, 1728–1740.

Andrews, L.S., Park, D.L., and Chen, Y.P. (2000). Low temperature pasteurization to reduce the risk of vibrio infections from raw shell-stock oysters. Food Addit. Contam. *17*, 787–791.

Andrews, S.C. (1998). Iron storage in bacteria. Adv. Microb. Physiol. *40*, 281–351.

Audia, J.P., Webb, C.C., and Foster, J.W. (2001). Breaking through the acid barrier: an orchestrated response to proton stress by enteric bacteria. Int. J. Med. Microbiol. *291*, 97–106.

Baba, K., Shirai, H., Terai, A., Kumagai, K., Takeda, Y., and Nishibuchi, M. (1991). Similarity of the *tdh* gene-bearing plasmids of *Vibrio cholerae* non-O1 and *Vibrio parahaemolyticus*. Microb. Pathog. *10*, 61–70.

Baffone, W., Citterio, B., Vittoria, E., Casaroli, A., Campana, R., Falzano, L., and Donelli, G. (2003). Retention of virulence in viable but non-culturable halophilic *Vibrio* spp. Int. J. Food Microbiol. *89*, 31–39.

Balcazar, J.L., Vendrell, D., de Blas, I., Ruiz-Zarzuela, I., Girones, O., and Muzquiz, J.L. (2007). *In vitro* competitive adhesion and production of antagonistic compounds by lactic acid bacteria against fish pathogens. Vet. Microbiol. *122*, 373–380.

Baross, J., and Liston, J. (1968). Isolation of *Vibrio parahaemolyticus* from the Northwest Pacific. Nature *217*, 1263–1264.

Baross, J., and Liston, J. (1970). Occurrence of *Vibrio parahaemolyticus* and related hemolytic vibrios in marine environments of Washington State. Appl. Microbiol. *20*, 179–186.

Barrow, G.I., and Miller, D.C. (1974). Growth studies on *Vibrio parahaemolyticus* in relation to pathogenicity. In International Symposium of *Vibrio parahaemolyticus*, Fujino, T., Sakaguchi, G., Sakazaki, R., and Takeda, Y., eds. (Saikon Publishing Co. Ltd, Tokyo), pp. 205–210.

Bates, T.C., and Oliver, J.D. (2004). The viable but nonculturable state of Kanagawa positive and negative strains of *Vibrio parahaemolyticus*. J. Microbiol. *42*, 74–79.

Bej, A.K., Patterson, D.P., Brasher, C.W., Vickery, M.C., Jones, D.D., and Kaysner, C.A. (1999). Detection of total and hemolysin-producing *Vibrio parahaemolyticus* in shellfish using multiplex PCR amplification of *tl*, *tdh* and *trh*. J. Microbiol. Methods *36*, 215–225.

Bhattacharjee, R.N., Park, K.S., Kumagai, Y., Okada, K., Yamamoto, M., Uematsu, S., Matsui, K., Kumar, H., Kawai, T., Iida, T., *et al.* (2006). VP1686, a *Vibrio* type III secretion protein, induces toll-like receptor-independent apoptosis in macrophage through NF-kappaB inhibition. J. Biol. Chem. *281*, 36897–36904.

Blake, P.A., Weaver, R.E., and Hollis, D.G. (1980). Diseases of humans (other than cholera) caused by vibrios. Annu. Rev. Microbiol. *34*, 341–367.

Broberg, C.A., Zhang, L., Gonzalez, H., Laskowski-Arce, M.A., and Orth, K. (2010). A *Vibrio* effector protein is an inositol phosphatase and disrupts host cell membrane integrity. Science *329*, 1660–1662.

Burdette, D.L., Yarbrough, M.L., Orvedahl, A., Gilpin, C.J., and Orth, K. (2008). *Vibrio parahaemolyticus* orchestrates a multifaceted host cell infection by induction of autophagy, cell rounding, and then cell lysis. Proc. Natl. Acad. Sci. U.S.A. *105*, 12497–12502.

Burdette, D.L., Yarbrough, M.L., and Orth, K. (2009). Not without cause: *Vibrio parahaemolyticus* induces acute autophagy and cell death. Autophagy *5*, 100–102.

Canizalez-Roman, A., Flores-Villaseñor, H., Zazueta-Beltran, J., Muro-Amador, S., and Leon-Sicairos, N. (2011). Comparative evaluation of a chromogenic agar medium-PCR protocol with a conventional method for isolation of *Vibrio parahaemolyticus* strains from environmental and clinical samples. Can. J. Microbiol. *57*, 136–142.

CDC (2010a). Preliminary FoodNet data on the incidence of infection with pathogens transmitted commonly through food –10 States, 2009. MMWR Morb. Mortal. Wkly Rep. April 16, 2010. *59*, 418–422.

CDC (2010b). Surveillance for Foodborne Disease Outbreaks – United States, 2007. MMWR Morb. Mortal. Wkly. Rep. August 13, 2010. *59*, 973–977.

Chan, S.F., and Chan, Z.C.Y. (2008). A review of foodborne disease outbreaks from 1996 to 2005 in Hong Kong and its implication son food safety promotion. J. Food Saf. *28*, 276–299.

Chao, G., Jiao, X., Zhou, X., Yang, Z., Pan, Z., Huang, J., Zhou, L., and Qian, X. (2009). Systematic functional pandemic strain-specific genes, three genomic islands, two T3SSs in foodborne, and clinical *Vibrio parahaemolyticus* isolates in China. Foodborne Pathog. Dis. *6*, 689–698.

Chao, Y., and Vogel, J. (2010). The role of Hfq in bacterial pathogens. Curr. Opin. Microbiol. *13*, 24–33.

Chauhan, B.S., Shaikh, N.M., De, S.P., Sil, S., Ganguly, P.K., and Kumar, R. (1997). Arsenate resistance as a possible marker in the differentiation of environmental and clinical isolates of *Vibrio parahaemolyticus*. Zentralbl Bakteriol. *285*, 486–490.

Chen, H.C., and Stern, N.J. (2001). Competitive exclusion of heterologous *Campylobacter* spp. in chicks. Appl. Environ. Microbiol. *67*, 848–851.

Chen, S., and Ge, B. (2010). Development of a *toxR*-based loop-mediated isothermal amplification assay for detecting *Vibrio parahaemolyticus*. BMC Microbiol. *10*, 41.

Chen, Y., Liu, X.M., Yan, J.W., Li, X.G., Mei, L.L., Mao, Q.F., and Ma, Y. (2010). Foodborne pathogens in retail oysters in south China. Biomed. Environ. Sci. *23*, 32–36.

Chiu, S.W., Chen, S.Y., and Wong, H.C. (2008). Localization and expression of MreB in *Vibrio parahaemolyticus* under different stresses. Appl. Environ. Microbiol. *74*, 7016–7022.

Chowdhury, N.R., Chakraborty, S., Ramamurthy, T., Nishibuchi, M., Yamasaki, S., Takeda, Y., and Nair, G.B. (2000). Molecular evidence of clonal *Vibrio parahaemolyticus* pandemic strains. Emerg. Infect. Dis. *6*, 631–636.

Clarke, S.C., Haigh, R.D., Freestone, P.P., and Williams, P.H. (2003). Virulence of enteropathogenic *Escherichia coli*, a global pathogen. Clin. Microbiol. Rev. *16*, 365–378.

Cleveland, J., Montville, T.J., Nes, I.F., and Chikindas, M.L. (2001). Bacteriocins: safe, natural antimicrobials for food preservation. Int. J. Food. Microbiol. *71*, 1–20.

Colwell, R.R., West, A.P.A., Maneval, D., Remmers, E.F., Elliot, E.L., and Carlson, N.E. (1984). Ecology of pathogenic vibrios in Chesapeake Bay. In Vibrios in the environment, Colwell, R.R., ed. (John Wiley & Sons, Inc., New York), pp. 367–387.

Cook, D.W. (2003). Sensitivity of *Vibrio* species in phosphate-buffered saline and in oysters to high-pressure processing. J. Food Prot. *66*, 2276–2282.

Cook, D.W., Oleary, P., Hunsucker, J.C., Sloan, E.M., Bowers, J.C., Blodgett, R.J., and Depaola, A. (2002). *Vibrio vulnificus* and *Vibrio parahaemolyticus* in U.S. retail shell oysters: a national survey from June 1998 to July 1999. J. Food Prot. *65*, 79–87.

Coutard, F., Crassous, P., Droguet, M., Gobin, E., Colwell, R.R., Pommepuy, M., and Hervio-Heath, D. (2007a). Recovery in culture of viable but nonculturable *Vibrio parahaemolyticus*: regrowth or resuscitation? ISME J. *1*, 111–120.

Coutard, F., Lozach, S., Pommepuy, M., and Hervio-Heath, D. (2007b). Real-time reverse transcription-PCR for transcriptional expression analysis of virulence and housekeeping genes in viable but nonculturable *Vibrio parahaemolyticus* after recovery of culturability. Appl. Environ. Microbiol. 73, 5183–5189.

Croci, L., Suffredini, E., Cozzi, L., and Toti, L. (2002). Effects of depuration of molluscs experimentally contaminated with *Escherichia coli*, *Vibrio cholerae* O1 and *Vibrio parahaemolyticus*. J. Appl. Microbiol. *92*, 460–465.

Croci, L., Suffredini, E., Cozzi, L., Toti, L., Ottaviani, D., Pruzzo, C., Serratore, P., Fischetti, R., Goffredo, E., Loffredo, G., *et al.* (2007). Comparison of different biochemical and molecular methods for the identification of *Vibrio parahaemolyticus*. J. Appl. Microbiol. *102*, 229–237.

Daniels, N.A., MacKinnon, L., Bishop, R., Altekruse, S., Ray, B., Hammond, R.M., Thompson, S., Wilson, S., Bean, N.H., Griffin, P.M., *et al.* (2000a). *Vibrio parahaemolyticus* infections in the United States, 1973–1998. J. Infect. Dis. *181*, 1661–1666.

Daniels, N.A., Ray, B., Easton, A., Marano, N., Kahn, E., McShan, A.L., 2nd, Del Rosario, L., Baldwin, T., Kingsley, M.A., Puhr, N.D., *et al.* (2000b). Emergence of a new *Vibrio parahaemolyticus* serotype in raw oysters: A prevention quandary. JAMA *284*, 1541–1545.

Defoirdt, T., Crab, R., Wood, T.K., Sorgeloos, P., Verstraete, W., and Bossier, P. (2006). Quorum sensing-disrupting brominated furanones protect the gnotobiotic brine shrimp *Artemia franciscana* from pathogenic *Vibrio harveyi*, *Vibrio campbellii*, and *Vibrio parahaemolyticus* isolates. Appl. Environ. Microbiol. 72, 6419–6423.

DePaola, A., Kaysner, C.A., Bowers, J., and Cook, D.W. (2000). Environmental investigations of *Vibrio parahaemolyticus* in oysters after outbreaks in Washington, Texas, and New York (1997 and 1998). Appl. Environ. Microbiol. *66*, 4649–4654.

DePaola, A., Nordstrom, J.L., Bowers, J.C., Wells, J.G., and Cook, D.W. (2003a). Seasonal abundance of total and pathogenic *Vibrio parahaemolyticus* in Alabama oysters. Appl. Environ. Microbiol. *69*, 1521–1526.

DePaola, A., Ulaszek, J., Kaysner, C.A., Tenge, B.J., Nordstrom, J.L., Wells, J., Puhr, N., and Gendel, S.M. (2003b). Molecular, serological, and virulence characteristics of *Vibrio parahaemolyticus* isolated from environmental, food, and clinical sources in North America and Asia. Appl. Environ. Microbiol. *69*, 3999–4005.

DePaola, A., Jones, J.L., Woods, J., Burkhardt, W., 3rd, Calci, K.R., Krantz, J.A., Bowers, J.C., Kasturi, K., Byars, R.H., Jacobs, E., *et al.* (2010). Bacterial and viral pathogens in live oysters: 2007 United States market survey. Appl. Environ. Microbiol. *76*, 2754–2768.

DeVinney, R., Gauthier, A., Abe, A., and Finlay, B.B. (1999). Enteropathogenic *Escherichia coli*: a pathogen that inserts its own receptor into host cells. Cell Mol. Life. Sci. *55*, 961–976.

Dion, P., Charbonneau, R., and Thibault, C. (1994). Effect of ionizing dose rate on the radioresistance of some food pathogenic bacteria. Can. J. Microbiol. *40*, 369–374.

Dore, W.J., and Lees, D.N. (1995). Behavior of *Escherichia coli* and male-specific bacteriophage in environmentally contaminated bivalve molluscs before and after depuration. Appl. Environ. Microbiol. *61*, 2830–2834.

Dziejman, M., Serruto, D., Tam, V.C., Sturtevant, D., Diraphat, P., Faruque, S.M., Rahman, M.H., Heidelberg, J.F., Decker, J., Li, L., *et al.* (2005). Genomic characterization of non-O1, non-O139 *Vibrio cholerae* reveals genes for a type III secretion system. Proc. Natl. Acad. Sci. U.S.A. *102*, 3465–3470.

Eddabra, R., Piemont, Y., and Scheftel, J.M. (2011). Evaluation of a new chromogenic medium, chromID Vibrio, for the isolation and presumptive identification of *Vibrio cholerae* and *Vibrio parahaemolyticus* from human clinical specimens. Eur. J. Clin. Microbiol. Infect. Dis. *30*, 733–737.

Fabbri, A., Falzano, L., Frank, C., Donelli, G., Matarrese, P., Raimondi, F., Fasano, A., and Fiorentini, C. (1999). *Vibrio parahaemolyticus* thermostable direct hemolysin modulates cytoskeletal organization and calcium homeostasis in intestinal cultured cells. Infect. Immun. *67*, 1139–1148.

Fabbro, C., Cataletto, B., and Del Negro, P. (2010). Detection of pathogenic *Vibrio parahaemolyticus* through biochemical and molecular-based methodologies in coastal waters of the Gulf of Trieste (North Adriatic Sea). FEMS Microbiol. Lett. *307*, 158–164.

Falcioni, T., Papa, S., Campana, R., Manti, A., Battistelli, M., and Baffone, W. (2008). State transitions of *Vibrio parahaemolyticus* VBNC cells evaluated by flow cytometry. Cytometry B. Clin. Cytom. *74*, 272–281.

Faruque, S.M., Biswas, K., Udden, S.N., Ahmad, Q.S., Sack, D.A., Nair, G.B. and Mekalanos, J.J. (2006). Transmissibility of cholera: *in vivo*-formed biofilms and their relationship to infectivity and persistence in the environment. Proc. Natl. Acad. Sci. U.S.A. *103*, 6350–6355.

Fishbein, M., Wentz, B., Landry, W.L., and MacEachern, B. (1974). *Vibrio parahaemolyticus* isolates in the U.S.: 1696–1972. In International Symposium of *Vibrio parahaemolyticus*, Fujino, T., Sakaguchi, G., Sakazaki, R., and Takeda, Y., eds. (Saikon Publishing Co. Ltd, Tokyo), pp. 53–58.

Food and Drug Administration. (2004). Bacteriological analytical manual. Available at: http://www.fda.gov/Food/ScienceResearch/LaboratoryMethods/BacteriologicalAnalyticalManualBAM/UCM070830

Food and Drug Administration. (2005). Quantitative risk assessment on the public health impact of *Vibrio parahaemolyticus* in raw oysters. Available at: http://www.fda.gov/Food/ScienceResearch/ResearchAreas/RiskAssessmentSafetyAssessment/ucm050421.htm

Franzon, V.L., Barker, A., and Manning, P.A. (1993). Nucleotide sequence encoding the mannose-fucose-resistant hemagglutinin of *Vibrio cholerae* O1 and construction of a mutant. Infect. Immun. *61*, 3032–3037.

Fujino, T. (1974). Discovery of *Vibrio parahaemolyticus*. In International Symposium of *Vibrio parahaemolyticus*, Fujino, T., Sakaguchi, G., Sakazaki, R., and Takeda, Y., eds. (Saikon Publishing Co. Ltd, Tokyo), pp. 1–4.

Gardel, C.L., and Mekalanos, J.J. (1996). Alterations in *Vibrio cholerae* motility phenotypes correlate with changes in virulence factor expression. Infect. Immun. *64*, 2246–2255.

Gelli, D.S., Tachibana, T., Sakuma, H. (1979). Ocorrência de *Vibrio parahaemolyticus, Escherichia coli* e de bactérias mesófilas em ostras. Revista do Instituto Adolfo Lutz.. *39*, 61–66.

Gode-Potratz, C.J., Chodur, D.M., and McCarter, L.L. (2010). Calcium and iron regulate swarming and type III secretion in *Vibrio parahaemolyticus*. J. Bacteriol. *192*, 6025–6038.

Gode-Potratz, C.J., Kustusch, R.J., Breheny, P.J., Weiss, D.S., and McCarter, L.L. (2011). Surface sensing in *Vibrio parahaemolyticus* triggers a programme of gene expression that promotes colonization and virulence. Mol. Microbiol. *79*, 240–263.

Gooch, J.A., DePaola, A., Kaysner, C.A., and Marshall, D.L. (2001). Evaluation of two direct plating methods using nonradioactive probes for enumeration of *Vibrio parahaemolyticus* in oysters. Appl. Environ. Microbiol. *67*, 721–724.

Gooch, J.A., DePaola, A., Bowers, J., and Marshall, D.L. (2002). Growth and survival of *Vibrio parahaemolyticus* in postharvest American oysters. J. Food Prot. *65*, 970–974.

Gorden, J., and Small, P.L. (1993). Acid resistance in enteric bacteria. Infect. Immun. *61*, 364–367.

Gram, L., Melchiorsen, J., Spanggaard, B., Huber, I., and Nielsen, T.F. (1999). Inhibition of *Vibrio anguillarum* by *Pseudomonas fluorescens* AH2, a possible probiotic treatment of fish. Appl. Environ. Microbiol. *65*, 969–973.

Griffitt, K.J., Noriea, N.F., 3rd, Johnson, C.N., and Grimes, D.J. (2011). Enumeration of *Vibrio parahaemolyticus* in the viable but nonculturable state using direct plate counts and recognition of individual gene fluorescence in situ hybridization. J. Microbiol. Methods *85*, 114–118.

Hackney, C.R., Kleeman, E.G. Ray, B., and Speck, M.L. (1980). Adherence as a method of differentiating virulent and avirulent strains of *Vibrio parahaemolyticus*. Appl. Environ. Microbiol. *40*, 652–658.

Hara-Kudo, Y., Nishina, T., Nakagawa, H., Konuma, H., Hasegawa, J., and Kumagai, S. (2001). Improved method for detection of *Vibrio parahaemolyticus* in seafood. Appl. Environ. Microbiol. *67*, 5819–5823.

Henke, J.M., and Bassler, B.L. (2004). Quorum sensing regulates type III secretion in *Vibrio harveyi* and *Vibrio parahaemolyticus*. J. Bacteriol. *186*, 3794–3805.

Hiyoshi, H., Kodama, T., Iida, T., and Honda, T. (2010). Contribution of *Vibrio parahaemolyticus* virulence factors to cytotoxicity, enterotoxicity, and lethality in mice. Infect. Immun. *78*, 1772–1780.

Hlady, W.G. (1997) *Vibrio* infections associated with raw oyster consumption in Florida, 1981–1994. J. Food Protect. *60*, 353–357.

Hlady, W.G., and Klontz, K.C. (1996). The epidemiology of *Vibrio* infections in Florida, 1981–1993. J. Infect. Dis. *173*, 1176–1183.

Hoashi, K., Ogata, K., Taniguchi, H., Yamashita, H., Tsuji, K., Mizuguchi, Y., and Ohtomo, N. (1990). Pathogenesis of *Vibrio parahaemolyticus*: intraperitoneal and orogastric challenge experiments in mice. Microbiol. Immunol. *34*, 355–366.

Honda, T., Ni, Y.X., and Miwatani, T. (1988). Purification and characterization of a hemolysin produced by a clinical isolate of Kanagawa phenomenon-negative *Vibrio parahaemolyticus* and related to the thermostable direct hemolysin. Infect. Immun. *56*, 961–965.

Honda, T., Abad-Lapuebla, M.A., Ni, Y.X., Yamamoto, K., and Miwatani, T. (1991). Characterization of a new thermostable direct haemolysin produced by a Kanagawa-phenomenon-negative clinical isolate of *Vibrio parahaemolyticus*. J. Gen. Microbiol. *137*, 253–259.

Honda, T., Ni, Y., Miwatani, T., Adachi, T., and Kim, J. (1992). The thermostable direct hemolysin of *Vibrio parahaemolyticus* is a pore-forming toxin. Can. J. Microbiol. *38*, 1175–1180.

Hondo, S., Goto, I., Minematsu, I., Ikeda, N., Asano, N., Ishibashi, M., Kinoshita, Y., Nishibuchi, N., Honda, T., and Miwatani, T. (1987). Gastroenteritis due to Kanagawa negative *Vibrio parahaemolyticus*. Lancet *1*, 331–332.

Hongping, W., Jilun, Z., Ting, J., Yixi, B., and Xiaoming, Z. (2011). Insufficiency of the Kanagawa hemolytic test for detecting pathogenic *Vibrio parahaemolyticus* in Shanghai, China. Diagn. Microbiol. Infect. Dis. *69*, 7–11.

Iida, T., and Yamamoto, K. (1990). Cloning and expression of two genes encoding highly homologous hemolysins from a Kanagawa phenomenon-positive *Vibrio parahaemolyticus* T4750 strain. Gene *93*, 9–15.

Iida, T., Park, K.S., Suthienkul, O., Kozawa, J., Yamaichi, Y., Yamamoto, K., and Honda, T. (1998). Close proximity of the *tdh, trh* and *ure* genes on the chromosome of *Vibrio parahaemolyticus*. Microbiology *144*, 2517–2523.

Iijima, Y., Yamada, H., and Shinoda, S. (1981). Adherence of *Vibrio parahaemolyticus* and its relation to pathogenicity. Can. J. Microbiol. *27*, 1252–1259.

Izutsu, K., Kurokawa, K., Tashiro, K., Kuhara, S., Hayashi, T., Honda, T., and Iida, T. (2008). Comparative genomic analysis using microarray demonstrates a strong correlation between the presence of the 80-kilobase pathogenicity island and pathogenicity in Kanagawa phenomenon-positive *Vibrio parahaemolyticus* strains. Infect. Immun. *76*, 1016–1023.

Jackson, H. (1974). Temperature relationships of *Vibrio parahaemolyticus*. In International Symposium of *Vibrio parahaemolyticus*, Fujino, T., Sakaguchi, G., Sakazaki, R., and Takeda, Y., eds. (Saikon Publishing Co. Ltd, Tokyo), pp. 139–145.

Jakabi, M., Gelli, D.S., Torre, J.C., Rodas, M.A., Franco, B.D., Destro, M.T., and Landgrafi, M. (2003). Inactivation by ionizing radiation of *Salmonella* enteritidis, *Salmonella* infantis, and *Vibrio parahaemolyticus* in oysters (*Crassostrea brasiliana*). J. Food Prot. *66*, 1025–1029.

Jiang, X., and Chai, T.J. (1996). Survival of *Vibrio parahaemolyticus* at low temperatures under starvation conditions and subsequent resuscitation of viable, nonculturable cells. Appl. Environ. Microbiol. *62*, 1300–1305.

Johnston, M.D., and Brown, M.H. (2002). An investigation into the changed physiological state of *Vibrio* bacteria as a survival mechanism in response to cold temperatures and studies on their sensitivity to heating and freezing. J. Appl. Microbiol. *92*, 1066–1077.

Ju, C.H., Yeung, P.S., Oesterling, J., Seigerman, D.A., and Boor, K.J. (2006). *Vibrio parahaemolyticus* growth under low-iron conditions and survival under high-magnesium conditions. J. Food Prot. *69*, 1040–1045.

Kamruzzaman, M., Udden, S.M., Cameron, D.E., Calderwood, S.B., Nair, G.B., Mekalanos, J.J., and Faruque, S.M. (2010) Quorum-regulated biofilms enhance the development of conditionally viable, environmental *Vibrio cholerae*. Proc. Natl. Acad. Sci. U.S.A. *107*, 1588–93.

Kang, M.H., Kim, I.W., Lee, D.W., Yoo, M.S., Han, S.H., and Yoon, B.S. (2011). Development of a rapid detection method to detect *tdh* gene in *Vibrio parahaemolyticus* using 2-step ultrarapid real-time polymerase chain reaction. Diagn. Microbiol. Infect. Dis. *69*, 21–29.

Kaufman, G.E., Bej, A.K., Bowers, J., and DePaola, A. (2003). Oyster-to-oyster variability in levels of *Vibrio parahaemolyticus*. J. Food Prot. *66*, 125–129.

Kaysner, C.A., Abeyta, C., Jr., Trost, P.A., Wetherington, J.H., Jinneman, K.C., Hill, W.E., and Wekell, M.M. (1994). Urea hydrolysis can predict the potential pathogenicity of *Vibrio parahaemolyticus* strains isolated in the Pacific Northwest. Appl. Environ. Microbiol. *60*, 3020–3022.

Kim, D.H., Lee, B.K., Kim, Y.D., Rhee, S.K., and Kim, Y.C. (2010). Detection of representative enteropathogenic bacteria, *Vibrio* spp., pathogenic *Escherichia coli*, *Salmonella* spp., *Shigella* spp., and *Yersinia enterocolitica*, using a virulence factor gene-based oligonucleotide microarray. J. Microbiol. *48*, 682–688.

Kim, Y.B., Okuda, J., Matsumoto, C., Takahashi, N., Hashimoto, S., and Nishibuchi, M. (1999). Identification of *Vibrio parahaemolyticus* strains at the species level by PCR targeted to the *toxR* gene. J. Clin. Microbiol. *37*, 1173–1177.

Kirov, S.M. (2003). Bacteria that express lateral flagella enable dissection of the multifunctional roles of flagella in pathogenesis. FEMS Microbiol. Lett. *224*, 151–159.

Kirs, M., Depaola, A., Fyfe, R., Jones, J.L., Krantz, J., Van Laanen, A., Cotton, D., and Castle, M. (2011). A survey of oysters (*Crassostrea gigas*) in New Zealand for *Vibrio parahaemolyticus* and *Vibrio vulnificus*. Int. J. Food Microbiol. *147*, 149–153

Kishishita, M., Matsuoka, N., Kumagai, K., Yamasaki, S., Takeda, Y., and Nishibuchi, M. (1992). Sequence variation in the thermostable direct hemolysin-related hemolysin (*trh*) gene of *Vibrio parahaemolyticus*. Appl. Environ. Microbiol. *58*, 2449–2457.

Kodaka, H., Teramura, H., Mizuochi, S., Saito, M., and Matsuoka, H. (2009). Evaluation of the Compact Dry VP method for screening raw seafood for total *Vibrio parahaemolyticus*. J. Food Prot. 72, 169–173.

Kodama, T., Hiyoshi, H., Gotoh, K., Akeda, Y., Matsuda, S., Park, K.S., Cantarelli, V.V., Iida, T., and Honda, T. (2008). Identification of two translocon proteins of *Vibrio parahaemolyticus* type III secretion system 2. Infect. Immun. *76*, 4282–4289.

Kodama, T., Gotoh, K., Hiyoshi, H., Morita, M., Izutsu, K., Akeda, Y., Park, K.S., Cantarelli, V.V., Dryselius, R., Iida, T., *et al.* (2010). Two regulators of *Vibrio parahaemolyticus* play important roles in enterotoxicity by controlling the expression of genes in the Vp-PAI region. PLoS One *5*, e8678.

Koga, T., Sakamoto, F., Yamoto, A., and Takumi, K. (1999). Acid adaptation induces cross-protection against some environmental stresses in *Vibrio parahaemolyticus*. J. Gen. Appl. Microbiol. *45*, 155–161.

Kogure, K., Simidu, U., and Taga, N. (1979). A tentative direct microscopic method for counting living marine bacteria. Can. J. Microbiol. *25*, 415–420.

Kueh, C.S., and Chan, K.Y. (1985). Bacteria in bivalve shellfish with special reference to the oyster. J. Appl. Bacteriol. *59*, 41–47.

Lassus, P., Bardouil, M., Massselin, P., Naviner, M., and Truquet, P. (2000). Comparative efficiencies of different non-toxic microalgal diets in detoxification of PSP-contaminated oysters (*Crassostrea gigas* Thunberg). J. Nat. Toxins. *9*, 1–12.

Le Guyader, F., Loisy, F., Atmar, R.L., Hutson, A.M., Estes, M.K., Ruvoen-Clouet, N., Pommepuy, M., and Le Pendu, J. (2006). Norwalk virus-specific binding to oyster digestive tissues. Emerg. Infect. Dis *12*, 931–936.

Lesmana, M., Subekti, D., Simanjuntak, C.H., Tjaniadi, P., Campbell, J.R., and Oyofo, B.A. (2001). *Vibrio parahaemolyticus* associated with cholera-like diarrhea among patients in North Jakarta, Indonesia. Diagn Microbiol. Infect. Dis. *39*, 71–75.

Lin, Z., Kumagai, K., Baba, K., Mekalanos, J.J., and Nishibuchi, M. (1993). *Vibrio parahaemolyticus* has a homolog of the *Vibrio cholerae toxRS* operon that mediates environmentally induced regulation of the thermostable direct hemolysin gene. J Bacteriol. *175*, 3844–3855.

Liverman, A.D., Cheng, H.C., Trosky, J.E., Leung, D.W., Yarbrough, M.L., Burdette, D.L., Rosen, M.K., and Orth, K. (2007). Arp2/3-independent assembly of actin by *Vibrio* type III effector VopL. Proc. Natl. Acad. Sci. U.S.A. *104*, 17117–17122.

Ma, L., and Su, Y.C. (2011). Validation of high pressure processing for inactivating *Vibrio parahaemolyticus* in Pacific oysters (*Crassostrea gigas*). Int. J. Food. Microbiol. *144*, 469–474.

Mahmoud, B.S., and Burrage, D.D. (2009). Inactivation of *Vibrio parahaemolyticus* in pure culture, whole live and half shell oysters (*Crassostrea virginica*) by X-ray. Lett. Appl. Microbiol. *48*, 572–578.

Mahoney, J.C., Gerding, M.J., Jones, S.H., and Whistler, C.A. (2010). Comparison of the pathogenic potentials of environmental and clinical *Vibrio parahaemolyticus* strains indicates a role for temperature regulation in virulence. Appl. Environ. Microbiol. *76*, 7459–7465.

Makino, K., Oshima, K., Kurokawa, K., Yokoyama, K., Uda, T., Tagomori, K., Iijima, Y., Najima, M., Nakano, M., Yamashita, A., *et al.* (2003). Genome sequence of *Vibrio parahaemolyticus*: a pathogenic mechanism distinct from that of *V. cholerae*. Lancet *361*, 743–749.

Marlovits, T.C., and Stebbins, C.E. (2010). Type III secretion systems shape up as they ship out. Curr. Opin. Microbiol. *13*, 47–52.

Marshall, S., Clark, C.G., Wang, G., Mulvey, M., Kelly, M.T., and Johnson, W.M. (1999). Comparison of molecular methods for typing *Vibrio parahaemolyticus*. J. Clin. Microbiol. *37*, 2473–2478.

Martinez, O., Rodriguez-Calleja, J.M., Santos, J.A., Otero, A., and Garcia-Lopez, M.L. (2009). Foodborne and indicator bacteria in farmed molluscan shellfish before and after depuration. J. Food. Prot. *72*, 1443–1449.

Matlawska-Wasowska, K., Finn, R., Mustel, A., O'Byrne, C.P., Baird, A.W., Coffey, E.T., and Boyd, A. (2010). The *Vibrio parahaemolyticus* Type III Secretion Systems manipulate host cell MAPK for critical steps in pathogenesis. BMC Microbiol. *10*, 329.

Matsuda, S., Kodama, T., Okada, N., Okayama, K., Honda, T., and Iida, T. (2010). Association of *Vibrio parahaemolyticus* thermostable direct hemolysin with lipid rafts is essential for cytotoxicity but not hemolytic activity. Infect. Immun. *78*, 603–610.

Matsumoto, C., Okuda, J., Ishibashi, M., Iwanaga, M., Garg, P., Rammamurthy, T., Wong, H.C., Depaola, A., Kim, Y.B., Albert, M.J., *et al.* (2000). Pandemic spread of an O3:K6 clone of *Vibrio parahaemolyticus* and emergence of related strains evidenced by arbitrarily primed PCR and *toxRS* sequence analyses. J. Clin. Microbiol. *38*, 578–585.

McLaughlin, J.B., DePaola, A., Bopp, C.A., Martinek, K.A., Napolilli, N.P., Allison, C.G., Murray, S.L., Thompson, E.C., Bird, M.M., and Middaugh, J.P. (2005). Outbreak of *Vibrio parahaemolyticus* gastroenteritis associated with Alaskan oysters. N. Engl. J. Med. *353*, 1463–1470.

Mead, P.S., Slutsker, L., Dietz, V., McCaig, L.F., Bresee, J.S., Shapiro, C., Griffin, P.M., and Tauxe, R.V. (1999). Food-related illness and death in the United States. Emerg. Infect. Dis. *5*, 607–625.

Miyamoto, Y., Kato, T., Obara, Y., Akiyama, S., Takizawa, K., and Yamai, S. (1969). *In vitro* hemolytic characteristic of *Vibrio parahaemolyticus*: its close correlation with human pathogenicity. J. Bacteriol. *100*, 1147–1149.

Mizunoe, Y., Wai, S.N., Ishikawa, T., Takade, A., and Yoshida, S. (2000). Resuscitation of viable but nonculturable cells of *Vibrio parahaemolyticus* induced at low temperature under starvation. FEMS Microbiol. Lett. *186*, 115–120.

Molenda, J.R., Johnson, W.G., Fishbein, M., Wentz, B., Mehlman, I.J., and Dadisman, T.A., Jr. (1972). *Vibrio parahaemolyticus* gastroenteritis in Maryland: laboratory aspects. Appl. Microbiol. *24*, 444–448.

Monnet, D., Lafay, D., Desmonceaux, M., Boeufgras, J.M., Allard, F., and Freney, J. (1994). Evaluation of a semi-automated 24-hour commercial system for identification of Enterobacteriaceae and other Gram-negative bacteria. Eur. J. Clin. Microbiol. Infect. Dis. *13*, 424–430.

Montieri, S., Suffredini, E., Ciccozzi, M., and Croci, L. (2010). Phylogenetic and evolutionary analysis of *Vibrio parahaemolyticus* and *Vibrio alginolyticus* isolates based on *toxR* gene sequence. New Microbiol. *33*, 359–372.

Motes, M.L., and DePaola, A. (1996). Offshore suspension relaying to reduce levels of *Vibrio vulnificus* in oysters (*Crassostrea virginica*). Appl. Environ. Microbiol. *62*, 3875–3877.

Nagayama, K., Yamamoto, K., Mitawani, T., and Honda, T. (1995). Characterisation of a haemolysin related to Vp-TDH produced by a Kanagawa phenomenon-negative clinical isolate of *Vibrio parahaemolyticus*. J. Med. Microbiol. *42*, 83–90.

Nakaguchi, Y., and Nishibuchi, M. (2005). The promoter region rather than its downstream inverted repeat sequence is responsible for low-level transcription of the thermostable direct hemolysin-related hemolysin (*trh*) gene of *Vibrio parahaemolyticus*. J. Bacteriol. *187*, 1849–1855.

Nakano, M., Takahashi, A., Su, Z., Harada, N., Mawatari, K., and Nakaya, Y. (2008). Hfq regulates the expression of the thermostable direct hemolysin gene in *Vibrio parahaemolyticus*. BMC Microbiol. *8*, 155.

Nasser, W., and Reverchon, S. (2007). New insights into the regulatory mechanisms of the LuxR family of quorum sensing regulators. Anal. Bioanal. Chem. *387*, 381–390.

Nasu, H., Iida, T., Sugahara, T., Yamaichi, Y., Park, K.S., Yokoyama, K., Makino, K., Shinagawa, H., and Honda, T. (2000). A filamentous phage associated with recent pandemic *Vibrio parahaemolyticus* O3:K6 strains. J. Clin. Microbiol. *38*, 2156–2161.

National Shellfish Sanitation Program. (2009). NSSP Guide for the control of molluscan Shellfish. Section II Chapter VIII Control of Shellfish Harvesting. Available at: http://www.fda.gov/Food/FoodSafety/Product-SpecificInformation/Seafood/FederalStatePrograms/NationalShellfishSanitationProgram/UCM047104.

Neogi, S.B., Chowdhury, N., Asakura, M., Hinenoya, A., Haldar, S., Saidi, S.M., Kogure, K., Lara, R.J., and Yamasaki, S. (2010). A highly sensitive and specific multiplex PCR assay for simultaneous detection of *Vibrio cholerae*, *Vibrio parahaemolyticus* and *Vibrio vulnificus*. Lett Appl. Microbiol. *51*, 293–300.

Nishibuchi, M., and Kaper, J.B. (1985). Nucleotide sequence of the thermostable direct hemolysin gene of *Vibrio parahaemolyticus*. J. Bacteriol. *162*, 558–564.

Nishibuchi, M., and Kaper, J.B. (1990). Duplication and variation of the thermostable direct haemolysin (*tdh*) gene in *Vibrio parahaemolyticus*. Mol. Microbiol. *4*, 87–99.

Nishibuchi, M., and Kaper, J.B. (1995). Thermostable direct hemolysin gene of *Vibrio parahaemolyticus*: a virulence gene acquired by a marine bacterium. Infect. Immun. *63*, 2093–2099.

Nishibuchi, M., Taniguchi, T., Misawa, T., Khaeomanee-Iam, V., Honda, T., and Miwatani, T. (1989). Cloning and nucleotide sequence of the gene (*trh*) encoding the hemolysin related to the thermostable direct hemolysin of *Vibrio parahaemolyticus*. Infect. Immun. *57*, 2691–2697.

Nishibuchi, M., Kumagai, K., and Kaper, J.B. (1991). Contribution of the *tdh1* gene of Kanagawa phenomenon-positive *Vibrio parahaemolyticus* to production of extracellular thermostable direct hemolysin. Microb. Pathog. *11*, 453–460.

Nishikawa, Y., Ogasawara, J., and Kimura, T. (1993). Heat and acid sensitivity of motile *Aeromonas*: a comparison with other food-poisoning bacteria. Int. J. Food. Microbiol *18*, 271–278.

Nizet, V. (2002). Streptococcal beta-hemolysins: genetics and role in disease pathogenesis. Trends. Microbiol. *10*, 575–580.

Nolan, C.M., Ballard, J., Kaysner, C.A., Lilja, J.L., Williams, L.P., Jr., and Tenover, F.C. (1984). *Vibrio parahaemolyticus* gastroenteritis. An outbreak associated with raw oysters in the Pacific northwest. Diagn Microbiol. Infect. Dis. *2*, 119–128.

Normanno, G., Parisi, A., Addante, N., Quaglia, N.C., Dambrosio, A., Montagna, C., and Chiocco, D. (2006). *Vibrio parahaemolyticus*, *Vibrio vulnificus* and microorganisms of fecal origin in mussels (*Mytilus galloprovincialis*) sold in the Puglia region (Italy). Int. J. Food Microbiol. *106*, 219–222.

Nurmi, E., and Rantala, M. (1973). New aspects of *Salmonella* infection in broiler production. Nature *241*, 210–211.

Oberbeckmann, S., Wichels, A., Maier, T., Kostrzewa, M., Raffelberg, S., and Gerdts, G. (2011). A polyphasic approach for the differentiation of environmental *Vibrio* isolates from temperate waters. FEMS Microbiol. Ecol. *75*, 145–162.

Okada, K., Iida, T., Kita-Tsukamoto, K., and Honda, T. (2005). Vibrios commonly possess two chromosomes. J. Bacteriol. *187*, 752–757.

Okada, N., Iida, T., Park, K.S., Goto, N., Yasunaga, T., Hiyoshi, H., Matsuda, S., Kodama, T., and Honda, T. (2009). Identification and characterization of a novel type III secretion system in *trh*-positive *Vibrio parahaemolyticus* strain TH3996 reveal genetic lineage and diversity of pathogenic machinery beyond the species level. Infect. Immun. *77*, 904–913.

Okuda, J., Ishibashi, M., Abbott, S.L., Janda, J.M., and Nishibuchi, M. (1997). Analysis of the thermostable direct hemolysin (*tdh*) gene and the *tdh*-related hemolysin (*trh*) genes in urease-positive strains of *Vibrio parahaemolyticus* isolated on the West Coast of the United States. J. Clin. Microbiol. *35*, 1965–1971.

Okuda, J., and Nishibuchi, M. (1998). Manifestation of the Kanagawa phenomenon, the virulence-associated phenotype, of *Vibrio parahaemolyticus* depends on a particular single base change in the promoter of the thermostable direct haemolysin gene. Mol. Microbiol. *30*, 499–511.

Okura, M., Osawa, R., Iguchi, A., Arakawa, E., Terajima, J., and Watanabe, H. (2003). Genotypic analyses of *Vibrio parahaemolyticus* and development of a pandemic group-specific multiplex PCR assay. J. Clin. Microbiol. *41*, 4676–4682.

Okura, M., Osawa, R., Iguchi, A., Takagi, M., Arakawa, E., Terajima, J., and Watanabe, H. (2004). PCR-based identification of pandemic group *Vibrio parahaemolyticus* with a novel group-specific primer pair. Microbiol. Immunol. *48*, 787–790.

Oliver, J.D. (1995). The viable but non-culturable state in the human pathogen *Vibrio vulnificus*. FEMS Microbiol. Lett. *133*, 203–208.

Oliver, J.D. (2010). Recent findings on the viable but nonculturable state in pathogenic bacteria. FEMS Microbiol. Rev. *34*, 415–425.

Oliver, J.D., and Kaper, J.B. (2007). *Vibrio* species. In Food Microbiology: Fundamentals and Frontiers 3rd Ed., Doyle M.P. and Beuchat, L.R., eds. (ASM Press, Washington, DC), pp. 343–379.

Osawa, R., Okitsu, T., Morozumi, H., and Yamai, S. (1996). Occurrence of urease-positive *Vibrio parahaemolyticus* in Kanagawa, Japan, with specific reference to presence of thermostable direct hemolysin (TDH) and the TDH-related-hemolysin genes. Appl. Environ. Microbiol. *62*, 725–727.

Osawa, R., and Yamai, S. (1996). Production of thermostable direct hemolysin by *Vibrio parahaemolyticus* enhanced by conjugated bile acids. Appl. Environ. Microbiol. *62*, 3023–3025.

Pan, T.M., Wang, T.K., Lee, C.L., Chien, S.W., and Horng, C.B. (1997). Food-borne disease outbreaks due to bacteria in Taiwan, 1986 to 1995. J. Clin. Microbiol. *35*, 1260–1262.

Park, K.S., Ono, T., Rokuda, M., Jang, M.H., Okada, K., Iida, T., and Honda, T. (2004). Functional characterization of two type III secretion systems of *Vibrio parahaemolyticus*. Infect. Immun. *72*, 6659–6665.

Parveen, S., Hettiarachchi, K.A., Bowers, J.C., Jones, J.L., Tamplin, M.L., McKay, R., Beatty, W., Brohawn, K., Dasilva, L.V., and Depaola, A. (2008). Seasonal distribution of total and pathogenic *Vibrio parahaemolyticus* in Chesapeake Bay oysters and waters. Int. J. Food. Microbiol. *128*, 354–361.

Pineyro, P., Zhou, X., Orfe, L.H., Friel, P.J., Lahmers, K., and Call, D.R. (2010). Development of two animal models to study the function of *Vibrio parahaemolyticus* type III secretion systems. Infect. Immun. *78*, 4551–4559.

ProMED-mail. (2006a). *Vibrio parahaemolyticus*, North America: background. ProMED-mail 2006; 17 Aug: 20060717.1963. Available at: http://www.promedmail.org

ProMED-mail. (2006b). *Vibrio parahaemolyticus*, shellfish –USA (Washington, New York). ProMED-mail 2006; 5 Aug: 20060805.2175. Available at: http://www.promedmail.org

ProMED-mail. (2006c). *Vibrio parahaemolyticus*, shellfish –USA (Multistate). ProMED-mail 2006; 12 Aug: 20060812.2264. Available at: http://www.promedmail.org

ProMED-mail. (2006d). *Vibrio parahaemolyticus*, shellfish –USA (Washington, New York). ProMED-mail 2006; 21 Aug: 20060821.2347. Available at: http://www.promedmail.org.

ProMED-mail. (2007). *Vibrio parahaemolyticus*, oysters – USA (Washington): international alert. ProMED-mail 2007; 13 Aug: 20070813.2637. Available at: http://www. promedmail.org

ProMED-mail. (2008). *Vibrio parahaemolyticus*, oysters – USA (Maryland ex New Jersey). ProMED-mail 2008; 21 Aug: 20080821.2608. Available at: http://www.promedmail.org

ProMED-mail. (2010). *Vibrio parahaemolyticus*, oysters –USA (Washington). ProMED-mail 2010; 3 Aug: 20100803.2604. Available at: http://www.promedmail.org

Qadri, F., Alam, M.S., Nishibuchi, M., Rahman, T., Alam, N.H., Chisti, J., Kondo, S., Sugiyama, J., Bhuiyan, N.A., Mathan, M.M., *et al.* (2003). Adaptive and inflammatory immune responses in patients infected with strains of *Vibrio parahaemolyticus*. J. Infect. Dis. *187*, 1085–1096.

Raimondi, F., Kao, J.P., Fiorentini, C., Fabbri, A., Donelli, G., Gasparini, N., Rubino, A., and Fasano, A. (2000). Enterotoxicity and cytotoxicity of *Vibrio parahaemolyticus* thermostable direct hemolysin in *in vitro* systems. Infect. Immun. *68*, 3180–3185.

Ren, T., and Su, Y.C. (2006). Effects of electrolyzed oxidizing water treatment on reducing *Vibrio parahaemolyticus* and *Vibrio vulnificus* in raw oysters J. Food Prot. *69*, 1829–1834.

Rico-Mora, R., Voltolina, D., and Villaescusa-Celaya, J.A. (1998) Biological control of *Vibrio alginolyticus* in *Skeletonema costatum* (Bacillariophyceae) cultures. Aquacult. Eng. *19*, 1–6.

Ristori, C.A., Iaria, S.T., Gelli, D.S., and Rivera, I.N. (2007). Pathogenic bacteria associated with oysters (*Crassostrea brasiliana*) and estuarine water along the south coast of Brazil. Int. J Environ. Health. Res. *17*, 259–269.

Rizvi, A.V., and Bej, A.K. (2010). Multiplexed real-time PCR amplification of *tlh*, *tdh* and *trh* genes in *Vibrio parahaemolyticus* and its rapid detection in shellfish and Gulf of Mexico water. Antonie Van Leeuwenhoek *98*, 279–290.

Robert-Pillot, A., Copin, S., Gay, M., Malle, P., and Quilici, M.L. (2010). Total and pathogenic *Vibrio parahaemolyticus* in shrimp: fast and reliable quantification by real-time PCR. Int. J. Food Microbiol. *143*, 190–197.

Sakurai, J., Matsuzaki, A., Takeda, Y., and Miwatani, T. (1974). Existence of two distinct hemolysins in *Vibrio parahaemolyticus*. Infect. Immun. *9*, 777–780.

Sanyal, S.C., and Sen, P.C. (1974). Human volunteer study on the pathogenicity of *Vibrio parahaemolyticus*. In International Symposium of *Vibrio parahaemolyticus*, Fujino, T., Sakaguchi, G., Sakazaki, R., and Takeda, Y., eds. (Saikon Publishing Co. Ltd, Tokyo), pp. 227–230.

Scallan, E., Hoekstra, R.M., Angulo, F.J., Tauxe, R.V., Widdowson, M.A., Roy, S.L., Jones, J.L., and Griffin, P.M. (2011). Foodborne illness acquired in the United States – major pathogens. Emerg. Infect. Dis. *17*, 7–15.

Schroeder, G.N., and Hilbi, H. (2008). Molecular pathogenesis of *Shigella* spp.: controlling host cell signaling, invasion, and death by type III secretion. Clin Microbiol Rev *21*, 134–156.

Schuch, R., Sandlin, R.C., and Maurelli, A.T. (1999). A system for identifying post-invasion functions of invasion genes: requirements for the Mxi-Spa type III secretion pathway of *Shigella flexneri* in intercellular dissemination. Mol. Microbiol. *34*, 675–689.

Shiaris, M.P., Rex, A.C., Pettibone, G.W., Keay, K., McManus, P., Rex, M.A., Ebersole, J., and Gallagher, E. (1987). Distribution of indicator bacteria and *Vibrio parahaemolyticus* in sewage-polluted intertidal sediments. Appl. Environ. Microbiol. *53*, 1756–1761.

Shimohata, T., and Takahashi, A. (2010). Diarrhea induced by infection of *Vibrio parahaemolyticus*. J. Med. Invest. *57*, 179–182.

Shimohata, T., Nakano, M., Lian, X., Shigeyama, T., Iba, H., Hamamoto, A., Yoshida, M., Harada, N., Yamamoto, H., Yamato, M., *et al.* (2011). *Vibrio parahaemolyticus* infection induces modulation of IL-8 secretion through dual pathway via VP1680 in Caco-2 cells. J. Infect. Dis. *203*, 537–544.

Shirai, H., Ito, H., Hirayama, T., Nakamoto, Y., Nakabayashi, N., Kumagai, K., Takeda, Y., and Nishibuchi, M. (1990). Molecular epidemiologic evidence for association of thermostable direct hemolysin (TDH) and TDH-related hemolysin of *Vibrio parahaemolyticus* with gastroenteritis. Infect. Immun. *58*, 3568–3573.

Sobrinho Pde, S., Destro, M.T., Franco, B.D., and Landgraf, M. (2011). Occurrence and distribution of *Vibrio parahaemolyticus* in retail oysters in Sao Paulo State, Brazil. Food. Microbiol. *28*, 137–140.

Sochard, M.R., and Colwell, R.R. (1977). Toxin isolation from a Kanagawa-phenomenon negative strain of *Vibrio parahaemolyticus*. Microbiol. Immunol. *21*, 243–254.

Staley, C., and Harwood, V.J. (2010). The use of genetic typing methods to discriminate among strains of *Vibrio cholerae*, *V. parahaemolyticus*, and *V. vulnificus*. J. AOAC Int. *93*, 1553–1569.

Steele-Mortimer, O., Brumell, J.H., Knodler, L.A., Meresse, S., Lopez, A., and Finlay, B.B. (2002). The invasion-associated type III secretion system of *Salmonella enterica* serovar Typhimurium is necessary for intracellular proliferation and vacuole biogenesis in epithelial cells. Cell Microbiol. *4*, 43–54.

Stevens, M.P., Wood, M.W., Taylor, L.A., Monaghan, P., Hawes, P., Jones, P.W., Wallis, T.S., and Galyov, E.E. (2002). An Inv/Mxi-Spa-like type III protein secretion system in *Burkholderia pseudomallei* modulates intracellular behaviour of the pathogen. Mol. Microbiol. *46*, 649–659.

Su, Y.C., Duan, J., and Wu, W.H. (2005). Selectivity and specificity of a chromogenic medium for detecting *Vibrio parahaemolyticust*. J. Food Prot. *68*, 1454–1456.

Su, Y.C., Yang, Q., and Hase, C. (2010). Refrigerated seawater depuration for reducing *Vibrio parahaemolyticus* contamination in pacific oyster (*Crassostrea gigas*). J. Food Prot. *73*, 1111–1115.

Sugiyama, T., Iida, T., Izutsu, K., Park, K.S., and Honda, T. (2008). Precise region and the character of the pathogenicity island in clinical *Vibrio parahaemolyticus* strains. J. Bacteriol. *190*, 1835–1837.

Suthienkul, O., Ishibashi, M., Iida, T., Nettip, N., Supavej, S., Eampokalap, B., Makino, M., and Honda, T. (1995). Urease production correlates with possession of the *trh* gene in *Vibrio parahaemolyticus* strains isolated in Thailand. J. Infect. Dis. *172*, 1405–1408.

Takahashi, A., Kenjyo, N., Imura, K., Myonsun, Y., and Honda, T. (2000a). Cl(-) secretion in colonic epithelial cells induced by the *Vibrio parahaemolyticus* hemolytic toxin related to thermostable direct hemolysin. Infect. Immun. *68*, 5435–5438.

Takahashi, A., Sato, Y., Shiomi, Y., Cantarelli, V.V., Iida, T., Lee, M., and Honda, T. (2000b). Mechanisms of chloride secretion induced by thermostable direct haemolysin of *Vibrio parahaemolyticus* in human colonic tissue and a human intestinal epithelial cell line. J. Med. Microbiol. *49*, 801–810.

Takeda, Y. (1983). Thermostable direct hemolysin of *Vibrio parahaemolyticus*. Pharmacol. Ther. *19*, 123–146.

Takeda, Y. (2011). *Vibrio parahaemolyticus*, enterotoxigenic *Escherichia coli*, enterohemorrhagic *Escherichia coli* and *Vibrio cholerae*. Proc. Jpn. Acad. Ser. B. Phys. Biol. Sci. *87*, 1–12.

Takeda, Y., Takeda, T., Honda, T., and Miwatani, T. (1976). Inactivation of the biological activities of the thermostable direct hemolysin of *Vibrio parahaemolyticus* by ganglioside Gt1. Infect. Immun. *14*, 1–5.

Taniguchi, H., Ohta, H., Ogawa, M., and Mizuguchi, Y. (1985). Cloning and expression in *Escherichia coli* of *Vibrio parahaemolyticus* thermostable direct hemolysin and thermolabile hemolysin genes. J. Bacteriol. *162*, 510–515.

Taniguchi, H., Hirano, H., Kubomura, S., Higashi, K., and Mizuguchi, Y. (1986). Comparison of the nucleotide sequences of the genes for the thermostable direct hemolysin and the thermolabile hemolysin from *Vibrio parahaemolyticus*. Microb. Pathog. *1*, 425–432.

Toro, H., Price, S.B., McKee, A.S., Hoerr, F.J., Krehling, J., Perdue, M., and Bauermeister, L. (2005). Use of bacteriophages in combination with competitive exclusion to reduce *Salmonella* from infected chickens. Avian. Dis. *49*, 118–124.

Troisfontaines, P., and Cornelis, G.R. (2005). Type III secretion: more systems than you think. Physiology (Bethesda) *20*, 326–339.

Trosky, J.E., Mukherjee, S., Burdette, D.L., Roberts, M., McCarter, L., Siegel, R.M., and Orth, K. (2004). Inhibition of MAPK signaling pathways by VopA from *Vibrio parahaemolyticus*. J. Biol. Chem. *279*, 51953–51957.

Trosky, J.E., Li, Y., Mukherjee, S., Keitany, G., Ball, H., and Orth, K. (2007). VopA inhibits ATP binding by acetylating the catalytic loop of MAPK kinases. J. Biol. Chem. *282*, 34299–34305.

Tsunasawa, S., Sugihara, A., Masaki, T., Sakiyama, F., Takeda, Y., Miwatani, T., and Narita, K. (1987). Amino acid sequence of thermostable direct hemolysin produced by *Vibrio parahaemolyticus*. J. Biochem. *101*, 111–121.

Wang, D., Yu, S., Chen, W., Zhang, D., and Shi, X. (2010). Enumeration of *Vibrio parahaemolyticus* in oyster tissues following artificial contamination and depuration. Lett. Appl. Microbiol. *51*, 104–108.

Wang, H.Z., Wong, M.M., O'Toole, D., Mak, M.M., Wu, R.S., and Kong, R.Y. (2006). Identification of a DNA methyltransferase gene carried on a pathogenicity island-like element (VPAI) in *Vibrio parahaemolyticus* and its prevalence among clinical and environmental isolates. Appl. Environ. Microbiol. *72*, 4455–4460.

Wang, R., Huang, J., Zhang, W., Lin, G., Lian, J., Jiang, L., Lin, H., and Wang, S. (2011). Detection and identification of *Vibrio parahaemolyticus* by multiplex PCR and DNA–DNA hybridization on a microarray. J. Genet. Genomics *38*, 129–135.

Waterman, S.R., and Small, P.L. (1996). Identification of sigma S-dependent genes associated with the stationary-phase acid-resistance phenotype of *Shigella flexneri*. Mol. Microbiol. *21*, 925–940.

Wiles, T.J., Kulesus, R.R., and Mulvey, M.A. (2008). Origins and virulence mechanisms of uropathogenic *Escherichia coli*. Exp. Mol. Pathol. *85*, 11–19.

Williams, T.L., Musser, S.M., Nordstrom, J.L., DePaola, A., and Monday, S.R. (2004). Identification of a protein biomarker unique to the pandemic O3:K6 clone of *Vibrio parahaemolyticus*. J. Clin. Microbiol. *42*, 1657–1665.

Wong, H.-C., Liu, S.-H., Ku, L.-W., Lee, I.-Y., Wang, T.-K., Lee, Y.-S., Lee, C.-L., Kuo, L.-P., and Shih, D.Y.-C. (2000). Characterization of *Vibrio parahaemolyticus* isolates obtained from foodborne illness outbreaks during 1992 through 1995 in Taiwan. J Food Prot. *63*, 900–906.

Wong, H.C., and Lin, C.H. (2001). Evaluation of typing of *Vibrio parahaemolyticus* by three PCR methods using specific primers. J. Clin. Microbiol. *39*, 4233–4240.

Wong, H.C., Peng, P.Y., Han, J.M., Chang, C.Y., and Lan, S.L. (1998). Effect of mild acid treatment on the survival, enteropathogenicity, and protein production in *Vibrio parahaemolyticus*. Infect. Immun. *66*, 3066–3071.

Wong, H.C., Peng, P.Y., Lan, S.L., Chen, Y.C., Lu, K.H., Shen, C.T., and Lan, S.F. (2002). Effects of heat shock on the thermotolerance, protein composition, and toxin production of *Vibrio parahaemolyticus*. J. Food Prot. *65*, 499–507.

Wong, H.C., Wang, P., Chen, S.Y., and Chiu, S.W. (2004). Resuscitation of viable but non-culturable *Vibrio parahaemolyticus* in a minimum salt medium. FEMS Microbiol. Lett. *233*, 269–275.

Xu, H.S., Roberts, N., Singleton, F.L., Attwell, R.W., Grimes, D. and Colwel, R.R. (1982) Survival and viability of nonculturable *Escherichia coli* and *Vibrio cholerae* in the estuarine and marine environment. Microb. Ecol. *8*, 313–323.

Xu, M., Iida, T., Yamamoto, K., Takarada, Y., Miwatani, T., and Honda, T. (1994a). Demonstration and characterization of simultaneous production of a thermostable direct hemolysin (TDH/I) and a TDH-related hemolysin (TRHx) by a clinically isolated *Vibrio parahaemolyticus* strain, TH3766. Infect. Immun. *62*, 166–171.

Xu, M., Yamamoto, K., Honda, T., and Ming, X. (1994b). Construction and characterization of an isogenic mutant of *Vibrio parahaemolyticus* having a deletion in the thermostable direct hemolysin- related hemolysin gene (*trh*). J. Bacteriol. *176*, 4757–4760.

Yamamoto, S., Okujo, N., Matsuura, S., Fujiwara, I., Fujita, Y., and Shinoda, S. (1994). Siderophore-mediated utilization of iron bound to transferrin by *Vibrio parahaemolyticus*. Microbiol. Immunol. *38*, 637–693.

Yamamoto, S., Okujo, N., Miyoshi, S., Shinoda, S., and Narimatsu, S. (1999). Siderophore production of *Vibrio parahaemolyticus* strains from different sources. Microbiol. Immunol. *43*, 909–912.

Yamazaki, W., Kumeda, Y., Misawa, N., Nakaguchi, Y., and Nishibuchi, M. (2010). Development of a loop-mediated isothermal amplification assay for sensitive and rapid detection of the *tdh* and *trh* genes of *Vibrio parahaemolyticus* and related *Vibrio* species. Appl. Environ. Microbiol. *76*, 820–828.

Yan, Y., Cui, Y., Han, H., Xiao, X., Wong, H.C., Tan, Y., Guo, Z., Liu, X., Yang, R., and Zhou, D. (2011). Extended MLST-based population genetics and phylogeny of *Vibrio parahaemolyticus* with high levels of recombination. Int. J. Food Microbiol. *145*, 106–112.

Yarbrough, M.L., Li, Y., Kinch, L.N., Grishin, N.V., Ball, H.L., and Orth, K. (2009). AMPylation of Rho GTPases by *Vibrio* VopS disrupts effector binding and downstream signaling. Science *323*, 269–272.

Yeung, P.S., and Boor, K.J. (2004). Effects of acid stress on *Vibrio parahaemolyticus* survival and cytotoxicity. J. Food Prot. *67*, 1328–1334.

Yeung, P.S., Hayes, M.C., DePaola, A., Kaysner, C.A., Kornstein, L., and Boor, K.J. (2002). Comparative phenotypic, molecular, and virulence characterization of *Vibrio parahaemolyticus* O3:K6 isolates. Appl. Environ. Microbiol. *68*, 2901–2909.

Yeung, P.S., Wiedmann, M., and Boor, K.J. (2007). Evaluation of a tissue culture-based approach for differentiating between virulent and avirulent *Vibrio parahaemolyticus* strains based on cytotoxicity. J. Food Prot. *70*, 348–354.

Yeung, P.S.M., DePaola, A., Kaysner, C.A., and Boor, K.J. (2003). A PCR assay for specific detection of the pandemic *Vibrio parahaemolyticus* O3:K6 clonal group from shellfish. J. Food Sci. *68*, 1459–1466.

Yoh, M., Tang, G.Q., Iida, T., Morinaga, N., Noda, M., and Honda, T. (1996). Phosphorylation of a 25 kDa protein is induced by thermostable direct hemolysin of *Vibrio parahaemolyticus*. Int. J. Biochem. Cell Biol. *28*, 1365–1369.

Yokomaku, D., Yamaguchi, N., and Nasu, M (2000). Improved direct viable count procedure for quantitative estimation of bacterial viability in freshwater environments. Appl. Environ. Microbiol. *66*, 5544–5548.

Yu, S., Chen, W., Wang, D., He, X., Zhu, X., and Shi, X. (2010). Species-specific PCR detection of the food-borne pathogen *Vibrio parahaemolyticus* using the *irgB* gene identified by comparative genomic analysis. FEMS Microbiol. Lett. *307*, 65–71.

Zhang, L.H., and Dong, Y.H. (2004). Quorum sensing and signal interference: diverse implications. Mol. Microbiol. *53*, 1563–1571.

Zhao, T., Tkalcic, S., Doyle, M.P., Harmon, B.G., Brown, C.A., and Zhao, P. (2003). Pathogenicity of enterohemorrhagic *Escherichia coli* in neonatal calves and evaluation of fecal shedding by treatment with probiotic *Escherichia coli*. J. Food Prot. *66*, 924–930.

Zhou, X., Konkel, M.E., and Call, D.R. (2010a). Regulation of type III secretion system 1 gene expression in *Vibrio parahaemolyticus* is dependent on interactions between ExsA, ExsC, and ExsD. Virulence *1*, 260–272.

Zhou, X., Konkel, M.E., and Call, D.R. (2010b). Vp1659 is a *Vibrio parahaemolyticus* type III secretion system 1 protein that contributes to translocation of effector proteins needed to induce cytolysis, autophagy, and disruption of actin structure in HeLa cells. J. Bacteriol. *192*, 3491–3502.

Epidemiology and Genetics of the Pandemic Clone of *Vibrio parahaemolyticus*

8

Indrani Karunasagar, Ballamoole Krishna Kumar and G. Balakrish Nair

Abstract

Members of the *Vibrionaceae* family occurring in marine environment are responsible for many of the reported cases of infection worldwide. Among these, *Vibrio parahaemolyticus* is an important foodborne pathogen transmitted through contaminated seafood. Historically, food poisoning due to *V. parahaemolyticus* occurred as sporadic cases caused by different serotypes without the clustering of one particular serotype. With the emergence of the pandemic clone belonging to O3:K6 serotype in Kolkata, India, in 1996, the epidemiology of this organism changed abruptly causing large outbreaks and rapid hospitalizations. This new highly virulent strain is now globally disseminated. This chapter traces the epidemiology of the pandemic strain of *V. parahaemolyticus*, its emergence, molecular characteristics and clonal dissemination.

Characteristics of the pandemic clone of *V. parahaemolyticus*

Till the mid 1990s, *V. parahaemolyticus* gastroenteritis were considered sporadic with occasional local outbreaks involving diverse serotypes. Analysis of strains isolated during active surveillance of diarrheal aetiologies among hospitalized patients in Calcutta, India, in 1996 revealed that 50–80% of *V. parahaemolyticus* infections recorded after February 1996 were due to strains belonging to O3:K6 serotype (Okuda *et al.*, 1997). Subsequently, infections caused by this serovar were detected in different parts of the world and the outbreaks were termed 'pandemic'. But as pointed out by Nair *et al.* (2007), in the epidemiological sense, there was no outbreak that could fit the Webster dictionary definition of pandemic – 'occurring over wide geographical areas and afflicting an exceptionally high proportion of population'. Though outbreaks were reported over wide geographical areas, it had not affected exceptionally high proportion of the population and mortality has been low. However, the term pandemic continues to be used in literature to describe the strains. Further, the outbreaks are not caused by a single strain of O3:K6 serotype; other serovariants have been involved, but the isolates share certain genetic and phenotypic characters that support the view that the isolates are clonally related. All the isolates in what is termed as the pandemic group carry *tdh* but not *trh* gene, are urease negative and belong to the same biochemical phenotype according to biochemical fingerprinting by Pheneplate system (Php-48; Pheneplate microplate Techniques, Stockholm, Sweden), while non-pandemic isolates are heterogeneous (Rahman *et al.*, 2006). Currently, more than 20 serovariants including O3:K6, O4:K68, O1:K25, O1: KUT (untypeable) and other serotypes are known to belong to the pandemic strains (Chen *et al.*, 2011).

The *V. parahaemolyticus* O3:K6 isolates of 1996 from Calcutta differ from strains of the same serotype isolated prior to 1996. The global regulatory gene *toxR* involved in the regulation of many virulence-associated genes is conserved within the members of the genus *Vibrio*. A Group Specific PCR (GS PCR) that specifically detects *toxRS* sequence of the pandemic O3:K6 strain was established to differentiate it from non pandemic strains of *V. parahaemolyticus* (Matsumoto *et al.*, 2000). This PCR was developed on the basis of the differences in the *toxRS* sequences of the pandemic and the pre-pandemic O3:K6 isolates. The former were invariably found to differ from the latter in seven base regions within the 1364 bp *toxRS* region. Also, pandemic strains representing serovars other than O3:K6, e.g. O4:K68 and O1: K untypeable (KUT) serovars give positive GS-PCR results thus providing additional evidence for the clonality of these isolates (Chowdhury *et al.*, 2000a,b; Matsumoto *et al.*, 2000). The pandemic strains belonging to serovars O3:K6, O4:K68 and O1: KUT from different countries show identical ribotypes and pulse field gel electrophoresis (PFGE) patterns (Nair *et al.*, 2007). However, exceptional strains of pandemic O3:K6 group that lack the *tdh* gene, and exhibit arbitrarily primed PCR (AP-PCR) profiles slightly different from the typical profiles of the pandemic group have been reported (Matsumoto *et al.*, 2000). Multilocus sequence typing (MLST) indicates genetic diversity among pre-pandemic O3:K6 isolates and clonality among pandemic isolates (Chowdhury *et al.*, 2004). The presence in the genome of the pandemic isolates of a filamentous phage f237 is another important discriminating feature and *orf8* of f237 has been used as marker for identification of the pandemic strain by PCR (Nasu *et al.*, 2000). However, some of the O3:K6 strains from Bangladesh isolated between 1997 and 2000 (Bhuiyan *et al.*, 2002) and from Vietnam (Chowdhury *et al.*, 2004) were negative for *orf8*. Four genomic islands (VPI1, VPI4, VPI5 and VPI6) have been found to be unique to pandemic strains of *V. parahaemolyticus* (Hurley *et al.*, 2006), but screening of 91 pandemic isolates showed the presence of VPI1 in all and VPI5 in 90 isolates (Chao *et al.*, 2009).

Epidemiology

Though O3:K6 pandemic isolates were detected during active surveillance of diarrhoea in Calcutta in 1996, the first reported isolate of this serotype was from a traveller returning from Indonesia to Japan in 1995 (Okuda *et al.*, 1997). Detection of *toxRS* sequences identical to that of pandemic strains in four *tdh* negative O3:K6 isolates from Japan obtained between 1983 and 1988 led Okura *et al.* (2003) to hypothesize that the pandemic O3:K6 clone could have originated from these non-pathogenic strains by acquisition of *tdh* gene in the environs of Japan. In India, though O3:K6 serotype accounted for 50–80% of infections in 1996, its proportion declined subsequently. Analysis of 258 isolates of the pandemic group during 1996–2004 showed fluctuation with peaks in 1996 and 2004 (Fig. 8.1). The 1996 peak was due to O3:K6 serotype and the 2004 peak was due to O1:K25 serotype that accounted for most isolates in 2000 and 2004 (Nair *et al.*, 2007). As shown in Fig. 8.2, the pandemic clones of *V. parahaemolyticus* have been reported from Asia, Africa, Europe, North and South America. The appearance of different serotypes of the pandemic strains in different regions are summarized in Table 8.1. The number of *V. parahaemolyticus* cases in Japan in 1996 were 5241 which increased to 6786 in 1997, and almost doubled to 12,346 cases in 1998 (WHO, 1999). Serotype O4:K8 was the dominant serotype in Japan till 1996 after which O3:K6 replaced this. In Taiwan the dominance of O3:K6 serotype in *V. parahaemolyticus* infections was noted almost at the same period as in India. While this serotype accounted for only 0.6% of infections in 1995 in Taiwan, it accounted for 50.1% infections in 1996 and 83.7% infections in 1997 (Chiou *et al.*, 2000). Isolation of the pandemic strain of *V. parahaemolyticus* in Bangladesh,

	1994	1995	1996	1997	1998	1999	2000	2001	2002	2003	2004
O3:K6	0	0	70	33	12	20	4	1	2	9	1
O1:K25	0	0	0	0	7	5	0	1	2	0	0
Total	0	0	70	39	24	26	29	7	15	12	36

Figure 8.1 Prevalence of pandemic *Vibrio parahaemolyticus* in Kolkata, India from 1994 to 2004 (reprinted with modification from Nair *et al.*, 2007).

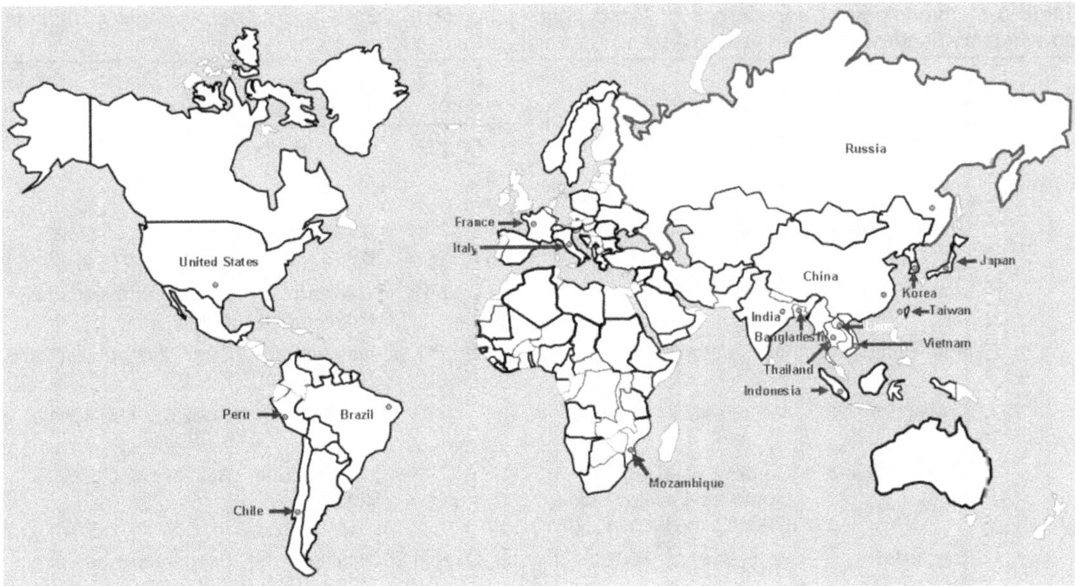

Figure 8.2 Global dissemination of the unique O3:K6 isolate of *Vibrio parahaemolyticus* and its serovariants.

Laos, Vietnam, China, Thailand, Indonesia, Korea and Japan has been reported subsequently (Nair *et al.*, 2007). *V. parahaemolyticus* O3:K6 isolates that resembled the pandemic strains in PFGE were isolated from 28 stool samples of patients involved in an oyster (harvested from Galveston Bay) associated outbreak of gastroenteritis in Texas, United States in 1998 (Daniels *et al.*, 2000a,b). During the same year, *V. parahaemolyticus* O3:K6 infections associated with consumption of oysters and clams harvested from Long Island Sound occurred among residents of Connecticut, New Jersey and New York (CDC, 1999).

A retrospective analysis of isolates of *V. parahaemolyticus* in Peru showed that the earliest O3:K6 isolate was seen in 1996. Other serovariants of the pandemic clone were also found in Peru (Gil *et al.*, 2007). In 1998, about 300 cases of *V. parahaemolyticus* gastroenteritis were reported from Antofagasta, Chile (Cordova *et al.*, 2002). Most of the tested isolates of this outbreak belonged to O3:K6 serotype and had molecular features of the pandemic clone (Gonzales-Escalona *et al.*, 2005). During 2004–2007, there were large outbreaks in Puerto Montt, southern Chile, with a peak in 2005 (about 1500 cases in 2004, 3600 in 2005, 900 in 2006 and 475 in 2007). Even during 2004, a high proportion of the strains were found to be of the pandemic group. The diversity in strains seemed to increase during later years. Pandemic strains accounted for 88% of isolates in 2005 and decreased to 66% in 2007. In 2005, nine different PFGE profiles were identified and this increased to 16 profiles in 2007 (Dauros *et al.*, 2011). Though there was a marginal increase (1153 cases) in 2008, the number of cases reduced to 441 in 2009. During 2009, 64% of clinical isolates and 24% of shellfish isolates tested belonged to the pandemic type (Garcia *et al.*, 2009).

Molecular studies on the *V. parahaemolyticus* isolates from outbreaks and sporadic cases that occurred in northeast Brazil during 2001–2002 indicated that the 2001 outbreak involved O3:KUT serotype and the 2002 outbreak involved O3:K6 serotypes showing characteristics of the pandemic strain (Leal *et al.*, 2008). During a surveillance of diarrhoea in Beira, Mozambique 42 stool samples in 2004 and 16 in 2005 yielded *V. parahaemolyticus*. In 2004, 32/42 belonged to O3:K6 serotype, and in 2005, 6/16 belonged to this serotype. Two isolates each during the two years belonged to O4:K68 serotype. Thus, 72% of the isolates belonged to pandemic serotype. In PFGE studies, these isolates clustered with

Table 8.1 Chronology of appearance of *Vibrio parahaemolyticus* O3:K6 and its serovariants in different countries (Expanded from Nair *et al.*, 2007)

Serotype	Country (year of isolation)	Reference(s)	Serotype	Country (year of isolation)	Reference(s)
O3:K6	Taiwan (1995) India (1996)	Chiou *et al.* (2000) Okuda *et al.* (1997), Chowdhury *et al.* (2000)	O1:KUT	India (1998)	Chowdhury *et al.* (2000b)
	Vietnam (1997)	Chowdhury *et al.* (2004)		Bangladesh (1998, 2000)	Bhuiyan *et al.* (2002), Matsumoto *et al.* (2000)
	Laos (1997)	Matsumoto *et al.* (2000)	O4:K12	Thailand (1998–1999)	Laohaprertthisan *et al.* (2000)
	Indonesia (1997)	Okuda *et al.* (1997)		Vietnam (1998–1999)	Chowdhury *et al.* (2004)
	Korea (1997–98)	Matsumoto *et al.* (2000)		Chile (2004)	González-Escalona *et al.* (2005)
	Chile (1998 and 2004)	Cordova *et al.* (2002), Gonza´lez-Escalona *et al.* (2005), Cabello *et al.* (2007)		United states (2006)	Balter *et al.* (2006)
	Bangladesh (1996–2000)	Bhuiyan *et al.* (2002)	O1:K41	Thailand (1998–1999)	Laohaprertthisan *et al.* (2000)
	Japan (1998)	WHO (1999)		Vietnam (1998–1999)	Chowdhury *et al.* (2004)
	Thailand (1999)	Chowdhury *et al.* (2000), Laohaprertthisan *et al.* (2000)	O1:K56	Vietnam (1998–1999)	Chowdhury *et al.* (2004)
	Russia (2001)	Smolikova *et al.* (2001)	O3:K75	Vietnam (1998–1999)	Chowdhury *et al.* (2004)
	France (2004)	Quilici *et al.* (2005)	O4:K8	Vietnam (1998–1999)	Chowdhury *et al.* (2004)
	Mozambique (2004)	Ansaruzzaman *et al.* (2005)	O4:KUT	Vietnam (1998–1999)	Chowdhury *et al.* (2004)
	Europe (2004)	Martinez-Urtaza *et al.* (2005)	O5:KUT	Vietnam (1998–1999)	Chowdhury *et al.* (2004)
	Peru (1996)	Gil *et al.* (2007)		India (2004)	Nair *et al.* (2007)
	Italy (2007–2008)	Ottaviani *et al.* (2008, 2010)	O5:K17	India (2002)	Nair *et al.* (2007)
	Northeast Brazil (2008)	Leal *et al.* (2008)	O5:K25	India (2002)	Nair *et al.* (2007)
	China (2005–2008)	Chao *et al.* (2009)	O1:K33	India (2002)	Nair *et al.* (2007)
	Thailand (2001–2002, 2000–2003)	Bhoopong *et al.* (2007), Serichantalergs *et al.* (2007)	O2:K3	India (2002)	Nair *et al.* (2007)
O4:K68	India (1998)	Chowdhury *et al.* (2000a)	OUT:KUT	India (2003–2004)	Nair *et al.* (2007)
	Thailand (1999)	Chowdhury *et al.* (2000b)	O3:KUT	India (2003–2004)	Nair *et al.* (2007)
	Bangladesh (1998 and 2000)	Bhuiyan *et al.* (2002)	O3:K5	India (2004)	Nair *et al.* (2007)
	Vietnam (1998)	Chowdhury *et al.* (2004)	O4:K4	India (2004)	Nair *et al.* (2007)
	Mozambique (2004)	Ansaruzzaman *et al.* (2005)	O4:K10	India (2004)	Nair *et al.* (2007)

Serotype	Country (year of isolation)	Reference(s)	Serotype	Country (year of isolation)	Reference(s)
O1:K25	India (1998, 2002)	Chowdhury *et al.* (2000b), Matsumoto *et al.* (2000), Nair *et al.* (2007)	O6:K18	Taiwan (2005)	Nair *et al.* (2007)
	Thailand (1999, 2001–2002)	Laohaprertthisan *et al.* (2000), Serichantalergs *et al.* (2007)	O1:K26	China (2005–2008)	Chao *et al.* (2009)
	Vietnam (1998–1999)	Chowdhury *et al.* (2004)	O4:K68	China (2005–2008)	Chao *et al.* (2009)
	Bangladesh (1999–2000)	Bhuiyan *et al.* (2002)	O1:K36,	China (2005–2008)	Chao *et al.* (2009)
	China (2005–2008)	Chao *et al.* (2009)	O3:K25	China (2005–2008)	Chao *et al.* (2009)
O1:KUT	India (1998)	Chowdhury *et al.* (2000b)	O3:K68	China (2005–2008)	Chao *et al.* (2009)
	Bangladesh (1998 and 2000)	Matsumoto *et al.* (2000), Bhuiyan *et al.* (2002)	O3: K46	Thailand (2001–2002)	Serichantalergs *et al.* (2007)
	Italy (2008)	Ottaviani *et al.* (2010)	–	–	–
	China (2005–2008)	Chao *et al.* (2009)	–	–	–
	Thailand (2001–2002)	Serichantalergs *et al.* (2007)	–	–	–

the Asian pandemic clone (Ansaruzzaman *et al.*, 2005, 2008).

Analysis of 13 clinical isolates of *V. parahaemolyticus* obtained during 1997–2004 in France showed that five isolates (one isolated during 1997, 1998, 1999 and two during 2003) belonged to O3:K6 serotype and had molecular features of the pandemic clone (Quilici *et al.*, 2005). It is interesting to note that the presence of the pandemic strain in clinical cases in France came to light owing to retrospective analysis of clinical isolates, and not owing to any noticeable outbreak occurring. Though isolates were obtained as early as 1997, there were only one isolate per year except in 2003, when there were two. This suggests that though pandemic strains were found in clinical cases in France, they did not cause any noticeable spike in hospitalizations. This should be viewed in the context that France is one of the largest consumers of bivalve molluscs within the European Union (Erwan and Paquotte, 1998). In Spain, there were two clinical isolates of *V. parahaemolyticus* O3:K6 and one of O3: KUT that showed molecular features of the pandemic clone during an oyster-associated outbreak in 2004 (Martinez-Urtaza *et al.*, 2005). One case of *V. parahaemolyticus* O3:K6 gastroenteritis was reported from Italy in 2007 and one in 2008 (Ottaviani *et al.*, 2008, 2010). The pandemic clone of *V. parahaemolyticus* has also been isolated from Russia (Smolikova *et al.*, 2001).

Ecology, virulence and mechanism of global spread

Often in the environment, pathogenic *V. parahaemolyticus* (*tdh*$^+$) constitutes only a small proportion of the total population; hence isolation of pandemic strains from environment requires special effort. Hara-Kudo *et al.* (2003) isolated pandemic strains belonging to O3:K6, O4:K68 and O1:K25 serotypes from environs around Japan and since pre-pandemic *tdh*$^-$ O3:K6 serotypes have also been isolated in Japan, suggested that serovar conversion might have taken place in the environs of Japan. These investigators used most probable number method combined with chromogenic agar isolation and *tdh* PCR for detection of pathogenic strains. Similar observations have been made by Mahmud *et al.* (2006) based on their studies in Kii Channel in Japan. Studies in India, Bangladesh, Thailand and China also showed the presence of pandemic O3:K6 in the environment, suggesting that this serovar has established an ecological niche in Asia (Arakawa *et al.*, 1999; Vuddhakul *et al.*, 2000; Islam *et al.*, 2004; Deepanjali *et al.*, 2005; Vongxay *et al.*, 2008). Studies in Chile in 2005 and 2007 indicated the presence of the pandemic clone in the environment (Dauros *et al.*, 2011). However, studies in the United States could not detect O3:K6 serotype in oysters in Alabama (DePaola *et al.*, 2003).

There are several theories to explain the global dissemination of pandemic *V. parahaemolyticus*. Ballast water of ships has been suspected to be involved in dissemination of pathogens like *V. cholerae* O1 (McCarthy and Khambaty, 1994) and this theory has been proposed to explain the spread of pandemic clone of *V. parahaemolyticus*. Multiple loci tandem repeat (VNTR) analysis using seven polymorphic loci of representative strains from Asia, Peru and Chile showed that strains from Peru and Chile have differentiated recently from strains in Asia which show higher diversity (Ansede-Bermejo *et al.*, 2010). It has been argued that the genetic diversity of O3:K6 subpopulation that arrived in Peru in 1997 is suggestive of events of higher magnitude than single ballast water discharge. The arrival of this population coincides with the El Nino phenomenon characterized by the arrival of warm equatorial waters in a sequence of invasive waves that lasted for over 6 months in 1997. It has also been suggested that the outbreak in Chile in 2004 and 2005 are related to currents and movement of warm waters along the coast of South America. While Humboldt current flows northwards along the west coast of South America ushering in low-saline cold waters, the Peru–Chile countercurrent has been described as a conduit for transport of equatorial warm waters from Peru to Chile and this is supported by evidence for transport of subtropical zooplankton to Chilean areas south of Puerto Montt (Ansede-Bermejo *et al.*, 2010).

Currently there is not enough molecular explanation for the higher virulence of the pandemic strains. Clinical symptoms of patients infected with O3:K6 were not different from those infected with non-pandemic strains (Nair *et al.*, 2007). There is no significant difference in the amounts of TDH produced by pandemic and non-pandemic strains. One amino acid polymorphism has been observed in the TDH produced by O3:K6 (Gly109) and non-O3:K6 (Asp109) serotypes and sequence polymorphism has been observed in putative *tdh* promoter region of these isolates (Okuda *et al.*, 1997; Wong *et al.*, 2000), but the significance of these on the expression of TDH and its function is not known. Pandemic strains of *V. parahaemolyticus* have a filamentous phage f237 that is similar to the CTX filamentous phage of *V. cholerae* O1 and the ORF8 that is present in this filamentous phage has been investigated for its role in virulence (Nair *et al.*, 2007). The predicted amino acid sequence of ORF8 has similarities with a putative adhesion molecule encoded by *plx* gene of *Drosophila* and *in vitro* cytotoxicity and adherence studies with human epithelial cells show that O3:K6 strains show statistically significant higher activity than non-O3:K6 strains (Yeung *et al.*, 2002). But the absence of ORF8 in some pandemic strains makes definitive conclusions about its role in virulence difficult.

Bioinformatic analysis of the whole genome sequence of an O3:K6 pandemic clone (RIMD2210633) indicated the presence of four genomic islands, but later studies showed that only two are found in most pandemic strains (Chao *et al.*, 2009). However, the role of genes carried by these genomic islands is not known. Pandemic strains carry a 23-kb pathogenicity island that contains a methyltransferase gene (Wang *et al.*, 2006), but the role of this pathogenicity island is not yet understood. *V. parahaemolyticus* pandemic strains isolated after a decade after the pandemic started seems to have lower virulence potential, being less involved in clinical cases (Nair *et al.*, unpublished). Genomic sequencing of one of these late pandemic strains shows no major gene deletions, but several insertions including cassettes added to the strain's super integron and 23 ORFs containing a DNA phosphorothioation (dnd) system (Chen *et al.*, 2011). However, it is not clear whether these insertions have led to reduced virulence in these strains.

Genome and evolution of pandemic strains

Complete genome sequencing of the two chromosomes, chromosome I and chromosome II, of *V. parahaemolyticus* strains has been conducted. A comparison of *V. parahaemolyticus* genome with that of *V. cholerae* indicates that their chromosome I does not differ much in size (3.3 Mb of *V. parahaemolyticus* vs. 3.0 Mb of *V. cholerae*) but chromosome II of *V. parahaemolyticus* (1.9 Mb) is much larger than that of *V. cholerae* (1.1 Mb). This difference in size could have arisen during evolution through acquisition of genes, gene duplication or by horizontal transfer as in *V. parahaemolyticus* or through frequent gene decay or deletion as in *V. cholerae*. The genome sequencing also revealed the presence of genes encoding type III secretion systems (T3SS) in *V. parahaemolyticus* and consists of two non-identical T3SS encoded within a pathogenicity island (Makino *et al.*, 2003). Both T3SS are located in pathogenicity islands, one on the larger chromosome I (T3SS-1) and the other on the smaller chromosome II (T3SS-2). T3SS-1 is known to be present in *V. parahaemolyticus* isolated from clinical and environmental sources regardless of their pathogenicity and has a G+C content similar to the rest of the genome indicating that this region is ancestral to the species. T3SS-2 is located predominantly in chromosome II of highly virulent strains of *V. parahaemolyticus* isolated after 1995, whereas most clinical isolates recovered before 1995 do not encode T3SS-2 indicating that the region is not essential for virulence (Makino *et al.*, 2003; Hurley *et al.*, 2006). The pathogenicity island of *V. parahaemolyticus* located on chromosome I encode a T3SS that is highly homologous to the T3SS found in *Yersinia* spp. The locus is in the reverse order of the locus of *Yersinia* spp. and there is a set of twelve open reading frames (ORFs) that are predicted to encode hypothetical proteins. The pathogenicity island located on chromosome II exhibits many of the characteristics of a 'classic' pathogenicity island. The island has insertion elements located at both ends and the G+C content is lower (39.8%) than the rest of the genome (45.4%) suggesting the acquisition of these genes through a recent lateral transfer event (Schmidt and Hensel *et al.*, 2004). This island encodes the *tdh* genes and the *toxRS* genes that are known to be involved in virulence. In addition, the island also contains the necessary structural genes for a second T3SS. There are also other genes located in the island that are candidates for virulence factors including VPA1321, a homologue of the *E. coli* protein cytotoxic necrotizing factor-1. The T3SS-2 has been demonstrated in rabbit ileal loop model to be involved in cytotoxicity and enterotoxicity. This data along with the presence of this system in all clinical isolates implies that it is required for virulence in humans. The gene product of T3SS-1 has been found to induce cytotoxicity in cell culture but does not appear to affect enterotoxicity in a rabbit ileal loop model (Park *et al.*, 2004). In 2009, a second type of T3SS-2 was discovered in *trh*$^+$ *V. parahaemolyticus* strain TH3996 (Okada *et al.*, 2009) which showed significant homology to the T3SS-2 of KP-positive *V. parahaemolyticus* RIMD2210633. After the report of novel type III secretion system in *trh*-positive *V. parahaemolyticus*, it has been classified into two distinct phylogroups, T3SS2α (in KP-positive *V. parahaemolyticus* RIMD2210633) and T3SS2β (in *trh*$^+$ *V. parahaemolyticus* TH3996). These are reportedly also found in other human pathogenic vibrios such as *V. cholerae* non-O1/non-O139 strains and *V. mimicus*. These findings demonstrate that these two distinct types are distributed not only within a species but also beyond the species level (Henke and Bassler, 2004; Okada *et al.*, 2010).

Genomic islands (GI) are gene cassettes in the chromosome of *V. parahaemolyticus* that used to be mobile and are now fixed. It can code for many functions like symbiosis or pathogenesis and may help an organism's adaptation. Genomic islands have unusual base composition compared with the core genome of the organism. Hurley *et al.* (2006) identified seven genomic islands (Vp-PAI-1 to Vp-PAI-7) in the genome of *V. parahaemolyticus* strain RIMD2210633, O3:K6 serotype isolated in Japan in 1996. Five GI were present on chromosome 1 and two on chromosome 2. The GIs, namely Vp-PAI-1, Vp-PAI-2, Vp-PAI-3 and Vp-PAI-4, are inserted adjacent to tRNA genes. The size of each GI was found to be in the range of 10 kb to 81 kb and had an unusual base composition as compared with the core genome of *V. parahaemolyticus* RIMD2210633. GIs of *V. parahaemolyticus* RIMD2210633 are flanked by direct repeats and have lower G+C content (ranging from 38% to 43%) than the overall genome G+C content of 45%. These GIs may increase the fitness of the bacteria either for survival in the aquatic environment or in their ability to infect humans. (Hurley *et al.*, 2006; Boyd *et al.*, 2008; Chao *et al.*, 2009, 2011). All these GIs were found to encode an integrase gene with the exception of Vp-PAI-7, which contain a number of transposase genes. Vp-PAI-7 is an 81 kb region that encodes a T3SS and the *tdh* gene and was previously identified as a potential pathogenicity island. The study on genomic analysis of pandemic strain of KP-positive *V. parahaemolyticus* (RIMD2210633) demonstrated that more than 86% of the RIMD2210633 genes are conserved and genes acquired through lateral transfer formed a cluster of genes. Among the genes that were variably present, 11 loci were found to be specifically present in the pandemic strains. This suggests that difference between pandemic and non-pandemic strains is not due to a simple genetic event (Izutsu *et al.*, 2008). The gene cluster from VPA1310 to VPA1396 was found to be exclusively conserved in KP-positive pathogenic strains but not in KP-negative strains. This cluster is described as the 'pathogenicity island Vp-PAI' (Makino *et al.*, 2003; Izutsu *et al.*, 2008) and found to be unique and common to all KP-positive strains suggesting a strong correlation with pathogenicity. The Vp-PAI is known to be absent in the KP-negative strains. It is observed that *tdh* genes and other putative virulence genes, including T3SS-2, are often located in the 80 kb region of Vp-PAI. From the literature, it seems that not only the *tdh* gene but the whole region of Vp-PAI is required for the pathogenicity of KP-positive *V. parahaemolyticus* strains.

Molecular analyses of worldwide clinical isolates of pandemic and non-pandemic *V. parahaemolyticus* belonging to pre-1996 and post-1995 demonstrated the presence of a 24 kb region which was named as *V. parahaemolyticus* island-1 (Vp-PAI-1) in O3:K6 and related strains recovered after 1995. Further analysis of the other genomic islands showed that Vp-PAI-4, Vp-PAI-5 and Vp-PAI-6 are also highly unique to the virulent clonal complexes of pandemic strains recovered after 1995 (Hurley *et al.*, 2006). GIs specific to the pandemic strains are Vp-PAI-1 (24 kb), Vp-PAI-4 (17 kb), Vp-PAI-5 (12 kb) and Vp-PAI-6 (27 kb) and among them, Vp-PAI-1 is suggested to be one of the markers of pandemicity owing to the presence of a virulence-associated gene (Wang *et al.*, 2006; Nishioka *et al.*, 2008). Chao *et al.* (2009) determined the presence of functional pandemic strain specific markers (*tdh, toxRS* new, GS-PCR, orf8, Vp-PAI-1 Vp-PAI-5, Vp-PAI-7 including T3SS-2). They noted that all the pandemic strains harboured Vp-PAI-1 and Vp-PAI-5 except for one pandemic strain belonging to O4:K68 isolated from an outbreak of food poisoning. This strain lacks Vp-PAI-5, possibly owing to chromosomal rearrangement.

From several studies it is apparent that the O3:K6 strains post 1995 isolated from different countries had probably evolved from a common ancestor. The arbitrarily primed PCR, ribotyping, and pulsed-field gel electrophoresis (PFGE) on post-1995 O3:K6 strains showed that, they were genetically similar but significantly different from the genetically variable pre-1995 O3:K6 strains. In addition, the newly emerged O3:K6 clone has diversified into various other serotypes such as O1: KUT, O4:K68, O1:K25, O4:K12 etc. since its initial isolation. These strains are postulated to be derivatives of the O3:K6 serotype because they are genetically similar in group-specific (GS)-PCR for *toxRS* and ORF8 PCR for detection of the f237 filamentous phage as well as in the epidemiological techniques of arbitrarily primed PCR, PFGE, and ribotyping. DNA restriction patterns of O3:K6 strains isolated in Peru and Chile have shown profiles closely related to strains from Asian countries obtained from the first epidemic in 1996–1997 (Gonzalez-Escalona *et al.*, 2005; Martinez-Urtaza *et al.*, 2008). Similar relationships were obtained through MLST using seven housekeeping genes (Gonzalez-Escalona *et al.*, 2008).

The study of Han *et al.* (2008) showed that the pandemic clone post-1996 (new O3:K6) and its serovariants might have emerged or evolved from the old-O3:K6 clone, which was promoted by stepwise acquisition of genomic islands, *toxRS*/new sequence and differentiation of O/K antigen genes. The study also provides direct evidence for the concept that the post-1996 strains of pandemic serovariants evolved from the new O3:K6 clone (Han *et al.*, 2008). They also demonstrated that, the acquisition of *toxRS*/new sequence led to the phylogenesis of the intermediate-O3:K6 clade from the old-O3:K6 clone, and that the post-1996 new O3:K6 strain stemmed from this intermediate clade after the acquisition of *tdh*, Vp-PAI-5 and other unidentified genes. It is concluded that, intermediate-O3:K6 clade served as the phylogenetic intermediate between new-O3:K6 and old-O3:K6, and the differentiation of O/K antigen genes promoted the derivation of new-O3:K6 serovariants from new-O3:K6. Several published reports highlight the point that the new pandemic O3:K6 and its serovariants originated from the non-pathogenic environmental strain of O3:K6 serotype by lateral transfer of virulence genes from other vibrios (Nair *et al.*, 2007; Boyd *et al.*, 2008; Chao *et al.*, 2011) and that the acquisition of additional serotypes may be a selection response to host immunological pressure. Chowdhury *et al.* (2004) surmised that the transition of major serovars occurred among the pandemic strains and change in serovar was possibly related to the change in the environmental conditions.

Multilocus sequence typing (MLST) is a procedure to characterize isolates of bacterial species using the DNA sequences of internal fragments of multiple housekeeping genes. It plays a role in studying epidemiology of infectious diseases, generating the information necessary for identifying, tracking, and intervening against disease outbreaks (Maiden *et al.*, 1998; Urwin *et al.*, 2003). MLST provided strong molecular evidence for the clonal origin of *V. parahaemolyticus* O3:K6 and revealed that isolates within such a clonal group may acquire previously identified serotypes of *V. parahaemolyticus*. The MLST study also confirmed genetic diversity among the *V. parahaemolyticus* strains that prevailed before the emergence of the pandemic O3:K6 strain and genetic uniformity between O3:K6 and its serovariants in spite of their serotype diversity.

The sequence analysis of the amplified housekeeping genes (*gyrB*, *recA*, *dnaE* and *gnd*) of different serovars of *V. parahaemolyticus* showed that prepandemic strains were highly variable and the pandemic O3:K6 isolates shared two alleles (Chowdhury *et al.*, 2004). The study on molecular analysis of the emergence of pandemic *V. parahaemolyticus*, clustered the diverse panel of 42 *V. parahaemolyticus* belonging to ten different serotypes into two closely related but distinct groups based on the MLST data (Boyd *et al.*, 2008). The former contained highly virulent isolates whereas the latter comprised mainly environmental isolates recovered in early 2000. González-Escalona *et al.*, (2008) found that the O3:K6 pandemic clone belonged to a clonal complex (CC) and confirmed the first reported pandemic spread of *V. parahaemolyticus*. CC corresponded to the pandemic strains with four different sequence types, ST-3, ST-42, ST-27 and ST-51. ST-3 was defined as the ancestral type or founder of the clonal complex and unequivocally established the clonal relationship of the pandemic complex. The common feature between non-pathogenic and pandemic strains is the presence of Vp-PAI-3. Further it has been shown that, the acquisition of Vp-PAI-3 was the first step to form pandemic strains. The ancestor of O3:K6 pandemic clone is believed to be O3:K6, ST-3 environmental non-pathogenic strain (Chao *et al.*, 2011). Literature suggests that the origin of pandemic clones was not from *tdh*$^+$ (embedded in Vp-PAI-7) or *trh*$^+$ strains, and that the Vp-PAI-7 was not the island acquired after Vp-PAI-3 and Vp-PAI-2. As Vp-PAI-1, Vp-PAI-5 play an important role in forming pandemic clone, Vp-PAI-7 was acquired after Vp-PAI-1 and Vp-PAI-5; previous studies showed that Vp-PAI-4 may be the last genomic island acquired by pandemic clones (Chao *et al.*, 2011). In conclusion, based on the environmental strain (O3:K6, ST-3), the pandemic O3:K6 clone emerged around 1996 by lateral transfer of large regions of DNA to acquire virulence genes and genomic islands.

Acknowledgements

The financial support from the Department of Biotechnology, Government of India, and the Indian Council of Medical Research towards some of the research results presented here is gratefully acknowledged.

References

Ansaruzzaman, M., Chowdhury, A., Bhuiyan, N.A., Sultana, M., Safa, A., Lucas, M., Von Seidlein, L., Barreto, A., Chaignat, C.L., Sack, D.A., *et al.* (2008). Characteristics of a pandemic clone of O3: K6 and O4: K68 *Vibrio parahaemolyticus* isolated in Beira, Mozambique. J. Med. Microbiol. *57*, 1502–1507.

Ansaruzzaman, M., Lucas, M., Deen, J.L., Bhuiyan, N.A., Wang, X.Y., Safa, A., Sultana, M., Chowdhury, A., Nair, G.B, Sack, D.A., *et al.* (2005). Pandemic serovars (O3:K6 and O4:K68) of *Vibrio parahaemolyticus* associated with diarrhea in Mozambique: spread of the pandemic into the African continent. J. Clin. Microbiol. *43*, 2559–2562.

Ansede-Bermejo, J., Gavilan, R.G., Triñanes, J., Espejo, R.T., and Martinez-Urtaza, J. (2010). Origins and colonization history of pandemic *Vibrio parahaemolyticus* in South America. Mol. Ecol. *19*, 3924–3937.

Arakawa, E., Murase, T., Shimada, T., Okitsu, T., Yamai, S., and Watanabe, H. (1999). Emergence and prevalence of a novel *Vibrio parahaemolyticus* O3:K6 clone in Japan. Jpn. J. Infect. Dis. *52*, 246–247.

Balter, S., Hanson, H., Kornstein, L., Lee, L., Reddy, V., Sahl, S., Stavinsky, F., Fage, M., Johnson, G., Bancroft, J., *et al.* (2006). *Vibrio parahaemolyticus* infections associated with consumption of raw shellfish – three states, 2006. Morb. Mortal. Wkly Rep. *55*, 1–2.

Bhoopong, P., Palittapongarnpim, P., Pomwised, R., Kiatkittipong, A., Kamruzzaman, M., Nakaguchi, Y., Nishibuchi, M., Ishibashi, M., and Vuddhakul, V. (2007). Variability of properties of *Vibrio parahaemolyticus* strains isolated from individual patients. J. Clin. Microbiol. *45*, 1544–1550.

Bhuiyan, N.A., Ansaruzzaman, M., Kamruzzaman, M., Alam, K., Chowdhury, N.R., Nishibuchi, M., Faruque, S.M., Sack, D.A, Takeda, Y., and Nair, G.B. (2002). Prevalence of the pandemic genotype of *Vibrio parahaemolyticus* in Dhaka, Bangladesh, and significance of its distribution across different serotypes. J. Clin. Microbiol. *40*, 284–286.

Boyd, E.F., Cohen, A.L., Naughton, L.M., Ussery, D.W., Binnewies, T.T., Stine, O.C., and Parent, M.A. (2008). Molecular analysis of the emergence of pandemic *Vibrio parahaemolyticus*. BMC Microbiol. *8*, 110.

Cabello, F.C., Espejo, R.T., Hernandez, M.C., Rioseco, M.L., Ulloa, J., and Vergara, J.A. (2007). *Vibrio parahaemolyticus* O3:K6 epidemic diarrhea, Chile, 2005. Emerg. Infect. Dis. *13*, 655–656.

Centers for Disease Control and Prevention (CDC). (1999). Outbreak of *Vibrio parahaemolyticus* infection associated with eating raw oysters and clams harvested from Long Island Sound – Connecticut, New Jersey, and New York, 1998. MMWR Morb. Mortal. Wkly Rep. *48*, 48–51.

Chao, G., Jiao, X., Zhou, X., Yang, Z., Huang, J., Pan, Z., Zhou, L., and Qian, X. (2009). Serodiversity, pandemic O3:K6 clone, molecular typing, and antibiotic susceptibility of foodborne and clinical *Vibrio parahaemolyticus* isolates in Jiangsu, China. Foodborne Pathog. Dis. *6*, 1021–1028.

Chao, G., Wang, F., Zhou, X., Jiao, X., Huang, J., Pan, Z., Zhou, L., and Qian, X. (2011). Origin of *Vibrio parahaemolyticus* O3:K6 pandemic clone. Int. J. Food Microbiol. *145*, 459–463.

Chen, Y., Stine, O.C., Badger, J.H., Gil, A.I., Nair, G.B., Nishibuchi, M., and Fouts, D.E. (2011). Comparative genomic analysis of *Vibrio parahaemolyticus*: serotype conversion and virulence. BMC Genomics *12*, 294.

Chiou, C.S., Hsu, S.Y., Chiu, S.I., Wang, T.K., and Chao, C.S. (2000). *Vibrio parahaemolyticus* serovar O3:K6 as cause of unusually high incidence of food-borne disease outbreaks in Taiwan from 1996 to 1999. J. Clin. Microbiol. *38*, 4621–4625.

Chowdhury, A., Ishibashi, M., Thiem, V.D., Tuyet, D.T., Tung, T.V., Chien, B.T., Seidlein, L.V., Canh, D.G., Clemens, J., Trach, D.D., *et al.* (2004). Emergence and serovar transition of *Vibrio parahaemolyticus* pandemic strains isolated during a diarrhea outbreak in Vietnam between 1997 and 1999. Microbiol. Immunol. *48*, 319–327.

Chowdhury, N.R., Chakraborty, S., Eampokalap, B., Chaicumpa, W., Chongsa-Nguan, M., Moolasart, P., Mitra, R., Ramamurthy, T., Bhattacharya, S.K., Nishibuchi, M., *et al.* (2000a). Clonal dissemination of *Vibrio parahaemolyticus* displaying similar DNA fingerprint but belonging to two different serovars (O3:K6 and O4:K68) in Thailand and India. Epidemiol. Infect. *125*, 17–25.

Chowdhury, N.R., Chakraborty, S., Ramamurthy, T., Nishibuchi, M., Yamasaki, S., Takeda, Y., and Nair, G.B. (2000b). Molecular evidence of clonal *Vibrio parahaemolyticus* pandemic strains. Emerg. Infect. Dis. *6*, 631–636.

Chowdhury, N.R., Stine, O.C., Morris, J.G., and Nair, G.B. (2004). Assessment of evolution of pandemic *Vibrio parahaemolyticus* by multilocus sequence typing. J. Clin. Microbiol. *42*, 1280–1282.

Córdova, J.L., Astorga, J., Silva, W., and Riquelme, C. (2002). Characterization by PCR of *Vibrio parahaemolyticus* isolates collected during the 1997-1998 Chilean outbreak. Biol. Res. *35*, 433–440.

Daniels, N.A., MacKinnon, L., and Bishop, R. (2000a). *Vibrio parahaemolyticus* infections in the United States, 1973–1998. J. Infect. Dis.*181*, 1661–1666.

Daniels, N.A., Ray, B., Easton, A., Marano, N., Kahn, E., McShan, A.L., Rosario, L.D., Baldwin, T., Kingsley, M.A., Puhr, N.D., *et al.* (2000b). Emergence of a new O3:K6 *Vibrio parahaemolyticus* serotype in raw oysters. JAMA *284*, 1541–1545.

Dauros, P., Bello, H., Domínguez, M., Hormazábal, J.C, and González, G. (2011). Characterization of *Vibrio parahaemolyticus* strains isolated in Chile in 2005 and in 2007. J. Infect. Dev. Ctries *27*, 502–510.

Deepanjali, A., Kumar, H.S., Karunasagar, I., and Karunasagar, I. (2005). Seasonal variation in abundance of total and pathogenic *Vibrio parahaemolyticus* bacteria in oysters along the southwest coast of India. Appl. Envirion. Microbiol. *71*, 3575–3580.

DePaola, A., Ulaszek, J., Kaysner, C.A., Tenge, B.J., Nordstrom, J.L., Wells, J., Puhr, N., and Gendel, S.M. (2003). Molecular, serological, and virulence characteristics of *Vibrio parahaemolyticus* isolated from environmental, food, and clinical sources in North America and Asia. Appl. Environ. Microbiol. *69*, 3999–4005.

Erwan, C. and Paquotte, P. (1998). Product differentiation and quality approach in the French market for oysters and mussels. Communication at the IIFET Conference Tromso 1998. Available at: http://archimer.ifremer.fr/doc/00000/2523/

García, K., Torres, R., Uribe, P., Hernández, C., Rioseco, M.L., Romero, J., and Espejo, R.T. (2009). Dynamics of clinical and environmental *Vibrio parahaemolyticus* strains during seafood-related summer diarrhea outbreaks in southern Chile. Appl. Environ. Microbiol. *75*, 7482–7487.

Gil, A.I., Miranda, H., Lanata, C.F., Prada, A., Hall, E.R., Barreno, C.M., Nusrin, S., Bhuiyan, N.A., Sack, D.A., and Nair, G.B. (2007). O3:K6 Serotype of *Vibrio parahaemolyticus* identical to the global pandemic clone associated with diarrhea in Peru. Int. J. Infect. Diseases. *11*, 324–328.

González-Escalona, N., Cachicas, V., Acevedo, C., Rioseco, M.L., Vergara, J.A., Cabello, F., Romero, J.. andEspejo. R.T. (2005). *Vibrio parahaemolyticus* diarrhea, Chile, 1998 and 2004. Emerg. Infect. Dis. *11*, 129–131.

González-Escalona, N., Martinez-Urtaza, J., Romero, J., Espejo, R.T., Jaykus, L.A., and DePaola, A. (2008). Determination of molecular phylogenetics of *Vibrio parahaemolyticus* strains by multilocus sequence typing. J. Bacteriol. *190*, 2831–2840.

Han, H., Wong. H.C., Kan, B., Guo, Z., Zeng, X , Yin, S., Liu, X., Yang, R., and Zhou, D. (2008). Genome plasticity of *Vibrio parahaemolyticus*: microevolution of the pandemic group. BMC Genomics *28*, 570.s

Hara-Kudo, Y., Sugiyama, K., Nishibuchi, M., Chowdhury, A., Yatsuyanagi, J., Ohtomo, Y., Saito, A., Nagano, H., Nishina, T., Nakagawa, H., *et al.* (2003). Prevalence of pandemic thermostable direct hemolysin-producing *Vibrio parahaemolyticus* O3:K6 in seafood and the coastal environment in Japan. Appl. Environ. Microbiol. *69*, 3883–3891.

Henke, J.M., and Bassler, B.L. (2004). Quorum sensing regulates type III secretion in *Vibrio harveyi* and *Vibrio parahaemolyticus*. J. Bacteriol. *186*, 3794–3805.

Hurley, C.C., Quirke, A., Reen, F.J., and Boyd, E.F. (2006). Four genomic islands that mark post-1995 pandemic *Vibrio parahaemolyticus* isolates. BMC Genomics *7*, 104–122.

Islam, M.S., Tasmin, R., Khan, S.I., Bakht, H.B., Mahmood, Z.H., Rahman, M.Z., Bhuiyan, N.A., Nishibuchi, M., Nair, G.B., Sack, R.B., *et al.* (2004). Pandemic strains of O3:K6 *Vibrio parahaemolyticus* in the aquatic environment of Bangladesh. Can. J. Microbiol. *50*, 827–834.

Izutsu, K., Kurokawa, K., Tashiro, K., Kuhara, S., Hayashi, T., Honda, T., and Iida, T. (2008). Comparative genomic analysis using microarray demonstrates strong correlation between presence of Vp-PAI and pathogenicity in Kanagawa phenomenon-positive *Vibrio parahaemolyticus*. Infect Immun. *76*, 1016–1023.

Laohaprertthisan, V., Chowdhury, A., Kongmuang, U., Kalnauwakul, S., Ishibashi, M., Matsumoto, C.. and Nishibuchi. M. (2000). Prevalence of serodiversity of the pandemic clone among the clinical strains of *Vibrio parahaemolyticus* isolated from Thailand. Epidemiol. Infect. *130*, 395–406.

Leal, N.C., da Silva, S.C., Cavalcanti, V.O., Figueiroa, A.C., Nunes, V.V., Miralles, I.S., and Hofer, E. (2008). *Vibrio parahaemolyticus* serovar O3:K6 gastroenteritis in northeast Brazil. J. Appl. Microbiol. *105*, 691–697.

McCarthy, S.A., and Khambaty, F.M. (1994). International dissemination of epidemic *Vibrio cholerae* by cargo ship ballast and other nonpotable waters. Appl. Environ. Microbiol. *60*, 2597–2601.

Mahmud, Z.H., Kassu, A., Mohammad, A., Yamato, M., Bhuiyan, N.A., Nair, G.B., and Ota, F. (2006). Isolation and molecular characterization of toxigenic *Vibrio parahaemolyticus* from the Kii Channel, Japan. Microbiol. Res. *161*, 25–37.

Maiden, M.C., Bygraves, J.A., and Feil, E. (1998). Multilocus sequence typing: a portable approach to the identification of clones within populations of pathogenic microorganisms. Proc. Natl. Acad. Sci. U.S.A. *95*, 3140–3145.

Makino, K., Oshima, K., Kurokawa, K., Yokoyama, K., Uda, T., Tagomori, K., Iijima, Y., Najima, M., Nakano, M., Yamashita, A., *et al.* (2003). Genome sequence of *Vibrio parahaemolyticus*: a pathogenic mechanism distinct from that of *V. cholerae*. Lancet *361*, 743–749.

Martinez-Urtaza, J., Simental, L., Velasco, D., DePaola, A., Ishibashi, M., Nakaguchi,Y., Nishibuchi, M., Carrera-Flores, D., Rey-Alvarez, C., and Pousa, A. (2005). Pandemic *Vibrio parahaemolyticus* O3:K6, Europe. Emerg. Infect. Dis. *11*, 319–320.

Matsumoto, C., Okuda, J., Ishibashi, M., Iwanaga, M., Garg, P., Ramamurthy, T., Wong, H.C., Depaola, A., Kim, Y.B., Albert, M.J., *et al.* (2000). Pandemic spread of an O3:K6 clone of *Vibrio parahaemolyticus* and emergence of related strains evidenced by arbitrarily primed PCR and toxRS sequence analyses. J. Clin. Microbiol. *38*, 578–585.

Nair, G.B., Ramamurthy, T., Bhattacharya, S.K., Dutta, B., Takeda, Y., and Sack, D.A. (2007). Global dissemination of *Vibrio parahaemolyticus* serotype O3:K6 and its serovariants. Clin. Microbiol. Rev. *20*, 39–48.

Nasu, H., Lida, T., Sughara, T., Yamaichi, Y, Park, K.S., Yokoyama, K., Makino, K., Shinagawa, H. and Honda, T. (2000). A filamentous phage associated with recent pandemic *Vibrio parahaemolyticus* 03:K6 strains. J. Clin. Microbiol. *38*, 2156–2161.

Nishioka, T., Kamruzzaman, M., Nishibuchi, M., and Satta, Y. (2008). On the origin and function of an insertion element Vp-PAI-1 specific to post-1995 pandemic *Vibrio parahaemolyticus* strains. Genes Genet. Syst. *83*, 101–110.

Okada, N., Iida, T., Park, K.S., Goto, N., Yasunaga, T., Hiyoshi, H., Matsuda, S., Kodama, T., and Honda, T. (2009). Identification and characterization of a novel type III secretion system in trh-positive *Vibrio parahaemolyticus* strain TH3996 reveal genetic lineage and diversity of pathogenic machinery beyond the species level. Infect. Immun. *77*, 904–913.

Okada, N., Matsuda, S., Matsuyama, J., Park, K.S., de los Reyes, C., Kogure, K., Honda, T., and Iida, T. (2010). resence of genes for type III secretion system 2 in *Vibrio mimicus* strains. BMC Microbiol. *10*, 302.

Okuda, J., Ishibashi, M., Hayakawa, E., Nishino, T., Takeda, Y., and Mukhopadhyay, A.K. (1997). Emergence of a unique O3:K6 clone of *Vibrio parahaemolyticus* in Calcutta, India and isolation of strains from the same clonal group from Southeast Asian travelers arriving in Japan. J. Clin. Microbiol. *35*, 3150–3155.

Okura, M., Osawa, R., Iguchi, A., Arakawa, E., Terajima, J., and Watanabe, H. (2003). Genotypic analyses of *Vibrio parahaemolyticus* and development of a pandemic group-specific multiplex PCR assay. J. Clin. Microbiol. *41*, 4676–4682.

Ottaviani, D., Leonim, F., Rocchegiani, E., Santarelli, S., Canonico, C., Masini, L., Ditrani, V., and Carraturo, A. (2008). First clinical report of pandemic *Vibrio parahaemolyticus* O3:K6 infection in Italy. J. Clin. Microbiol. *46*, 2144–2145.

Ottaviani, D., Leoni, F., Rocchegiani, E., Canonico, C., Potenziani, S., Santarelli, S., Masini, L., Scuota, S., and Carraturo, A. (2010). *Vibrio parahaemolyticus*-associated gastroenteritis in Italy: persistent occurrence of O3:K6 pandemic clone and emergence of O1:KUT serotype. Diagn. Microbiol. Infect. Dis. *66*, 452–455.

Park, K.S, Ono, T, Rokuda, M., Jang, M.H., and Okada. K. (2004) Functional characterization of two type III secretion systems of *Vibrio parahaemolyticus*. Infect Immun. 72, 6659–6665.

Quilici, M.L., Robert-Pillot, A., Picart, J., and Fournier, J.M. (2005). Pandemic *Vibrio parahaemolyticus* O3:K6 spread, France. Emerg. Infect. Dis. *11*, 1148–1149.

Rahman, M., Bhuiyan, N.A., Kuhn, I., Ramamurthy, T., Rahman, M., Mollby, R., and Nair, G.B. (2006) Biochemical fingerprinting of *Vibrio parahaemolyticus* by the PhenePlate system: comparison between pandemic and non-pandemic serotypes. Epidemiol. Infect. *134*, 985–989.

Schmidt, H., and Hensel. M. (2004). Pathogenicity islands in bacterial pathogenesis. Clin. Microbiol. Rev. *17*, 14–56.

Serichantalergs, O., Bhuiyan, N.A., Nair, G.B., Chivaratanond, O., Srijan, A., Bodhidatta, L., Anuras, S., and Mason, C J. (2007). The dominance of pandemic serovars of *Vibrio parahaemolyticus* in expatriates and sporadic cases of diarrhoea in Thailand, and a new emergent serovar (O3: K46) with pandemic traits. J. Med. Microbiol. *56*, 608–13.

Smolikova, L.M., Lomov, IuM., Khomenko, T.V., Murnachev, G.P., Kudriakova, T.A., Fetsaĭlova, O.P., Sanamiants, E.M., Makedonova, L.D., Kachkina, G.V., and Golenishcheva, E.N. (2001). Studies on halophilic vibrios causing a food poisoning outbreak in the city of Vladivostok. Zh. Mikrobiol. Epidemiol. Immunobiol. *6*, 3–7.

Urwin, R., and Maiden, M.C.J. (2003). Multi-locus sequence typing: a tool for global epidemiology. Trends Microbiol. *11*, 479–487.

Vongxay, K., Wang, S., Zhang, X., Wu, B., Hu, H., Pan, Z., Chen, S., and Fang, W. (2008). Pathogenetic characterization of *Vibrio parahaemolyticus* isolates from clinical and seafood sources. Int. J. Food Microbiol. *126*, 71–75.

Vuddhakul, V., Chowdhury, A., Laohaprertthisan, V.,. Pungrasamee, P., Patararungrong, N., Thianmontri, P., Ishibashi, M., Matsumoto, C., and Nishibuchi, M. (2000). Isolation of a pandemic O3:K6 clone of a *Vibrio parahaemolyticus* strain from environmental and clinical sources in Thailand. Appl. Environ. Microbiol. *66*, 2685–2689.

Wang, H., Wong, M.M., O'Toole, D., Mak, M.M., Wu, R.S., and Kong, R.Y. (2006). Identification of a DNA methyltransferase gene carried on a pathogenicity island-like element (VP-PAI) in *Vibrio parahaemolyticus* and its prevalence among clinical and environmental isolates. Appl. Environ. Microbiol. *72*, 4455–4460.

Wong, H.C., Liu, S.H., Wang, T.K., Lee, C.L., Chiou, C.S., Liu, D.P., Nishibuchi, M., and Lee, B.K. (2000). Characteristics of *Vibrio parahaemolyticus* O3:K6 from Asia. Appl. Environ. Microbiol. *66*, 3981–3986.

World Health Organization. (1999). *Vibrio parahaemolyticus*, Japan, 1996–1998. Wkly Epidemiol. Rec. *74*, 361–363.

Yeung, P.S., Hayes, M.C., DePaola, A., Kaysner, C.A., Kornstein, L., and Boor, K.J. (2002). Comparative phenotypic, molecular and virulence characterization of *Vibrio parahaemolyticus* O3:K6 isolates. Appl. Environ. Microbiol. *68*, 2901–2909.

Epidemiology and Molecular Pathogenesis of *Vibrio vulnificus*

9

Iddya Karunasagar and Anusha Rohit

Abstract

Vibrio vulnificus is a normal inhabitant of warm estuarine environments all over the world and may be associated with a wide variety of seafood. In susceptible individuals with underlying liver disease, diabetes or other immunocompromised condition and consuming raw seafood, the organism can cause primary septicaemia with a mortality rate of over 50%. A number of putative virulence factors such as capsule, cytotoxic factors, iron acquisition factors and factors responsible for evading the immune system of the host have been described and multiple factors seem to be involved in causing disease symptoms. The organism can be isolated, identified and enumerated by traditional microbiological methods as well as molecular methods such as polymerase chain reaction (PCR) and real time PCR. Clinical strains can be generally distinguished from most environmental strains by genetic fingerprinting techniques. The organism does not grow at temperatures below 13°C and therefore an important control measure is to cool the seafood to temperatures below this within a few hours of harvest. The organism is sensitive to mild heat, which can be used as a postharvest treatment method to minimize the risk of infection. Relaying shellfish to waters with salinity of > 30 ppt has also been found to be an important control measure.

Vibrio vulnificus as a foodborne pathogen

V. vulnificus is a natural inhabitant of warm estuarine and coastal environments throughout the world. The name *Vibrio vulnificus* (wound inflicting) was given in 1980 to a group of halophilic lactose-fermenting vibrios isolated from blood cultures (Farmer, 1980). *V. vulnificus* has been associated with three distinct clinical syndromes: primary septicaemia, wound infections and gastroenteritis. *V. vulnificus* is a Gram-negative curved rod shaped organism motile by polar flagellum. The bacterium is characterized by oxidase positive reaction, sensitivity to 10 μg O129, ability to ferment lactose, and cellobiose; inability to ferment sucrose and arabinose; positive for ornithine and lysine decarboxylase, negative for arginine decarboxylase, and capable of growth at 6% NaCl, but no growth at 8% NaCl. However, phenotypic characters may vary among isolates including lactose and sucrose fermentation that are often considered important for identification of this species (Oliver and Kaper, 2007). Currently three biotypes of *V. vulnificus* are recognized based on a combination of phenotypic, serological and host range characters (Drake *et al.*, 2007). Biotype 1 strains are indole positive, serologically diverse and are associated with human infections. Biotype 2 strains are indole negative and considered mainly as eel pathogens, but may also be opportunistic human pathogens, being associated with infections in eel handlers. This biotype has three serotypes and strains associated with eel and human infections belong to serotype E (Sanjuan and Amaro, 2004). Biotype 3 has five atypical biochemical characters, is genetically clonal and has been isolated from 62 Israeli patients with wound infection or septicaemia. This biotype has not been associated with foodborne infections and reported infections are limited to persons handling tilapia fish (Drake *et al.*, 2007).

V. vulnificus has been isolated from waters where the temperatures range from 9 to 31°C and proliferates in waters, where temperature exceeds 18°C (Kasper and Tamplin, 1993; Strom and Paranjpaye, 2000). Though this organism has worldwide distribution and has been isolated from coastal marine and estuarine waters, sediment, plankton, various shellfish (both molluscan and crustacean) and finfish species (FAO/WHO, 2005; Drake *et al.*, 2007), detailed ecological studies have been done only from a few countries. The abundance varies considerably and is greatly influenced by temperature and salinity. In North America, higher densities are observed in mid-Atlantic, Chesapeake Bay and Gulf coast waters, where temperatures are higher throughout the year, while densities are lower in Pacific, Canadian and North Atlantic waters (Kaysner *et al.*, 1987; O'Neil *et al.*, 1992; DePaola *et al.*, 1994; Wrights *et al.*, 1996; Motes *et al.*, 1998). The lowest temperature at which *V. vulnificus* has been isolated varies geographically, being 8°C at Chesapeake Bay (Wright *et al.*, 1996) and < 12.5°C in Gulf coast (Simonson and Siebeling, 1986). The organism survives in sediment during winter. In tropical waters, where temperature does not drop below 18°C, abundance of *V. vulnificus* is influenced by salinity (Parvathi *et al.*, 2004). In south India, highest *V. vulnificus* levels were found during monsoon season when the salinities were less than 5 ppt and at salinities exceeding 25 ppt, these organisms were not detectable (Parvathi *et al.*, 2004). Salinity has a significant effect on the abundance of the organism even in temperate waters. In US waters, numbers of *V. vulnificus* were high at salinity between 5–25 ppt, but dropped by 58–88% at salinities over 30 ppt (Kasper and Tamplin, 1993; Motes *et al.*, 1998). *V. vulnifucus* can colonize plankton and fish gut (DePaola *et al.*, 1994; Wright *et al.*, 1996; FAO/WHO, 2005). *V. vulnificus* produces chitinase, which might help the organism to colonize zooplankton (Strom and Paranjpaye, 2000). Through fish, the organism even reaches gut of birds since Miyasaka *et al.* (2005) found 14.1% aquatic birds in Japan to be positive for *V. vulnificus*.

Levels of *V. vulnificus* in oysters could be 100 times higher than in water surrounding them. In US Gulf coast, the levels in oysters may reach 10^4 cfu/g during summer months (Drake *et al.*, 2007) and in tropical waters of India, similar levels were found in oysters when salinities were less than 10 ppt (Parvathi *et al.*, 2004). *V. vulnificus* counts exceeding 10^6/g have been reported from the intestines of benthic fish inhabiting oyster reefs (DePaola *et al.*, 1994). There is no correlation between the prevalence or occurrence of *V. vulnificus* and faecal contamination of waters (Tamplin *et al.*, 1982; Parvathi *et al.*, 2004), hence faecal coliforms/*Escherichia coli* cannot be used as indicator organism for this pathogen. In the US, raw oysters seem to be the main source of foodborne outbreaks, since 96% of those developing primary septicaemia are linked to this commodity (Strom and Paranjapye, 2000; Drake *et al.*, 2007). However, in other regions, where raw seafood consumption is common, these may also be vehicles of infection. Three cases in Kumamoto Prefecture, Japan were attributed to consumption of raw flathead fish that had been marinated in soy sauce for one day (Ono *et al.*, 2001).

The disease rarely (< 5%) occurs in healthy individuals and liver disease including cirrhosis due to alcohol consumption is a risk factor for *V. vulnificus* infection. Other predisposing factors are diabetes, gastrointestinal disorders (ulcer, surgery), haematological conditions, and immunocompromised condition associated with cancer and therapy with immunosuppressive drugs. The fatality rate (about 50%), is the highest among foodborne pathogens (FAO/WHO, 2005). However, the attack rate is low with one illness occurring per 10,000 meals of raw US Gulf coast oysters (containing *V. vulnificus*) served to the highest risk population, i.e. people with liver diseases (FAO/WHO, 2005). The incubation period ranges from 7 hours to 10 days, with symptoms appearing in 36 hours in most cases (Oliver and Kaper, 2007). Most patients present with sudden onset of fever and chills, generally accompanied with nausea, vomiting, abdominal pain, hypotension (systolic pressure < 85 mm). In over 60% cases, secondary lesions appear, mostly on the legs that often develop necrotizing fasciitis or vasculitis that may require surgical debridement or amputation (Strom and Paranjapaye, 2000; Oliver and Kaper, 2007). *V. vulnificus* can be isolated from blood and cutaneous lesions. Epidemiological data suggests that men are more susceptible than women to *V. vulnificus* infection (Sharpio *et al.*, 1989; Merkel *et al.*, 2001). Experimental studies with rats suggest the oestrogen may play a role in protection, since mortality rate was 82% in normal male rats, 21% in normal female rats and 75% in ovariectomized female rats (Merkel *et al.*, 2001). Rare cases of atypical infections have been reported and these include septic arthritis (Johnson and Arnett, 2001), meningoencephalitis (Kim *et al.*, 2003) and ocular infection (Jung *et al.*, 2005) following consumption of raw oysters or raw fish.

Apart from primary septicaemia, *V. vulnificus* may be associated with wound infections. Strom and Paranjapye (2000) noted that 69% of such infections were associated with occupational exposures among oyster shuckers and commercial fishermen. Wound infections may progress to ecchymoses, cellulitis, bullae and necrotizing fasciitis, but mortality rate (25%) is much lower than in case of primary septicaemia (Jones and Oliver, 2009). However, 50% of cases may require surgical debridement or amputation. *V. vulnificus* infection may also result in mild gastroenteritis with vomiting, diarrhoea and abdominal cramps.

Antibiotic therapy is important for both primary septicaemia and wound infections. Tetracycline has been the most effective and in some cases, this antibiotic has been used in combination with third generation cephalosporin or gentamycin or chloramphenicol (Strom and Paranjapye, 2000; Liu *et al.*, 2006).

Epidemiology

In US and Korea, *V. vulnificus* infection is a reportable disease, hence most epidemiological information is available from there. It is estimated that about 100 cases of primary septicaemia due to *V. vulnificus* occur per year in the United States (Drake *et al.*, 2007) and nearly all the cases have been associated with consumption of raw oysters harvested from Gulf Coast with 90% of cases occurring during April to November (FAO/WHO, 2005). A retrospective study of cases reported to the US Centers for disease control and Prevention during 1988–1996 indicate that wound infections were almost as frequent as primary septicaemia (Storm and Paranjapye, 2000). In this study, primary septicaemia was defined as a systemic illness characterized as fever and shock where *V. vulnificus* was isolated from blood or other sterile site, and with a history of raw shellfish consumption but no wound infection preceding illness. Wound infections were defined as those where a patient incurred a wound before or during exposure to seawater or seafood drippings, and where *V. vulnificus* was subsequently cultured from that wound, blood, or otherwise normally sterile site. Gastroenteritis was characterized by illness with abdominal cramps, vomiting or diarrhoea, no evidence of wound infection, and where *V. vulnificus* was isolated from stool alone. Using these criteria, out of 422 infections reported between 1988 and 1996, 45% were wound infections, 43% primary septicaemia, 5% gastroenteritis, and the remaining 7% from undetermined exposure.

The Korean CDC estimates 40–70 confirmed cases per year and this high rate is suspected to be due to consumption of raw seafood or higher prevalence of predisposing factors (Drake *et al.*, 2007). However, in Japan, Inoue *et al* (2008) estimated 12–24 cases per year and in Taiwan, there was a peak occurrence in 2000 with 26 cases per million population and these included both septicaemia and wound infections (Hush *et al.*, 2004). A study of cases in Taiwan during 1995–200 revealed that over 80% had underlying conditions like liver disease, diabetes or steroid use, and 23.8% cases had primary septicaemia (Hush *et al.*, 2004). Interestingly, in Japan, oysters are not the primary source, since raw oysters are eaten only in winter and most infections occur during June-November with a peak in July. A mud shrimp *Upogebia major* was the common agent associated with *V. vulnificus* infections (Inoue *et al.*, 2008). Most cases occurred in western Japan with about 50% of cases occurring in Kyushu. In Japan, 72.3% of infections had septicaemia and mortality rate was 75%. Most patients (86.5%) had liver function impairment with 56.9% having liver cirrhosis and 10.1% liver cancer (Inoue *et al.*, 2008). In Europe, *V. vulnificus* infections are rare and mostly wound infections (Baker-Austin *et al.*, 2010). Rare cases of septicaemia have been reported from Thailand (Thamlikitkul, 1990) and India (Saraswathi *et al.*, 1989).

Virulence factors and pathogenesis

The virulence of *V. vulnificus* seems to be related to multiple factors such as presence of a polysaccharide capsule, ability to obtain iron from transferrin, ability to produce extracellular enzymes and exotoxin (Drake *et al.*, 2007). Pathogens ingested through food encounter highly acidic gastric environment as the first host defence. Exposure of *V. vulnificus* to low acid environments leads to increased expression of *cadBA* operon encoded lysine decarboxylase, involved in the breakdown of lysine to cadavarine, which functions as acid neutraliser and as a superoxide radical scavenger that contributes to oxidative stress tolerance (Jones and Oliver, 2009). A variety of bacteria use manganese superoxide dismutase (SOD) for acid neutralization and in *V. vulnificus*, mutation of *sodA* encoding manganese SOD increases the LD_{50} by two log units indicating a role in virulence (Jones and Oliver, 2000). The capsule has been reported to play an important role in resistance to host defences like phagocytosis and non-encapsulated transposon mutants have four times higher lethal dose compared with the wild type strains. Microbial capsules are known to provide resistance to opsonisation by complement and avoid phagocytosis by macrophages. Capsules are also known to confer resistance to bactericidal effect of serum and capsules are hypothesized to mask the immunogenic structures that would trigger non-specific immune response in the host system (Jones and Oliver, 2009). Mutants unable to express capsules have increased LD_{50}, suggesting the role of capsule in the virulence of *V. vulnificus*. Antibodies specific to capsule are detected in patients with *V. vulnificus* infections (Foire *et al.*, 1992). However, most freshly isolated environmental strains are capsulated, irrespective of their virulence (Strom and Paranjpye, 2000) and *V. vulnificus* strains may show phase variable expression of capsule as indicated by opaque (capsulated) and translucent (reduced capsule expression or acapsulate) colony morphologies. Macrophages are known to be important for clearing *V. vulnificus* entering the human system and clinical strains of this organism have been shown to be capable of inducing apoptosis in macrophages suggesting that ability to kill leucocytes involved in defence system would be important virulence factor of this organism (Jones and Oliver, 2009).

Iron is a limiting factor for the growth of microorganisms in host systems and ability to acquire iron in the host body plays an important role in microbial virulence. Lethal doses of *V. vulnificus* are 10^5-fold lower in iron dextran treated mice compared with normal mice (Starks *et al.*, 2000). Individuals with underlying chronic liver diseases have a higher susceptibility to *V. vulnificus* infections and generally such conditions are associated with impaired iron metabolism (FAO/WHO, 2005). Most iron in human serum is bound to transferrin making it unavailable to invading pathogens and *V. vulnificus* has multiple systems of iron acquisition with ability to produce both catechol (vulnibactin) and hydroxamate siderophores. Using mutational analysis of vulnibactin synthesis and transport, it has been demonstrated that this siderophore would be required for acquisition of iron from transferrin and holotransferrin (Jones and Oliver, 2009). It has been suggested that a haem receptor HupA is involved in non-transferrin bound iron uptake and this receptor could be involved in growth of *V. vulnificus* after entering the host (Litwin and Byrrne, 1998). Haemolytic factors contribute to iron availability through release of iron from haemoglobin. The *V. vulnificus* haemolysin not only contributes to iron acquisition, but also has cytotoxicity to a variety of mammalian cell lines and has 60–65% amino acid homology with non-O1 *V. cholerae* haemolysin and *V. cholerae* El Tor haemolysin (Strom and Paranjpye, 2000). Mice injected with *V. vulnificus* haemolysin display lesions similar to that noticed in infected humans (Gray and Kreger, 2000). *V. vulnificus* haemolysin is a pore forming toxin and it has been suggested that its activity contributes to vascular permeability and hypotension that is characteristic disease caused by this organism. However, inactivation of *vvhA* gene encoding haemolysin does not significantly reduce LD_{50} in iron-overloaded mice and haemolysin negative mutants induce necrosis and tissue damage in mice indicating that this toxin is not solely responsible for tissue damage in *V. vulnificus* disease. A gene homologous to *rtxA* toxin gene of *V. cholerae* known as *rtxA1* has been identified in *V. vulnificus* and RTX toxins consist of repeating subunits that form pores in cell membranes. Isogenic *rtxA1* mutants of *V. vulnificus* showed reduced cellular damage, and *in vivo*, significantly reduced spread to liver suggesting that this toxin could play a role in causing cell necrosis and facilitating invasion of *V. vulnificus* into bloodstream by crossing the intestinal epithelium (Jones and Oliver, 2009). Studies on pilin (structural protein encoded by *pilA* and pilin peptidase encoded by *pilD*), outer membrane protein ompU and flagella suggest that they may be involved in attachment of *V. vulnificus* to host cells that seems to be important for expression of cytotoxicity (Jones and Oliver, 2009). The mechanisms by which *V. vulnificus* overcomes host defence factors are summarized in Table 9.1.

Considering that expression of several factors is required for *V. vulnificus* to bring about tissue destruction and lethality, it can be expected that the organism depends on global regulators of genes to control pathogenesis. Quorum sensing through LuxR/LuxS type system seems to be involved in regulation of virulence gene expression. Mutational inactivation of *luxS* or *luxR* (called *smcR*) in *V. vulnificus* leads to reduced cytotoxicity and increased LD_{50} (Jones and Oliver, 2009). Another global regulator of genes in microorganisms is cyclic AMP (cAMP) generated by cleavage of ATP by adenyl cyclase (encoded by *cya* in *V. vulnificus*) and binds to cAMP receptor protein (CRP encoded by *crp*), which then binds to DNA to regulate gene expression. *V. vulnificus cya* mutant showed 2 log unit higher LD_{50} compared with wild type, suggesting a role for this regulator in virulence gene expression (Kim *et al.*, 2005). A transcriptional regulator AphB encoded by *aphB* gene has been shown to regulate a variety of functions such as acid tolerance, adhesion, motility, cytotoxicity and lethality in *V. vulnificus* (Jeong and Choi, 2008). A regulatory protein HlyU has been shown to regulate expression of both vvhA and rtx1, and hlyU mutants have reduced cytotoxicity and increased LD50 (Liu *et al.*, 2007). Thus, regulation of virulence gene expression seems to be complicated in *V. vulnificus* involving several global gene regulators.

Most of the virulence associated factors are present in over 95% of environmental strains. Considering the worldwide distribution of this organism in warm coastal waters, *V. vulnificus* infections are rare in most countries. Even in countries, where *V. vulnificus* is a reportable disease e.g. in the USA, it has been observed that only a small proportion of susceptible people contract the disease. For example in Florida, where there are estimated 70,000 at risk population (with liver disease) eating raw oysters, only 5–10 infections occur every year (Jackson *et al.*, 1997). This has led to the suggestion that though *V. vulnificus* can be commonly found in oysters, only certain strains are involved in human infections (Jackson *et al.*, 1997). Further studies using various molecular typing methods indicate that

Table 9.1 Mechanisms by which *V. vulnificus* overcomes host defence factors

Stage of entry into human system	Host defence	Mechanism by which *V. vulnificus* overcomes the host defence	Reference
Stomach	Acidic pH	Acid neutralization through lysine decarboxylase, manganese superoxide dismutase	Jones and Oliver (2000)
Intestine	Epithelial cell layer – physical barrier	RTX1 toxin induces apoptosis in epithelial cells	Jones and Oliver (2000), Lee *et al.* (2008)
Blood stream	Phagocytic cells – macrophages	Capsule reduces susceptibility to phagocytosis, induction of apoptosis in macrophages	Jones and Oliver (2000), Strom and Paranjpye (2000)
	Iron limitation	Vulnibactin and hydroxamate siderophores, haemolytic factors releasing iron from haemoglobin, haem receptor, HupA	Litwin and Byrrne (1998), Jones and Oliver (2000), Strom and Paranjpye (2000)
	Bactericidal factors in serum	Capsule, proteases	Jones and Oliver (2000)

it is possible to discriminate environmental and clinical strains. Nilsson *et al.* (2003) recorded differences in the sequence of 492 bp fragment of 16S rRNA gene of *V. vulnificus* strains, which has an *Alu*I cleavage site after nucleotides 202 and 244, and a *Hae*III cleavage site after nucleotides 168 and 372. Most environmental strains have an additional *Alu*I cleavage site after nucleotide 140 and most clinical strains have an additional *Hae*III cleavage site after nucleotide 147 (Nilsson *et al.*, 2003). Analysis of clinical and environmental strains using multilocus sequence typing has revealed that two main lineages can be recognized with Lineage I, consisting exclusively of Biotype 1 isolates, and Lineage II, consisting of isolates belonging to Biotype 1 and all isolates of Biotype 2. The proportion of clinical isolates in Lineage 1 was much higher than those in Lineage II and a 33-kb genomic island was unique to Lineage I (Cohen *et al.*, 2007). Rosche *et al.* (2005) using nucleotide sequence analysis showed that Biotype 1 strains can be distinguished into two types that strongly correlate with clinical (C) or environmental (E) origin. C-genotypes showed greater resistance to human serum than E-genotypes and had lower LD_{50}, suggesting that C-genotype strains may be more virulent (Bogard and Oliver, 2007; Rosche *et al.*, 2010). While similar levels of C- and E-genotypes were found in estuarine waters, oysters had 85% E-genotypes (Warner and Oliver, 2008). Thus it is possible that total *V. vulnificus* levels in shellfish may not give a true picture of the level of exposure of consumers to C-genotype strains that are mostly the cause of illness.

Detection and quantification in seafood

Methods used for isolation and identification of *V. vulnificus* from foods and the environment vary widely and both traditional culture based methods as well as molecular methods like polymerase chain reaction (PCR) and colony hybridization have been described (Harwood *et al.*, 2004; Drake *et al.*, 2007). The method described in US FDA Bacteriological Analytical Manual (BAM) for enumeration of *V. vulnificus* involves Most probable number (MPN) analysis using alkaline peptone water, plating the tubes showing growth onto modified cellobiose–polymyxin B–colistin (mCPC) agar, identification of typical colonies by biochemical tests or by colony hybridization using enzyme labelled DNA probe (Kaysner and DePaola, 2004). Though traditionally, TCBS agar has been used for isolation of pathogenic *Vibrio* spp., a number of studies have indicated that the selectivity of this medium is not adequate for isolation of *V. vulnificus* from environmental samples (Harwood *et al.*, 2004). CPC agar, developed by Massad and Oliver (1987), has colistin and polymyxin as selective agent- and cellobiose to differentiate *V. vulnificus* from non-cellobiose fermenting vibrios such as *V. cholerae*. Selectivity is further improved by incubation at 40°C. This was modified by Tamplin *et al.* (1991) by reducing the concentration of colistin. However, some investigators (Hori *et al.*, 1998; Parvathi *et al.*, 2005) reported higher isolation rate using alkaline peptone water with polymyxin for enrichment and cellobiose–colistin (CC) agar for selective plating. Other selective agars described for *V. vulnificus* include *Vibrio vulnificus* agar (VV agar), sodium dodecyl sulfate–polymyxin B–sucrose (SPS) agar, *Vibrio vulnificus* medium (VVM) and *Vibrio vulnificus* enumeration (VVE) agar (Harwood *et al.*, 2004).

Identification of colonies coming up on selective agars by biochemical tests could be a challenge due phenotypic plasticity, but systems such as API20E or Biolog have been used (Harwood *et al.*, 2004). Immunoassay using monoclonal antibody has been used for identification (Tamplin *et al.*, 1991). The US FDA BAM recommends use of DNA probe binding to the cytolysin gene *vvhA* or PCR amplification of *vvhA* gene for identification of *V. vulnificus* colonies (Kaysner and DePaola, 2004).

Alternate method described for isolation/enumeration of *V. vulnificus* involves direct plating of samples on non-selective agar, *Vibrio vulnificus* agar (VVA), followed by colony lifts on Whatman filter paper, lysis of colonies, fixing of DNA and hybridization with alkaline phosphatase labelled probes binding to *V. vulnificus* cytolysin/haemolysin (*vvhA*) gene. (Kaysner and DePaola, 2004). Digoxigenin-labelled *vvhA* probes can also be used (Gooch *et al.*, 2001). NBT/BCIP substrate is used for colour development. For isolation of colonies reacting with the probe, the filter is aligned with the plate the colonies from the area reacting with probe are picked up, isolated by streaking on mCPC or CC agars and confirmed by PCR or probe hybridization. Results with MPN and direct probe hybridization methods have been found to be comparable capable of detecting as few as 10 viable cells per 100 g (Kaysner and DePaola, 2004). Digoxigenin-labelled probes binding to *vvhA* gene or 16S rRNA gene in conjunction with selective differential medium VVM could also be used (Cerda-Cuellar *et al.*, 2000).

A number of PCR based assays have been used for detection of *V. vulnificus* in seafood and the target for PCR amplification has been various regions of *vvhA* gene or 16S rRNA/23S rRNA gene or *gyrB* gene (Harwood *et al.*, 2004; Kumar *et al.*, 2006; Drake *et al.*, 2007). One of the problems in applying PCR for detection of *V. vulnificus* in seafood directly is the inhibition of reaction by seafood matrix. Generally, DNA extraction protocols are used to minimize this problem. Another problem is the inability to distinguish between dead and viable cells. Nested PCR using 23S rRNA gene universal external primers and *V. vulnificus* specific internal primers had a sensitivity of detecting 120 cells of *V. vulnificus* in artificially inoculated samples without enrichment in 24 hours (Arias *et al.*, 1995). However, these primers might miss some strains that are positive by *vvhA* PCR (Parvathi *et al.*, 2005). Nested PCR amplifying *vvhA* gene could detect as low as one colony forming unit per gram of seafood, when coupled with direct extraction of template DNA (Lee *at al.*, 1999). PCR can be used in conjunction with MPN technique for enumeration of *V. vulnificus* (Kaysner and DePaola, 2004). Multiplex PCR for detection of *V. vulnificus* with other pathogens such as *Salmonella* Typhimurium, *V. cholerae, V. parahaemolyticus* and *Escherichia coli* has been described. In oyster homogenates seeded with pathogens, 10–100 cfu/g could be detected with two rounds of PCR (Brasher *et al.*, 1998). Combination of PCR and microtitre plate sandwich hybridization has been used to detect multiple pathogens like *V. vulnificus, V. cholerae, V. parahaemolyticus* and *S.* Typhimurium. Amplified target DNA was hybridized with capture probes in microtitre plates and detection was achieved by hybridization with a biotinylated oligonucleotide probe internal to amplified DNA. In oyster homogenates seeded with *V. vulnificus*, detection limits were 100 cfu/g (Lee *et al.*, 2003).

Recently, there have been several reports on the application of real time PCR for quantification of *V. vulnificus* associated with seafood and assay based on TaqMan probe as well as SYBR green dye have been used (Drake *et al.*, 2007). Sensitivity of the assay based on vvhA gene fragment amplification was 100 cells/g in oyster homogenates or seawater. Enrichment for 5 hours increased the sensitivity of detection to one cell (Campbell and Wright, 2003). Results of enumeration of *V. vulnificus* by real-time PCR performed on MPN tubes after 6-hour enrichment showed 93% agreement with that obtained by plating the MPN enrichments followed by confirmation of colonies by colony hybridization but real-time PCR gave faster results (Wright *et al.*, 2007). An SYBR green-based assay by Panicker *et al.* (2004), targeting the cytolysin gene, had a detection limit of 10^2 cfu/g oyster homogenate and 10^2 cfu/ml seawater and the sensitivity could be increased to 1 cfu/g with 5 hours' enrichment. TaqMan real-time PCR assay that discriminated between viable and non-viable cells of *V. vulnificus* using the DNA intercalating agent ethidium bromide has been described by Wang and Levin (2006).

Loop mediated isothermal amplification (LAMP) is another technology that is gaining popularity in food safety testing laboratories, since this can be performed and the results read without expensive equipment. LAMP assay using primers binding to *vvhA* of *V. vulnificus*, had a detection limit of 10^7 cfu/g with oysters and 7 cfu/g after 5 hours enrichment (Han and Ge, 2008). Surasilp *et al.* (2011) combined LAMP targeting *rpoS* gene with amplicon detection by chromatographic lateral flow dipstick assay (LFD). In spiked oyster samples without enrichment, the detection limit was 1.2×10^4 cfu/g or equivalent to 11 cells per reaction. Specific detection of potentially virulent strain in LAMP assay using virulence-coregulated gene (*vcg*) was reported by Han *et al.* (2011). In spiked raw oysters, the assay was capable of detecting 2.5×10^3 cfu/g and the sensitivity could be increased to 1 cfu/g using 6 hour enrichment.

Survival and growth in seafood

V. vulnificus resides inside various tissues of oysters, hence depuration is ineffective in elimination of this pathogen, but relaying oysters in high salinity (> 30 ppt) waters for 17–49 days caused a decrease in population of *V. vulnificus* from 10^3 cfu/g to < 10 MPN/g (Motes and Depaola, 1996). *V. vulnificus* does not grow in oysters at temperatures below 13°C and prolonged refrigeration could lead to reduction in numbers (Cook and Ruple, 1992; Cook, 1994). While Cook and Ruple (1992) noted that levels in refrigerated shellfish became non-detectable (< 3/g) in 14–21 days, Kaysner *et al.* (1989) observed survival in artificially contaminated oysters for 14 days at 2°C, suggesting that refrigeration cannot be relied upon for elimination of this pathogen in oysters. The rate of decline in refrigerated oyster shellstock has been estimated to be 0.041 log unit per day (Cook *et al.*, 2002). In fact, if the temperature is not controlled immediately after harvest, growth of *V. vulnificus* in oyster could occur. Cook (1997) demonstrated that *V. vulnificus* levels in oyster shell stock held without refrigeration for 3.5, 7, 10.5 and 14 hours increased respectively by 0.75, 1,3, 1.74 and 1.94 log units. It has also been reported that *V. vulnificus* levels in retail oysters originating from Gulf of Mexico were 1–2 log units greater than at harvest (Cook *et al.*, 2002). It has been estimated that commercial cooling of oyster stocks could take an average of 5.5 hours (CFSAN/FDA, 2005) and therefore the time shell stock is kept unrefrigerated on boat deck is an issue in control plans.

Four to five $\log_{10}$ reduction in numbers of natural *V. vulnificus* population in oysters occur when frozen to −40°C and stored for 3 weeks (Cook and Ruple, 1992). However cold adaptation at 15°C may reduce the effectiveness of freezing (Bryan *et al.*, 1999). A combination of vacuum packaging and freezing can bring down *V. vulnificus* counts by 3–4 $\log_{10}$ units in 7 days and though numbers continue to decline till day 7, complete elimination cannot be achieved (Parker *et al.*, 1994). *V. vulnificus* is sensitive to temperature with 6 $\log_{10}$ reduction in numbers occurring when subjected to 50°C for 5 minutes in shucked oyster meat (Cook and Ruple, 1992). Natural population of *V. vulnificus* (4.3×10^3 cfu/g) could be reduced to non-detectable levels by exposing them to 50°C for 10 minutes (Cook and Ruple, 1992). D-values at 47°C were 3.44 to 3.66 minutes for opaque colonies and 3.18 to 3.38 minutes for translucent colonies (Kim *et al.*, 1997). In North and South Carolina, commercial shellstock is subjected to heat shock by submerging about 70 chilled oysters in wire baskets into a heat-shock tank containing about 850 l of potable water at a temperature of 67°C for about 5 minutes depending on oyster size and condition (Darke *et al.*, 2007). This process has been shown to reduce *V. vulnificus* levels by 2–4 $\log_{10}$ units (Hesselman *et al.*, 1999). The bacteria were also found to be inactivated at pH 2.0 (Koo *et al.*, 2000). *V. vulnificus* is sensitive to ionizing radiation and irradiation doses of 1.0 kGy applied on whole shell oysters can reduce the bacterial numbers from 10^7 cfu/g to undetectable levels (Andrews *et al.*, 2003). Hydrostatic pressure of 250 MPa for 120 seconds also reduced *V. vulnificus* > 5 $\log_{10}$ units in oyster (Cook, 2003).

V. vulnificus has been reported to enter into a viable but non-culturable (VBNC) phase under conditions of low nutrient or temperature. In the VBNC state, *V. vulnificus* cells are small (0.3 µm) cocci and on resuscitation, they regain rod shaped (3 µm × 0.7 µm) morphology (Linder and Oliver, 1989; Nilsson *et al.*, 1991). Changes in fatty acid composition of cell membrane in *V. vulnificus* as a response to lower temperatures have been recorded, with proportional increase in unsaturated fatty acids (Linder and Oliver, 1989). Temperature stress also impacted on the shift of *V. vulnificus* from culturable to VBNC state. Temperature shift from 35°C to 5°C resulted in VBNC state, however, when cells were subjected to 15°C before downwards shift, *V. vulnificus* cells remained viable (Bryan *et al.*, 1999). *V. vulnificus* cells pre-starved for 24 hours before exposure to 5°C remained culturable, but cells starved for the same period at 5°C entered into the VBNC state (Oliver *et al.*, 1991). VBNC *V. vulnificus* cannot be resuscitated after temperature upshift to 22°C in nutrient-rich broth, but can be revived in media supplemented with pyruvate or catalase, and this has been attributed to increased sensitivity of cells in VBNC state to hydrogen peroxide (Bogosian *et al.*, 2000). Notably, 10^5 VBNC cells injected intraperitoneally were lethal to mice (Oliver and Bockain, 1995).

Risk assessment and control measures

A quantitative risk assessment for *V. vulnificus* in raw oysters was documented by FAO/WHO (2005) and this study modified the US FDA *V. parahaemolyticus* risk assessment model to assess the risk of *V. vulnificus* primary septicaemia in United States. The geographical coverage was limited because quantitative data for *V. vulnificus* levels in oysters, at the point of consumption and the data for the susceptible population was available only for the US (FAO/WHO, 2005). Data on *V. vulnificus* levels in oysters were based on weekly analysis of oysters from four Gulf States conducted during 1994–95 (Motes *et al.*, 1998, FAO/WHO, 2005) and all strains were considered equally virulent. Though association of certain genotypes with clinical cases has been reported (Nilsson *et al.*, 2003), data on seasonal and regional distribution of such strains or on the ability of such strains to grow/survive in oysters under typical industry practices was not available. The model used for determining exposure assessment is illustrated in Fig. 9.1. Harvest and post-harvest module was based on post harvest practices (duration of oysters in harvest vessel in water, time to first refrigeration, cool down time) based on surveys conducted in the Gulf Coast. *V. vulnificus* growth in oysters, survival during refrigeration and levels at consumption were estimated based on data from studies in the US Gulf Coast (Cook, 1997; Cook *et al.*, 2002). The model predicted that the mean *V. vulnificus* levels in oysters would be 5.7×10^4/g in summer and 8.0×10^1/g in winter. At a serving size of 196 g, the ingested dose would be 1.1×10^7 *V. vulnificus* in summer and 1.6×10^4 in winter. The US FDA data on prevalence of risk factors in US population and oyster consumption data from the surveys were used in the model (FAO/WHO, 2005). The dose response relationship was modelled by estimating the exposure per eating occasion and number of eating occasions for oyster associated *V. vulnificus* cases reported to US Centers for disease control

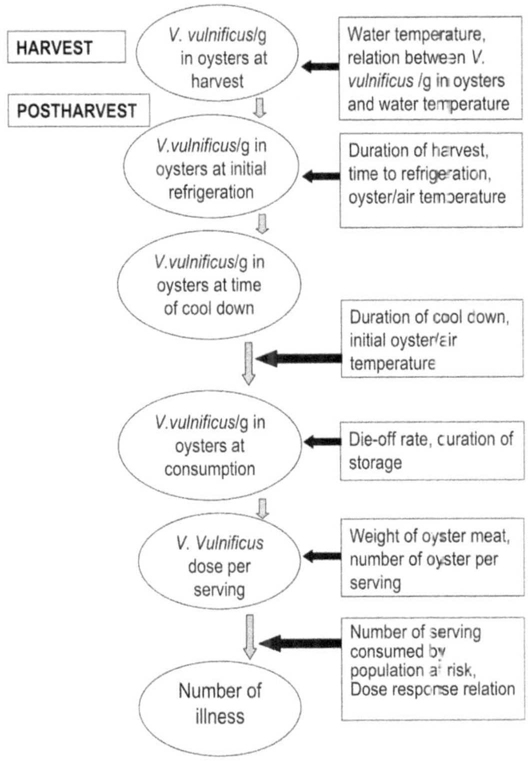

Figure 9.1 Harvest to consumption model for risk assessment of *V. vulnificus* (modified from FAO/WHO, 2005)

during 1995–2001. The model predicted 0.5, 11.5, 12.2 and 8 illnesses for winter (January to March), spring (April to June), summer (July to September) and autumn (October to December) seasons respectively. When compared with epidemiological data, the number of reported cases (average for 1995–2001 was 0.6 in winter, 9.6 in spring, 13.5 in summer and 7.4 in autumn) were within the 90% confidence limit predicted by the model (FAO/WHO, 2005).

The risk assessment also predicted the reductions in illness that could be achieved by post harvest treatments to reduce *V. vulnificus* levels to target values such as 3/g or 30/g or 300/g. In the US, there are three validated methods to achieve end-point criterion of <3 MPN/g *V. vulnificus* and these include mild heat treatment (50°C), freezing with extended frozen storage and high hydrostatic pressure. If all oysters are treated to achieve target level of 3/g, the model predicted that the number of cases could be reduced from current 32 reported cases per year to one case every 6 years. If the target is shifted to 30 or 300/g, then the predicted cases would increase to 1.2 and 7.7 cases per year respectively (FAO/WHO, 2005). At a time to refrigeration range of 0–20 hours, the predicted illness ranged from 17.7 to 59.3 cases, suggesting that immediate cooling of oysters alone is not adequate to achieve substantial reduction in the number of *V. vulnificus* illnesses. Since *V. vulnificus* levels in oysters harvested from waters with a salinity of > 30 ppt is greatly reduced, it is predicted that if all oysters are harvested from waters at salinity of > 30 ppt, irrespective of the water temperature, *V. vulnificus* illness would be < 1 case per year (FAO/WHO, 2005). Relaying oysters to high-salinity waters (> 32 ppt) has been shown to reduce *V. vulnificus* levels by 3–4 log units (< 10/g) within 2 weeks. Based on FAO/WHO risk assessment, Codex Committee on Food Hygiene developed Code of Hygienic practice for control of *Vibrio* spp. in seafood with an annex on control measures for *V. parahaemolyticus* and *V. vulnificus* in bivalve molluscs. This Code recommends assessment of the need for control measures based on (a) number of sporadic illness associated with bivalve molluscs in the area, (b) water temperature at harvest, air temperature, and harvest and post-harvest practices, and (c) water salinity at harvest. Since there is wide geographical variation in prevalence and levels of *V. vulnificus* in bivalves, control measures that have been validated and appropriate for the region may be adopted by the appropriate authority having jurisdiction and implemented under HACCP system. Validation of control measure should be carried out in accordance with the Codex Guidelines for the validation of food safety control measures (CAC/GL 69–2008).

The US National Shellfish Sanitation Programme (NSSP) guide (2009) includes the following strategies for minimizing the risk due to *V. vulnificus* in molluscan shellfish in states reporting two or more cases of *V. vulnificus* illness per year: (a) increased educational efforts targeted towards the population at risk to improve their awareness of the risks of eating raw molluscan shellfish and to change their eating behaviour to reduce or stop eating raw or untreated molluscan shellfish, (b) limited harvest restrictions on areas incriminated in outbreaks, (c) requirement for the temperature of shellstock to be brought down to 10°C or less by using ice, mechanical refrigeration or other means within specified period (10 h when water temperature is > 28°C; 12 h when water temperature is > 23°C up to 28°C; 14 hours when water temperature is 18–23°C and 36 h when water temperature is < 18°C) and (d) phased-in postharvest treatment requirements or other controls.

References

Andrews, L.S., Jahncke, M., Millikarjunan, K. (2003). Low-dose gamma irradiation to reduce pathogenic *Vibrios* in live oysters (*Crassostrea virginica*). J. Aquat. Food Prod. Tech *12*, 71–82.

Arias, C.R., Garay, E., and Aznar, R. (1995). Nested PCR method for rapid and sensitive detection of *Vibrio vulnificus* in fish, sediments, and water. Appl. Environ. Microbiol. *61*, 3476–3478.

Baker-Austin, C., Stockley, L., Rangdale, R. and Martinez-Urtaza, J. (2010). Environmental occurrence and clinical impact of *Vibrio vulnificus* and *Vibrio parahaemolyticus*: a European perspective. Environ. Microbiol. Rep. 2, 7–18.

Bogard, R.W., and Oliver, J.D. (2007). Role of iron in human serum resistance of the clinical and environmental *Vibrio vulnificus* genotypes. Appl. Environ. Microbiol. *73*, 7501–7505.

Bogosian, G., Aardema, N.D., Bourneuf, E.V., Morris, P.J., O'Neil, J.P. (2000). Recovery of hydrogen peroxide-sensitive culturable cells of *Vibrio vulnificus* gives the appearance of resuscitation from a viable but nonculturable state. J. Bacteriol. *182*, 5070–5075.

Brasher, C.W., DePaola, A., Jones, D.D, and Bej, A.K. (1998). Detection of microbial pathogens in shellfish with multiplex PCR. Curr. Microbiol. *37*, 101–107.

Bryan, P.J., Steffan, R.J., DePaola, A., Foster, J.W., Bej, A.K. (1999). Adaptive response to cold temperatures in *Vibrio vulnificus*. Curr Microbiol *38*, 168–175.

Campbell, M.S., and Wright, A.C. (2003). Real-time PCR analysis of *Vibrio vulnificus* from oysters. Appl. Environ. Microbiol. *69*, 7137–44.

Cerda-Cuellar, M., Jofre, J., Blanch, A.R., (2000). A selective medium and a specific probe for detection of *Vibrio vulnificus*. Appl. Environ. Microbiol. *66*, 855–859.

CFSAN/FDA, (2005). Quantitative risk assessment on the public health impact of pathogenic *Vibrio parahaemolyticus* in raw oysters. Available at: http://www.fda.gov/Food/ScienceResearch/ResearchAreas/RiskAssessmentSafetyAssessment/ucm050421.htm

Cohen, A.V., Oliver, J.D., DePaola, A., Feil, E.J., and Boyd, E.F. (2007). Emergence of a virulent clade of *Vibrio vulnificus* and correlation with the presence of a 33-kilobase genomic island. Appl. Environ Microbiol. *73*, 5553–5565.

Cook, D.W. (1994). Effect of time and temperature on multiplication of *Vibrio vulnificus* in postharvest Gulf coast shellstock oysters. Appl. Environ. Microbiol. *60*, 3483–3484.

Cook, D.W. (1997). Refrigeration of oyster shellstock: conditions which minimize the outgrowth of *Vibrio vulnificus*. J. Food Prot. *60*, 349–352.

Cook, D.W. (2003). Sensitivity of *Vibrio* species in phosphate-buffered saline and in oysters to high-pressure processing. J. Food Prot. *66*, 2276–2282.

Cook, D.W. and Ruple, A.D. (1992). Cold storage and mild heat treatment as processing aids to reduce the numbers of *Vibrio vulnificus* in raw oysters. J. Food Prot.*55*, 985–989.

Cook, D.W, O'Leary, P., Hunsucher, J.C., Sloan, E.M., Bowers, J.C., Blodgett, B.J., DePaola, A. (2002). *Vibrio vulnificus* and *Vibrio parahaemolyticus* in U.S. retail shell oysters: a national survey from June 1998 to July 1999. J. Food Prot. *65*, 79–87.

DePaola, A., Capers, G.M., Alexander, D. (1994). Densities of *Vibrio vulnificus* in the intestines of fish from the U.S. Gulf Coast. Appl. Environ. Microbiol. *60*, 984–988.

Drake, S.L., DePaola, A. and Jaykus, L. (2007). An overview of *Vibrio vulnificus* and *Vibrio parahaemolyticus*. Comprehensive Rev. Food Sci. Food Safety *6*, 120–144.

FAO/WHO (2005). Risk assessment of *Vibrio vulnificus* in raw oysters. Microbiological Risk Assessment Series 8, pp. 90.

Farmer, J.J., III. (1980). Revival of the name *Vibrio vulnificus*. Int. J. Syst. Bacteriol. *30*, 656.

Foire, A., Hayat, U, Wright, A.C., Wasserman, S.S., Morris, J.G., Jr. (1992). Infection with *Vibrio vulnificus* elicits an antibody response to capsular polysaccharide. Clin. Res. *40*, 428–431.

Gooch, J.A., DePaola, A., Kaysner, C.A., Marshall, D.L., (2001). Evaluation of two direct plating methods using nonradioactive probes for enumeration of *Vibrio parahaemolyticus* in oysters. Appl. Environ. Microbiol. *67*, 721–724.

Han, F. and Ge, B. (2008). Evaluation of a loop mediated isothermal amplification assay for detecting *Vibrio vulnificus* in oysters. Foodborne Pathog. Dis. *5*, 311–320.

Han, F., Wang, F., and Ge, B. (2011). Detecting potentially virulent strains of *Vibrio vulnificus* in raw oysters by quantitative loop-mediated isothermal amplification. Appl. Environ. Microbiol. *77*, 2589–2595.

Harwood, V.J., Gandhi, J.P., Wright, A.C. (2004). Methods for isolation and confirmation of *V. vulnificus* from oysters and environmental sources: A review. J. Microbiol. Methods *59*, 301–316.

Hesselman, D.M., Motes, M.L. and Lewis, L.P. (1999). Effects of a commercial heatshock process on *Vibrio vulnificus* in the American oyster *Crassostrea virginica*, harvested from the Gulf Coast. J. Food Prot. *62*, 1266–1269.

Hoi, L., Dalsgaard, I., and Dalsgaard, A. (1998). Improved isolation of *Vibrio vulnificus* from seawater and sediment with cellobiose-colistin agar. Appl. Environ. Microbiol. *64*, 1721–1724.

Hsueh, P.R., Lin, C.Y., Tang, H.J., Lee, H., Liu, J., Liu, Y., and Chuang, Y. (2004). *Vibrio vulnificus* in Taiwan. Emerg. Infect. Dis. *10*, 1363- 1369.

Inoue, Y., Ono, T., Matsui, T., Miyasaka, J., Kinoshita, Y., and Ihn, H. (2008). Epidemiological survey of *Vibrio vulnificus* infection in Japan, between 1999–2003. J. Dermatol. *25*, 129–139.

Jackson, J.K., Murphree, R.L., and Tamplin, M.L. (1997). Evidence that mortality from *Vibrio vulnificus* infection results from single strains among heterogeneous populations in shellfish, J. Clin. Microbiol. *35*, 2098–2101.

Jeong, H.G., and S.H.Choi. (2008). Evidence that AphB, essential for the virulence of *Vibrio vulnificus*, is a global regulator. J. Bacteriol. *190*, 3768–3773.

Johnson, R.W., and Arnett, F.C. (2001). A fatal case of *Vibrio vulnificus* presenting as septic arthritis. Arch. Intern. Med. *161*, 2616–2618.

Jones, M.K., and Oliver, J.D. (2009). *Vibrio vulnificus*: Disease and pathogenesis. Infect. Immun. *77*, 1723–1733.

Jung, S.I., Shin, D.H., Park, K.H., Shin, J.H. Seo, M.S. (2005). *Vibrio vulnificus* endophthalmitis occurring after ingestion of raw seafood. J. Inf. *51*, 281–283.

Kaspar, C.W., and Tamplin, M.L. (1993). Effects of temperature and salinity on the survival of *Vibrio vulnificus* in seawater and shellfish. Appl. Environ. Microbiol. *59*, 2425–2429.

Kaysner, C.A. and DePaola, A., Jr. (2004). *Vibrio*. Bacteriological Analytical Manual, 8th Edition, Chapter 9. Substantially rewritten and revised May 2004. Available at: http://www.fda.gov/Food/ScienceResearch/LaboratoryMethods/BacteriologicalAnalyticalManualBAM/ucm070830.htm

Kaysner, C.A., Abeyta, C., Wekell, M.M., DePaola, A., Stott, R.F., and Leitch, J.M. (1987). Virulent strains of *Vibrio vulnificus* isolated from estuaries of the United States west coast. Appl. Environ. Microbiol. *53*,1349–1351.

Kaysner, C.A., Tamplin, M.L.,Wkell, M.M. Stott, R.F., Colburn, K.G. (1989). Survival of *Vibrio vulnificus* in shellstock and shucked oysters (*Crassostrea gigas* and *Crassostrea virginica*) and effects of isolation medium on recovery. Appl. Environ. Microbiol. *55*, 3072–3079.

Kim, C.M., Jeong, K.C, Rhee, J.H., Choi, S.H. (1997). Thermal-death times of opaque and translucent morphotypes of *Vibrio vulnificus*. Appl. Environ. Microbiol. *63*, 3308–3310.

Kim, S.J., Kim, B.C., Kim, D.C., Kim, M.K., Cho, K.H., Seo, J.J., Shin, J.H. (2003). A fatal case of *Vibrio vulnificus* meningoencephalitis. Clin. Micro. Infect. *9*, 568–571.

Kim, Y.R., Kim, S.Y., Kim, C.M., Lee, S.E., and Rhee, J.H.. (2005). Essential role of an adenylate cyclase in regulating *Vibrio vulnificus* virulence. FEMS Microbiol. Lett. *243*, 497–503.

Kumar, S.H., Parvathi, A., Karunasagar, I. and Karunasagar, I. (2006). A gyrB-based PCR for detection of *Vibrio vulnificus* and its application for direct detection of this pathogen in oyster enrichment broths. Int. J. Food Microbiol. *111*, 216–220.

Koo, J., DePaola, A., and Marshall, D.L. (2000). Impact of acid on survival of *Vibrio vulnificus* and *Vibrio vulnificus* phage. J. Food Prot. *63*, 1049–52.

Lee, B.C., Choi, S.H., and Kim, T.S. (2008). *Vibrio vulnificus* RTX toxin plays an important role in the apoptotic death of human intestinal epithelial cells exposed to *Vibrio vulnificus*. Microbes Infect. *10*, 1504–1513.

Lee, C.Y., Panicker, G., and Bej, A.K. (2003). Detection of pathogenic bacteria in shellfish using multiplex PCR followed by CovaLink NH microwell plate sandwich hybridization. J. Microbiol. Methods 53, 199–209.

Lee, J.Y., Bang, Y.B., Rhee, J.H. and Choi, S.H. (1999). Two stage nested PCR effectiveness for direct detection of *Vibrio vulnificus* in natural samples. J. Food Sci. *64*, 157–162.

Linder, K., and Oliver, J.D. (1989). Membrane fatty acid and virulence changes in the viable but nonculturable state of *Vibrio vulnificus*. Appl. Environ. Microbiol. *55*, 2837–2842.

Litwin, C.M., and Byrne, B.L. (1998). Cloning and characterization of an outer membrane protein of *Vibrio vulnificus* required for heme utilization: regulation of expression and determination of the gene sequence. Infect. Immun. *66*, 3134–3141.

Liu, J.W., Lee, I.K., Tang, H.J., Ko, W.C., Lee, H.C., Liu, Y.C., Hsueh, P.R., and Chang, Y.C. (2006). Prognostic factors and antibiotics in *Vibrio vulnificus* septicaemia. Arch. Intern. Med. *166*, 2117–2123.

Liu, M., Alice, A.F., Naka, H., and Crosa, J.H. (2007). The HlyU protein is a positive regulator of *rtxA1*, a gene responsible for cytotoxicity and virulence in the human pathogen *Vibrio vulnificus*. Infect. Immun. *75*, 3282–3289.

Massad, G., and Oliver, J.D. (1987). New selective and differential medium for *Vibrio cholerae* and *Vibrio vulnificus*. Appl. Environ. Microbiol. *53*, 2262–2264.

Merkel, S.M., Alexander, S., Zufall, E., Oliver, J.D., and Huet-Hudson, Y.M. (2001). Essential role for estrogen in protection against *Vibrio vulnificus*-induced endotoxic shock. Infect. Immun. *69*, 6119–6122.

Miyasaka, J., Yahiro, S., Arahira, Y., Tokunaga, H., Katsuki, K., and Hara-Kudo, Y. (2005). Isolation of *Vibrio parahaemolyticus* and *Vibrio vulnificus* from wild aquatic birds in Japan. Epidemiol. Infect. *134*, 780–785.

Motes, M.L., and DePaola, A. (1996). Offshore suspension relaying to reduce levels of *Vibrio vulnificus* in oysters (*Crassostrea virginica*). Appl. Environ. Microbiol. *62*, 3875–3877.

Motes, M.L., DePaola, A., Cook, D.W., Yeazey, J.W., Hunsucker, J.C., Garthright, W.E., Blodgett, R.J., Chirtel, S.J. (1998). Influence of water temperature and salinity on *Vibrio vulnificus* in northern Gulf and Atlantic coast oysters (*Crassostrea virginica*). Appl. Environ. Microbiol. *64*, 1459–1465.

Nilsson, L., Oliver, J.D., and Kjelleberg, S. (1991). Resuscitation of *Vibrio vulnificus* from the viable but nonculturable state. J. Bacteriol. *173*, 5054–5059.

Nilsson, W.B., Parajpye, R.N., DePaola, A., Strom, M. (2003). Sequence polymorphisms of the 16S rRNA gene of *Vibrio vulnificus* is a possible indicator of strain virulence. J. Clin. Microbiol. *41*, 442–6.

NSSP. (2009). National Shellfish Sanitation Programme Guide for the Control of Molluscan shellfish 2009 Revision. Available at: http://www.fda.gov/Food/FoodSafety/Product-SpecificInformation/Seafood/FederalStatePrograms/NationalShellfishSanitationProgram/ucm046353.htm

Oliver, J.D., and Bockian, R. (1995). *In vivo* resuscitation, and virulence towards mice, of viable but nonculturable cells of *Vibrio vulnificus*. Appl. Environ. Microbiol. *61*, 2620–2623.

Oliver, J.D., and Kaper, J.B. (2007). *Vibrio* spp. In Food Microbiology : Fundamentals and Frontiers 3rd Ed., Doyle, M.P. and Beuchat, L.R., eds. (ASM Press, Washington, DC), pp. 343–379.

Oliver, J.D., Nilsson, L., Kjelleberg, S. (1991). Formation of nonculturable *Vibrio vulnificus* cells and its relationship to the starvation state. Appl. Environ. Microbiol. *57*, 2640–2644.

O'Neil, K.R., Jones, S.H., Grimes, D.J. (1992). Seasonal incidence of *Vibrio vulnificus* in the Great Bay estuary of New Hampshire and Maine. Appl Environ Microbiol. *58*, 3257–3262.

Ono, M., Inoue, Y., and Yokoyama, M., (2001). A cluster of *Vibrio vulnificus* infection in Kumamoto Prefecture (in Japanese). Infectious Agents Surveillance Reports, No 22.

Panicker, G., Meyers, M.L., Bej, A.K. (2004). Rapid detection of *Vibrio vulnificus* in shellfish and Gulf of Mexico water by real-time PCR. Appl. Environ. Microbiol. *70*, 498–507.

Parker, R.W., Maurer, E.M., Childers, A.B., Lewis, D.H. (1994). Effect of frozen storage and vacuum packaging on survival of *Vibrio vulnificus* in Gulf Coast oysters (*Crassostrea virginica*). J. Food Prot. *57*, 604–606.

Parvathi, A., Kumar, H.S., Karunasagar, I., and Karunasagar, I. (2004). Detection and enumeration of *Vibrio vulnificus* in oysters from two estuaries along the Southwest coast of India using molecular methods. Appl. Environ. Microbiol. *70*, 6909–6913.

Parvathi, A., Kumar, H.S., Karunasagar, I., Karunasagar, I. (2005). Study of the occurrence of *Vibrio vulnificus* in oysters in India by polymerase chain reaction (PCR) and heterogeneity among *V. vulnificus* by randomly amplified polymorphic DNA PCR and gyrB sequence analysis. Environ. Microbiol. *7*, 995–1002.

Rosche, T.M., Yano, Y., and Oliver, J.D. (2005). A rapid and simple PCR analysis indicates there are two subgroups of *Vibrio vulnificus* which correlate with clinical or environmental isolation. Microbiol. Immunol. *49*, 381–389.

Rosche, T.M., Binder, E.A., and Oliver, J.D. (2010). *Vibrio vulnificus* genome suggests two distinct ecotypes. Environ. Microbiol. Rep. *2*, 128–132.

Saraswathi, K., Barve, S.M. and Deodhar, L.P. (1989). Septicaemia due to *Vibrio vulnificus*. Trans. Royal Soc. Trop. Med. Hyg. *83*, 714.

Sanjuan, E., and Amaro, C. (2004). Protocol for specific isolation of virulent strains of *Vibrio vulnificus* Serovar E (Biotype 2) from environmental samples. Appl. Environ. Microbiol. *70*, 7024–7032.

Simonson, J., and Siebeling, J. (1986). Rapid serological identification of *Vibrio vulnificus* by anti-H coagglutination. Appl. Environ. Microbiol. *52*, 1299–1304.

Shapiro, R.L., Altekruse, S., Hutwagner, L., Bishop, R., Hammond, R., Wilson, S., Ray, B., Thompson, S., Tauxe, R.V., and Griffin, P.M. (1998). The role of Gulf Coast oysters harvested in warmer months in *Vibrio vulnificus* infections in the United States, 1988–1996. J. Infect. Dis. *178*, 754–759.

Starks, A.M., Schoeb, T.R., Tamplin, M.L., Parveen, S., Doyle, T.J., Bomeisl, P.E., Escudero, G.M., and Gulig, P.A. (2000). Pathogenesis of infection by clinical and environmental strains of *Vibrio vulnificus* in iron-dextran-treated mice. Infect. Immun. *38*, 5785–5793.

Strom, M.S., and Paranjpye, R.N. (2000). Epidemiology and pathogenesis of *Vibrio vulnificus* infection. Microb. Inf. 73, 177–188.

Suraslip, T., Longyant, S., Rukprataporn, S., Sridulyakul, P., Sithigorngul, P. and Chiavisuthangkura, P. (2011). Rapid and sensitive detection of *Vibrio vulnificus* by loop-mediated isothermal amplification combined with lateral flow dipstick targeted to rpoS gene. Mol. Cell. Probes *25*, 158–163.

Tamplin, M., Rodrick, G.E., Blake, N.J., and Cuba, T. (1982). Isolation and characterization of *Vibrio vulnificus* from two Florida estuaries. Appl. Environ. Microbiol. *44*, 1466–1470.

Tamplin, M.L., Martin, A.L., Ruple, A.D., Cook, W.D., Kaspar, and C.W. (1991). Enzyme immuno assay for identification of *Vibrio vulnificus* in seawater, sediment, and oysters. Appl. Environ. Microbiol. *57*, 1235–1240.

Thamlikitkul, V. (1990). *Vibrio* bacteraemia in Siriraj hospital. J. Med. Assoc. Thailand 73, 136–138.

Wang, S., and Levin, R.F. (2006). Discrimination of viable *Vibrio vulnificus* cells from dead cells in real-time PCR. J. Microbiol. Methods *64*, 1–8.

Warner, E., and Oliver, J.D. (2008). Population structures of two genotypes of *Vibrio vulnificus* in oysters (*Crassostrea virginica*) and seawater. Appl. Environ. Microbiol. 74, 80–85.

Wright, A.C., Hill, R.T., Johnson, J.A., Roghman, M.C., Colwell, R.R., Morris, J.G., Jr. (1996). Distribution of *Vibrio vulnificus* in the Chesapeake Bay. Appl. Environ. Microbiol. *62*, 717–724.

Wright, A.C., Garrido, V., Debuex, G., Farrel-Evans, M., Mudbidri, A.A., and Otwell, W.S. (2007). Evaluation of postharvest processed oysters by using PCR-based Most-Probable-Number enumeration of *Vibrio vulnificus* bacteria. Appl. Environ. Microbiol. 73, 7477–7481.

Epidemiology and Molecular Biology of *Vibrio cholerae* 10

Shah M. Faruque and John J. Mekalanos

Abstract

Vibrio cholerae belonging to O1 and O139 serogroups cause cholera, a life-threatening diarrhoeal disease that spreads through consumption of water and food contaminated with the pathogen. Other serogroups of *V. cholerae* are also occasionally associated with mild to moderate enteric infections. Although *V. cholerae* is a human pathogen, the bacteria are part of the normal aquatic flora in estuarine and brackish waters, and thus are able to persist in the environment outside the human host. The ability of *V. cholerae* strains to cause disease in humans depends on their virulence gene content, which varies between pathogenic and non-pathogenic strains. Horizontal transfer of critical virulence genes among different *V. cholerae* strains, as well as microevolution of bacterial genes contribute significantly to the emergence of *V. cholerae* strains with altered antigenic and pathobiological properties. Seasonal cholera epidemics may selectively enrich genetic variants with unique properties that promote transmission or environmental persistence. The ecosystem comprising *V. cholerae*, the aquatic environment and the human host offers an understanding of the complex relationship between pathogenesis and the evolution of a typical waterborne bacterial pathogen.

Introduction

Vibrio cholerae as a species is a normal flora of the aquatic environment and is well-defined on the basis of biochemical tests and DNA homology studies. However, the species is not homogeneous with regard to pathogenic potential, and the vast majority of *V. cholerae* strains in the environment are thought to be non-pathogenic to humans. Of more than 200 known serogroups of *V. cholerae*, two serogroups, O1 and O139, account for the cholera cases both in endemic and epidemic settings (Faruque *et al.*, 1998c). Other serogroups of *V. cholerae* occasionally cause gastrointestinal or extra-intestinal illness utilizing similar or distinctly different virulence mechanisms (Morris, 1994; Mukhopadhyay, 1995, 2001). Even within the O1 and O139 serogroups, there are genetic variations that determine antibiotic resistance, phage resistance, and other phenotypes that may be linked to environmental fitness (Guidolin, 1987; Kaper *et al.*, 1995; Faruque *et al.*, 1998c). However, the ability of *V. cholerae* strains to cause severe enteric infection in humans depends primarily on their virulence gene content. The major virulence associated genes which encode colonization factors, and the cholera enterotoxin are part of larger genetic elements comprised clusters of genes (Faruque *et al.*, 1998c). Pathogenic *V. cholerae* appear to have evolved from non-pathogenic environmental strains fairly recently in the evolutionary history of the pathogen by acquisition of virulence genes.

Pathogenesis of *V. cholerae* relies on the synergistic effect of a number of pathogenic factors produced by *V. cholerae*. Strains capable of causing cholera invariably produce cholera toxin (CT) and a pilus colonization factor known as toxin coregulated pilus (TCP). All pathogenic strains may also produce a number of putative accessory virulence associated factors, e.g. the mannose sensitive haemagglutinin (MSHA) pilus, haemolysins, and the RTX toxin. In addition certain non-O1 non-O139 strains may also produce other toxins or virulence-associated factors including effectors of a type III secretion system. However, the roles of the accessory virulence factors in cholera pathogenesis are not well established, and recent studies are beginning to reveal that at least some of these putative virulence factors also play a role in environmental fitness of the pathogen and horizontal transfer of DNA (Watnick *et al.*, 1999; Chiavelli *et al.*, 2001). The major virulence factors themselves might also be involved in environmental persistence given that it was recently shown that TCP pili are involved in biofilm formation on chitin surfaces (Reguera and Kolter, 2005).

A distinctive epidemiological feature of cholera is its appearance in a seasonal pattern in endemic areas, in explosive outbreaks and often appear to begin in several distinct foci simultaneously (Glass *et al.*, 1982; Kaper *et al.*, 1995). This observation indicates a possible role of environmental factors in triggering the epidemic process as well as very rapid spread of the disease through water contaminated with the pathogen. The epidemiological success of a pathogenic *V. cholerae* strain may depend on its environmental fitness in addition to the ability to produce major virulence factors (Rahman *et al.*, 2008).

Many aspects of the pathogen and the disease are being further studied, particularly molecular aspects of the ecology of *V. cholerae*, the inter-epidemic reservoirs, and the role of virulence factors in the natural selection and possible ecological association of the organism with the aquatic environment as well as with the human host. Recent studies are also beginning to reveal the origin of the major virulence genes and the mechanism of their horizontal transfer among strains.

Nomenclature and classification of *Vibrio cholerae*

Vibrio cholerae is a Gram-negative bacterium belonging to the γ-subdivision of the family *Proteobacteriaceae*. Taxonomic studies on related Gram-negative rod-shaped bacteria indicated a close relationship between three genera, *Vibrio*, *Aeromonas* and *Plesiomonas* (Kaper *et al.*, 1995; Faruque *et al.*, 1998c). Members of the genus *Vibrio* can be differentiated from those of allied genera on the basis of biochemical characteristics (West and Colwell, 1984). The genus *Vibrio* contains several species, of which *V. cholerae*, *V. parahaemolyticus* and *V. vulnificus* are the most important pathogens of humans. *V. cholerae* is classified on the basis of its somatic antigens (O-antigens) into serovars or serogroups, and there are more than 200 serogroups of *V. cholerae*. Until 1993, the serogroup O:1 was supposed to include all the strains responsible for epidemic and endemic cholera. The O1 serogroup has two major serotypes, Ogawa and Inaba; a Hikojima serotype has been reported rarely. These serotypes can be further distinguished into two biotypes, classical and El Tor based on some biochemical properties, and susceptibility to bacteriophages (Sakazaki, 1970).

In the conventional tests, classical strains of *V. cholerae* O1 are negative for chicken red blood cell agglutination (CCA), haemolysis of sheep erythrocytes, Voges–Proskauer (VP) test and are sensitive to polymyxin B and to phage IV but resistant to phage 5. Conversely, El Tor strains are positive for haemolysis of sheep erythrocytes, CCA, and VP test but resistant to polymyxin B and phage IV but sensitive to phage 5. However, over period of time, most of these tests have proven to be inadequate in classifying strains of *V. cholerae* O1 into their biotype. Currently, haemolysis of sheep erythrocytes is of limited use since, apart from isolates in the early 1960s, most isolates of *V. cholerae* O1 El Tor are negative for haemolysis of sheep erythrocytes and therefore non-haemolytic (Barrett and Blake, 1981). On the other hand, the CCA and VP tests are more consistent in distinguishing the two biotypes. Using molecular techniques biotype specific alleles of some critical virulence genes of *V. cholerae* have been identified. Although these alleles encode identical function, they differ partly in their DNA sequence. Two major virulence genes of *V. cholerae* O1 strains, *tcpA* and *hlyA* possess classical- and El Tor-specific alleles. Other genes which can be useful genetic markers to distinguish between the classical and El Tor biotypes include the repressor gene (*rstR*) of the CTX prophage (Waldor and Mekalanos, 1996) which carry the cholera toxin genes (*ctxAB*) as well as the *ctxB* gene with sequence slightly different between the two biotypes. Therefore, the current biotyping scheme based on phenotypic traits could be augmented by genotypic tests. For example, *ctxB* genotyping largely matches with the conventional biotyping scheme, except for the US Gulf Coast El Tor strains. Since, typically classical strains possess classical type *rstR* and *ctxB* genes and El Tor strains possess El Tor type *rstR* and *ctxB* genes, the possession of classical *ctxB* gene by the US Gulf Coast strains despite being El Tor is an anomaly (Nair *et al.*, 2008).

There are over 200 known serogroups of non-01 *V. cholerae*. These vibrios possess biochemical and morphological characteristics very similar to those of *V. cholerae* O1 but are non-agglutinable with polyvalent O:1 antiserum. Such vibrios are agglutinable by their own antisera. Non-O1 serogroups of *V. cholerae* had been mostly associated with sporadic cases of diarrhoea and extraintestinal infections (Janda *et al.*, 1998) until 1993, when a large cholera-like outbreak in Bangladesh and India was caused by a *V. cholerae* non-O1 strain (Albert *et al.*, 1993). This organism did not belong to any of the 138 O serogroups for *V. cholerae* described until then, but belonged to a new serogroup, which was later designated as O139 (Shimada *et al.*, 1993). Since then *V. cholerae* O139 has persisted as a second etiologic agent of cholera. The incidence of cholera due to *V. cholerae* O139 is rare now, and the last large outbreak of *V. cholerae* O139 infection occurred in Bangladesh in 2002 (Faruque *et al.*, 2003). However, there are now two recognized serogroups, O1 and O139, that have been associated with epidemic disease, although there are also strains of these serogroups which do not produce cholera toxin, and are not involved in epidemics.

The hybrid biotypes

A remarkable event in recent years has been the emergence of strains of *V. cholerae* O1 which possess traits of both the classical and El Tor biotypes. Nair *et al.* (2002) reported polymyxin B-sensitive El Tor strains of *V. cholerae* O1 from southern parts of Bangladesh. A retrospective analysis of *V. cholerae* O1 strains isolated from hospitalized cases of diarrhoea in Matlab, a rural area of Bangladesh, revealed that some strains could not be biotyped as El Tor or classical using the conventional biotyping tests, and these strains were designated as the 'Matlab variants' (Nair *et al.*, 2002). Genotypically, the Matlab variants carried the set of typical virulence genes. However, in these strains, the classical and El Tor alleles of *tcpA* and the CTX prophage repressor gene (*rstR*) were found in different combination (Nair *et al.*, 2002). Furthermore, the nucleotide-sequence-derived CtxB amino acid sequence of the Matlab variants was of the classical biotype (Safa *et al.*, 2006). Ribotyping was used as a phylogenetic tool for assessing similarity between the hybrid strains versus standard strains of classical and El Tor biotypes. Ribotyping demonstrated that Matlab variants were similar to the El Tor reference strain and were completely different from the classical reference strain suggesting that these strains originated from an El Tor progenitor. Therefore, overall the Matlab strains were of the El Tor lineage displaying some traits of the classical biotype Similar strains were later isolated from cholera epidemics in Mozambique (Ansaruzzaman *et al.*, 2004). Extensive analysis showed that the Mozambique strain carries two copies of CTX prophage of the classical type located on the small chromosome in a tandem array (Faruque *et al.*, 2007).

Recently another variant of the El Tor strains has been identified. This variant strain harbours a hybrid CTX prophage which has an El Tor type rstR gene (rstR2) but the ctxB gene is of the classical type (Nair *et al.*, 2006). This strain has been classified as 'altered El Tor,' which refers to an atypical El Tor strain containing CTX prophage with an El Tor type *rstR* gene (*rstR2*) but produces ctxB of the classical type. On the other hand, the 'hybrid El Tor,' refers to an atypical El Tor strain which carries a typical CTX^{Class} prophage with a classical type rstR gene (*rstR1*) and also produces ctxB of the classical type.

Epidemiology of cholera

Hallmarks of the epidemiology of cholera include (a) a high degree of clustering of cases by location and season, (b) highest rates of infection in children 1–5 years of age in endemic areas, (c) antibiotic resistance patterns of the predominant *V. cholerae* strain that frequently change from year to year, (d) clonal diversity of epidemic strains and (e) protection against the disease being afforded by pre-existing immunity and improved sanitation/hygienic conditions.

Cholera is an ancient scourge, and at least seven distinct pandemics of cholera have occurred since the first recorded pandemic began in 1817 (Kaper *et al.*, 1995; Pollitzer, 1959). The seventh pandemic originated in Indonesia, but the other six arose in the Indian subcontinent (usually in the Ganges Delta region) and spread from there to other continents affecting many countries and extending over many years (Snow 1855; Pollitzer 1959; Laval 1989). The sixth pandemic and presumably the fifth pandemic were caused by *V. cholerae* O1 of the classical biotype. The current seventh pandemic is the most extensive in geographic spread and duration, and the causative agent is *V. cholerae* O1 of the El Tor biotype. The seventh pandemic began in 1961 from the island of Sulawesi, in Indonesia and spread to other nearby islands and then to the Philippines, Sabah, and Taiwan (Kamal, 1974). During 1963 to 1969, the pandemic spread to Malaysia, Thailand, Burma, Cambodia, Vietnam, India, Bangladesh and Pakistan. Soon after El Tor cholera reached Afghanistan, Iran, Iraq, and nearby republics within the Soviet Union (Kamal, 1974). By 1970, El Tor cholera invaded the Arabian Peninsula, Syria, Jordan, and Israel (Cohen *et al.*, 1971). The seventh pandemic reached the sub-Saharan West Africa in the early seventies in explosive outbreaks causing over 400,000 cases with a high case fatality (Goodgame and Greenough, 1975). In this epidemic, cholera spread along the coast and into the interior through waterways, and further disseminated into the interior of the Sahelian states. In 1970, cholera outbreaks were reported in 36 countries, of which 28 were newly affected countries and 16 were in Africa (Kaper *et al.*, 1995).

In January 1991, the seventh pandemic reached South America and caused an epidemic that began in Peru, and thus caused the return of cholera to the continent after more than a century (Levine, 1991; Pan American Health Organization, 1991; Ries *et al.*, 1992). Subsequently epidemic cholera was reported in neighbouring Ecuador, and then in Colombia. In each of these countries low socioeconomic populations lacking proper drinking water and sanitation facilities were mostly affected (Pan American Health Organization, 1991). By April 1991, the pandemic reached Chile (Levine, 1991), and cholera began to spread along the Pacific coast of South America to progressively enter more countries in South and Central America. According to an estimate of the Pan American Health Organization during 1991–1992 there were 750,000 cases of cholera in the Americas with 6500 deaths.

In July 1994, one of the worst cholera outbreaks occurred in Goma, eastern Zaire, in (Siddique *et al.*, 1995) with an estimated 12,000 deaths among Rwandan refugees to Zaire. Conflicts between tribes in Rwanda had displaced nearly a million people to neighbouring Zaire and were sheltered in refugee camps. The seventh pandemic continues and the El Tor biotype of *V. cholerae* O1 cause seasonal outbreaks in many developing countries, especially in Bangladesh and India. However, in 1992, *V. cholerae* belonging to a non-O1 serogroup (now referred to as O139) caused large epidemics of cholera in India and Bangladesh, and spread to some other countries as described later.

Surveillance of cholera in Bangladesh, has provided important information regarding the epidemiology of the disease, and temporal changes in the properties of *V. cholerae* strains isolated from different epidemics. Bangladesh is situated in the Ganges delta, the region from where all the cholera pandemics except for the seventh pandemic started. Systematic surveillance of cholera has been carried out in Bangladesh by the International Centre for Diarrhoeal Disease Research (ICDDR,B) and the former Pakistan Seato Cholera Research Laboratory during a period of over 35 years (McCormack *et al.*, 1969; Martin *et al.*, 1969; Samadi *et al.*, 1983; Siddique *et al.*, 1991, 1992). A number of studies have shown that in Bangladesh epidemic outbreaks usually occur twice during a year, with the highest number of cases during September to December, and somewhat smaller peak of cholera cases between March and May (reviewed by Faruque *et al.*, 1998c). Cholera surveillance in Matlab, a rural subdistrict of Bangladesh, showed that cholera outbreaks occur every year without exception, with yearly attack rates varying from 0.5 to 8.0 per 1000 population. Until 1970, more than 90% of cholera in Bangladesh was caused by *V. cholerae*

O1 of the classical Inaba serotype. However, by 1972, 85% of all cases were reported to be due to the classical strain of the Ogawa serotype (Khan *et al.*, 1986). The El Tor biotype of *V. cholerae* O1 appeared in Bangladesh during 1973 and replaced the classical biotype. However, in 1982, the classical biotype re-emerged as a predominant epidemic strain in Bangladesh (Khan *et al.*, 1983, 1986) and coexisted with the El Tor strain until 1992.

In late 1992 and 1993, a new strain of *V. cholerae*, later serotyped as O139 appeared in India and Bangladesh (cholera working group, 1993; Ramamurthy *et al.*, 1993) and was associated with large cholera epidemics. At first, this new strain totally displaced the existing O1 strains, but during 1994–95, the O139 vibrios were replaced in most northern and central areas of Bangladesh by a new El Tor clone (Faruque *et al.*, 1997). Since the second half of 1995, both O1 and O139 have been a significant cause of infection and morbidity in Dhaka, as well as in many rural districts of Bangladesh (Faruque *et al.*, 1997a,b). Cholera cases or outbreaks due to *V. cholerae* O139 were reported in Pakistan, Nepal, China, Thailand, Kazakhstan, Afghanistan and Malaysia (Chongsanguan, 1993; Swerdlow, 1993), and imported cases were reported from the United Kingdom and the United States. Until 2002, *V. cholerae* O139 continued to cause cholera outbreaks in India and Bangladesh and coexisted with *V. cholerae* of the El Tor biotype. Notably, *V. cholerae* O139 was the primary cause of cholera in Dhaka and adjoining areas of Bangladesh during March and April 2002, causing an estimated 35,000 cases (Faruque *et al.*, 2003). However, currently O1 strains dominate in this region.

Regional distribution of recent cholera outbreaks

During the decade 1995 to 2005, a total of 632 outbreaks of cholera were reported worldwide, of which 66% occurred in sub-Saharan Africa followed by 16.8% in South-East Asia (Griffith *et al.*, 2006). Sub-Saharan Africa accounted for 88% of the total number of reported cases. The highest number of multiple outbreak reports was in the Eastern Democratic Republic of Congo and western Uganda. Eastern and south-eastern Africa also had high incidences, along with parts of West Africa. In South Africa, the outbreak reports mostly came from its western provinces near the border with Mozambique, which reported cholera outbreaks throughout each province. The most commonly cited risk factor in West and Southern Africa, was heavy rainfall and flooding, which accounted for 33% and 39% of all the risk factor reports, respectively. However, in East Africa rainfall and flooding accounted for only 10% of reports; a further 36% of reports identified a refugee setting or internally displaced persons setting as the primary risk factor. In Western Africa, a refugee setting accounted for only 5%, and in Southern Africa, there were no outbreaks attributed to refugee settings. In West, Southern, and East Africa, contamination of water source was the second most common risk factor reported, and represented 32%, 30%, and 24% of the total, respectively. Water source contamination was most common In Central Africa, accounting for 30% of the reported risk factors. Globally, the three most common risk factors were water source contamination (29%), rainfall and flooding (25%), and refugee settings (13%). In South America and East Asia, the most commonly noted risk factor for cholera outbreaks was associated with contaminated food, and foodborne transmission accounted for 32% of outbreaks in South America and 71% in East Asia. Since 1999, the reported cases of cholera declined greatly in Latin America.

The annual cholera statistics released by the World Health Organization showed a dramatic rise in the number of cholera cases during 2006 reaching the level of the late 1990s. Cholera was notified from 52 countries in 2005 with the number of cases registering an overall increase of 79% compared with the number of cases reported in 2005, and the case fatality rate rising from 1.72% in 2005 to 2.66% in 2006 (Nair *et al.*, 2008).

In recent times, cholera epidemic in Zimbabwe began in August 2008, and spread across the country (Islam *et al.*, 2011) to Botswana, Mozambique, South Africa and Zambia. By January 2010 there had been 98,741 reported cases and 4293 deaths making it the deadliest African cholera outbreak in the last 15 years. The principal cause of the outbreak was reported as lack of access to safe water in urban areas and communities. The *V. cholerae* strain in Zimbabwe was confirmed as belonging to serogroup O1, serotype Inaba. Of the 88 isolates that were analysed in details, 83 were classified as 'altered El Tor' and 5 as 'hybrid El Tor' strains.

The Caribbean country Haiti experienced a powerful earthquake which devastated the country on 12 January 2010. Cholera appeared in the poverty-stricken country in an explosive outbreak that began in late October 2010 in the rural Artibonite Department, about 100 kilometres north of the capital, Port-au-Prince (Chin *et al.*, 2011). By March 2011 the epidemic killed 4672 people, with 252,640 reported cases. By the first 10 weeks of the epidemic, cholera spread to all of Haiti's ten departments or provinces. In November 2010, the first cases of cholera were reported in the Dominican Republic and a single case in Florida, United States, whereas in January 2011 a few cases were reported in Venezuela. As of late September, 2011, some 6435 deaths were reported. The neighbouring Dominican Republic reported 135 cholera deaths as well.

Molecular epidemiology

The continually changing relative prevalence of different clones of pathogenic *V. cholerae* has prompted numerous studies of the molecular basis of these changes. Improvement in DNA analysis techniques has enhanced molecular epidemiological studies through the introduction of several new typing methods. Multilocus enzyme electrophoresis analysis and restriction fragment length polymorphism (RFLP) analysis of different genes have facilitated the study of *V. cholerae* epidemiology (Chen *et al.*, 1991; Faruque *et al.*, 1992, 1993, 2003; Cook *et al.*, 1993). RFLPs of conserved rRNA genes have been used to distinguish *V. cholerae* strains into different clonal groups. For example, ribotyping indicates that toxigenic US Gulf coast isolates are clonal, and that they differ from other seventh pandemic isolates (Wachsmuth *et al.*, 1994). On the other hand, analysis of isolates from the 1991 Latin American epidemic showed that these were clearly related to seventh pandemic isolates from other parts of the world (Faruque and Albert, 1992; Wachsmuth *et al.*, 1993, 1994), and hence the Latin American cholera epidemic was an extension of the seventh pandemic. Analysis of toxigenic El Tor strains by multilocus enzyme electrophoresis has also been used to group the El Tor strains into major clonal groups. The clones seem to reflect broad geographical and epidemiological associations (Wachsmuth *et al.*, 1994).

Based on heterogeneity within the *ctxB* gene encoding the B subunit of cholera toxin, three *ctxB* genotypes of the *V. cholerae* O1 strains have been identified. Amino acid residue substitutions at positions 39, 46, and 68, allowed all classical and US Gulf Coast El Tor strains to be categorized as genotype 1, the Australian El Tor strains as genotype 2, and the El Tor strains of the seventh pandemic and the Latin American epidemic as genotype 3. The ctxB gene in classical strains is not identical to the ctxB gene carried by the El Tor strains except for the US Gulf Coast El Tor clone (Olsvik *et al.*, 1993). The US Gulf Coast hybrid El Tor strains that harbour the classical ctxB gene have been associated with sporadic outbreaks in the United States (Blake, 1994) and, until 2004, had not been reported anywhere else in the world. In 2004, hybrid El Tor strains that carry classical ctxB gene were isolated from cholera patients in Matlab, Bangladesh (Safa *et al.*, 2006), and in Beira, Mozambique (Ansaruzzaman *et al.*, 2004). In 2006, Nair *et al.* reported that the prototype seventh pandemic El Tor strains had been replaced by hybrid El Tor strains in Bangladesh (Nair *et al.*, 2006). Latest reports showed that the hybrid El Tor strains have spread more widely in Asia and Africa (Safa *et al.*, 2008).

Microarray-based comparative genomic analysis of *V. cholerae* isolates (Dziejman *et al.*, 2002) identified genes that are shared by all pandemic strains, as well as genes specific to seventh pandemic O1 El Tor and O139 strains. The latter genes were contained on two chromosomal 'islands' namely Vibrio seventh pandemic island-1 and -2 (VSP-1 and VSP-2) and may define genes involved in the endemic as well as global success of the seventh pandemic clone. Notably, characterization of ten *V. cholerae* O1 strains of Australian origin revealed the absence of genes that make up the VSP islands in all of the strains indicating their pre-seventh pandemic ancestry (Safa *et al.*, 2009).

Comparative molecular analysis of O1 El Tor strains and the epidemic O139 strains suggested that the O139 strains were derived from the El Tor strains by serotype-specific genetic changes (Faruque *et al.*, 1994, Wachsmuth *et al.*, 1994).

Antibiotic resistance of toxigenic *V. cholerae*

Although the main course of treatment in cholera is rehydration therapy, administration of an effective antibiotic can shorten the duration of diarrhoea. Temporal changes in the prevalence of different clones of *V. cholerae* is often reflected in shifts in antibiotic susceptibility patterns of the clinical isolates. Pattern of antimicrobial resistance in *V. cholerae* has been changing frequently. In a worldwide survey carried out in 1976, only 3% of the randomly collected strains were found to be resistant to commonly used antibiotics. This scenario however, changed rapidly primarily owing to the indiscriminate use of antibiotics, and 3 years later in another survey conducted in Bangladesh, about 18% of the strains were found to be resistant to a number of commonly used drugs (Ghosh and Ramamurthy, 2011).

On several occasions multiple antibiotic-resistant (MAR) strains have been associated with epidemics. In 1979, an outbreak of cholera in Matlab, a rural sub-district of Bangladesh was caused by MAR strains (Glass *et al.*, 1980, 1983). Screening of isolates from the outbreak showed that 16.7% of the isolates were resistant to five antibiotics – tetracycline, ampicillin, kanamycin, streptomycin and trimethoprim–sulfamethoxazole – while another 10% of the isolates were resistant to any four of these antibiotics including tetracycline. Epidemiological assessment of the outbreak suggested that the outbreak began from the introduction into the area of a single multiple drug-resistant strain of *V. cholerae* O1 which carried a conjugative plasmid encoding resistance to multiple antibiotics (Glass *et al.*, 1983). By 1986, the drug resistance pattern changed again, and the predominant strains were found to be susceptible to tetracycline, streptomycin, chloramphenicol, amoxicillin or nalidixic acid (Nakasone, 1987). However, during 1988 and 1989, nearly all classical *V. cholerae* isolated in Bangladesh were resistant to tetracycline, whereas El Tor strains were sensitive to the drug (Siddique *et al.*, 1989). Tetracycline-resistant El Tor biotype strains re-emerged during the 1991 epidemic in Bangladesh (Siddique *et al.*, 1992). In October 1995, emergence of nalidixic acid-resistant *V. cholerae* O1 was observed in southern India (Jesudason and Saaya, 1997).

The susceptibility of *V. cholerae* O1 strains to certain antibiotics changed in regard to the isolation time and geographical location. Between March 1994 and December 1996, 80–100% of *V. cholerae* O1 isolates in Kenya and south Sudan, and 65–90% of isolates in Somalia were sensitive to tetracycline (Materu *et al.*, 1997), whereas all isolates in Tanzania and Rwanda were resistant to the drug. In Kenya and Somalia, the percentage of isolates resistant to chloramphenicol and cotrimoxazole markedly increased from 15% in 1994 to more than 90% in 1996 (Materu *et al.*, 1997). The O139 serogroup of *V. cholerae* which emerged during 1992–1993, were sensitive to tetracycline (cholera working group, 1993). Although *V. cholerae* O139 showed a trend of increased resistance to trimethoprim–sulfamethoxazole, it was more susceptible to ampicillin and tetracycline than the O1 serogroup of *V. cholerae* (Sciortino *et al.*, 1996). A self-transmissible transposon-like element (SXT element) was found to encode resistance to sulfamethoxazole, trimethoprim and streptomycin in *V. cholerae* O139 (Waldor *et al.*, 1996). Comparison of the antibiotic resistance patterns between the O139 strains isolated during 1992 and 1993 and those isolated in 1996 and 1997 in India showed that the later strains were susceptible to co-trimoxazole, unlike the O139 strains from 1992 and 1993. In Kolkata, India a trend of reduced susceptibility to fluoroquinolones amongst *V. cholerae* isolates was reported during 1998 (Garg *et al.*, 2001). During 2004–2006, the Inaba serotype of *V. cholerae* O1 emerged as the predominant serotype in Delhi, India, and all isolates were resistant to nalidixic acid, furazolidone and co-trimoxazole (Das *et al.*, 2008). Multiple antibiotic-resistant *V. cholerae* O1 Inaba isolates exhibited increased resistance to ciprofloxacin with MIC > 4 μg/ml, but largely remained susceptible to other antibiotics such as gentamicin, tetracycline and chloramphenicol.

In recent times cholera outbreaks were reported in sub-Saharan African countries and in Caribbean Haiti. In both these regions the isolates were resistant to one or more antibiotics. Isolates of *V. cholerae* obtained from the epidemic in Zimbabwe were susceptible to tetracycline, doxycycline, ciprofloxacin and azithromycin, but susceptibility to tetracycline and azithromycin was diminished on the basis of minimal inhibitory concentration (MIC). Isolates from the Haitian epidemic were determined to be susceptible to tetracycline and erythromycin but resistant to trimethoprim–sulfamethoxazole and nalidixic acid (Chin *et al.*, 2011).

The *V. cholerae* genome

The *V. cholerae* genome comprises two unique circular chromosomes (Trucksis *et al.*, 1998). The first complete genome sequence of a representative *V. cholerae* O1 El Tor biotype strain N16961 was reported by Heidelberg *et al.* (2000). Since then a number of *V. cholerae* isolates have been fully sequenced. Based on the genome sequence of N16961, the large chromosome designated as ChrI and the small chromosome designated as ChrII consist of 2,961,146 and 1,072,314 bp, respectively. These chromosomes together encode 3885 open reading frames. ChrI was found to contain most of the recognizable genes required for essential cell functions. These genes include those involved in DNA replication, transcription, translation, and cell-wall biosynthesis. Genes involved in pathogenesis including those encoding toxins, surface antigens, and adhesins are also located on the large chromosome. In contrast, the small chromosome contains a larger fraction (59%) of hypothetical genes compared with the large chromosome (42%). The small chromosome also carries the integron island, a gene capture system. The GC content of the two chromosomes is almost the same, suggesting that the two chromosomes have been cohabitants for a considerably long evolutionary time. Therefore, the origination of the small chromosome possibly occurred by excision from a single large ancestral chromosome. Alternatively, the small chromosome could originally have been a megaplasmid captured by an ancestral *Vibrio* species.

Virulence genes in *V. cholerae* and evolution of pathogenic strains

The ability of *V. cholerae* strains to cause severe enteric infection in humans depends on their virulence gene content, which varies between pathogenic and non-pathogenic strains. Strains capable of causing cholera invariably carry genes encoding a potent enterotoxin, cholera toxin (CT) and a pilus colonization factor called toxin co-regulated pilus (TCP) (Herrington *et al.*, 1988; Faruque *et al.*, 1998c). Biochemical and genetic analysis, as well as studies in experimental animals and human volunteers, have revealed many details of cholera pathogenesis. In brief, orally ingested bacteria survive passage through the stomach and then use motility and chemotaxis functions to adhere to, and penetrate the mucus coat of the upper intestinal epithelium. Vibrios then coordinately express two critical sets of genes encoded by the *tcp* and *ctx* operons (Faruque *et al.*, 1998c). TCP is used to colonize the intestinal epithelium whereas the enterotoxin CT disrupts ion transport by intestinal epithelial cells leading to severe diarrhoea. These major virulence genes exist as part of larger gene clusters. These include the TCP pathogenicity island (Kovach *et al.*, 1996) which encodes TCP, and the CTX prophage which encodes CT (Waldor and Mekalanos, 1996). Since the natural environmental reservoir for *V. cholerae* is aquatic, ingestion by a human, passage through the stomach acid barrier, and subsequent colonization of the intestine understandably involves the expression of many additional genes that are crucial for survival and adaptation of the bacterium in the gastrointestinal environment.

Horizontally acquired gene clusters

The pathogenesis of cholera involves virulence factors encoded by a number of genes that exist in clusters. The *ctxAB* genes encoding cholera toxin reside in the genome of a lysogenic filamentous phage (CTXΦ) (Waldor and Mekalanos, 1996), whereas genes encoding the major colonization factor, TCP, are part of a gene cluster also referred to as the TCP pathogenicity island (Ogierman *et al.*, 1993; Kovach *et al.*, 1996).The major subunit of TCP is encoded by the *tcpA* gene, a gene that undergoes major sequence variation within the species (Nandi *et al.*, 2000; Karaolis *et al.*, 2001; Boyd and Waldor, 2002). Immediately downstream of the *tcp* gene cluster is the *acf* gene cluster (Harkey *et al.*, 1994; Everiss *et al.*, 1996), which also encodes an accessory colonization factor although the exact nature of this determinant is not clear. A putative integrase gene (*int*) is located downstream of and adjacent to the TCP/ACF gene cluster. A region of nearly 40 kb, including the TCP/ACF clusters flanked on both sides by a putative 20-bp *att*-like attachment sequence, an integrase, and a transposase gene, constitutes the TCP island alternatively designated as the VPI (*Vibrio* pathogenicity island). The structure, GC content, and codon usage within the VPI suggests that the acquisition of VPI by *V. cholerae* was a recent event in its evolutionary history.

The CTXΦ genome comprises a number of open reading frames (ORFs) other than *ctxAB*, including those required for CTXΦ morphogenesis (Waldor and Mekalanos, 1996; Waldor *et al.*, 1997) located on a 4.5 kb 'core region'. The ORFs in the RS2 region (*rstR*, *rstA2* and *rstB2*) encode products required for the integration, replication and regulatory functions of CTXΦ. Under appropriate conditions, toxigenic *V. cholerae* strains produce extracellular CTXΦ particles (Waldor and Mekalanos, 1996; Faruque *et al.*, 1998a,b) that can infect recipient non-toxigenic strains of *V. cholerae* and convert them to toxigenic strains. Remarkably, the CTXΦ uses TCP pili as its receptor for infecting recipient cells. Thus, it appears that acquisition of the TCP pathogenicity island is the most likely initial genetic event required for the evolution of epidemic strains. However, the mechanism of horizontal transfer of the TCP pathogenicity island is not clear.

Other clusters of genes having putative additional roles in pathogenesis include the RTX toxin gene cluster (Lin *et al.*, 1999) as well as genes for a new type IV pilus (Fullner and Mekalanos, 1999), The structures of these gene clusters also suggest that these genes are transferable horizontally, and have been recently acquired by *V. cholerae*. A mobile gene cluster which carries genes for multiple antibiotic resistance in *V. cholerae* is the SXT constin (Hochhut *et al.*, 2002). It is a self-transmissible transposon-like element encoding resistance to sulfamethoxazole, trimethoprim and streptomycin, the resistance genes being clustered together in a 9.4 kb region. Although the SXT element is not directly involved in pathogenesis, alteration in antibiotic sensitivity of epidemic strains can significantly influence the epidemiology of cholera.

V. cholerae DNA microarrays have been recently used to perform a comparative genomic analysis of *V. cholerae* isolates (Dziejman *et al.*, 2002). These studies have identified genes specific to the seventh pandemic clone of *V. cholerae*, and these genes are grouped into two chromosomal islands, the VSP-1 and VSP-2 whose GC content suggests they were also recently acquired by *V. cholerae*. A 57 kb chromosomal insert encoding genes for neuraminidase (*nanH*) and amino sugar metabolism (*nan*) was proposed to correspond to a pathogenicity island designated VPI-2 (Jermyn and Boyd, 2002). Based on microarray analysis that detected variability of the *nan-nanH* gene cluster between non-toxigenic and pre-seventh pandemic TCP positive *V. cholerae* strains, it has been suggested that VPI-2 must have been acquired by these strains shortly after or just before the emergence of toxigenic classical and seventh pandemic El Tor O1 strains. Thus, the *V. cholerae* genome appears to contain considerable number of gene clusters that are potentially mobile horizontally, although the mechanisms involved in their efficient lateral transfer are not well known. In addition, the *V. cholerae* genome has also been found to contain a distinctive class of integrons, which constitutes an efficient gene capture system (Mazel *et al.*, 1998). Nearly 12% of the small chromosome (Heidelberg *et al.*, 2000) of *V. cholerae* consists of a large gene cluster termed the 'integron island'. Integrons are gene expression elements that acquire open reading frames and convert them to functional genes. This region is hypervariable between different strains of *V. cholerae*. The contribution of integrons in the evolutionary biology of *V. cholerae* is yet to be fully explored.

Genomic variation and *V. cholerae* recombination

As noted earlier, horizontal transfer of virulence-associated genes by phages is believed to play a major role in the evolution of *V. cholerae*. The best understood example in *V. cholerae* is the CTX . The emergence of the O139 serogroup is attributed to the acquisition of O139 lipopolysaccharide antigen genes by an El Tor strain (Bik *et al.*, 1995). However the mechanism how these LPS genes, or other gene clusters and pathogenicity islands, are acquired naturally by *V. cholerae* is yet to be fully elucidated. There have been frequent changes in the drug resistance pattern of clinical *V. cholerae* strains (Glass *et al.*, 1980, 1983; Faruque *et al.*, 1998c). The rapidly changing pattern of antibiotic resistance observed among toxigenic *V. cholerae* isolated from different outbreaks, suggests that there is also a substantial mobility of genetic elements encoding antibiotic resistance in *V. cholerae*. While conjugative plasmids and a single generalized transducing phage for *V. cholerae* are known to exist (Ogg *et al.*, 1981), other studies are beginning to reveal the role of additional vibriophages in the horizontal transfer of chromosomal segments as well as selection of different clones of *V. cholerae* (Faruque *et al.*, 2005). *V. cholerae* has also been reported to become naturally competent to uptake exogenous DNA, while growing on a chitin substrate (Meibom *et al.*, 2005). More recently, it has been suggested that DNA released in the aquatic environment by phage mediated lysis of *V. cholerae* strains can be harnessed and assimilated by phage-resistant strains (Udden *et al.*, 2008).

Many studies have focused on gross genetic differences among *V. cholerae* strains, such as the presence or absence of individual genes or gene clusters, but research is now turning towards characterizing the role of finer genetic variations in the biology of *V. cholerae*. For example, different alleles of *tcpA* exist in vibrios of various serogroups, including toxigenic and non-toxigenic strains from clinical and environmental sources (Nandi *et al.*, 2000; Karaolis *et al.*, 2001; Boyd and Waldor, 2002), and it has been postulated that the different *tcpA* alleles may have evolved in response to selective pressures that vary between the environment and the host. Similarly, considerable genetic mosaicism in both the CTXΦ and the *tcp* gene cluster has been found in non-O1, non-O139 environmental isolates (Mukhopadhyay *et al.*, 2001). Some of the isolates were found to carry new alleles of the *tcpA*, *toxT*, and *tcpF* genes. Interestingly, some of the non-O1, non-O139 environmental strains possessing novel *tcpA*, *toxT*, and *tcpF* alleles were able to colonize mice indicating that non-canonical virulence alleles may still have virulence function. Colonization capacity has also been demonstrated for other non-canonical *tcpA* alleles (Boyd and Waldor, 2002).

Single nucleotide variations (SNVs) with biologic and epidemiologic significance have accumulated in the CTX prophage region. The gene encoding cholera toxin B subunit (*ctxB*) in the Haitian strain carries three non-synonymous substitutions relative to strain N16961. Two of these changes are characteristic of *ctxB* in classical strains of the sixth pandemic, and they have been detected in recent El Tor O1 strains from South Asia. http://www.ncbi.nlm.nih.gov/pubmed/20348258

Taken together, these studies imply that (1) genetic variation among *V. cholerae* strains may be quite extensive; (2) the fitness of strains in the environment may be favoured by different alleles of virulence genes other than those that increase fitness in the human host; and (3) the interactions among the various virulence alleles and elements encoding them are clearly important in the evolution of pathogenic *V. cholerae* strains.

Coordinate regulation of *V. cholerae* genes

Several critical virulence genes in *V. cholerae* are coordinately regulated by a cascading system of regulatory factors in response to specific environmental conditions such as temperature, osmolarity, bile salts, and pH (DiRita, 1992; Skorupski and Taylor, 1997). Most of these genes are controlled by the ultimate regulatory gene in the cascade ToxT, a member of the AraC family of bacterial transcription activators, but it is unclear how environmental signals are integrated in the control of virulence. Other regulatory genes and processes (e.g. motility, iron concentration, and quorum sensing) are also known to control virulence gene expression or growth (Sigel and Payne, 1982; Hase and Mekalanos, 1998; Zhu *et al.*, 2002). 'Quorum sensing' is a regulatory response thought to occur in bacterial communities through the sensing of extracellular signal molecules called autoinducers that are produced by members of the community (Camara *et al.*, 2002; Hammer and Bassler, 2003). Quorum-sensing systems in *Vibrio cholerae* have been shown to regulate certain phenotypes including biofilm-formation and virulence.

As mentioned above, the *tcp* and *ctx* operons which encode the major virulence factors are expressed in a coordinated fashion (Faruque *et al.*, 1998a). Both these operons are regulated by two membrane proteins, ToxR and ToxS (DiRita, 1992). ToxR can directly affect gene expression by binding to promoters, including those for the *ctx* operon, the genes encoding two outer membrane proteins OmpU and OmpT, and the *toxT* gene. ToxT is a transcriptional regulator belonging to the AraC family of proteins (Higgins *et al.*, 1992) and is the most downstream regulator of the ToxR regulon. ToxT can activate *ctx* and *tcp* promoters independent of ToxR once ToxT expression has occurred (DiRita *et al.*, 1991, 1996). However, two other membrane proteins, TcpP and TcpH, are required to act synergistically with ToxR and ToxS to activate *toxT* transcription (Hase and Mekalanos, 1998) (Fig. 10.1). Several other factors, including AphAB (Kovacikova and Skorupski, 1999), CRP (Skorupski and Taylor, 1997), Nqr (Hase and Mekalanos, 1999), HapR (Kovacikova and Skorupski, 2002), and PepA (Behari *et al.*, 2001), fine tune this regulatory hierarchy by controlling expression of *tcpPH* in response to various environmental signals (Fig. 10.1). These signals include temperature and pH as well as physiological signals such as cell density, growth phase, and motility. There are additional genes with putative functions in determining pathogenic and epidemiological properties of *V. cholerae*. These include genes encoding accessory colonization factors (Peterson and Mekalanos, 1988), iron acquisition genes (Goldberg *et al.*, 1990a,b; Henderson and Payne, 1994), phosphate transfer genes, and biotin synthesis genes (Chiang and Mekalanos, 1998). Despite various studies

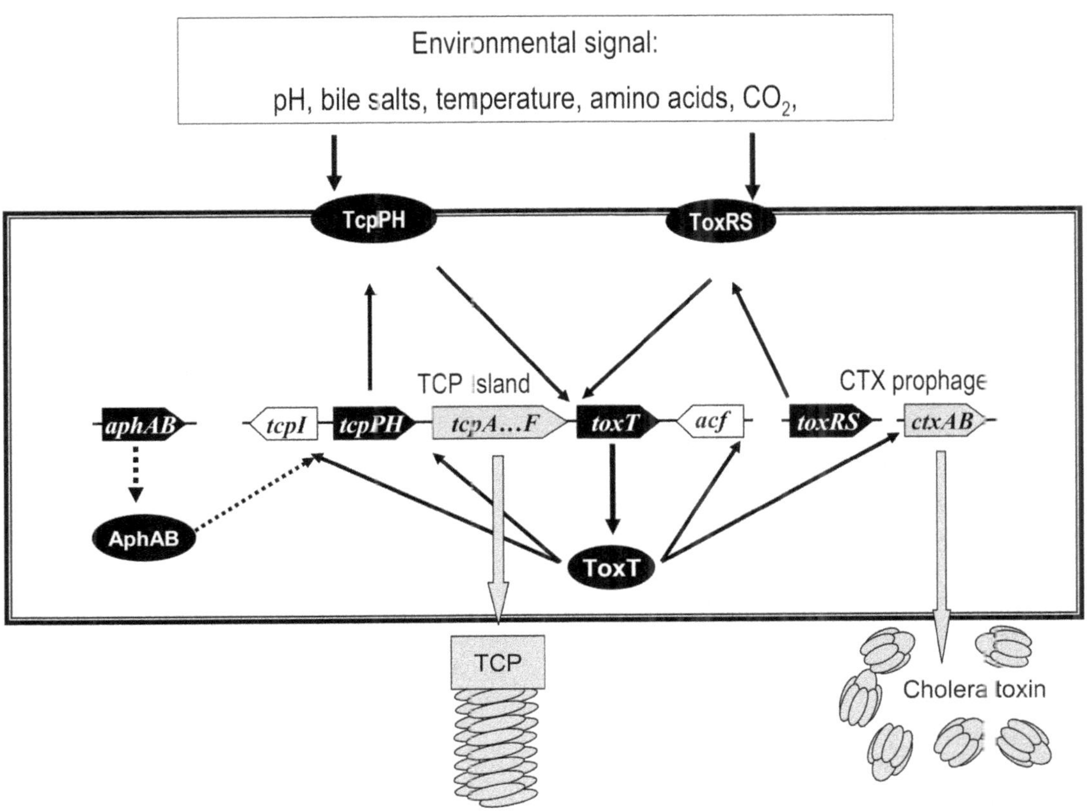

Figure 10.1 Coordinate regulation of virulence genes in *Vibrio cholerae* in response to environmental signals. Filled arrows denote positive regulation whereas broken arrows denote inhibition. See text for details.

our understanding of *V. cholerae* gene expression patterns during infection is still poor.

Recently expression profiling with microarrays has been used to study virulence regulation in *V. cholerae* on the genomic scale. For example, a custom *V. cholerae* DNA microarray was used to study gene expression in various quorum-sensing mutants (Zhu and Mekalanos, 2003) and found that quorum sensing is associated with the repression of ToxT-regulated genes and induction of genes involved in proteolysis and biofilm formation. Transcription profiling of *V. cholerae* shed in cholera stools, or after passage in animal models is beginning to reveal the nature of metabolic changes that occur in the pathogen under *in vivo* conditions during natural or experimental infection of a host (Merrel *et al.*, 2002; Bina *et al.*, 2003).

Molecular ecology of *V. cholerae*

V. cholerae persists in estuarine and other brackish waters, and the physicochemical conditions required for the survival of *V. cholerae* have been investigated (Colwell and Huq, 1994). The survival may depend on several factors, such as occurrence of particular physicochemical conditions, association of the bacteria with aquatic plants or animals, and/or the existence of specific ecological conditions involving several components of the aquatic environment including unique carbon sources such as chitin. However, further studies are required to establish whether the association between *V. cholerae* and aquatic organisms constitutes a mechanism for selective enrichment of *V. cholerae* with pathogenic potential in natural environments, or whether it is a phenomenon common to all *V. cholerae* including non-pathogenic strains.

In contrast, studies based on clinical and environmental monitoring and molecular characterization of *V. cholerae* strains, have proposed a model according to which seasonal factors cause a bloom of diverse *V. cholerae* in the environment, with epidemics occurring only after amplification of pathogenic strains in human victims who consume these environmental surface waters (Faruque *et al.*, 2003). Furthermore, phage mediated killing of the epidemic strain has been found to be responsible for the collapse of epidemics, and hence cholera epidemics are self-limiting in nature (Faruque *et al.*, 1995). This model involves both environmental and host factors in the epidemic cycle of cholera and also explains the limited clonality apparent in the strains causing these epidemics.

Merrell *et al.* (2002) reported that human colonization creates a hyperinfectious bacterial state that is maintained after dissemination into the aquatic environment and that this too may contribute to the epidemic spread of cholera. It was later found that stools of cholera victims contain *V. cholerae* in more than one morphological state including free-swimming planktonic cells, and biofilm-like cells attached to debris (Faruque *et al.*, 2006a). Estimation of relative infectivity of these different population of *V. cholerae* cells suggested that the enhanced infectivity of *V. cholerae* shed in human stools is largely due to presence of biofilm-like clumps of cells which disperse *in vivo* providing a high dose of the pathogen. However, owing to differences in their physiology, these biofilm associated cells may also be innately more infectious than planktonic cells. The various morphological forms of *V. cholerae* excreted in stools may display different transcriptional programmes, which in turn may affect their infectivity. Thus, stools from cholera patients contain complex mixtures of bacteria that have not been fully characterized.

It has been shown that a significant proportion of epidemic *V. cholerae* cells in the environment exists in a 'conditionally viable' state requiring special culture techniques for their recovery, and these cells were termed CVEC for 'conditionally viable environmental cells' (Faruque *et al.*, 2005, 2006a,b). CVEC appeared to be clumps of partially dormant cells that could be revived under appropriate *in vivo* and *in vitro* conditions. Recent studies are beginning to reveal that CVEC originate from clumps of cells excreted in human cholera stools (Faruque *et al.*, 2006a; Kamruzzaman *et al.*, 2010).

Genetics of CVEC formation

Recently, using strains with defined insertion mutations in genes involving quorum sensing and biofilm formation it has been confirmed that biofilm formation is a prerequisite to CVEC formation (Kamruzzaman *et al.*, 2010). These data further suggested that a temporary loss of quorum sensing due to dilution of extracellular autoinducers confers a selective advantage to communities of *V. cholerae* by blocking quorum-mediated regulatory responses that would break down biofilms.

CVEC are derived from quorum-regulated biofilms. V. cholerae quorum sensing systems have been found to influence biofilm formation *in vitro* as well as intestinal colonization *in vivo* (Zhu *et al.*, 2002; Hammer *et al.*, 2003; Zhu and Mekalanos, 2003). In this process, the information from the sensory circuits converges to the response regulator protein LuxO, which negatively controls the transcriptional regulator HapR that in turn controls certain genes involved in the formation or dissolution of biofilms, and in regulating virulence genes involved in intestinal colonization (Fig. 10.2). El Tor biotype strain C6706 and its derivatives

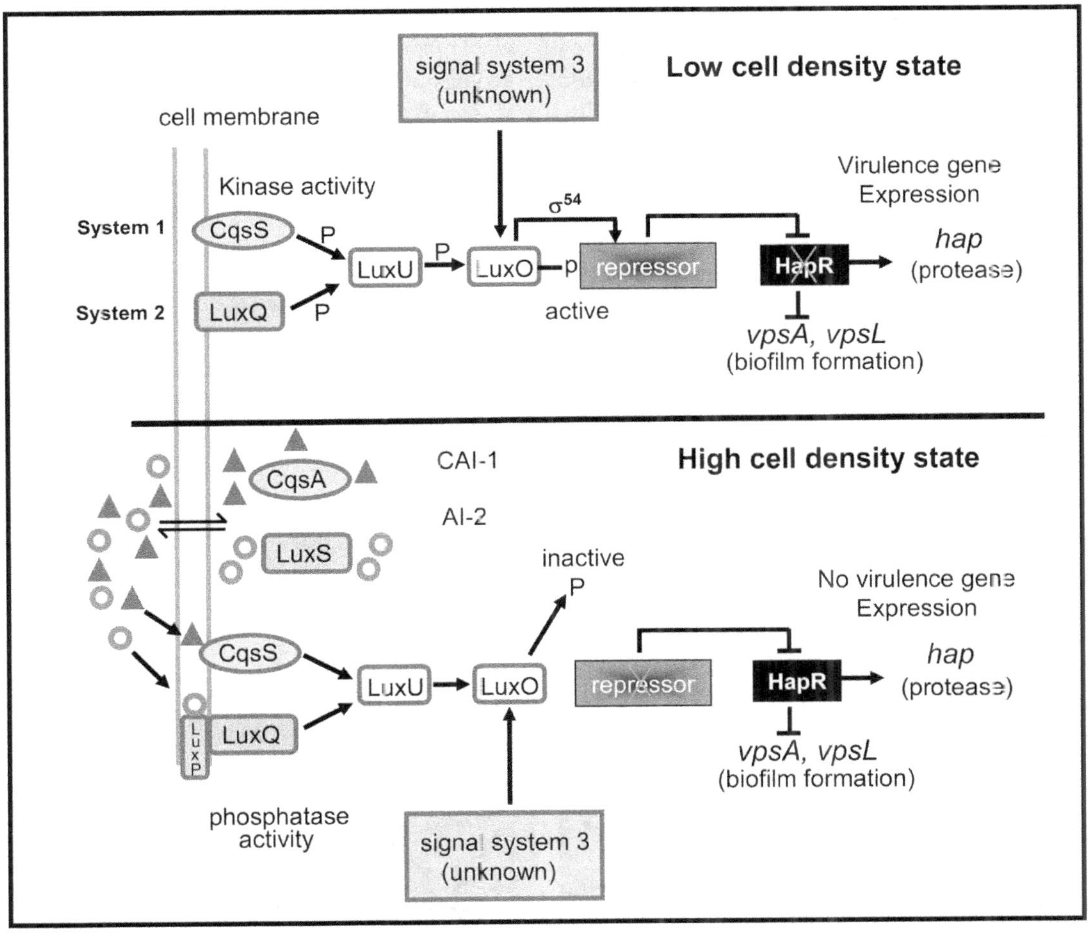

Figure 10.2 Quorum sensing circuits of *Vibrio cholerae*, and regulators of biofilm formation. Three sensory circuits converge through LuxO, the central regulator. Gene regulation occurs in response to the presence or absence of signal molecules via a phosphorylation–dephosphorylation cascade. At low cell density, when autoinducer concentrations are low, phosphate is transferred from the CqsS and LuxP/LuxQ sensors to the LuxU protein, which then transfers the phosphate to LuxO. The phosphorylated LuxO in association with σ^{54} activates a repressor that represses HapR expression. HapR activates or represses a number of target genes. Two of the sensory circuits, system 1 and system 2 have been defined, whereas the third circuit, system 3 remains to be characterized but is hypothesized to respond to an intracellular signal, and does not act through LuxU, but transmits its signal directly to LuxO.

carrying defined mutations, were analysed for their ability to develop biofilms and CVEC under different conditions (Fig. 10.3). These experiments showed that an isogenic strain carrying a mutation in *hapR* produced more robust biofilm-like clumps than the wild type strain, whereas the *luxO* mutant produced very little biofilm. Additionally mutants carrying insertions in genes encoding enzymes required for *Vibrio* exopolysaccharide synthesis (EPS) were also very poor biofilm producers (Fig. 10.3).

To determine if CVEC formation is induced following intestinal infection, *V. cholerae* present in the ileal loop fluids of rabbits pre-challenged with the strain were introduced into samples of environmental water and monitored for the formation of CVEC-like cells. Generally, the clumped cells survived in water for a significantly longer duration than the planktonic cells. Once cells were no longer viable by conventional culture, these biofilm associated cells remained as CVEC for minimally 60 days (Kamruzzaman *et al.*, 2010). In contrast, when planktonic cells were added to water, they did not form CVEC and the cell counts dropped rapidly with no live cells left within a few days (< 10 days). These results suggested that biofilm development is a prerequisite to CVEC formation. Similarly, when *V. cholerae* cells excreted in cholera stools were added to samples of water, these cells entered into the CVEC state. Furthermore, in this study *V. cholerae* isolated from environmental CVEC were compared with the predominant epidemic strain by genetic fingerprinting, and these results indicated that naturally occurring CVEC are derived from stools of cholera victims. These and other studies have led to propose a model to explain enhanced transmissibility and persistence of *V. cholerae* in water leading to rapid spread of cholera epidemics (Fig. 10.4).

There is much to elucidate regarding *V. cholerae* CVEC, including their metabolic status, innate infectivity, phage sensitivity, and the changes in gene regulation that occur in during formation of CVEC and their reactivation as viable planktonic cells. Thus, it appears that the ecology of toxigenic *V. cholerae* that supports the seasonal pattern of cholera epidemics involve a multitude of environmental and biological factors, most of which are yet to be fully understood.

Acknowledgements

Research activities of SMF and JJM are funded in part by National Institutes of Health grants 2RO1-GM068851-5, AI070963–01A1, and U01-AI058935 under different sub-agreements between the Harvard Medical School, Massachusetts General Hospital and the ICDDR,B. The ICDDR,B is supported by countries and agencies which share its concern for the health problems of developing countries.

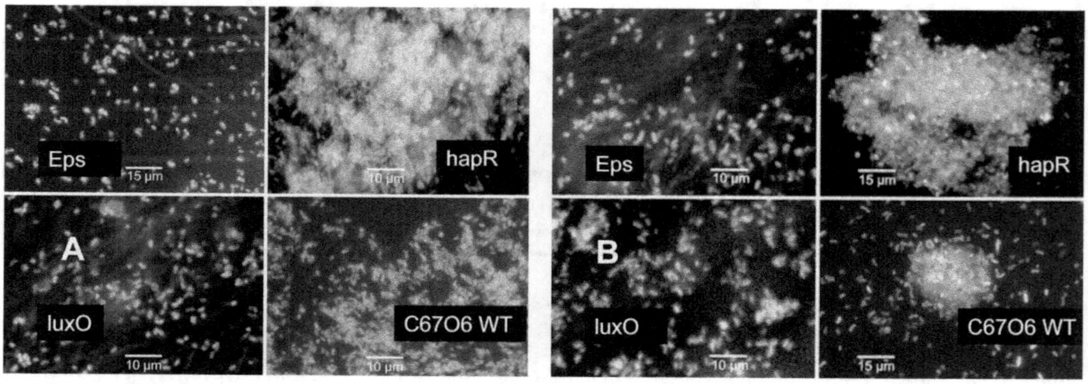

Figure 10.3 Biofilm formation by strain C6706 and its mutants carrying insertions in different genes under laboratory conditions (A) and in the ileal loops of adult rabbits (B). Formation of clumps of cells under in vivo conditions are down regulated in the *luxO* mutant but up-regulated in the *hapR* mutant.

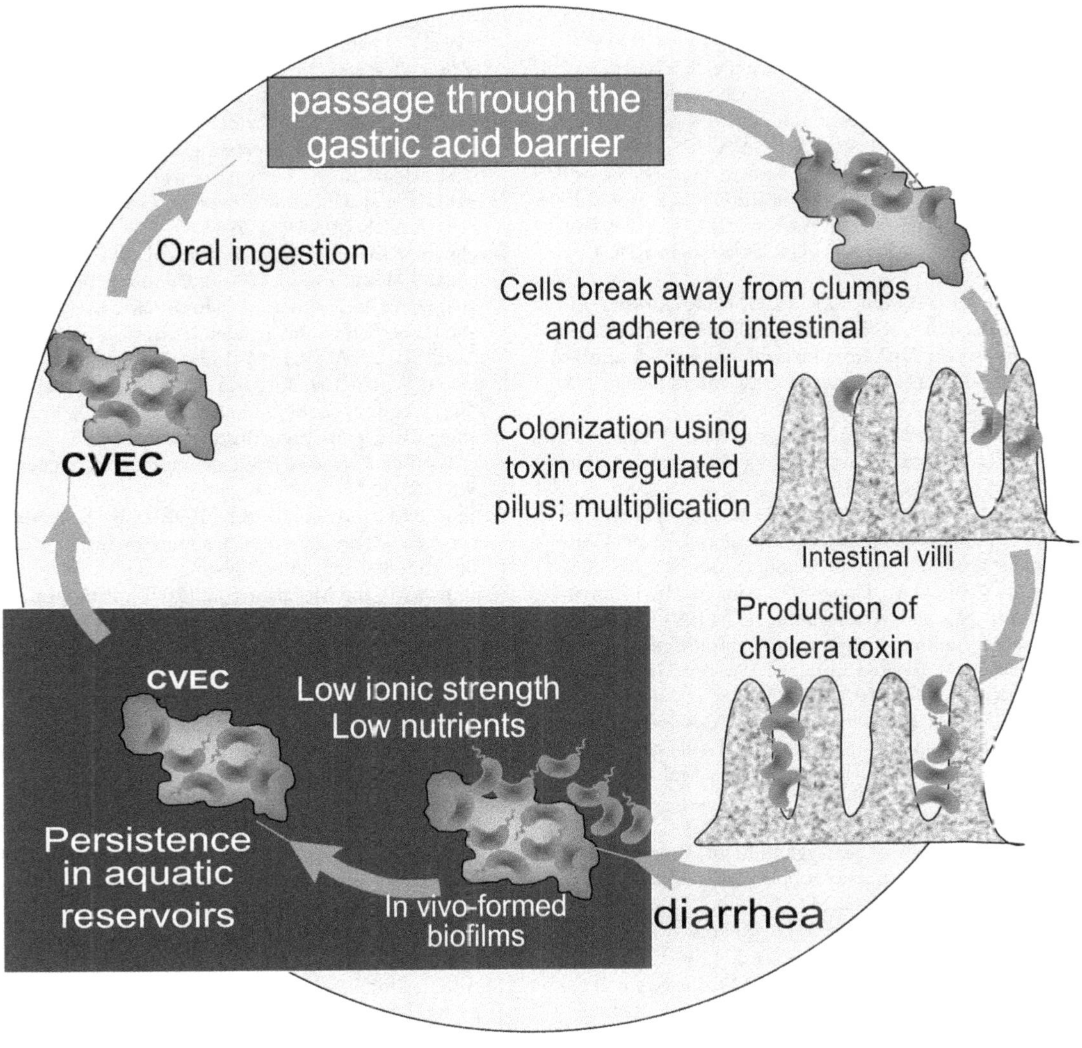

Figure 10.4 Model for enhanced transmissibility and persistence of *V. cholerae* in water leading to rapid spread of cholera epidemics. *V. cholerae* derived from cholera patients persist in the aquatic environment as conditionally viable environmental cells (CVEC) which are derived from in vivo-formed biofilms. As cholera stools containing *V. cholerae* contaminate water reservoirs, a temporary loss of quorum sensing occurs due to dilution of extracellular autoinducers leading to block in quorum-mediated regulatory responses that would break down biofilms. These biofilms instead convert to CVEC and persist in the aquatic environment. The clumped structure allows delivery of a high dose of the pathogen to a potential victim and facilitates survival of the bacteria while passing through the gastric acid barrier. See text for details.

References

Albert, M.J., Siddique, A.K., Islam, M.S., Faruque, A.S.G., Ansaruzzaman, M., Faruque, S.M., and Sack, R.B. (1993). A large outbreak of clinical cholera due to *Vibrio cholerae* non-01 in Bangladesh. Lancet *341*, 704.

Ansaruzzaman, M., Bhuiyan, N.A., Nair, G.B., Sack, D.A., Lucas, M., and Deen, J.L. (2004). Cholera in Mozambique, variant of *Vibrio cholerae*. Emerg. Infect. Dis. 10, 2057–2059.

Barrett, T.J., and Blake, P.A. (1981). Epidemiological usefulness of changes in hemolytic activity of *Vibrio cholerae* biotype El Tor during the seventh pandemic. J. Clin. Microbiol. *13*, 126–129.

Behari, J., Stagon, L., and Calderwood, S.B. (2001). *pepA*, a gene mediating pH regulation of virulence genes in *Vibrio cholerae*. J. Bacteriol. *183*, 178–188.

Bik, E.M., Bunschoten, A.E., Gouw, R.D., and Mooi, F. (1995). Genesis of the novel epidemic *Vibrio cholerae* O139 strain: evidence for horizontal transfer of genes involved in polysaccharide synthesis. EMBO J. *14*, 209–216

Bina, J., Zhu, J., Dziejman, M., Faruque, S., Calderwood, S.B., Mekalanos, J.J. (2003). ToxR regulon of *Vibrio cholerae* and its expression in vibrios shed by cholera patients. Proc. Natl. Acad. Sci. U.S.A. *100*, 2801–06.

Blake, P.A. (1994). Endemic cholera in Australia and United States. In *Vibrio cholerae* and Cholera: Molecular to Global Perspectives, Wachsmuth I.K., Blake P.A., and Olsvik Ø., eds. (ASM Press, Washington, DC), pp. 309–319.

Boyd, E.F., and Waldor, M.K. (2002). Evolutionary and functional analyses of variants of the toxin- coregulated pilus protein TcpA from toxigenic *Vibrio cholerae* non-O1/non-O139 serogroup isolates. Microbiology *148*, 1655–66.

Camara, M., Hardman, A., Williams, P., and Milton, D. (2002). Quorum sensing in *Vibrio cholerae*. Nature Genetics 32, 217–218.

Cholera Working Group. (1993). Large epidemic of cholera-like disease in Bangladesh caused by *Vibrio cholerae* O139 synonym Bengal. Lancet *342*, 387–390.

Chen, F., Evins, G.M., Cook, W.L., Almeida, R., Hargrett-Bean, N., and Wachsmuth, K. (1991). Genetic diversity among toxigenic and nontoxigenic *Vibrio cholerae* O1 isolated from the Western Hemisphere. Epidemiol. Infect. *107*, 225–233.

Chiang, S.L., and Mekalanos, J.J. (1998). Use of signature-tagged transposon mutagenesis to identify *Vibrio cholerae* genes critical for colonization. Mol. Microbiol. *27*, 797–805.

Chiavelli, D.A., Marsh, J.W. and Taylor, R.K. (2001). The mannose-sensitive hemagglutinin of *Vibrio cholerae* promotes adherence to zooplankton. Appl. Environ. Microbiol. *67*, 3220–3225.

Chin, C.S., Sorenson, J., Harris, J.B., Robins, W.P., Charles, R.C., Jean-Charles, R.R., Bullard, J., Webster, D.R., Kasarskis, A., Peluso, P., *et al.* (2011). The origin of the Haitian cholera outbreak strain. N. Engl. J. Med. *364*, 33–42.

Chongsanguan, M., Chaicumpa, W., Moolasart, P., Kandhasingha, P., Shimada, T., Kurazono, H., and Takeda, Y. (1993). *Vibrio cholerae* O139 Bengal in Bangkok. Lancet *342*, 430–431.

Cohen, J., Schwartz, T., Klasmer, R., Pridan, D., Ghalayini, H., and Davies, A.M. (1971). Epidemiological aspects of cholera El Tor outbreak in a non-endemic area. Lancet *2(7715)*, 86–89.

Colwell, R.R., and A.Huq. (1994). Vibrios in the environment: viable but nonculturable *Vibrio cholerae*, In *Vibrio cholerae* and Cholera: Molecular to Global Perspectives, Wachsmuth, I.K., Blake, P.A., and Olsvik, Ø. eds. (ASM Press, Washington, DC), pp. 117–133.

Cook, W.L., Wachsmuth, K., Feeley, J. and Huq, I. (1983). The question of classical cholera. Lancet *1*, 879–80.

Das, S., Saha, R., Kaur, I.R. (2008). Trend of antibiotic resistance of *Vibrio cholerae* strains from East Delhi. Indian J. Med. Res. *127*, 478–82.

DiRita, V.J. (1992). Co-ordinate expression of virulence genes by ToxR in *Vibrio cholerae*. Mol. Microbiol. *6*, 451–458.

DiRita, V.J., Parsot, C., Jander, G., and Mekalanos, J.J. (1991). Regulatory cascade controls virulence in *Vibrio cholerae*. Proc. Natl. Acad. Sci. U.S.A. *88*, 5403–5407.

DiRita, V.J., Neely, M., Taylor, R.K., and Bruss, P.M. (1996). Differential expression of ToxR regulon in classical and El Tor biotypes of *Vibrio cholerae* is due to biotype-specific control over ToxT expression. Proc. Natl. Acad. Sci. *93*, 7991–7995.

Dziejman, M., Balon, E., Boyd, D., Fraser, C.M., Heidelberg, J.F. and Mekalanos, J.J. (2002). Comparative genomic analysis of *Vibrio cholerae*: genes that correlate with cholera endemic and pandemic disease. Proc. Natl. Acad. Sci. U.S.A. *99*, 1556–1561.

Everiss, K.D., Hughes, K.J., and Peterson, K.M. (1996). The accessory colonization factor and the toxin-coregulated pilus gene clusters are physically linked in the *Vibrio cholerae* O395 genome. DNA Sequence *5*, 51–55.

Faruque, S.M., and Albert, M.J. (1992). Genetic relation between *Vibrio cholerae* O1 strains in Ecuador and Bangladesh. Lancet *339*, 740–741.

Faruque, S.M., and Mekalanos, J.J. (2003). Pathogenicity islands and phages in *Vibrio cholerae* evolution. Trends. Microbiology *11*, 505–510.

Faruque, S.M., Adbul Alim, A.R., Rahman, M.M., Siddique, A.K., Sack, R.B., and Albert, M.J. (1993). Clonal relationships among classical *Vibrio cholerae* O1 strains isolated between 1961 and 1992 in Bangladesh. J. Clin. Microbiol. *31*, 2513–2516.

Faruque, S.M., Adbul Alim, A.R., Roy, S.K., Khan, F., Nair, G.B., Sack, R.B. and Albert, M.J. (1994). Molecular analysis of rRNA and cholera toxin genes carried by the new epidemic strain of toxigenic *Vibrio cholerae* O139 synonym Bengal. J. Clin. Microbiol. *32*, 1050–1053.

Faruque, S.M., Ahmed, K.M., Abdul Alim, A.R., Siddique, A.K., Qadri, F. and Albert, M.J. (1997a). Emergence of a new clone of toxigenic *Vibrio cholerae* O1 biotype El Tor displacing *V. cholerae* O139 Bengal in Bangladesh. J. Clin. Microbiol. 35, 624–30.

Faruque, S.M., Ahmed, K.M., Siddique, A.K., Zaman, K., Alim, A.R., and Albert, M.J. (1997b). Molecular analysis of toxigenic *Vibrio cholerae* O139 Bengal strains isolated in Bangladesh between 1993 and 1996: evidence for emergence of a new clone of the Bengal vibrios. J. Clin. Microbiol. *35*, 2299–2306.

Faruque, S.M., Asadulghani, Alim, A.R.Albert, M.J.Islam, K.M. and Mekalanos, J.J. (1998a). Induction of the lysogenic phage encoding cholera toxin in naturally occurring strains of toxigenic *Vibrio cholerae* O1 and O139. Infect. Immun. *66*, 3752–3757.

Faruque, S.M., Asadulghani, Saha, M.N., Alim, A.R., Albert, M.J., Islam, K.M., and Mekalanos, J.J. (1998b). Analysis of clinical and environmental strains of nontoxigenic *Vibrio cholerae* for susceptibility to CTXΦ: molecular basis for origination of new strains with epidemic potential. Infect Immun. *66*, 5819–5825.

Faruque, S.M., Albert, M.J., and Mekalanos, J.J. (1998c). Epidemiology, genetics, and ecology of toxigenic *Vibrio cholerae*. Microbiol. Mol. Biol. Rev. *62*, 1301–1314.

Faruque, S.M., Asadulghani, Kamruzzaman, M., Nandi, R.K., Ghosh, A.N., Nair, G.B., Mekalanos, J.J., and Sack, D.A. (2002). RS1 element of *Vibrio cholerae* can propagate horizontally as a filamentous phage exploiting the morphogenesis genes of CTXphi. Infect Immun. *70*, 163–170.

Faruque, S.M., Chowdhury, N., Kamruzzaman, M., Ahmad, Q.S., Faruque, A.S., Salam, M.A., Ramamurthy, T., Nair, G.B., Weintraub, A., and Sack, D.A. (2003). Reemergence of epidemic *Vibrio cholerae* O139, Bangladesh. Emerg. Infect. Dis. *9*, 1116–1122.

Faruque, S.M., Chowdhury, N., Kamruzzaman, M., Dziejman, M., Rahman, M.H., Sack, D.A., Nair, G.B., and Mekalanos, J.J. (2004). Genetic diversity and virulence potential of environmental *Vibrio cholerae* population in a cholera-endemic area. Proc. Natl. Acad. Sci. U.S.A. *101*, 2123–2128.

Faruque, S.M., Iftekar, B.N., Kazukata, F., Diraphat, P., Chowdhury, N., Kamruzzaman, M., Qadri, F., Yamasaki, S., and Mekalanos, J.J. (2005a). Genomic sequence and receptor for the *Vibrio cholerae* phage KSF-1φ: Evolutionary divergence among filamentous vibriophages mediating lateral gene transfer. J. Bacteriol. *187*, 4095–4103.

Faruque, S.M., Islam, M.J., Ahmad, Q.S., Faruque, A.S.G., Sack, D.A., Nair, G B., and Mekalanos, J.J. (2005b). Self-limiting nature of seasonal cholera epidemics: Role of host-mediated amplification of phage. Proc. Natl. Acad. Sci. U.S.A. *102*, 6119–24.

Faruque, S.M., Biswas, K., Udden, S.M., Ahmad, Q.S., Sack, D.A., Nair, G.B., and Mekalanos, J.J. (2006a). Transmissibility of cholera: *in vivo*-formed biofilms and their relationship to infectivity and persistence in the environment. Proc. Natl. Acad. Sci. U.S.A. *103*, 6350–6355.

Faruque, S.M., Islam, M.J., Ahmad, Q.S., Biswas, K., Faruque, A.S.G., Nair, G.B., Sack, R.B., Sack, D.A., and Mekalanos. J.J. (2006b). An improved technique for isolation of environmental *Vibrio cholerae* with epidemic potential: monitoring the emergence of a multiple-antibiotic-resistant epidemic strain in Bangladesh. J. Infect. Dis. *193*, 1029–1036.

Faruque, S.M., Tam, V.C., Chowdhury, N., Diraphat, P., Dziejman, M., Heidelberg, J.F., Clemens, J.D., Mekalanos, J.J. and Nair, G.B. (2007). Genomic analysis of the Mozambique strain of *Vibrio cholerae* O1 reveals the origin of El Tor strains carrying classical CTX prophage. Proc. Natl. Acad. Sci. U.S.A. *104*, 5151–5156.

Fullner, K.J., and Mekalanos, J.J. (1999). Genetic characterization of a new type IV-A pilus gene cluster found in both classical and El Tor biotypes of *Vibrio cholerae*. Infect Immun. *67*,1393–1404.

Garg, P., Sinha, S., Chakraborty, R., Bhattacharya, S.K., Nair, G.B., and Ramamurthy, T. (2001). Emergence of fluoroquinolone-resistant strains of *Vibrio cholerae* O1 biotype El Tor among hospitalized patients with cholera in Calcutta, India. Antimicrob. Agents Chemother. *45*, 1605–1606.

Ghosh, A., and Ramamurthy, T. (2011). Antimicrobials & cholera: are we stranded?. Indian J. Med. Res. *133*, 225–31.

Glass, R.I., Huq, I., Adbul Alim, A.R., and Yunus, M. (1980). Emergence of multiple antibiotic-resistant *Vibrio cholerae* in Bangladesh. J. Infect Dis. *142*, 939–942.

Glass, R.I., Huq, M.I., Lee, J.V., Threlfall, E.J., Khan, M.R., Alim, A.R.M.A., Rowe, B. and Gross, R.J (1983). Plasmid-borne multiple drug resistance in *Vibrio cholerae* serogroup O1, biotype El Tor: evidence of a point-source outbreak in Bangladesh. J. Infect. Dis. *147*, 204–209.

Goodgame, R.W., and Greenough, W.B., III. (1975). Cholera in Africa: a message for the West. Ann. Intern. Med. *82*, 101–106.

Griffith, D.C., Kelly-Hope, L.A., and Miller, M.A. (2006). Review of reported cholera outbreaks, 1995–2005. Am. J. Trop. Med. Hyg. *75*, 973–977.

Goldberg, M.B., Boyko, S.A., and Calderwood, S.B. (1990a). Transcripional regulation by iron of a *Vibrio cholerae* virulence gene and homology of the gene to the *Escherichia coli* Fur system. J. Bacteriol. *172*, 6863–6870.

Goldberg, M.B., DiRita, V.J. and Calderwood, S.B. (1990b). Identification of an iron-regulated virulence determinant in *Vibrio cholerae*, using Tn*phoA* mutagenesis. Infect. Immun. *58*, 55–60.

Guidolin, A. and Manning, P.A. (1987). Genetics of *Vibrio cholerae* and its bacteriophages. Micriobiol. Rev. *51*, 285–298.

Hammer, B.K., and Bassler, B.L. (2003) Quorum sensing controls biofilm formation in *Vibrio cholerae*. Mol. Microbiol. *50*, 101–104.

Harkey, C.W., Everiss, K.D., and Peterson, K.M. (1994). The *Vibrio cholerae* toxin-coregulated-pilus gene *tcpI* encodes a homolog of methyl-accepting chemotaxis proteins. Infect. Immun. *62*, 2669–2678.

Hase, C.C., and Mekalanos, J.J. (1998). TcpP protein is a positive regulator of virulence gene expression in *Vibrio cholerae*. Proc. Natl. Acad. Sci. U.S.A. *95*,730–734.

Häse, C.C., and Mekalanos, J.J. (1999). Effects of changes in membrane sodium flux on virulence gene expression in *Vibrio cholerae*. Proc. Natl. Acad. Sci. U.S.A. *96*, 3183–3187.

Heidelberg, J.F., Eisen, J.A., Nelson, W.C., Clayton, R.A., Gwinn, M.L., Dodson, R.J., Haft, D.H., Hickey, E.K., Peterson, J.D., Umayam, L., *et al.* (2000). DNA sequence of both chromosomes of the cholera pathogen *Vibrio cholerae*. Nature *406*, 477–483.

Henderson, D.P., and Payne, S.M. (1994). *Vibrio cholerae* iron transport systems: roles of heme and siderophore iron transport in virulence and identification of a gene associated with multiple iron transport systems. Infect Immun. *62*, 5120–5125.

Higgins, D.E., and DiRita, V.J. (1996). Genetic analysis of the interaction between *Vibrio cholerae* transcription activator ToxR and ToxT promoter DNA. J Bacteriol. *178*, 1080–1087.

Hochhut, B., Lotfi, Y., Mazel, D., Faruque, S.M., Woodgate, R., and Waldor, M.K. (2002). Molecular analysis of antibiotic resistance gene clusters in *Vibrio cholerae* O139 and O1 SXT constins. Antimicrob. Agents. Chemother. *45*, 2991–3000.

Islam, M.S., Mahmud, Z.H., Ansaruzzaman, M., Faruque, S.M., Talukder, K.A., Qadri, F., Alam, M., Islam, S., Bardhan, P.K., Mazumder, R.N., *et al.* (2011). Phenotypic, genotypic, and antibiotic sensitivity patterns of strains isolated from the cholera epidemic in Zimbabwe. J Clin. Microbiol. *49*, 2325–2327.

Janda, J.M., Powers, C., Bryant, R.G., and Abbott, S.L. (1988). Current perspective on the epidemiology and pathogenesis of clinically significant *Vibrio* spp. Clin. Microbiol. Rev. *1*, 245–267.

Jermyn, W.S., and Boyd, E.F. (2002). Characterization of a novel *Vibrio* pathogenicity island (VPI-2) encoding neuraminidase (nanH) among toxigenic *Vibrio cholerae* isolates. Microbiology *148*, 3681–93.

Jesudason, M.V. and Saaya, R. (1997). Resistance of *Vibrio cholerae* O1 to nalidixic acid. Indian J. Med. Res. *105*, 153–154.

Kamal, A.M. (1974). The seventh pandemic of cholera. In Cholera, Barua, D., and Burrows, W., eds. (Saunders, Philadelphia, PA), pp. 1–14.

Kamruzzaman, M., Udden, S.M., Cameron, D.E., Calderwood, S.B., Nair, G.B., Mekalanos, J.J., and Faruque, S.M. (2010). Quorum-regulated biofilms enhance the development of conditionally viable, environmental *Vibrio cholerae*. Proc. Natl. Acad. Sci. U.S.A. *107*, 1588–93

Kaper, J.B., Morris, J.G., Jr, and Levine, M.M. (1995). Cholera. Clin. Microbiol. Rev. *8*, 48–86.

Karaolis, D.K., Lan, R., Kaper, J.B., and Reeves, P.R. (2001). Comparison of *Vibrio cholerae* pathogenicity islands in sixth and seventh pandemic strains. Infect. Immun. *69*, 1947–1952.

Kovach, M.E., Shaffer, M.D., and Peterson, K.M. (1996). A putative integrase gene defines the distal end of a large cluster of ToxR-regulated colonization genes in *Vibrio cholerae*. Microbiology *142*, 2165–2174.

Kovacikova, G., and Skorupski, K. (1999). A *Vibrio cholerae* LysR homolog, AphB, cooperates with AphA at the tcpPH promoter to activate expression of the ToxR virulence cascade. J. Bacteriol. *181*, 4250–4256.

Kovacikova, G. and Skorupski, K. (2002). Regulation of virulence gene expression in *Vibrio cholerae* by quorum sensing: HapR functions at the aphA promoter. Mol. Microbiol. *46*, 1135–1147.

Laval, E. (1989). El cholera en, Chile (1886–1888). Rev. Chil. Infect. *6*, 96–99.

Levine, M.M. (1991). South America: the return of cholera. Lancet 338,45–46.

Lin, W., Fullner, K.J., Clayton, R., Sexton, J.A., Rogers, M.B., Calia, K.E., Calderwood, S.B., Fraser, C., and Mekalanos, J.J. (1999). Identification of a *Vibrio cholerae* RTX toxin gene cluster that is tightly linked to the cholera toxin prophage. Proc. Natl. Acad. Sci. U.S.A. *96*, 1071–1076.

Materu, S.F., Lema, O.E., Mukunza, H.M., Adhiambo, C.G., and Carter, J.Y. (1997). Antibiotic resistance pattern of *Vibrio cholerae* and *Shigella* causing diarrhoea outbreaks in the eastern Africa region: 1994–1996. East Afr. Med. J. *74*, 193–197.

Mazel, D., Dychinco, B., Webb, V.A., and Davies, J. (1998). A distinctive class of integron in the *Vibrio cholerae* genome. Science. *280*, 605–608.

Meibom, K.L., Blokesch, M., Dolganov, N.A., Wu, C., and Schoolnik, G.K. (2005). Chitin Induces Natural Competence in *Vibrio cholerae*. Science *310*, 1824–1827.

Merrell, D.S., Butler, S.M., Qadri, F., Dolganov, N.A., Alam, A., Cohen, M.B., Calderwood, S.B., Schoolnik, G.K., and Camilli, A. (2002). Host-induced epidemic spread of the cholera bacterium. Nature *417*, 642–645.

Morris, J.G. (1994). Non-O1 group 1 *Vibrio cholerae* strains not associated with epidemic disease. In *Vibrio cholerae* and Cholera: Molecular to Global Perspectives, Wachsmuth, I.K., Blake, P.A., and Olsvik, O., eds. (ASM Press, Washington, DC), pp. 103–116.

Mukhopadhyay, A.K., Saha, P.K., Garg, S., Bhattacharya, S.K., Shimada, T., Takeda, T., Takeda, Y., and Nair, G.B. (1995). Distribution and virulence of *Vibrio cholerae* belonging to serogroups other than O1 and O139: a nation survey. Epidemiol. Infect. *114*, 65–70.

Mukhopadhyay, A.K., Chakraborty, S., Takeda, Y., Nair, G.B., and Berg, D.E. (2001). Characterization of VPI pathogenicity island and CTXphi prophage in environmental strains of *Vibrio cholerae*. J. Bacteriol. *183*, 4737–4746.

Nair, G.B., Faruque, S.M., Bhuiyan, N.A., Kamruzzaman, M., Siddique, A.K. and Sack, D.A. (2002). New variants of *Vibrio cholerae* O1 biotype El Tor with attributes of the classical biotype from hospitalized patients with acute diarrhea in Bangladesh. J. Clin. Microbiol. *40*, 3296–3299.

Nair, G.B., Qadri, F., Holmgren, J., Svennerholm, A.M., Safa, A., and Bhuiyan, N.A. (2006) Cholera due to altered El Tor strains of *Vibrio cholerae* O1 in Bangladesh. J. Clin. Microbiol. *44*, 4211–4213.

Nair, G.B., Mukhopadhyay, A.K., Safa, A. and Takeda, Y. (2008). Emerging hybrid variants of *Vibrio cholerae* O1. In *Vibrio cholerae*: Genomics and Molecular Biology, Faruque S.M., Nair, G.B. eds. (Horizon Scientific Press Ltd, Norfolk, UK), pp. 179–190.

Nakasone, N., Iwanaga, M., and Eeckels, R. (1987). Characterization of *Vibrio cholerae* O1 recently isolated in Bangladesh. Trans. Roy. Soc. Trop. Med. Hyg. *81*, 876–878.

Nandi, B., Nandy, R.K., Vicente, A.C., and Ghose, A.C. (2000). Molecular characterization of a new variant of toxin-coregulated pilus protein (TcpA) in a toxigenic non-O1/non-O139 strain of *Vibrio cholerae*. Infect. Immun. *68*, 948–952.

Ogg, J.E., Timme, T.L., and Alemohammad, M.M. (1981). General transduction in *Vibrio cholerae*. Infect Immun. *31*, 737–41

Ogierman, M.A., Zabihi, S., Mourtzios, L., and Manning, P.A. (1993). Genetic organization and sequence of the promoter-distal region of the *tcp* gene cluster of *Vibrio cholerae*. Gene *126*, 51–60.

Olsvik, Ø., Wahlberg, J., Petterson, B., Uhlen, M., Popovic, T., Wachsmuth, I.K., (1993). Use of automated sequencing of polymerase chain reaction generated amplicons to identify three types of cholera toxin subunit B in *Vibrio cholerae* O1 strains. J. Clin. Microbiol. *31*, 22–25.

Pan American Health Organization. (1991). Cholera situation in the Americas. Epidemiol. Bull. *12*, 1–4.

Pollitzer, R. (1959). Cholera. World Health Organization Monograph 42, Geneva, Switzerland.

Peterson, K.M., and Mekalanos, J.J. (1988). Characterization of the *Vibrio cholerae* ToxR regulon: identification of novel genes involved in intestinal colonization. Infect. Immun. *56*, 2822–2829.

Rahman, M.H., Biwas, K., Hossain, M.A., Sack, R.B., Mekalanos, J.J., and Faruque, S.M. (2008). Distribution of genes for virulence and ecological fitness among diverse *Vibrio cholerae* population in a cholera endemic area: tracking the evolution of pathogenic strains. DNA Cell Biol. *27*, 347–355.

Ramamurthy, T., Garg, S., Sharma, R., Bhattacharya, S.K., Nair, G.B., Shimada, T., Takeda, T., Karasawa, T., Kurazano, H., Pal, A. *et al.* (1993). Emergence of a novel strain of *Vibrio cholerae* with epidemic potential in southern and eastern India. Lancet *341*, 703–704.

Reguera, G. and Kolter, R. (2005). Virulence and the environment: a novel role for *Vibrio cholerae* toxin-coregulated pili in biofilm formation on chitin. J. Bacteriol. *187*, 3551–3555.

Ries, A.A., Vugia, D.J., Beingolea, L., Palacios, A.M., Vasquez, E., Wells, J.G., Baca, N.G., Swerdlow, D.L., Pollack, M., Bean, N.H., *et al.* (1992). Cholera in Piura, Peru: a modern urban epidemic. J. Infect. Dis. *166*, 1429–1433.

Safa, A., Bhuyian, N.A., Nusrin, S., Ansaruzzaman, M., Alam, M., and Hamabata, T. (2006). Genetic characteristics of Matlab variants of *Vibrio cholerae* O1 that are hybrids between classical and El Tor biotypes. J. Med. Microbiol. *55*, 1563–1569.

Safa, A., Sultana, J., Cam, P.D., Mwansa, J.C., and Kong, R.Y.C. (2008). *Vibrio cholerae* O1 hybrid El Tor strains, Asia and Africa. Emerg. Infect. Dis. *14*, 987–988.

Safa, A., Bhuiyan, N.A., Murphy, D., Bates, J., Nusrin, S., Kong, R.Y.C., Chongsanguan, M., Chaicupma, W., and Nair, G.B. (2009). Multilocus genetic analysis reveals that the Australian strains of *Vibrio cholerae* O1 are similar to the pre-seventh pandemic strains of the El Tor biotype. J. Med. Microbiol. *58*, 105–111.

Safa, A., Nair, G.B., and Kong, R.Y. (2010). Evolution of new variants of *Vibrio cholerae* O1. Trends Microbiol. *18*, 46–54.

Sakajaki, R. (1970). Classification and characteristics of vibrios. Public Health Papers 40. Geneva WHO: 33.

Sciortino, C.V., Johnson, J.A., and Hamad, A. (1996). Vitek system antimicrobial susceptibility testing of O1, O139, and non-O1 *Vibrio cholerae*. J. Clin. Microbiol. *34*, 897–900.

Skorupski, K., and Taylor, R.K. (1997). Control of the ToxR virulence regulon in *Vibrio cholerae* by environmental stimuli. Mol. Microbiol. *25*, 1003–1009.

Siddique, A.K., Zaman, K., Mazumder, Y., Islam, Q., Bashir, I., Mutsuddy, P., and Eusof, A. (1989). Simultaneous outbreaks of contrasting drug resistant classic and El Tor *Vibrio cholerae* O1 in Bangladesh. Lancet *2(8659)*, 396.

Siddique, A.K., Zaman, K., Baqui, A.H., Akram, K., Mutsuddy, P., Eusof, A., Haider, K., Islam, S. and Sack, R.B. (1992). Cholera epidemics in Bangladesh: 1985–1991. J. Diarrhoeal Dis. Res. *10*, 79–86.

Siddique, A.K., Salam, A., Islam, M.S., Akram, K., Majumdar, R.N., Zaman, K., Fronczak, N., and Laston, S. (1995). Why treatment centres failed to prevent cholera deaths among Rwandan refugees in Goma, Zaire. Lancet *345*, 359–361.

Sigel, S.P., and Payne, S.M. (1982). Effect of iron limitation on growth, siderophore production, and expression of outer membrane proteins of *Vibrio cholerae*. J. Bacteriol. *150*, 148–55.

Shimada, T., Nair, G.B., Deb, B.C., Albert, M.J., Sack, R.B. and Takeda, Y. (1993). Outbreak of *Vibrio cholerae* non-O1 in India and Bangladesh. Lancet *341*, 1347.

Snow, J. (1855). On the mode of communication of cholera, 2nd ed (J.Churchill, London), pp. 1–162.

Swerdlow, D.L., and Ries, A.A. (1993). *Vibrio cholerae* non-O1-the eighth pandemic? Lancet *342*, 382–3.

Trucksis, M., Michalski, J., Deng, Y.K., and Kaper, J.B. (1998). The *Vibrio cholerae* genome contains two unique circular chromosomes. Proc. Natl Acad. Sci. U.S.A. *95*, 14464–14469.

Udden, S.M.N., Zahid, M.S.H., Biswas, K., Ahmad, Q.S., Cravioto, A., Nair, G.B., Mekalanos, J.J., and Faruque, S.M. (2008). Acquisition of classical CTX prophage from *Vibrio cholerae* O141 by El Tor strains aided by lytic phages and chitin-induced competence. Proc. Natl. Acad. Sci U.S.A. 23, 11951–11958.

Wachsmuth, I.K., Evins, G.M., Fields, P.I., Olsvik, Ø., Popovic, T., Bopp, C.A., Wells, J.G., Carrillo, C., and Blake, P.A. (1993). The molecular epidemiology of cholera in Latin America. J. Infect. Dis. *167*, 621–626.

Wachsmuth, I.K., Olsvik, Ø., Evins, G.M., and Popovic, T. (1994). Molecular epidemiology of cholera. In *Vibrio cholerae* and Cholera: Molecular to Global Perspectives, Wachsmuth, I.K., Blake, P.A., and Olsvik, Ø., eds. (ASM Press, Microbiology, Washington, DC), pp. 357–370.

Waldor, M.K., Tschape, H., and Mekalanos, J.J. (1996). A new type of conjugative transposon encodes resistance to sulfamethoxazole, trimethoprim, and streptomycin in *Vibrio cholerae* O139. J. Bacteriol. *178*, 4157–4165.

Waldor, M.K., and Mekalanos, J.J. (1996). Lysogenic conversion by a filamentous bacteriophage encoding cholera toxin. Science. *272*, 1910–1914.

Waldor, M.K., Rubin, E.J., Pearson, G.D., Kimsey, H., and Mekalanos, J.J. (1997). Regulation, replication, and integration functions of the *Vibrio cholerae* CTXΦ are encoded by region RS2. Mol. Microbiol. *24*, 917–26.

Watnick, P.I., Fullner, K.J., and Kolter, R. (1999). A role for the mannose sensitive hemagglutinin in biofilm formation by *Vibrio cholerae* El Tor. J. Bacteriol. *181*, 3606–3609.

West, P.A., and Colwell, R.R. (1984). Identification and classification of *Vibrionaceae* – an overview. In Vibrios in the Environment, Colwell, R.R., ed. (Willey, New York), pp. 285–363.

Zhu, J., and Mekalanos, J.J. (2003). Quorum sensing-dependent biofilms enhance colonization in *Vibrio cholerae*. Dev. Cell 5, 647–656.

Zhu, J., Miller, M.B., Vance, R.E., Dziejman, M., Bassler, B.L., and Mekalanos, J.J. (2002). Quorum-sensing regulators control virulence gene expression in *Vibrio cholerae*. Proc. Natl. Acad. Sci. U.S.A. *99*, 3129–3134.

Biology of *Campylobacter* Infection

11

Lieneke I. Bouwman and Jos P.M. van Putten

Abstract

Campylobacter species are one of the leading causes of bacterial foodborne disease. *Campylobacter* survives and flourishes in a variety of environmental niches but causes pathology mainly in humans. The pathogenesis of the disease is still poorly understood, and currently only a few *Campylobacter* virulence determinants have been proposed. Recent studies indicate that *Campylobacter* displays extensive genome plasticity and a range of environmental adaptations, that probably contribute to the success of the pathogen. In this chapter, we will discuss the state-of-the-art of the epidemiology, molecular biology and pathogenesis of *Campylobacter* infection.

Introduction

The discovery of *Campylobacter* probably goes back to 1886 when Theodor Escherich described the presence of *Vibrio*-like organisms in the colon of infants who died from enteritis (Escherich, 1886). Early in the twentieth century, the bacterium was for the first time isolated from the infected uterus of sheep and intestines of cattle and pigs (Jones *et al.*, 1931). Awareness of *Campylobacter* as a principal bacterial enteric pathogen in humans first became evident in the 1970s when improved methods of *Campylobacter* isolation from faeces (Dekeyser *et al.*, 1972) and selective culture media (Skirrow, 1977) facilitated routine diagnostics. Nowadays, campylobacteriosis cases are estimated to reach incidences of up to 5–20% for the general population in developing countries (Coker *et al.*, 2002). Yet, the pathogenic mechanism of the bacteria is still an enigma. In this review, we will discuss the current knowledge on key aspects of *Campylobacter* epidemiology, molecular biology and virulence.

Campylobacter – the microbe

Phylogenetic and genomewide analyses indicate that the genus *Campylobacter* comprises at least 20 species and six subspecies, with several novel subspecies being proposed (Debruyne *et al.*, 2008). The genus *Campylobacter* belongs to the family *Campylobacteraceae* together with the genera *Arcobacter* and *Sulfurospirillum*. The *Campylobacteraceae* family belongs to the class of Epsilon-proteobacteria (Debruyne *et al.*, 2008). Members of the genus *Campylobacter* are mostly Gram-negative spiral-shaped rods that are 0.5 to 5 µm long and 0.2–0.8 µm wide. They carry a single unsheathed uni- or bipolar flagellum. *Campylobacter* generally requires microaerobic growth (5% O_2 is optimal) and capnophilic (3–5% CO_2) conditions. The thermophilic *Campylobacters* do not grow below 30°C and grow optimally at 42°C, although the bacteria may survive at low temperatures for weeks. *Campylobacter* genus is considered non-saccharolytic and utilize amino acids and tricarboxylic acid cycle intermediates as energy sources. Under unfavourable conditions including nutrient limitation, high temperature (60°C) or aeration, *Campylobacter* may change their morphology into coccoid forms, a state in which the bacteria remain viable but can not be cultured (Boucher *et al.*, 1994). However, controversy exists whether the coccoid form is a degenerative state or a dormant state.

Campylobacter species are mainly distinguished based on genetic, biochemical and/or growth characteristics. A phylogenetic tree of *Campylobacter* species based on 16S rRNA sequence similarity is shown in Fig. 11.1. Although members of most *Campylobacter* species have zoonotic potential (Humphrey *et al.*, 2007), *Campylobacter jejuni* subsp. *jejuni*, commonly referred to as *C. jejuni*, causes up to 90% of human campylobacteriosis (Debruyne *et al.*, 2008). *Campylobacter coli*, the closest genetic relative of *C. jejuni*, is the second most common *Campylobacter* species causing gastrointestinal disease. *C. upsaliensis* and *C. consisus* are recognized as pathogens for children at least in some countries (Lastovica and Allos, 2008). Reportedly, non-culturable *Campylobacter* species may also contribute to disease (Bullman *et al.*, 2011).

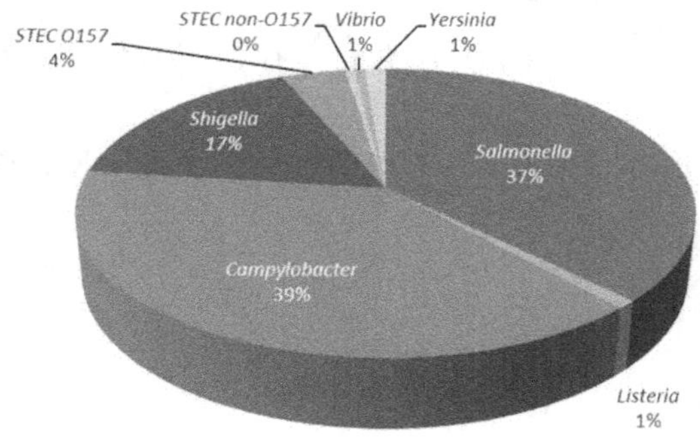

Figure 11.2 Incidence of laboratory-confirmed bacterial foodborne infections (1996–2009) in the USA. (http://www.cdc.gov/foodnet/factsandfigures/2009/Table1b_all_incidence_96–09.pdf).

Epidemiology of *Campylobacter*

Campylobacter is a leading cause of bacterial enterocolitis worldwide, surpassing *Salmonella* and *Shigella* in many regions (Fig. 11.2) (Allos, 2001; Girard *et al.*, 2006). In developed countries *Campylobacter* infection usually affects children under the age of four and adults aged from 15 to 44 years (Olson *et al.*, 2008). Infection generally results in enteritis with manifestations ranging from watery to severe bloody diarrhoea, severe abdominal pain and fever. The incidence of the disease approximates 20 per 100,000 population in the United Sates to 396 per 100,000 individuals in New Zealand (Allos, 2001) but reported cases probably represent only a small fraction of the actual number. In industrialized countries, *Campylobacter* is estimated to infect up to 1% of the human population annually (Humphrey *et al.*, 2007; Havelaar *et al.*, 2009). Hospitalization occurs in ~10–15% of diagnosed cases and *C. jejuni* is lethal in about 0.1% of culture-confirmed cases (Olson *et al.*, 2008).

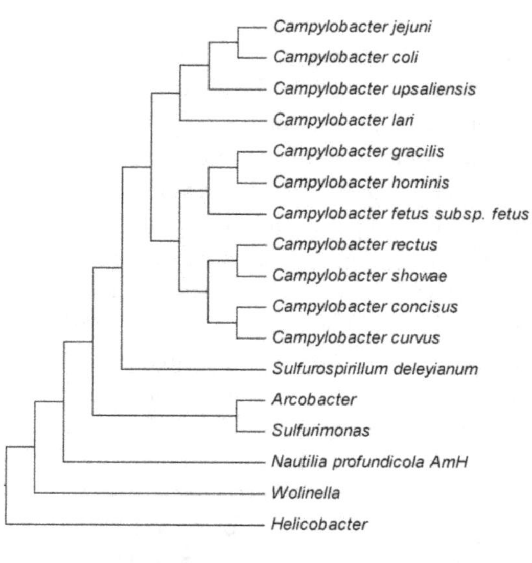

Figure 11.1 Phylogenetic tree of the genus *Campylobacter* based on 16S rRNA sequence similarity.

In developing countries the incidence of clinical manifestations due to *C. jejuni* infection is several orders of magnitude higher than in the industrialized world. Approximately, 40–60% of young children are estimated to become infected every year and high numbers of asymptomatic carriage in older children and adults are reported (Coker *et al.*, 2002; Samuel *et al.*, 2004). Experiments with human volunteers indicate that ingestion of a minor dose of 500–800 *C. jejuni* is sufficient to cause disease (Robinson, 1981; Black *et al.*, 1988).

Although the symptoms usually disappear within ten days, *Campylobacter* species affect the health of millions of people worldwide with an estimated annual economic burden of up to several billion dollars in the United States alone (Buzby *et al.*, 1997). This can partly be attributed to severe complications which can follow infection such as reactive arthritis and the paralysing autoimmune neuropathies Guillain-Barré syndrome (GBS) and Miller–Fisher syndrome (Allos, 2001).

Reservoirs and transmission of *Campylobacter*

Campylobacters are isolated from a variety of animals (e.g. cattle, swine, cats, dogs, reptiles, and birds) as well as from contaminated surface water. Different species appear to have favourite niches. *C. jejuni* and *C. coli* are frequently present in cattle, sheep, swine and birds, *C. upsaliensis* is often found in domestic dogs and cats, while *C. fetus* is primarily isolated from livestock. Although *C. fetus* is a major cause of septic abortion in cattle and sheep, *Campylobacter* colonization of the digestive tract is generally not associated with animal disease. Chicken caecum, for example, may easily contain up to 10^9 cfu of *C. jejuni* per gram of content without apparent pathology.

One significant source of infection in developed countries is the handling and consumption of contaminated food products, especially poultry meat (Lee and Newell, 2006; Sheppard *et al.*, 2009). Other identified risk factors are exposure to surface water, consumption of raw milk, contact with pets and farm animals, and foreign travel (Allos, 2001; Gascon, 2006; Girard *et al.*, 2006; Humphrey *et al.*, 2007; Havelaar *et al.*, 2009). In developing countries environmental contamination, for example the use of contaminated water and close contact with farm animals are major sources of infection (Ashbolt, 2004; Havelaar *et al.*, 2009). Recent studies indicate that *Campylobacter* can survive for prolonged periods inside amoebae (Axelsson-Olsson *et al.*, 2005), which may facilitate waterborne spread to humans and animals. The majority of human *C. jejuni* infections are sporadic cases. Community outbreaks are usually associated with the ingestion of contaminated water or raw milk. The sensitivity to atmospheric oxygen and the fastidious growth requirements of *Campylobacter* probably limits survival and growth on food products during processing and/or storage, reducing the risk of outbreaks.

Molecular typing and diversity of *Campylobacter*

Successful tracking of *Campylobacter* infection routes requires reliable, robust and stable typing methods that have sufficient discriminatory power. *Campylobacter* serotyping can distinguish more than 60 Penner serotypes (heat-stable antigens) (Frost *et al.*, 1998) and over 100 Lior serotypes (heat-labile antigens) (Alm *et al.*, 1991). Phagetyping has been applied as a complementary method (Evans *et al.*, 1998). Both typing methods are valuable but have a moderate resolution and are vulnerable to phenotype variation. These methods are increasingly being replaced by genotyping (Nachamkin *et al.*, 1993; Dingle *et al.*, 2001; Quinones *et al.*, 2007; Sheppard *et al.*, 2009; Behringer *et al.*, 2011). Applied genotyping methods include PCR-based methods, band-based methods e.g. pulsed-field gel electrophoresis (PFGE) and amplified-fragment-length polymorphism (AFLP), multilocus sequence typing (MLST), microarray based methods and pangenome sequence analysis.

Genotyping methods indicate that *C. jejuni* is genetically highly diverse with only a few clonal lineages (Wassenaar and Newell, 2000; Dingle *et al.*, 2001; Suerbaum *et al.*, 2001). The genetic plasticity and rapid exchange of DNA are apparent even after a single *in vivo* passage in chicken (Wassenaar *et al.*, 1998; Hanninen *et al.*, 1999; de Boer *et al.*, 2002). Most *C. jejuni* strains are naturally competent to acquire exogenous DNA. Identified factors indispensable or influencing the efficiency of natural transformation are the type II secretion system and the genes Cj1211, *galE*, *dprA*, and Cj0011c. Surface polysaccharides (capsule, LOS) and the variable presence of restriction -modification systems also influence natural transformation (Jeon *et al.*, 2009). Acquired DNA fragments integrate via RecA-dependent homologous recombination into the genome (Guerry *et al.*, 1994). Other processes involved in the generation of genome diversity are conjugation and transduction. Conjugative plasmids are present in 19–56% of clinical *C. jejuni* isolates (Batchelor *et al.*, 2004;

Schmidt-Ott *et al.*, 2005), while prophages are present in the genomes of a large panel of *C. jejuni* isolates (Fouts *et al.*, 2005; Parker *et al.*, 2006; Clark and Ng, 2008). The frequency of horizontal gene transfer and intragenomic rearrangements often depends on the activity of the available DNA repair mechanisms. *C. jejuni* appears to have functional excision (NER, BER) and recombinational DNA repair pathways, but lack methyl-directed mismatch (MMR) repair (Gaasbeek *et al.*, 2009). The latter probably explains the high frequency of on- and off-switching of phase-variable genes.

The clonality of certain *C. jejuni* lineages has been attributed in part to the presence of insertion elements that contain genes encoding exogenous DNases that limit natural transformation and uptake of DNA (Gaasbeek *et al.*, 2009, 2010). Overall, the *C. jejuni* population diversity and genome plasticity which may result in multiple *C. jejuni* genotypes present in a single specimen complicate global epidemiological surveillance with low resolution methods (e.g. MLST). Whole genome sequencing may resolve this problem.

Molecular biology of *Campylobacter*

The *Campylobacter* genome

In recent years, genome sequences of several *Campylobacter* species have been determined. The genomes typically are 1.5–2.0 Mb in size, have a G+C content between 29% and 34%, and are predicted to contain approximately 1500–1850 genes. About 45% of the genes have no assigned function. Sequenced strains have a variable number (0–3) of prophages, genomic islands, CRISP and IS elements, and plasmids. Many of the sequenced strains (*C. jejuni* subsp. *jejuni*, *C. coli*, *C. upsaliensis*, *C lari* and *C. fetus* subsp. *fetus*, but not *C. jejuni* subsp. *doylei* and *C. concisus*) have genes encoding the cytolethal distending toxin (CDT), one of the few proposed virulence determinants of *Campylobacter*. All genomes have an extensive number (9–75) of homopolymeric G:C tracts, which are short sequence DNA repeats that are prone to slipped strand mispairing. This feature results in high frequency on–off switching of gene function. The G:C tracts are mainly present within genes encoding surface structures (capsule, LOS, flagella) and restriction-modification systems. The large number of genes (7–12) encoding putative two-component regulatory systems probably reflects the ability of *Campylobacter* to adapt to a variety of environmental niches (Parkhill *et al.*, 2000; Miller, 2008). The genomes of the different *Campylobacter* species display relative low overall colinearity. However, the genome contains clusters of genes that remain linked in the various species. Nevertheless, the majority of identified proteins are conserved among the sequenced *Campylobacter* species (Fouts *et al.*, 2005; Miller, 2008).

Genome-wide analysis of multiple *C. jejuni* subsp. *jejuni* isolates indicates that the genomes are syntenic with hot spots for intra- and intergenomic recombination (Champion *et al.*, 2008). The genome contains approximately 1650 genes that mainly are involved in metabolic, biosynthetic, cellular and regulatory processes (Parkhill *et al.*, 2000). The limited number (< 5%) of strain specific (variable) genes are located on the genomes mainly in clusters, named hypervariable plasticity regions. These variable genes are mostly involved in processes like respiration, LOS and capsule variation, and the biosynthesis and modification of the flagella (Pearson *et al.*, 2003). Some of the variable genes contain homopolymeric nucleotide tracts (Parkhill *et al.*, 2000) enabling variable gene expression, resulting in heterogeneity of phenotypes within the bacterial population of a single strain. Several *C. jejuni* strains contain one or more integrated elements and/or plasmids (Fouts *et al.*, 2005). The *C. jejuni* genome appears to lack typical pathogenicity islands, although some strains carry a type VI secretion system (Bleumink and van Putten, unpublished results).

Genetics of environmental adaptation

Campylobacter survives in a variety of ecological niches and can adapt to changes in nutrients, oxygen, pH, temperature and the complex environments encountered in the various hosts. This adaptation is regulated at different levels (transcription, translation, metabolism), although knowledge of the mechanisms is still fragmentary (Wosten *et al.*, 2008). The genome of *C. jejuni* encodes up to 37 transcriptional regulators and contains three sigma factors (RpoD, FliA, and RpoN). About 20% of the *C. jejuni* genes are estimated to be under the control of 7–10 two-component signal transduction systems that respond to different environmental cues. The FliA (σ28) and RpoN (σ54) sigma factors and the FlgS/FlgR two-component system are required for the regulation of flagella assembly (Carrillo *et al.*, 2004; Wosten *et al.*, 2004), the glycosylation of flagellin subunits (Logan *et al.*, 2002), and/or the secretion of virulence-associated proteins (FspA, Cj0977) through the flagellum (Goon *et al.*, 2006; Poly *et al.*, 2007). Several other two-component systems (PhosR/PhosS, DccS/DccR, RacS/RacR) enable metabolic adaptation in response to changes in nutrient- and oxygen availability (MacKichan *et al.*, 2004; Wosten *et al.*, 2004, 2006, 2010).

As non-saccharolytic bacterium *Campylobacter* mainly utilizes amino acids and Krebs cycle intermediates. These molecules generate the essential carbon molecules and drive the electron flow through the highly branched electron transport chain, providing energy. Preferred amino acids utilized by *C. jejuni* are serine, aspartate, glutamate and proline, which all are abundantly present in the chicken caecum. Serine is mainly converted to pyruvate, while the other amino acids are converted via aspartate to fumarate (Guccione *et al.*, 2008). The preferred electron acceptor of *C. jejuni* is oxygen (Hoffman and Goodman, 1982). In a low-oxygen environment, i.e. in the gut, organic and inorganic compounds can serve as alternative electron donors (Myers and Kelly, 2005; Pajaniappan *et al.*, 2008; Weerakoon *et al.*, 2009) and electron acceptors (Sellars *et al.*, 2002; Pittman *et al.*, 2007; Weingarten *et al.*, 2008). The hierarchical order in which the different redox couples of electron donors and acceptors are used in an oxygen-limited environment is not known for *Campylobacter*.

Campylobacter motility and chemotaxis

To reach its favourable niche *Campylobacter* is equipped with a single unsheathed flagellum at one or both poles. The polar flagella are essential for bacterial motility. The spiral shape of the bacterium supposedly aids in its movement in viscous environments e.g. mucus. The genes that contribute to the corkscrew morphology of *Campylobacter* have not been identified, but differ from those present in *Helicobacter* (Bonis *et al.*, 2010; Sycuro *et al.*, 2010). The flagellum of *C. jejuni* basically consists of three structural elements: the basal body, the hook, and the filament (Fig. 11.3). The flagellar filament is composed of the major subunits FlaA and FlaB. Expression of *flaA* and *flaB* is regulated via different sigma factors (σ28 and σ54) and the two-component system FlgS/FlgR (Wosten *et al.*, 2004; Joslin and Hendrixson, 2009). The ratio between the two types of flagellin varies with the growth temperature. The anti-sigma factor FlgM represses both sigma factors in a temperature-dependent manner and thus limits flagella length (Wosten *et al*, 2010). Each flagellin consists of a conserved N and C terminal domain interspaced with a surface exposed domain that is highly variable in protein sequence among strains. The surface-exposed domain is heavily decorated with *O*-linked pseudaminic acid and pseudaminic acid derivatives (Thibault *et al.*, 2001). The glycosylation is essential for flagellin transport through the flagellar secretion apparatus and/or for the actual assembly of the flagellum, as mutants unable to glycosylate flagellin do not produce functional flagella. The flagellar filaments from *C. jejuni* have seven protofilaments rather than the 11 in *S. typhimurium* (Galkin *et al.*, 2008).

Campylobacter is equipped with several taxis systems that enable directional motility. The architecture of the *C. jejuni* chemotaxis system grossly follows the classical chemotaxis systems in prokaryotes. Alterations in concentrations of attractants or repellents are first sensed by different methyl accepting sensory receptors (MCPs) in the periplasm. The resulting conformational change in the receptors starts a phosphorelay signalling cascade that requires the CheA histidine kinase and the CheY response regulator. The phosphorylation status of CheY influences the direction of the rotation and/or speed of the flagellar rotor. The molecular details behind *C. jejuni* chemotaxis have not yet been resolved. *In silico* analysis indicates the presence in *C. jejuni* of up to ten possible methyl accepting receptors

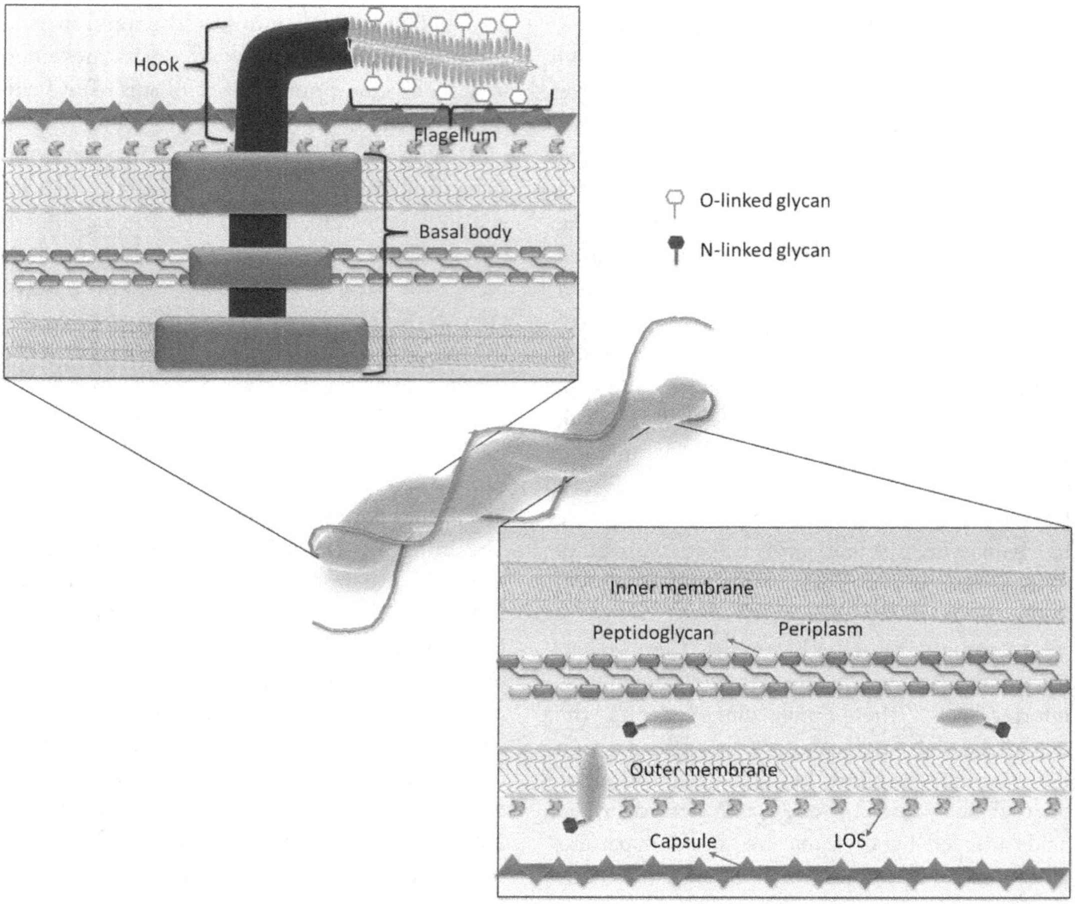

Figure 11.3 *Campylobacter* membrane architecture. The spiral shape and single polar flagella are typical of *Campylobacter*. Glycan structures are present in the peptidogycan, LOS, capsule and attached to *N*-glycoproteins. The flagellum consists of the basal body, the flagellar hook, and the protruding flagellar filament which is heavily O-glycosylated. Besides for motility, the *C. jejuni* flagellum acts as a type III secretion system.

(MCPs) and the proteins CheA, CheW and CheY as well as CheB and CheR proteins involved in receptor adaptation (Korolik and Ketely, 2008; Kanungpean *et al.*, 2011). In contrast to *E. coli*, *C. jejuni* also has a CheV protein and three chemotaxis proteins with a CheY domain (CheA, CheV and CheY). *C. jejuni* displays chemotaxis towards a variety of substances, including serine, fucose, pyruvate and succinate (Hugdahl *et al.*, 1988).

An additional taxis system of *C. jejuni* is the aerotaxis or energy taxis which involves the proteins CetA and CetB. This system probably senses internal bacterial signals in particular the redox state of the bacterium. The CetA/B system is most active under conditions of nutrient starvation. The *C. jejuni* taxis systems are important in bacterial colonization and virulence (Takata *et al.*, 1992; Yao *et al.*, 1997; Hendrixson and DiRita, 2004).

Surface architecture of *Campylobacter*

The surface characteristics of *Campylobacter* have been investigated mostly for the major foodborne pathogen *C. jejuni*. The outer surface of *C. jejuni* is almost entirely decorated with sugars. Outer membrane protein components include the major outer membrane porin proteins (MOMP, OMP50), the adhesin CadF, several minor proteins and, in some strains, the needle structure of the type VI secretion system. Fimbriae appear to be absent. The glycan repertoire of *C. jejuni* consists of capsular polysaccharides (CPS), lipooligosaccharide (LOS), and *O*-glycosylated and *N*-glycosylated proteins (Fig. 11.3). Approximately 10% of the *C. jejuni* genome is predicted to be involved in carbohydrate biosynthesis. Most of the genes group on the genome in one of four gene clusters. The locus containing the genes coding for *N*-glycosylation is conserved among the species. The other three loci coding for the CPS, LOS and *O*-glycosylation biosynthetic pathways respectively, are hypervariable among strains (Dorrell *et al.*, 2001; Pearson *et al.*, 2003). This genetic variation provides the species with the ability to decorate its population with a huge diversity in surface glycans. The large variation in glycan structures probably contributes to *C. jejuni* survival by preventing attacks by the immune system and bacteriophages (Karlyshev *et al.*, 2005a; Guerry and Szymanski, 2008).

Capsule

The presence of *Campylobacter* capsular polysaccharides was appreciated after analysis of the whole genome sequence (Karlyshev *et al.*, 2000; Bacon *et al.*, 2001). The *C. jejuni* capsule consists of repeating oligosaccharide units (Corcoran *et al.*, 2006) and is only loosely attached to the bacterial cell surface. The polysaccharides are biosynthesized in the cytoplasm where glycosyltransferases attach nucleotide diphosphate sugars to the undecaprenyl pyrophosphate carrier. The whole complex is then flipped over the membrane via an ABC transporter complex consisting of the proteins KpsM, KpsT, and three other Kps proteins. The polysaccharide is anchored to the membrane via the lipid dipalmitoyl-glycerolphosphate (Corcoran *et al.*, 2006).

The composition of the *C. jejuni* CPS is highly variable among strains. These differences turned out to be the basis of the Penner serotype, which has been used to distinguish *C. jejuni* isolates for many years (Karlyshev *et al.*, 2000). The *cps* locus varies between strains and ranges in size from 15 kb to 34 kb (Karlyshev *et al.*, 2005b). The hypervariability is caused by genetic exchange of the capsular genes, horizontal transfer of entire clusters, gene duplication, deletion, and fusion, but also by contingency gene variation. Variable attachment of methyl-, ethanolamine- and aminoglycerol groups to the carbohydrates further increases the plasticity of the polysaccharide.

The *C. jejuni* capsule is important for bacterial colonization of the intestine and confers serum resistance (Bacon *et al.*, 2001). A capsule deficient *kpsM* mutant displays reduced invasion of INT-407 cells and colonization of chickens and is attenuated in a ferret infection model (Bacon *et al.*, 2001). Capsule production is phase variable (Bacon *et al.*, 2001) and may vary depending on culture conditions including the presence of human epithelial cells (Corcionivoschi *et al.*, 2009).

Lipooligosaccharide (LOS)

The major constituent of the surface outer membrane leaflet of *C. jejuni* is LOS. *C. jejuni* LOS consists of a non-repeating oligosaccharide that is attached via an inner core oligosaccharide to a hydrophobic lipid A anchor (Karlyshev *et al.*, 2005a; Corcoran *et al.*, 2006). The LOS inner core consists of a single 3-deoxy-D-manno-octulosonic residue (KDO) and two attached glycero-D-mannose heptose residues. The composition and length of the outer core oligosaccharide are variable and show both inter- and intrastrain variation. Each strain has a defined LOS repertoire. The variation between strains is due to differences in specific gene content. Intrastrain LOS heterogeneity is mainly caused by slipped strand mispairing in contingency genes (Parkhill *et al.*, 2000) and mutations leading to gene inactivation or different acceptor specificities of glycosyltransferases (Gilbert *et al.*, 2002), but is also influenced by environmental factors such as temperature (Semchenko *et al.*, 2010).

Based on the gene content and organization of *C. jejuni* LOS loci, up to 19 different LOS classes (A-S) have been identified. Eighty per cent of strains belong to class A-H (Parker *et al.*, 2008). Some strains, mainly the strains from class A-C, are able to incorporate sialic acid into the LOS. The LOS of these strains resembles gangliosides. Auto-reactive antibodies directed to these LOS forms are linked to Guillain–Barré syndrome and Miller–Fisher syndrome (Yuki *et al.*, 2004). The genes responsible for sialic acid synthesis are located in the *cgt* cluster. Phase variation of the *cgtA* gene determines whether the LOS mimics GM1, GM2, GM3, or GD2 gangliosides. Besides gangliosides, *C. jejuni* LOS can mimic a range of host cell glycolipid/glycoprotein antigens including P blood group antigen, paragloboside (lacto-*N*-neotetraose), lacto-*N*-biose, and sialyl-Lewis c units. The enzymes sialyltransferase CstII and galactosyltransferase CgtD play a key role in the synthesis of these mimic types (Houliston *et al.*, 2011).

The impact of *C. jejuni* LOS variation on bacterial virulence is increasingly appreciated. *C. jejuni* LOS influences the susceptibility of the pathogen to complement-mediated killing and antibiotics (Jeon *et al.*, 2009; van Mourik *et al.*, 2010), and affects DNA uptake (Guerry *et al.*, 2002; Jeon *et al.*, 2009). Distinct LOS isoforms increase bacterial invasion of eukaryotic cells and influences cell tropism (Louwen *et al.*, 2008; Heikema *et al.*, 2010). Mammalian receptors recognizing distinct *C. jejuni* LOS types are sialoadhesin (siglec-1) found on a subset of macrophages, and siglec-7 and MGL lectin receptors present on dendritic cells (van Sorge *et al.*, 2009; Heikema *et al.*, 2010; Bax *et al.*, 2011). The lipid A part of *C. jejuni* LOS activates the innate immune receptor TLR4 and stimulate the release of pro-inflammatory mediators (Hu *et al.*, 2006a; de Zoete *et al.*, 2010a).

O-glycosylation

The major protein structure extruding from the *C. jejuni* surface are the polar flagella. Both *C. jejuni* and *C. coli* flagella are heavily decorated with *O*-linked carbohydrates. Flagellar glycosylation mainly involves the attachment of 5,7-diacetamideo-3,5,7,9-tetradeoxy-L-*glycero*-α-L-*manno*-nonulosonic acid (pseudaminic acid, Pse5Ac7Ac) or 5,7-diacetamido-D-*glycero*-β-D-*galacto*-nonulosonic acid (legionaminic acid, Leg5Am7Ac) derivatives to hydroxyl groups of serine or threonine residues located on the surface-exposed variable domain of flagellin (Thibault *et al.*, 2001). Glycans are bound to up to 19 serine or threonine residues per flagellin subunit and contribute up to 10% of the flagellin weight. The genes encoding the *O*-glycosylation machinery are located in a hypervariable region of the genome in close vicinity but independently regulated from the flagellin genes (Parkhill *et al.*, 2000). The biological function of flagellin glycosylation remains unclear. Flagellin glycosylation is required for flagellin export and/or flagella assembly, while mutants with altered post-translation modification of the *O*-glycosyl moiety exhibit loss of auto-agglutination, and/or reduced adherence and invasion of human epithelial cells (Guerry *et al.*, 2006; van Alphen *et al.*, 2008b; Ewing *et al.*, 2009). The post-translational modification of the flagella also contributes to serospecificity (Alm *et al.*, 1991). These results suggest that the variation in O-glycans contributes to immune evasion from antibodies developed during infection and perhaps attack by bacteriophages.

N-glycosylation

C. jejuni has been instrumental in the discovery of *N*-linked protein glycosylation in prokaryotes (Szymanski *et al.*, 1999). At present, more than 50 *N*-glycosylated proteins have been identified in *C. jejuni*. *N*-glycosylation involves the transfer *en bloc* of the heptasaccharide GalNAc-α1,4-GalNAc-α1,4-[Glc-β1,3]GalNAc-α1,4-GalNac-α1,4-GalNAc-α1,3-di-*N*-acetyl bacillosamine from the lipid carrier (most likely undecaprenyl pyrophosphate) to asparagine residues in the specific glycosylation consensus sequence, Asp/Glu-Y-Asn-X-Ser/Thr, where Y and X is any amino acid except proline (Kowarik *et al.*, 2006). The key enzyme in the transfer of the oligosaccharides to proteins is PglB (Wacker *et al.*, 2002). The genes encoding the *N*-glycosylation pathway are all located in the protein glycosylation (*pgl*) locus. The gene composition of this locus is highly conserved among *C. jejuni* and *C. coli* suggesting restricted functional flexibility of the *N*-glycosylation and/or the absence of environmental pressure to modify the glycosyl group. As most *N*-glycoproteins of *C. jejuni* are predicted to be located in the periplasm, it can be imagined that they are not a natural target of selective pressure by the immune system or bacteriophages.

The function of the *Campylobacter* *N*-linked protein glycosylation is still largely unknown. Mutants defective in *N*-glycosylation display reduced natural competence, reduced immunoreactivity with human sera, reduced adherence and invasion of host cells, and reduced colonization of chicken and mice (Szymanski *et al.*, 1999,

2002; Larsen *et al.*, 2004; Linton *et al.*, 2005). The *N*-linked heptasaccharide, which is also present at high concentrations as free oligosaccharide in the periplasm, has recently been reported to interact with the MGL lectin receptor on human dendritic cells (van Sorge *et al.*, 2009), suggesting a possible immunomodulatory function.

Virulence of *Campylobacter*

C. jejuni and *C. coli* are major pathogens causing > 100 million cases of enteritis each year (Girard *et al.*, 2006). However, the pathogenesis of the disease remains elusive. Both pathogens frequently colonize a large variety of animal hosts usually without apparent pathology, suggesting the existence of human specific virulence determinants. Thus far several putative virulence determinants have been proposed (Table 11.1). The contribution of these factors to the development of pathology remains difficult to assess in the absence of a suitable animal disease model. Present knowledge of colonization and virulence factors is largely based on the results of *in vitro* assays and experiments with (often immunocompromised) animals. In general, ingestion of *C. jejuni* or *C. coli* by an individual is believed to result in colonization of the intestine, in particular, the intestinal crypts. At some point the epithelial monolayer is damaged or crossed and inflammation is observed. These events clinically manifest as an acute watery or bloody diarrhoea, fever and abdominal pain. The symptoms usually disappear after 5–7 days. The molecular events that cause the acute intestinal inflammation remain to be defined. Details of the assumed key steps in the development of campylobacteriosis are outlined below.

Colonization of the mucus layer

After ingestion of contaminated food or water *C. jejuni* efficiently colonizes the mucus of the small and large bowel. Mucin glycoproteins contain fucose and serine, which are strong chemoattractants and major carbon source of *C. jejuni* (Hugdahl *et al.*, 1988; Stahl *et al.*, 2011). The spiral shape and polar flagella movement probably facilitate the efficient penetration of the mucus layer (Ferrero and Lee, 1988; Szymanski *et al.*, 1995). In human biopsies, large number of *C. jejuni* are found deep in the intestinal crypts. *C. jejuni* contains several putative mucin-degrading enzymes but their functionality remains to be validated. *Ex vivo* studies with human ileal tissue indicated that at prolonged infection *C. jejuni* forms microcolonies that adhere to the tissue via flagella (Haddock *et al.*, 2010). Many studies indicate that *C. jejuni* that lack flagella or carry a short flagellum consisting of only FlaB show reduced levels of colonization of animals and reduced invasion of epithelial cells (Wassenaar *et al.*, 1991; Szymanski

Table 11.1 Putative virulence factors of *C. jejuni*

Virulence factor	Functions in	Reference
Flagella	Colonization, adhesion, invasion	Grant *et al.* (1993)
Chemotaxis	Colonization, invasion	van Alphen *et al.* (2008a)
Capsule	Colonization, invasion, evasion	Bacon *et al.* (2001)
LOS	Colonization, invasion, evasion	Louwen *et al.* (2008)
N-glycosylation	Colonization, immunemodulation	Karlyshev *et al.* (2004)
CDT	Apoptosis, G2/M cell cycle arrest	Lara-Tejero and Galan (2000)
FspA2	Apoptosis	Poly *et al.* (2007)
Cia	Invasion, intracellular survival	Konkel *et al.* (1999)
CadF	Adhesion, invasion	Konkel *et al.* (1997)
CapA	Adhesion	Ashgar *et al.* (2007)
JlpA	Adhesion	Jin *et al.* (2001)
PEB1	Adhesion	Pei *et al.* (1998)
FlaC	Adhesion, invasion	Song *et al.* (2004)

et al., 1995). The most common human mucin MUC2, increases the expression of the flagellin major subunit FlaA (but not flaB) as well as putative virulence factors cytolethal distending toxin (CDT), the assumed adhesin JlpA, and the *Campylobacter* invasion associated protein CiaB (Tu *et al.*, 2008). Together, these findings indicate a major role of intestinal mucus and bacterial motility and chemotaxis in *C. jejuni* colonization.

Adhesion to host cells

Once within the crypts *C. jejuni* is supposed to adhere to the mucosal tissue. The specific mechanisms of cellular adhesion have not been fully elucidated. *Campylobacter* lacks fimbriae-mediated adhesion. Instead an array of putative surface adhesins have been proposed including surface polysaccharides, flagella, CadF, JlpA, CapA, Cj1349, FlpA and several PEB proteins. Corresponding receptors present on the apical site of intestinal cells, however, have not been identified for any of these proteins and systematic studies that evaluate the relative significance of the various proposed adhesins are lacking. One class of potential adhesins that include proteins CadF, Cj1349 and FlpA, is predicted to contain a fibronectin binding domain (Konkel *et al.*, 1997; Flanagan *et al.*, 2009). CadF and FlpA are located on the bacterial surface and promote bacterial adhesion in cell culture systems as well as chicken colonization (Konkel *et al.*, 1997; Flanagan *et al.*, 2009; Konkel *et al.*, 2010). Another putative adhesin is the surface exposed lipoprotein JlpA that binds to Hsp90 on HEp-2 cells and triggers the activation of NF-κB and p38 MAP kinases (Jin *et al.*, 2001). Several studies however, failed to confirm JlpA-dependent adhesion (van Alphen *et al.*, 2008a; Novik *et al.*, 2010) and a role of JlpA in the adhesion to chicken epithelial cells or in chicken colonization could not be demonstrated (Flanagan *et al.*, 2009). The surface-exposed autotransporter protein Cap confers moderate adhesion to human intestinal Caco-2 cells (Ashgar *et al.*, 2007) but CapA is not conserved in *Campylobacter* and its role in the *C. jejuni* colonization of chicken is controversial (Ashgar *et al.*, 2007; Flanagan *et al.*, 2009). For a long time, the proteins PEB1, PEB3 and PEB4 were considered to act as adhesins (Pei *et al.*, 1998). Recent studies indicate that most of these proteins reside in the periplasm and are involved in amino acid transport. PEB1 has a high binding affinity for L-aspartate and L-glutamate. Inactivation of PEB4 reduces *C. jejuni* adhesion but also reduces the amount of CadF protein (Rathbun and Thompson, 2009). At this time, the function of PEB proteins as *C. jejuni* adhesins is still under debate. Overall, it can be concluded that in comparison to other enteropathogens, *Campylobacter* appears to have a variable repertoire of rather weak bacterial surface adhesins.

Cellular uptake of *C. jejuni*

Several studies on human intestinal biopsies and infected macaques suggest that *C. jejuni* is endocytosed by epithelial cells (van Spreeuwel *et al.*, 1985; Russell *et al.*, 1993). *In vitro* studies confirm that *C. jejuni* is internalized by non-professional phagocytes, although the efficiency of uptake varies between laboratories perhaps owing to the use of different strains and methods. One mechanism of bacterial uptake appears to be dependent on the so-called *Campylobacter* invasion antigens (especially CiaB) that are secreted from the flagellar export apparatus (Konkel *et al.*, 1999; Konkel *et al.*, 2004). Several groups, however, reported unaltered invasive phenotypes for CiaB insertional mutants (Goon *et al.*, 2006; van Alphen *et al.*, 2008a; Novik *et al.*, 2010). Recently, a novel mechanism of *C. jejuni* invasion has been identified that is active under conditions of nutrient starvation. This mechanism involved rapid migration of *C. jejuni* underneath epithelial cells (a process termed subvasion), followed by unprecedented efficient entry (up to 50 bacteria per cell at 2 hours of infection) from the basal cell surface (van Alphen *et al.*, 2008a) (Fig. 11.4). The molecular basis of this process remains to be elucidated.

Cellular uptake of *C. jejuni* has been reported to require polymerized host cell microtubules and/or microfilaments or none of these cytoskeletal structures (Oelschlaeger *et al.*, 1993; Hu and Kopecko, 1999; Biswas *et al.*, 2003; Monteville *et al.*, 2003). These results probably reflect the diversity of existing uptake pathways under different environmental conditions (Hu and Kopecko, 2008). Once inside cells *C. jejuni* resides in a

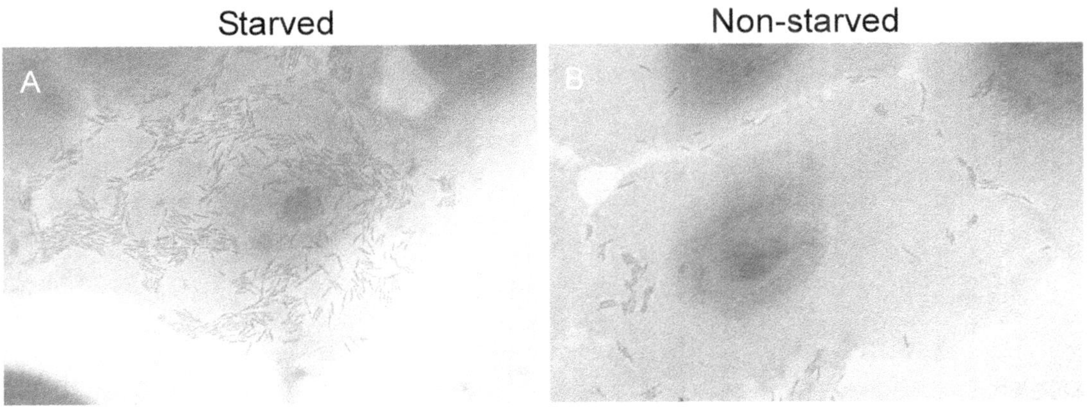

Figure 11.4 *C. jejuni* infection (1 h) of primary cultures of epithelial cells. (A) Nutrient-starved *C. jejuni* subvade epithelial cells and then enter the cells at the basal cell surface in high numbers. (B) In nutrient-rich conditions the efficiency of cellular infection is low.

membrane-bound cellular compartment (Watson and Galan, 2008). Some studies indicate the formation of a unique *C. jejuni*-containing vacuole (CCV) where the bacteria can survive and even replicate, possibly in a CiaI protein-dependent fashion (Watson and Galan, 2008; Buelow *et al.*, 2011). Other suggest that *C. jejuni* cannot survive intracellularly (De Melo *et al.*, 1989). Whether the conflicting reports are solely caused by the poor recovery from bacteria that are adapted to the intracellular environment to the extracellular conditions (Watson and Galan, 2008), awaits further study.

Bacterial translocation

Enteropathogens including *C. jejuni* have been proposed to translocate across the epithelial barrier via three possible routes (Fig. 11.5). The first pathway is via M-cells that sample antigens and particles from the lumen and transport them across the epithelium. This is a major infection route utilized by, for example, *Salmonella*, but may also be employed by *Campylobacter* (Walker *et al.*, 1988). A second route is transcellular migration. In this scenario, *C. jejuni* is endocytosed by the epithelial cells, transported towards the basolateral cell surface, and released into the subepithelial space (Grant *et al.*, 1993; Bras and Ketley, 1999; Monteville and Konkel, 2002). The third mechanism is paracellular transport during which the bacteria pass intact, weakened or disrupted tight junctions between epithelial cells (MacCallum *et al.*, 2005; Chen *et al.*, 2006; Wine *et al.*, 2008). Which pathway(s) are utilized by *Campylobacter* during the natural infection is unknown. *In vitro* disruption of the epithelial monolayer is observed after 24 to 48 hours depending on start inoculum, strain and the cell type, while bacterial translocation across the monolayer appears a much more rapid event (Kalischuk *et al.*, 2009). *In vivo* the production of the cytolethal distending toxin by *C. jejuni* may limit cell renewal and thus facilitate bacterial passage of the epithelium.

Secretion of virulence factors

Many bacterial enteropathogens secrete virulence factors to damage the mucosal barrier and induce pathology. For a long time, the only toxic factor reported for *C. jejuni* was the cytolethal distending toxin (CDT). More recently, other factors have been identified including the proteins FspA, FlaC and *C. jejuni* invasion antigens (Cia). Typical type III secretion systems as present in certain other bacteria e.g. *Salmonella* spp. are absent in *Campylobacter*. Some *Campylobacter* strains, however, express a functional type VI secretion system.

CDT consists of three membrane-associated subunits CdtA, CdtB and CdtC. All three proteins are needed for biological activity. CdtA and CdtC bind to host cell and forms a pore which enables CdtB to enter the cell and cause cytoxicity (Lara-Tejero and Galan, 2001; Lee *et al.*, 2003). CDT causes cell cycle arrest at the G2/M phase, resulting in cellular distension and eventually apoptotic

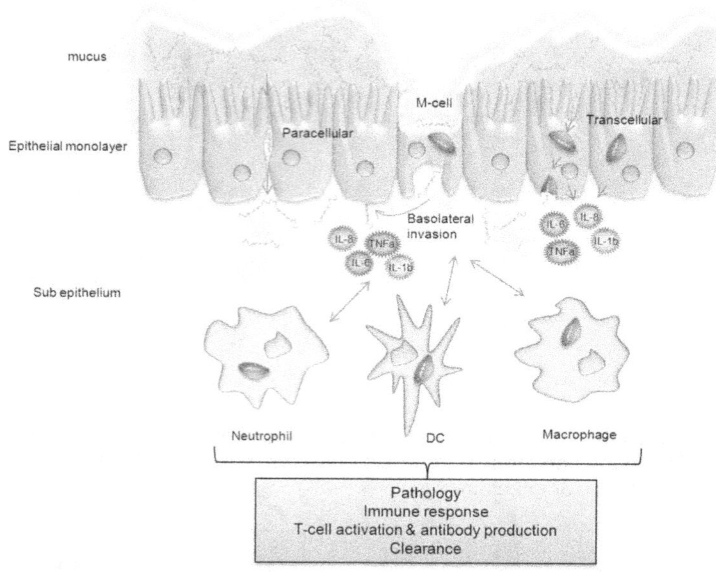

Figure 11.5 Possible translocation routes of *C. jejuni*. After penetration of intestinal mucus bacteria can migrate between epithelial cells (paracellular translocation), and/or invade the epithelial cells at the apical site and exit the cells at the basolateral site (transcellular translocation). *C. jejuni* may also cross the epithelial barrier via M-cells and then invade the epithelial cells at the basolateral surface. Once across the barrier, *C. jejuni* encounters professional immune cells. Infection of the human mucosa induces an inflammatory response that contributes to pathology.

cell death (Whitehouse *et al.*, 1998). In addition, CDT induces NF-κB activation and IL-8 production (Hickey *et al.*, 2005). The toxin is secreted into the supernatant via outer membrane vesicles (Lindmark *et al.*, 2009).

The FspA and FlaC proteins and members of the Cia protein family are secreted through the flagellar export system (Konkel *et al.*, 2004; Song *et al.*, 2004; Poly *et al.*, 2007). There are two isoforms of FspA (FspA1 and FspA2), and both require only a minimal flagella structure for secretion. Recombinant FspA2 causes apoptosis in INT-407 epithelial cells (Poly *et al.*, 2007). As mentioned above, the secretion of the Cia proteins may facilitate bacterial uptake into host cells and influence intracellular survival by preventing fusion of the endosome with the lysosome (Buelow *et al.*, 2011). The FlaC protein structurally resembles the flagellin subunits FlaA and FlaB, but is not involved in flagellar assembly. Purified FlaC binds to HEp-2 cells (Song *et al.*, 2004), but its function as a putative virulence determinant remains elusive.

Analysis of the culture supernatants of *C. jejuni* suggests the presence of several other secreted factors that are possibly involved in activation of the immune system (Mellits *et al.*, 2002; Al-Sayeqh *et al.*, 2010). The structure and significance of these factors await further investigation.

Interaction of *C. jejuni* with the immune system

The contact between *C. jejuni* and the intestinal tissues eventually induces inflammation with an influx of neutrophils and high numbers of leucocytes present in the faeces (Black *et al.*, 1988). The molecular signals that drive the inflammation are beginning to be explored. Important receptors present on the intestinal cells and immune cells that initiate the innate immune response are the Toll-like receptors (TLR) and NOD-like receptors (NLRs). These receptors recognize conserved structural motifs from microbes, including DNA (TLR9), lipoproteins (TLR2 complex), LOS (TLR4), and flagellin (TLR5). Stimulation of the receptors activates NF-κB, which ultimately drives the production of pro-inflammatory cytokines and chemokines. Recent studies by our group indicate that *C. jejuni* bacterial lysates activate human TLR2 and TLR4. Intact *C. jejuni*, however, did not activate the innate immune receptors, probably owing to shielding of the TLR ligands in the native bacteria (de Zoete *et al.*, 2010a). Furthermore, *C. jejuni* flagellin and DNA were unable to activate human TLR5 and TLR9, suggesting that live *C. jejuni* can evade immune recognition by these TLRs. Exchange of *C. jejuni* flagellins domains for homologues regions from *Salmonella* flagellin restored activation of human TLR5 (de Zoete *et al.*, 2010b).

Human and chicken differ in TLR repertoire and function. Chicken have a larger variety of TLR2 complexes, a TLR4 complex that is unable to activate the MyD88-independent signalling pathway (unlike mammalian TLR4), and lack TLR9 (Keestra *et al.*, 2008). Instead chickens express TLR15 and TLR21 which are absent in mammals. TLR15 is activated by microbial proteases (de Zoete *et al.*, 2011), while TLR21 recognizes bacterial DNA (Keestra *et al.*, 2010). Comparison of the *C. jejuni* activation of human and chicken TLRs revealed that *C. jejuni* flagellin is also unable to activate chicken TLR5, but that *C. jejuni* DNA stimulates chicken TLR21 (de Zoete *et al.*, 2010a). The possible significance of these findings for the different course of *C. jejuni* infection in humans and chickens remains to be elucidated.

C. jejuni has also been reported to activate human NLRC1 (also known as NOD1). This member of the NLR family responds to the peptidoglycan component D-glutamyl-meso-diaminopimelic acid (iE-DAP). NLRC1 stimulation activates NF-κB and the inflammasome (Zilbauer *et al.*, 2007). The activation of the TLRs and NLR by *C. jejuni* molecules stimulate the production of among others IL-1β, IL-6, IL-8, and TNFα (Hu and Hickey, 2005; Hu *et al.*, 2006b; Zheng *et al.*, 2008). These factors contribute to the influx and activation of professional immune cells to the site of infection. The professional phagocytes will internalize *C. jejuni*. Whether the pathogen can survive within monocytes or macrophages is still under debate. Both intracellular replication and efficient bacterial killing by monocytes have been reported (Kiehlbauch *et al.*, 1985; Hickey *et al.*, 2005; Watson and Galan, 2008). In some cases, survival in primary monocytes was only observed in a subset of human donors (Wassenaar *et al.*, 1997). The molecular basis and significance of these findings for the course of infection are unknown.

Campylobacter – commensal or pathogen?

C. jejuni and *C. coli* are major foodborne pathogens but rarely cause animal disease. The molecular basis for this different behaviour remains to be elucidated. The pathogenic *Campylobacter* appear to have only a few virulence determinants, of which CDT may show host specificity. Other factors that may contribute to pathogenicity are host-specific receptors and environmental factors, including the composition of the mucus, the composition of the intestinal flora, the availability of nutrients, and the innate host defence. In developing countries, *Campylobacter* is a pathogen early in life but shows mainly commensal behaviour at older ages (Havelaar *et al.*, 2009), suggesting that changes in intestinal flora and/or development of immunity cause a transition from pathogen to commensal behaviour.

Besides optimal hygiene, future preventative measures can focus on prevention of pathology and/or the reduction of (non-symptomatic) intestinal colonization and spread of *Campylobacter*. Prevention of disease requires neutralization of critical virulence determinants, which are not completely known. Fundamental research on *Campylobacter* pathogenesis is needed to enable rational vaccine design. Reduction of *Campylobacter* colonization and commensal behaviour in humans and/or animals seems much more complex and requires detailed knowledge of the factors that enable *Campylobacter* survival in the intestine and spread into the environment. Better characterization of the bacterial adaptation mechanisms and metabolism are important first steps in this direction. One ongoing approach is to develop *Campylobacter* vaccines for use in poultry (de Zoete *et al.*, 2007). Here a reduction of bacterial load may already be sufficient to achieve the urgently needed reduction of *Campylobacter* load in the food chain (Havelaar *et al.*, 2009). At this time, only hygienic measures can be advocated as the simple solution to the problem of *Campylobacter*. The occurrence of > 100 millions of cases of campylobacteriosis each year indicates that the preventative measures are not sufficient and that more innovative strategies are required to combat *Campylobacter* infections.

Acknowledgements

The van Putten laboratory acknowledges the financial support for the development of a *Campylobacter* vaccine by the Dutch Ministry of Economics, Agriculture and Innovation, the Danish Agency for Science, Technology and Innovation (CamVac grant number: 09-067131) and the European Union (CamCon grant number: 244547).

References

Allos, B.M. (2001). *Campylobacter jejuni* infections: update on emerging issues and trends. Clin. Infect. Dis. *32*, 1201–1206.

Alm, R.A., Guerry, P., Power, M.E., Lior, H., and Trust, T.J. (1991). Analysis of the role of flagella in the heat-labile Lior serotyping scheme of thermophilic *Campylobacters* by mutant allele exchange. J. Clin. Microbiol. *29*, 2438–2445.

van Alphen, L.B., Bleumink-Pluym, N.M., Rochat, K.D., van Balkom, B.W., Wosten, M.M., and van Putten, J.P. (2008a). Active migration into the subcellular space precedes *Campylobacter jejuni* invasion of epithelial cells. Cell. Microbiol. *10*, 53–66.

van Alphen, L.B., Wuhrer, M., Bleumink-Pluym, N.M., Hensbergen, P.J., Deelder, A.M., and van Putten, J.P. (2008b). A functional *Campylobacter jejuni maf4* gene results in novel glycoforms on flagellin and altered autoagglutination behaviour. Microbiology *154*, 3385–3397.

Al-Sayeqh, A.F., Loughlin, M.F., Dillon, E., Mellits, K.H., and Connerton, I.F. (2010). *Campylobacter jejuni* activates NF-kappaB independently of TLR2, TLR4, Nod1 and Nod2 receptors. Microb. Pathog. *49*, 294–304.

Ashbolt, N.J. (2004). Microbial contamination of drinking water and disease outcomes in developing regions. Toxicology *198*, 229–238.

Ashgar, S.S., Oldfield, N.J., Wooldridge, K.G., Jones, M.A., Irving, G.J., Turner, D.P., and Ala'Aldeen, D.A. (2007). CapA, an autotransporter protein of *Campylobacter jejuni*, mediates association with human epithelial cells and colonization of the chicken gut. J. Bacteriol. *189*, 1856–1865.

Axelsson-Olsson, D., Waldenstrom, J., Broman, T., Olsen, B., and Holmberg, M. (2005). Protozoan *Acanthamoeba polyphaga* as a potential reservoir for *Campylobacter jejuni*. Appl. Environ. Microbiol. *71*, 987–992.

Bacon, D.J., Szymanski, C.M., Burr, D.H., Silver, R.P., Alm, R.A., and Guerry, P. (2001). A phase-variable capsule is involved in virulence of *Campylobacter jejuni* 81–176. Mol. Microbiol. *40*, 769–777.

Batchelor, R.A., Pearson, B.M., Friis, L.M., Guerry, P., and Wells, J.M. (2004). Nucleotide sequences and comparison of two large conjugative plasmids from different *Campylobacter* species. Microbiology *150*, 3507–3517.

Bax, M., Kuijf, M.L., Heikema, A.P., van Rijs, W., Bruijns, S.C., Garcia-Vallejo, J.J., Crocker, P.R., Jacobs, B.C., van Vliet, S.J., and van Kooyk, Y. (2011). *Campylobacter jejuni* lipo-oligosaccharides modulate DC-mediated T cell polarization in a sialic acid linkage dependent manner. Infect. Immun. *79*, 2681–2689.

Behringer, M., Miller, W.G., and Oyarzabal, O.A. (2011). Typing of *Campylobacter jejuni* and *Campylobacter coli* isolated from live broilers and retail broiler meat by *flaA*-RFLP, MLST, PFGE and REP-PCR. J. Microbiol. Methods *84*, 194–201.

Biswas, D., Itoh, K., and Sasakawa, C. (2003). Role of microfilaments and microtubules in the invasion of INT-407 cells by *Campylobacter jejuni*. Microbiol. Immunol. *47*, 469–473.

Black, R.E., Levine, M.M., Clements, M.L., Hughes, T.P., and Blaser, M.J. (1988). Experimental *Campylobacter jejuni* infection in humans. J. Infect. Dis. *157*, 472–479.

de Boer, P., Wagenaar, J.A., Achterberg, R.P., van Putten, J.P., Schouls, L.M., and Duim, B. (2002). Generation of *Campylobacter jejuni* genetic diversity *in vivo*. Mol. Microbiol. *44*, 351–359.

Bonis, M., Ecobichon, C., Guadagnini, S., Prevost, M.C., and Boneca, I.G. (2010). A M23B family metallopeptidase of *Helicobacter pylori* required for cell shape, pole formation and virulence. Mol. Microbiol. *78*, 809–819.

Boucher, S.N., Slater, E.R., Chamberlain, A.H., and Adams, M.R. (1994). Production and viability of coccoid forms of *Campylobacter jejuni*. J. Appl. Bacteriol. *77*, 303–307.

Bras, A.M., and Ketley, J.M. (1999). Transcellular translocation of *Campylobacter jejuni* across human polarised epithelial monolayers. FEMS Microbiol. Lett. *179*, 209–215.

Buelow, D.R., Christensen, J.E., Neal-McKinney, J.M., and Konkel, M.E. (2011). *Campylobacter jejuni* survival within human epithelial cells is enhanced by the secreted protein CiaI. Mol. Microbiol. *80*, 1296–1312.

Bullman, S., O'Leary, J., Corcoran, D., Sleator, R.D., and Lucey, B. (2011). Molecular-based detection of non-culturable and emerging *Campylobacteria* in patients presenting with gastroenteritis. Epidemiol. Infect. 1–5.

Buzby, J.C., Allos, B.M., and Roberts, T. (1997). The economic burden of *Campylobacter*-associated Guillain-Barré syndrome. J. Infect. Dis. *176*, S192–7.

Carrillo, C.D., Taboada, E., Nash, J.H., Lanthier, P., Kelly, J., Lau, P.C., Verhulp, R., Mykytczuk, O., Sy, J., Findlay, W.A., *et al.* (2004). Genome-wide expression analyses of *Campylobacter jejuni* NCTC11168 reveals coordinate regulation of motility and virulence by flhA. J. Biol. Chem. *279*, 20327–20338.

Champion, O.L., Al-Jaberi, S., Stabler, R.A., and and Wren, B.W. (2008). Comparative genomics of *Campylobacter jejuni*. In *Campylobacter*, 3rd ed, Nachamkin, I., Szymanski, C.M., and Blaser, M.J., eds. (ASM Press, Washington, DC), pp. 63–71.

Chen, M.L., Ge, Z., Fox, J.G., and Schauer, D.B. (2006). Disruption of tight junctions and induction of proinflammatory cytokine responses in colonic epithelial cells by *Campylobacter jejuni*. Infect. Immun. *74*, 6581–6589.

Clark, C.G., and Ng, L.K. (2008). Sequence variability of *Campylobacter* temperate bacteriophages. BMC Microbiol. *8*, 49.

Coker, A.O., Isokpehi, R.D., Thomas, B.N., Amisu, K.O., and Obi, C.L. (2002). Human *Campylobacter*iosis in developing countries. Emerg. Infect. Dis. *8*, 237–244.

Corcionivoschi, N., Clyne, M., Lyons, A., Elmi, A., Gundogdu, O., Wren, B.W., Dorrell, N., Karlyshev, A.V., and Bourke, B. (2009). *Campylobacter jejuni* cocultured with epithelial cells reduces surface capsular polysaccharide expression. Infect. Immun. *77*, 1959–1967.

Corcoran, A.T., Annuk, H., and Moran, A.P. (2006). The structure of the lipid anchor of *Campylobacter jejuni* polysaccharide. FEMS Microbiol. Lett. *257*, 228–235.

Debruyne, L., Gevers, D., and Vandamme, P. (2008). Taxonomy of the family *Campylobacteraceae*. In *Campylobacter*, Nachamkin, I., Szymanski, C.M., and Blaser M.J., eds. (ASM Press, Washington DC), pp. 3–25.

Dekeyser, P., Gossuin-Detrain, M., Butzler, J.P., and Sternon, J. (1972). Acute enteritis due to related *vibrio*: first positive stool cultures. J. Infect. Dis. *125*, 390–392.

Dingle, K.E., Colles, F.M., Wareing, D.R., Ure, R., Fox, A.J., Bolton, F.E., Bootsma, H.J., Willems, R.J., Urwin, R., and Maiden, M.C. (2001). Multilocus sequence typing system for *Campylobacter jejuni*. J. Clin. Microbiol. *39*, 14–23.

Dorrell, N., Mangan, J.A., Laing, K.G., Hinds, J., Linton, D., Al-Ghusein, H., Barrell, B.G., Parkhill, J., Stoker, N.G., Karlyshev, A.V., *et al.* (2001). Whole genome comparison of *Campylobacter jejuni* human isolates using a low-cost microarray reveals extensive genetic diversity. Genome Res. *11*, 1706–1715.

Escherich, T. (1886). Beitrage zur kenntnis der darmbacterien. Munch. Med. Wschr. 33, 815–833.

Evans, M.R., Lane, W., Frost, J.A., and Nylen, G. (1998). A *Campylobacter* outbreak associated with stir-fried food. Epidemiol. Infect. *121*, 275–279.

Ewing, C.P., Andreishcheva, E., and Guerry, P. (2009). Functional characterization of flagellin glycosylation in *Campylobacter jejuni* 81–176. J. Bacteriol. *191*, 7086–7093.

Ferrero, R.L., and Lee, A. (1988). Motility of *Campylobacter jejuni* in a viscous environment: comparison with conventional rod-shaped bacteria. J. Gen. Microbiol. *134*, 53–59.

Flanagan, R.C., Neal-McKinney, J.M., Dhillon, A.S., Miller, W.G., and Konkel, M.E. (2009). Examination of *Campylobacter jejuni* putative adhesins leads to the identification of a new protein, designated FlpA, required for chicken colonization. Infect. Immun. *77*, 2399–2407.

Fouts, D.E., Mongodin, E.F., Mandrell, R.E., Miller, W.G., Rasko, D.A., Ravel, J., Brinkac, L.M., DeBoy, R.T., Parker, C.T., Daugherty, S.C., *et al.* (2005). Major structural differences and novel potential virulence mechanisms from the genomes of multiple *Campylobacter* species. PLoS Biol. *3*, e15.

Frost, J.A., Oza, A.N., Thwaites, R.T., and Rowe, B. (1998). Serotyping scheme for *Campylobacter jejuni* and *Campylobacter coli* based on direct agglutination of heat-stable antigens. J. Clin. Microbiol. *36*, 335–339.

Gaasbeek, E.J., van der Wal, F.J., van Putten, J.P., de Boer, P., van der Graaf-van Bloois, L., de Boer, A.G., Vermaning, B.J., and Wagenaar, J.A. (2009). Functional characterization of excision repair and RecA-dependent recombinational DNA repair in *Campylobacter jejuni*. J. Bacteriol. *191*, 3785–3793.

Gaasbeek, E.J., Wagenaar, J.A., Guilhabert, M.R., van, Putten, J.P., Parker, C.T., and van der, Wal, F.J. (2010). Nucleases encoded by the integrated elements CJIE2 and CJIE4 inhibit natural transformation of *Campylobacter jejuni*. J. Bacteriol. *192*, 936–941.

Gaasbeek, E.J., Wagenaar, J.A., Guilhabert, M.R., Wosten, M.M., van Putten, J.P., van der Graaf-van Bloois, L., Parker, C.T., and van der Wal, F.J. (2009). A DNase encoded by integrated element CJIE1 inhibits natural transformation of *Campylobacter jejuni*. J. Bacteriol. *191*, 2296–2306.

Galkin, V.E., Yu, X., Bielnicki, J., Heuser, J., Ewing, C.P., Guerry, P., and Egelman, E.H. (2008). Divergence of quaternary structures among bacterial flagellar filaments. Science *320*, 382–385.

Gascon, J. (2006). Epidemiology, etiology and pathophysiology of traveler's diarrhea. Digestion 73, 102–108.

Gilbert, M., Karwaski, M.F., Bernatchez, S., Young, N.M., Taboada, E., Michniewicz, J., Cunningham, A.M., and Wakarchuk, W.W. (2002). The genetic bases for the variation in the lipo-oligosaccharide of the mucosal pathogen, *Campylobacter jejuni*. Biosynthesis of sialylated ganglioside mimics in the core oligosaccharide. J. Biol. Chem. *277*, 327–337.

Girard, M.P., Steele, D., Chaignat, C.L., and Kieny, M.P. (2006). A review of vaccine research and development: human enteric infections. Vaccine *24*, 2732–2750.

Goon, S., Ewing, C.P., Lorenzo, M., Pattarini, D., Majam, G., and Guerry, P. (2006). A sigma28-regulated nonflagella gene contributes to virulence of *Campylobacter jejuni* 81–176. Infect. Immun. *74*, 769–772.

Grant, C.C., Konkel, M.E., Cieplak, W., Jr, and Tompkins, L.S. (1993). Role of flagella in adherence, internalization, and translocation of *Campylobacter jejuni* in nonpolarized and polarized epithelial cell cultures. Infect. Immun. *61*, 1764–1771.

Guccione, E., Leon-Kempis Mdel, R., Pearson, B.M., Hitchin, E., Mulholland F., van Diemen, P.M., Stevens, M.P., and Kelly, D.J. (2008). Amino acid-dependent growth of *Campylobacter jejuni*: key roles for aspartase (AspA) under microaerobic and oxygen-limited conditions and identification of AspB (Cj0762), essential for growth on glutamate. Mol. Microbiol. *69*, 77–93.

Guerry, P., and Szymanski, C.M. (2008). *Campylobacter* sugars sticking out. Trends Microbiol. *16*, 428–435.

Guerry, P., Pope, P.M., Burr, D.H., Leifer, J., Joseph, S.W., and Bourgeois, A.L. (1994). Development and characterization of recA mutants of *Campylobacter jejuni* for inclusion in attenuated vaccines. Infect. Immun. *62*, 426–432.

Guerry, P., Szymanski, C.M., Prendergast, M.M., Hickey, T.E., Ewing, C.P., Pattarini, D.L., and Moran, A.P. (2002). Phase variation of *Campylobacter jejuni* 81–176 lipooligosaccharide affects ganglioside mimicry and invasiveness *in vitro*. Infect. Immun. *70*, 787–793.

Guerry, P., Ewing, C.P., Schirm, M., Lorenzo, M., Kelly, J., Pattarini, D., Majam, G., Thibault, P., and Logan, S. (2006). Changes in flagellin glycosylation affect *Campylobacter* autoagglutination and virulence. Mol. Microbiol. *60*, 299–311.

Haddock, G., Mullin, M., MacCallum, A., Sherry, A., Tetley, L., Watson, E., Dagleish, M., Smith, D.G., and Everest, P. (2010). *Campylobacter jejuni* 81–176 forms distinct microcolonies on *in vitro*-infected human small intestinal tissue prior to biofilm formation. Microbiology *156*, 3079–3084.

Hanninen, M.L., Hakkinen, M., and Rautelin, H. (1999). Stability of related human and chicken *Campylobacter jejuni* genotypes after passage through chick intestine studied by pulsed-field gel electrophoresis. Appl. Environ. Microbiol. *65*, 2272–2275.

Havelaar, A.H., van Pelt, W., Ang, C.W., Wagenaar, J.A., van Putten, J.P., Gross, U., and Newell, D.G. (2009). Immunity to *Campylobacter*: its role in risk assessment and epidemiology. Crit. Rev. Microbiol. *35*, 1–22.

Heikema, A.P., Bergman, M.P., Richards, H., Crocker, P.R., Gilbert, M., Samsom, J.N., van Wamel, W.J., Endtz, H.P., and van Belkum, A. (2010). Characterization of the specific interaction between sialoadhesin and sialylated *Campylobacter jejuni* lipooligosaccharides. Infect. Immun. *78*, 3237–3246.

Hendrixson, D.R., and DiRita, V.J. (2004). Identification of *Campylobacter jejuni* genes involved in commensal colonization of the chick gastrointestinal tract. Mol. Microbiol. *52*, 471–484.

Hickey, T.E., Majam, G., and Guerry, P. (2005). Intracellular survival of *Campylobacter jejuni* in human monocytic cells and induction of apoptotic death by cytholethal distending toxin. Infect. Immun. *73*, 5194–5197.

Hoffman, P.S., and Goodman, T.G. (1982). Respiratory physiology and energy conservation efficiency of *Campylobacter jejuni*. J. Bacteriol. *150*, 319–326.

Houliston, R.S., Vinogradov, E., Dzieciatkowska, M., Li, J., St Michael, F., Karwaski, M.F., Brochu, D., Jarrell, H.C., Parker, C.T., Yuki, N., *et al.* (2011). Lipooligosaccharide of *Campylobacter jejuni*: similarity with multiple types of mammalian glycans beyond gangliosides. J. Biol. Chem. *286*, 12361–12370.

Hu, L., and Hickey, T.E. (2005). *Campylobacter jejuni* induces secretion of proinflammatory chemokines from human intestinal epithelial cells. Infect. Immun. *73*, 4437–4440.

Hu, L., and Kopecko, D.J. (1999). *Campylobacter jejuni* 81–176 associates with microtubules and dynein during invasion of human intestinal cells. Infect. Immun. *67*, 4171–4182.

Hu, L., and Kopecko, D.J. (2008). Cell Biology of Human Host Cell Entry by *Campylobacter jejuni*. In *Campylobacter* 3rd ed., Nachamkin, I., Szymanski, C. M., and Blaser, M.J., eds. (ASM Press, Washington, DC), pp. 297–314.

Hu, L., McDaniel, J.P., and Kopecko, D.J. (2006a). Signal transduction events involved in human epithelial cell invasion by *Campylobacter jejuni* 81–176. Microb. Pathog. *40*, 91–100.

Hu, L., Bray, M.D., Osorio, M., and Kopecko, D.J. (2006b). *Campylobacter jejuni* induces maturation and cytokine production in human dendritic cells. Infect. Immun. *74*, 2697–2705.

Hugdahl, M.B., Beery, J.T., and Doyle, M.P. (1988). Chemotactic behavior of *Campylobacter jejuni*. Infect. Immun. *56*, 1560–1566.

Humphrey, T., O'Brien, S., and Madsen, M. (2007). *Campylobacters* as zoonotic pathogens: a food production perspective. Int. J. Food Microbiol. *117*, 237–257.

Jeon, B., Muraoka, W., Scupham, A., and Zhang, Q. (2009). Roles of lipooligosaccharide and capsular polysaccharide in antimicrobial resistance and natural transformation of *Campylobacter jejuni*. J. Antimicrob. Chemother. *63*, 462–468.

Jin, S., Joe, A., Lynett, J., Hani, E.K., Sherman, P., and Chan, V.L. (2001). JlpA, a novel surface-exposed lipoprotein specific to *Campylobacter jejuni*, mediates adherence to host epithelial cells. Mol. Microbiol. *39*, 1225–1236.

Jones, F.S., Orcutt, M., and Little, R.B. (1931). *Vibrios* (*Vibrio jejuni* N.S.P) associated with intestinal disorders of cows and calves. J. Exp. Med. *53*, 853–867.

Joslin, S.N., and Hendrixson, D.R. (2009). Activation of the *Campylobacter jejuni* FlgSR two-component system is linked to the flagellar export apparatus. J. Bacteriol. *191*, 2656–2667.

Kalischuk, L.D., Inglis, G.D., and Buret, A.G. (2009). *Campylobacter jejuni* induces transcellular translocation of commensal bacteria via lipid rafts. Gut Pathog. *1*, 2.

Kanungpean, D., Kakuda, T., and Takai, S. (2011). Participation of CheR and CheB in the chemosensory response of *Campylobacter jejuni*. Microbiology *157*, 1279–1289.

Karlyshev, A.V., Linton, D., Gregson, N.A., Lastovica, A.J., and Wren, B.W. (2000). Genetic and biochemical evidence of a *Campylobacter jejuni* capsular polysaccharide that accounts for Penner serotype specificity. Mol. Microbiol. *35*, 529–541.

Karlyshev, A.V., Everest, P., Linton, D., Cawthraw, S., Newell, D.G., and Wren, B.W. (2004). The *Campylobacter jejuni* general glycosylation system is important for attachment to human epithelial cells and in the colonization of chicks. Microbiology *150*, 1957–1964.

Karlyshev, A.V., Ketley, J.M., and Wren, B.W. (2005a). The *Campylobacter jejuni* glycome. FEMS Microbiol. Rev. *29*, 377–390.

Karlyshev, A.V., Champion, O.L., Churcher, C., Brisson, J.R., Jarrell, H.C., Gilbert, M., Brochu, D., St Michael, F., Li, J., Wakarchuk, W.W., *et al.* (2005b). Analysis of *Campylobacter jejuni* capsular loci reveals multiple mechanisms for the generation of structural diversity and the ability to form complex heptoses. Mol. Microbiol. *55*, 90–103.

Keestra, A.M., de Zoete, M.R., van Aubel, R.A., and van Putten, J.P. (2008). Functional characterization of chicken TLR5 reveals species-specific recognition of flagellin. Mol. Immunol. *45*, 1298–1307.

Keestra, A.M., de Zoete, M.R., Bouwman, L.I., and van Putten, J.P. (2010). Chicken TLR21 is an innate CpG DNA receptor distinct from mammalian TLR9. J. Immunol. *185*, 460–467.

Kiehlbauch, J.A., Albach, R.A., Baum, L.L., and Chang, K.P. (1985). Phagocytosis of *Campylobacter jejuni* and its intracellular survival in mononuclear phagocytes. Infect. Immun. *48*, 446–451.

Konkel, M.E., Garvis, S.G., Tipton, S.L., Anderson, D.E., Jr, and Cieplak, W., Jr. (1997). Identification and molecular cloning of a gene encoding a fibronectin-binding protein (CadF) from *Campylobacter jejuni*. Mol. Microbiol. *24*, 953–963.

Konkel, M.E., Kim, B.J., Rivera-Amill, V., and Garvis, S.G. (1999). Bacterial secreted proteins are required for the internaliztion of *Campylobacter jejuni* into cultured mammalian cells. Mol. Microbiol. *32*, 691–701.

Konkel, M.E., Klena, J.D., Rivera-Amill, V., Monteville, M.R., Biswas, D., Raphael, B., and Mickelson, J. (2004). Secretion of virulence proteins from *Campylobacter jejuni* is dependent on a functional flagellar export apparatus. J. Bacteriol. *186*, 3296–3303.

Konkel, M.E., Larson, C.L., and Flanagan, R.C. (2010). *Campylobacter jejuni* FlpA binds fibronectin and is required for maximal host cell adherence. J. Bacteriol. *192*, 68–76.

Korolik, V., and Ketely, J. (2008). Chemosensory signal transduction pathway of *Campylobacter jejuni* In *Campylobacter*, 3rd ed, Nachamkin, I., Szymanski, C.M., Blaser, M.J. eds. (ASM Press, Washington, DC), pp. 351–366.

Kowarik, M., Numao, S., Feldman, M.F., Schulz, B.L., Callewaert, N., Kiermaier, E., Catrein, I., and Aebi, M. (2006). *N-linked* glycosylation of folded proteins by the bacterial oligosaccharyltransferase. Science *314*, 1148–1150.

Lara-Tejero, M., and Galan, J.E. (2000). A bacterial toxin that controls cell cycle progression as a deoxyribonuclease I-like protein. Science *290*, 354–357.

Lara-Tejero, M., and Galan, J.E. (2001). CdtA, CdtB, and CdtC form a tripartite complex that is required for cytolethal distending toxin activity. Infect. Immun. *69*, 4358–4365.

Larsen, J.C., Szymanski, C., and Guerry, P. (2004). *N*-linked protein glycosylation is required for full competence in *Campylobacter jejuni* 81–176. J. Bacteriol. *186*, 6508–6514.

Lastovica, A.J., and Allos, B.M. (2008). Clinical significance of *Campylobacter* and related species other than *Campylobacter jejuni* and *Campylobacter coli*. In *Campylobacter*, 3rd ed, Nachamkin, I., Szymanski, C.M., Blaser, M.J. eds. (ASM Press, Washington, DC), pp. 123–150.

Lee, M.D., and Newell, D.G. (2006). *Campylobacter* in poultry: filling an ecological niche. Avian Dis. *50*, 1–9.

Lee, R.B., Hassane, D.C., Cottle, D.L., and Pickett, C.L. (2003). Interactions of *Campylobacter jejuni* cytolethal distending toxin subunits CdtA and CdtC with HeLa cells. Infect. Immun. *71*, 4883–4890.

Lindmark, B., Rompikuntal, P.K., Vaitkevicius, K., Song, T., Mizunoe, Y., Uhlin, B.E., Guerry, P., and Wai, S.N. (2009). Outer membrane vesicle-mediated release of cytolethal distending toxin (CDT) from *Campylobacter jejuni*. BMC Microbiol. *9*, 220.

Linton, D., Dorrell, N., Hitchen, P.G., Amber, S., Karlyshev, A.V., Morris, H.R., Dell, A., Valvano, M.A., Aebi, M., and Wren, B.W. (2005). Functional analysis of the *Campylobacter jejuni N*-linked protein glycosylation pathway. Mol. Microbiol. *55*, 1695–1703.

Logan, S.M., Kelly, J.F., Thibault, P., Ewing, C.P., and Guerry, P. (2002). Structural heterogeneity of carbohydrate modifications affects serospecificity of *Campylobacter* flagellins. Mol. Microbiol. *46*, 587–597.

Louwen, R., Heikema, A., van Belkum, A., Ott, A., Gilbert, M., Ang, W., Endtz, H.P., Bergman, M.P., and Nieuwenhuis, E.E. (2008). The sialylated lipooligosaccharide outer core in *Campylobacter jejuni* is an important determinant for epithelial cell invasion. Infect. Immun. *76*, 4431–4438.

MacCallum, A., Hardy, S.P., and Everest, P.H. (2005). *Campylobacter jejuni* inhibits the absorptive transport functions of Caco-2 cells and disrupts cellular tight junctions. Microbiology *151*, 2451–2458.

MacKichan, J.K., Gaynor, E.C., Chang, C., Cawthraw, S., Newell, D.G., Miller, J.F., and Falkow, S. (2004). The *Campylobacter jejuni* dccRS two-component system is required for optimal *in vivo* colonization but is dispensable for *in vitro* growth. Mol. Microbiol. *54*, 1269–1286.

Mellits, K.H., Mullen, J., Wand, M., Armbruster, G., Patel, A., Connerton, P.L., Skelly, M., and Connerton, I.F. (2002). Activation of the transcription factor NF-kappaB by *Campylobacter jejuni*. Microbiology *148*, 2753–2763.

De Melo, M.A., Gabbiani, G., and Pechere, J.C. (1989). Cellular events and intracellular survival of *Campylobacter jejuni* during infection of HEp-2 cells. Infect. Immun. *57*, 2214–2222.

Miller, W.G. (2008). Comparative Genomics of *Campylobacter* Species Other Than *Campylobacter jejuni*. In *Campylobacter* 3rd ed., Nachamkin, I., Szymanski, C.M., and Blaser, M.J., eds. (ASM Press, Washington, DC), pp. 73–95.

Monteville, M.R., and Konkel, M.E. (2002). Fibronectin-facilitated invasion of T84 eukaryotic cells by *Campylobacter jejuni* occurs preferentially at the basolateral cell surface. Infect. Immun. *70*, 6665–6671.

Monteville, M.R., Yoon, J.E., and Konkel, M.E. (2003). Maximal adherence and invasion of INT 407 cells by *Campylobacter jejuni* requires the CadF outer-membrane protein and microfilament reorganization. Microbiology *149*, 153–165.

van Mourik, A., Steeghs, L., van Laar, J., Meiring, H.D., Hamstra, H.J., van Putten, J.P., and Wosten, M.M. (2010). Altered linkage of hydroxyacyl chains in lipid A of *Campylobacter jejuni* reduces TLR4 activation and antimicrobial resistance. J. Biol. Chem. *285*, 15828–15836.

Myers, J.D., and Kelly, D.J. (2005). A sulphite respiration system in the chemoheterotrophic human pathogen *Campylobacter jejuni*. Microbiology *151*, 233–242.

Nachamkin, I., Bohachick, K., and Patton, C.M. (1993). Flagellin gene typing of *Campylobacter jejuni* by restriction fragment length polymorphism analysis. J. Clin. Microbiol. *31*, 1531–1536.

Novik, V., Hofreuter, D., and Galan, J.E. (2010). Identification of *Campylobacter jejuni* genes involved in its interaction with epithelial cells. Infect. Immun. *78*, 3540–3553.

Oelschlaeger, T.A., Guerry, P., and Kopecko, D.J. (1993). Unusual microtubule-dependent endocytosis mechanisms triggered by *Campylobacter jejuni* and *Citrobacter freundii*. Proc. Natl. Acad. Sci. U.S.A. *90*, 6884–6888.

Olson, K.C., Ethelberg, S., van Pelt, W., and and Tauxe, R.V. (2008). Epidemiology of *Campylobacter jejuni* infections in industrialized nations. In *Campylobacter* 3rd ed., Nachamkin, I., Szymanski, C.M., and Blaser, M.J., eds. (ASM Press, Washington, DC), pp. 163–189.

Pajaniappan, M., Hall, J.E., Cawthraw, S.A., Newell, D.G., Gaynor, E.C., Fields, J.A., Rathbun, K.M., Agee, W.A., Burns, C.M., Hall, S.J., *et al.* (2008). A temperature-regulated *Campylobacter jejuni* gluconate dehydrogenase is involved in respiration-dependent energy conservation and chicken colonization. Mol. Microbiol. *68*, 474–491.

Parker, C.T., Quinones, B., Miller, W.G., Horn, S.T., and Mandrell, R.E. (2006). Comparative genomic analysis of *Campylobacter jejuni* strains reveals diversity due to genomic elements similar to those present in *C. jejuni* strain RM1221. J. Clin. Microbiol. *44*, 4125–4135.

Parker, C.T., Gilbert, M., Yuki, N., Endtz, H.P., and Mandrell, R.E. (2008). Characterization of lipooligosaccharide-biosynthetic loci of *Campylobacter jejuni* reveals new lipooligosaccharide classes: evidence of mosaic organizations. J. Bacteriol. *190*, 5681–5689.

Parkhill, J., Wren, B.W., Mungall, K., Ketley, J.M., Churcher, C., Basham, D., Chillingworth, T., Davies, R.M., Feltwell, T., Holroyd, S., *et al.* (2000). The genome sequence of the food-borne pathogen *Campylobacter jejuni* reveals hypervariable sequences. Nature *403*, 665–668.

Pearson, B.M., Pin, C., Wright, J., I'Anson, K., Humphrey, T., and Wells, J.M. (2003). Comparative genome analysis of *Campylobacter jejuni* using whole genome DNA microarrays. FEBS Lett. *554*, 224–230.

Pei, Z., Burucoa, C., Grignon, B., Baqar, S., Huang, X.Z., Kopecko, D.J., Bourgeois, A.L., Fauchere, J.L., and Blaser, M.J. (1998). Mutation in the *peb1A* locus of *Campylobacter jejuni* reduces interactions with epithelial cells and intestinal colonization of mice. Infect. Immun. *66*, 938–943.

Pittman, M.S., Elvers, K.T., Lee, L., Jones, M.A., Poole, R.K., Park, S.F., and Kelly, D.J. (2007). Growth of *Campylobacter jejuni* on nitrate and nitrite: electron transport to NapA and NrfA via NrfH and distinct roles for NrfA and the globin Cgb in protection against nitrosative stress. Mol. Microbiol. *63*, 575–590.

Poly, F., Ewing, C., Goon, S., Hickey, T.E., Rockabrand, D., Majam, G., Lee, L., Phan, J., Savarino, N.J., and Guerry, P. (2007). Heterogeneity of a *Campylobacter jejuni* protein that is secreted through the flagellar filament. Infect. Immun. *75*, 3859–3867.

Quinones, B., Parker, C.T., Janda, J.M., Jr, Miller, W.G., and Mandrell, R.E. (2007). Detection and genotyping of Arcobacter and *Campylobacter* isolates from retail chicken samples by use of DNA oligonucleotide arrays. Appl. Environ. Microbiol. 73, 3645–3655.

Rathbun, K.M., and Thompson, S.A. (2009). Mutation of PEB4 alters the outer membrane protein profile of *Campylobacter jejuni*. FEMS Microbiol. Lett. *300*, 188–194.

Robinson, D.A. (1981). Infective dose of *Campylobacter jejuni* in milk. Br. Med. J. (Clin. Res. Ed) *282*, 1584.

Russell, R.G., O'Donnoghue, M., Blake, D.C., Jr, Zulty, J., and DeTolla, L.J. (1993). Early colonic damage and invasion of *Campylobacter jejuni* in experimentally challenged infant *Macaca mulatta*. J. Infect. Dis. *168*, 210–215.

Samuel, M.C., Vugia, D.J., Shallow, S., Marcus, R., Segler, S., McGivern, T., Kassenborg, H., Reilly, K., Kennedy, M., Angulo, F., *et al.* (2004). Epidemiology of sporadic *Campylobacter* infection in the United States and declining trend in incidence, FoodNet 1996–1999. Clin. Infect. Dis. *38*, S165–74.

Schmidt-Ott, R., Pohl, S., Burghard, S., Weig, M., and Gross, U. (2005). Identification and characterization of a major subgroup of conjugative *Campylobacter jejuni* plasmids. J. Infect. *50*, 12–21.

Sellars, M.J., Hall, S.J., and Kelly, D.J. (2002). Growth of *Campylobacter jejuni* supported by respiration of fumarate, nitrate, nitrite, trimethylamine-*N*-oxide, or dimethyl sulfoxide requires oxygen. J. Bacteriol. *184*, 4187–4196.

Semchenko, E.A., Day, C.J., Wilson, J.C., Grice, I.D., Moran, A.P., and Korolik, V. (2010). Temperature-dependent phenotypic variation of *Campylobacter jejuni* lipooligosaccharides. BMC Microbiol. *10*, 305.

Sheppard, S.K., Dallas, J.F., Strachan, N.J., MacRae, M., McCarthy, N.D., Wilson, D.J., Gormley, F.J., Falush, D., Ogden, I.D., Maiden, M.C., *et al.* (2009). *Campylobacter* genotyping to determine the source of human infection. Clin. Infect. Dis. *48*, 1072–1078.

Skirrow, M.B. (1977). *Campylobacter* enteritis: a "new" disease. Br. Med. J. *2*, 9–11.

Song, Y.C., Jin, S., Louie, H., Ng, D., Lau, R., Zhang, Y., Weerasekera, R., Al Rashid, S., Ward, L.A., Der, S.D., *et al.* (2004). FlaC, a protein of *Campylobacter jejuni* TGH9011 (ATCC43431) secreted through the flagellar apparatus, binds epithelial cells and influences cell invasion. Mol. Microbiol. 53, 541–553.

van Sorge, N.M., Bleumink, N.M., van Vliet, S.J., Saeland, E., van der Pol, W.L., van Kooyk, Y., and van Putten, J.P. (2009). *N*-glycosylated proteins and distinct lipooligosaccharide glycoforms of *Campylobacter jejuni* target the human C-type lectin receptor MGL. Cell. Microbiol. *11*, 1768–1781.

van Spreeuwel, J.P., Duursma, G.C., Meijer, C.J., Bax, R., Rosekrans, P.C., and Lindeman, J. (1985). *Campylobacter* colitis: histological immunohistochemical and ultrastructural findings. Gut *26*, 945–951.

Stahl, M., Friis, L.M., Nothaft, H., Liu, X., Li, J., Szymanski, C.M., and Stintzi, A. (2011). L-Fucose utilization provides *Campylobacter jejuni* with a competitive advantage. Proc. Natl. Acad. Sci. U.S A. *108*, 7194–7199.

Suerbaum, S., Lohrengel, M., Sonnevend, A., Ruberg, F., and Kist, M. (2001). Allelic diversity and recombination in *Campylobacter jejuni*. J. Bacteriol. *183*, 2553–2559.

Sycuro, L.K., Pincus, Z., Gutierrez, K.D., Biboy, J., Stern, C.A., Vollmer, W., and Salama, N.R. (2010). Peptidoglycan crosslinking relaxation promotes *Helicobacter pylori's* helical shape and stomach colonization. Cell *141*, 822–833.

Szymanski, C.M., King, M., Haardt, M., and Armstrong, G.D. (1995). *Campylobacter jejuni* motility and invasion of Caco-2 cells. Infect. Immun. *63*, 4295–4300.

Szymanski, C.M., Yao, R., Ewing, C.P., Trust, T.J., and Guerry, P. (1999). Evidence for a system of general protein glycosylation in *Campylobacter jejuni*. Mol. Microbiol. *32*, 1022–1030.

Szymanski, C.M., Burr, D.H., and Guerry, P. (2002). *Campylobacter* protein glycosylation affects host cell interactions. Infect. Immun. *70*, 2242–2244.

Takata, T., Fujimoto, S., and Amako, K. (1992). Isolation of nonchemotactic mutants of *Campylobacter jejuni* and their colonization of the mouse intestinal tract. Infect. Immun. *60*, 3596–3600.

Thibault, P., Logan, S.M., Kelly, J.F., Brisson, J.R., Ewing, C.P., Trust, T.J., and Guerry, P. (2001). Identification of the carbohydrate moieties and glycosylation motifs in *Campylobacter jejuni* flagellin. J. Biol. Chem. *276*, 34862–34870.

Tu, Q.V., McGuckin, M.A., and Mendz, G.L. (2008). *Campylobacter jejuni* response to human mucin MUC2: modulation of colonization and pathogenicity determinants. J. Med. Microbiol. *57*, 795–802.

Wacker, M., Linton, D., Hitchen, P.G., Nita-Lazar, M., Haslam, S.M., North, S.J., Panico, M., Morris, H.R., Dell, A., Wren, B.W., *et al.* (2002). *N*-linked glycosylation in *Campylobacter jejuni* and its functional transfer into *E. coli*. Science *298*, 1790–1793.

Walker, R.I., Schmauder-Chock, E.A., Parker, J.L., and Burr, D. (1988). Selective association and transport of *Campylobacter jejuni* through M cells of rabbit Peyer's patches. Can. J. Microbiol. *34*, 1142–1147.

Wassenaar, T.M., and Newell, D.G. (2000). Genotyping of *Campylobacter* spp. Appl. Environ. Microbiol. *66*, 1–9.

Wassenaar, T.M., Bleumink-Pluym, N.M., and van der Zeijst, B.A. (1991). Inactivation of *Campylobacter jejuni* flagellin genes by homologous recombination demonstrates that *flaA* but not *flaB* is required for invasion. EMBO J. *10*, 2055–2061.

Wassenaar, T.M., Engelskirchen, M., Park, S., and Lastovica, A. (1997). Differential uptake and killing potential of *Campylobacter jejuni* by human peripheral monocytes/macrophages. Med. Microbiol. Immunol. *186*, 139–144.

Wassenaar, T.M., Geilhausen, B., and Newell, D.G. (1998). Evidence of genomic instability in *Campylobacter jejuni* isolated from poultry. Appl. Environ. Microbiol. *64*, 1816–1821.

Watson, R.O., and Galan, J.E. (2008). *Campylobacter jejuni* survives within epithelial cells by avoiding delivery to lysosomes. PLoS Pathog. *4*, e14.

Weerakoon, D.R., Borden, N.J., Goodson, C.M., Grimes, J., and Olson, J.W. (2009). The role of respiratory donor enzymes in *Campylobacter jejuni* host colonization and physiology. Microb. Pathog. *47*, 8–15.

Weingarten, R.A., Grimes, J.L., and Olson, J.W. (2008). Role of *Campylobacter jejuni* respiratory oxidases and reductases in host colonization. Appl. Environ. Microbiol. *74*, 1367–1375.

Whitehouse, C.A., Balbo, P.B., Pesci, E.C., Cottle, D.L., Mirabito, P.M., and Pickett, C.L. (1998). *Campylobacter jejuni* cytolethal distending toxin causes a G2-phase cell cycle block. Infect. Immun. *66*, 1934–1940.

Wine, E., Chan, V.L., and Sherman, P.M (2008). *Campylobacter jejuni* mediated disruption of polarized epithelial monolayers is cell-type specific, time dependent, and correlates with bacterial invasion. Pediatr. Res. *64*, 599–604.

Wosten, M.M., Parker, C.T., van, Mourik, A., Guilhabert, M.R., van, Dijk, L., and van, Putten, J.P. (2006). The *Campylobacter jejuni* PhosS/PhosR operon represents a non-classical phosphate-sensitive two-component system. Mol. Microbiol. *62*, 278–291.

Wosten, M.M., van, Dijk, L., Veenendaal, A.K., de, Zoete, M.R., Bleumink-Pluijm, N.M., and van, Putten, J.P. (2010). Temperature-dependent FlgM/FliA complex formation regulates *Campylobacter jejuni* flagella length. Mol. Microbiol. *75*, 1577–1591.

Wosten, M.M., Wagenaar, J.A., and van, Putten, J.P. (2004). The FlgS/FlgR two-component signal transduction system regulates the *fla* regulon in *Campylobacter jejuni*. J. Biol. Chem. *279*, 16214–16222.

Wosten, M.M., van Mourik, A., and van Putten, J.P. (2008). Regulation of Genes in *Campylobacter jejuni*. In *Campylobacter* 3rd ed., Nachamkin, I., Szymanski, C.M., and Blaser, M.J., eds. (ASM Press, Washington, DC), pp. 611–624.

Yao, R., Burr, D.H., and Guerry, P. (1997). CheY-mediated modulation of *Campylobacter jejuni* virulence. Mol. Microbiol. *23*, 1021–1031.

Yuki, N., Susuki, K., Koga, M., Nishimoto, Y., Odaka, M., Hirata, K., Taguchi, K., Miyatake, T., Furukawa, K., Kobata, T., *et al.* (2004). Carbohydrate mimicry between human ganglioside GM1 and *Campylobacter jejuni* lipooligosaccharide causes Guillain-Barré syndrome. Proc. Natl. Acad. Sci. U.S.A. *101*, 11404–11409.

Zheng, J., Meng, J., Zhao, S., Singh, R., and Song, W. (2008). *Campylobacter*-induced interleukin-8 secretion in polarized human intestinal epithelial cells requires *Campylobacter*-secreted cytolethal distending toxin- and Toll-like receptor-mediated activation of NF-kappaB. Infect. Immun. *76*, 4498–4508.

Zilbauer, M., Dorrell, N., Elmi, A., Lindley, K.J., Schuller, S., Jones, H.E., Klein, N.J., Nunez, G., Wren, B.W., and Bajaj-Elliott, M. (2007). A major role for intestinal epithelial nucleotide oligomerization domain 1 (NOD1) in eliciting host bactericidal immune responses to *Campylobacter jejuni*. Cell. Microbiol. *9*, 2404–2416.

de Zoete, M.R., van Putten, J.P., and Wagenaar, J.A. (2007). Vaccination of chickens against *Campylobacter*. Vaccine *25*, 5548–5557.

de Zoete, M.R., Keestra, A.M., Roszczenko, P., and van Putten, J.P. (2010a). Activation of human and chicken toll-like receptors by *Campylobacter* spp. Infect. Immun. *78*, 1229–1238.

de Zoete, M.R., Keestra, A.M., Wagenaar, J.A., and van Putten, J.P. (2010b). Reconstitution of a functional Toll-like receptor 5 binding site in *Campylobacter jejuni* flagellin. J. Biol. Chem. *285*, 12149–12158.

de Zoete, M.R., Bouwman, L.I., Keestra, A.M., and van Putten, J.P. (2011). Cleavage and activation of a Toll-like receptor by microbial proteases. Proc. Natl. Acad. Sci. U.S.A. *108*, 4968–4973.

Epidemiology, Pathogenesis, Ecology and Genetics of *Listeria monocytogenes*

12

Sangmi Lee, Robin M. Siletzky and Sophia Kathariou

Abstract

Listeria monocytogenes is a Gram-positive foodborne pathogen that causes a severe, potentially fatal illness (listeriosis) in animals and humans. The only human pathogen within the genus *Listeria*, this bacterium is equipped with sophisticated mechanisms to invade mammalian cells and proliferate intracellularly. Population genetics data indicate that some groups of *L. monocytogenes* are more frequently associated with human listeriosis. However, the ecology and potentially unique characteristics of such groups remain to be elucidated. This chapter discusses recent advances in the epidemiology of listeriosis and epidemic-associated clonal groups; the characterization of pathogenicity of *L. monocytogenes*; the evolution of *Listeria* species and *L. monocytogenes*; and the increasingly recognized importance of *Listeria* phages in the ecology of *L. monocytogenes* in food processing facilities.

Introduction

Listeria monocytogenes is a rod-shaped, facultative anaerobic, non-spore-forming Gram-positive bacterium responsible for listeriosis in humans and animals. Human listeriosis is transmitted exclusively via foods and its clinical manifestations range from mild flu-like symptoms to life-threatening meningitis, encephalitis, sepsis, abortions and stillbirths (Mead *et al.*, 1999; Swaminathan and Gerner-Smidt, 2007; Varma *et al.*, 2007; Voetsch *et al.*, 2007; Drevets and Bronze, 2008; Wagner and MacLauchlin, 2008; Warriner and Namvar, 2009; Scallan *et al.*, 2011). The number of listeriosis cases is relatively low in comparison to other foodborne pathogens such as *Salmonella* and *Campylobacter*; however, the mortality rate remains high (approximately 16%) (Warriner and Namvar, 2009; Scallan *et al.*, 2011). *L. monocytogenes* is often found in the environment including water, soil, vegetation and food processing facilities (Kornacki and Gurtler, 2007; Wagner and MacLauchlin, 2008; Warriner and Namvar, 2009). The high mortality and morbidity associated with listeriosis and the ability of the bacterium to contaminate food processing and other environments continue to render *L. monocytogenes* a considerable public health concern.

Epidemiology of listeriosis

The first conclusive evidence for foodborne transmission of listeriosis was obtained from the investigation of an outbreak of perinatal listeriosis in the Maritime Provinces, Canada, attributed to contaminated coleslaw (Schlech *et al.*, 1983). Subsequently, North America and Europe witnessed a number of listeriosis outbreaks primarily associated with unpasteurized milk, pâté, soft cheese, hot dogs and deli meats, while a major multistate outbreak in the United States in 2011 involved cantaloupe (Kathariou, 2003; Swaminathan and Gerner-Smidt, 2007; Varma *et al.*,2007; Wagner and MacLauchlin, 2008; Warriner and Namvar, 2009; Centers for Disease Control and Prevention, 2011). These outbreaks triggered the introduction of rigorous regulations to reduce food contamination by *L. monocytogenes*. In the United States, *L. monocytogenes* has been designated a zero-tolerance bacterium in all ready-to-eat (RTE) foods, while the European Union allows up to 100 colony-forming units (cfu) per g in those RTE foods that do not permit growth of *L. monocytogenes* (Gerner-Smidt and Whichard, 2008; Wagner and MacLauchlin, 2008; Warriner and Namvar, 2009).

Listeriosis occurs primarily among specific at-risk populations that include the elderly, pregnant women and neonates, and those immunocompromised due to cancer and chemotherapy, HIV infection, organ transplants and other conditions and treatments (Swaminathan and Gerner-Smidt, 2007; Drevets and Bronze, 2008; Wagner and MacLauchlin, 2008). In the United States, Hispanics have been identified as another at-risk population since listeriosis, in particular pregnancy-associated cases, has been observed at a higher rate among Hispanics than among other ethnic groups, probably because of consumption of Mexican-style soft cheeses made from raw milk (MacDonald *et al.*, 2005; Voetsch *et al.*, 2007; Jackson *et al.*, 2011).

The number of cases of listeriosis exhibits a seasonal peak during the late summer into the fall (Voetsch *et al.*, 2007; Bennion *et al.*, 2008; Wagner and MacLauchlin, 2008). An estimated 37% of listeriosis cases among persons over 50 years of age in the US between 1996 and 2003 occurred between August and October (Voetsch *et al.*, 2007). Mortality data from 1990 to 2005 also showed an increase of listeriosis-related deaths during this period with the highest level in September (Bennion *et al.*, 2008). Similar seasonality was reported in France and among non-pregnancy-associated listeriosis cases in England and the Wales (Gillespie *et al.*, 2006; Goulet *et al.*, 2008). The reasons for this seasonal peak remain unclear although it has been suggested that this tendency may be linked to the increased consumption of RTE meats in the summer through picnics or barbecues (Voetsch *et al.*, 2007; Bennion *et al.*, 2008).

Even though 13 serotypes of *L. monocytogenes* have been identified, over 95% of clinical cases involve strains of three serotypes: 1/2a, 1/2b and 4b (Kathariou, 2002; Swaminathan and Gerner-Smidt, 2007; Voetsch *et al.*, 2007; Wagner and MacLauchlin, 2008). Serotype 4b has been associated with the majority of listeriosis outbreaks in spite of the fact that serotypes 1/2a and 1/2b are more frequently isolated from foods or food processing facilities (Kathariou, 2003; Swaminathan and Gerner-Smidt, 2007). This is one indication that serotype 4b may be more virulent for humans than strains of other serotypes; however, conclusive evidence for such relatively high virulence of serotype 4b is currently lacking (Swaminathan and Gerner-Smidt, 2007).

Recent epidemiological trends of listeriosis

In the United States, listeriosis is currently estimated to result in approximately 1600 cases and 250 deaths annually (Scallan *et al.*, 2011). Following implementation of food safety policies against *L. monocytogenes*, the incidence of listeriosis decreased significantly, from 7.9 cases per million in 1989 to 4.2 cases per million in 1993 (Voetsch *et al.*, 2007; Wagner and MacLauchlin, 2008). Analysis of FoodNet data collected from 1996 through 2003 indicated that the overall incidence continued to decline (from 4.1 to 3.1 cases per million) (Voetsch *et al.*, 2007). Interestingly, the incidence of listeriosis declined until 2001 but has remained at a relatively stable level since then (Gerner-Smidt and Whichard, 2008; Goulet *et al.*, 2008; Centers for Disease Control and Prevention, 2010). Possible reasons for the decreased incidence include strict food safety regulations, enhanced industry compliance and test-and-hold systems, and improved outreach with at-risk groups (Swaminathan and Gerner-Smidt, 2007; Voetsch *et al.*, 2007).

In contrast to the decreasing trends in the United States, listeriosis has recently increased significantly in several European countries. In England and the Wales, annual listeriosis incidence increased from 2.13 cases per million between 1990 and 2000 to 3.47 cases per million during 2001–2004 (Gillespie *et al.*, 2006). During 2001–2007, the incidence doubled in comparison to that during 1990–1999 (Gillespie *et al.*, 2009). This rise was mainly due to increases in sporadic cases affecting patients aged over 60 with most of the increases corresponding to cases with septicaemia but without central nervous system involvement (Gillespie *et al.*, 2006, 2009). In France, listeriosis decreased between 1987 and 1997 after extensive control strategies were implemented by the food industry in response to large outbreaks (Goulet *et al.*, 2006). Subsequent to the implementation of mandatory reporting, listeriosis incidence was further reduced from 4.5 cases per million in 1999–2000 to 3.5 cases per million in 2001–2003 (Goulet *et al.*, 2006). This level was maintained until 2005; however, listeriosis incidence rose to 4.7 cases per million in 2006 and was estimated at 5.6 cases per million in 2007, suggesting a 46% increase in incidence during 2006–2007 in comparison with that during 2001–2004 (Goulet *et al.*, 2008). As observed in England and Wales, this increase was accounted for by a rise in listeriosis patients over the age of 60 years and with septicaemia (Goulet *et al.*, 2008). In Germany, listeriosis increased substantially during 2001–2005, especially among the elderly population (Koch and Stark, 2006). Elevated listeriosis incidence was also observed in other European countries such as the Netherlands, Switzerland, Denmark, Finland and Belgium (Doorduyn *et al.*, 2006; Goulet *et al.*, 2008).

The reasons for the increased incidence in Europe and the upsurge in septicaemia cases remain unknown. It has been hypothesized, however, that the higher incidence of listeriosis might be related to higher exposure to *L. monocytogenes* (Gillespie *et al.*, 2006; Goulet *et al.*, 2008). As a scenario that could lead to increased exposure to this bacterium, Goulet *et al.* proposed that, in compliance to recommendations against hypertension, less salt was used in some RTE foods, thus permitting growth of *L. monocytogenes* and resulting in higher probability of listeriosis in at-risk groups (Goulet *et al.*, 2008). Increasing use of acid inhibitors was also postulated to contribute to increased vulnerability to *L. monocytogenes* infection (Gillespie *et al.*, 2009).

Another topic of note is the emergence of a new type of listeriosis outbreak first documented in the 1990s and 2000s and characterized by febrile gastroenteritis (Swaminathan and Gerner-Smidt, 2007). Unlike invasive listeriosis, patients in febrile gastroenteritis outbreaks typically were immunocompetent adults (Swaminathan and Gerner-Smidt, 2007). These outbreaks have involved a variety of products including rice salad, chocolate milk, cold maize and tuna salad, RTE deli meats, cold-smoked fish and cheese (Salamina *et al.*, 1996; Dalton *et al.*, 1997; Aureli *et al.*, 2000; Frye *et al.*, 2002; Swaminathan and Gerner-Smidt, 2007). The implicated foods were typically contaminated with large numbers of *L. monocytogenes*; for instance, ~ 10^9 cfu/ml in the 1994 chocolate milk outbreak (Dalton *et al.*, 1997; Swaminathan and Gerner-Smidt, 2007). Many of the febrile gastroenteritis outbreaks were caused by strains of serotypes 1/2a and 1/2b although serotype 4b was also involved (Kathariou, 2002; Swaminathan and Gerner-Smidt, 2007; Cheng *et al.*, 2008). Further research is needed to determine whether these non-invasive listeriosis outbreaks stem from potentially unusual characteristics of the strains involved or from exposure to high doses of the pathogen.

Populations susceptible to invasive listeriosis may change with time. For instance, the percentage of HIV patients among deaths from listeriosis decreased from 4.9% during 1990–1995 to 3.4% during 1996–2005 in the US, possibly due to the introduction of effective treatments against diseases of AIDS patients caused by opportunistic organisms including *L. monocytogenes* (Bennion *et al.*, 2008). In the case of the Hispanic population in the US, listeriosis occurred at a disproportionately high incidence in 1997–2001; however, incidence was considerably diminished in 2002 and 2003 (Voetsch *et al.*, 2007). This change may originate from various public education campaigns or decreased availability of high-risk foods from the market (Voetsch *et al.*, 2007).

Epidemiological studies have been performed to identify foods posing high risk for sporadic cases of listeriosis, which represent the majority of human listeriosis cases (Varma *et al.*, 2007; Wagner and MacLauchlin, 2008). Several studies in the 1980s and 1990s linked sporadic listeriosis with frankfurters, chicken, milk, pâté, soft cheese and deli foods (Schwartz *et al.*, 1988; Schuchat *et al.*, 1992; Jensen *et al.*, 1994). A case–control study involving 169 sporadic listeriosis cases indicated an association between sporadic listeriosis and the consumption of melons and hummus from commercial establishments (Varma *et al.*, 2007). Given that foods posing high risk for listeriosis are often ready-to-eat, refrigerated and subjected to storage for a long period, Goulet suggested that hummus was more likely to belong to this category of foods than melons (Goulet, 2007).

Such assessments are of interest in light of a major, multistate outbreak of listeriosis in the Unite States in 2011, which involved cantaloupe (Centers for Disease Control and Prevention, 2011)

Pathogenicity of *L. monocytogenes*

Intracellular life of *L. monocytogenes*

Following the intake of contaminated foods, infection by *L. monocytogenes* is initiated with this bacterium penetrating through the epithelial lining of the intestine via junctions at the villus tips in which cell extrusion occurs (Pentecost *et al.*, 2006). *L. monocytogenes* spreads via the blood circulation from the intestine into the liver and spleen, and grows in these organs (Orndorff *et al.*, 2006). At this stage, *L. monocytogenes* is either eliminated by the immune system or proceeds into other organs such as the central nervous system and fetoplacental unit (Orndorff *et al.*, 2006). This infection route is mainly due to the capability of *L. monocytogenes* to invade and proliferate within diverse non-phagocytic cells such as enterocytes, endothelial cells, hepatocytes and fibroblasts (Gray *et al.*, 2006; Drevets and Bronze, 2008; Bonazzi *et al.*, 2009; Corr and O'Neill, 2009). Mechanisms underlying this ability have been extensively studied, resulting in the elucidation of key virulence factors (Dussurget *et al.*, 2004, 2008).

Entry into mammalian cells is mediated by surface invasion proteins such as internalin A (InlA) and internalin B (InlB). At the carboxyl terminus, InlA and InlB possess LPXTG and GW motifs, respectively, responsible for anchoring the protein to the peptidoglycan layer (Cabanes *et al.*, 2002; Bierne and Cossart, 2007). InlA plays an indispensable role in the translocation into the intestinal tract by interacting with E-cadherin, an adhesion molecule located at adherens junctions between epithelial cells (Dussurget *et al.*, 2004; Orndorff *et al.*, 2006; Drevets and Bronze, 2008; Bonazzi *et al.*, 2009). InlA binds to E-cadherin in such a species-specific manner that E-cadherins of rats and mice do not function as receptor for InlA whereas those of other mammalian species including guinea pigs and gerbils do. This attribute is dictated by the amino acid occupying position 16 in the first extracellular domain of E-cadherin, which is glutamic acid in rats and mice but proline in susceptible species (Dussurget, 2008; Bonazzi *et al.*, 2009).

InlB contributes to invasion into a wider range of cell types such as endothelial cells, hepatocytes and fibroblasts owing to the ubiquitous nature of its receptor, the hepatocyte growth factor receptor Met. InlB also interacts with the receptor of the globular part of the complement component Clq (gClq-R)/p32 (Dussurget *et al.*, 2004, 2008; Drevets and Bronze, 2008; Bonazzi *et al.*, 2009).

Following attachment on mammalian cells *L. monocytogenes* is internalized within a vacuole which is subsequently lysed via the synergistic actions of listeriolysin O (LLO), a cholesterol-dependent, pore-generating cytolysin and two phospholipases C (PLCs), PC-PLC (PlcA) and PI-PLC (PlcB) (Tilney and Portnoy, 1989; Dussurget *et al.*, 2004; Orndorff *et al.*, 2006; Drevets and Bronze, 2008). After vacuole lysis, *L. monocytogenes* is released into the cytosol that is permissible for replication (Tilney and Portnoy, 1989). For multiplication in the intracellular environment *L. monocytogenes* relies on nutrients available in the cytosol such as glucose-1-phosphate through Hpt, a hexose phosphate transporter (Dussurget *et al.*, 2004; Gray *et al.*, 2006; Orndorff *et al.*, 2006).

In parallel to cytosolic replication, *L. monocytogenes* disseminates into adjacent host cells by actin-mediated motility that has been reviewed extensively (Tilney and Portnoy, 1989; Cabanes *et al.*, 2002; Dussurget *et al.*, 2004; Hain *et al.*, 2006a; Orndorff *et al.*, 2006; Bierne and Cossart, 2007; Drevets and Bronze, 2008). *L. monocytogenes* moves at the extremity of the growing actin tails and creates a pseudopod-like protrusion into neighbouring cells resulting in cell-to-cell transfer (Tilney and Portnoy, 1989; Dussurget *et al.*, 2004; Hain *et al.*, 2006a; Orndorff *et al.*, 2006). Internalin C (InlC) contributes to cell-to-cell dissemination between polarized epithelial cells by decreasing cortical tension at apical junctions (Rajabian *et al.*, 2009). The pseudopod-like protrusion is engulfed by the adjacent cell resulting in a double-membrane secondary vacuole (Tilney and

Portnoy, 1989; Dussurget *et al.*, 2004; Orndorff *et al.*, 2006). *L. monocytogenes* escapes from this vacuole through the actions of LLO and the two PLCs repeating the cycle of cytosolic replication and actin-based movement into the neighbouring cells (Tilney and Portnoy, 1989; Dussurget *et al.*, 2004; Orndorff *et al.*, 2006).

Listeria pathogenicity islands

In *Listeria* species three virulence gene clusters have been identified to date and termed *Listeria* pathogenicity island 1, 2 and 3 (LIPI-1, LIPI-2 and LIPI-3). LIPI-2 is found exclusively in *L. ivanovii*, a species associated with animal listeriosis (Cotter *et al.*, 2008; Wagner and MacLauchlin, 2008). Here we will discuss in further detail LIPI-1 and LIPI-3, both harboured by *L. monocytogenes* (Cotter *et al.*, 2008; Wagner and MacLauchlin, 2008).

LIPI-1 refers to a genomic region flanked by *prs* and *ldh* and harbouring several well-known virulence genes (Fig. 12.1). This region is identified in three different *Listeria* species, *L. monocytogenes*, *L. seeligeri* and *L. ivanovii* (Schmid *et al.*, 2005; Wagner and MacLauchlin, 2008). Virulence genes within LIPI-1 are encoded in a 9-kb cluster of six genes, *prfA-plcA-hly-mpl-actA-plcB*. Among these genes, *prfA* encodes the positive regulatory factor A (PrfA) that regulates the expression of itself and most of the key virulence genes identified so far, including *hly*, *mpl*, *actA*, *plcA*, *plcB*, *inlA*, *inlB*, *inlC* and *hpt* (Scortti *et al.*, 2007).

Genes regulated by PrfA are transcribed from promoters with a 'PrfA box' to which the PrfA dimer is bound. PrfA boxes are located ~200 bp upstream of the start codon and characterized with the consensus sequence TTAACANNTGTAAA with seven conserved nucleotides (underlined) (Scortti *et al.*, 2007). In addition to auto-regulation, PrfA expression is dependent on the sigma factor σ^B involved in the general stress response (Scortti *et al.*, 2007; de las Heras *et al.*, 2011). A 5′ untranslated thermo-sensor region whose secondary structure blocks the ribosomal binding site at ≤ 30°C inhibits translation of the *prfA* message at low temperatures, accounting for the well-known temperature-regulated expression of *hly* and other genes under PrfA control; these genes are expressed at 37°C but poorly or not at all below 30°C (Johansson *et al.*, 2002; Scortti *et al.*, 2007; Freitag *et al.*, 2009; de las Heras *et al.*, 2011). PrfA regulation of virulence genes is mediated by transforming PrfA into an active form. PrfA activation is enhanced when the cells are inside the cytosol or grown in medium supplemented with activated charcoal. On the other hand, PrfA activation is repressed by sugars such as glucose and cellobiose that are transported by the phosphoenolpyruvate–sugar phosphotransferase system (PTS) and abundant in the environment; repression is not conferred by non-PTS sugar hexose phosphates that are rich in the cytosol (Scortti *et al.*, 2007; Freitag *et al.*, 2009; de las Heras *et al.*, 2011).

LIPI-3 is the most recently identified

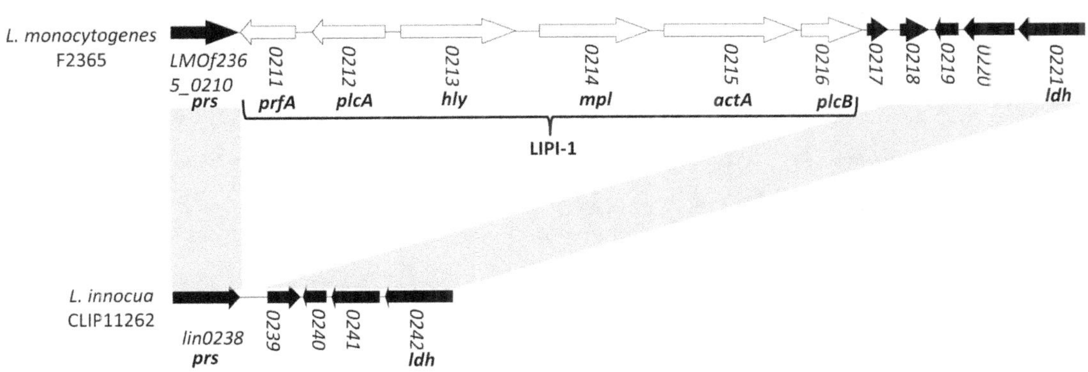

Figure 12.1 Comparison of the genetic region flanked by *prs* and *ldh* between *L. monocytogenes* F2365 and *L. innocua* CLIP11262. Genes comprising LIPI-1 are marked in white and other genes are in black. The grey shadow indicates region of homology, identified with WebACT (http://www.webact.org/WebACT/home) (Carver *et al.*, 2005).

pathogenicity island, which is located between *LMOf2365_1111* and *LMOf2365_1120* genes in *L. monocytogenes* F2365 (Fig. 12.2). This pathogenicity island contains the eight gene cluster (*llsA, llsG, llsH, llsX, llsB, llsY, llsD,* and *llsP*) encoding listeriolysin S (LLS), a bacteriocin-like modified peptide exhibiting haemolytic and cytotoxic activities (Cotter *et al.*, 2008; Clayton *et al.*, 2011). LLS contributes to the virulence and intracellular survival of the bacterium since an LLS-negative mutant (lacking *llsB*) exhibited reduced bacterial loads in a mouse model and within human polymorphonuclear neutrophils (PMNs). Deletion mutants of other genes in *llsG-llsD* exhibited loss of LLS-derived haemolysis, suggesting that these genes also play indispensable roles for LLS synthesis. In fact, *llsX* was employed as a genetic marker in a real-time PCR scheme to detect *L. monocytogenes* isolates with LIPI-3 (Clayton *et al.*, 2011). In contrast to LIPI-1 which is found in all strains of *L. monocytogenes* as well as in other *Listeria* species, LIPI-3 is possessed only by a subset of *L. monocytogenes* lineage I. GC content analysis suggested that LIPI-3 may have been acquired via horizontal gene transfer by an ancestral strain of lineage I, and was subsequently maintained by a subset of lineage I strains (Cotter *et al.*, 2008).

Premature stop codon mutations in *inlA*

Many *L. monocytogenes* strains harbour premature stop codon (PMSC) mutations in *inlA* that result in truncated, secreted InlA proteins that lack the C-terminal part with the membrane-anchoring LPXTG motif. To date, at least 18 such PMSC mutations in *inlA* have been identified (Olier *et al.*, 2003; Jacquet *et al.*, 2004; Handa-Miya *et al.*, 2007; Ragon *et al.*, 2008; Ward *et al.*, 2008, 2010; Van Stelten *et al.*, 2010). Strains harbouring PMSCs in *inlA* exhibited reduced invasion efficiency in human intestinal epithelial cells (Olier *et al.*, 2003; Handa-Miya *et al.*, 2007; Ward *et al.*, 2010). Conclusive evidence was provided by characterizing isogenic mutants with and without PMSCs. The PMSCs in *inlA* caused decrease in invasion of human hepatocyte and intestinal epithelial cells as well as attenuated virulence in the guinea pig model (Nightingale *et al.*, 2008). Surveys of isolates with PMSCs in *inlA* revealed that these mutations were mostly identified in food isolates and were rarely encountered among clinical isolates (Jacquet *et al.*, 2004; Van Stelten *et al.*, 2010). PMSCs were harboured by an estimated 48.5% of isolates from RTE foods and processing facilities in the United States (Ward *et al.*, 2010), suggesting that a large proportion of *L. monocytogenes* food isolates might be attenuated in virulence.

PMSCs in *inlA* have been identified primarily among isolates of serotypes 1/2a, 1/2b and 1/2c, i.e. serotypes often found among food and environmental isolates. Interestingly, *inlA* PMSC mutations were absent among serotype 4b isolates (Jacquet *et al.*, 2004; Ward *et al.*, 2010).

Truncated InlA has also been encountered among faecal carriage *L. monocytogenes* isolates (Olier *et al.*, 2003). Severity of *L. monocytogenes* infection was lessened in guinea pigs previously exposed to *L. monocytogenes* harbouring a PMSC in *inlA*, indicating that virulence-attenuated strains with truncated InlA might function as a natural vaccine against *L. monocytogenes* with full-length InlA (Nightingale *et al.*, 2008).

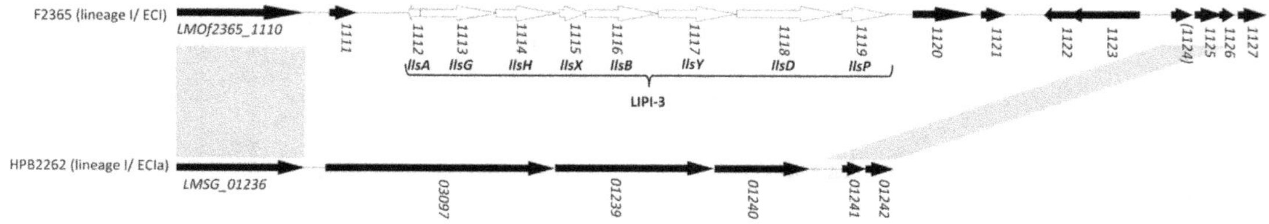

Figure 12.2 Comparison of the genetic region harbouring LIPI-3 in *L. monocytogenes* F2365 (lineage I/ECI) and the genomically equivalent region in HPB2262 (*L. monocytogenes* lineage I/ECIa), which is deficient of LIPI-3. White genes represent LIPI-3 whereas all other genes are marked in black. The name of a pseudogene (*LMOf2365_1124* in F2365) is surrounded by parentheses. The grey shadow indicates region of homology, identified with WebACT

Phylogeny and genomic comparisons of *Listeria* species

Phylogeny of *Listeria* species

The genus *Listeria* is composed of eight species: *L. monocytogenes*, *L. seeligeri*, *L. innocua*, *L. welshimeri*, *L. ivanovii*, *L. grayi*, *L marthii* and *L. rocourtiae* (Schmid *et al.*, 2005; Graves *et al.*, 2010; Leclercq *et al.*, 2010). *L. marthii* and *L. rocourtiae* have been recently identified from the natural environment of the Finger Lakes region of New York and a lettuce from Salzburg (Austria), respectively (Graves *et al.*, 2010; Leclercq *et al.*, 2010). Neither of these species exhibited haemolysis on blood agar or *in vitro* virulence, suggesting that they are non-pathogenic (Graves *et al.*, 2010; Leclercq *et al.*, 2010). For *L. marthii*, the absence of LIPI-1 and inlA was confirmed with PCR and sequencing (den Bakker *et al.*, 2010b; Graves *et al.*, 2010). Among the eight *Listeria* species, only two are considered to be pathogenic: *L. monocytogenes* and *L. ivanovii* (Schmid *et al.*, 2005). *L. monocytogenes* causes listeriosis in humans and animals while *L. ivanovii* is mainly responsible for bovine abortions (Schmid *et al.*, 2005).

The phylogenetic relatedness and evolution of *Listeria* species have been investigated by various approaches (Boerlin *et al.*, 1991; Vaneechoutte *et al.*, 1998; Schmid *et al.*, 2005; den Bakker *et al.*, 2010a,b). These studies consistently showed that

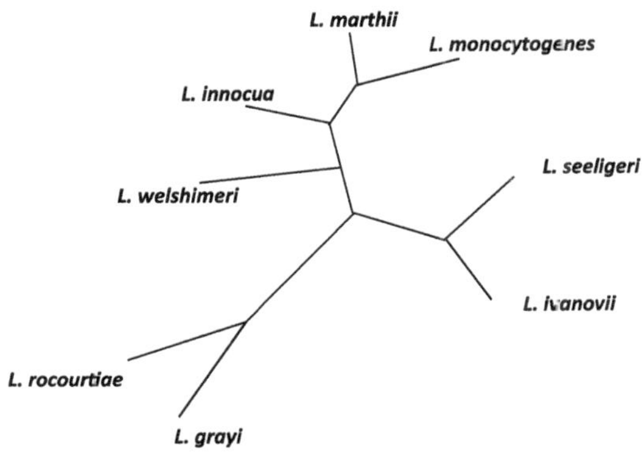

Figure 12.3 Schematic of putative phylogenetic relationships among *Listeria* species based on data provided by previous studies (Schmidt *et al.*, 2005; den Bakker *et al.*, 2010a,b; Graves *et al.*, 2010; Leclerq *et al.*, 2010).

L. grayi is separated from the other *Listeria* species (Fig. 12.3); however, different relationships have been proposed among the remaining species (Boerlin *et al.*, 1991; Vaneechoutte *et al.*, 1998; Schmid *et al.*, 2005; den Bakker *et al.*, 2010a,b). Boerlin *et al.* proposed that *L. monocytogenes*, *L. innocua*, and *L. welshimeri* form one cluster in which *L. innocua* and *L. welshimeri* are grouped as sister branches, whereas *L. seeligeri* and *L. ivanovii* compose the other (Boerlin *et al.*, 1991). Phylogenetic relationships based on the 16S rRNA gene sequences divided *Listeria* species into the same two clusters; however, this analysis placed *L. innocua* and *L. monocytogenes* as sister species instead (Vaneechoutte *et al.*, 1998). A phylogenetic tree based on the concatenated DNA sequence of 16S and 23S rRNA genes, LIPI-1 flanking genes (*prs*, *ldh* and *vclB*) and *iap* placed *L. monocytogenes* and *L. innocua* in one group while *L. welshimeri*, *L. seeligeri* and *L. ivanovii* were in a separate cluster, with *L. welshimeri* being located in the deepest branch (Fig. 12.3) (Schmid *et al.*, 2005).

Two recent studies employed concatenated DNA sequences of 10 and 100 select genes, respectively, to determine the phylogenetic relationships among *Listeria* species, including the recently identified *L. marthii* (den Bakker *et al.*, 2010a,b). These examinations identified two clades of *Listeria* species, one consisting of *L. seeligeri* and *L. ivanovii* and the other composed of *L. welshimeri*, *L. innocua*, *L. marthii* and *L. monocytogenes* (den Bakker *et al.*, 2010a,b). In the latter clade, *L. welshimeri* was placed as a lineage basal to the other species in the same group and *L. marthii* was a sister group to *L. monocytogenes* (Fig. 12.3) (den Bakker *et al.*, 2010a,b). The close relationship between *L. marthii* and *L. monocytogenes* was also observed in the phylogeny based on concatenated partial *sigB*, *gap* and *prs* sequences (Graves *et al.*, 2010). The other recently described *Listeria* species, *L. rocourtiae*, was placed separately from other *Listeria* species in the phylogeny inferred from 16S rRNA gene sequences (Fig. 12.3) (Leclercq *et al.*, 2010).

Excluding *L. grayi*, the common ancestor of *Listeria* species seems to possess pathogenic potential (Schmid *et al.*, 2005; den Bakker *et al.*, 2010a,b). After comparing the DNA sequences of LIPI-1, Schmid *et al.* proposed that this pathogenicity island was inherited from a common progenitor and hypothesized that LIPI-1 was lost through two deletion events leading to the current *L. innocua* and *L. welshimeri*, respectively (Schmid *et al.*, 2005). Similar conclusions have been reached based on multilocus sequence typing (MLST) data and absence/presence of LIPI-1. This analysis suggested that LIPI-1 might have been deleted at least five times during the evolution of *Listeria* species, resulting in pathogenic and non-pathogenic *Listeria* species and atypical *Listeria* strains, e.g. *L. innocua* strains with LIPI-1 and *L. seeligeri* strains without LIPI-1 (den Bakker *et al.*, 2010a). This scenario of virulence gene deletion is consistent with the large-scale deletions observed in *L. innocua* and *L. welshimeri* genomes in comparison with *L. monocytogenes* (Hain *et al.*, 2006b), which will be discussed further in the next section.

Genomic comparisons of different *Listeria* species

Listeria species possess circular genomes of similar sizes, ranging from 2.8 to 3.2 Mbp, that encode approximately 2700 to 3000 genes (Glaser *et al.*, 2001; Hain *et al.*, 2006b; Buchrieser, 2007; den Bakker *et al.*, 2010b; Steinweg *et al.*, 2010). The GC content varied somewhat among different species: 38% for *L. monocytogenes*, 37% for *L. innocua*, 37.4% for *L. seeligeri* and 36.4% for *L. welshimeri* (Glaser *et al.*, 2001; Hain *et al.*, 2006b; Buchrieser, 2007; Steinweg *et al.*, 2010).The genome encodes numerous surface proteins, transporter proteins, and transcriptional regulators, a finding compatible with the ability of *Listeria* to inhabit various environments (Buchrieser, 2007). The *Listeria* pan-genome is estimated to consist of 6494 genes with 1994 commonly shared genes and 4500 accessory genes (den Bakker *et al.*, 2010b). *Listeria* genomes are characterized by pronounced synteny which is disrupted mainly by prophages, monocins, strain-specific genes, and to a lesser degree by translocation (Buchrieser, 2007). Notably, *comK* encoding a natural competence protein has often been disrupted by prophage integration (Buchrieser, 2007; den Bakker *et al.*, 2010b).

Strain-specific genes are dispersed in various genomic locations and tend to deviate from the average GC content of the genome, suggesting acquisition through horizontal gene transfer (Buchrieser, 2007; Nelson *et al.*, 2004). Certain genes of *L. monocytogenes* seem to be absent from the non-pathogenic *L. innocua* and *L. welshimeri* (Glaser *et al.*, 2001; Hain *et al.*, 2006b). Most of the deletion events were observed in the same genetic location in these two non-pathogenic species and the deletion was usually more extensive in *L. welshimeri* relative to *L. innocua*, thus contributing to the smaller genome size of the former. Genes absent in *L. welshimeri* and *L. innocua* included several among those encoding proteins possessing LPXTG- and LRR motifs, which often are involved in adhesion and invasion into mammalian cells (Hain *et al.*, 2006b). Although rarely observed, translocations also interrupt the synteny of *Listeria* genomes. An example of translocation was identified in *L. welshimeri* where a series of genes encoding the F_0F_1-ATP synthase were found in a different location than in the *L. monocytogenes* or *L. innocua* genomes (Hain *et al.*, 2006b; Buchrieser, 2007).

Serotyping and population structure of *L. monocytogenes*

L. monocytogenes has been classified into 13 serotypes according to reactivity to somatic and flagellar antigens: serotypes 1/2a, 1/2b, 1/2c, 3a, 3b, 3c, 4a, 4ab, 4b, 4c, 4d, 4e and 7 (Gasanov *et al.*, 2005). Our understanding of the relationship between *L. monocytogenes* lineages and serotypes has grown over time through the use of various typing schemes. Early studies identified two genomic divisions (lineages) by means of a variety of genotyping tools. An additional lineage, lineage III, was first discovered by Rasmussen *et al.* and has been further supported by several other studies (Rasmussen *et al.*, 1995; Wiedmann *et al.*, 1997; Meinersmann *et al.*, 2004; Ward *et al.*, 2004; Doumith *et al.*, 2004b; Ragon *et al.*, 2008). More recently, four lineages of *L. monocytogenes* were proposed by dividing lineage III into two separate evolutionary groups (den Bakker *et al.*, 2008; den Bakker *et al.*, 2010a; Orsi *et al.*, 2008b; Ward *et al.*, 2008). In this review, we will describe the population structure of *L. monocytogenes* on the basis of the four-lineage classification.

Lineage I

Lineage I includes strains of serotypes 4b, 4d, 4e, 1/2b and 3b (Doumith *et al.*, 2004b). Serotype 4b is implicated in many outbreaks and sporadic cases resulting in lineage I being overrepresented among clinical samples (Cheng *et al.*, 2008). Factors contributing to the strong association of serotype 4b in lineage I to human listeriosis have not been elucidated yet; however, the pathogenicity island LIPI-3 was exclusively found in a subset of lineage I isolates and PMSC mutations in *inlA* have not been identified among serotype 4b isolates (Jacquet *et al.*, 2004; Cotter *et al.*, 2008; Van Stelten *et al.*, 2010; Ward *et al.*, 2010).

Lineage I exhibits a high degree of clonality in comparison to the other *L. monocytogenes* lineages, primarily owing to low rate of recombination and positive selection (Meinersmann *et al.*, 2004; Ward *et al.*, 2004; den Bakker *et al.*, 2008; Orsi *et al.*, 2008b; Ragon *et al.*, 2008). Two lineage I subgroups were identified through DNA–DNA hybridization between the strain-specific *Listeria* DNA array and genomic DNA from various isolates: one consisting of serotypes 4b, 4d and 4e (serotype 4b complex) and the other consisting of serotypes 1/2b and 3b (Doumith *et al.*, 2004b).

Population genetic analysis has led to the hypothesis that this lineage experienced a bottleneck either separately or along with lineage II, estimated to have occurred between one to one-half million years ago (Meinersmann *et al.*, 2004; Ward *et al.*, 2004; den Bakker *et al.*, 2008). The population bottleneck may be connected with reduced hosts and lower temperatures during the first glaciation of the Pleistocene (den Bakker *et al.*, 2008).

Lineage II

Lineage II encompasses serotypes 1/2a, 1/2c, 3a and 3c (Ward *et al.*, 2004; Roberts *et al.*, 2006). Isolates belonging to this lineage are frequently found in foods and natural environments although serotype 1/2a is also frequently identified among clinical isolates and sometimes implicated in outbreaks (Jacquet *et al.*, 2004; Handa-Miya *et al.*, 2007; Handa-Orsi *et al.*, 2008a; Gilmour *et al.*, 2010; Ward *et al.*, 2010).

Studies on the population structure of *L. monocytogenes* suggest that lineage II differs considerably from the other two lineages (Doumith *et al.*, 2004b; Ward *et al.*, 2004; Ragon *et al.*, 2008). For instance, when Doumith *et al.* compared the partial genome sequence of a serotype 4b strain (lineage I) with the complete sequence of the serotype 1/2a EGDe (lineage II), 8% of the serotype 4b genes were not found in EGDe. Since there were 9.5% of EGDe- specific genes in comparison with *L. innocua*, they concluded that lineage II differs from lineage I to a degree equivalent to the interspecies level between *L. monocytogenes* and *L. innocua* (Doumith *et al.*, 2004b). MLST analysis with seven housekeeping genes suggested that the nucleotide divergence of lineage II amounts to 4.99% relative to lineage I and to 7.57% in comparison to lineage III (Ragon *et al.*, 2008).

Genes specific to or highly conserved in lineage II are mainly related to sugar transport and metabolism (Doumith *et al.*, 2004b). According to a study using the species-specific *Listeria* array, 12 out of 19 lineage II-specific genes were putatively involved in sugar metabolism. Most of the PTS genes conserved among lineage II strains were absent in serotypes 4b and 1/2b, which belong to lineage I (Doumith *et al.*, 2004b).

Lineage II represents a more heterogeneous group of strains in comparison to lineage I (Meinersmann *et al.*, 2004; den Bakker *et al.*, 2008; Orsi *et al.*, 2008b; Ragon *et al.*, 2008). The diversity within lineage II was computed to be 0.61%, which was higher than that of lineage I (0.33%) (Ragon *et al.*, 2008). It is possible that this heterogeneity results from more active recombination in addition to positive selection (Meinersmann *et al.*, 2004; den Bakker *et al.*, 2008; Orsi *et al.*, 2008b). Owing to the diversity present in lineage II, multiple subgroups have been identified. Based on the genetic content revealed by the DNA array composed of strain-specific genes, Doumith *et al.* proposed two lineage II subgroups, in which serotypes 1/2a and 3a constitute one subgroup whereas serotypes 1/2c and 3c belong to the other (Doumith *et al.*, 2004b).

Lineages III and IV

Lineage III consists of serotypes 4a, 4c, and certain strains of serotype 4b (Roberts *et al.*, 2006). Even though this lineage has been implicated in occasional sporadic cases, it is markedly less common in human listeriosis than strains of lineage I or II (Ward *et al.*, 2004; Swaminathan and Gerner-Smidt, 2007). However, DNA arrays showed that all *L. monocytogenes* strains tested harboured all known virulence genes regardless of phylogenetic lineage (Doumith *et al.*, 2004b). Also, lineage III isolates were capable of forming plaques in a cell pathogenicity test and exhibited *in vivo* virulence in an intraperitoneal mouse infection (Liu *et al.*, 2006; Roberts *et al.*, 2006). In addition, no premature stop codon mutations in *inlA* have been observed among the lineage III (and IV) isolates (Ward *et al.*, 2008).

Since lineage III seems to have the potential to cause infection, an alternative explanation for its low clinical incidence could be lack of human exposure to these strains. In fact, lineage III has rarely been found among food isolates (Ward *et al.*, 2004, 2008, 2010). It has been suggested that lineage III might not be well equipped against environmental stresses encountered in foods or food-processing plants. This was based on the observations that lineage III strains had a long lag period at refrigeration temperature and relatively high vulnerability to thermal treatment, exhibited low propensity for biofilm formation, and harboured a premature stop codon in the stress response regulator *sigB* (Roberts *et al.*, 2006). Interestingly, lineage III is a major contributor to animal listeriosis, (Roberts *et al.*, 2006).

Analysis of a set of 46 lineage III isolates utilizing partial *actA* and *sigB* sequences identified lineage III as the one with the most pronounced diversity (Roberts *et al.*, 2006). This was further supported by MLST data of seven housekeeping genes: genetic diversity within lineage III was 1.25%, conspicuously higher than either lineage I (0.33%) or lineage II (0.61%). However, recombination in lineage III was found to be highly restricted (Ragon *et al.*, 2008).

Several subgroups have been identified within lineage III. A *Listeria* DNA array composed of strain-specific genes divided lineage III into two subgroups according to serotypes, one subgroup with strains of serotype 4a and the other with strains of serotype 4c (Doumith *et al.*, 2004b). On the other hand, three subgroups (IIIA, IIIB, and IIIC) without associations with serotype were identified based on partial *actA* and *sigB* sequence analysis, with subgroup IIIA found to be the most heterogeneous. In terms of rhamnose fermentation often used in biochemical tests for *L. monocytogenes*, subgroup IIIA utilized rhamnose as observed for typical *L. monocytogenes* isolates, whereas subgroups IIIB and IIIC did not (Roberts *et al.*, 2006). Phylogenetic studies revealed that subgroup IIIB differed from other *L. monocytogenes* lineages including subgroups IIIA and IIIC that formed one cluster. Hence, lineage IIIB was proposed to constitute a distinct lineage (lineage IV) (den Bakker *et al.*, 2008; Orsi *et al.*, 2008b; Ward *et al.*, 2008).

Evolution of different *L. monocytogenes* serotypes

Several efforts have been made to elucidate the evolutionary relationships among *L. monocytogenes* serotypes. Findings from *Listeria* DNA arrays suggested that divergence began as serotypes 1/2b and 1/2c were separated from a common ancestor. Divergence of serotype 1/2c resulted in lineage II while serotypes 4b, 4a, and 4c evolved from serotype 1/2b; *L. innocua* branched off this group (serotypes 4b, 4a, and 4c) mainly through gene loss. (Doumith *et al.*, 2004b). MLST data also suggested that serotype 1/2b, the ancestor of lineage I, evolved into serotype 4b but did not support emergence of *L. innocua* from *L. monocytogenes* serotype 4b; the study also led to the proposal that serotype 1/2a was the likely ancestor of lineage II (Ragon *et al.*, 2008).

PCR-based serotyping of *L. monocytogenes*

Traditionally, serotypes have been determined by reactivity with antisera (Gasanov *et al.*, 2005); however, this method requires specialized reagents and can only be performed in a small number of reference laboratories. In an attempt to overcome these limitations PCR-based serotyping methods were developed (Doumith *et al.*, 2004a; Chen and Knabel, 2007). For instance, Doumith *et al.* described a multiplex PCR that divides *L. monocytogenes* isolates into four groups employing primers annealing to *Listeria* genus-specific *prs* and genes specific to serotype-associated phylogenetic lineages of *L. monocytogenes*. This multiplex PCR can differentiate between strains of serotype 1/2a or 3a, 1/2c or 3c, 1/2b or 3b and the serotype 4b complex (4b, 4d, 4e) (Doumith *et al.*, 2004a). Serotype of lineage III strains cannot be identified via this multiplex PCR scheme. In the case of the two *Listeria* species identified after this scheme was designed, *L. marthii* generated only the *prs* amplicon whereas *L. rocourtiae* failed to amplify any of the target genes (Graves *et al.*, 2010; Leclercq *et al.*, 2010). Indeed, *L. rocourtiae* did not react with any of the antisera against known somatic and flagella antigens of *Listeria* species, suggesting that strains of this species have novel serotypes (Leclercq *et al.*, 2010).

Several investigations have identified atypical serotype 4b isolates that yield the multiplex PCR results expected for serotype 4b as well as the band expected with primers for *lmo0737* (lineage II-specific) (Huang *et al.*, 2011; Leclercq *et al.*, 2011; Lee *et al.*, 2012). This atypical multiplex PCR profile was designated profile IVb-v1 (Leclerq *et al.*, 2011) and was also detected among certain serotype 4b isolates characterized in our laboratory. Analysis of 23 such isolates from the United States revealed that they constituted three distinct clonal groups. One of these groups had a novel MLGT haplotype and appeared to be restricted in its regional distribution: all isolates of this group were from one state, North Carolina. In contrast, the other two groups consisted of human clinical isolates from various states in the United States (Lee *et al.*, 2011).

Epidemic clones

The comparison of the strains implicated in outbreaks showed that a few clonal groups of *L. monocytogenes* have been involved with numerous geographically and temporally unlinked listeriosis outbreaks (Kathariou, 2002; Kathariou, 2003; Cheng *et al.*, 2008). Initially discovered by means of multilocus enzyme electrophoresis, this finding has been also supported by numerous other subtyping tools including pulsed-field gel electrophoresis (PFGE), ribotyping, multiple-locus variable-number tandem-repeat analysis (MLVA), multivirulence-locus sequence typing (MVLST) and MLST (Kathariou, 2003; den Bakker *et al.*, 2008; Chen *et al.*, 2007; Cheng *et al.*, 2008; Lomonaco *et al.*, 2008; Sperry *et al.*, 2008). These groups of epidemic-associated strains are referred to as epidemic clones. To date, three major epidemic clones of serotype 4b have been identified and characterized: epidemic clone (EC) I, II, and Ia (also designated ECIV) (Cheng *et al.*, 2008). Recent major outbreaks have involved strains of serotype 1/2a and 1/2b (Gilmour *et al.*, 2010; Centers for Disease Control and Prevention, 2011) and such strains may be members of novel epidemic clones. ECI, II and Ia have been extensively described in a previous review (Cheng *et al.*, 2008).

Increased exposure or higher pathogenicity?

The recurrent association of epidemic clones with outbreaks can reflect enhanced virulence or high prevalence of these strains in food processing environments and foods. To address these two hypotheses, several studies examined the prevalence of ECI, Ia and II among serotype 4b isolates (Yildirim *et al.*, 2004; Eifert *et al.*, 2005; Franciosa *et al.*, 2007; Ward *et al.*, 2010). ECI, ECII and ECIa isolates have been frequently encountered among serotype 4b isolates from foods and food processing environments. In a survey of food and environmental isolates of serotype 4b *L. monocytogenes*, Yildirim *et al.* discovered that 20 out of 34 isolates (59%) belonged to ECI (Yildirim *et al.*, 2004). Another study of 18 serotype 4b isolates from turkey processing facilities in the US revealed that 22 and 56% of the isolates were of ECI and ECII, respectively (Eifert *et al.*, 2005). ECI and ECII were also common among food isolates in a study in Italy that employed PCR targeting ECI- and ECII-specific markers (Franciosa *et al.*, 2007). MLGT analysis of 34 serotype 4b isolates from ready to eat (RTE) foods and processing facilities in the United States revealed that ECI, ECIa and ECII accounted for 26.5, 41.2 and 26.5%, respectively, of the isolates (Ward *et al.*, 2010). Such findings suggest that ECI, Ia and II may have higher fitness than other serotype 4b strains in food processing facilities and foods. This could increase human exposures and, along with the absence of truncated Internalin in serotype 4b strains, it may contribute to the relatively high prevalence of these strains in outbreaks (as well as sporadic listeriosis).

Certain unique adaptations of ECs may promote survival in food processing facilities. For instance, ECII strains are characteristically resistant to *Listeria* phages at low temperatures; the genes mediating this phenotype are unique to ECII strains and can in fact be employed as ECII-specific probes (Kim and Kathariou, 2009; Cheng *et al.*, 2010). Further analysis has revealed that the genes constitute a novel restriction-modification system with a thermoregulated restriction endonuclease (Kim *et al.*, 2012). Also, the ECII strains from the 1998–1999 hotdog outbreak of listeriosis displayed resistance to quaternary ammonium disinfectants, with the genes being harboured on a large plasmid (pLM80) (Nelson *et al.*, 2004; Elhanafi *et al.*, 2010).

Increased pathogenicity cannot be excluded in explaining the frequent association of ECs with outbreaks. For instance, strains harbouring LIPI-3 included those of ECI, Ia and II (Cotter *et al.*, 2008). Although not exclusively harboured by the isolates of these epidemic clonal groups, presence of this additional pathogenicity island might enhance virulence potential of these ECs. Further investigations, including rigorous analysis of epidemiological data, are needed for quantitative assessments of virulence of strains associated with outbreaks (and also significantly contributing to the burden of sporadic listeriosis).

Listeria phages

Bacteriophages specific for *Listeria* species have been employed as typing and cloning tools and for biocontrol against *L. monocytogenes* (Carlton *et al.*, 2005; Loessner and Rees, 2005). Currently, no less than 400 *Listeria* phages have been isolated from various sources. All *Listeria* phages described to date possess double-stranded DNA genomes of various sizes with GC contests ranging from 34.7% to 40.8% (Loessner and Rees, 2005). The majority of *Listeria* phages are members of the families *Siphoviridae* (isometric capsids and long, non-contractile tails) and *Myoviridae* (isometric capsids and long, contractile tails). Lytic phages have wide host range, being able to infect different *Listeria* serotypes and species while temperate phages are specific for certain serotypes and can often mediate generalized transduction (Hodgson, 2000; Loessner and Rees, 2005).

The lysogenic state is common in *Listeria*, conferring resistance to infection by phages of similar immunity groups (Loessner *et al.*, 2000; Loessner and Rees, 2005; Dorscht *et al.*, 2009). However, infectious phages are estimated to be produced only by 6 to 37% of the strains. Of the five phage-like elements identified in the genome of *L. innocua* CLIP11262 only one was induced into infective virions. Although prophages can be induced with UV light or mitomycin, many seem to be cryptic (Loessner and Rees, 2005). Cryptic prophages result in the production of substances (monocins) inhibitory to other *Listeria* strains. The lethality of monocins derives from intact lysis genes encoded by the cryptic prophages (Loessner and Rees, 2005).

Influence of *Listeria* phages on *L. monocytogenes* ecology and evolution in the food processing environment

The ecological and evolutionary significance of *Listeria* phages has been suggested by several studies (Chen and Knabel, 2008; Kim *et al.*, 2008; Orsi *et al.*, 2008a; Verghese *et al.*, 2011). Two *Listeria* phages isolated from different turkey processing facilities in the US had identical wide-host-range and partial genome sequence data suggested >96% identity with homologous sequences in the wide host range virulent phages P100 and A511 (Kim *et al.*, 2008), indicating the presence of a closely related group of virulent *Listeria* phages that frequently inhabit food processing facilities. Resistance to these phages was significantly more common among environmental *L. monocytogenes* of serotype 1/2a (or 3a) and 1/2b (or 3b) than among those of the serotype 4b complex. This resistance differential was speculated to be associated with the extensively documented predominance of serotypes 1/2a and 1/2b among environmental and food isolates of *L. monocytogenes* (Kim *et al.*, 2008).

Phage resistance may be a major contributor to environmental fitness of *L. monocytogenes* ECII. Strains of this epidemic clone are uniformly resistant to all tested phages when grown at temperatures < 30°C but are phage-susceptible following growth at 37°C (Kim and Kathariou, 2009). Since low temperatures prevail in foods and food processing facilities, such strains would have a competitive advantage in the presence of phage. In fact, use of phage as biocontrol may inadvertently select for these strains, should they be present together with other strains of *Listeria*. As described earlier, temperature-dependent phage resistance of ECII is mediated by a gene cassette unique to this clonal group.

Comparative genome sequence analysis of *L. monocytogenes* strains that originated from the same processing plant but were isolated twelve years apart revealed that the genomic differences between these isolates resided mainly in two prophage sequences incorporated into *comK* (extensive diversification) and tRNA-Thr-4 (one single SNP), suggesting that phages might actively contribute to genomic diversification of *L. monocytogenes* in this environment (Orsi *et al.*, 2008a).

Genetic changes in *comK* prophages among closely related *L. monocytogenes* strains were also found in two studies that characterized strains belonging to the same epidemic clone based on DNA sequences from either the internal portion or the junction fragments of *comK* prophages (Chen and Knabel, 2008; Verghese *et al.*, 2011). Defective *comK* phages were shown to be capable of spontaneous induction and release of infectious particles suggesting potential for transduction-mediated diversification of the *comK* prophages (Verghese *et al.*, 2011). Another intriguing finding was that strains without *comK* prophages (e.g. ECI) exhibited lower growth on meat- and poultry-conditioned biofilms whereas strains harbouring *comK* prophages (e.g. ECII, and the 2008 Canadian outbreak strain) grew better under these conditions. This observation, in addition to sequence diversity in *comK* prophages prompted the proposal that the defective *comK* prophages constituted 'rapid adaptation islands (RAI)' encoding genes ('adaptons') that enhance the ability of *L. monocytogenes* to adapt to and persist in individual food processing facilities by promoting attachment, growth and biofilm formation (Verghese *et al.*, 2011).

As reflected in these studies, the role of phages in the evolution and ecology of *L. monocytogenes* has only recently begun to be appreciated and more studies are warranted to enhance our understanding on this subject. Such research will be particularly relevant now that the influence of *Listeria* phages is expected to increase owing to *Listeria* phages having been approved as biocontrol agent against *L. monocytogenes* in foods (Mahony *et al.*, 2011).

Summary and conclusions

Despite extensive efforts to reduce contamination of the food supply by *L. monocytogenes*, this pathogen remains a serious public health concern as indicated by the recent increase in incidence of listeriosis in Europe (Hedberg, 2006; Gerner-Smidt and Whichard, 2008; Goulet *et al.*, 2008; Wagner and MacLauchlin, 2008; Warriner and Namvar, 2009) and a major, unprecedented outbreak in the United States involving cantaloupe (Centers for Disease Control and Prevention, 2011). Our understanding of listerial pathogenesis at the cellular and immunological level continues to grow, and a wealth of genomic data and omics applications are contributing to the elucidation of the organism's genomic landscape, evolution and adaptations. Such advances have facilitated identification of genes specific to species, serotypes, and clonal groups frequently involved in listeriosis outbreaks. They also led to the development of novel subtyping tools (MLST, MLVA, MLGT, etc.) with significant applications in efforts to elucidate the epidemiology, ecology and evolution of *L. monocytogenes*. Further advances in genome sequencing and omics-based analyses are expected to lead to new insights on strain-specific genetic attributes that may be associated with enhanced virulence or with persistence in a particular environmental niche.

References

Aureli, P., Fiorucci, G.C., Caroli, D., Marchiaro, G., Novara, O., Leone, L., and Salmaso, S. (2000). An outbreak of febrile gastroenteritis associated with corn contaminated by *Listeria monocytogenes*. N. Engl. J. Med. 342, 1236–1241.

den Bakker, H.C., Bundrant, B.N., Fortes, E.D., Orsi, R.H., and Wiedmann, M. (2010a). A population genetics-based and phylogenetic approach to understanding the evolution of virulence in the genus *Listeria*. Appl. Environ. Microbiol. 76, 6085–6100.

den Bakker, H.C., Cummings, C.A., Ferreira, V., Vatta, P., Orsi, R.H., Degoricija, L., Barker, M., Petrauskene, O., Furtado, M.R., and Wiedmann, M. (2010b). Comparative genomics of the bacterial genus *Listeria*: Genome evolution is characterized by limited gene acquisition and limited gene loss. BMC Genomics *11*, 688.

den Bakker, H.C., Didelot, X., Fortes, E.D., Nightingale, K.K., and Wiedmann, M. (2008). Lineage specific recombination rates and microevolution in *Listeria monocytogenes*. BMC Evol. Biol. *8*, 277.

Bennion, J.R., Sorvillo, F., Wise, M.E., Krishna, S., and Mascola, L. (2008). Decreasing listeriosis mortality in the United States, 1990–2005. Clin. Infect. Dis. *47*, 867–874.

Bierne, H., and Cossart, P. (2007). *Listeria monocytogenes* surface proteins: from genome predictions to function. Microbiol. Mol. Biol. Rev. *71*, 377–397.

Boerlin, P., Rocourt, J., and Piffaretti, J.C. (1991). Taxonomy of the genus *Listeria* by using multilocus enzyme electrophoresis. Int. J. Syst. Bacteriol. *41*, 59–64.

Bonazzi, M., Lecuit, M., and Cossart, P. (2009). *Listeria monocytogenes* internalin and E-cadherin: from structure to pathogenesis. Cell. Microbiol. *11*, 693–702.

Buchrieser, C. (2007). Biodiversity of the species *Listeria monocytogenes* and the genus *Listeria*. Microbes Infect. *9*, 1147–1155.

Cabanes, D., Dehoux, P., Dussurget, O., Frangeul, L., and Cossart, P. (2002). Surface proteins and the pathogenic potential of *Listeria monocytogenes*. Trends Microbiol. *10*, 238–245.

Carlton, R.M., Noordman, W.H., Biswas, B., de Meester, E.D., and Loessner, M.J. (2005). Bacteriophage P100 for control of *Listeria monocytogenes* in foods: genome sequence, bioinformatic analyses, oral toxicity study, and application. Regul. Toxicol. Pharmacol. *43*, 301–312.

Carver, T.J., Rutherford, K.M., Berriman, M., Rajandream, M.A., Barrell, B.G., and Parkhill, J. (2005). ACT: the Artemis Comparison Tool. Bioinformatics *21*, 3422–3423.

Centers for Disease Control and Prevention (CDC). (2010). Preliminary FoodNet data on the incidence of infection with pathogens transmitted commonly through food –10 states, 2009. MMWR Morb. Mortal. Wkly Rep. *59*, 418–422.

Centers for Disease Control and Prevention (CDC). (2011). Multistate outbreak of listeriosis linked to whole cantaloupes from Jensen Farms, Colorado-United States, 2011. Available at: http://www.cdc.gov/listeria/outbreaks/cantaloupes-jensen-farms/120811/index.html

Chen, Y., and Knabel, S.J. (2007). Multiplex PCR for simultaneous detection of bacteria of the genus *Listeria, Listeria monocytogenes,* and major serotypes and epidemic clones of *L. monocytogenes*. Appl. Environ. Microbiol. *73*, 6299–6304.

Chen, Y., and Knabel, S.J. (2008). Prophages in *Listeria monocytogenes* contain single-nucleotide polymorphisms that differentiate outbreak clones within epidemic clones. J. Clin. Microbiol. *46*, 1478–1484.

Chen, Y., Zhang, W., and Knabel, S.J. (2007). Multi-virulence-locus sequence typing identifies single nucleotide polymorphisms which differentiate epidemic clones and outbreak strains of *Listeria monocytogenes*. J. Clin. Microbiol. *45*, 835–846.

Cheng, Y., Siletzky, R.M., and Kathariou, S. (2008). Genomic divisions/lineages, epidemic clones, and population structure. In Handbook of *Listeria monocytogenes*, Liu, D., ed. (CRC Press, Boca Raton, FL), pp. 337–358.

Cheng, Y., Kim, J.W., Lee, S., Siletzky, R.M., and Kathariou, S. (2010). DNA probes for unambiguous identification of *Listeria monocytogenes* epidemic clone II strains. Appl. Environ. Microbiol. *76*, 3061–3068.

Clayton, E.M., Hill, C., Cotter, P.D., and Ross, R.P. (2011). Real-time PCR assay to differentiate Listeriolysin S-positive and -negative strains of *Listeria monocytogenes*. Appl. Environ. Microbiol. *77*, 163–171.

Corr, S.C., and O'Neill, L.A. (2009). *Listeria monocytogenes* infection in the face of innate immunity. Cell. Microbiol. *11*, 703–709.

Cotter, P.D., Draper, L.A., Lawton, E.M., Daly, K.M., Groeger, D.S., Casey, P.G., Ross, R.P., and Hill, C. (2008). Listeriolysin S, a novel peptide haemolysin associated with a subset of lineage I *Listeria monocytogenes*. PLoS Pathog. *4*, e1000144.

Dalton, C.B., Austin, C.C., Sobel, J., Hayes, P.S., Bibb, W.F., Graves, L.M., Swaminathan, B., Proctor, M.E., and Griffin, P.M. (1997). An outbreak of gastroenteritis and fever due to *Listeria monocytogenes* in milk. N. Engl. J. Med. *336*, 100–105.

Doorduyn, Y., de Jager, C.M., van der Zwaluw, W.K., Wannet, W.J., van der Ende, A., Spanjaard, L., and van Duynhoven, Y.T. (2006). First results of the active surveillance of *Listeria monocytogenes* infections in the Netherlands reveal higher than expected incidence. Euro Surveill. *11*, E060420.4.

Dorscht, J., Klumpp, J., Bielmann, R., Schmelcher, M., Born, Y., Zimmer, M., Calendar, R., and Loessner, M.J. (2009). Comparative genome analysis of *Listeria* bacteriophages reveals extensive mosaicism, programmed translational frameshifting, and a novel prophage insertion site. J. Bacteriol. *191*, 7206–7215.

Doumith, M., Buchrieser, C., Glaser, P., Jacquet, C., and Martin, P. (2004a). Differentiation of the major *Listeria monocytogenes* serovars by multiplex PCR. J. Clin. Microbiol. *42*, 3819–3822.

Doumith, M., Cazalet, C., Simoes, N., Frangeul, L., Jacquet, C., Kunst, F., Martin, P., Cossart, P., Glaser, P., and Buchrieser, C. (2004b). New aspects regarding evolution and virulence of *Listeria monocytogenes* revealed by comparative genomics and DNA arrays. Infect. Immun. *72*, 1072–1083.

Drevets, D.A., and Bronze, M.S. (2008). *Listeria monocytogenes*: epidemiology, human disease, and mechanisms of brain invasion. FEMS Immunol. Med. Microbiol. *53*, 151–165.

Dussurget, O. (2008). New insights into determinants of *Listeria monocytogenes* virulence. Int. Rev. Cell. Mol. Biol. *270*, 1–38.

Dussurget, O., Pizarro-Cerda, J., and Cossart, P. (2004). Molecular determinants of *Listeria monocytogenes* virulence. Annu. Rev. Microbiol. *58*, 587–610.

Eifert, J.D., Curtis, P.A., Bazaco, M.C., Meinersmann, R.J., Berrang, M.E., Kernodle, S., Stam, C., Jaykus, L.A., and Kathariou, S. (2005). Molecular characterization of *Listeria monocytogenes* of the serotype 4b complex (4b, 4d, 4e) from two turkey processing plants. Foodborne Pathog. Dis. *2*, 192–200.

Elhanafi, D., Dutta, V., and Kathariou, S. (2010). Genetic characterization of plasmid-associated benzalkonium chloride resistance determinants in a *Listeria monocytogenes* strain from the 1998–1999 outbreak. Appl. Environ. Microbiol. *76*, 8231–8238.

Franciosa, G., Scalfaro, C., Maugliani, A., Floridi, F., Gattuso, A., Hodzic, S., and Aureli, P. (2007). Distribution of epidemic clonal genetic markers among *Listeria monocytogenes* 4b isolates. J. Food Prot. *70*, 574–581.

Freitag, N.E., Port, G.C., and Miner, M.D. (2009). *Listeria monocytogenes* – from saprophyte to intracellular pathogen. Nat. Rev. Microbiol. *7*, 623–628.

Frye, D.M., Zweig, R., Sturgeon, J., Tormey, M., LeCavalier, M., Lee, I., Lawani, L., and Mascola, L. (2002). An outbreak of febrile gastroenteritis associated with delicatessen meat contaminated with *Listeria monocytogenes*. Clin. Infect. Dis. *35*, 943–949.

Gasanov, U., Hughes, D., and Hansbro, P.M. (2005). Methods for the isolation and identification of *Listeria* spp. and *Listeria monocytogenes*: a review. FEMS Microbiol. Rev. *29*, 851–875.

Gerner-Smidt, P., and Whichard, J.M. (2008). Foodborne disease trends and reports. Foodborne Pathog. Dis. *5*, 365–367.

Gillespie, I.A., McLauchlin, J., Grant, K.A., Little, C.L., Mithani, V., Penman, C., Lane, C., and Regan, M. (2006). Changing pattern of human listeriosis, England and Wales, 2001–2004. Emerg. Infect. Dis. *12*, 1361–1366.

Gillespie, I.A., McLauchlin, J., Little, C.L., Penman, C., Mook, P., Grant, K., and O'Brien, S.J. (2009). Disease presentation in relation to infection foci for non-pregnancy-associated human listeriosis in England and Wales, 2001 to 2007. J. Clin. Microbiol. *47*, 3301–3307.

Gilmour, M.W., Graham, M., Van Domselaar, G., Tyler, S., Kent, H., Trout-Yakel, K.M., Larios, O., Allen, V., Lee, B., and Nadon, C. (2010). High-throughput genome sequencing of two *Listeria monocytogenes* clinical isolates during a large foodborne outbreak. BMC Genomics *11*, 120.

Glaser, P., Frangeul, L., Buchrieser, C., Rusniok, C., Amend, A., Baquero, F., Berche, P., Bloecker, H., Brandt, P., Chakraborty, T., *et al.* (2001). Comparative genomics of *Listeria* species. Science *294*, 849–852.

Goulet, V. (2007). What can we do to prevent listeriosis in 2006? Clin. Infect. Dis. *44*, 529–530.

Goulet, V., Jacquet, C., Martin, P., Vaillant, V., Laurent, E., and de Valk, H. (2006). Surveillance of human listeriosis in France, 2001–2003. Euro Surveill. *11*, 79–81.

Goulet, V., Hedberg, C., Le Monnier, A., and de Valk, H. (2008). Increasing incidence of listeriosis in France and other European countries. Emerg. Infect. Dis. *14*, 734–740.

Graves, L.M., Helsel, L.O., Steigerwalt, A.G., Morey, R.E., Daneshvar, M.I., Roof, S.E., Orsi, R.H., Fortes, E.D., Milillo, S.R., den Bakker, H.C., *et al.* (2010). *Listeria marthii* sp. nov., isolated from the natural environment, Finger Lakes National Forest. Int. J. Syst. Evol. Microbiol. *60*, 1280–1288.

Gray, M.J., Freitag, N.E., and Boor, K.J. (2006). How the bacterial pathogen *Listeria monocytogenes* mediates the switch from environmental Dr. Jekyll to pathogenic Mr. Hyde. Infect. Immun. *74*, 2505–2512.

Hain, T., Steinweg, C., and Chakraborty, T. (2006a). Comparative and functional genomics of *Listeria* spp. J. Biotechnol. *126*, 37–51.

Hain, T., Steinweg, C., Kuenne, C.T., Billion, A., Ghai, R., Chatterjee, S.S., Domann, E., Karst, U., Goesmann, A., Bekel, T., *et al.* (2006b). Whole-genome sequence of *Listeria welshimeri* reveals common steps in genome reduction with *Listeria innocua* as compared with *Listeria monocytogenes*. J. Bacteriol. *188*, 7405–7415.

Handa-Miya, S., Kimura, B., Takahashi, H. Sato, M., Ishikawa, T., Igarashi, K., and Fujii, T. (2007). Nonsense-mutated *inlA* and *prfA* not widely distributed in *Listeria monocytogenes* isolates from ready-to-eat seafood products in Japan. Int. J. Food Microbiol. *117*, 312–318.

Hedberg, C. (2006). *Listeria* in Europe: the need for a European surveillance network is growing. Euro Surveill. *11*, 75–76.

de las Heras, A., Cain, R.J., Bielecka, M.K., and Vazquez-Boland J.A. (2011). Regulation of *Listeria* virulence: PrfA master and commander. Curr. Opin. Microbiol. *14*, 118–127.

Hodgson, D.A. (2000). Generalized transduction of serotype 1/2 and serotype 4b strains of *Listeria monocytogenes*. Mol. Microbiol. *35*, 312–323.

Huang, B., Fang, N., Dimovski, K., Wang, X., Hogg, G., and Bates, J. (2011). Observation of a new pattern in serogroup-related PCR typing of *Listeria monocytogenes* 4b isolates. J. Clin. Microbiol. *49*, 426–429.

Jackson, K.A., Biggerstaff, M., Tobin-D'Angelo, M., Sweat, D., Klos, R., Nosari, J., Garrison, O., Boothe, E., Saathoff-Huber, L., Hainstock, L., *et al.* (2011). Multistate outbreak of *Listeria monocytogenes* associated with Mexican-style cheese made from pasteurized milk among pregnant, Hispanic women. J. Food Prot. *74*, 949–953.

Jacquet, C., Doumith, M., Gordon, J.I., Martin, P.M., Cossart, P., and Lecuit, M. (2004). A molecular marker for evaluating the pathogenic potential of foodborne *Listeria monocytogenes*. J. Infect. Dis. *189*, 2094–2100.

Jensen, A., Frederiksen, W., and Gerner-Smidt, P. (1994). Risk factors for listeriosis in Denmark, 1989–1990. Scand. J. Infect. Dis. *26*, 171–178.

Johansson, J., Mandin, P., Renzoni, A., Chiaruttini, C., Springer, M., and Cossart, P. (2002). An RNA thermosensor controls expression of virulence genes in *Listeria monocytogenes*. Cell *110*, 551–561.

Kathariou, S. (2002). *Listeria monocytogenes* virulence and pathogenicity, a food safety perspective. J. Food Prot. *65*, 1811–1829.

Kathariou, S. (2003). Foodborne outbreaks of listeriosis and epidemic-associated lineages of *Listeria monocytogenes*. In Microbial Food Safety in Animal Agriculture-current Topics, Torrence, M. E., Isaacson, R.E., eds. (Iowa State Press, Ames, IA), pp. 243–256.

Kim, J.W., and Kathariou, S. (2009). Temperature-dependent phage resistance of *Listeria monocytogenes* epidemic clone II. Appl. Environ. Microbiol. *75*, 2433–2438.

Kim, J.W., Siletzky, R.M., and Kathariou, S. (2008). Host ranges of *Listeria*-specific bacteriophages from the turkey processing plant environment in the United States. Appl. Environ. Microbiol. *74*, 6623–6630.

Kim, J.W., Dutta, V., Elhanafi, D., Lee, S., Osborne, J.A., and Kathariou, S. (2012). A novel restriction-modification system is responsible for temperature-dependent phage resistance in *Listeria monocytogenes* ECII. Appl. Environ. Microbiol. *78*, 1995–2004.

Koch, J., and Stark, K. (2006). Significant increase of listeriosis in Germany – epidemiological patterns 2001–2005. Euro Surveill. *11*, 85–88.

Kornacki, J.L., and Gurtler, J.B. (2007). Incidence and control of *Listeria* in food processing facilities. In *Listeria*, Listeriosis and Food Safety, Ryser, E. T., and Marth, E. H. eds., (Boca Raton, FL: CRC Press) pp. 681–766.

Leclercq, A., Clermont, D., Bizet, C., Grimont, P.A., Le Fleche-Mateos, A., Roche, S.M., Buchrieser, C., Cadet-Daniel, V., Le Monnier, A., Lecuit, M., *et al.* (2010). *Listeria rocourtiae* sp. nov. Int. J. Syst. Evol. Microbiol. *60*, 2210–2214.

Leclercq, A., Chenal-Francisque, V., Dieye, H., Cantinelli, T., Drali, R., Brisse, S., and Lecuit, M. (2011). Characterization of the novel *Listeria monocytogenes* PCR serogrouping profile IVb-v1. Int. J. Food Microbiol. *147*, 74–77.

Lee, S., Ward, T.J., Graves, L.M., Wolf, L.A., Sperry, K., Siletzky, R.M., and Kathariou, S. (2012). Characterization of atypical *Listeria monocytogenes* serotype 4b strains harboring a lineage II-specific gene cassette. Appl. Environ. Microbiol. *78*, 660–667.

Liu, D., Lawrence, M.L., Wiedmann, M., Gorski, L., Mandrell, R.E., Ainsworth, A.J., and Austin, F.W. (2006). *Listeria monocytogenes* subgroups IIIA, IIIB, and IIIC delineate genetically distinct populations with varied pathogenic potential. J. Clin. Microbiol. *44*, 4229–4233.

Loessner, M.J., and Rees, C.E.D. (2005). *Listeria* phages: basics and applications. In Phages: Their Role in Bacterial Pathogenesis and Biotechnology, Waldor, K.M., Friedman, D.I., and Adhya, S.L., eds. (ASM Press, Washington, DC), pp. 362–379.

Loessner, M.J., Inman, R.B., Lauer, P., and Calendar, R. (2000). Complete nucleotide sequence, molecular analysis and genome structure of bacteriophage A118 of *Listeria monocytogenes*: implications for phage evolution. Mol. Microbiol. *35*, 324–340.

Lomonaco, S., Chen, Y., and Knabel, S.J. (2008). Analysis of additional virulence genes and virulence gene regions in *Listeria monocytogenes* confirms the epidemiologic relevance of multi-virulence-locus sequence typing. J. Food Prot. *71*, 2559–2566.

MacDonald, P.D., Whitwam, R.E., Boggs, J.D., MacCormack, J.N., Anderson, K.L., Reardon, J.W., Saah, J.R., Graves, L.M., Hunter, S.B., and Sobel, J. (2005). Outbreak of listeriosis among Mexican immigrants as a result of consumption of illicitly produced Mexican-style cheese. Clin. Infect. Dis. *40*, 677–682.

Mahony, J., McAuliffe, O., Ross, R.P., and van Sinderen, D. (2011). Bacteriophages as biocontrol agents of food pathogens. Curr. Opin. Biotechnol. *22*, 157–163.

Mead, P.S., Slutsker, L., Dietz, V., McCaig, L.F., Bresee, J.S., Shapiro, C., Griffin, P.M., and Tauxe, R.V. (1999). Food-related illness and death in the United States. Emerg. Infect. Dis. *5*, 607–625.

Meinersmann, R.J., Phillips, R.W., Wiedmann, M., and Berrang, M.E. (2004). Multilocus sequence typing of *Listeria monocytogenes* by use of hypervariable genes reveals clonal and recombination histories of three lineages. Appl. Environ. Microbiol. *70*, 2193–2203.

Nelson, K.E., Fouts, D.E., Mongodin, E.F., Ravel, J., DeBoy, R.T., Kolonay, J.F., Rasko, D.A., Angiuoli, S.V., Gill, S.R., Paulsen, I.T., *et al.* (2004). Whole genome comparisons of serotype 4b and 1/2a strains of the food-borne pathogen *Listeria monocytogenes* reveal new insights into the core genome components of this species. Nucleic Acids Res. *32*, 2386–2395.

Nightingale, K.K., Ivy, R.A., Ho, A.J., Fortes, E.D., Njaa, B.L., Peters, R.M., and Wiedmann, M. (2008). *inlA* premature stop codons are common among *Listeria monocytogenes* isolates from foods and yield virulence-attenuated strains that confer protection against fully virulent strains. Appl. Environ. Microbiol. *74*, 6570–6583.

Olier, M., Pierre, F., Rousseaux, S., Lemaitre, J.P., Rousset, A., Piveteau, P., and Guzzo, J. (2003). Expression of truncated Internalin A is involved in impaired internalization of some *Listeria monocytogenes* isolates carried asymptomatically by humans. Infect. Immun. *71*, 1217–1224.

Orndorff, P.E., Hamrick, T.S., Smoak, I.W., and Havell, E.A. (2006). Host and bacterial factors in listeriosis pathogenesis. Vet. Microbiol. *114*, 1–15.

Orsi, R.H., Borowsky, M.L., Lauer, P., Young, S.K., Nusbaum, C., Galagan, J.E., Birren, B.W., Ivy, R.A., Sun, Q., Graves, L.M., *et al.* (2008a). Short-term genome evolution of *Listeria monocytogenes* in a non-controlled environment. BMC Genomics 9, 539.

Orsi, R.H., Sun, Q., and Wiedmann, M. (2008b). Genome-wide analyses reveal lineage specific contributions of positive selection and recombination to the evolution of *Listeria monocytogenes*. BMC Evol. Biol. *8*, 233.

Pentecost, M., Otto, G., Theriot, J.A., and Amieva, M.R. (2006). *Listeria monocytogenes* invades the epithelial junctions at sites of cell extrusion. PLoS Pathog. *2*, e3.

Ragon, M., Wirth, T., Hollandt, F., Lavenir, R., Lecuit, M., Le Monnier, A., and Brisse, S. (2008). A new perspective on *Listeria monocytogenes* evolution. PLoS Pathog. *4*, e1000146.

Rajabian, T., Gavicherla, B., Heisig, M., Muller-Altrock, S., Goebel, W., Gray-Owen, S.D., and Ireton, K. (2009). The bacterial virulence factor InlC perturbs apical cell junctions and promotes cell-to-cell spread of *Listeria*. Nat. Cell Biol. *11*, 1212–1218.

Rasmussen, O.F., Skouboe, P., Dons, L., Rossen, L., and Olsen, J.E. (1995). *Listeria monocytogenes* exists in at least three evolutionary lines: evidence from flagellin, invasive associated protein and listeriolysin O genes. Microbiology *141*, 2053–2061.

Roberts, A., Nightingale, K., Jeffers, G., Fortes, E., Kongo, J.M., and Wiedmann, M. (2006). Genetic and phenotypic characterization of *Listeria monocytogenes* lineage III. Microbiology *152*, 685–693.

Salamina, G., Dalle Donne, E., Niccolini, A., Poda, G., Cesaroni, D., Bucci, M., Fini, R., Maldini, M., Schuchat, A., Swaminathan, B., *et al.* (1996) A foodborne outbreak of gastroenteritis involving *Listeria monocytogenes*. Epidemiol. Infect. *117*, 429–436.

Scallan, E., Hoekstra, R.M., Angulo, F.J., Tauxe, R.V., Widdowson, M.A., Roy, S.L., Jones, J.L., and Griffin, P.M. (2011). Foodborne illness acquired in the United States-major pathogens. Emerg. Infect. Dis. *17*, 7–15.

Schlech, I., W.F., Lavigne, P.M., Bortolussi, R.A., Allen, A.C., Haldane, E.V., Wort, A.J., Hightower, A.W., Johnson, S.E., King, S.H., Nicholls, E.S., *et al.* (1983). Epidemic listeriosis –evidence for transmission by food. N. Engl. J. Med. *308*, 203–206.

Schmid, M.W., Ng, E.Y., Lampidis, R., Emmerth, M., Walcher, M., Kreft, J., Goebel, W., Wagner, M., and Schleifer, K.H. (2005). Evolutionary history of the genus *Listeria* and its virulence genes. Syst. Appl. Microbiol. *28*, 1–18.

Schuchat, A., Deaver, K.A., Wenger, J.D., Plikaytis, B.D., Mascola, L., Pinner, R.W., Reingold, A.L., and Broome, C.V. (1992). Role of foods in sporadic listeriosis. I. Case–control study of dietary risk factors. The *Listeria* Study Group. JAMA *267*, 2041–2045.

Schwartz, B., Ciesielski, C.A., Broome, C.V., Gaventa, S., Brown, G.R., Gellin, B.G., Hightower, A.W., and Mascola, L. (1988). Association of sporadic listeriosis with consumption of uncooked hot dogs and undercooked chicken. Lancet *2*, 779–782.

Scortti, M., Monzo, H.J., Lacharme-Lora, L., Lewis, D.A., and Vazquez-Boland J.A. (2007). The PrfA virulence regulon. Microbes Infect. *9*, 1196–1207.

Sperry, K.E., Kathariou, S., Edwards, J.S., and Wolf, L.A. (2008). Multiple-locus variable-number tandem-repeat analysis as a tool for subtyping *Listeria monocytogenes* strains. J. Clin. Microbiol. *46*, 1435–1450.

Steinweg, C., Kuenne, C.T., Billion, A., Mraheil, M.A., Domann, E., Ghai, R., Barbuddhe, S.B., Karst, U., Goesmann, A., Puhler, A., *et al.* (2010). Complete genome sequence of *Listeria seeligeri*, a nonpathogenic member of the genus *Listeria*. J. Bacteriol. *192*, 1473–1474.

Swaminathan, B., and Gerner-Smidt, P. (2007). The epidemiology of human listeriosis. Microbes Infect. *9*, 1236–1243.

Tilney, L.G., and Portnoy, D.A. (1989). Actin filaments and the growth, movement, and spread of the intracellular bacterial parasite, *Listeria monocytogenes*. J. Cell Biol. *109*, 1597–1608.

Van Stelten, A., Simpson, J.M., Ward, T.J., and Nightingale, K.K. (2010). Revelation by single-nucleotide polymorphism genotyping that mutations leading to a premature stop codon in *inlA* are common among *Listeria monocytogenes* isolates from ready-to-eat foods but not human listeriosis cases. Appl. Environ. Microbiol. *76*, 2783–2790.

Vaneechoutte, M., Boerlin, P., Tichy, H.V., Bannerman, E., Jager, B., and Bille, J. (1998). Comparison of PCR-based DNA fingerprinting techniques for the identification of *Listeria* species and their use for atypical *Listeria* isolates. Int. J. Syst. Bacteriol. *48*, 127–139.

Varma, J.K., Samuel, M.C., Marcus, R., Hoekstra, R.M., Medus, C., Segler, S., Anderson, B.J., Jones, T.F., Shiferaw, B., Haubert, N., *et al.* (2007). *Listeria monocytogenes* infection from foods prepared in a commercial establishment: a case–control study of potential sources of sporadic illness in the United States. Clin. Infect. Dis. *44*, 521–528.

Verghese, B., Lok, M., Wen, J., Alessandria, V., Chen, Y., Kathariou, S., and Knabel, S. (2011). *comK* prophage junction fragments as markers for *Listeria monocytogenes* genotypes unique to individual meat and poultry processing plants and a model for rapid niche-specific adaptation, biofilm formation, and persistence. Appl. Environ. Microbiol. *77*, 3279–3292.

Voetsch, A.C., Angulo, F.J., Jones, T.F., Moore, M.R., Nadon, C., McCarthy, P., Shiferaw, B., Megginson, M.B., Hurd, S., Anderson, B.J., *et al.* (2007). Reduction in the incidence of invasive listeriosis in foodborne diseases active surveillance network sites, 1996–2003. Clin. Infect. Dis. *44*, 513–520.

Wagner, M., and MacLauchlin, J. (2008). Biology. In Handbook of *Listeria monocytogenes*, Liu, D., ed. (CRC Press, Boca Raton, FL), pp. 3–25.

Ward, T.J., Gorski, L., Borucki, M.K., Mandrell, R.E., Hutchins, J., and Pupedis, K. (2004). Intraspecific phylogeny and lineage group identification based on the *prfA* virulence gene cluster of *Listeria monocytogenes*. J. Bacteriol. *186*, 4994–5002.

Ward, T.J., Ducey, T.F., Usgaard, T., Dunn, K.A., and Bielawski, J.P. (2008). Multilocus genotyping assays for single nucleotide polymorphism-based subtyping of *Listeria monocytogenes* isolates. Appl. Environ. Microbiol. *74*, 7629–7642.

Ward, T.J., Evans, P., Wiedmann, M., Usgaard, T., Roof, S.E., Stroika, S.G., and Hise, K. (2010). Molecular and phenotypic characterization of *Listeria monocytogenes* from U.S. Department of Agriculture Food Safety and Inspection Service surveillance of ready-to-eat foods and processing facilities. J. Food Prot. *73*, 861–869.

Warriner, K., and Namvar, A. (2009). What is the hysteria with *Listeria*? Trends Food Sci. Tech. *20*, 245–254.

Wiedmann, M., Bruce, J.L., Keating, C., Johnson, A.E., McDonough, P.L., and Batt, C.A. (1997). Ribotypes and virulence gene polymorphisms suggest three distinct *Listeria monocytogenes* lineages with differences in pathogenic potential. Infect. Immun. *65*, 2707–2716.

Yildirim, S., Lin, W., Hitchins, A.D., Jaykus, L.A., Altermann, E., Klaenhammer, T.R., and Kathariou, S. (2004). Epidemic clone I-specific genetic markers in strains of *Listeria monocytogenes* serotype 4b from foods. Appl. Environ. Microbiol. *70*, 4158–4164.

Epidemiology, Virulence Genes, and Reservoirs of Enteropathogenic *Yersinia* Species 13

Riikka Laukkanen-Ninios and Maria Fredriksson-Ahomaa

Abstract

Enteropathogenic yersiniosis is caused due to foodborne infection with *Yersinia enterocolitica* and *Y. pseudotuberculosis*. Several virulence factors have been identified that are common to these two pathogens even though *Y. pseudotuberculosis* is genetically more related to *Y. pestis*, which is typically transmitted by fleas and not through foods. Diarrhoea and abdominal pain are the most dominant symptoms for *Y. enterocolitica* and *Y. pseudotuberculosis* infections. Occasionally, complications such as joint pain and skin rash may occur, typically among adults. Most human cases are caused by *Y. enterocolitica*. The reported cases are mainly sporadic and outbreaks are uncommon. However, outbreaks of *Y. pseudotuberculosis* infection, often in school children, have occurred in Finland, Russia and Japan. The most important transmission route is proposed to be via contaminated foods even though these pathogens have seldom been isolated from foods. The low isolation rates are probably due to the low sensitivity of the culture methods. *Y. enterocolitica* infections have been linked to raw or undercooked pork and pork products while *Y. pseudotuberculosis* infections have been linked to raw fresh produce and surface water. Pigs are so far the most important reservoir for human pathogenic *Y. enterocolitica*. The principal reservoir of *Y. pseudotuberculosis* is believed to be wild animals, especially rodents and birds. Transmission routes of these pathogens from animals to humans are mostly unknown.

Taxonomy, species and subtypes

The genus *Yersinia* currently comprises three human pathogens: *Y. enterocolitica, Y. pseudotuberculosis* and *Y. pestis*. The genomes of these pathogens are very similar in size (4.6–4.7 Mb) (Thomson *et al.,* 2006). The G + C contents of *Y. enterocolitica* and of *Y. pseudotuberculosis* and *Y. pestis* are 47.3 and 47.6, respectively, which is lower than that of most members of the *Enterobacteriaceae* family. Multilocus sequence analysis (MLSA) and DNA–DNA hybridization studies suggest that *Y. enterocolitica* has diverged from *Y. pseudotuberculosis* 0.4–1.9 million years ago and that *Y. pestis* is a non-recombining monomorphic clone of *Y. pseudotuberculosis* that has emerged within the last 1500–20,000 years (Achtman *et al.,* 1999; Wren, 2003; Morelli *et al.,* 2010). Genome sequence data confirm that *Y. pestis* and *Y. pseudotuberculosis* are closely related, with DNA sequence homology of nearly 97%, whereas both share only ~ 50% homology with *Y. enterocolitica* (Wren, 2003).

Neubauer *et al.* (2000) subdivided *Y. enterocolitica* into two subspecies based on the 16S rRNA sequences and DNA–DNA re-association values. *Y. enterocolitica* subsp. *enterocolitica* comprises the *Y. enterocolitica* type strain ATCC 9610^T and strains from high pathogenic biotype 1B, whereas *Y. enterocolitica* subsp. *palearctica* comprises the strains from low-pathogenic biotypes 2–5 and non-pathogenic biotype 1A (Neubauer *et al.,* 2000; Howard *et al.,* 2006). Additionally, the genus *Yersinia* comprises *Y. ruckeri*, which is a fish pathogen causing enteric red mouth disease in fish (Ewing *et al.,* 1978), and at least 13 other species, which are considered non-pathogenic. These are *Y. aldovae*, *Y. bercovieri*, *Y. frederiksenii*, *Y. intermedia*, *Y. kristensenii*, *Y. mollaretii*, *Y. rohdei*, *Y. aleksiciae* (Sprague and Neubauer, 2005), *Y. similis* (Sprague *et al.*, 2008), *Y. massiliensis* (Merhej *et al.*, 2008), *Y. entomophaga* (Hurst *et al.*, 2010), *Y. nurmii* (Murros-Kontiainen *et al.*, 2011a) and *Y. pekkanenii* (Murros-Kontiainen *et al.*, 2011b).

Y. enterocolitica

Y. enterocolitica is a heterogeneous group of bacteria based on biochemical, antigenic and virulence properties. Six different biotypes (1A, 1B, 2–5) and about 30 serotypes of *Y. enterocolitica* strains have been described (Wauters *et al.*, 1987, 1991). Strains belonging to biotypes 1B and 2–5 have been associated with diseases in humans and animals while strains belonging to biotype 1A are generally regarded as non-pathogenic (Bhagat and Virdi, 2011). However, strains of biotype 1A represent a large fraction of strains from patients with enteritis and may have been opportunistic pathogens (Tennant *et al.*, 2003; Huovinen *et al.*, 2010). *Y. enterocolitica* strains can further be divided into high-pathogenic (biotype 1B), low-pathogenic (biotypes 2–5) and non-pathogenic (biotype 1A) strains (Carniel, 2001). Recently, Lamps *et al.* designed an assay for molecular biotyping of pathogenic *Y. enterocolitica* strains based on a single PCR targeting the 23S rRNA gene (Lamps *et al.*, 2006). With this method high-pathogenic biotype 1B could be discriminated from the low-pathogenic biotypes (2–5).

Serotype O:3, which has a world-wide distribution, is the most common serotype associated with human yersiniosis in Europe followed by O:9 and O:5,27. Geographical distribution of *Y. enterocolitica* O:8 is generally restricted to North America and has been found only seldom in Japan and Europe. Recently, this serotype has been isolated from human clinical samples in Poland (Gierczyński *et al.*, 2009; Rastawicki *et al.*, 2009). The serotypes associated with human yersiniosis are also shared by non-pathogenic *Y. enterocolitica* strains and other *Yersinia* spp. and, thus, more accurate tests for pathogenic strains are needed in addition to serological typing. Slide agglutination using commercial antisera O:3, O:5, O:8, O:9 and O:27 is still the most commonly used method for serotyping but there are also PCR assays available for identification of serotype O:3 (Weynants *et al.*, 1996) and O:9 (Jacobsen *et al.*, 2005).

Y. pseudotuberculosis

There is little variation in biochemical reactions among *Y. pseudotuberculosis* strains. However, a simple biotyping scheme using melibiose, raffinose, and citrate has been suggested for *Y. pseudotuberculosis* (Tsubokura and Aleksić, 1995). Using this scheme, *Y. pseudotuberculosis* strains can be divided into four biotypes (Table 13.1) but without any clear clinical relevance it is seldom used. Interestingly, the inability to ferment melibiose has been associated with lowered pathogenicity, mainly due to lack of virulence for mice or guinea pigs, although the strains harbour chromosomal virulence genes, such as *inv*, and the virulence plasmid pYV (Tsubokura *et al.*, 1984;

Table 13.1 Standard biochemical reactions for biotyping of *Y. enterocolitica* and *Y. pseudotuberculosis* strains

	Y. enterocolitica						*Y. pseudotuberculosis*		
Biotypes	PYZ	TWE	IND	XYL	TRE	Biotypes	MEL	CIT	RAF
1A	+	+	+	+	+	1	+	–	–
1B	–	+	+	+	+				
2	–	–	+	+	+	2	–	–	–
3	–	–	–	+	+	3	–	+	–
4	–	–	–	–	+	4	+	–	+
5	–	–	–	–	–				

PYZ, pyrazinamidase; TWE, tween; IND, indole; XYL, xylose; TRE, trehalose; MEL, melibiose; CIT, citrate; RAF, raffinose.

Nagano *et al.*, 1997; Fukushima *et al.*, 2001). However, melibiose-negative O:3 strains have been isolated from both sick and healthy animals, and from humans with yersiniosis (Tsubokura *et al.*, 1984; Aleksić *et al.*, 1995; Martins *et al.*, 1998), suggesting that the strains are probably pathogenic.

Serotyping of *Y. pseudotuberculosis* strains has proven to be useful for diagnosis and epidemiological purpose. The serotyping scheme of *Y. pseudotuberculosis* comprises 15 different O serotypes (O:1–O:15) and 10 subtypes (O:1a, O:1b, O:1c, O:2a, O:2b, O:2c, O:4a, O:4b, O:5a, O:5b) (Tsubokura and Aleksić, 1995). The serotypes differ in their geographical distribution and ecological niches. The serotypes O:1–O:3 are widely distributed, while serotypes O:4–O:15 mainly occur in Asia. Serotypes O:1 and O:3 dominates in humans in Europe (Aleksić *et al.*, 1995; Jalava *et al.*, 2004) and serotypes O:4 and O:5 in Japan (Fukushima *et al.*, 2001). Serotypes O:6–O:15 have so far been isolated only from non-human sources (Fukushima *et al.*, 2001). Serotyping by commercial antisera has been used extensively for *Y. pseudotuberculosis*. However, the slide agglutination test is unable to identify rough strains and commercial antisera are available only for serotypes O:1–O:6. An alternative is to use a multiplex-PCR assay to identify *Y. pseudotuberculosis* serotypes and subserotypes (Bogdanovich *et al.*, 2003).

Virulence genes and pathogenesis of enteropathogenic *Yersinia*

Enteropathogenic *Yersinia* strains have several plasmid and chromosome-encoded virulence factors (Table 13.2). All pathogenic *Yersinia* strains need a 70-kb virulence plasmid called pYV (plasmid for *Yersinia* virulence) to survive and multiply in the host (Cornelis *et al.*, 1998). The pYV codes for an outer membrane protein YadA (*Yersinia* adhesin A), a set of secreted proteins called Yops (*Yersinia* outer membrane protein), and their secretion apparatus called Ysc (Yop secretion) which is a type III secretion system (Cornelis, 2002). With the Ysc, extracellularly located *Yersinia* strains, which are in close contact with the eukaryotic cell, deliver Yops into the cytosol of the target cell. Yops protect *Yersinia* strains from the macrophage by destroying the phagocytic and signalling capacities of the macrophage, and finally, inducing apoptosis. The *virF* gene is an important transcriptional activator of the production of YOPs.

Enteropathogenic *Yersinia* strains need also chromosomally encoded virulence factors for full virulence. At least two chromosomal genes, *inv* and *ail*, are needed for mammalian cell invasion. Attachment invasion locus (Ail) is involved in the attachment and invasion of the host cell by *Y. enterocolitica* and in the serum resistance of both *Y. enterocolitica* and *Y. pseudotuberculosis*

Table 13.2 Virulence factors of *Y. enterocolitica* (YE) and *Y. pseudotuberculosis* (YP)

Location	Gene	Determinant	Function	YE	YP
Plasmid	*ysc/yop*	Yops	Resistance to phagocytosis	+	+
	yadA	YadA	Attachment, invasion, serum resistance	+	+[a]
Chromosome	*inv*	Invasin	Attachment, invasion	+	+
	ail	Ail	Attachment, invasion	+	–
			Serum resistance	+	+
	ystA	YstA	Fluid secretion in intestine	+	–
	ypm	YPM	Stimulation of T cells	–	+[c]
	ybt	Ybt	Iron uptake	+[b]	+[c]
	myfA	MyfA	Unknown	+	–
	psa	pH6	Haemagglutination, adhesion	–	+

[a]Only serum resistance.
[b]Strains belonging to biotype 1B.
[c]Variable.

(Miller *et al.*, 2001). Invasin (Inv) promotes the entry of enteropathogenic *Yersinia* strains into epithelial cells of the ileum during the initial stage of infections (Isberg and Barnes, 2001). Additionally, thermostable enterotoxin (Yst) and mucoid Yersinia factor (MyfA) of *Y. enterocolitica* and pH 6 antigen of *Y. pseudotuberculosis* are encoded in the chromosome. The pH 6 antigen is a surface (fimbrial) structure that mediates agglutination of erythrocytes and is responsible for adhesion to cultured mammal cells (Yang and Isberg, 1997). MyfA of *Y. enterocolitica* is the homologue of pH 6 antigen of *Y. pseudotuberculosis* because of the common features of environmental regulation and the high sequence similarity, however, the function of MyfA is unknown.

The high-pathogenic *Y. enterocolitica* strains (biotype 1B) and *Y. pseudotuberculosis* strains (serotypes O:1 and O:3) harbour the high-pathogenicity island (HPI), a cluster of chromosomal genes intended for iron uptake (Carniel, 2001). HPI encodes for the yersiniabactin (Ybt) siderophore, which is responsible for iron uptake of the bacterium and therefore systemic dissemination of the bacteria in the host. The complete HPI is present in most *Y. pseudotuberculosis* strains of serotype O:1, while melibiose non-fermenting serotype O:3 strains commonly have an incomplete HPI (Fukushima *et al.*, 2001). Additionally, highly pathogenic *Y. enterocolitica* 1B strains carry Ysa pathogenicity island (Ysa-PI) encoding a type III secretion system that is different from the plasmid-encoded Ysc (Matsumoto and Young, 2009). Some *Y. pseudotuberculosis* strains produce a super-antigen toxin, *Y. pseudotuberculosis*-derived mitogen (YPM), which is known to play an important role in causing severe systemic infections (Iwata *et al.*, 2008). The presence of *ypmA* gene in human *Y. pseudotuberculosis* strains is associated with strains of Far East origin and scarlatinoid fever syndrome (Fukushima *et al.*, 2001).

Biotype 1A strains are considered non-pathogenic owing to the lack of major chromosomal virulence genes and the virulence plasmid (pYV) (Bhagat and Virdi, 2011). *Y. enterocolitica* strains belonging to biotype 1A usually carry the *ystB* but not *ail* and *ystA*, which are found in biotypes 1B and 2–5 (Bonardi *et al.*, 2010). However, some strains of biotype 1A have recently been reported to carry *ail* (Sihvonen *et al.*, 2011a).

Identification and characterization

Accurate identification is necessary to discriminate between harmless environmental *Yersinia* spp. and enteropathogenic *Y. enterocolitica* or *Y. pseudotuberculosis*. Traditionally, biochemical and serological characteristics are used to identify *Yersinia* spp. However, atypical biochemical reactions are not rare among *Yersinia* spp., making identification difficult (Hallanvuo *et al.*, 2006; Sihvonen *et al.*, 2009). For example, *Y. similis* and *Y. pekkanenii* have similar biochemical reactions to *Y. pseudotuberculosis* (Sprague *et al.*, 2008; Niskanen *et al.*, 2009; Murros-Kontiainen *et al.* 2011b) and the species cannot be differentiated from each other using API 20E system which is widely used to identify *Yersinia* spp. Matrix-assisted laser desorption/ionization time-of-flight (MALDI-TOF) protein profiling has been used for accurate and rapid identification of *Yersinia* spp. (Ayyadurai *et al.*, 2010; Lasch *et al.*, 2010). *Yersinia* spp. can also be identified at species level using Fourier transform infrared (FT-IR) spectroscopy (Kuhm *et al.*, 2009). The FT-IR also allows identification of different biotypes and serotypes of *Y. enterocolitica* strains. However, PCR assays targeting the plasmid (*vir*F/*lcr*F and *yad*A) and chromosomal (*ail, inv* and *yst*) virulence genes are so far the most widely used methods for rapid and accurate identification of enteropathogenic *Yersinia* strains (Fredriksson-Ahomaa *et al.*, 2010).

Multiple DNA-based typing methods have been used for epidemiological studies to determine the infection sources and transmission routes of enteropathogenic *Yersinia* (Table 13.3). Pulsed-field gel electrophoresis (PFGE) is so far the most widely used method for both *Y. enterocolitica* and *Y. pseudotuberculosis* (Virdi and Sachdeva, 2005; Fredriksson-Ahomaa *et al.*, 2006a; Fredriksson-Ahomaa, 2009).

PFGE has proved to be highly discriminatory for molecular typing of *Yersinia* spp., and thus this method is a commonly used technique in epidemiological studies of *Y. enterocolitica* and *Y. pseudotuberculosis*. It allows subtyping of strains

Table 13.3 DNA-based method for subtyping of *Y. enterocolitica* and *Y. pseudotuberculosis*

Typing method	Species	References
AFLP	*Y. enterocolitica*	Fearnley *et al.* (2005), Kuehni-Boghenbor *et al.* (2006)
MLVA	*Y. enterocolitica*	Gierczyński *et al.* (2007), Gulati *et al.* (2009)
MLST	*Y. enterocolitica*	Kotetishvili *et al.* (2005)
	Y. pseudotuberculosis	Ch'Ng *et al.* (2011)
RAPD	*Y. enterocolitica*	Rasmussen *et al.* (1994)
	Y. pseudotuberculosis	Makino *et al.* (1994), Kageyama *et al.* (2002)
REAC	*Y. enterocolitica*	Kapperud *et al.* (1990)
REAP	*Y. enterocolitica*	Kapperud *et al.* (1990), Iteman *et al.* (1996)
	Y. pseudotuberculosis	Fukushima *et al.* (1994), Fukushima *et al.* (1998), Han *et al.* (2003)
Rep-PCR	*Y. enterocolitica*	Odinot *et al.* (1995)
	Y. pseudotuberculosis	Kim *et al.* (2003)
Ribotyping	*Y. enterocolitica*	Kapperud *et al.* (1990), Iteman *et al.* (1996)
	Y. pseudotuberculosis	Voskressenskaya *et al.* (2005)
PFGE	*Y. enterocolitica*	Najdenski *et al.* (1994), Fredriksson-Ahomaa *et al.* (1999), Fredriksson-Ahomaa *et al.* (2006b), Laukkanen *et al.* (2009)
	Y. pseudotuberculosis	Niskanen *et al.* (2002), Nuorti *et al.* (2004), Laukkanen *et al.* (2008)

AFLP, amplified-fragment-length polymorphism; MLVA, multiple locus VNTR (variable number of tandem repeat regions) analysis; MLST, multilocus sequence typing; RAPD, randomly amplified polymorphic DNA; REAC, restriction endonuclease analysis of the chromosome; REAP, restriction endonuclease analysis of the plasmid; Rep-PCR, repetitive element PCR; PFGE, pulsed-field gel electrophoresis.

belonging to the same bioserotype (Iteman *et al.*, 1996). However, sometimes digestion with two or three enzymes is needed to discriminate strains belonging to the same bioserotype (Fredriksson-Ahomaa *et al.*, 1999), and the complexity of the profiles can make comparison of large numbers of restriction patterns of enteropathogenic *Yersinia* strains difficult. Furthermore, the method is very laborious and time-consuming.

Ribotyping, random amplification of polymorphic DNA (RAPD) assay and restriction endonuclease analysis of the plasmid (REAP) have also frequently been applied to characterize enteropathogenic *Yersinia* strains. Ribotyping has been shown to be a useful tool for molecular typing, especially to characterize *Y. pseudotuberculosis* strains (Voskressenskaya *et al.*, 2005). REAP has been used to demonstrate differences among *Y. pseudotuberculosis* strains belonging to same serotype isolated in different geographical location (Fukushima *et al.*, 1994, 1998). RAPD assay, which is simple and quick to perform but may possess a low reproducibility (Blixt *et al.*, 2003), has been used to characterize *Y. enterocolitica* (Rasmussen *et al.*, 1994; Odinot *et al.*, 1995) and *Y. pseudotuberculosis* strains at the sub-serotype level in some studies including outbreak investigations (Kageyama *et al.*, 2002; Makino *et al.*, 1994).

Multilocus variable-number tandem-repeat analysis (MLVA) is a PCR-based typing method which is based on variable-number tandem repeats (VNTR) present in most bacterial genomes including *Yersinia* (de Benito *et al.*, 2004). This method is increasingly used for typing pathogenic bacteria. The MLVA is easy to perform and it seems to be more discriminatory than other genotyping methods for characterization of *Y. enterocolitica* strains (Gulati *et al.*, 2009). Recently, Sihvonen *et al.* (2011b) characterized 104 sporadic and outbreak-related *Y. enterocolitica* 4/O:3 strains with MLVA using six VNTR markers and showed that this method had a clearly better discriminatory power than PFGE with *NotI* enzyme. Furthermore, MLVA is less labour intensive and the results are easier to analyse than PFGE.

Multilocus sequence typing (MLST) has been

used to assess the genetic relationship, phylogenetic traits and clustering of the *Yersinia* spp. (Kotetishvili *et al.*, 2005). Two MLST schemes for *Y. pseudotuberculosis* have been developed to study population structure of the species: schemes from Ch'Ng *et al.* (2011) and from Laukkanen-Ninios *et al.* (2011). The seven housekeeping genes included for MLST were *mhd, recA, sucA, fumC, aroC, pgi* and *gyrB* in the scheme of Ch'Ng *et al.*, whereas *adk, argA, aroA, glnA, thrA, tmk* and *trpE* genes were selected for MLST in the scheme of Laukkanen-Ninios *et al.* (2011).

Some other molecular methods have also been used for typing of enteropathogenic *Yersinia*, e.g. repetitive element PCR (rep-PCR) (Kim *et al.*, 2003; Wojciech *et al.*, 2004; Souza *et al.*, 2010), PCR-ribotyping (ITC-profiling) (Wojciech *et al.*, 2004; Lucero Estrada *et al.*, 2011), insertion sequence (IS) typing (Odaert *et al.*, 1996) and amplified-fragment-length polymorphism (AFLP), a recently adopted PCR-based typing method, which allows differentiation of *Y. enterocolitica* strains within serotype-related clusters (Fearnley *et al.*, 2005; Kuehni-Boghenbor *et al.*, 2006).

Epidemiology and clinical manifestations

Human enteric yersiniosis

Yersiniosis is an infectious disease caused by *Yersinia*, and enteric yersiniosis is due to *Y. enterocolitica* or *Y. pseudotuberculosis*. Enteric yersiniosis ranks third after campylobacteriosis and salmonellosis among the notifiable bacterial zoonoses in Europe. Most of the reported cases are caused by *Y. enterocolitica*. However, *Y. pseudotuberculosis* infections are probably under-recognized because this pathogen is not routinely tested in many countries and the clinical diagnosis can be challenging (Long *et al.*, 2010). Most reported cases of yersiniosis are sporadic and outbreaks are uncommon. However, in recent years, *Y. pseudotuberculosis* has emerged as an outbreak associated organism (Table 13.4). Outbreaks of *Y. pseudotuberculosis*

Table 13.4 Reported *Y. enterocolitica* (YE) and *Y. pseudotuberculosis* (YP) foodborne outbreaks during the past decade

Country	Year	Species	Serotype	Source	Reference
Canada	1998	YP	O:1	Milk	Nowgesic *et al.* (1999)
Finland	1998–9	YP	O:3	Iceberg lettuce	Nuorti *et al.* (2004)
	2001	YP	O:1, O:3	Iceberg lettuce	Jalava *et al.* (2004)
	2003	YP	O:1	Grated carrots	Jalava *et al.* (2006)
	2004	YP	O:1	Grated carrots	Kangas *et al.* (2008)
	2006	YP	O:1	Grated carrots	Rimhanen-Finne *et al.* (2009)
	2008	YP	O:1	Grated carrots	THL (2009)
Korea	2001	YP	O:4	Spring water	Han *et al.* (2003)
Japan	1998	YP	O:5	Well water	Sunahara *et al.* (2000)
	2004	YE	O:8	Salads	Sakai *et al.* (2005)
	2006	YE	O:9	Food?	Moriki *et al.* (2010)
Norway	2005–6	YE	O:9	Brawn/pork shops	Grahek-Ogden *et al.* (2007)
Russia	2002	YE	NR	Vegetables	ProMED: 20020206.3400
	2005	YP	NR	Cabbage, onion	ProMED: 2005427.1169
	2005	YP	NR	Vegetables	ProMED: 2005216.3617
	2007	YP	NR	Vegetables	ProMED: 20071001.3240
	2008	YP	NR	Vegetables	ProMED: 20080718.2184
USA	2001–2	YE	O:3	Chitterlings	Jones *et al.* (2003)

ProMED (www.promedmail.org) archive number.

infection, often in school children, have occurred in Finland, Russia and Japan (Tsubokura *et al.*, 1989; Voskressenskaya *et al.*, 2003; Jalava *et al.*, 2004).

Infection with *Y. enterocolitica* or *Y. pseudotuberculosis* can cause a variety of symptoms depending on the age of the infected person (Cover and Aber, 1989). Common symptoms are fever, abdominal pain and diarrhoea. Symptoms typically develop 4–7 days after exposure and may last 5–14 days, although diarrhoea may persist for several weeks. In older children and young adults, especially those with *Y. pseudotuberculosis* infection, abdominal pain in the right lower abdomen may be mistaken for appendicitis (pseudoappendicitis). Linking of unnecessary appendectomies can help to identify a *Yersinia* outbreak (Nuorti *et al.*, 2004). The most common post-infectious sequelae of yersiniosis are joint pain (reactive arthritis) and skin rash (erythema nodosum), occurring usually a few weeks after the initial infection (Hannu *et al.*, 2003; Jalava *et al.*, 2006; Rosner *et al.*, 2010). The reactive arthritis is associated with the presence of HLA-B27 antigen. The joint pains, most commonly in the knees, ankles or wrists, usually develop within 1 week to 1 month after the initial infection and generally resolve after 1–6 months (Granfors *et al.*, 1989; Hannu *et al.*, 2003).

Most of the yersiniosis cases resolve on their own without any antimicrobial treatment. In more severe infections, like systemic infection and bacteraemia, antimicrobials may be useful. Treatment of *Yersinia* enteritis should also be considered for patients who are immune-compromised and patients with iron overload. In contrast to *Y. pseudotuberculosis*, which lacks ß-lactamase activity and is hence susceptible to penicillins, *Y. enterocolitica* does possess β-lactamase and is therefore generally resistant to penicillins and first-generation cephalosporins (Stock and Wiedemann, 1999). Aminoglycosides, third-generation cephalosporins, co-trimoxazole, tetracyclines, and fluoroquinolones are used for therapy. Although antimicrobial resistance among enteropathogenic *Yersinia* strains is rare, multiresistant *Y. enterocolitica* strains have been reported (Capilla *et al.*, 2003; Sihvonen *et al.*, 2011b). The resistance of *Y. enterocolitica* to nalidixic acid is based on point mutations in the fluoroquinolone resistance-determining regions of *gyrA* (Capilla *et al.*, 2004; Sihvonen *et al.* 2011b).

Enteropathogenic Yersinia in animals

Animals have long been suspected of being reservoirs for pathogenic *Y. enterocolitica* and *Y. pseudotuberculosis* and, hence, sources of human infections. These bacteria have been recovered from diverse animal sources ranging from farm animals, domestic pets and experimental animals to wild and captive animals. However, enteropathogenic *Yersinia* strains, especially *Y. enterocolitica* strains, have frequently been isolated only from pigs (Fredriksson-Ahomaa *et al.*, 2010). The host range of *Y. pseudotuberculosis* is broad but the principal reservoir hosts are believed to be wild animals, especially birds, rodents and their predators (Fredriksson-Ahomaa, 2009).

Y. enterocolitica of bioserotype 4/O:3 has a wide distribution in the pig population and it has shown to be the most common type in Europe and also in the US (Bhaduri and Wesley, 2006) (Table 13.5). Bioserotype 2/O:5,27 of *Y. enterocolitica* has frequently been isolated from pigs in the US and Canada (Bhaduri and Wesley, 2006; Poljak *et al.*, 2010). In England, bioserotypes 2/O:9 and 2/O:5,27 (Ortiz Martínez *et al.*, 2010) and in China, bioserotypes 2/O:9 and 3/O:3 (Wang *et al.*, 2009) are the most common ones. *Y. pseudotuberculosis* belonging to bioserotype 2/O:3 is widely distributed in pig population and the most common type found in pigs at slaughter (Niskanen *et al.*, 2002; Martínez *et al.*, 2009, 2011), however, in England, several different bioserotypes of *Y. pseudotuberculosis* strains among pigs have been reported (Ortiz Martínez *et al.*, 2010).

Enteropathogenic *Yersinia* strains have seldom been isolated from other farm animals. In the UK, bioserotype 3/O:5,27 has sporadically been found in sheep (Milnes *et al.*, 2008) and in France, *Y. enterocolitica* O:9 has been isolated from bovine and ovine faeces (Gourdon *et al.*, 1999). *Y. pseudotuberculosis* infections have been reported among Australian sheep flocks owing to serotype O:3 (Slee and Skilbeck, 1992) and among farmed deer owing to serotypes O:1,

Table 13.5 Isolation rate of *Y. enterocolitica* (YE) and *Y. pseudotuberculosis* (YP) in pigs at slaughter

Country	Number of pigs	Species	Biotype/serotype	Number of positives (%) in		Reference
				Tonsils	Faeces	
Belgium	201	YE	3/O:9 (8), 4/O:3 (81)	89 (44)	–	Martinez *et al.* (2011)
		YP	1/O:1 (1), 1/O:2 (1), 2/O:3 (3)	5 (2)	–	
	139	YE	4/O:3	52 (37)	–	Van Damme *et al.* (2010)
England	630	YE	2/O:5 (97), 2/O:9 (124), 4/O:3 (39)	278 (44)	–	Ortiz Martínez *et al.* (2010)
		YP	1/O:1 (32), 1/O:4 (29), 2/O:3 (41)	114 (18)	–	
Estonia	151	YE	4/O:3	135 (89)	–	Martínez *et al.* (2009)
		YP	2/O:3	2 (1)	–	
Finland	210	YE	4/O:3	109 (52)	–	Korte *et al.* (2004)
		YP	2/O:3	8 (4)	–	Niskanen *et al.* (2002)
	350	YE	4/O:3	124 (35)	–	Laukkanen *et al.* (2009)
	358	YE	4/O:3		27 (8)	
	350	YP	O:3	34 (10)	–	Laukkanen *et al.* (2008)
	358	YP	O:3		24 (7)	
	301	YE	4/O:3	177 (59)	90 (30)	Laukkanen *et al.* (2010)
		YP	O:3	8 (3)	13 (4)	
Germany	372	YE	4/O:3 (142), O:9 (1)	143 (38)	–	Gürtler *et al.* (2005)
	379	YE	4/O:3		2 (1)	
	164	YE	4/O:3	101 (62)	17 (10)	Bucher *et al.* (2008)
Italy	428	YE	2/O:5 (1), 4/O:3 (136)	137 (32)	–	Martínez *et al.* (2011)
		YP	1/O:1 (3), 2/O:2 (1), NT (2)	5 (1)	–	
Latvia	109	YE	4/O:3	70 (64)	–	Martínez *et al.* (2009)
		YP	2/O:3	5 (5)	–	
	404	YE	4/O:3	143 (35)	–	Terentjeva and Berzins (2010)
		YP	NR	12 (3)	–	
Russia	197	YE	4/O:3	66 (34)	–	Martínez *et al.* (2009)
		YP	2/O:3	13 (7)	–	
Spain	200	YE	4/O:3	185 (93)	–	Martínez *et al.* (2011)
Switzerland	212	YE	2/O:5,27 (6), 2/O:9 (2), 4/O:3 (69)	72 (34)	–	Fredriksson-Ahomaa *et al.* (2007)

NR, not reported.

O:2 and O:3 (Jerrett *et al.*, 1990). In Brazil, *Y. pseudotuberculosis* infections caused by bioserotypes 1/O:1 and 2/O:3 have been reported in buffalo and cattle, respectively (Martins *et al.*, 1998). *Y. pseudotuberculosis* has more frequently been isolated from wild and captive animals than *Y. enterocolitica* (Fredriksson-Ahomaa *et al.*, 2010). Wild rodents have been shown to excrete both pathogens in the faeces (Bachans *et al.*, 2010). Recently, *Y. enterocolitica* of bioserotypes 2/O:5,27, 2/O:9 and 4/O:3 and *Y. pseudotuberculosis* of bioserotypes 1–2/O:1 and 1/O:2 have been isolated from wild boars in Switzerland (Fredriksson-Ahomaa *et al.*, 2009). Dogs and cats have been reported to excrete enteropathogenic *Yersinia* sporadically in the faeces

(Fukushima *et al.*, 1985; Fredriksson-Ahomaa *et al.*, 2001; Wang *et al.*, 2010).

Enteropathogenic *Yersinia* in pigs

Pathogenic *Y. enterocolitica* is frequently isolated from fattening or finishing pigs, but not usually from piglets or farrowing sows (Fukushima *et al.*, 1983; Korte *et al.*, 2004; Gürtler *et al.*, 2005; Bowman *et al.*, 2007; Farzan *et al.*, 2010). Surprisingly, Bowman *et al.* (2007) detected *Y. enterocolitica* frequently from gestating sows. In fattening pigs, the occurrence of pathogenic *Y. enterocolitica* is dependent on the age of the pigs and sample source: the pathogen can first be isolated from faeces and tonsils before antibodies can be detected by serology (Fukushima *et al.*, 1983; Nielsen *et al.*, 1996; Nesbakken *et al.*, 2006). They excrete a high amount of this pathogen into faeces usually only around 12–21 weeks of age, after which the excretion decreases. Repeated sampling on a farm shows that *Y. enterocolitica* persists on the farm, and both environmental and pig-to-pig transmissions have been suggested (Fukushima *et al.*, 1983; Skjerve *et al.*, 1998; Pilon *et al.*, 2000; Poljak *et al.*, 2010). This pathogen has been isolated from pig house structures such as pen floors and separators, stairs and passage, suckling devices, water tap, and piping (Fukushima *et al.*, 1983; Pilon *et al.*, 2000). Working equipment and clothing have also been found to be contaminated (Pilon *et al.*, 2000; von Altrock *et al.*, 2010).

High prevalence of *Y. enterocolitica* has been associated with conventional production, where as *Y. pseudotuberculosis* has been more prevalent in organic production (Nowak *et al.*, 2006; Laukkanen *et al.*, 2008; Ortiz Martínez *et al.*, 2010; Virtanen *et al.*, 2011) but no clear difference between farrow-to-finish farms and fattening farms or production capacity has been seen (Wesley *et al.*, 2008; Laukkanen *et al.*, 2010; Virtanen *et al.*, 2011). Lower health status and snout contact of the pigs, as well as use of untreated water and industrial by-products in feed are factors associated with high prevalence of *Y. enterocolitica* (Wesley *et al.*, 2008; Virtanen *et al.*, 2011). It has been suggested that feed containing prebiotic component designed to prevent intestinal colonization of enteropathogenic bacteria may reduce carriage and shedding of *Y. enterocolitica* (Virtanen *et al.*, 2011). Outdoor facilities for fattening pigs have been associated with high prevalence of *Y. pseudotuberculosis* in one study (Laukkanen *et al.*, 2008). Furthermore, rodents can act as carriers of enteropathogenic *Yersinia* on pig farms (Laukkanen *et al.*, 2008; Bachans *et al.*, 2010). More information regarding the sources of contamination is needed to reduce the prevalence of enteropathogenic *Yersinia* in pigs.

Enteropathogenic *Yersinia* has frequently been isolated from pig at slaughter (Table 13.5). The isolation rates of human pathogenic *Y. enterocolitica* strains in pig tonsils are very high (between 30 and 90%) in Europe. The isolation rates of *Y. pseudotuberculosis* are usually low (between 1–10%). However, in England, *Y. pseudotuberculosis* was isolated from 18% of the slaughtered pigs (Ortiz Martínez *et al.*, 2010). The prevalence of *Y. enterocolitica* and *Y. pseudotuberculosis* has been shown to be clearly lower in sows than in fattening pigs at slaughter (Niskanen *et al.*, 2002; Korte *et al.*, 2004) and pathogenic *Y. enterocolitica* can more often be isolated from tonsils than from faeces of pigs at slaughter (Table 13.5). In southern Germany, isolation rates between 56% and 67% in tonsils and between 8% and 12% in faeces samples have been demonstrated (Bucher *et al.*, 2008). In Finland, isolation rates of 35% and 8% in tonsils and faeces, respectively have been reported (Laukkanen *et al.*, 2003). In Belgium, the pig tonsils at slaughter were shown to be highly contaminated (average concentration of 4.5 log cfu/g) with *Y. enterocolitica* 4/O:3 (Van Damme *et al.*, 2010).

Pigs infected with enteropathogenic *Yersinia* on farms have been identified as a major source of enteropathogenic *Yersinia* contamination of carcasses and high prevalence of enteropathogenic *Yersinia* in live pigs predispose carcasses and pluck set to *Yersinia* contamination (Laukkanen *et al.*, 2008, 2009, 2010). Pigs are mostly asymptomatic carriers of enteropathogenic *Yersinia* that cannot be identified during meat inspection. During the slaughter process, *Y. enterocolitica* and *Y. pseudotuberculosis* can easily be transmitted from tonsils and intestinal content to other parts of the carcass and offal, and to the slaughterhouse environment (Fredriksson-Ahomaa *et al.*, 2006a). Pathogenic *Y. enterocolitica* can frequently be detected on the pig carcasses, whereas the prevalence of *Y.*

pseudotuberculosis is low (Fredriksson-Ahomaa *et al.*, 2000; Gürtler *et al.*, 2005; Lindblad *et al.*, 2007; Laukkanen *et al.*, 2010). Removal of the head with the tonsils and tongue prior to evisceration may probably reduce the *Yersinia* contamination of carcasses. Additionally, bagging of the rectum has shown to reduce the carcass contamination (Nesbakken *et al.*, 1994; Laukkanen *et al.*, 2010). Based on modelling, the combined prevalence of *Listeria monocytogenes* and enteropathogenic *Yersinia* in Finnish pork was estimated to be reduced from 1–11% to 0–2% if the head was removed intact and the rectum sealed off (Ranta *et al.*, 2010).

Prevalence of enteropathogenic *Yersinia* in food

Foodstuffs have often been suggested to be the main source of *Y. enterocolitica* and *Y. pseudotuberculosis*, even though these pathogens have seldom been isolated from food samples. The low isolation rate of enteropathogenic *Yersinia* is thought to be due to the low sensitivity of available culture methods (Fredriksson-Ahomaa and Korkeala, 2003; Fredriksson-Ahomaa, 2009). Using PCR assays, however, pathogenic *Y. enterocolitica* has been detected frequently, especially on pig tongues, and on the surface of freshly slaughtered pig and poultry carcasses, as well as in minced pork and fermented sausages (Lindblad *et al.*, 2007; Fredriksson-Ahomaa *et al.*, 2010) (Table 13.6).

In Northern Europe and Germany, the only bioserotype associated with human disease isolated from pork is 4/O:3 (Bucher *et al.*, 2008; Messelhäusser *et al.*, 2011). Bioserotypes 2/O:5,27 and 2/O:9 have sporadically been recovered from pork in Ireland (Logue *et al.*, 1996) and bioserotypes 3/O:3 and 3/O:5,27 from that in Japan (Fukushima *et al.*, 1997). Recently, bioserotypes 2/O:9 and 3/O:5 carrying the common virulence genes *virF, ail, ystA* and *myfA* were reported in pork products in Argentina (Lucero Estrada *et al.*, 2011). Enteropathogenic *Y. enterocolitica* 2/O:9 was isolated from a minced pork sample, raw sausage and from an uncooked chicken sample in Italy (Bonardi et al., 2010; Lucero Estrada *et al.*, 2011). The level of contaminating bacteria in the raw chicken sample was shown to be low with a most probable number (MPN) value of 0.30–0.90/g. In southern Germany, a high (over 10^5 cfu/g) contamination level of enteropathogenic (*ail*-positive) *Y. enterocolitica* on raw pig tongues has recently been reported (Messelhäusser *et al.*, 2011). In a study in New Zealand, enteropathogenic (*virF* and *ail*-positive) *Y. enterocolitica* was detected in 32% and isolated from 18% of the pork samples, using a combination of PCR and MPN methods (Hudson *et al.*, 2008), and the number of *Y. enterocolitica* was found to be within a range of 0.30 to 5.42 MPN/cm^2.

Enteropathogenic *Yersinia* has rarely been detected in fresh vegetables. In two studies on retail fresh produce, pathogenic *Y. enterocolitica* was detected by PCR in 3% of lettuce samples, but none were positive by culture methods (Fredriksson-Ahomaa, 2001; Johannessen *et al.*, 2002). *Y. pseudotuberculosis* has not been isolated from fresh carrots unrelated to outbreaks in Europe (Thisted Lambertz *et al.*, 2008). In India, *Y. pseudotuberculosis* has been isolated from lettuce on one occasion (Pingulkar *et al.*, 2001). In Finland, *Y. pseudotuberculosis* has been isolated from iceberg lettuce and carrot-peeling line as well as from spoiled carrots, and from fluid draining from spoiled carrot in outbreak investigations (Nuorti *et al.*, 2004; Jalava *et al.*, 2006; Kangas *et al.*, 2008; Rimhanen-Finne *et al.*, 2009).

Enteropathogenic *Yersinia* in the environment

Pathogenic strains of *Y. enterocolitica* have very seldom been isolated from environmental samples including surface water. One reason can be the low detection levels and insensitive isolation methods (Cheyne *et al.*, 2010). Sandery *et al.* (1996) detected pathogenic *Y. enterocolitica* in environmental water by PCR.

Pathogenic *Y. pseudotuberculosis* are widespread in the soil and water where they survive for a long time. The environment can also be contaminated with faeces of infected wild animals (Fukushima *et al.*, 1994). The prevalent serotypes found in Japan in wild animals and environmental samples have been the same suggesting that there is a link between wild animals carrying *Y. pseudotuberculosis* and environmental contamination by the organism (Fukushima and Gomyoda, 1991).

Table 13.6 Prevalence of enteropathogenic *Y. enterocolitica* in raw foods

Sample	No. of samples	No. of positives by				Reference
		Culture	(%)	PCR	(%)	
Pig tongues	15	7	(47)	10	(67)	Vishnubhatla *et al*. (2001)
	99	79	(80)	82	(83)	Fredriksson-Ahomaa and Korkeala (2003)
	129	34	(26)	58	(45)	Messelhäusser *et al*. (2011)
Pig offal[a]	110	38	(35)	77	(70)	Fredriksson-Ahomaa and Korkeala (2003)
Broiler carcass	98	0	–	9	(10)	Lindblad *et al*. (2006)
Pig carcass	518	0	–	81	(16)	Lindblad *et al*. (2007)
Chitterling	350	8	(2)	278	(79)	Boyapalle *et al*. (2001)
Ground pork	350	0	–	133	(38)	Boyapalle *et al*. (2001)
	100	32	(32)	47	(47)	Vishnubhatla *et al*. (2001)
Ground beef	100	23	(23)	31	(31)	Vishnubhatla *et al*. (2001)
Minced pork	255	4	(2)	63	(25)	Fredriksson-Ahomaa and Korkeala (2003)
	100	5	(5)	35	(35)	Lambertz *et al*. (2007)
	102	1	(1)	5	(5)	Messelhäusser *et al*. (2011)
Pork[b]	300	6	(2)	50	(17)	Johannessen *et al*. (2000)
	91	6	(7)	9	(10)	Thisted Lambertz and Danielsson-Tham (2005)
	150	0	–	9	(6)	Fredriksson-Ahomaa *et al*. (2007)
	97	0	–	11	(11)	Lambertz *et al*. (2007)
	34	6	(18)	11	(32)	Hudson *et al*. (2008)
	125	2	(2)	ND[c]	–	Bonardi *et al*. (2010)
	201	11	(5)	18	(9)	Messelhäusser *et al*. (2011)
Game	60	0	–	23	(38)	Bucher *et al*. (2008)
	51	0	–	3	(6)	Messelhäusser *et al*. (2011)
Chicken	43	0	–	0	–	Fredriksson-Ahomaa and Korkeala (2003)
	102	0	–	1	(1)	Bucher *et al*. (2008)
	80	1	(1)	ND	–	Bonardi *et al*. (2010)
Fish	200	0	–	0	–	Fredriksson-Ahomaa and Korkeala (2003)
Lettuce	200	0	–	6	(3)	Johannessen *et al*. (2002)
	101	0	–	3	(3)	Fredriksson-Ahomaa and Korkeala (2003)
Milk	61	0	–	0	–	Messelhäusser *et al*. (2011)
Tofu	50	0	–	6	(12)	Vishnubhatla *et al*. (2001)

[a]Heart, kidney, liver.
[b]Except pig tongues and offal.
[c]Not done.

Transmission of *Y. enterocolitica* and *Y. pseudotuberculosis* to humans

Human enteric yersiniosis is thought to be primarily transmitted to humans faecal–orally by food, especially by raw or undercooked pork and pork products and untreated water (Tauxe *et al*., 1987; Ostroff *et al*., 1994; Satterthwaite *et al*., 1999; Jones *et al*., 2003; Fredriksson-Ahomaa *et al*., 2006a; Grahek-Ogden *et al*., 2007; Boqvist *et al*., 2009). Transmission is likely to occur more often via cross-contamination of cooked pork or foods not normally harbouring *Yersinia* because consumption of raw pork does not play an important role in most countries. Another transmission route may be from person to person, which can occur directly or indirectly. Direct person-to-person contact has not been demonstrated, but Lee *et al.* (1990) reported *Y. enterocolitica* O:3 infections in infants who were probably exposed to infection by their caretakers. Indirect person-to-person transmission has apparently occurred in several instances by transfusion of contaminated blood products (Sen, 2000). In these cases, the most likely source of *Yersinia* has been blood donors with subclinical bacteraemia. Yersiniosis can also occur by direct contact with pigs and dogs (Seuri and Granfors, 1992; Fredriksson-Ahomaa *et al.*, 2001; Wang *et al.*, 2010).

In Finland, *Y. pseudotuberculosis* has caused many outbreaks via domestically grown carrots and iceberg lettuce (Table 13.4). Wild animals and irrigation water have been suspected as contamination sources for the vegetables (Nuorti *et al.*, 2004; Jalava *et al.*, 2006; Kangas *et al.*, 2008). A combination of direct contact with wildlife faeces during the storage and cross-contamination of the equipment are the most likely contributing factors. Un-chlorinated water from wells, springs and streams contaminated with faeces of wild animals has been a suspected cause of *Y. pseudotuberculosis* infections particularly in Japan and Korea (Fukushima *et al.*, 1988; Sunahara *et al.*, 2000; Han *et al.*, 2003).

Several studies using different DNA-based typing methods have been conducted to compare human strains with animal, especially pig strains. Most of the reports support the hypothesis that pigs are an important source of human *Y. enterocolitica* infections (Lee *et al.*, 1990; Wojciech *et al.*, 2004; Fearnley *et al.*, 2005; Fredriksson-Ahomaa *et al.*, 2006a,b; Wang *et al.*, 2009). Thisted Lamberz and Danielsson-Tham (2005) compared *Y. enterocolitica* 4/O:3 strains recovered from patients and pork products in Sweden using PFGE and found indistinguishable profiles. Furthermore, Fredriksson-Ahomaa *et al.* (2006b) analysed human and retailed pork strains of *Y. enterocolitica* 4/O:3 using three different restriction enzymes, and demonstrated indistinguishable genotypes among the strains. This finding further reinforces the assumption that contaminated pork is an important source of human infection with *Y. enterocolitica* 4/O:3.

Falcão *et al.* (2006) reported a high similarity between human and animal *Y. enterocolitica* 4/O:3 strains using PFGE and ERIC-PCR suggesting that pigs, dogs and rats are an important infection source for humans in Brazil. In China, the PFGE patterns of the pathogenic (*ail*- and *virF*-positive) *Y. enterocolitica* 3/O:3 and 2/O:9 strains isolated from diarrhoea patients were indistinguishable from that of the animal isolates, especially from pigs, rodents and dogs, in the same area, thus, indicating that humans infections originated from the animals (Wang *et al.*, 2009). These investigators have also demonstrated using PFGE that dogs belonging to farmers may be an important carrier of *Y. enterocolitica* 3/O:3 and a potential source of human infections (Wang *et al.*, 2010). In Nigeria, *Y. enterocolitica* 2/O:9 and *Y. pseudotuberculosis* 1/O:1 strains have been isolated from different human and non-human sources, and the dominating PFGE types of both pathogens were found in humans, animals (pigs and sheep) and foods (milk and soap) indicating a link between animal reservoirs and humans, and a transmission via contaminated foods (Okwori *et al.*, 2009).

Conclusion

Our knowledge about the epidemiology of enteropathogenic *Yersinia* is inadequate. Molecular methods including PCR have a better chance to yield useful information on the prevalence of enteropathogenic *Y. enterocolitica* and *Y. pseudotuberculosis* in non-human sources like animals, foods and environment. However, live isolates are also needed for further characterization to identify infection sources and transmission routes, and hence more effective culture method should be developed. Although several studies on the epidemiology of enteropathogenic *Yersinia* have been conducted, a lot more studies using discriminating identification and characterization methods will be required to address questions which remain to be solved.

References

Achtman, M., Zurth, K., Morelli, G., Torrea, G., Guiyoule, A., and Carniel, E. (1999). *Yersinia pestis*, the cause of plague, is a recently emerged clone of *Yersinia pseudotuberculosis*. Proc. Natl. Acad. Sci. U.S.A. *96*, 14043–14048.

Aleksić, S., Bockemühl, J., and Wuthe, H.H. (1995). Epidemiology of Y. *pseudotuberculosis* in Germany, 1983–1993. Contrib. Microbiol. Immunol. *13*, 55–58.

von Altrock, A., Roesler, U., Merle, R., and Waldmann, K.H. (2010). Prevalence of pathogenic *Yersinia enterocolitica* strains on liver surfaces of pigs and their antimicrobial susceptibility. J. Food Prot. *73*, 1680–1683.

Ayyadurai, S., Flaudrops, C., Raoult, D., and Drancourt, M. (2010). Rapid identification and typing of *Yersinia pestis* and other *Yersinia* species by matrix-assisted laser desorption/ionization time-of-flight (MALDI-TOF) mass spectrometry. BMC Microbiol. *10*, 285.

Bachans, A., Fellström, C., and Lambertz, S.T. (2010). Occurrence of pathogenic *Yersinia enterocolitica* and *Yersinia pseudotuberculosis* in small wild rodents. Epidemiol. Infect. *139*, 1230–1238.

de Benito, I., Cano, M.E., Agüero, J., and García Lobo, J.M. (2004). A polymorphic tandem repeat potentially useful for typing in the chromosome of *Yersinia enterocolitica*. Microbiol. *150*, 199–204.

Bhaduri, S., and Wesley, I. (2006). Isolation and characterization of *Yersinia enterocolitica* from swine feces recovered during the National Animal Health Monitoring System Swine 2000 study. J. Food Protection *69*, 2107–2112.

Bhagat, N., and Virdi, J.S. (2011). The enigma of *Yersinia enterocolitica* biovar 1A. Crit. Rev. Microbiol. *37*, 25–39.

Blixt, Y., Knutsson, R., Borch, E., and Rådström, P. (2003). Interlaboratory random amplified polymorphic DNA typing of *Yersinia enterocolitica* and *Y. enterocolitica*-like bacteria. Int. J. Food Microbiol. *83*, 15–26.

Bogdanovich, T., Carniel, E., Fukushima, H., and Skurnik, M. (2003). Use of O-antigen gene cluster-specific PCRs for the identification and O-genotyping of *Yersinia pseudotuberculosis* and *Yersinia pestis*. J. Clin. Microbiol. *41*, 5103–5112.

Bonardi, S., Paris, A., Bassi, L., Salmi, F., Bacci, C., Riboldi, E., Boni, E., D'incau, M., Tagliabue, S., and Brindani, F. (2010). Detection, semiquantitative enumeration, and antimicrobial susceptibility of *Yersinia enterocolitica* in pork and chicken meats in Italy. J. Food Prot. *73*, 1785–1792.

Boqvist, S., Pettersson, H., Svensson, Å., and Andersson, Y. (2009). Sources of sporadic *Yersinia enterocolitica* infection in children in Sweden, 2004: A case–control study. Epidemiol. Infect. *137*, 897–905.

Bowman, A.S., Glendening, C., Wittum, T.E., LeJeune, J.T., Stich, R.W., and Funk, J.A. (2007). Prevalence of *Yersinia enterocolitica* in different phases of production on swine farms. J. Food Prot. *70*, 11–16.

Boyapalle, S., Wesley, I.V., Hurd, H.S., and Reddy, P.G. (2001). Comparison of culture, multiplex, and 5′ nuclease polymerase chain reaction assays for the rapid detection of *Yersinia enterocolitica* in swine and pork products. J. Food Prot. *64*, 1352–1361.

Bucher, M., Meyer, C., Grötzbach, B., Wacheck, S., Stolle, A., and Fredriksson-Ahomaa, M. (2008). Epidemiological data on pathogenic *Yersinia enterocolitica* in Southern Germany during 2000–2006. Foodborne Pathog. Dis. *5*, 273–280.

Capilla, S., Goñi, P., Rubio, M.C., Castillo, J., Millán, L., Cerdá, P., Sahagún, J., Pitart, C., Beltrán, A., and Gómez-Lus, R. (2003). Epidemiological study of resistance to nalidixic acid and other antibiotics in clinical *Yersinia enterocolitica* O:3 isolates. J. Clin. Microbiol. *41*, 4876–4878.

Capilla, S., Ruiz, J., Goñi, P., Rubio, M.C., Jiménez de Anta, M.T., Gómez-Lus, R., and Vila, J. (2004). Characterization of the molecular mechanisms of quinolone resistance in *Yersinia enterocolitica* O:3 clinical isolates. J. Antimicrob. Chemother. *53*, 1068–1071.

Carniel, E. (2001). The *Yersinia* high-pathogenicity island: An iron-uptake island. Microb. Infect. *3*, 561–569.

Cheyne, B.M., Van Dyke, M.I., Anderson, W.B., and Huck, P.M. (2010). The detection of *Yersinia enterocolitica* in surface water by quantitative PCR amplification of the *ail* and *yadA* genes. J. Water Health *8*, 487–499.

Ch'Ng, S.L., Octavia, S., Xia, Q., Duong, A., Tanaka, M.M., Fukushima, H., and Lan, R. (2011). Population structure and evolution of pathogenicity of *Yersinia pseudotuberculosis*. Appl. Environ. Microbiol. *77*, 768–775.

Cornelis, G.R. (2002). *Yersinia* type III secretion: Send in the effectors. J. Cell Biol. *158*, 401–408.

Cornelis, G.R., Boland A., Boyd, A.P., Geuijen, C., Iriarte, M., Neyt, C., Sory, M.P., and Stainier, I. (1998). The virulence plasmid of *Yersinia*, an antihost genome. Microbiol. Mol. Biol. Rev. *62*, 1315.

Cover, T.L., and Aber, R.C. (1989). *Yersinia enterocolitica*. N. Engl. J. Med. *321*, 16–24.

Ewing, W.H., Ross, A.J., Brenner, D.J., and Fanning, G.R. (1978). *Yersinia ruckeri* sp. nov, redmouth (Rm) bacterium. Int. J. Syst. Bacteriol. *28*, 37–44.

Falcão, J.P., Falcão, D.P., Pitondo-Silva, A., Malaspina, A.C., and Brocchi, M. (2006). Molecular typing and virulence markers of *Yersinia enterocolitica* strains from human, animal and food origins isolated between 1968 and 2000 in Brazil. J. Med. Microbiol. *55*, 1539–1548.

Farzan, A., Friendship, R.M., Cook, A., and Pollari, F. (2010). Occurrence of *Salmonella*, *Campylobacter*, *Yersinia enterocolitica*, *Escherichia coli* O157 and *Listeria monocytogenes* in swine. Zoonoses Public Health *57*, 388–396.

Fearnley, C., On, S.L.W., Kokotovic, B., Manning, G., Cheasty, T., and Newell, D.G. (2005). Application of fluorescent amplified fragment length polymorphism for comparison of human and animal isolates of *Yersinia enterocolitica*. Appl. Environ. Microbiol. *71*, 4960–4965.

Fredriksson-Ahomaa, M. (2001). Prevalence of pathogenic *Yersinia enterocolitica* in food samples at retail level in Finland. Arch. Lebensmittelhyg. *52*, 66–68.

Fredriksson-Ahomaa, M. (2009). Epidemiology of human *Yersinia pseudotuberculosis* infection. Arch. Lebensmittelhyg. *60*, 82–87.

Fredriksson-Ahomaa, M., and Korkeala, H. (2003). Low occurrence of pathogenic *Yersinia enterocolitica* in clinical, food, and environmental samples: A methodological problem. Clin. Microbiol. Rev. *16*, 220–229.

Fredriksson-Ahomaa, M., Autio, T., and Korkeala, H. (1999). Efficient subtyping of *Yersinia enterocolitica* bioserotype 4/O:3 with pulsed-field gel electrophoresis. Lett. Appl. Microbiol. *29*, 308–312.

Fredriksson-Ahomaa, M., Korte, T., and Korkeala, H. (2000). Contamination of carcasses, offals, and the environment with *yadA*-positive *Yersinia enterocolitica* in a pig slaughterhouse. J. Food Prot. *63*, 31–35.

Fredriksson-Ahomaa, M., Korte, T., and Korkeala, H. (2001). Transmission of *Yersinia enterocolitica* 4/O:3 to pets via contaminated pork. Lett. Appl. Microbiol. *32*, 375–378.

Fredriksson-Ahomaa, M., Stolle, A., and Korkeala, H. (2006a). Molecular epidemiology of *Yersinia enterocolitica* infections. FEMS Immunol. Med. Microbiol. *47*, 315–329.

Fredriksson-Ahomaa, M., Stolle, A., Siitonen, A., and Korkeala, H. (2006b). Sporadic human *Yersinia enterocolitica* infections caused by bioserotype 4/O:3 originate mainly from pigs. J. Med. Microbiol. *55*, 747–749.

Fredriksson-Ahomaa, M., Stolle, A., and Stephan, R. (2007). Prevalence of pathogenic *Yersinia enterocolitica* in pigs slaughtered at a Swiss abattoir. Int. J. Food Microbiol. *119*, 207–212.

Fredriksson-Ahomaa, M., Wacheck, S., Koenig, M., Stolle, A., and Stephan, R. (2009). Prevalence of pathogenic *Yersinia enterocolitica* and *Yersinia pseudotuberculosis* in wild boars in Switzerland. Int. J. Food Microbiol. *135*, 199–202.

Fredriksson-Ahomaa, M., Lindström, M., and Korkeala, H. (2010). *Yersinia enterocolitica* and *Yersinia pseudotuberculosis*. In Pathogens and Toxins in Foods: Challenger and Interventions, Juneja, V.K., and Sofos, N.J., eds, (ASM Press, Washington, DC), pp. 164–180.

Fukushima, H., and Gomyoda, M. (1991). Intestinal carriage of *Yersinia pseudotuberculosis* by wild birds and mammals in Japan. Appl. Environ. Microbiol. *57*, 1152–1155.

Fukushima, H., Nakamura, R., and Ito, Y. (1983). Ecological studies of *Yersinia enterocolitica*. I. Dissemination of *Y. enterocolitica* in pigs. Vet. Microbiol. *8*, 469–483.

Fukushima, H., Nakamura, R., and Iitsuka, S. (1985). Presence of zoonotic pathogens (*Yersinia* spp., *Campylobacter jejuni*, *Salmonella* spp., and *Leptospira* spp.) simultaneously in dogs and cats. Zentralbl. Bakteriol. Mikrobiol. Hyg. *181*, 430–440.

Fukushima, H., Gomyoda, M., Shiozawa, K., Kaneko, S., and Tsubokura, M. (1988). *Yersinia pseudotuberculosis* infection contracted through water contaminated by a wild animal. J. Clin. Microbiol. *26*, 584–585.

Fukushima, H., Gomyoda, M., Kaneko, S., Tsubokura, M., Takeda, N., Hongo, T., and Shubin, F.N. (1994). Restriction endonuclease analysis of virulence plasmids for molecular epidemiology of *Yersinia pseudotuberculosis* infections. J. Clin. Microbiol. *32*, 1410–1413.

Fukushima, H., Hoshina, K., Itogawa, H., and Gomyoda, M. (1997). Introduction into Japan of pathogenic *Yersinia* through imported pork, beef and fowl. Int. J. Food Microbiol. *35*, 205–212.

Fukushima, H., Gomyoda, M., Hashimoto, N., Takashima, I., Shubin, F.N., Isachikova, L.M., Paik, I.K., and Zheng, X.B. (1998). Putative origin of *Yersinia pseudotuberculosis* in Western and Eastern countries. A comparison of restriction endonuclease analysis of virulence plasmids. Zentralbl. Bakteriol. *288*, 93–102.

Fukushima, H., Matsuda, Y., Seki, R., Tsubokura, M., Takeda, N., Shubin, F.N., Paik, I.K., and Zheng, X.B. (2001). Geographical heterogeneity between Far Eastern and western countries in prevalence of the virulence plasmid, the superantigen *Yersinia pseudotuberculosis*-derived mitogen, and the high-pathogenicity island among *Yersinia pseudotuberculosis* strains. J. Clin. Microbiol. *39*, 3541–3547.

Gierczyński, R., Golubov, A., Neubauer, H., Pham, J.N., and Rakin, A. (2007). Development of multiple-locus variable-number tandem-repeat analysis for *Yersinia enterocolitica* subsp. *palearctica* and its application to bioserogroup 4/O3 subtyping. J. Clin. Microbiol. *45*, 2508–2515.

Gierczyński, R., Szych, J., Rastawicki, W., Wardak, S., and Jagielski, M. (2009). Molecular characterization of human clinical isolates of *Yersinia enterocolitica* bioserotype 1B/O8 in Poland: emergence and dissemination of three highly related clones. J. Clin. Microbiol. *47*, 1225–1228.

Gourdon, F., Beytout, J., Reynaud, A., Romaszko, J., Perre, D., Theodore, P., Soubelet, H., and Sirot, J. (1999). Human and animal epidemic of *Yersinia enterocolitica* O:9, 1989–1997, Auvergne, France. Emerg. Inf. Dis. *5*, 719–721.

Grahek-Ogden, D., Schimmer, B., Cudjoe, K.S., Nygård, K., and Kapperud, G. (2007). Outbreak of *Yersinia enterocolitica* serogroup O:9 infection and processed pork, Norway. Emerg. Infect. Dis. *13*, 754–756.

Granfors, K., Jalkanen, S., Von Essen, R., Lahesmaa-Rantala, R., Isomaki, O., Pekkola-Heino, K., Merilahti-Palo, R., Saario, R., Isomaki, H., and Toivanen, A. (1989). *Yersinia* antigens in synovial-fluid cells from patients with reactive arthritis. N. Engl. J. Med. *320*, 216–221.

Gulati, P., Varshney, R.K., and Virdi, J.S. (2009). Multilocus variable number tandem repeat analysis as a tool to discern genetic relationships among strains of *Yersinia enterocolitica* biovar 1A. J. Appl. Microbiol. *107*, 875–884.

Gürtler, M., Alter, T., Kasimir, S., Linnebur, M., and Fehlhaber, K. (2005). Prevalence of *Yersinia enterocolitica* in fattening pigs. J. Food Prot. *68*, 850–854.

Hallanvuo, S., Peltola, J., Heiskanen, T., and Siitonen, A. (2006). Simplified phenotypic scheme evaluated by 16S rRNA sequencing for differentiation between *Yersinia enterocolitica* and *Y. enterocolitica*-like species. J. Clin. Microbiol. *44*, 1077–1080.

Han, T.H., Paik, I.K., and Kim, S.J. (2003). Molecular relatedness between isolates *Yersinia pseudotuberculosis* from a patient and an isolate from mountain spring water. J. Korean Med. Sci. *18*, 425–428.

Hannu, T., Mattila, L., Nuorti, J.P., Ruutu, P., Mikkola, J., Siitonen, A., and Leirisalo-Repo, M. (2003). Reactive arthritis after an outbreak of *Yersinia pseudotuberculosis* serotype O:3 infection. Ann. Rheum. Dis. *62*, 866–869.

Howard, S.L., Gaunt, M.W., Hinds, J., Witney, A.A., Stabler, R., and Wren, B.W. (2006). Application of comparative phylogenomics to study the evolution of *Yersinia enterocolitica* and to identify genetic differences relating to pathogenicity. J. Bacteriol. *188*, 3645–3653.

Hudson, J.A., King, N.J., Cornelius, A.J., Bigwood, T., Thom, K., and Monson, S. (2008). Detection, isolation and enumeration of *Yersinia enterocolitica* from raw pork. Int. J. Food Microbiol. *123*, 25–31.

Huovinen, E., Sihvonen, L.M., Virtanen, M.J., Haukka, K., Siitonen, A., and Kuusi, M. (2010). Symptoms and sources of *Yersinia enterocolitica*-infection: A case–control study. BMC Inf. Dis. *10*, 122.

Hurst, M.R., Becher, S.A., Young, S.D., Nelson, T.L., and Glare, T.R. (2011). *Yersinia entomophaga* sp. nov. isolated from the New Zealand grass grub *Costelytra zealanfica*. Int. J. Syst. Evol. Microbiol. 61, 844–849.

Isberg, R.R., and Barnes, P. (2001). Subversion of integrins by enteropathogenic *Yersinia*. J. Cell. Sci. *114*, 21–28.

Iteman, I., Guiyoule, A., and Carniel, E. (1996). Comparison of three molecular methods for typing and subtyping pathogenic *Yersinia enterocolitica* strains. J. Med. Microbiol. 45, 48–56.

Iwata, T., Une, Y., Okatani, A.T., Kato, Y., Nakadai, A., Lee, K., Watanabe, M., Taniguchi, T., Elhelaly, A.E., Hirota, Y., *et al.* (2008). Virulence characteristics of *Yersinia pseudotuberculosis* isolated from breeding monkeys in Japan. Vet. Microbiol. *129*, 404–409.

Jacobsen, N.R., Bogdanovich, T., Skurnik, M., Lübeck, P.S., Ahrens, P., and Hoorfar, J. (2005). A real-time PCR assay for the specific identification of serotype O:9 of Yersinia enterocolitica. J. Microbiol. Methods *63*, 151–156.

Jalava, K., Hallanvuo, S., Nakari, U., Ruutu, P., Kela, E., Heinäsmäki, T., Siitonen, A., and Nuorti, J.P. (2004). Multiple outbreaks of *Yersinia pseudotuberculosis* infections in Finland. J. Clin. Microbiol. *42*, 2789–2791.

Jalava, K., Hakkinen, M., Valkonen, M., Nakari, U., Palo, T., Hallanvuo, S., Ollgren, J., Siitonen, A., and Nuorti, J.P. (2006). An outbreak of gastrointestinal illness and erythema nodosum from grated carrots contaminated with *Yersinia pseudotuberculosis*. J. Infect. Dis. *194*, 1209–1216.

Jerrett, I.V., Slee, K.J., and Robertson, B.I. (1990). Yersiniosis in farmed deer. Aust. Vet. J. *67*, 212–214.

Johannessen, G.S., Kapperud, G., and Kruse, H. (2000). Occurrence of pathogenic *Yersinia enterocolitica* in Norwegian pork products determined by a PCR method and a traditional culturing method. Int. J. Food Microbiol. *54*, 75–80.

Johannessen, G.S., Loncarevic, S., and Kruse, H. (2002). Bacteriological analysis of fresh produce in Norway. Int. J. Food Microbiol. 77, 199–204.

Jones, T.F., Buckingham, S.C., Bopp, C.A., Ribot, E., and Schaffner, W. (2003). From pig to pacifier: Chitterling-associated Yersiniosis outbreak among black infants. Emerg. Inf. Dis. *9*, 1007–1009.

Kageyama, T., Ogasawara, A., Fukuhara, R., Narita, Y., Miwa, N., Kamanaka, Y., Abe, M., Kumazaki, K., Maeda, N., Suzuki, J., *et al.* (2002). *Yersinia pseudotuberculosis* infection in breeding monkeys: Detection and analysis of strain diversity by PCR. J. Med. Primatol. *31*, 129–135.

Kangas, S., Takkinen, J., Hakkinen, M., Nakari, U., Johansson, T., Henttonen, H., Virtaluoto, L., Siitonen, A., Ollgren, J., and Kuusi, M. (2008). *Yersinia pseudotuberculosis* O:1 traced to raw carrots, Finland. Emerg. Inf. Dis. *14*, 1959–1961.

Kapperud, G., Nesbakken, T., Aleksic, S., and Mollaret, H.H. (1990). Comparison of restriction endonuclease analysis and phenotypic typing methods of differentiation of *Yersinia enterocolitica* isolates. J. Clin. Microbiol. 28, 1125–1131.

Kim, W., Song, M., Song, W., Kim, K., Chung, S., Choi, C., and Park, Y. (2003). Comparison of 16S rDNA analysis and rep-PCR genomic fingerprinting for molecular identification of *Yersinia pseudotuberculosis*. Int. J. Gen. Mol. Microbiol. 83, 125–133.

Korte, T., Fredriksson-Ahomaa, M., Niskanen, T., and Korkeala, H. (2004). Low prevalence of *yadA*-positive *Yersinia enterocolitica* in sows. Foodborne Pathog. Dis. *1*, 45–52.

Kotetishvili, M., Kreger, A., Wauters, G., Morris, J.G., Sulakvelidze, A., and Stine, O.C. (2005) Multilocus sequence typing for studying genetic relationships among *Yersinia* species. J. Clin. Microbiol. *43*, 2674–2684.

Kuehni-Boghenbor, K., On, S.L.W., Kokotovic, B., Baumgartner, A., Wassenaar, T.M., Wittwer, M., Bissig-Choisat, B., and Frey, J. (2006). Genotyping of human and porcine *Yersinia enterocolitica*, *Yersinia intermedia*, and *Yersinia bercovieri* strains from Switzerland by amplified fragment length polymorphism analysis. Appl. Environ. Microbiol. 72, 4061–4066.

Kuhm, A.E., Suter, D., Felleisen, R., and Rau, J. (2009). Identification of *Yersinia enterocolitica* at the species and subspecies levels by fourier transform infrared spectroscopy. Appl. Environ. Microbiol. *75*, 5809–5813.

Lambertz, S.T., Granath, K., Fredriksson-Ahomaa, M., Johansson, K., and Danielsson-Tham, M. (2007). Evaluation of a combined culture and PCR method (NMKL-163A) for detection of presumptive pathogenic *Yersinia enterocolitica* in pork products. J. Food Prot. 70, 335–340.

Lamps, L.W., Havens, J.M., Gilbrech, L.J., Dube, P.H., and Scott, M.A. (2006). Molecular biogrouping of pathogenic *Yersinia enterocolitica*: Development of a diagnostic PCR assay with histologic correlation. Am. J. Clin. Pathol. *125*, 658–664.

Lasch, P., Drevinek, M., Nattermann, H., Grunow, R., Stämmler, M., Dieckmann, R., Schwecke, T., and Naumann, D. (2010). Characterization of *Yersinia* using MALDI-TOF mass spectrometry and chemometrics. Anal. Chem. *82*, 8464–8475.

Laukkanen, R., Niskanen, T., Fredriksson-Ahomaa, M., and Korkeala, H. (2003). *Yersinia pseudotuberculosis* in pigs and pig houses in Finland. Adv. Exp. Med. Biol. *529*, 371–373.

Laukkanen, R., Martinez, P.O., Siekkinen, K., Ranta, J., Maijala, R., and Korkeala, H. (2008). Transmission of *Yersinia pseudotuberculosis* in the pork production chain from farm to slaughterhouse. Appl. Environ. Microbiol. *74*, 5444–5450.

Laukkanen, R., Martinez, P.O., Siekkinen, K.M., Ranta, J., Maijala, R., and Korkeala, H. (2009). Contamination of carcasses with human pathogenic *Yersinia enterocolitica* 4/O:3 originates from pigs infected on farms. Foodborne Pathog. Dis. *6*, 681–688.

Laukkanen, R., Ranta, J., Dong, X., Hakkinen, M., Martínez, P.O., Lundén, J., Johansson, T., and Korkeala, H. (2010). Reduction of enteropathogenic *Yersinia* in the pig slaughterhouse by using bagging of the rectum. J. Food Prot. *73*, 2161–2168.

Laukkenen-Ninios, R., Didelot, X., Jolley, K.A., Morelli, G., Sangal, V., Kristo, P., Brehony, C., Imori, P.F., Fukushima, H., Siitonen, A., *et al.* (2011). Population structure of the *Yersinia pseudotuberculosis* complex according to multilocus sequence typing. Environ. Microbiol. *13*, 3114–3127.

Lee, L.A., Gerber, A.R., Lonsway, D.R., Smith, J.D., Carter, G.P., Puhr, N.D., Parrish, C.M., Sikes, R.K., Finton, R.J., and Tauxe, R.V. (1990). *Yersinia enterocolitica* O:3 infections in infants and children, associated with the household preparation of chitterlings. N. Engl. J. Med. *322*, 984–987.

Lindblad, M., Lindmark, H., Lambertz, S.T., and Lindqvist, R. (2006). Microbiological baseline study of broiler chickens at Swedish slaughterhouses. J. Food Prot. *69*, 2875–2882.

Lindblad, M., Lindmark, H., Thisted, Lambertz, S., and Lindqvist, R. (2007). Microbiological baseline study of swine carcasses at Swedish slaughterhouses. J. Food Prot. *70*, 1790–1797.

Logue, C.M., Sheridan, J.J., Wauters, G., Mc Dowell, D.A., and Blair, I.S. (1996). *Yersinia* spp. and numbers, with particular reference to *Y. enterocolitica* bio/serotypes, occurring on Irish meat and meat products, and the influence of alkali treatment on their isolation. Int. J. Food Microbiol. *33*, 257–274.

Long, C., Jones, T.F., Vugia, D.J., Scheftel, J., Strockbine, N., Ryan, P., Shiferaw, B., Tauxe, R.V., and Gould, L.H. (2010). *Yersinia pseudotuberculosis* and *Y. enterocolitica* infections, FoodNet, 1996-2007. Emerg. Inf. Dis. *16*, 566–567.

Lucero Estrada, C.S.M., del Carmen Velazquez, L., Esther Escudero, M., Isabel Favier, G., Lazarte, V., and Maria Stefanini de Guzman, A. (2011). Pulsed field, PCR ribotyping and multiplex PCR analysis of *Yersinia enterocolitica* strains isolated from meat food in San Luis Argentina. Food Microbiol. *28*, 21–28.

Makino, S., Okada, Y., Maruyama, T., Kaneko, S., and Sasakawa, C. (1994). PCR-based random amplified polymorphic DNA fingerprinting of *Yersinia pseudotuberculosis* and its practical applications. J. Clin. Microbiol. *32*, 65–69.

Martínez, P.O., Fredriksson-Ahomaa, M., Sokolova, Y., Roasto, M., Berzins, A., and Korkeala, H. (2009). Prevalence of enteropathogenic *Yersinia* in Estonian, Latvian, and Russian (Leningrad Region) pigs. Foodborne Pathog. Dis. *6*, 719–724.

Martínez, P.O., Fredriksson-Ahomaa, M., Pallotti, A., Rosmini, R., Houf, K., and Korkeala, H. (2011). Variation in the prevalence of enteropathogenic *Yersinia* in slaughter pigs from Belgium, Italy and Spain. Foodborne Pathog. Dis. *8*, 445–450.

Martins, C.H.G., Bauab, T.M., and Falcao, D.P. (1998). Characteristics of *Yersinia pseudotuberculosis* isolated from animals in Brazil. J. Appl. Microbiol. *85*, 703–707.

Matsumoto, H., and Young, G.M. (2009). Translocated effectors of *Yersinia*. Curr. Opin. Microbiol. *12*, 94–100.

Merhej, V., Adekambi, T., Pagnier, I., Raoult, D., and Drancourt, M. (2008). *Yersinia massiliensis* sp nov., isolated from fresh water. Int. J. Syst. Evol. Microbiol. *58*, 779–784.

Messelhäusser, U., Kämpf, P., Colditz, J., Bauer, H., Schreiner, H., Höller, C., and Busch, U. (2011). Qualitative and quantitative detection of human pathogenic *Yersinia enterocolitica* in different food matrices at retail level in Bavaria. Foodborne Pathog. Dis. *8*, 39–44.

Miller, V.L., Beer, K.B., Heusipp, G., Young, B.M., and Wachtel, M.R. (2001). Identification of regions of Ail required for the invasion and serum resistance phenotypes. Mol. Microbiol. *41*, 1053–1062.

Milnes, A.S., Stewart, I., Clifton-Hadley, F.A., Davies, R.H., Newell, D.G., Sayers, A.R., Cheasty, T., Cassar, C., Ridley, A., Cook, A.J.C., *et al.* (2008). Intestinal carriage of verocytotoxigenic *Escherichia coli* O157, *Salmonella*, thermophilic *Campylobacter* and *Yersinia enterocolitica*, in cattle, sheep and pigs at slaughter in Great Britain during 2003. Epidemiol. Infect. *136*, 739–751.

Morelli, G., Song, Y., Mazzoni, C.J., Eppinger, M., Roumagnac, P., Wagner, D.M., Feldkamp, M., Kusecek, B., Vogler, A.J., Li, Y., *et al.* (2010). *Yersinia pestis* genome sequencing identifies patterns of global phylogenetic diversity. Nat. Genet. *42*, 1140–1143.

Moriki, S., Nobata, A., Shibata, H., Nagai, A., Minami, N., Taketani, T., and Fukushima, H. (2010). Familial outbreak of *Yersinia enterocolitica* serotype O9 biotype 2. J. Inf. Chemother. *16*, 56–58.

Murros-Kontiainen, A.E., Fredriksson-Ahomaa, M., Korkeala, H., Johansson, P., Rahkila, R., and Björkroth, J. (2011a) *Yersinia nurmii* sp. nov. Int. J. Syst. Evol. Microbiol. *61*, 2368–2372.

Murros-Kontiainen, A.E., Johansson, P., Niskanen, T., Fredriksson-Ahomaa, M., Korkeala, H., and Björkroth, J. (2011b) *Yersinia pekkanenii* sp. nov. Int. J. Syst. Evol. Microbiol. *61*, 2363–2367.

Nagano, T., Ichimura, K., Haji, N., Nagao, K., Someya, K., Kiyohara, T., Suzuki, K., Tsubokura, M., and Otsuki, K. (1997). Characteristics and pathogenicity of non-melibiose-fermenting strains of *Yersinia pseudotuberculosis* O3. Microbiol. Immunol. *41*, 175–183.

Najdenski, H., Iteman, I., and Carniel, E. (1994). Efficient subtyping of pathogenic *Yersinia enterocolitica* strains by pulsed-field gel electrophoresis. J. Clin. Microbiol. *32*, 2913–2920.

Nesbakken, T., Nerbrink, E., Rotterud, O., and Borch, E. (1994). Reduction of *Yersinia entercolitica* and *Listeria* spp. on pig carcasses by enclosure of the rectum during slaughter. Int. J. Food Microbiol. 23, 197–208.

Nesbakken, T., Iversen, T., Eckner, K., and Lium, B. (2006). Testing of pathogenic *Yersinia enterocolitica* in pig herds based on the natural dynamic of infection. Int. J. Food Microbiol. *111*, 99–104.

Neubauer, H., Aleksic, S., Hensel, A., Finke, E.J., and Meyer, H. (2000). *Yersinia enterocolitica* 16S rRNA gene types belong to the same genospecies but form three homology groups. Int. J. Med. Microbiol. *290*, 61–64.

Nielsen, B., Heisel, C., and Wingstrand, A. (1996). Time course of the serological response to *Yersinia enterocolitica* O:3 in experimentally infected pigs. Vet. Microbiol. *48*, 293–303.

Niskanen, T., Fredriksson-Ahomaa, M., and Korkeala, H. (2002). *Yersinia pseudotuberculosis* with limited genetic diversity is a common finding in tonsils of fattening pigs. J. Food Prot. *65*, 540–545.

Niskanen, T., Laukkanen, R., Murros, A., Björkroth, J., Skurnik, M., Korkeala, H., and Fredriksson-Ahomaa, M. (2009). Characterisation of non-pathogenic *Yersinia pseudotuberculosis*-like strains isolated from food and environmental samples. Int. J. Food Microbiol. *129*, 150–156.

Nowak, B., Mueffling, T.V., Caspari, K., and Hartung, J. (2006). Validation of a method for the detection of virulent *Yersinia enterocolitica* and their distribution in slaughter pigs from conventional and alternative housing systems. Vet. Microbiol. *117*, 219–228.

Nowgesic, E., Fyfe, M., Hockin, J., King, A., Ng, H., Paccagnella, A., Trinidad, A., Wilcott, L., Smith, R., Denney, A., *et al.* (1999). Outbreak of *Yersinia pseudotuberculosis* in British Columbia – November 1998. Can. Commun. Dis. Rep. *25*, 97–100.

Nuorti, J.P., Niskanen, T., Hallanvuo, S., Mikkola, J., Kela, E., Hatakka, M., Fredriksson-Ahomaa, M., Lyytikäinen, O., Siitonen, A., Korkeala, H., *et al.* (2004). A widespread outbreak of *Yersinia pseudotuberculosis* O:3 infection from iceberg lettuce. J. Infect. Dis. *189*, 766–774.

Odaert, M., Berche, P., and Simonet, M. (1996). Molecular typing of *Yersinia pseudotuberculosis* by using an IS200-like element. J. Clin. Microbiol. *34*, 2231–2235.

Odinot, P.T., Meis, J.F.G.M., Van den Hurk, P.J.J.C., Hoogkamp-Korstanje, J.A.A., and Melchers, W.J.G. (1995). PCR-based characterization of *Yersinia enterocolitica*: Comparison with biotyping and serotyping. Epidemiol. Infect. *115*, 269–277.

Okwori, A.E.J., Martínez, P.O., Fredriksson-Ahomaa, M., Agina, S.E., and Korkeala, H. (2009). Pathogenic *Yersinia enterocolitica* 2/O:9 and *Yersinia pseudotuberculosis* 1/O:1 strains isolated from human and non-human sources in the Plateau State of Nigeria. Food Microbiol. *26*, 872–875.

Ortiz Martínez, P., Mylona, S., Drake, I., Fredriksson-Ahomaa, M., Korkeala, H., and Corry, J.E.L (2010). Wide variety of bioserotypes of enteropathogenic *Yersinia* in tonsils of English pigs at slaughter. Int. J. Food Microbiol. *139*, 64–69.

Ostroff, S.M., Kapperud, G., Hutwagner, L.C., Nesbakken, T., Bean, N.H., Lassen, J., and Tauxe, R.V. (1994). Sources of sporadic *Yersinia enterocolitica* infections in Norway: A prospective case–control study. Epidemiol. Infect. *112*, 133–141.

Pilon, J., Higgins, R., and Quessy, S. (2000). Epidemiological study of *Yersinia enterocolitica* in swine herds in Québec. Can. Vet. J. *41*, 383–387.

Pingulkar, K., Kamat, A., and Bongirwar, D. (2001). Microbiological quality of fresh leafy vegetables, salad components and ready-to-eat salads: An evidence of inhibition of *Listeria monocytogenes* in tomatoes. Int. J. Food Sci. Nutr. *52*, 15–23.

Poljak, Z., Dewey, C.E., Martin, S.W., Rosendal, T., Christensen, J., Ciebin, B., and Friendship, R.M. (2010). Prevalence of *Yersinia enterocolitica* shedding and bioserotype distribution in Ontario finisher pig herds in 2001, 2002, and 2004. Prev. Vet. Med. *93*, 110–120.

Ranta, J., Siekkinen, K., Nuotio, L., Laukkanen, R., Hellström, S., Korkeala, H., and Maijala, R. (2010). Causal hidden variable model of pathogenic contamination from pig to pork. Stat. Model. *10*, 69–87.

Rasmussen, H.N., Olsen, J.E., and Rasmussen, O.F. (1994). RAPD analysis of *Yersinia enterocolitica*. Lett. Appl. Microbiol. *19*, 359–362.

Rastawicki, W., Szych, J., Gierczyński, R., and Rokosz, N. (2009). A dramatic increase of *Yersinia enterocolitica* serogroup O:8 infections in Poland. Eur. J. Clin. Microbiol. Inf. Dis. *28*, 535–537.

Rimhanen-Finne, R., Niskanen, T., Hallanvuo, S., Makary, P., Haukka, K., Pajunen, S., Siitonen, A., Ristolainen, R., Pöyry, H., Ollgren, J., *et al.* (2009). *Yersinia pseudotuberculosis* causing a large outbreak associated with carrots in Finland, 2006. Epidemiol. Infect. *137*, 342–347.

Rosner, B.M., Stark, K., and Werber, D. (2010). Epidemiology of reported *Yersinia enterocolitica* infections in Germany, 2001–2008. BMC Public Health *10*, 337.

Sakai, T., Nakayama, A., Hashida, M., Yamamoto, Y., Takebe, H., and Imai, S. (2005). Outbreak of food poisoning by *Yersinia enterocolitica* serotype O8 in Nara Prefecture: The first case report in Japan. Jpn. J. Infect. Dis. *58*, 257–258.

Sandery, M., Stinear, T., and Kaucner, C. (1996). Detection of pathogenic *Yersinia enterocolitica* in environmental waters by PCR. J. Appl. Bacteriol. *80*, 327–332.

Satterthwaite, P., Pritchard, K., Floyd, D., and Law, B. (1999). A case–control study of *Yersinia enterocolitica* infections in Auckland. Aust. N.Z. J. Public Health 23, 482–485.

Sen, K. (2000). Rapid identification of *Yersinia enterocolitica* in blood by the 5′ nuclease PCR assay. J. Clin. Microbiol. 38, 1953–1958.

Seuri, M., and Granfors, K. (1992). Possible confounders of the relationship between occupational swine contact and *Yersinia enterocolitica* 0:3 and 0:9 antibodies Eur. J. Epidemiol. *8*, 532–538.

Sihvonen, L.M., Haukka, K., Kuusi, M., Virtanen, M.J., and Siitonen, A. (2009). *Yersinia enterocolitica* and *Y. enterocolitica*-like species in clinical stool specimens of humans: Identification and prevalence of bio/

serotypes in Finland. Eur. J. Clin. Microbiol. Inf. Dis. *28*, 757–765.

Sihvonen, L.M., Hallanvuo, S., Haukka, K., Skurnik, M., and Siitonen, A. (2011a). The *ail* gene is present in some *Yersinia enterocolitica* biotupe 1A strains. Foodborne Pathog. Dis. *8*, 455–457.

Sihvonen, L.M., Toivonen, S., Haukka, K., Kuusi, M., Skurnik, M., and Siitonen, A. (2011b). Multilocus variable-number tandem-repeat analysis, pulsed-field gel electrophoresis, and antimicrobial susceptibility patterns in discrimination of sporadic and outbreak-related strains of *Yersinia enterocolitica*. BMC Microbiol. *11*, 42.

Skjerve, E., Lium, B., Nielsen, B., and Nesbakken, T. (1998). Control of *Yersinia enterocolitica* in pigs at herd level. Int. J. Food Microbiol. *45*, 195–203.

Slee, K.J., and Skilbeck, N.W. (1992). Epidemiology of *Yersinia pseudotuberculosis* and *Y. enterocolitica* infections in sheep in Australia. J. Clin. Microbiol. *30*, 712–715.

Souza, R.A., Pitondo-Silva, A., Falcão, D.P., and Falcão, J.P. (2010). Evaluation of four molecular typing methodologies as tools for determining taxonomy relations and for identifying species among *Yersinia* isolates. J. Microbiol. Meth. *82*, 141–150.

Sprague, L.D., and Neubauer, H. (2005). *Yersinia aleksiciae* sp nov. Int. J. Syst. Evol. Microbiol. *55*, 831–835.

Sprague, L.D., Scholz, H.C., Amann, S., Busse, H.J., and Neubauer, H. (2008). *Yersinia similis* sp. nov. Int. J. Syst. Evol. Microbiol. *58*, 952–958.

Stock, I., and Wiedemann, B. (1999). An in-vitro study of the antimicrobial susceptibilities of *Yersinia enterocolitica* and the definition of a database. J. Antimicrob. Chemother. *43*, 37–45.

Sunahara, C., Yamanaka, Y., and Yamanishi, S. (2000). Sporadic cases of *Yersinia pseudotuberculosis* serotype 5 infection in Shodo Island, Kagawa Prefecture. Jpn. J. Infect. Dis. *53*, 74–75.

Tauxe, R.V., Vandepitte, J., and Wauters, G. (1987). *Yersinia enterocolitica* infections and pork: The missing link. Lancet *1*, 1129–1132.

Tennant, S.M., Grant, T.H., and Robins-Browne, R.M. (2003). Pathogenicity of *Yersinia enterocolitica* biotype 1A. FEMS Immunol. Med. Microbiol. *38*, 127–137.

Terentjeva, M., and Berzins, A. (2010). Prevalence and antimicrobial resistance of *Yersinia enterocolitica* and *Yersinia pseudotuberculosis* in slaughter pigs in Latvia. J. Food Prot. *73*, 1335–1338.

Thisted Lambertz, S., and Danielsson-Tham, M.L. (2005). Identification and characterization of pathogenic *Yersinia enterocolitica* isolates by PCR and pulsed-field gel electrophoresis. Appl. Environ. Microbiol. *71*, 3674–3681.

Thisted Lambertz, S., Nilsson, C., and Hallanvuo, S. (2008). TaqMan-based real-time PCR method for detection of *Yersinia pseudotuberculosis* in food. Appl. Environ. Microbiol. *74*, 6465–6469.

THL (National Institute for Health and Welfare). (2009). Outbreak of *Yersinia pseudotuberculosis* in Kajaani-grated carrots suspected as a vehicle of infections. Infectious diseases in Finland 2008. THL Report. *12*, 19.

Thomson, N.R., Howard, S., Wren, B.W., Holden, M.T.G., Crossman, L., Challis, G.L., Churcher, C., Mungall, K., Brooks, K., Chillingworth, T., *et al.* (2006). The complete genome sequence and comparative genome analysis of the high pathogenicity *Yersinia enterocolitica* strain 8081. PLoS Genetics 2, 2039–2051.

Tsubokura, M., and Aleksić, S. (1995). A simplified antigenic scheme for serotyping of *Yersinia pseudotuberculosis*: phenotypic characterization of reference strains and preparation of O and H factor sera. Contrib. Microbiol. Immunol. *13*, 99–105.

Tsubokura, M., Otsuki, K., Kawaoka, Y., and Maruyama, T. (1984). Characterization of pathogenicity of *Yersinia pseudotuberculosis* isolated from swine and other animals. J. Clin. Microbiol. *19*, 754–756.

Tsubokura, M., Otsuki, K., Sato, K., Tanaka, M., Hongo, T., Fukushima, H., Maruyama, T., and Inoue, M. (1989). Special features of distribution of *Yersinia pseudotuberculosis* in Japan. J. Clin. Microbiol. *27*, 790–791.

Van Damme, I., Habib, I., and De Zutter, L. (2010). *Yersinia enterocolitica* in slaughter pig tonsils: Enumeration and detection by enrichment versus direct plating culture. Food Microbiol. *27*, 158–161.

Virdi, J.S., and Sachdeva, P. (2005). Molecular heterogeneity in *Yersinia enterocolitica* and '*Y. enterocolitica*-like' species –Implications for epidemiology, typing and taxonomy. FEMS Immunol. Med. Microbiol. *45*, 1–10.

Virtanen, S.E., Salonen, L.K., Laukkanen, R., Hakkinen, M., and Korkeala, H. (2011). Factors related to the prevalence of pathogenic *Yersinia enterocolitica* on pig farms. Epidemiol. Infect. *139*, 1919–1927.

Vishnubhatla, A., Oberst, R.D., Fung, D.Y.C., Wonglumsom, W., Hays, M.P., and Nagaraja, T.G. (2001). Evaluation of a 5′-nuclease (TaqMan) assay for the detection of virulent strains of *Yersinia enterocolitica* in raw meat and tofu samples. J. Food Prot. *64*, 355–360.

Voskressenskaya, E., Leclercq, A., Tseneva, G., and Carniel, E. (2005). Evaluation of ribotyping as a tool for molecular typing of *Yersinia pseudotuberculosis* strains of worldwide origin. J. Clin. Microbiol. *43*, 6155–6160.

Voskressenskaya, E.A., Klimov, V.T., Tseneva, G.Y., Carniel, E., Foulon, J., and Chesnokova, M.V. (2003). Molecular epidemiological characterization of *Yersinia pseudotuberculosis* circulating in different geographic areas of the Russian Federation. Adv. Exp. Med. Biol. *529*, 391–394.

Wang, X., Cui, Z., Jin, D., Tang, L., Xia, S., Wang, H., Xiao, Y., Qiu, H., Hao, Q., Kan, B., *et al.* (2009). Distribution of pathogenic *Yersinia enterocolitica* in China. Eur. J. Clin. Microbiol. Inf. Dis. *28*, 1237–1244.

Wang, X., Cui, Z., Wang, H., Tang, L., Yang, J., Gu, L., Jin, D., Luo, L., Qiu, H., Xiao, Y., *et al.* (2010). Pathogenic strains of *Yersinia enterocolitica* isolated from domestic dogs (*Canis familiaris*) belonging to farmers are of the same subtype as pathogenic *Y. enterocolitica* strains isolated from humans and may be a source of human infection in Jiangsu Province, China. J. Clin. Microbiol. *48*, 1604–1610.

Wauters, G., Kandolo, K., and Janssens, M. (1987). Revised biogrouping scheme of *Yersinia enterocolitica*. Contrib. Microbiol. Immunol. *9*, 14–21.

Wauters, G., Aleksic, S., Charlier, J., and Schulze, G. (1991). Somatic and flagellar antigens of *Yersinia enterocolitica* and related species. Contrib. Microbiol. Immunol. *12*, 239–243.

Wesley, I.V., Bhaduri, S., and Bush, E. (2008). Prevalence of *Yersinia enterocolitica* in market weight hogs in the United States. J. Food Prot. *71*, 1162–1168.

Weynants, V., Jadot, V., Denoel, P.A., Tibor, A., and Letesson, J.J. (1996). Detection of *Yersinia enterocolitica* serogroup O:3 by a PCR method. J. Clin. Microbiol. *34*, 1224–1227.

Wojciech, Ł., Staroniewicz, Z., Jakubczak, A., and Ugorski, M. (2004). Typing of *Yersinia enterocolitica* isolates by ITS profiling, REP- and ERIC-PCR. J. Vet. Med. Ser. B Infect. Dis. Vet. Public Health *51*, 238–244.

Wren, B.W. (2003). The yersiniae – a model genus to study the rapid evolution of bacterial pathogens. Nat. Rev. Microbiol. *1*, 55–64.

Yang, Y.X., and Isberg, R.R. (1997). Transcriptional regulation of the *Yersinia pseudotuberculosis* pH 6 antigen adhesin by two envelope-associated components. Mol. Microbiol. *24*, 499–510.

Bacterial Factors Encoded by Mobile and Integrative Genetic Elements in Enteric Pathogens

14

E. Fidelma Boyd

Abstract

The acquisition of new phenotypes by bacteria is largely driven by horizontal gene transfer (HGT), a process that is ubiquitous among bacteria and universally present among enteric pathogens. The common vectors of HGT in enteric pathogens include phages, pathogenicity islands and plasmids, all genetic elements that can encode virulence factors essential for host colonization and infection. In this review, *Salmonella enterica*, *Escherichia coli*, *Vibrio cholerae* and *V. parahaemolyticus* are discussed in terms of their virulence genes encoded within mobile and integrative genetic elements (MIGE) and their role in pathogenesis.

Enteric pathogens and mobile and integrative genetic elements (MIGEs)

Within the *Gamma-Proteobacteria* are a large number of waterborne and foodborne enteric pathogens, the most well studied of which are *Escherichia coli* and *Salmonella enterica* and to a much lesser extent *Vibrio cholerae* and *V. parahaemolyticus*, even though these organisms affect a larger number of individuals world-wide. Each of these species contains isolates that can cause gastroenteritis; the key symptoms of which are diarrhoea and/or vomiting. The diarrhoea can be secretory, either bloody or watery, or inflammatory, which is clinically also associated with a high grade fever that is generally absent in patients with secretory diarrhoea. *Salmonella enterica* serovar Typhimurium exemplifies the causative agent of inflammatory diarrhoea and is an intracellular pathogen whereas *V. cholerae* serogroup O1 infection causes a profuse watery diarrhoea characterized by rice water stool and is the prototypical extracellular pathogen. *E. coli*, depending on the isolate, can cause a range of intestinal and extra-intestinal infections; to date eight different pathovars have been identified. These include the intestinal pathovars Enterohaemorrhagic *E. coli* (EHEC), enteropathogenic *E. coli* (EPEC), enteroinvasive *E. coli* (EIEC), enterotoxigenic *E. coli* (ETEC), enteroaggregative *E. coli* (EAEC) and diffusely adherent *E. coli* (DAEC) and the two extraintestinal pathovars that cause Neonatal Meningitis (NMEC) and urinary tract infections (UroPathogenicEC). EHEC, EPEC, ETEC, and EAEC are all extracellular pathogens that cause secretory diarrhoea in an infected host whereas EIEC similar to *Shigella* species is an intracellular pathogen that causes inflammatory diarrhoea. EHEC, EIEC and EAEC are associated with large intestinal infections whereas EPEC, ETEC, DAEC and EAEC are associated with infection of the small intestine. *E. coli* also is a commensal of the human gut without which we would not survive. So the question arises: how does an organism that is essential for proper intestinal function transform into one of the most deadly pathogens known to human kind? The answer to this question came before genome sequencing when researchers examined genome size variation among *E. coli* natural isolates and found that some *E. coli* isolates have approximately 1 million base pairs of DNA less than pathogenic isolates based on pulse field gel electrophoresis analysis (Bergthorsson and Ochman, 1995; Ochman and Bergthorsson, 1995). Subsequently, whole bacterial genome sequencing confirmed this initial finding and identified the reason for the discrepancies in genome size in these otherwise highly related strains (Blattner *et al.*, 1997; Hayashi *et al.*, 2001; Perna *et al.*, 2001; Welch *et al.*, 2002). Each *E. coli* pathovar encodes within its genome, large regions of chromosomal and extrachromosomal DNA that have been acquired by horizontal gene transfer, in most cases from an unknown source (Dobrindt *et al.*, 2010). This includes mobile and integrative genetic elements (MIGEs) such as plasmids, pathogenicity islands, phages, conjugative transposons and integrons (Fig. 14.1) (Stokes and Hall, 1989; Hacker *et al.*, 1990, 1997; Recchia and Hall, 1995; Pembroke *et al.*, 2002). Some of these MIGEs are unique to a particular pathovar while others are present in several different pathovars. In *S. enterica*, a bacterium that is prey for a number of single celled eukaryotes, contains numerous MIGEs important for intracellular survival and these tend to be present in most strains of the species. In *V. cholerae* MIGEs play an essential role in human pathogenesis and all the virulence factors required for cholera are encoded within MIGEs present in a limited number of isolates (Faruque and Mekalanos, 2003). The subject of this review is to discuss the virulence genes that have been acquired by subsets of strains that make them pathogenic.

Salmonella enterica: the causative agent of inflammatory diarrhoea

Salmonella enterica serotypes are associated with a wide range of acute infections, ranging from gastroenteritis (*S. enterica* serovar Typhimurium) to enteric fever (*S. enterica* serovar Typhi) (Tsolis

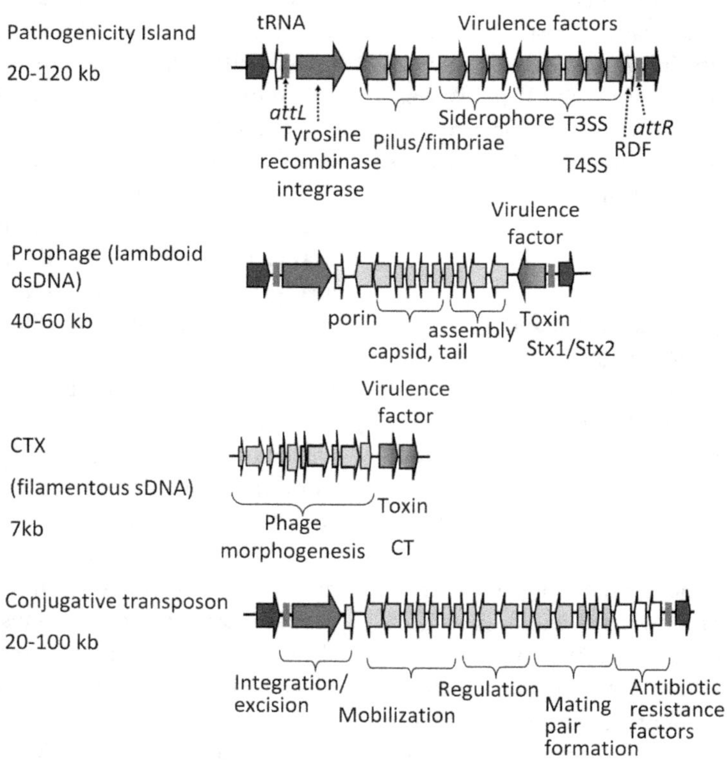

Figure 14.1 Schematic representation of MIGEs; Pathogenicity island, prophage (lambdoid like), CTX (filamentous phage), and a conjugative transposon or ICE (integrative and conjugative element). Block arrows represent genes encoding proteins. Black arrows represent core proteins presents in all strains of a species, these core genes flank MIGE regions acquired by a subset of strains within a species. Stripe arrows represent the integrase gene required for site specific recombination, *att*= attachment site, RDF=recombination directionality factor. Different type of genes are carried on each element.

et al., 2008). *S. enterica* is a facultative intracellular pathogen that is transmitted to humans via the food chain, leading to morbidity and mortality as well as cause severe economic losses owing to food recalls (Miller *et al.*, 2008). Nearly 1.5 million cases of *S. enterica* infections are recorded each year and cost billions in losses in the United States (Miller *et al.*, 2008).

The *Salmonella* genus contains two species, *Salmonella enterica* and *S. bongori* (formerly subspecies V), which are found in birds, reptiles and mammals (Reeves *et al.*, 1989; Popoff *et al.*, 2004; McQuiston *et al.*, 2008; Lan *et al.*, 2009). The species *S. enterica* is subdivided into seven subspecies: I, II, IIIa, IIIb, IV, VI and VII (Thampapillai *et al.*, 1994; Boyd *et al.*, 1996; McQuiston *et al.*, 2008); however, infections in warm-blooded animals are generally caused by isolates from subspecies I, encompassing more than 2300 serovars (Le Minor *et al.*, 1990; Popoff *et al.*, 2004; Porwollik *et al.*, 2004). Specific serovars cause variable disease symptoms in different hosts, which range from gastroenteritis to highly invasive diseases. For example *S. enterica* serovar Choleraesuis causes sepsis or extra-intestinal focal infection in humans and paratyphoid in swine. Whereas *S. enterica* serovars Typhi and Paratyphi C are highly adapted to humans causing typhoid, and a typhoid-like disease and recurrent intra-abdominal abscesses, respectively. *S. enterica* serovar Typhimurium, one of the most common causes of foodborne infections, is the causative agent of inflammatory diarrhoea.

The pathogenic mechanism of inflammatory diarrhoea differs from that of secretory bloody diarrhoea caused by pathovars of *E. coli* and secretory watery diarrhoea cause by *V. cholerae* reviewed later. In *S. enterica* serovar Typhimurium many of the key virulence genes are encoded on MIGE acquired by horizontal gene transfer and many of which have incorporated themselves into the host chromosome (Tables 14.1 and 14.2 and Fig. 14.2) (Groisman and Ochman, 1997; Porwollik *et al.*, 2002, 2004; Lan *et al.*, 2009). Five pathogenicity islands name *Salmonella* pathogenicity island 1 (SPI-1) to SPI-5 have been identified within the genome of Typhimurium isolates and range in size from 40 kb to 7 kb (Fig. 14.2) (Hensel *et al.*, 1995; Mills *et al.*, 1995; Groisman and Ochman, 1997; Wood *et al.*, 1998; Blanc-Potard *et al.*, 1999a; Porwollik *et al.*, 2002). In addition most Typhimurium isolates encode five different prophages named Gifsy-1, Gifsy-2, Fels-1, Fels-2 and sopEphi (Fig. 14.2) (Fang *et al.*, 1999; Figueroa-Bossi and Bossi, 1999; Figueroa-Bossi *et al.*, 2001; Porwollik *et al.*, 2002).

Pathogenicity islands (PAIs)

Pathogenicity Islands (PAIs) are a group of integrative elements that encode multiple virulence factors that confer an advantage to the bacterium within its target eukaryotic host (Hacker *et al.*, 1997; Hacker and Carniel, 2001). PAIs were first described in an *E. coli* uropathogenic strain 536 where two unstable regions termed PAI-1 and PAI-2 encoded multiple virulence factors (Blum *et al.*, 1994). Genome sequence data has revealed seven PAIs in *E. coli* 536 and several more PAIs among diarrhoeagenic strains Sakai, and EDL933 (Hayashi *et al.*, 2001; Perna *et al.*, 2001; Welch *et al.*, 2002; Dobrindt *et al.*, 2010). The key features of a PAI are that it consists of a large region (> 10 kb) of contiguous DNA that encodes multiple virulence factors and is present in pathogenic isolates and absent from non-pathogenic isolates. The DNA region encompassing the PAI should have a per cent guanine plus cytosine (G+C) content and/or codon bias that is significantly different from the host chromosome, a feature that is indicative of horizontal gene transfer. The PAI encodes an integrase or transposase that allows for integration into the host chromosome usually at a tRNA locus. The five PAIs that have been identified in serovar Typhimurium contain many of these features (Schmidt and Hensel, 2004). SPI-1 is a 40 kb region that is present in all *S. enterica* subspecies. The region is inserted at the *fhlA* gene and has a per cent GC content of 42% that is significantly different from the 52% GC content of the chromosome. The SPI-1 region appears to be conserved among all *S. enterica* and *S. bongori* isolates (Groisman and Ochman, 1996, 1997). SPI-1 encodes multiple genes involved in invasion and enteropathogenesis including a type III secretion system (T3SS) and effector proteins discussed below (Table 14.1) (Mills *et al.*, 1995; Galan, 1996). SPI-2 is also a 40-kb region present in all *S. enterica* subspecies and is inserted at the

Table 14.1 Pathogenicity island-encoded bacterial virulence factors among *S. enterica*, *E. coli*, *V. cholerae* and *V. parahaemolyticus*

Bacterial species	PAI	Proteins
E. coli EHEC O157:H7	PAI-I	Urease, tellurium resistance
E. coli EHEC O157:H7	PAI-II	Enterotoxin, VrmP, Tnp
E. coli EHEC O157:H7	PAI-III	LEE T3SS
E. coli EHEC O157:H7	PAI-IV	Resolvase, helicases, Tnp
S. enterica Typhimurium	SPI-1	Effectors, T3SS
S. enterica Typhimurium	SPI-2	Effectors, T3SS
S. enterica Typhimurium	SPI-3	Effectors, invasion
S. enterica Typhimurium	SPI-4	Effectors, invasion
S. enterica Typhimurium	SPI-5	Effectors, intracellular survival
Vibrio cholerae O1 serogroup	VPI	TCP, ACF, ToxT, TcpPH
V. cholerae O1 serogroup	VPI-2	SAC, sialidase, restriction modification
V. cholerae O1 serogroup	VSP-I	Unknown
V. cholerae O1 serogroup	VSP-II	Unknown
V. cholerae O39 serogroup	VPI-2V	SAC, sialidase, T3SS
V. parahaemolyticus O3:K6	VPaI-1	Haemagglutinin, restriction modification
V. parahaemolyticus O3:K6	VPaI-2	Omp, resolvase, RadC
V. parahaemolyticus O3:K6	VPaI-3	Histidine kinase, helicase, chemotaxis protein
V. parahaemolyticus O3:K6	VPaI-4	Cytotoxic integrase, M protein
V. parahaemolyticus O3:K6	VPaI-5	Unknown
V. parahaemolyticus O3:K6	VPaI-6	DHBP synthase, Colicin, hydrolase
V. parahaemolyticus O3:K6	VPaI-7	T3SS, TDH

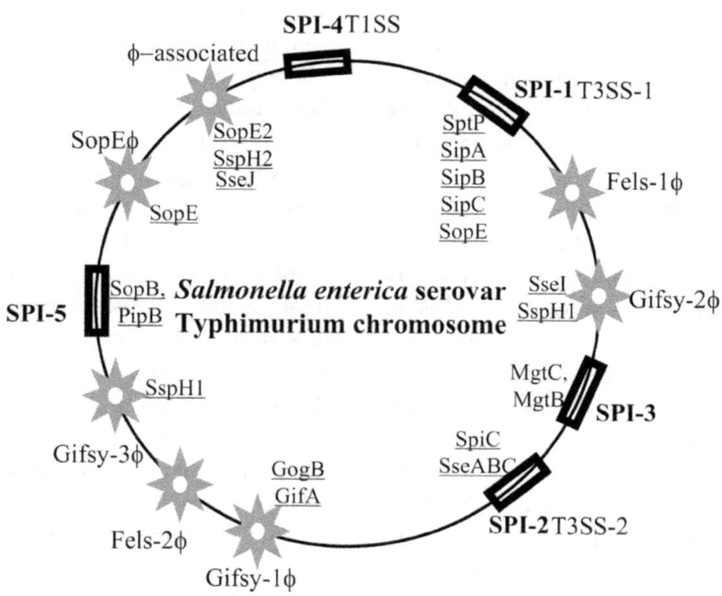

Figure 14.2 Schematic representation of prophages and PAIs within the *S. enterica* chromosome. Effector proteins and their designated pathogenicity island encoded type 3 secretion system translocator in *S. enterica* genome are shown. Stars represent the different prophages in *S. enterica* and their effector proteins and the open rectangular boxes represent pathogenicity islands. Effector proteins are underlined.

Table 14.2 Phage-encoded bacterial virulence factors among enteric pathogens *E. coli*, *S. enterica*, *V. parahaemolyticus* and *V. cholerae*

Bacterial species	Phage	Gene
E. coli EHEC	H-19B	*stx1,2*/Shiga toxins
E. coli EHEC	lambdoid	*tccP*/effector invasion
E. coli EHEC	lambdoid	*ecp*/effector invasion
E. coli EHEC	lambdoid	*nleA-F*/effector
E. coli EHEC	Sp4, 10	*sodC*/intracellular survival
Salmonella enterica	SopEphi	*sopE*/effector invasion
S. enterica	*Gifsy*-1	*gogB*/effector invasion
S. enterica	*Gifsy*-1	*gipA*/invasion
S. enterica	*Gifsy-2*	*sseI(gtgB)*/effector invasion
S. enterica	*Gifsy-3*	*sspH1*/ç
S. enterica	*Gifsy-2*	*sodCI*/intracellular survival
S. enterica	*Fels-1*	*sodCIII*/intracellular survival
S. enterica	*Gifsy*-1,2	*ailT,ailF*/serum resistance
S. enterica	eplison34	*rfb*/glucosylation antigenicity
S. enterica	P22	*gtr*/glucosylation antigenicity
S. enterica	*Fels-1*	*nanH*/neuraminidase
S. enterica	*Gifsy*	*grvA*/antivirulence gene
Vibrio cholerae	CTXphi	*ctxAB*/cholera enterotoxin
V. cholerae	CTXphi	*ace,zot*/enterotoxin
V. cholerae	K139	*glo*/G-protein-like
V. parahaemolyticus	f237	Unknown

valine tRNA locus. The region encodes genes involved in intracellular proliferation and similar to SPI-1 encodes a T3SS and several effector proteins translocated by this system directly into the eukaryotic host cell (Shea *et al.*, 1996). SPI-2 is present in all *S. enterica* isolates but is absent from *S. bongori* isolates suggesting that it was acquired after the split between the two species (Groisman and Ochman, 1996). SPI-3 is a 17 kb region that is inserted at the *selC* tRNA locus and has a GC content of 47%. The SPI-3 region encodes genes required for intracellular proliferation and magnesium uptake (Blanc-Potard and Groisman, 1997; Blanc-Potard *et al.*, 1999b). The SPI-3 region is present in all *S. enterica* isolates with some minor differences among isolates (Blanc-Potard and Groisman, 1997; Blanc-Potard *et al.*, 1999b). SPI-4 is a 25 kb region inserted at the *ssb/soxR* loci and has an aberrant GC content compared with the entire chromosome. The SPI-4 region encodes a type I secretion system and may be important in toxin secretion and apoptosis (Wong *et al.*, 1998). The SPI-5 region is a 7 kb region that encodes a number of effector proteins that are secreted by SPI-1 and SPI-2 (Wood *et al.*, 1998). The region appears to have lost some of its functions owing to its composite structure since the region containing effector protein PipB is missing from *S. bongori* isolates. SPI-5 is inserted at the tRNA-serine locus and has a GC content of 44% and was first identified in *S. enterica* serovar Dublin but is present among all *S. enterica* isolates (Wood *et al.*, 1998).

Type III secretion systems

S. enterica serovar Typhimurium isolates contain two T3SSs as stated above. T3SSs are multicomponent structures that transverse the bacterial cell membrane and cell wall and into the eukaryotic cell membrane to deliver bacterial proteins directly into host cell cytosol (Marlovits and Stebbins, 2010). The translocated proteins are known as effectors or effector toxins and to date well over a hundred different effector proteins have been described (see Dean, 2011, for a comprehensive review). These proteins are multifunctional containing several distinct domains or modules that appear to have arisen through recombination and reassortment events (Stavrinides *et al.*, 2006). Effector proteins can function by targeting a range of structures within the host eukaryotic cell such as actin filaments, microtubules, mitochondria, lysosomes, the nucleus, inner cell membrane, and tight junctions, in addition to disrupting a range of host cell signalling pathways such as (de)activating cyclic AMP and GTPases Rho, Rac and Cdc42 (Dean and Kenny, 2009; Marlovits and Stebbins, 2010).

In *S. enterica*, the two T3SSs are unrelated to one another but do share homology with T3SSs in other species. In addition to the diversity of T3SSs present within *S. enterica*, the number and type of effector proteins translocated by the T3SSs can differ from strain to strain (Dean, 2011). In *S. enterica*, one of the most well characterized systems, 30 different effector proteins have been identified (McGhie *et al.*, 2009). The effectors may be encoded within the same region that contains its cognate T3SS or they can be dispersed on the chromosome within prophages or other PAIs (Tables 14.1 and 14.2, Fig. 14.2). Given the large number of effectors identified in *S. enterica* and the multifunctional nature and structure of these proteins, redundancy in functions encoded by these proteins is well documented (McGhie *et al.*, 2009; Dean, 2011).

In *S. enterica* serovar Typhimurium the effectors secreted by T3SS-1 encoded on SPI-1 are required for host cell invasion by triggering membrane ruffling and internalization within a vacuole. The effector proteins activate cytoskeleton rearrangement by causing actin nucleation, which disrupts the microvilli on the apical cell surface and aid in bacterial uptake into a vacuole. The *Salmonella* containing vacuole (SCV) within the host cell cytosol then secretes effector proteins into the cytosol to modulate intracellular survival, firstly by preventing lysosome fusion with the SCV (Ibarra and Steele-Mortimer, 2009). Both SPI-2 encoded T3SS-2 and SPI-1 encoded T3SS-1 are involved in translocating proteins from the SCV into the host cell cytosol (Ibarra and Steele-Mortimer, 2009). In Fig. 14.2, some examples of effector proteins identified to date are listed and the regions that encode these genes.

Enterohaemorrhagic *E. coli* (EHEC): the causative agent of bloody diarrhoea

In the genome of EHEC strain O157:H7 Sakai, 18 integrases are present, 13 associated with prophages and five associated with PAIs (Hayashi *et al.*, 2001). The five PAIs range in size from 86 kb to 23 kb and all have a GC content below 48% and inserted within a tRNA loci (Table 14.1) (Napolitano *et al.*, 2011). The Locus of enterocyte effacement (LEE) pathogenicity island present in EHEC isolates is a 35 kb region that encodes a T3SS that is required for translocation of effector proteins into the host cell that is required for EHEC internalization (Table 14.1) (Schmidt, 2010). EHEC strains can inject an array of effectors into epithelial cells using the T3SS, primarily to promote cell attachment and colonization (Ogura *et al.*, 2006; Tobe *et al.*, 2006; Schmidt, 2010; Wong *et al.*, 2011). Over 39 different effector proteins are known to be transported by EHEC T3SS (Tobe *et al.*, 2006). Seven effector proteins are encoded within the LEE pathogenicity island that contains the T3SS structural genes. The additional effector proteins are dispersed throughout the chromosome and many of these are encoded within prophage genomes (Table 14.2) (Tobe *et al.*, 2006). For example, 9 lambdoid prophages are proposed to encode 2–8 different effector proteins (Ogura *et al.*, 2006, 2009). Depending on the EHEC strain examined and the prophage content, the repertoire of effector proteins can be very different, which could explain differences in pathogenesis and cell tropism (Ogura *et al.*, 2009).

Vibrio cholerae serogroup O1: the causative agent of secretory diarrhoea

Choleragenic *V. cholerae* is spread via contaminated water and in infected individuals causes a profuse and watery diarrhoea characterized by 'rice water' stools that can lead to dehydration, circulatory collapse and death within 24 hours. The main symptom of cholera, profuse diarrhoea, is caused by the action of cholera toxin (CT) (De, 1959). CT is an A-B_5 type exotoxin that consists of five binding B subunits and one active A subunit (Gill, 1976). The B subunits bind the active A subunit to the glycoprotein receptor the GM_1 ganglioside in the small intestine (Holmgren *et al.*, 1975). CT is internalized within the intestinal cell by endocytosis and the CT containing endosome is retrograde trafficked via the Golgi to the endoplasmic reticulum where the active A1 subunit is cleaved and released into the cytosol (Fig. 14.3). In the cytosol CT activates adenylate cyclase resulting in production of cyclic adenosine monophosphate (cAMP), which leads to hyper-secretion of water and electrolytes into the gut via the activation of the CFTR transporter (Fig. 14.3). The vast quantity of liquid lost exceeds the capacity of absorption of the intestine and therefore is excreted as diarrhoea (Field *et al.*, 1972). *Vibrio cholerae* choleragenic strains also encode sialidase, a glycohydrolase that cleaves higher order gangliosides to unmask the receptor for CT with the release of the nine carbon amino sugar sialic acid (Galen *et al.*, 1992). *V. cholerae* colonizes the small intestine using a type IV bundle forming pilus known as the toxin-coregulated pilus (TCP) (Taylor *et al.*, 1987).

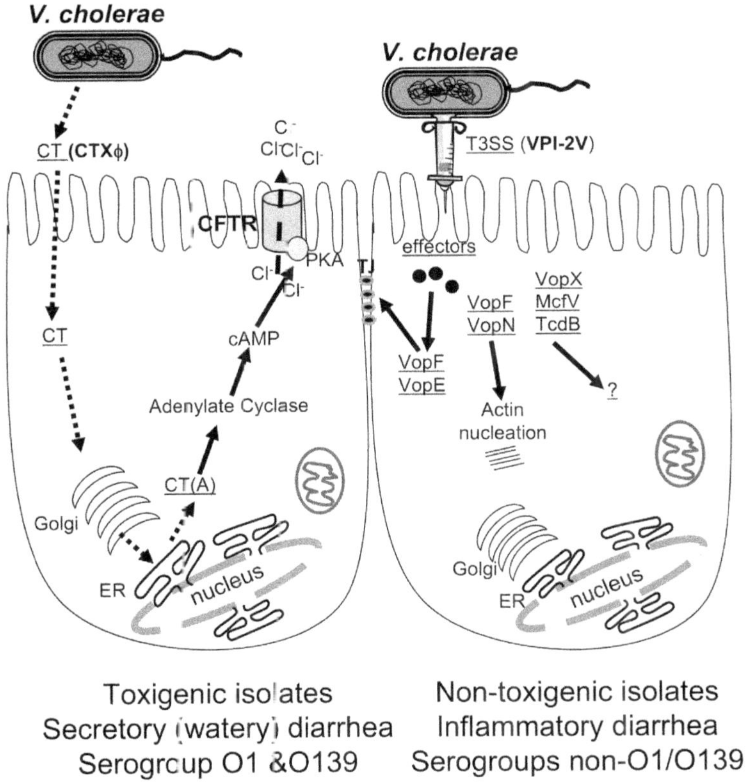

Figure 14.3 A simplified cartoon of the mechanism of pathogenesis of toxigenic and non-toxigenic *V. cholerae* strains. *V. cholerae* strains that encode CTX produce and secrete cholera toxin. CT is internalized within the target cell and retrograde trafficked to the endoplasmic reticulum (ER) where it is activated and then trafficked to its site of activity, adenylate cyclase. Activation of adenylate cyclase activates cAMP, which in turn results in CFTR activation with subsequent chloride loss. In pathogenic *V. cholerae* strains that do not contain CT, a T3SS is present which secretes effector proteins directly from bacterial cytosol into target cell cytosol. In a number of pathogens effectors induce actin microfilament rearrangement. Some effectors are believed to prevent lysosome fusion and a number of effectors are known to activate and deactivate the host cell immune responses. Few effectors have been identified in *V. cholerae*.

CTXΦ: the cholera toxin encoding phage

Lambdoid phages are double-stranded DNA phages generally 40–80 kb in size and can integrate and excise from the host chromosome catalysed by a phage integrase. Lambdoid phages can replicate within the bacterial cell, lyse the cell and can then infect new host cells. Many lambdoid phages have been shown to encode bacterial virulence factors, particularly bacterial toxins such as the A-B_5 Shiga toxins (Stx1, stx2) encoded within *E. coli* pathovars as well as a range of other toxins in Gram-positive and Gram-negative species (Boyd and Brussow, 2002; Brussow *et al.*, 2004; Boyd, 2011). CT is also encoded on a prophage, a single stranded DNA filamentous phage named CTXΦ. CT is the only known example of a bacterial virulence factor encoded within a filamentous phage genome. The CTXΦ genome similar to other characterized filamentous phages is 7 kb and encodes structural, assembly and regulation genes as well as the *ctxAB* genes (Fig. 14.1). CTXΦ does not encode an integrase but does contain an attachment site (*att*) and is integrated at the *dif* sites on the host chromosomes in *V. cholerae*, and the integration is catalysed by the host XerC/D recombinase (Huber and Waldor, 2002; McLeod and Waldor, 2004). The CTXΦ genome is divided into two functional regions; the core region encodes zonula occludens toxin (*zot*), accessory cholera enterotoxin (*ace*), and core-encoded pilin (*cep*), which are involved in the phage morphogenesis, as well as the genes for CT (*ctxAB*) (Waldor and Mekalanos, 1996). The genes required for phage replication, integration, and regulation are encoded within the RS2 region (Fig. 14.1). The receptor for the CTXΦ on the *V. cholerae* cell is the bundle-forming type IV pilus TCP and is a required for the uptake of CTXΦ (Waldor and Mekalanos, 1996).

Vibrio pathogenicity island (VPI)

In addition to the presence of the CTX prophage, in *V. cholerae* O1 serogroup isolates at the tmRNA locus encoded by *ssrAB* is the *Vibrio* pathogenicity island (VPI), the region that encodes TCP (Karaolis *et al.*, 1998). Similar to CTXΦ, VPI is only present in choleragenic isolates and absent from all other isolates. VPI is 40 kb in size and has a G+C content of 35% compared with 47% for the entire *V. cholerae* genome. VPI encodes an integrase that belongs to the tyrosine recombinase family, *att* site, and the genes required for the synthesis and assembly of TCP, the accessory colonization factor (ACF), and several virulence regulators such as ToxT and TcpPH (Karaolis *et al.*, 1998). Although, VPI is highly stable within the *V. cholerae* chromosome I, it can excise from the chromosome and form a circularized intermediate at very low rates (Rajanna *et al.*, 2003). The VPI region is non-self mobile but has been shown to be transmissible. O'Shea and Boyd demonstrated that VPI can be transferred via generalized transduction between O1 strains of *V. cholerae* (O'Shea and Boyd, 2002).

VPI-2 or sialic acid catabolism (SAC) island

The *Vibrio* pathogenicity island-2 (VPI-2) is a 57.3 kb region consisting of 52 open reading frames (ORFs) VC1758-VC1809 on chromosome I of *V. cholerae* N16961 (Jermyn and Boyd, 2002). VPI-2 is integrated at a tRNA-serine locus and encodes an integrase from the tyrosine recombinase family (Jermyn and Boyd, 2002; Murphy and Boyd, 2008). VPI-2 has an aberrant G+C content of 42%. Similar to VPI, VPI-2 is highly stable on the chromosome but can excise and form a circular intermediate that is non-replicative and non-self-mobilizable (Murphy and Boyd, 2008). Under induced stress conditions the rate of VPI-2 excision is increased (Almagro-Moreno *et al.*, 2010).

VPI-2 is divided into distinct functional regions (Fig. 14.4), which can differ between strains (Jermyn and Boyd, 2002; Murphy and Boyd, 2008). The first of these regions (in the 5′ to 3′ direction) is a type-1 restriction modification system, comprising of ORFs VC1764 to VC1769 (*hsdR, hsdM* and *hsdS*) as well as several chemotaxis proteins. The second region (ORFs VC1773–1784) of VPI-2 encodes genes for sialic acid scavenging (*nanH*), transport into the cell (*siaPQM*) and sialic acid catabolism (SAC) encoded by *nanM, nanA, nanEK* and *nagA* (Fig. 14.4) (Jermyn and Boyd, 2002; Murphy and Boyd, 2008). The third major region of VPI-2 contains

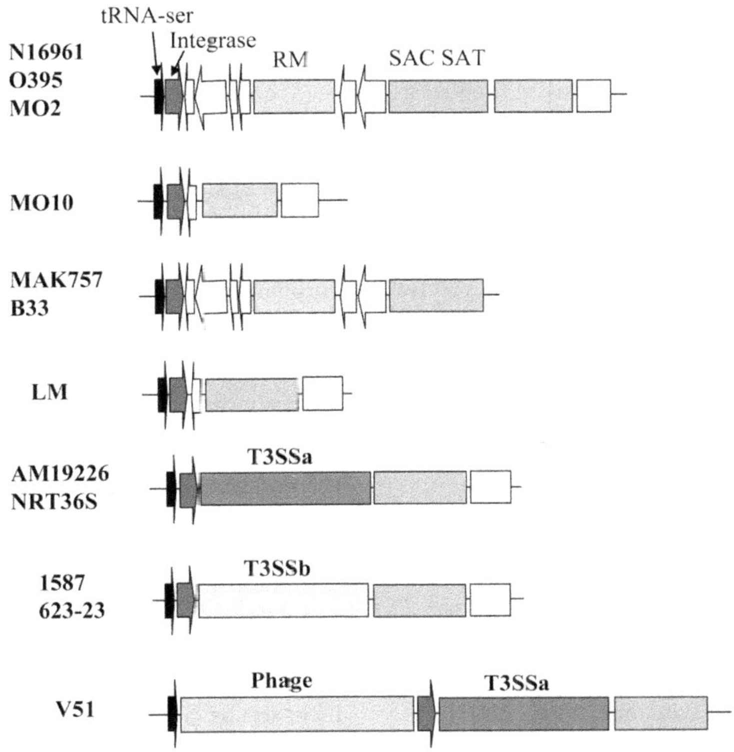

Figure 14.4 *Vibrio* pathogenicity island-2. The range of VPI-2 region identified to date among *V. cholerae* isolates. RM represents restriction modification gene cluster, SAC represents sialic acid transport and catabolism gene clusters, T3SS the type 3 secretion system cluster.

an IS element, two recombination directionality factors or excisionases and Mu-phage-like genes (Jermyn and Boyd, 2002; Almagro-Moreno *et al.*, 2010)

Sialic acid comprises a family of nine-carbon amino sugars (non-ulosonic acids) that are prevalent in all mucus rich environments in the human body. Interestingly the SAC gene cluster is present in only 12 *Gamma-Proteobacteria* and nine Firmicutes genera, all known pathogens of mammals or human gut commensals (Almagro-Moreno and Boyd, 2009a, 2010). Among enteric species, the SAC cluster is found in *Vibrio* species, *Yersinia* species, *S. enterica* and *E. coli* isolates. Reconstruction of the phylogenetic history of all NanA homologues in the databases demonstrated that NanA from *Vibrio* pathogenic species all group together with all *Yersinia* species separate from all other Gamma-Proteobacteria representatives examined. The data suggested that the SAC gene have been horizontally acquired in some genera from multiple origins (Almagro-Moreno and Boyd, 2009a, 2010).

Sialidase/neuraminidase (NanH) converts higher order gangliosides found in the intestinal mucus into GM_1 gangliosides, the receptors of the cholera toxin with the release of free sialic acid (Galen *et al.*, 1992). Within VPI-2, as mentioned above, are the genes for sialic acid uptake into the cell, the *siaPQM* cluster which encodes a tripartite ATP-independent periplasmic C4-dicarboxylate transport system (TRAP) (Fig. 14.4) (Almagro-Moreno and Boyd, 2009a, 2010). TRAP transporters have been shown to be high-affinity transporters for sialic acid and we recently found that in a *V. cholerae* knockout of this system, the mutant strain is no longer able to catabolize sialic acid (Chowdhury and Boyd, unpublished data). In the mucus-rich environment of the gut, the capacity to transport and catabolize sialic acid as a sole carbon and energy source should confer an advantage to *V. cholerae*.

We demonstrated that a *V. cholerae nanA* mutant is defective in the early stages of intestinal colonization compared with a wild-type strain (Almagro-Moreno and Boyd, 2009b). The *nanA* mutant demonstrated a decrease in the competitive index in colonization competition assays compared with the wild-type reinforcing the suggestion that catabolism of sialic acid may play a significant role in *V. cholerae* intestinal survival (Almagro-Moreno and Boyd, 2009b). Among O139 serogroup strains recovered after 1992, most of the VPI-2 region has been deleted, which may explain why O139 serogroup isolates are no longer a significant cause of cholera outbreaks (Fig. 14.4) (Jermyn and Boyd, 2005).

Vibrio seventh pandemic island-I (VSP-I)

The *Vibrio* seventh pandemic island-I (VSP-I) is confined to seventh pandemic strains (O1 El Tor and O139) and is a 16 kb region on the *V. cholerae* N16961 genome (Dziejman *et al.*, 2002). The 16-kb region has an atypical 40% G+C content and encodes a transposase gene rather than an integrase. VSP-I was identified by microarray analysis comparing *V. cholerae* isolates spanning ORFs VC0175 to VC0185. Of the eleven genes on VSP-I seven encode hypothetical proteins (Dziejman *et al.*, 2002). VC0175 encodes a protein which has similarity to a deoxycytidylate deaminase-related protein, VC0176 encodes an homologue of a transcriptional regulator whereas VC0178 encodes a patatin-related protein and VC0185 encodes a putative transposase (Dziejman *et al.*, 2002). VSP-I can excise from chromosome I and form a circular intermediate (Murphy and Boyd, 2008).

Vibrio seventh pandemic island-II (VSP-II)

The *Vibrio* seventh pandemic island-II (VSP-II) as the name implies was originally believed to be confined to seventh pandemic isolates. As with VSP-I, VSP-II was originally identified by DNA microarray analysis as a 7.5 kb region encompassing the ORFs VC0490-VC0497 (Dziejman *et al.*, 2002). However, O'Shea and colleagues showed that VSP-II is a 27 kb region encompassing ORFs VC0490-VC0516 (O'Shea *et al.*, 2004). The majority of VSP-II genes has yet to be assigned a function with 13 of 24 genes encoding conserved hypothetical proteins. VSP-II encodes an integrase from the tyrosine recombinase family and an RDF/excisionase, as well as methyl-accepting chemotaxis proteins and a type IV pilus protein (O'Shea *et al.*, 2004). The VSP-II region has also been found to excise from chromosome I (Murphy and Boyd, 2008). Its role in pathogenesis or fitness remains unknown.

V. cholerae non-O1/139: the causative agent of inflammatory diarrhoea

V. cholerae belonging to serogroup O1 and O139 are the causative agents of cholera, a secretory diarrhoea. However, many non-O1/O139 isolates are pathogenic and can cause an inflammatory diarrhoea. The mechanism of pathogenesis was unknown until recent genome sequencing of these strains (Dziejman *et al.*, 2005; Chen *et al.*, 2007). Genome sequence analysis of *V. cholerae* pathogenic non-choleragenic isolates that cause gastroenteritis demonstrated the presence of a T3SS encoded within VPI-2 (Chen *et al.*, 2007; Murphy and Boyd, 2008) (Fig. 14.4). In most of the strains that contain a T3SS, neither the CTX_ nor VPI regions are present. Studies have shown that two different T3SS can be present within VPI-2 depending on the strain examined and these different phylotypes show homology to the T3SS present on chromosome II in *V. parahaemolyticus* (Dziejman *et al.*, 2005; Chen *et al.*, 2007). In addition a small number of effector proteins including VopN, VopF and VopX have been identified, (Fig. 14.3) (Tam *et al.*, 2007; Alam *et al.*, 2011). Two of these effectors have homologues in *V. parahaemolyticus* but VopX appears to be unique (Tam *et al.*, 2007; Alam *et al.*, 2011). The role of these *V. cholerae* effectors is under investigation and it is thought that VopN and VopF are involved in actin nucleation resulting in cytoskeleton rearrangements but the function of VopX is unknown (Fig. 14.3) (Alam *et al.*, 2011). More recent data using an infant rabbit model of *V. cholerae* infection by a non-O1/O139 strain encoding a T3SS identified the effector proteins McfV and TcdB (Shin *et al.*, 2011).

Vibrio mimicus VPI-2 and VPI-2V

Vibrio mimicus, a species closely related to *V. cholerae*, can cause gastroenteritis. *Vibrio mimicus* isolates contain a 20-kb VPI-2 region, which harbours the SAC region and the integrase, and is inserted at the same tRNA but contains neither the 5′ or 3′ ends of the island (Jermyn and Boyd, 2002, 2005). In evolutionary terms, *V. cholerae* and *V. mimicus* strains cluster separately on phylogenetic trees based on housekeeping genes forming distinct lineages (Jermyn and Boyd, 2002, 2005). *V. mimicus* genome sequence analysis indicates that some *V. mimicus* strains also contain a T3SS encoded within VPI-2 suggesting a similar mechanism of pathogenesis as that recently shown for *V. cholerae* non-O1/O139 isolates (Okada *et al.*, 2010).

Vibrio parahaemolyticus: the causative agent of inflammatory diarrhoea

Disease caused by the ingestion of *V. parahaemolyticus* is typically self-limiting, resulting in a transient inflammatory diarrhoea and abdominal pain. Pathogenic strains cause haemolysis on Wagatsuma blood agar, known as the Kanagawa phenomenon (KP) and this is attributed to the production of the thermostable direct haemolysin (TDH) (Honda *et al.*, 1976). Gastrointestinal disease has also been associated with KP-negative strains from which a second haemolysin termed the TDH-related haemolysin (TRH) was identified, and the production of one or both of these haemolysins was initially believed to be the primary cause of virulence by *V. parahaemolyticus* (Honda *et al.*, 1989). TDH and TRH function within the target cell by activating calcium which leads to increased secretion of chloride via the calcium activated chloride channels (CLCAs) (Fig. 14.5).

Genome sequence of *V. parahaemolyticus* RIMD2210633, an O3:K6 serotype currently associated with world-wide outbreaks of seafood-borne gastroenteritis, identified genes that encode a T3SS on each chromosome (designated T3SS-1 and T3SS-2) (Makino *et al.*, 2003). In addition, RIMD2210633 encodes a class 1 integron on chromosome II and two f237 phage regions.

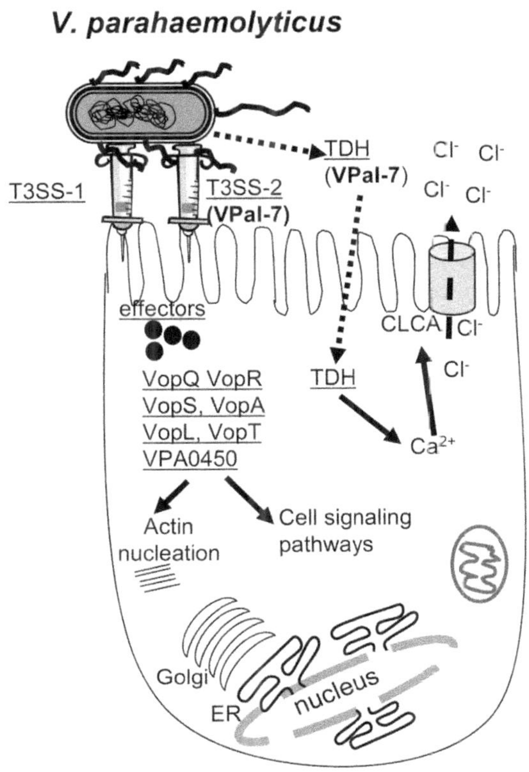

Figure 14.5 A Schematic representation of the mechanism of pathogenesis of *V. parahaemolyticus* and the MIGE encoded virulence genes identified to date. *V. parahaemolyticus* encodes two T3SSs and seven effector proteins are known which are underlined. This pathogen also encodes a toxin TDH that is secreted from the bacterial cell and taken up into the target cell, where it activates calcium that affects CLCA channels. The effectors' mode of action is not as well characterized, but they are known to affect actin formation and different cell signalling pathways.

T3SS-1 is encoded on chromosome I and is present in all isolates examined. Flanking the T3SS-2 gene cluster on chromosome II is a copy of the *tdh* gene (*tdhA* and *tdhS*) within an 80 kb pathogenicity island named the *Vibrio parahaemolyticus* island-7 (VPaI-7) encompassing ORFs VPA1312 to VPA1396, a region associated with the pandemic O3:K6 serogroup strains (Makino *et al.*, 2003; Hurley *et al.*, 2006; Boyd *et al.*, 2008). Both T3SSs in *V. parahaemolyticus* have been shown to be functional, though data suggests that the contribution of each to virulence differs markedly (Park *et al.*, 2004; Bhattacharjee *et al.*, 2006;

Hiyoshi *et al.*, 2010). Recent genome sequence analysis of a *trh*-positive *V. parahaemolyticus* strain identified a third gene cluster encoding a T3SS (Okada *et al.*, 2009).

Several functional T3SS-associated effector toxins have been identified from *in vitro* analysis (Fig. 14.5) (Bhattacharjee *et al.*, 2006; Ono *et al.*, 2006; Kodama *et al.*, 2007, 2008). The role of the T3SS effector proteins VopQ (*vp1680*), VopR (*vp1683*), VopS (*vp1686*) and VPA0450 on the host immune system also have been examined. The production of these effector proteins by *V. parahaemolyticus* induces autophagy, cell rounding, lysis and cell death *in vitro* (Trosky *et al.*, 2004; Liverman *et al.*, 2007; Burdette *et al.*, 2008, 2009; Yarbrough *et al.*, 2009; Broberg *et al.*, 2010).

By using bioinformatics and genome comparisons seven regions named VPaI-1 to VPaI-7 were identified as unique to *V. parahaemolyticus* compared with other sequenced *Vibrio* genomes (Hurley *et al.*, 2006; Boyd *et al.*, 2008). These seven regions had features suggestive of MIGE regions, i.e. each region has an aberrant base composition compared with the entire genome, encoded an integrase, inserted at a tRNA gene and were variably present among strains. A number of the VPaI regions in particular VPaI-4, VPaI-6 and VPaI-7 encoded putative virulence genes such as TDH, T3SS, M protein, hydrolases, cytotoxin integrase, and colicins. All the VPaI regions encoded a large proportion of hypothetical and unknown proteins. The VPaI-1, VPaI-4, VPaI-5, and VPaI-6 regions were unique to the pandemic group of *V. parahaemolyticus* strains isolated after 1995 suggesting that the acquisition of novel DNA by horizontal gene transfer may have played an important role in the emergence of these strains (Hurley *et al.*, 2006; Boyd *et al.*, 2008).

Conclusions

In this brief overview of MIGEs encoding virulence genes among various enteric pathogens a common theme has emerged. Within a specific species depending on its MIGE content, the mode and mechanism of pathogenesis can be very different. In *V. cholerae* the presence of CTXΦ and VPI can result in a strain that causes the potentially lethal cholera infection whereas in a strain that contains a T3SS within VPI-2, a self limiting diarrhoea results (Fig. 14.3). A similar scenario is found within *E. coli*; depending on the strain examined a very different set of MIGEs can be present each encoding a very different array of virulence genes. The added complication among *E. coli* isolates is that the repertoire of effector proteins present differs even among very closely related isolates. In many cases, this repertoire of effectors is driven by acquisition of different PAIs and phages by these isolates, indicating that these isolates are continuously evolving.

Acknowledgements

I thank members of my group for their enthusiasm and hard work; Megan Carpenter, Olivia Cosides, Sarah Gaffney, Brandy Haines, Nityananda Chowdhury, Joseph J. Kingston, Jean Bernard Lubin, Michael G. Napolitano, Serge Ongagna, W. Brian Whitaker, and Claire Yang. Work in the Boyd group is support by a National Science Foundation CAREER grant DEB-0844409. Literature citations have been limited in many cases to review articles owing to space consideration and I apologize to those authors whose primary research is not included or cited.

References

Alam, A., Miller, K.A., Chaand, M., Butler, J.S., and Dziejman, M. (2011). Identification of *Vibrio cholerae* type III secretion system effector proteins. Infect. Immun. 79, 1728–1740.

Almagro-Moreno, S., and Boyd, E.F. (2009a). Insights into the evolution of sialic acid catabolism among bacteria. BMC Evol. Biol. *9*, 118.

Almagro-Moreno, S., and Boyd, E.F. (2009b). Sialic acid catabolism confers a competitive advantage to pathogenic *Vibrio cholerae* in the mouse intestine. Infect. Immun. *77*, 3807–3816.

Almagro-Moreno, S., and Boyd, E.F. (2010). Bacterial catabolism of nonulosonic (sialic) acid and fitness in the gut. Gut Microbes *1*, 45–50.

Almagro-Moreno, S., Napolitano, M.G., and Boyd, E.F. (2010). Excision dynamics of *Vibrio* pathogenicity island-2 from *Vibrio cholerae*: role of a recombination directionality factor VefA. BMC Microbiol. *10*, 306.

Bergthorsson, U., and Ochman, H. (1995). Heterogeneity of genome sizes among natural isolates of *Escherichia coli*. J. Bacteriol. *177*, 5784–5789.

Bhattacharjee, R.N., Park, K.S., Kumagai, Y., Okada, K., Yamamoto, M., Uematsu, S., Matsui, K., Kumar, H., Kawai, T., Iida, T., *et al.* (2006). VP1686, a *Vibrio* type III secretion protein, induces toll-like receptor-independent apoptosis in macrophage through NF-kappaB inhibition. J. Biol. Chem. *281*, 36897–36904.

Blanc-Potard, A.B., and Groisman, E.A. (1997). The *Salmonella selC* locus contains a pathogenicity island mediating intramacrophage survival. EMBO J. *16*, 5376–5385.

Blanc-Potard, A., Solomon, F., Kayser, J., and Groisman, E.A. (1999a). The SPI-3 pathogenicity island of *Salmonella enterica*. J. Bacteriol. *181*, 998–1004.

Blanc-Potard, A.B., Solomon, F., Kayser, J., and. (1999b). The SPI-3 pathogenicity island of *Salmonella enterica*. J. Bacteriol. *181*, 998–1004.

Blattner, F.R., Plunkett, G., 3rd, Bloch, C.A., Perna, N.T., Burland, V., Riley, M., Collado-Vides, J., Glasner, J.D., Rode, C.K., Mayhew, G.F., *et al.* (1997). The complete genome sequence of *Escherichia coli* K-12. Science *277*, 1453–1462.

Blum, G., Ott, M., Lischewski, A., Ritter, A., Imrich, H., Tschape, H., and Hacker, J. (1994). Excision of large DNA regions termed pathogenicity islands from tRNA-specific loci in the chromosome of an *Escherichia coli* wild-type pathogen. Infect. Immun. *62*, 606–614.

Boyd, E.F. (2012). Phage-encoded bacterial virulence factors: phage-pathogenicity island and phage–phage interactions. Adv. Virus Res. *82*, 91–118.

Boyd, E.F., and Brussow, H. (2002). Common themes among bacteriophage-encoded virulence factors and diversity among the bacteriophages involved. Trends Microbiol. *10*, 521–529.

Boyd, E.F., Wang, F.S., Whittam, T.S., and Selander, R.K. (1996). Molecular genetic relationships of the salmonellae. Appl Environ Microbiol *62*, 804–808.

Boyd, E.F., Cohen, A.L.V., Naughton, L.M., Ussery, D.W., Binnewies, T.T., Stine, O.C., and Parent, M.A. (2008). Molecular analysis of the emergence of pandemic *Vibrio parahaemolyticus*. BMC Microbiol. *8*, 110.

Broberg, C.A., Zhang, L., Gonzalez, H., Laskowski-Arce, M.A., and Orth, K. (2010). A *Vibrio* effector protein is an inositol phosphatase and disrupts host cell membrane integrity. Science *329*, 1660–1662.

Brussow, H., Canchaya, C., and Hardt, W.D. (2004). Phages and the evolution of bacterial pathogens: from genomic rearrangements to lysogenic conversion. Microbiol. Mol. Biol. Rev. *68*, 560–602.

Burdette, D.L., Yarbrough, M.L., Orvedahl, A., Gilpin, C.J., and Orth, K. (2008). *Vibrio parahaemolyticus* orchestrates a multifaceted host cell infection by induction of autophagy, cell rounding, and then cell lysis. Proc. Natl. Acad. Sci. U.S.A. *105*, 12497–12502.

Burdette, D.L., Yarbrough, M.L., and Orth, K. (2009). Not without cause: *Vibrio parahaemolyticus* induces acute autophagy and cell death. Autophagy 5, 100–102.

Chen, Y., Johnson, J.A., Pusch, G.D., Morris, J.G., Jr., and Stine, O.C. (2007). The genome of non-O1 *Vibrio cholerae* NRT36S demonstrates the presence of pathogenic mechanisms that are distinct from O1 *Vibrio cholerae*. Infect Immun. 75, 2645–2647.

De, S.N. (1959). Enterotoxicity of bacteria-free culture-filtrate of *Vibrio cholerae*. Nature *183*, 1533–1534.

Dean, P. (2011). Functional domains and motifs of bacterial type III effector proteins and their roles in infection. FEMS Microbiol. Rev. *35*, 1100–1125.

Dean, P., and Kenny, B. (2009). The effector repertoire of enteropathogenic *E. coli*: ganging up on the host cell. Curr. Opin. Microbiol. *12*, 101–109.

Dobrindt, U., Chowdary, M., Krumbholz, G., and Hacker, J. (2010). Genome dynamics and its impact on evolution of *Escherichia coli*. Med Microbiol. Immunol. *199*, 145–154.

Dziejman, M., Balon, E., Boyd, D., Fraser, C.M., Heidelberg, J.F., and Mekalanos, J.J. (2002). Comparative genomic analysis of *Vibrio cholerae*: genes that correlate with cholera endemic and pandemic disease. Proc. Natl. Acad. Sci. U.S.A. *99*, 1556–1561.

Dziejman, M., Serruto, D., Tam, V.C., Sturtevant, D., Diraphat, P., Faruque, S.M., Rahman, M.H., Heidelberg, J.F., Decker, J., Li, L., *et al.* (2005). Genomic characterization of non-O1, non-O139 *Vibrio cholerae* reveals genes for a type III secretion system. Proc. Natl. Acad. Sci. U.S.A. *102*, 3465–3470.

Fang, F.C., DeGroote, M.A., Foster, J.W., Baumler, A.J., Ochsner, U., Testerman, T., Bearson, S., Giard, J.C., Xu, Y., Campbell, G., *et al.* (1999). Virulent *Salmonella* typhimurium has two periplasmic Cu, Zn-superoxide. Proc. Natl. Acad. Sci. U.S.A. *96*, 7502–7507.

Faruque, S.M., and Mekalanos, J.J. (2003). Pathogenicity islands and phages in *Vibrio cholerae* evolution. Trends Microbiol. *11*, 505–510.

Field, M., Fromm, D., al-Awqati, Q., and Greenough, W.B., 3rd. (1972). Effect of cholera enterotoxin on ion transport across isolated ileal mucosa. J. Clin. Invest. *51*, 796–804.

Figueroa-Bossi, N., and Bossi, L. (1999). Inducible prophages contribute to *Salmonella* virulence in mice. Mol. Microbiol. *33*, 167–176.

Figueroa-Bossi, N., Uzzau, S., Maloriol, D., and Bossi, L. (2001). Variable assortment of prophages provides a transferable repertoire of pathogenic determinants in *Salmonella*. Mol. Microbiol. *39*, 260–271.

Galan, J.E. (1996). Molecular and cellular bases of *Salmonella* entry into host cells. Curr. Top Microbiol. Immunol. *209*, 43–60.

Galen, J.E., Ketley, J.M., Fasano, A., Richardson, S.H., Wasserman, S.S., and Kaper, J.B. (1992). Role of *Vibrio cholerae* neuraminidase in the function of cholera toxin. Infect. Immun. *60*, 406–415.

Gill, D.M. (1976). The arrangement of subunits in cholera toxin. Biochemistry *15*, 1242–1248.

Groisman, E., and Ochman, H. (1997). How *Salmonella* became a pathogen. Trends Microbiol. *5*, 343–349.

Groisman, E.A., and Ochman, H. (1996). Pathogenicity islands: bacterial evolution in quantum leaps. Cell *87*, 791–794.

Hacker, J., and Carniel, E. (2001). Ecological fitness, genomic islands and bacterial pathogenicity. A Darwinian view of the evolution of microbes. EMBO Rep. 2, 376–381.

Hacker, J., Bender, L., Ott, M., Wingender, J., Lund, B., Marre, R., and Goebel, W. (1990). Deletions of chromosomal regions coding for fimbriae and hemolysins occur *in vitro* and *in vivo* in various extraintestinal *Escherichia coli* isolates. Microb. Pathog. *8*, 213–225.

Hacker, J., Blum-Oehler, G., Muhldorfer, I., and Tschape, H. (1997). Pathogenicity islands of virulent bacteria: structure, function and impact on microbial evolution. Mol. Microbiol. *23*, 1089–1097.

Hayashi, T., Makino, K., Ohnishi, M., Kurokawa, K., Ishii, K., Yokoyama, K., Han, C.G., Ohtsubo, E., Nakayama, K., Murata, T., *et al.* (2001). Complete genome sequence of enterohemorrhagic *Escherichia coli* O157:H7 and genomic comparison with a laboratory strain K-12. DNA Res *8*, 11–22.

Hensel, M., Shea, J.E., Gleeson, C., Jones, M.D., Dalton, E., and Holden, D.W. (1995). Simultaneous identification of bacterial virulence genes by negative selection. Science *269*, 400–403.

Hiyoshi, H., Kodama, T., Iida, T., and Honda, T. (2010). Contribution of *Vibrio parahaemolyticus* virulence factors to cytotoxicity, enterotoxicity, and lethality in mice. Infect. Immun. *78*, 1772–1780.

Holmgren, J., Lonnroth, I., Mansson, J., and Svennerholm, L. (1975). Interaction of cholera toxin and membrane GM1 ganglioside of small intestine. Proc. Natl. Acad. Sci. U.S.A. *72*, 2520–2524.

Honda, T., Taga, S., Takeda, T., Hasibuan, M.A., Takeda, Y., and Miwatani, T. (1976). Identification of lethal toxin with the thermostable direct hemolysin produced by *Vibrio parahaemolyticus*, and some physicochemical properties of the purified toxin. Infect. Immun. *13*, 133–139.

Honda, T., Ni, Y., and Miwatani, T. (1989). Purification of a TDH-related hemolysin produced by a Kanagawa phenomenon-negative clinical isolate of *Vibrio parahaemolyticus* 06: K46. FEMS Microbiol. Lett. *48*, 241–245.

Huber, K.E., and Waldor, M.K. (2002). Filamentous phage integration requires the host recombinases XerC and XerD. Nature *417*, 656–659.

Hurley, C.C., Quirke, A., Reen, F.J., and Boyd, E.F. (2006). Four genomic islands that mark post-1995 pandemic *Vibrio parahaemolyticus* isolates. BMC Genomics *7*, 104.

Ibarra, J., and Steele-Mortimer, O. (2009). *Salmonella* – the ultimate insider. *Salmonella* virulence factors that modulate intracellular survival. Cell Microbiol. *11*, 1579–1586.

Jermyn, W.S., and Boyd, E.F. (2002). Characterization of a novel *Vibrio* pathogenicity island (VPI-2) encoding neuraminidase (*nanH*) among toxigenic *Vibrio cholerae* isolates. Microbiology *148*, 3681–3693.

Jermyn, W.S., and Boyd, E.F. (2005). Molecular evolution of Vibrio pathogenicity island-2 (VPI-2): mosaic structure among *Vibrio cholerae* and *Vibrio mimicus* natural isolates. Microbiology *151*, 311–322.

Karaolis, D.K., Johnson, J.A., Bailey, C.C., Boedeker, E.C., Kaper, J.B., and Reeves, P.R. (1998). A *Vibrio cholerae* pathogenicity island associated with epidemic and pandemic strains. Proc. Natl. Acad. Sci. U.S.A. *95*, 3134–3139.

Kodama, T., Rokuda, M., Park, K.S., Cantarelli, V.V., Matsuda, S., Iida, T., and Honda, T. (2007). Identification and characterization of VopT, a novel ADP-ribosyltransferase effector protein secreted via the *Vibrio parahaemolyticus* type III secretion system 2. Cell Microbiol. *9*, 2598–2609.

Kodama, T., Hiyoshi, H., Gotoh, K., Akeda, Y., Matsuda, S., Park, K.S., Cantarelli, V.V., Iida, T., and Honda, T. (2008). Identification of two translocon proteins of *Vibrio parahaemolyticus* type III secretion system 2. Infect. Immun. *76*, 4282–4289.

Lan, R., Reeves, P.R., and Octavia, S. (2009). Population structure, origins and evolution of major *Salmonella enterica* clones. Infect. Genet. Evol. *9*, 996–1005.

Le Minor, L., Popoff, M.Y., and Bockemuhl, J. (1990). Supplement 1989 (n. 33) to the Kauffmann-White scheme. Res. Microbiol. *141*, 1173–1177.

Liverman, A.D., Cheng, H.C., Trosky, J.E., Leung, D.W., Yarbrough, M.L., Burdette, D.L., Rosen, M.K., and Orth, K. (2007). Arp2/3-independent assembly of actin by Vibrio type III effector VopL. Proc. Natl. Acad. Sci. U.S.A. *104*, 17117–17122.

McGhie, E., Brawn, L., Hume, P., Humphreys, D., and Koronakis, V. (2009). *Salmonella* takes control: effector-driven manipulation of the host. Curr. Opin. Microbiol. *12*, 117–124.

McLeod, S.M., and Waldor, M.K. (2004). Characterization of XerC- and XerD-dependent CTX phage integration in *Vibrio cholerae*. Mol. Microbiol. *54*, 935–947.

McQuiston, J.R., Herrera-Leon, S., Wertheim, B.C., Doyle, J., Fields, P.I., Tauxe, R.V., and Logsdon, J.M., Jr. (2008). Molecular phylogeny of the salmonellae: relationships among *Salmonella* species and subspecies determined from four housekeeping genes and evidence of lateral gene transfer events. J. Bacteriol. *190*, 7060–7067.

Makino, K., Oshima, K., Kurokawa, K., Yokoyama, K., Uda, T., Tagomori, K., Iijima, Y., Najima, M., Nakano, M., Yamashita, A., *et al.* (2003). Genome sequence of *Vibrio parahaemolyticus*: a pathogenic mechanism distinct from that of *V. cholerae*. Lancet *361*, 743–749.

Marlovits, T.C., and Stebbins, C.E. (2010). Type III secretion systems shape up as they ship out. Curr. Opin. Microbiol. *13*, 47–52.

Miller, M.A., Sentz, J., Rabaa, M.A., and Mintz, E.D. (2008). Global epidemiology of infections due to Shigella, *Salmonella* serotype Typhi, and enterotoxigenic *Escherichia coli*. Epidemiol. Infect. *136*, 433–435.

Mills, D., Bajaj, V, and Lee, C. (1995). A 40 kb chromosomal fragment encoding *Salmonella typhimurium* invasion genes is absent from the corresponding region of the *Escherichia coli* K-12 chromosome. Mol. Microbiol. *15*, 749–759.

Murphy, R., and Boyd, E. (2008). Three pathogenicity islands of *Vibrio cholerae* can excise from the chromosome and form circular intermediates. J. Bacteriol. *190*, 636–647

Napolitano, M.G., Almagro-Moreno, S., and Boyd, E.F. (2011). Dichotomy in the evolution of pathogenicity island and bacteriophage encoded integrases from pathogenic *Escherichia coli* strains. Infect. Genet. Evol. *11*, 423–436.

O'Shea, Y.A., and Boyd, E.F. (2002). Mobilization of the *Vibrio* pathogenicity island between *Vibrio cholerae* isolates mediated by CP-T1 generalized transduction. FEMS Microbiol. Lett. *214*, 153–157.

O'Shea, Y.A., Finnan, S., Reen, F.J., Morrissey, J.P., O'Gara, F., and Boyd, E.F. (2004). The *Vibrio* seventh pandemic island-II is a 26.9 kb genomic island present in *Vibrio cholerae* El Tor and O139 serogroup isolates that shows homology to a 43.4 kb genomic island in *V. vulnificus*. Microbiology *150*, 4053–4063.

Ochman, H., and Bergthorsson, U. (1995). Genome evolution in enteric bacteria. Curr. Opin. Genet. Dev. *5*, 734–738.

Ogura, Y., Kurokawa, K., Ooka, T., Tashiro, K., Tobe, T., Ohnishi, M., Nakayama, K., Morimoto, T., Terajima, J., Watanabe, H., *et al.* (2006). Complexity of the genomic diversity in enterohemorrhagic *Escherichia coli* O157 revealed by the combinational use of the O157 Sakai OligoDNA microarray and the Whole Genome PCR scanning. DNA Res. *13*, 3–14.

Ogura, Y., Ooka, T., Iguchi, A., Toh, H., Asadulghani, M., Oshima, K., Kodama, T., Abe, H., Nakayama, K., Kurokawa, K., *et al.* (2009). Comparative genomics reveal the mechanism of the parallel evolution of O157 and non-O157 enterohemorrhagic *Escherichia coli*. Proc. Natl. Acad. Sci. U.S.A. *106*, 17939–17944.

Okada, N., Iida, T., Park, K.S., Goto, N., Yasunaga, T., Hiyoshi, H., Matsuda, S., Kodama, T., and T., H. (2009). Identification and characterization of a novel type III secretion system in *trh*-positive *Vibrio parahaemolyticus* strain TH3996 reveal genetic lineage and diversity of pathogenic machinery beyond the species level. Infect. Immun. *77*, 904–913.

Okada, N., Matsuda, S., Matsuyama, J., Park, K.S., de los Reyes, C., Kogure, K., Honda, T., and Iida, T. (2010). Presence of genes for type III secretion system 2 in *Vibrio mimicus* strains. BMC Microbiol. *10*, 302.

Ono, T., Park, K.S., Ueta, M., Iida, T., and Honda, T. (2006). Identification of proteins secreted via *Vibrio parahaemolyticus* type III secretion system 1. Infect. Immun. *74*, 1032–1042.

Park, K.S., Ono, T., Rokuda, M., Jang, M.H., Okada, K., Iida, T., and Honda, T. (2004). Functional characterization of two type III secretion systems of *Vibrio parahaemolyticus*. Infect. Immun. *72*, 6659–6665.

Pembroke, J., MacMahon, C., and B., M. (2002). The role of conjugative transposons in the Enterobacteriaceae. Cell. Mol. Life Sci. *59*, 2055–2064.

Perna, N.T., Plunkett, G., 3rd, Burland, V., Mau, B., Glasner, J.D., Rose, D.J., Mayhew, G.F., Evans, P.S., Gregor, J., Kirkpatrick, H.A., *et al.* (2001). Genome sequence of enterohaemorrhagic *Escherichia coli* O157:H7. Nature *409*, 529–533.

Popoff, M.Y., Bockemuhl, J., and Gheesling, L.L. (2004). Supplement 2002 (no. 46) to the Kauffmann-White scheme. Res. Microbiol. *155*, 568–570.

Porwollik, S., Wong, R.M., and McClelland, M. (2002). Evolutionary genomics of Salmonella: gene acquisitions revealed by microarray analysis. Proc. Natl. Acad. Sci. U.S.A. *99*, 8956–8961.

Porwollik, S., Boyd, E.F., Choy, C., Cheng, P, Florea, L., Proctor, E., and McClelland, M. (2004). Characterization of *Salmonella enterica* subspecies I genovars by use of microarrays. J. Bacteriol. *186*, 5883–5898.

Rajanna, C., Wang, J., Zhang, D., Xu, Z., Ali, A., Hou, Y.M., and Karaolis, D.K. (2003). The *Vibrio* pathogenicity island of epidemic *Vibrio cholerae* forms precise extrachromosomal circular excision products. J. Bacteriol. *185*, 6893–6901.

Recchia, G.D., and Hall, R.M. (1995). Gene cassettes: a new class of mobile element. Microbiology *141*, 3015–3027.

Reeves, M.W., Evins, G.M., Heiba, A.A., Plikaytis, B.D., and Farmer, J.J., 3rd. (1989). Clonal nature of *Salmonella* typhi and its genetic relatedness to other salmonellae as shown by multilocus enzyme electrophoresis, and proposal of *Salmonella bongori* comb. nov. J. Clin. Microbiol. *27*, 313–320.

Schmidt, M. (2010). LEEways: tales of EPEC, ATEC and EHEC. Cell Microbiol. *12*, 1544–1552.

Schmidt, H., and Hensel, M. (2004). Pathogenicity islands in bacterial pathogenesis. Clin. Microbiol Rev. *17*, 14–56.

Shea, J.E., Hensel, M., Gleeson, C., and Holden, D.W. (1996). Identification of a virulence locus encoding a second type III secretion system in *Salmonella* typhimurium. Proc. Natl. Acad. Sci. U.S.A. *93*, 2593–2597.

Shin, O.S., Tam, V.C., Suzuki, M., Ritchie, J.M, Bronson, R.T., Waldor, M.K., and Mekalanos, J.J. (2011). Type III Secretion Is Essential for the Rapidly Fatal Diarrheal Disease Caused by Non-O1, Non-O139 *Vibrio cholerae*. Mbio. *2*, e00106–e00111.

Stavrinides, J., Ma, W., and Guttman, D.S. (2006). Terminal reassortment drives the quantum evolution of type III effectors in bacterial pathogens. PLoS Pathog. *2*, e104.

Stokes, H.W., and Hall, R.M. (1989). A novel family of potentially mobile DNA elements encoding site-specific gene-integration functions: integrons. Mol. Microbiol. *3*, 1669–1683.

Tam, V.C., Serruto, D., Dziejman, M., Brieher, W., and Mekalanos, J.J. (2007). A type III secretion system in *Vibrio cholerae* translocates a formin/spire hybrid-like actin nucleator to promote intestinal colonization. Cell Host Microbe *1*, 95–107.

Taylor, R.K., Miller, V.L., Furlong, D.B., and Mekalanos, J.J. (1987). Use of *phoA* gene fusions to identify a pilus colonization factor coordinately regulated with cholera toxin. Proc. Natl. Acad. Sci. U.S.A. *84*, 2833–2837.

Thampapillai, G., Lan, R., and Reeves, P.R. (1994). Molecular evolution in the gnd locus of *Salmonella enterica*. Mol. Biol. Evol. *11*, 813–828.

Tobe, T., Beatson, S.A., Taniguchi, H., Abe, H., Bailey, C.M., Fivian, A., Younis, R., Matthews, S., Marches, O., Frankel, G., *et al.* (2006). An extensive repertoire of type III secretion effectors in *Escherichia coli* O157 and the role of lambdoid phages in their dissemination. Proc. Natl. Acad. Sci. U.S.A. *103*, 14941–14946.

Trosky, J.E., Mukherjee, S., Burdette, D.L., Roberts, M., McCarter, L., Siegel, R.M., and Orth, K. (2004). Inhibition of MAPK signaling pathways by VopA from *Vibrio parahaemolyticus*. J. Biol. Chem. *279*, 51953–51957.

Tsolis, R.M., Young, G.M., Solnick, J.V., and Baumler, A.J. (2008). From bench to bedside: stealth of enteroinvasive pathogens. Nat. Rev. Microbiol. *6*, 883–892.

Waldor, M.K., and Mekalanos, J.J. (1996). Lysogenic conversion by a filamentous phage encoding cholera toxin. Science *272*, 1910–1914.

Welch, R.A., Burland, V., Plunkett, G., 3rd, Redford, P., Roesch, P., Rasko, D., Buckles, E.L., Liou, S.R., Boutin, A., Hackett, J., *et al.* (2002). Extensive mosaic structure revealed by the complete genome sequence of uropathogenic *Escherichia coli*. Proc. Natl. Acad. Sci. U.S.A. *99*, 17020–17024.

Wong, A.R., Pearson, J.S., Bright, M.D., Munera, D., Robinson, K.S., Lee, S.F., Frankel, G., and Hartland, E.L. (2011). Enteropathogenic and enterohaemorrhagic *Escherichia coli*: even more subversive elements. Mol. Microbiol. *80*, 1420–1438.

Wong, K.K., McClelland, M., Stillwell, L.C., Sisk, E.C., Thurston, S.J., and Saffer, J.D. (1998). Identification and sequence analysis of a 27-kilobase chromosomal fragment containing a *Salmonella* pathogenicity island located at 92 minutes on the chromosome map of *Salmonella enterica* serovar typhimurium LT2. Infect. Immun. *66*, 3365–3371.

Wood, M.W., Jones, M.A., Watson, P.R., Hedges, S., Wallis, T.S., and Galyov, E.E. (1998). Identification of a pathogenicity island required for *Salmonella* enteropathogenicity. Mol. Microbiol. *29*, 883–891.

Yarbrough, M.L., Li, Y., Kinch, L.N., Grishin, N.V., Ball, H.L., and Orth, K. (2009). AMPylation of Rho GTPases by *Vibrio* VopS disrupts effector binding and downstream signaling. Science *323*, 269–272.

Conclusion and Future Prospects 15

Shah M. Faruque

Abstract

Recent studies have provided remarkable insights to our general understanding of the epidemiology, genetics, and ecology of foodborne and waterborne bacterial pathogens. Some of these bacteria have recently been reclassified with updated nomenclature; identification methods have also improved substantially with more extensive use of molecular approaches. The bulk of information generated in different areas of research in these pathogenic organisms have been summarized to provide an over all impression of the progress made. Future directions for research into these organisms have also been discussed with a view to understanding general themes of bacterial pathogens, that cause foodborne and waterborne diseases. In addition, available preventative measures to reduce incidences of disease due to these organisms have been discussed.

Distribution of foodborne and waterborne bacterial pathogens

The major bulk of foodborne or waterborne diseases occur in developing countries and are associated with poor sanitation and hygienic conditions, and lack of adequate supply of clean water. The burden of the diseases in these countries are due mainly to diarrhoeagenic *Escherichia coli*, *Vibrio cholerae*, *Salmonella*, *Shigella*, and *Campylobacter* infections. Some of these diseases, e.g. cholera frequently occurs as epidemics which can spread across geographical borders in the form of pandemics affecting a large number of people over many years (Kaper *et al.*, 1995; Faruque *et al.*, 1998). The prevalence and distribution of different foodborne and waterborne pathogens differ between the developed and the developing worlds. For example, although all *Shigella* species, are pathogenic to humans, their epidemiology varies in different parts of the world with different species and subgroups of *Shigella* affecting different subsets of the population. *S. sonnei* and enterohaemorrhagic *E. coli* (EHEC) infections are more prevalent in developed countries including Japan, and countries of Europe and North America, whereas Enteropathogenic *Escherichia coli* (EPEC) and enterotoxigenic *E. coli* (ETEC) are important causes of infantile diarrhoea in developing countries. Within countries there have been changes in the seasonal prevalence, age groups affected and dominant strains of different species of foodborne and waterborne pathogens.

Interestingly, although EPEC are prevalent in developing countries, EHEC infections are rare (Rosa *et al.*, 1998; Albert *et al.*, 1999; Nguyen *et al.*, 2006), despite the fact that cattle in developing countries carry Stx-producing *E. coli* in their intestine (Wani *et al.*, 2003; Vu-Khac *et al.*, 2008). It is postulated that EPEC infection might immunize individuals in developing countries against EHEC infection, since these two pathogens share common antigens encoded by the LEE pathogenicity island. Campylobacter species are commonly reported bacterial cause of intestinal diseases in many industrialized countries as well as in developing countries with an estimated 400 million cases worldwide. In developing countries the incidence of *C. jejuni* infection is several orders of magnitude higher than in the industrialized world. Approximately, 40–60% of young children are estimated to become infected every year and high levels of asymptomatic carriage in older children and adults are reported (Coker *et al.*, 2002).

Pathogenic *V. parahaemolyticus* strains have been associated with epidemics and pandemics, and hence are considered to be major foodborne pathogens in many countries. Since *V. parahaemolyticus* is a natural inhabitant of marine waters and therefore commonly found in estuaries and seafood worldwide, human infections are common in regions where raw, or undercooked shellfish are widely consumed. *Vibrio* vulnificus is also an inhabitant of warm estuarine environments and have been associated with a wide variety of seafood. The organism can cause wound infections, and primary septicaemia in susceptible individuals with underlying liver disease, diabetes or other immunocompromised condition and consuming raw seafood (Strom and Paranjapye, 2000; Jones and Oliver, 2009).

Acute gastroenteritis due to *Yersinia enterocolitica* infection has been reported in many industrialized countries, e.g. Belgium, Canada, the Netherlands, Australia, New Zealand and Germany (Doyle, 1985; Cover and Aber, 1989; Wright *et al.*, 1995). Different serotypes of *Y. enterocolitica* are associated with infection in humans in different regions. In Europe, Canada and Japan, occasional infections caused by *Y. enterocolitica* serovar O:3 and to a lesser extent O:9 are predominant whereas in the United States sporadic cases, are caused by multiple serovars; however, outbreaks which are more common, mainly involve serovar O:8. The differential distribution of different species of foodborne and waterborne pathogens is a demonstration of the fact that these pathogens can adapt to changes in host susceptibility and environmental factors. Hence, the epidemiology of these pathogens varies in different parts of the world with different subgroups of strains affecting different subsets of the population.

Genomic diversity and pathogen evolution

Remarkable progress have been made in our understanding of the genetics and evolution of bacterial pathogens, and the factors associated with the ability of pathogenic bacteria to cause severe infection and rapidly spreading epidemics. Bacteria must acquire and possess a set of genes referred to as virulence associated genes, to be able to cause pathogenesis. Virulence factors encoded by these genes include cell surface proteins that mediate bacterial attachment, cell surface carbohydrates and proteins that protect a bacterium, bacterial toxins, and hydrolytic enzymes that may contribute to alter the metabolism of the target host cells infected by the bacterium. These virulence factors manipulate cellular processes inside the host cell, and often the host cells are forced to behave in ways that promote pathogen growth and make defence against infection more difficult. The availability of whole genome sequences and the tools to compare sequences between pathogenic and non-pathogenic strains within a species have provided opportunities to understand pathogen evolution.

As explained above all strains within a species are not homogeneous with regard to pathogenic potential. For example, of more than 200 serogroups of *V. cholerae* only two serogroups O1 and O139 can cause epidemic disease (Faruque *et al.*, 1998). Similarly *E. coli* as a species is a normal inhabitant of the intestine, yet a few of these can cause severe disease. Thus many established etiologic agents such as *Vibrio cholerae, E. coli, Salmonella, Yersinia,* which elicit various diseases in humans, have their harmless relatives. The difference is based mostly on the presence or absence of certain DNA segments, the virulence genes. Pathogenic bacteria possess such virulence genes, harmless bacteria do not. Bacterial diversification has been driven by a high rate of horizontal gene transfer, which introduces novel genes, confers beneficial phenotypic capabilities, and permits the rapid exploitation of competitive environment.

Identification of the virulence associated genes in pathogenic bacteria have also contributed to development of molecular diagnostics. In this approach DNA probes are used to detect the presence of different virulence genes specific to particular species of pathogenic bacteria. PCR assays have further enhanced detection by enabling genetic analysis of a small number of organisms or only a few copies of the relevant gene. Genetic fingerprinting based on restriction fragment length polymorphism has also been used as markers to track the origin of different clones of pathogenic strains in epidemiological investigations.

In different molecular epidemiological studies, it has been shown that pathogenic bacteria appear with different genetic backgrounds and with enormous diversity among isolates even within a particular species (Kahali *et al.*, 2004). For example, in multilocus sequence typing (MLST), ETEC isolates were distributed in all the phylogenetic lineages of *E. coli* thereby indicating that the acquisition of toxin genes is sufficient to generate an ETEC strain (Turner *et al.*, 2006). In addition, the polyphyletic distribution of ETEC strains indicated that they have arisen several times without any common clonal lineage. Studies indicated that there is extensive exchange of enterotoxin and colonization factor genes between different lineages. Sequence analysis of *mdh* gene using multilocus enzyme electrophoresis strongly indicated that certain pathogenic clones have originated from commensal *E. coli* strains by acquisition of suitable virulence genes (Pupo *et al.*, 1997). Phylogenetic analysis also confirmed genetic variability in the known virulence associated genes of ETEC (Sahl *et al.*, 2011).

Emerging antibiotic resistance

The emergence and rapid spread of antibiotic resistance determinants among pathogenic bacteria are matters of public health concern. The antibiotic resistance often spread among bacteria through mobile genetic elements, and the resistance is particularly a problem to mange infections by some of these pathogens. For example, diseases due to *Salmonella enterica* serovar Typhi is endemic in many countries, and treatment with an appropriate antibiotic should commence as soon as the clinical diagnosis is made, which may be well before the results of antimicrobial sensitivity tests, are available. Chloramphenicol introduced in 1948, was found to be particularly successful in treating typhoid. However plasmid-mediated chloramphenicol resistance in *S.* Typhi emerged in the 1970s, and spread rapidly in different parts of the world. Later trimethoprim and ampicillin resistance emerged, and outbreaks of multiresistant strain were reported in the Indian subcontinent (Rowe *et al.*, 1997). In the early 1990s, chromosomally encoded resistance to ampicillin, chloramphenicol, streptomycin, sulfonamides and tetracyclines was reported in *S.* Typhimurium in England and Wales (Threlfall, *et al.*,1994; Wall, *et al.*, 1994). In a similar way, multiple drug resistance has also emerged in *Shigella* species, and other foodborne pathogens. Often the resistance spread through plasmids and transposons, but chromosomal integration of these mobile elements encoding antibiotic resistance has also been reported. These developments have resulted in a progressive reduction in options for the management of invasive salmonellosis or shigellosis in humans.

Reservoirs and transmissibility

It remains a challenging task to fully explain the ecology, and transmissibility of foodborne and waterborne bacterial pathogens, and the factors that enhance infectivity of these pathogens. Transmission of these pathogens can occur through a variety of means including consumption of undercooked meat, unpasteurized milk, contaminated water, through person-to-person spread, and from animals to people. Both symptomatic and asymptomatic individuals can be reservoirs for various pathogenic species of *E. coli*, including EPEC, EIEC and ETEC. The enterohaemorrhagic (EHEC) group of *E coli* including *E. coli* O157 which produce Vero cytotoxin has its reservoir in cattle. *Campylobacter* spp. are part of the normal bacterial flora in the gastrointestinal tract of animals, particularly poultry, and hence transmission of infections from animal to human is common. *Salmonella* infection occurs because of ingestion of contaminated food, including dairy products, or water contaminated by excreta from patients or handled by infected persons (Parry, 2004).

Shigella species a cause of classic dysentery are less susceptible to gastric acid than many other bacterial pathogens, and as few as 10 to 100 organisms can cause disease (Bennish, 1991). Thus bacterial replication in contaminated food or water is not required to achieve the low infectious dose, and so *Shigella* transmission can occur through direct person-to-person spread as well as from contaminated food and water. *Salmonella* can adapt to different domestic and wild warm-blooded animals and take advantage of changing environmental factors. Contamination of eggs and poultry meat as well as an increased consumption of poultry meat have led to the predominance of *S.* Enteritidis as a cause of human salmonellosis in many countries.

Transmission of cholera typically occurs through water contaminated with the pathogenic organisms. Recent studies are beginning to reveal that transmissibility of cholera is enhanced because of *in vivo* formation of *V. cholerae* biofilms which are excreted by cholera patients (Faruque *et al.*, 2006a). The epidemiological cycle of *V. cholerae* may also be driven by additional, incompletely defined ecological and genetic factors. Stools from cholera patients have been shown to contain biofilm-like aggregates of *V. cholerae* and the enhanced infectivity of *V. cholerae* shed in human stools is largely due to the presence of these clumps of cells which disperse *in vivo* providing a high dose of the pathogen.

Recent studies also suggested that 'conditionally viable environmental cells (CVEC)' of pathogenic *V. cholerae* that exist in surface water but resist cultivation by conventional techniques are aggregates of cells which are possibly derived from cell clumps excreted in human cholera stools. Such CVEC can be recovered as fully virulent bacteria by inoculation of the water into rabbit intestines. A recently published method also allows conversion of CVEC from apparently non-culturable cells to normal metabolically active cells under *in vitro* conditions, and these cells can be isolated exploiting their antibiotic resistance properties (Faruque *et al.*, 2006b). In preliminary studies, this conversion process appeared to involve both dispersion of clumps and activation of partially dormant cells. However, the mechanisms supporting the formation of CVEC in the environment, and their dispersion to infectious bacteria once ingested by a victim are under further investigation.

Future trends in the epidemiology of foodborne and waterborne diseases

Demographic changes in parts of the world have led to a dramatic increase in certain population groups that are especially susceptible to foodborne and waterborne diseases. In addition climatic changes compounded with increased natural disasters are likely to cause increased frequency of infectious diseases, and emergence of new pathogens. The following points are to be especially considered in assessing the needs for future prevention measures against foodborne and waterborne diseases.

1. Many cities are developing so rapidly that the water supply and sewage treatment infrastructures are unable to keep pace with the expansion of population. Similarly in many developing countries, migration of people from rural to urban areas are causing high population densities at certain places without adequate water and sanitation facilities. This scenario is most likely to give rise to increased infections and high morbidity and mortality from infectious diseases.
2. Climatic changes resulting in increased frequency and severity of floods, droughts, famines, wars or civil unrest can disrupt routine food and water supplies and present opportunities for infectious diseases to spread rapidly through contaminated water and foodstuff.
3. The proportions of elderly population in many countries are increasing, and a growing number of people are becoming more susceptible to infections owing to many factors including immunodeficiency as a result of HIV infection and various therapies.
4. In addition to increased risk of foodborne and waterborne diseases, individuals with HIV infection are susceptible to infection by many opportunistic pathogens. Such infections can have devastating consequences for the vulnerable subgroup and will consume large amounts of health care resources.
5. Increasing international travel has been contributing to the introduction and rapid dissemination of new and existing pathogens.

For example, after an absence of a century, cholera caused an epidemic in Peru in January 1991 and spread rapidly through Latin America.

6 Owing to global climate changes, the incidence of water-borne enteric disease such as cholera is likely to increase since *V. cholerae* O1 infections are known to increase with warmer water temperatures. *V. cholerae* non-O1 and other *Vibrio* species including *V. vulnificus* and *V. parahaemolyticus* may also become more frequent agents of disease as a result of increasing ocean temperatures.

7 Natural disasters associated with climate change may also increase disease incidences and the spread of epidemics because of inadequate clean water and food in disaster-stricken areas. Such increases pose a risk not only to developing countries, but also to developed countries through imported diseases.

Future prospect and prevention strategies

Enhancement of basic scientific research

It is important that our basic science gives us sufficient understanding of the pathogen, before we can devise measures to efficiently control a particular disease. In order to understand the mode of action of virulence factors produced by invading pathogens, all kinds of information is needed. For example, with the help of the Type 3 secretion system, some of the foodborne or waterborne pathogens pump one or more virulence factors into the intestinal cells of the victim. The disease is ultimately based on the mode of action of this virulence factor cocktail. The molecular function of some of these virulence factors has already been clarified. It is essential to elucidate how the virulence factors cooperate (synergistically, antagonistically) and how they are coordinated with each other in their mode of action and dosage. This will explain why some of these pathogens can interact with its host so efficiently, and can provide new starting points for the development of therapeutic tools. Since the arsenal of virulence genes determines the kind of disease that a pathogenic bacterium can cause, understanding of the mode of action of the virulence factors is of great importance for the development of new strategies to fight against infectious diseases.

Multiple systems are involved in the regulation of virulence genes in pathogenic bacteria. Often critical genes are coordinately regulated so that they respond in a similar fashion to specific environmental condition. The availability of complete genomic sequence, has facilitated the use of genomic microarrays and other transcriptome analysis methods to study the expression profile of diverse bacteria under different conditions. These studies have great potential in revealing the role of multiple genes in the adaptation of the bacterium to different environmental conditions.

The rationale behind our efforts to examine evolutionary events in bacterial pathogens is to gain new knowledge, and eventually devise means to control diseases. Knowing how new pathogenic strains emerge is of immense scientific interest, and in foodborne and waterborne pathogens represents paradigms in the co-evolution of bacteria and the virulence associated genetic elements. Future studies should be aimed at explaining more about the emergence of pathogenic strains, and factors involved in the natural selection of the species to ensure its continued existence, as well as the means of combating these in an effort to improve protection and treatment against foodborne and waterborne diseases.

Conjugative plasmids and transducing phages exist for various bacterial pathogens. Further efforts should be made to establish whether these or other potentially mobile genetic elements can be efficiently mobilized. Besides, recognition of possible common themes among bacteriophage-encoded virulence factors and other mobile elements further studies should be sought to expand our understanding of the evolution of bacterial pathogens.

Our improved understanding of the molecular mechanisms of *E. coli* pathogenesis has uncovered new aspects of host response to infecting pathogens. However, our understanding of the interactions between virulence factors and host components remains incomplete. It will be interesting to dissect the interactions to uncover how they contribute to the progression of the disease. This knowledge could be exploited in better treatment and prevention strategies.

Global monitoring and epidemiological networks

Early warning systems to detect emerging infections in a timely fashion is required so that investigations and preventative measures can be rapidly initiated before they become major public health problems in particular regions. Recent developments in electronic communication should be exploited to set up Global public health networks for effective disease surveillance. To control the diseases associated with both food and water it is important that our monitoring systems provide an early warning. Surveillance systems such as Salm-Net (Fisher *et al.*, 1994), PulseNet (Swaminathan *et al.*, 2006) have already been set up. A number of other global and regional network exists, but there is a need to further develop the network so that these may successfully contribute to the educational and control measures.

Prevention of spread of pathogens

When there is a potential risk of contamination with organism that particularly require a low infectious dose to establish infections, such as EHEC or *Shigella,* extra care should be taken to cook foods properly. Cross-contamination of cooked food with raw food should be avoided. Appropriate hand washing after handling raw meat is also very important to reduce EHEC infection. Also, patients suffering from EHEC infection should be handled with attention to universal precautions to prevent person-to-person transmission of organisms. In addition, prevention of EHEC infection requires practices that reduce or eliminate faecal contamination of meat during slaughter and processing. An active surveillance system for timely identification of outbreaks should identify contaminated food items and appropriate handling of patients to prevent further spread of infection. Water hygiene is also an important aspect of prevention of outbreaks owing to drinking water and recreational water contamination.

Eating habits, especially in industrialized countries, are changing rapidly. There is an increasing tendency for people to eat more meals outside the home, to buy food in bulk (and less often) and to eat more fast foods, which require less preparation. Increasing mass production and global distribution of ready-to-eat foods present the opportunity for the dissemination of emerging or re-emerging infections and put large populations at risk. In addition, certain bacterial pathogens, e.g. *Y. enterocolitica and Listeria monocytogenes,* can grow at low temperatures, and the increase in prevalence of this organism in food may in part be due to the increased use of refrigeration in the food chain (Schlech *et al.*, 1983; Doyle, 1990).

Competition in the food industry may force manufacturers to put cost reduction before food safety; therefore it is important that retailers make food safety a top priority so that companies compete on that basis. The focus must be on limiting the growth, survival and spread of pathogens in foodstuffs, and, considering the susceptibility of the particular consumers, agreements should be reached on what levels of these pathogens are acceptable in which foods.

References

Albert, M.J., Faruque, A.S.G., Faruque, S.M., Sack, R.B., and Mahalanabis, D. (1999). Case–control study of enteropathogens associated with childhood diarrhea in Dhaka, Bangladesh. J. Clin. Microbiol. *37*, 3458–3464.

Bennish, M.L. (1991). Potentially lethal complications of shigellosis. Rev. Infect Dis. *13*, S319.

Coker, A.O., Isokpehi, R.D., Thomas, B.N., Amisu, K.O., and Obi, C.L. (2002). Human campylobacteriosis in developing countries. Emerg. Infect. Dis. *8*, 237–244.

Doyle, M.P. (1990). Pathogenic *Escherichia coli, Yersinia enterocolitica* and *Vibrio parahaemolyticus.* Lancet *336*, 1111–1115.

Faruque, S.M., Albert, M.J., and Mekalanos, J.J. (1998). Epidemiology, genetics, and ecology of toxigenic *Vibrio cholerae.* Microbiol. Mol. Biol. Rev. *62*,1301–14.

Faruque, S.M., Biswas, K., Udden, S.M., Ahmad, Q.S., Sack, D.A., Nair, G.B., and Mekalanos, J.J. (2006a). Transmissibility of cholera: *in vivo*-formed biofilms and their relationship to infectivity and persistence in the environment. Proc. Natl. Acad. Sci. U.S.A. *103*, 6350–6355.

Faruque, S.M., Islam, M.J., Ahmad, Q.S., Biswas, K., Faruque, A.S.G., Nair, G.B., Sack, R.B., Sack, D.A., and Mekalanos. J.J. (2006b). An improved technique for isolation of environmental *Vibrio cholerae* with epidemic potential: monitoring the emergence of a multiple-antibiotic-resistant epidemic strain in Bangladesh. J. Infect. Dis. *193*, 1029–1036.

Fisher, I.S.T., Rowe, B., Bartlett, C., and Gill, O.N. (1994). "Salm-Net" laboratory-based surveillance of human

Salmonella infections in Europe. PHLS digest *11*, 181–182.

Jones, M.K. and Oliver, J.D. (2009). *Vibrio vulnificus*: Disease and pathogenesis. Infect. Immun. 77, 1723–1733.

Kahali, S., Sarkar, B., Chakraborty, S., Macaden, R., Deokule, J.S., Ballal, M., Nandy, R.K., Bhattacharya., S.K., Takeda, Y., and Ramamurthy, T. (2004). Molecular epidemiology of diarrhoeagenic *Escherichia coli* associated with sporadic cases and outbreaks of diarrhoea between 2000 and 2001 in India. Eur. J. Epidemiol. *19*, 473–479.

Kaper, J.B., Morris, J.G., Jr, and Levine, M.M. (1995). Cholera. Clin. Microbiol. Rev. *8*, 48–86.

Nguyen, R.N., Taylor, L.S., Tauschek, M., and Robins-Browne, R. (2006). Atypical enteropathogenic *Escherichia coli* infection and prolonged diarrhcea in children. Emerg. Infect. Dis. 12, 597–603.

Parry, C.M. (2004). Typhoid Fever. Curr. Infect. Dis. Rep. *6*, 27–33.

Pupo, G.M., Karaolis, D.K., Lan, R., and Reeves, P.R. (1997). Evolutionary relationships among pathogenic and nonpathogenic *Escherichia coli* strains inferred from multilocus enzyme electrophoresis and *mdh* sequence studies. Infect. Immun. 65, 2685–2692.

Rosa, A.C., Mariano, A.T., Pereira, A.M., Tibana, A., Gomes, T.A., and Andrade, J.R. (1998). Enteropathogenicity markers in *Escherichia coli* isolated from infants with acute diarrhoea and healthy controls in Rio de Janeiro, Brazil. J. Med. Microbiol. *47*, 781–90.

Rowe, B., Ward, L.R., and Threlfall, E.J. (1997) Multidrug-resistant *Salmonella typhi*: a worldwide epidemic. Clin. Infect. Dis. *24*, S106–S109.

Sahl, J.W., Steinsland, H., Redman, J.C., Angiuoli, S.V., Nataro, J.P., Sommerfelt, H., and Rasko, D.A. (2011). A comparative genomic analysis of diverse clonal types of enterotoxigenic *Escherichia coli* reveals pathovar-specific conservation. Infect. Immun. 79, 950–960.

Schlech, W.F., 3rd, Lavigne, P.M., Bortolussi, R.A., Allen, A.C., Haldane, E.V., Wort, A.J., Hightower, A.W., Johnson, S.E., King, S.H., Nicholls, E.S., *et al.* (1983). Epidemic listeriosis-evidence for transmission by food. N. Engl. J. Med. *308*, 203–206.

Strom, M.S. and Paranjpye, R.N. (2000). Epidemiology and pathogenesis of *Vibrio vulnificus* infection. Microb Inf. 73, 177–188.

Swaminathan, B., Gerner-Smidt, P., Ng, L.K., Lukinmaa, S., Kam, K.M., Rolando, S., Gutierrez, E.P., and Binsztein, N. (2006). Building PulseNet International: an interconnected system of laboratory networks to facilitate timely public health recognition and response to foodborne disease outbreaks and emerging foodborne disease. Foodborne Pathog. Dis. 3, 36–50.

Threlfall, E.J., Frost, J.A., Ward, L.R., and Rowe, B. (1994). Epidemic in cattle of *S. typhimurium* DT104 with chromosomally-integrated multiple drug resistance. Vet. Rec. *134*, 577.

Turner, S.M., Chaudhuri, R.R., Jiang, Z.D., DuPont, H., Gyles, C., Penn, C.W., Pallen, M.J., and Henderson, I.R. (2006). Phylogenetic comparisons reveal multiple acquisitions of the toxin genes by enterotoxigenic *Escherichia coli* strains of different evolutionary lineages. J. Clin. Microbiol. 44, 4528–4536.

Vu-Khac, H., and Cornick, N.A. (2008). Prevalence and genetic profiles of Shiga toxin-producing *Escherichia coli* strains isolated from buffaloes, cattle, and goats in central Vietnam. Vet. Microbial. *126*, 356–363.

Wall, P.G., Morgan, D., Lamden, K., Ryan, M., Griffin, M., Threlfall, E.J., Ward, L.R., and Rowe, B. (1994). A case control study of infection with an epidemic strain of multiresistant *Salmonella* DT104 in England and Wales. Comm. Dis. Rep. Rev. *4*, R130–R135.

Wani, S.A., Bhat, M.A., Samanta, I., Nishikawa, Y., and Buchh, A.S. (2003). Isolation and characterization of Shiga toxin-producing *Escherichia coli* (STEC) and enteropathogenic *Escherichia coli* (EPEC) from calves and lambs with diarrhoea in India Lett. Appl. Microbiol. 37, 121–126.

Index

E

F

G

H

I

K

L

M

N

O

P

Q

R

U

V

Y

www.ingramcontent.com/pod-product-compliance
Lightning Source LLC
LaVergne TN
LVHW062116200926
848484LV00007B/1680
* 9 7 8 1 9 0 8 2 3 0 0 6 5 *